Chapter 4 Systems of Linear Equations and Inequalities

A system of linear equations may be solved: (a) graphically: (b) by the substitution method, (c) by the addition or elimination method, or (d) by determinants.

$$\begin{vmatrix} a_1 & b_1 \\ a_2 & b_2 \end{vmatrix} = a_1 b_2 - a_2 b_1$$

Cramer's Rule: Given a system of equations of the form

$$\begin{matrix} a_1x + b_1y = c_1 \\ a_2x + b_2y = c_2 \end{matrix} \text{ then } x = \frac{\begin{vmatrix} c_1 & b_1 \\ c_2 & b_2 \end{vmatrix}}{\begin{vmatrix} a_1 & b_1 \\ a_2 & b_2 \end{vmatrix}} \text{ and } y = \frac{\begin{vmatrix} a_1 & c_1 \\ a_2 & c_2 \end{vmatrix}}{\begin{vmatrix} a_1 & b_1 \\ a_2 & b_2 \end{vmatrix}}$$

Chapter 5 Polynomials

$a^m \cdot a^n = a^{m+n}$

$a^m/a^n = a^{m-n}, a \neq 0$

$a^{-m} = \dfrac{1}{a^m}, a \neq 0$

$a^0 = 1, a \neq 0$

$(a^m)^n = a^{mn}$

$(ab)^m = a^m b^m$

$\left(\dfrac{a}{b}\right)^m = \dfrac{a^m}{b^m}, b \neq 0$

FOIL method to multiply two binomials

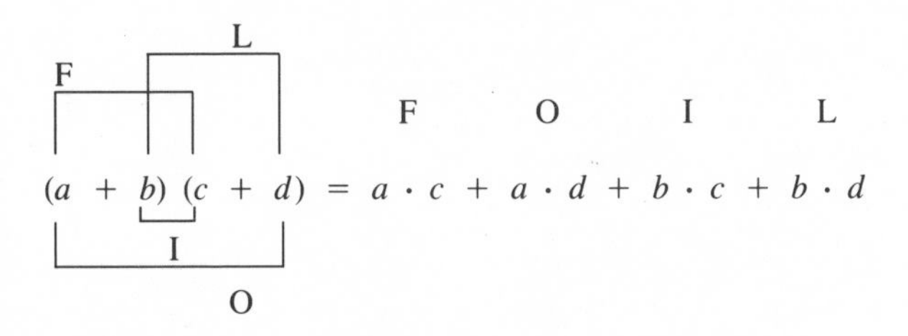

Square of a binomial:
$(a + b)^2 = a^2 + 2ab + b^2$, $(a - b)^2 = a^2 - 2ab + b^2$

Product of the sum and difference of the same two terms (also called the difference of two squares):
$(a + b)(a - b) = a^2 - b^2$

Polynomial function:
$f(x) = a_nx^n + a_{n-1}x^{n-1} + a_{n-2}x^{n-2} + \cdots + a_1x + a_0$

Quadratic function: $f(x) = ax^2 + bx + c, a \neq 0$

Chapter 6 Factoring

Difference of two squares: $a^2 - b^2 = (a + b)(a - b)$
Note: the sum of two squares $a^2 + b^2$ cannot be factored over the set of real numbers

Perfect square trinomials:
$a^2 + 2ab + b^2 = (a + b)^2$, $a^2 - 2ab + b^2 = (a - b)^2$

Sum of two cubes:
$a^3 + b^3 = (a + b)(a^2 - ab + b^2)$

Difference of two cubes:
$a^3 - b^3 = (a - b)(a^2 + ab + b^2)$

To Factor a Polynomial

1. Determine if the polynomial has a greatest common factor other than 1. If so factor out the GCF from every term in the polynomial.
2. If the polynomial has two terms (or is a binomial), determine if it is a difference of two squares or a sum or difference of two cubes. If so factor using the appropriate formula.
3. If the polynomial has 3 terms (or is a trinomial), determine if it is a perfect square trinomial. If so factor accordingly. If it is not, then factor the trinomial using the method discussed in Section 6.2.
4. If the polynomial has more than 3 terms, then try factoring by grouping.
5. As a final step examine your factored polynomial to see if any factors listed have a common factor and can be factored further. If you find a common factor, factor it out at this point.

Zero-factor property: If $a \cdot b = 0$, then either $a = 0$ or $b = 0$, or both a and $b = 0$.

Intermediate Algebra for College Students

Intermediate Algebra for College Students

SECOND EDITION

Allen R. Angel

Monroe Community College

Prentice Hall

Englewood Cliffs, New Jersey 07632

Library of Congress Cataloging-in-Publication Data

ANGEL. ALLEN R., date
Intermediate algebra for college students/Allen R. Angel-2nd ed.

p. cm.
Rev. ed. of: Intermediate Algebra. c1986.
Includes index.
ISBN 0-13-470055-4: ISBN 0-13-470063-5
(annotated instructor's ed.): Free
1. Algebra. I. Angel. Allen R., Intermediate algebra.
II. Title.
QA154.2.A53 1988
512.9—dc 19

Editorial/production supervision: Virginia Huebner
Interior and Cover design: Judith A. Matz-Coniglio
Manufacturing buyer: Paula Massenaro
Page layout: Meryl Poweski
Photo Editor: Lorinda Morris-Nantz
Photo Research: Teri Stratford
Cover Art: From original painting by Ronald Ghiz, courtesy of the artist

1, F.O.S., Inc. • *21*, Larry Mulvehill/*Photo Researchers* • *37*, Ed Landrock/*Rodale Press* • *55*, Courtesy of IBM •
89, Van Bucher/*Photo Researchers* • *103*, Boschung/*Leo de Wys, Inc.* • *134*, Guy Gillette/*Photo Researchers* •
157, Roger A. Clark, Jr./*Photo Researchers* • *158*, R.D. Ullmann/*Taurus Photos* • *177*, Allen R. Angel •
221, NASA • *225*, Norman R. Thompson/*Taurus Photos* • *252*, Kay Mori/*Taurus Photos* •
254, Scott Ransom/*Taurus Photos* • *255*, Richard Hutchings/*Photo Researchers* •
292, *303*, Mimi Forsyth/*Monkmeyer Press* • *304*, NASA • *336*, Bobby Maduro/*All Sport/Caryn Levy* •
352, G. Whitely/*Photo Researchers* • *367*, Lynn Pelham/*Leo de Wys. Inc.* • *384*, Allen R. Angel • *387*, Leo de Wys. Inc. •
389, St. *Louis Regional Commerce and Growth Association* • *425*, Francois Gohier/*Photo Researchers* •
462, Richard Megna/*Fundamental Photographs*

A Division of Simon & Schuster
Englewood Cliffs, New Jersey 07632
Previous edition published under the title of: Intermediate Algebra: A Practical Approach

Printed in the United States of America
10 9 8 7 6 5 4

ISBN 0-13-470055-4 01

Prentice-Hall International (UK) Limited, *London*
Prentice-Hall of Australia Pty. Limited, *Sydney*
Prentice-Hall Canada Inc., *Toronto*
Prentice-Hall Hispanoamericana, S.A., *Mexico*
Prentice-Hall of India Private Limited, *New Delhi*
Prentice-Hall of Japan, Inc., *Tokyo*
Prentice-Hall of Southeast Asia Pte. Ltd., *Singapore*
Editora Prentice-Hall do Brasil, Ltda., *Rio de Janeiro*

To my mother, Sylvia Angel-Baumgarten
and
To the memory of my father, Isaac Angel

Contents

10 Conic Sections 389

11 Exponential and Logarithmic Functions 425

12 Sequences, Series and the Binomial Theorem 462

Appendixes 489

Preface

This is the second book in a two-book algebra series. This book was written for college students who have successfully completed elementary algebra and wish to take a second course in algebra.

My primary goal in writing this book was to write a book that students can read, understand, and enjoy. To achieve this goal I have used short sentences, clear explanations, and many detailed, worked-out examples. I have tried to make the book relevant to college students by using practical applications of algebra throughout the text. For consistency in the series, I have used the same pedagogical features in this book as in the Elementary Algebra book. Some of these features are outlined below.

Features of the Text

Four-Color Format: The four colors are used pedagogically in the following ways:

Important definitions and procedures are color screened;

Color screening or color type is used to make other important items stand out;

Errors that students commonly make are given in colored boxes as warnings for students;

Artwork is enhanced and clarified with use of multiple colors;

Other important items such as the Helpful Hints, Just for Fun Problems, and Calculator Corners are enhanced with color;

The four color format allows for all these, and other features, to be presented in different forms and colors for easy identification by students;

The four color format helps make the text more appealing and interesting to students.

Practical Applications: Practical applications of algebra are stressed throughout the text. Students need to learn how to translate word problems into algebraic symbols. The problem solving approach used throughout this text gives students ample practice in setting up and solving word problems.

Detailed Worked-Out Examples: A wealth of examples have been worked out in a step by step detailed manner. Important steps are highlighted in color, and no steps are omitted until after the student has seen a sufficient number of similar examples.

Exercise Sets: Each exercise set is graded in difficulty. The early problems help develop the student's confidence, then they are eased gradually into the more difficult problems. A sufficient number and variety of examples are given in the section for the student to successfully complete even the more difficult exercises. The number of exercises in each section is more than ample for student assignments and practice.

Keyed Section Objectives: Each section opens with a list of skills that the student should learn in that section. The objectives are then keyed to the appropriate portions of the sections with symbols such as 1.

Common Student Errors: Errors that students often make are illustrated. The reasons these procedures are wrong are explained, and the correct procedure for working the problem is illustrated. These common student error boxes will help prevent your students from making those errors we see so often.

Helpful Hints: The helpful hint boxes offer useful suggestions for problem solving and other varied topics. They are set off in a special manner so that students will be sure to read them.

Just for Fun Problems: At the end of many exercise sets are Just for Fun problems. These problems offer more challenging problems for the bright students in your class who want something extra. These problems present additional applications of algebra, material to be presented later in the text, or material to be covered in a later mathematics course.

Calculator Corners: The Calculator Corners, placed at appropriate locations in the text, are written to reinforce the algebraic topics presented in the section.

Chapter Summaries: At the end of each chapter is a chapter summary which includes a glossary and important chapter facts.

Review Exercises: At the end of each chapter are review exercises that cover all types of exercises presented in the chapter. The review exercises are keyed to the sections where the material was first introduced.

Practice Tests: The comprehensive end of chapter practice test will enable the student to see how well they are prepared for the actual class test. The Instructor's Resource Manual includes 5 forms of each chapter test that are similar to the students practice test (multiple choice tests are also included in the Instructor's Resource Manual.)

Readability: One of the most important features of the text is its readability. The book is very readable even for those with weak reading skills. Short clear sentences are used and words that are more easily recognized and understood are used whenever possible. With so many of our students from different countries now taking algebra, this feature has become increasingly important.

Accuracy: Accuracy in a mathematics text is essential. To insure accuracy in this book, no fewer than five mathematicians from around the country have read the galleys carefully for typos, and have checked all the answers.

Prerequisite

The prerequisite for this course is a working knowledge of elementary algebra. Although some elementary algebra topics are briefly reviewed in the text, students should have a basic understanding of elementary algebra before taking this course.

Mode of Instruction

The format of this book—pedagogical use of four colors; short but complete sections; clear explanations; important points stressed in color; many detailed step-by-step worked-out examples; ample and graded exercise sets; Common Student Errors; Helpful Hints; chapter summaries, review exercises and practice test; answers to odd exercises, and all Just for Fun problems, review exercises and practice tests—makes this text suitable for many types of instructional modes including lecture, modified lecture, learning laboratory, and self-paced instruction. Many student supplements are available to assist the student in the learning process. Please see Available Supplements For Students which follows shortly.

Changes in the Second Edition

Many users of the first edition, and others, have indicated that they prefer a hardbound book for non-programmed textbooks. Therefore, the second edition of this algebra series has been written as a hardcover text.

(The first edition in softback will still remain in print for those who prefer the softcover edition).

Another major change is the new four color format. Many reviewers, and users of the text, felt that an increase from two to four colors would enhance the book in many ways, including increased clarity of the book's many distinguishing features.

A third major change is the content of the book. More intermediate algebra material is now covered and students get to this new material faster. There is now less overlap between the material covered in the elementary algebra book and the intermediate algebra book.

Other changes include: more precise definitions; increased number of worked-out examples and exercises; reordering of certain topics to allow for a smoother flow of material; rewriting of certain sections for greater clarity; earlier presentation of graphing of linear equations and introduction to functions (in Chapter 3); greater coverage of equations and inequalities containing absolute values; new sections on: systems of linear inequalities, a general review of factoring, polynomial functions, graphing polynomial functions, variation, the square root function (optional).

Available Supplements for the Instructor

Annotated Instructor's Edition: Contains the answers to all exercises clearly presented next to each exercise in the student's edition. This saves time and insures accuracy in classroom lectures. Answers in annotated edition are written in color for easy identification by the instructor.

Instructor's Solution Manual: Contains complete and detailed solutions to all even-numbered exercises in the text. Solutions to odd-numbered problems appear in the Student's Solution Manual.

Instructor's Resource Manual: Contains 8 forms of each chapter test. Five are open answer questions similar to the practice test in the text, and three are in multiple-choice format.

Software Testing Package for IBM and Apple PC's: This flexible package allows you to construct your own exams, or will generate any number of individualized exams to your specifications. Open answer *and* multiple choice items are both available.

Available Supplements for Students

Student's Study Guide: Includes additional worked out examples, additional drill problems and practice test and their answers. Important points are emphasized.

Student's Solution Manual: Includes detailed step-by-step solutions to every odd-numbered problem in the exercise sets, and all solutions to practice tests.

"How To Study Math": Designed to help your students overcome math anxiety and to offer helpful hints regarding study habits. This useful booklet is available free with each copy sold. To request copies for your students in quantity, contact your local Prentice Hall representative.

Video Tapes: Professionally done by qualified individuals, these tapes will enhance students' learning by providing further reinforcement of key concepts. Free with a qualified adoption; contact your local Prentice Hall representative for details.

Tutorial Software: More for your students. The program, available for both IBM and Apple pc's will give your students even more drill and practice. Free with a qualified adoption; contact your local Prentice Hall representative for details.

Acknowledgments

Writing a textbook is a long and time consuming project. Many people deserve thanks for encouraging and assisting me with this project. Most importantly I would like to thank my wife Kathy; and sons Robert and Steven. Without their constant encouragement and understanding, this project would not have become a reality. I would also like to thank each of them for contributing to this project in a variety of ways.

I would like to thank my colleagues at Monroe Community College for helping with this project, especially: Larry Clar, Gary Egan, Huebert Haefner, and Annette Leopard. Judy Conturo Karas did an excellent job of typing the manuscript.

I would like to thank my students, and students and faculty from around the country, for using the first edition and offering valuable suggestions for the second edition.

I would like to thank the following individuals for working with me on the various supplements for the book:

Instructor's Solution Manual, Student's Solution Manual

Helen Burrier *Kirkwood Community College*
Kathy Davis *Kirkwood Community College*
Leland Fry *Kirkwood Community College*
Robert D'addario *Brookdale Community College*
Maria Hall *Brookdale Community College*
Lucille Lubow *Brookdale Community College*
Linda Nelson *Brookdale Community College*
Susan Nusbaum *Brookdale Community College*

Instructor's Resource Manual

Ara Sullenberger *Tarrant County Junior College*

Student's Study Guide

Leonard Malinowski *Community College of the Finger Lakes*

Video Tapes

Roger Breen *Florida Community College at Jacksonville*
Margaret Greene *Florida Community College at Jacksonville*

I would like to thank my editor at Prentice-Hall, Priscilla McGeehon; executive editor, Robert Sickles; production editor, Virginia Huebner, and Judith A. Matz-Coniglio, designer.

I would like to thank the following reviewers and proofreaders for their valuable suggestions and their conscientiousness.

Helen Burrier *Kirkwood Community College*
Frank Cerrato *City College of San Francisco*
Peter Freedhand *New York University*
Peggy Greene *Florida Community College at Jacksonville*
Judy Kasabian *El Camino College*
Lois Miller *Golden West College*
Julie Monte *Daytona Beach Junior College*
Cathy Pace *Louisiana Tech University*
Ken Seydel *Skyline College*
Tommy Thompson *Brookhaven College*
John Wenger *Loop College*
Brenda Wood *Florida Community College at Jacksonville*

Finally, I would like to thank the following people who were helpful in advising me on issues of content and organization:

Ronald Bohuslov *Merritt College*
Francine Bortzel *Seton Hall University*
Dale Ewen *Parkland College*
Mark Gidney *Lees McRae College*
Robert Gesell *Oleary College*
Larry Hoehn *Austin Peay State University*
Herbert Kasube *Bradley University*
Melvin Kirkpatrick *Roane State Community College*
Adele LeGere *Oakton Community College*
Glenn Lipely *Malone College*
Charles Luttrell *Frederick Community College*
Merwin Lyng *Mayville State College*
P. William Magliaro *Buck County Community College*
Jack McCown *Central Oregon Community*
John Michaels *SUNY at Brockport*
Matthew Pickard *University of Puget Sound*
C. V. Peele *Marshall University*
James Perkins *Piedmont Community College*
Jon Plachy *Metropolitan State College*
Raymond Pluta *Castleton State College*
Dolores Schaffner *University of South Dakota*
Edith Silver *Mercer County Community College*
Fay Thames *Lamar University*
Karl Zilm *Lewis and Clark Community College*

To The Student

Algebra is a course that cannot be learned by observation. To learn algebra you must become an active participant. You must read the text, pay attention in class, and, most importantly, you must work the exercises. The more exercises you work the better.

If you purchased this text new then you should have received a complementary copy of "How to Study Math." I suggest you read, and follow, the instructions in the booklet very carefully. You will find them very helpful.

This text was written with you in mind. Short, clear sentences were used and many examples were given to illustrate specific points. The text stresses useful applications of algebra. Hopefully, as you progress through the course you will come to realize that algebra is not just another math course that you are required to take, but a course that offers a wealth of useful information and applications.

This text makes use of 4 different colors. The different colors are used to highlight important information. Important procedures, definitions and formulas are placed within colored boxes.

The boxes marked Common Student Errors should be studied carefully. These boxes point out errors that students commonly make and the correct procedures for doing these problems. The boxes marked Helpful Hints should also be studied carefully for they also stress important information.

At the end of many exercise sets are Just for Fun Problems. These exercises are not for everyone. They are for those students who are doing well in the course and are looking for more of a challenge. These exercises often present additional applications of algebra, material that will be presented in a later section, or material that will be presented in a later course.

At the end of each chapter is a chapter summary, a set of review exercises, and a chapter practice test. Before each examination you should review these sections carefully and take the practice test. If you do well on the practice test you should do well on the class test. The questions in the review exercises are marked to indicate the section in which that material was first introduced. If you have a problem with a review exercise question turn to and reread the section indicated.

In the back of the text there is an answer section which contains the answers to the odd-numbered exercises, Just for Fun Problems, all review exercises, and all practice tests. The answers should be used only to check your work.

Various supplements are available to help you achieve success in this course. They include: student's study guide, student's solution manual, video tapes, audio tapes, and tutorial software. Ask your instructor which of these are available for your use.

I have tried to make this text as clear and error free as possible. No text is perfect, however. If you find an error in the text, or an example or section that you believe can be improved, I would greatly appreciate hearing from you. If you enjoy the text, I would also appreciate hearing from you.

Allen R. Angel

1 Basic Concepts

See Section 1.4, Exercise 84.

1.1 Sets and the Real Number System

1. *Know the meaning of a set.*
2. *Identify subsets.*
3. *Perform set operations: union and intersection.*
4. *Identify various important sets of numbers.*
5. *Know the real numbers.*

1 Sets are used in many areas of mathematics, so an understanding of sets and set notation is important. A **set** is a collection of objects. The objects in a set are called **elements** of the sets. Sets are indicated by means of braces, { }, and are often named with capital letters. When the elements of a set are listed within the braces, as illustrated below, the set is said to be in **roster form.**

$$A = \{\text{a, b, c}\}$$
$$B = \{\text{yellow, green, blue, red}\}$$
$$C = \{1, 2, 3, 4, 5\}$$

Set A has three elements, set B has four elements, and set C has five elements. The symbol $\in$ is used to indicate that an item is an element of a set. Since 2 is an element of set C we may write $2 \in C$; this is read "2 is an element of set C." Note that 6 is not an element of set C. We may therefore write $6 \notin C$, which is read "6 is not an element of set C."

A set may be finite or infinite. Sets A, B, and C each have a finite number of elements and are therefore *finite sets.* In some sets it is impossible to list all the elements. These are *infinite sets.* The following set is an example of an infinite set.

$$N = \{1, 2, 3, 4, 5, \ldots\}$$

The three dots after the last comma indicate that the set continues in the same manner indefinitely. Set N is called the set of **natural numbers,** or **counting numbers.**

If we write

$$D = \{1, 2, 3, 4, 5, \ldots, 280\}$$

it means that the set continues in the same manner until the number 280. Set D is the set of the first 280 natural numbers.

A special set is the *null set,* or *empty set,* written { } or $\varnothing$. The null set is a set that contains no elements. For example, the set of students in your class over the age of 150 is the null or empty set.

A second method of writing a set is with *set builder notation.* An example of a set given in set builder notation is

$$E = \{x \mid x \text{ is a natural number greater than } 6\}$$

This is read "Set E is the set of all x such that x is a natural number greater than 6." In roster form this set would be

$$E = \{7, 8, 9, 10, 11, \ldots\}$$

The general form of set builder notation is

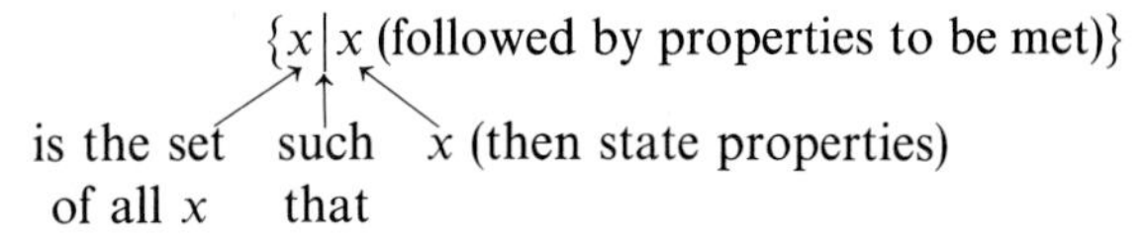

Consider

$$F = \{x \mid x \text{ is a natural number between 2 and 7}\}$$

This is read "Set F is the set of all x such that x is a natural number between 2 and 7." In roster form the set is

$$F = \{3, 4, 5, 6\}$$

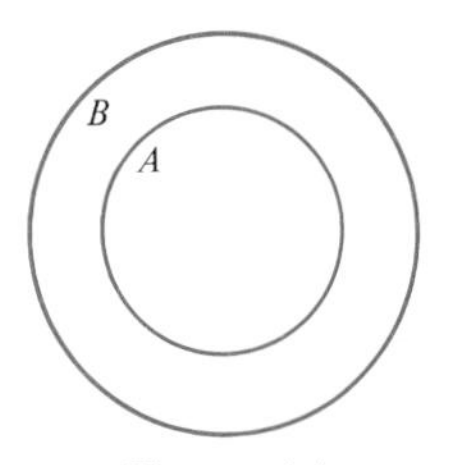

Figure 1.1

2 Set A is a *subset* of set B, written $A \subseteq B$, if every element of set A is also an element of set B. Figure 1.1 illustrates two sets A and B, where set A is a subset of set B. Note: Every element that is in set A must also be in set B.

EXAMPLE 1 If the first set is a subset of the second set, insert $\subseteq$ between the two sets. If the first set is not a subset of the second set, insert the symbol $\not\subseteq$ (read "is not a subset of") between the two sets.

(a) $\{1, 2, 3\} \quad \{1, 2, 3, 4\}$
(b) $\{a, b, c, d\} \quad \{b, c, d, e, f, g\}$

Solution: (a) $\{1, 2, 3\} \subseteq \{1, 2, 3, 4\}$
(b) $\{a, b, c, d\} \not\subseteq \{b, c, d, e, f, g\}$ Notice that a is in the first set but not the second set. ■

EXAMPLE 2 Determine if the set of natural numbers, $N = \{1, 2, 3, 4, 5, \ldots\}$, is a subset of the set of whole numbers, $W = \{0, 1, 2, 3, 4, 5, \ldots\}$.

Solution: Since every element in the set of natural numbers is also an element in the set of whole numbers, we may write

$$N \subseteq W$$

The natural numbers are a subset of the set of whole numbers. ■

3 Just as operations such as addition and multiplication are performed on numbers, operations are performed on sets. Two operations we will discuss are *union* and *intersection*. The **union** of set A and set B, written $A \cup B$, is the set of elements that belong to either set A or set B. The union is formed by combining, or joining together, the elements in set A with those in set B.

Examples of union of sets

$A = \{1, 2, 3, 4, 5\}$, $B = \{3, 4, 5, 6, 7\}$, $A \cup B = \{1, 2, 3, 4, 5, 6, 7\}$
$A = \{a, b, c, d, e\}$, $B = \{x, y, z\}$, $A \cup B = \{a, b, c, d, e, x, y, z\}$

In set builder notation we can express $A \cup B$ as

$$A \cup B = \{x \mid x \in A \quad or \quad x \in B\}$$

The **intersection** of set A and set B, written $A \cap B$, is the set of all elements that are common to both set A *and* set B.

Examples of intersection of sets

$$A = \{1, 2, 3, 4, 5\}, \quad B = \{3, 4, 5, 6, 7\}, \quad A \cap B = \{3, 4, 5\}$$
$$A = \{\text{a, b, c, d, e}\}, \quad B = \{\text{x, y, z}\}, \quad A \cap B = \{\ \}$$

Note that in the last example set A and B have no elements in common. Therefore, their intersection is the empty set. In set builder notation we can express $A \cap B$ as

$$A \cap B = \{x \mid x \in A \quad \textit{and} \quad x \in B\}$$

COMMON STUDENT ERROR

Students commonly make the following errors when working with sets.

Correct	Wrong
1. $\{\ \}$ or $\varnothing$ is the empty set.	$\{\varnothing\}$ is used to represent the empty set.
2. $3 \in \{1, 2, 3\}$	~~$3 \subseteq \{1, 2, 3\}$~~ Since there are no braces around the 3, it is not a set. 3 is an element of the set, not a subset.
3. $\{3\} \subseteq \{1, 2, 3\}$	~~$\{3\} \in \{1, 2, 3\}$~~ Since there are braces around the 3, $\{3\}$ is a set. Thus $\{3\}$ is a subset, not an element, of the set $\{1, 2, 3\}$.

4 The number line (Fig. 1.2) can be used to illustrate sets of numbers.

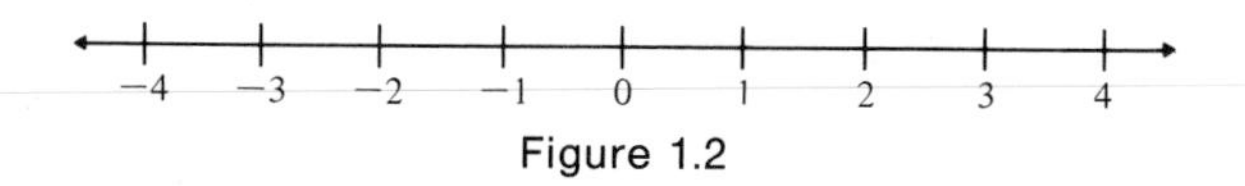

Figure 1.2

Some important sets of numbers are summarized below.

Important Sets of Numbers

Real Numbers	$\{x \mid x \text{ is a point on the number line}\}$
Natural or Counting Numbers	$\{1, 2, 3, 4, 5, \ldots\}$
Whole Numbers	$\{0, 1, 2, 3, 4, 5, \ldots\}$
Integers	$\{\ldots, -3, -2, -1, 0, 1, 2, 3, \ldots\}$
Rational Numbers	$\left\{\frac{p}{q} \mid p \text{ and } q \text{ are integers}, q \neq 0\right\}$
Irrational Numbers	$\{x \mid x \text{ is a real number that is not rational}\}$

Let us briefly look at the rational, irrational, and real numbers. A rational number is any number that can be represented as a quotient of two integers, with the denominator not zero.

Examples of rational numbers

$$\frac{3}{5}, \quad \frac{-2}{3}, \quad 0, \quad 1.63, \quad 7, \quad -12, \quad \sqrt{4}$$

Notice that 0, or any other integer, is also a rational number since it can be written as a fraction with a denominator of 1.

$$0 = \frac{0}{1} \qquad 7 = \frac{7}{1} \qquad -12 = \frac{-12}{1}$$

The number 1.63 can be written $\frac{163}{100}$, and is thus a quotient of two integers. Since $\sqrt{4} = 2$, and 2 is an integer, $\sqrt{4}$ is a rational number. *Every rational number when changed to a decimal number will be either a repeating or a terminating decimal number.*

Examples of repeating decimals	*Examples of terminating decimals*
$\frac{2}{3} = 0.6666\ldots$	$2 = 2.0$
$\frac{7}{3} = 2.3333\ldots$	$\frac{1}{2} = 0.5$
$\frac{1}{7} = 0.142857142857\ldots$ (the block 142857 repeats)	$\frac{7}{4} = 1.75$

Although $\sqrt{4}$ is a rational number, most square roots are not. Most square roots will be neither terminating nor repeating decimals when expressed as a decimal number, and are irrational numbers. Some irrational numbers are $\sqrt{2}, \sqrt{3}, \sqrt{5}, \sqrt{6}$, and so on. Another irrational number is pi, π. When we give a decimal value for an irrational number we are giving only an *approximation* of the value of the irrational number. The symbol $\approx$ means "is approximately equal to."

$$\sqrt{2} \approx 1.41$$

5 The **real numbers** are formed by taking the union of the rational numbers with the irrational numbers. Therefore, any real number must be either a rational number or an irrational number. The symbol $\mathbb{R}$ is often used to represent the set of real numbers. Figure 1.3 illustrates various real numbers on the number line.

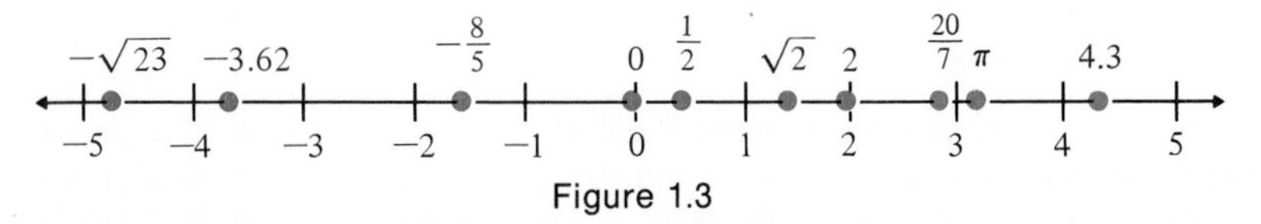

Figure 1.3

Figure 1.4 illustrates the relationship between the various sets of numbers. Earlier we stated that the natural numbers were a subset of the set of whole numbers.

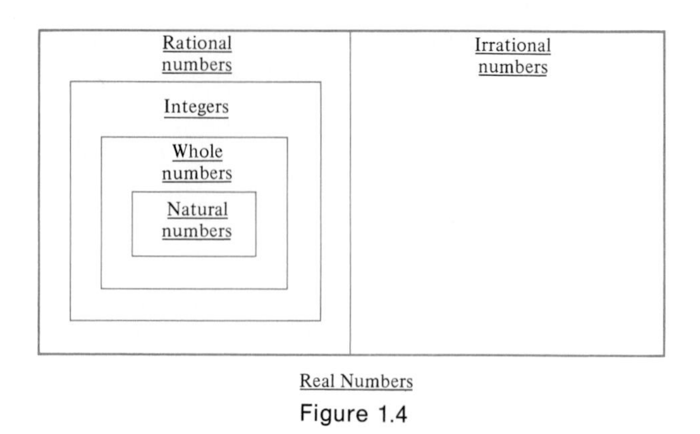

Figure 1.4

In Figure 1.4 you see that the set of natural numbers is a subset of the set of whole numbers, the set of integers, and the set of rational numbers. Therefore, every natural number must also be a whole number, integer, and rational number. Using the same reasoning, we can see that the set of whole numbers is a subset of the set of integers and the set of rational numbers; and the set of integers is a subset of the set of rational numbers.

EXAMPLE 3 Consider the following set of elements.

$$\left\{-4, 0, \frac{3}{5}, 1.8, \sqrt{3}, -\sqrt{5}, \frac{19}{6}, 18, 4.62, -23, \pi\right\}$$

List the elements of the set that are:

(a) Natural numbers. (b) Whole numbers. (c) Integers.
(d) Rational numbers. (e) Irrational numbers. (f) Real numbers.

Solution:
(a) 18
(b) 0, 18
(c) $-4, 0, 18, -23$
(d) $-4, 0, \frac{3}{5}, 1.8, \frac{19}{6}, 18, 4.62, -23$
(e) $\sqrt{3}, -\sqrt{5}, \pi$
(f) $-4, 0, \frac{3}{5}, 1.8, \sqrt{3}, -\sqrt{5}, \frac{19}{6}, 18, 4.62, -23, \pi$ ■

Not all numbers are real numbers. Some numbers that we discuss later in the text that are not real numbers include complex numbers and imaginary numbers.

Exercise Set 1.1

List each set in roster form.

1. $A = \{x|x$ is a natural number between 3 and $8\}$
2. $B = \{x|x$ is an even integer between 5 and $10\}$
3. $C = \{x|x$ is an even integer greater than or equal to 6 and less than or equal to $10\}$
4. $D = \{x|x$ is a natural number greater than $5\}$
5. $E = \{x|x$ is a whole number less than $7\}$
6. $F = \{x|x$ is a whole number less than or equal to $7\}$
7. $G = \{x|x$ is a natural number less than $0\}$
8. $H = \{x|x$ is a whole number multiple of $5\}$
9. $I = \{x|x$ is an integer greater than $-5\}$
10. $J = \{x|x$ is an integer between -6 and $-1\}$
11. $K = \{x|x$ is a whole number between 3 and $4\}$

Insert either $\in$ or $\notin$ in the space provided, to make the statement true.

12. $6 \quad \{1, 3, 6, 9\}$
13. $4 \quad \{1, 2, 3\}$
14. $0 \quad \{-1, 1, 3, 5\}$
15. $72 \quad \{1, 2, 3, 4, \ldots, 80\}$
16. $-12 \quad \{1, 2, 3, \ldots\}$
17. $\{3\} \quad \{1, 2, 3, 4\}$
18. $\{5\} \quad \{1, 2, 3, 4, 5, 6\}$
19. $\{0\} \quad \{-3, -2, -1, 0, 1, 2, 3\}$
20. $\{-5\} \quad \{-4, -3, -2\}$

Insert either $\subseteq$ or $\nsubseteq$ in the space provided, to make the statement true.

21. $\{3\} \quad \{1, 2, 3, 4\}$
22. $\{0\} \quad \{0, 1\}$
23. $\{5\} \quad \{-1, -2, -3\}$
24. $\{4\} \quad \{3, 4, 5, 6, \ldots\}$
25. $\{72\} \quad \{6, 7, 8, \ldots, 70\}$
26. $\{-1\} \quad \{-1, -2, -3\}$
27. $\{-1\} \quad \{-4, -3, -2\}$
28. $3 \quad \{1, 2, 3\}$
29. $0 \quad \{-1, 0, 1\}$
30. $5 \quad \{1, 2, 3, 4\}$

Let N = the set of natural numbers, W = the set of whole numbers, I = the set of integers, Q = the set of rational numbers, H = the set of irrational numbers, $\mathbb{R}$ = the set of real numbers. Insert either $\subseteq$ or $\nsubseteq$ to make the statement true.

31. $N \quad W$
32. $W \quad Q$
33. $I \quad Q$
34. $W \quad N$
35. $Q \quad H$
36. $Q \quad \mathbb{R}$
37. $H \quad \mathbb{R}$
38. $Q \quad I$
39. $Q \quad N$
40. $\mathbb{R} \quad Q$
41. $H \quad Q$
42. $I \quad N$
43. $N \quad I$
44. $N \quad \mathbb{R}$

Answer true or false.

45. 0 is a real number.
46. 0 is a rational number.
47. 0 is a natural number.
48. 0 is a whole number.
49. Some rational numbers are integers.
50. Some irrational numbers are rational numbers.
51. Some natural numbers are negative numbers.
52. Some whole numbers are natural numbers.
53. Every natural number is a whole number.
54. Every whole number is a natural number.
55. Every integer is a rational number.
56. Every rational number is an integer.
57. The union of the set of rational numbers with the set of irrational numbers forms the set of real numbers.
58. The intersection of the set of rational numbers with the set of irrational numbers is the empty set.
59. The set of natural numbers is a finite set.
60. The set of whole numbers is an infinite set.
61. The set of integers between 1 and 2 is the null set.
62. The set of rational numbers between 1 and 2 is an infinite set.

Consider the set of elements $\left\{-6, 4, \frac{1}{2}, \frac{5}{9}, 0, \sqrt{7}, \sqrt{5}, -1.23, \frac{99}{100}\right\}$

List the elements of the set that are:

63. Natural numbers
64. Whole numbers
65. Integers
66. Rational numbers
67. Irrational numbers
68. Real numbers

Consider the set of elements $\left\{2, 4, -5.33, \frac{9}{2}, \sqrt{7}, \sqrt{2}, -100, -7, 4.7\right\}$

List the elements of the set that are:

69. Whole numbers
70. Natural numbers
71. Rational numbers
72. Integers
73. Real numbers
74. Irrational numbers

Find $A \cup B$ and $A \cap B$ for each set A and B.

75. $A = \{1, 2, 3\}, B = \{2, 3, 4\}$
76. $A = \{2, 4, 6, 8\}, B = \{1, 3, 5, 7\}$
77. $A = \{-2, -4, -5\}, B = \{-1, -2, -4, -6\}$
78. $A = \{\ \}, B = \{1, 2, 3\}$
79. $A = \{\ \}, B = \{0, 1, 2, 3\}$
80. $A = \{-1, 0, 1\}, B = \{0, 2, 4, 6\}$

81. $A = \{2, 4, 6\}$, $B = \{2, 4, 6, 8, \ldots\}$

82. $A = \{1, 3, 5\}$, $B = \{1, 3, 5, 7, \ldots\}$

83. $A = \{0, 2, 4, 6, 8\}$, $B = \{1, 3, 5, 7\}$

84. $A = \{1, 2, 3, 4, \ldots\}$, $B = \{0, 1, 2, 3, 4, \ldots\}$

85. $A = \{1, 2, 3, 4, \ldots\}$, $B = \{2, 4, 6, 8, \ldots\}$

1.2 Properties of the Real Numbers

1 *Know the properties of the real numbers.*

2 *Know the multiplication property of 0 and the double negative property.*

The properties discussed in this section are important for an understanding of algebra. You should understand them thoroughly before moving on to the next section. When introducing the properties, letters called **variables** will be used to represent numbers. In this section the letters a, b, and c will be used. In algebra the letters x, y, and z are most often used to represent variables, as will be seen later in the text.

In algebra we do not use an $\times$ to indicate multiplication for it may be confused with the variable x. To indicate multiplication a dot may be used. When a number and a variable, or two variables are placed next to one another multiplication is also indicated. For example, both $2 \cdot x$ and $2x$ mean two times x, and both $x \cdot y$ and xy mean x times y.

The term **algebraic expression,** or simply **expression** will be used often in the text. An expression is any combination of numbers, variables, exponents, mathematical symbols, and mathematical operations.

1 Table 1.1 lists the basic properties for the operations of addition and multiplication on the real numbers.

Table 1.1 PROPERTIES OF REAL NUMBERS

For real numbers *a*, *b*, and *c*:	Addition	Multiplication
Commutative Property	$a + b = b + a$	$ab = ba$
Associative Property	$(a + b) + c = a + (b + c)$	$(ab)c = a(bc)$
Identity Property	$a + 0 = 0 + a = a$ (0 is called the **additive identity element**)	$a \cdot 1 = 1 \cdot a = a$ (1 is called the **multiplicative identity element**)
Inverse Property	$a + (-a) = (-a) + a = 0$ ($-a$ is called the **additive inverse** or **opposite** of a.)	$a \cdot \frac{1}{a} = \frac{1}{a} \cdot a = 1$ ($1/a$ is called the **multiplicative inverse** or **reciprocal** of a.)
Distributive Property (of multiplication over addition)	$a(b + c) = ab + ac$	

The distributive property applies when there are more than two numbers within the parentheses.

$$a(b + c + d + \cdots + n) = ab + ac + ad + \cdots + an$$

This expanded form of the distributive property is called the **extended distributive property.**

Note that the commutative property involves a change in *order*, and the associative property involves a change in *grouping*.

EXAMPLE 1 Name the properties illustrated.

(a) $6 \cdot x = x \cdot 6$ (b) $(x + 2) + 3 = x + (2 + 3)$
(c) $x + 3 = 3 + x$ (d) $3(x + 2) = 3x + 3(2) = 3x + 6$

Solution:
(a) Commutative property of multiplication.
(b) Associative property of addition.
(c) Commutative property of addition.
(d) Distributive property. ■

EXAMPLE 2 Name the properties illustrated.

(a) $4 \cdot 1 = 4$ (b) $x + 0 = x$
(c) $4 + (-4) = 0$ (d) $1(x + y) = x + y$

Solution:
(a) Identity property of multiplication.
(b) Identity property of addition.
(c) Inverse property of addition.
(d) Identity property of multiplication. ■

EXAMPLE 3 Write the additive inverse (or opposite) and multiplicative inverse (or reciprocal) of the following.

(a) -3 (b) $\frac{2}{3}$

Solution:
(a) Additive inverse: 3; multiplicative inverse: $\dfrac{1}{-3} = -\dfrac{1}{3}$

(b) Additive inverse: $\dfrac{-2}{3}$; multiplicative inverse: $\dfrac{1}{\frac{2}{3}} = \dfrac{3}{2}$. ■

2 Two other important properties are given below.

For any real number a:

Multiplication Property of 0

$$a \cdot 0 = 0 \cdot a = 0$$

Double negative Property

$$-(-a) = a$$

EXAMPLE 4 Name the following properties.

(a) $3 \cdot 0 = 0$
(b) $-(-4) = 4$
(c) $-(-x) = x$

Solution:
(a) Multiplication property of 0.
(b) Double negative property.
(c) Double negative property. ■

Exercise Set 1.2

Name the property.

1. $4 + 2 = 2 + 4$
2. $x + y = y + x$
3. $3(4 + 5) = 3 \cdot 4 + 3 \cdot 5$
4. $3(x + 2) = 3x + 6$
5. $(3 + 6) + 2 = 3 + (6 + 2)$
6. $(3 + x) + 5 = 3 + (x + 5)$
7. $x + 2 = 2 + x$
8. $(x + 3) + 6 = x + (3 + 6)$
9. $3 \cdot x = x \cdot 3$
10. $x = 1 \cdot x$
11. $x + 0 = x$
12. $2(x + 2) = 2x + 4$
13. $2 + (-4) = (-4) + 2$
14. $x(y + z) = xy + xz$
15. $3y + 4 = 4 + 3y$
16. $(2x \cdot 3y) \cdot 4y = 2x \cdot (3y \cdot 4y)$
17. $3x + 2y = 2y + 3x$
18. $3(x + y) = 3x + 3y$
19. $4(x + y + 2) = 4x + 4y + 8$
20. $3(2b) = (3 \cdot 2)b$
21. $-(-1) = 1$
22. $5 + 0 = 5$
23. $5 \cdot 1 = 5$
24. $4 \cdot \frac{1}{4} = 1$
25. $3 + (-3) = 0$
26. $6 \cdot 0 = 0$
27. $x + (-x) = 0$
28. $x \cdot \frac{1}{x} = 1$
29. $-4 + 4 = 0$
30. $(x + y) = 1(x + y)$
31. $(x + 2) = 1(x + 2)$
32. $\left(-\frac{1}{2}\right)(-2) = 1$
33. $3 \cdot \frac{1}{3} = 1$
34. $1(x + y) = 1x + 1y$
35. $-\left(-\frac{1}{2}\right) = \frac{1}{2}$
36. $-5 + 5 = 0$
37. $3 \cdot 0 = 0$
38. $x \cdot 0 = 0$
39. $-(-3) = 3$
40. $-(-x) = x$

Fill in the statement on the right side of the equal sign using the property indicated.

41. $x + 3 =$ commutative property of addition
42. $2 \cdot y =$ commutative property of multiplication
43. $(x + 2) + 3 =$ associative property of addition
44. $3(2x + 5) =$ distributive property
45. $x + 0 =$ identity property of addition
46. $3(x + y + 4) =$ distributive property

47. $1 \cdot x =$ identity property of multiplication

48. $(x \cdot 3) \cdot 4 =$ associative property of multiplication

49. $-(-x) =$ double negative property

50. $5x + (2y + 3x) =$ associative property of addition

51. $a \cdot 0 =$ multiplication property of zero

52. $5(x + y) =$ distributive property

53. $1(x + y) =$ distributive property

54. $0 + x =$ identity property of addition

55. $3 \cdot 1 =$ identity property of multiplication

56. $a \cdot \frac{1}{a} =$ inverse property of multiplication

57. $a + (-a) =$ inverse property of addition

58. $-(-2) =$ double negative property

List both the additive inverse and the multiplicative inverse for each of the following.

59. 4 **60.** 6 **61.** -3 **62.** -5 **63.** $\frac{2}{3}$

64. $\frac{5}{2}$ **65.** -6 **66.** -12 **67.** $-\frac{3}{7}$ **68.** $-\frac{2}{9}$

1.3 Inequalities and Absolute Value

1. *Identify and use inequality symbols.*
2. *Evaluate expressions containing absolute value.*

1 The symbols used to indicate an inequality are $>$, $\geq$, $<$, $\leq$, and $\neq$.

Inequality Symbols

$>$ is read "greater than."
$\geq$ is read "greater than or equal to."
$<$ is read "less than."
$\leq$ is read "less than or equal to."
$\neq$ is read "not equal to."

The number a is greater than the number b, $a > b$, when a is to the right of b on the number line (Fig. 1.5). We can also state that the number b is less than a, $b < a$, since b is to the left of a on the number line. The inequality $a \neq b$ means either $a < b$ or $a > b$.

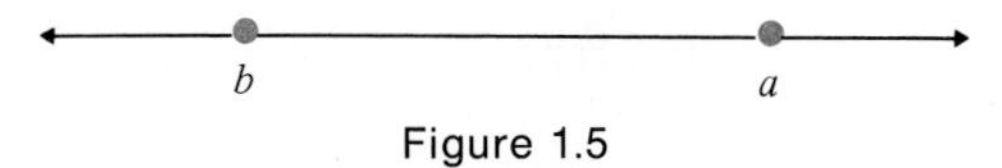

Figure 1.5

EXAMPLE 1 Insert either > or < to make the statement true.

(a) 7 3 (b) −4 −2 (c) 0 −2

Solution: Draw a number line illustrating the location of all the given points (Fig. 1.6).

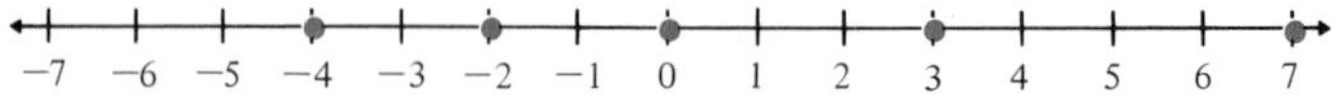

Figure 1.6

(a) $7 > 3$ since 7 is to the right of 3 on the number line
(b) $-4 < -2$ since −4 is to the left of −2 on the number line
(c) $0 > -2$ since 0 is to the right of −2 on the number line ■

Remember that the symbol used in an inequality, if it is true, always points to the smaller of the two numbers.

Absolute Value

2 The **absolute value** of a number is its distance from the number zero on the number line. The symbol $| \ |$ is used to indicate absolute value.

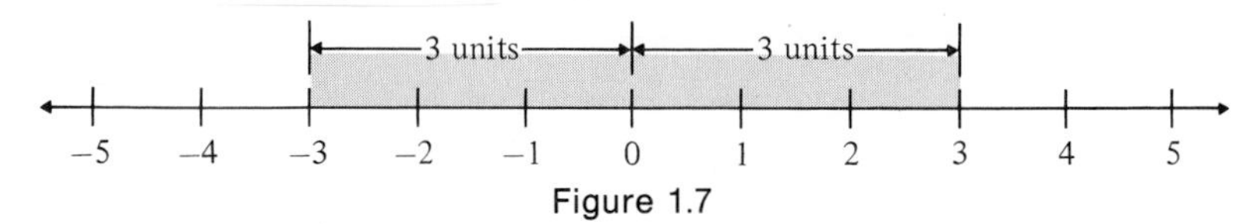

Figure 1.7

Consider the numbers 3 and −3 (see Fig. 1.7). Both numbers are 3 units from zero on the number line. Thus

$$|3| = 3$$
$$|-3| = 3$$

EXAMPLE 2 Evaluate each of the following.

(a) $|4|$ (b) $|-8|$ (c) $|0|$ (d) $\left|\frac{-5}{3}\right|$

Solution:
(a) $|4| = 4$ since 4 is 4 units from zero on the number line
(b) $|-8| = 8$ since −8 is 8 units from zero on the number line
(c) $|0| = 0$
(d) $\left|-\frac{5}{3}\right| = \frac{5}{3}$ ■

The absolute value of a number can also be found by the following definition.

Absolute value

If a represents any real number, then

$$|a| = \begin{cases} a & \text{if } a \geq 0 \\ -a & \text{if } a < 0 \end{cases}$$

The statement above says that the absolute value of any nonnegative number is the number itself, and the absolute value of any negative number is the additive inverse (or opposite) of the number. For example:

$\lvert 3\rvert = 3$	since 3 is greater than or equal to 0
$\lvert 8\rvert = 8$	since 8 is greater than or equal to 0
$\lvert 0\rvert = 0$	since 0 is greater than or equal to 0
$\lvert -4\rvert = -(-4) = 4$	since -4 is less than 0
$\lvert -12\rvert = -(-12) = 12$	since -12 is less than 0
$\lvert -5\rvert = -(-5) = 5$	since -5 is less than 0

The absolute value of any nonzero number will always be a positive number, and the absolute value of zero is zero.

EXAMPLE 3 Insert $>$, $<$, or $=$ between the two values to make the statement true.

(a) $\lvert -3\rvert \quad \lvert 3\rvert$ (b) $-3 \quad \lvert -4\rvert$ (c) $\lvert 6\rvert \quad \lvert 8\rvert$
(d) $-\lvert 4\rvert \quad \lvert -3\rvert$ (e) $-\lvert -6\rvert \quad \lvert -5\rvert$

Solution:
(a) $\lvert -3\rvert = \lvert 3\rvert$, since both $\lvert -3\rvert$ and $\lvert 3\rvert$ equal 3.
(b) $-3 < \lvert -4\rvert$, since $\lvert -4\rvert = 4$ and $-3 < 4$.
(c) $\lvert 6\rvert < \lvert 8\rvert$, since $\lvert 6\rvert = 6$ and $\lvert 8\rvert = 8$.
(d) $-\lvert 4\rvert < \lvert -3\rvert$, since $-\lvert 4\rvert = -4$ and $\lvert -3\rvert = 3$.
(e) $-\lvert -6\rvert < \lvert -5\rvert$, since $-\lvert -6\rvert = -6$ and $\lvert -5\rvert = 5$. ■

Exercise Set 1.3

Insert either $<$ or $>$ between the two numbers to make the statement true.

1. $4 \quad 2$ **2.** $6 \quad 8$ **3.** $5 \quad 3$ **4.** $-2 \quad 4$
5. $5 \quad -3$ **6.** $0 \quad -1$ **7.** $0 \quad 4$ **8.** $9 \quad -2$
9. $-1 \quad 1$ **10.** $-\frac{1}{2} \quad -1$ **11.** $-3 \quad -3.5$ **12.** $1 \quad -3$
13. $-4 \quad -2$ **14.** $-6 \quad -1$ **15.** $-2 \quad -5$ **16.** $-1 \quad -4$
17. $4 \quad -4$ **18.** $3 \quad -3.5$ **19.** $1.1 \quad 1.9$ **20.** $-1.1 \quad -1.9$
21. $15 \quad -12$ **22.** $-18 \quad -9$ **23.** $-36 \quad -42$ **24.** $-81 \quad -70$

Evaluate the absolute value expression.

25. $\lvert 6\rvert$ **26.** $\lvert 3\rvert$ **27.** $\lvert -4\rvert$ **28.** $\lvert -6\rvert$
29. $\lvert -2\rvert$ **30.** $\lvert 0.5\rvert$ **31.** $\left\lvert -\frac{1}{2}\right\rvert$ **32.** $\lvert -3\rvert$
33. $\lvert 0\rvert$ **34.** $\lvert -22\rvert$ **35.** $\lvert 45\rvert$ **36.** $\lvert 9.6\rvert$
37. $\lvert -13.84\rvert$ **38.** $\lvert -0.7\rvert$ **39.** $-\lvert 7\rvert$ **40.** $-\lvert -7\rvert$
41. $-\lvert -3\rvert$ **42.** $-\lvert -8\rvert$ **43.** $-\lvert 18\rvert$ **44.** $-\lvert -27\rvert$
45. $-\lvert 8\rvert$

Insert $<$, $>$, or $=$ between the two numbers to make the statement true.

46. $\lvert 6\rvert \quad \lvert -6\rvert$ **47.** $\lvert 9\rvert \quad \lvert 3\rvert$ **48.** $\lvert -9\rvert \quad \lvert 3\rvert$
49. $-4 \quad \lvert -4\rvert$ **50.** $\lvert -10\rvert \quad -5$ **51.** $-4 \quad -\lvert 4\rvert$

52. $|-3|\quad|-2|$
53. $|-4|\quad 6$
54. $|-20|\quad -|24|$
55. $|-16|\quad -|30|$
56. $-|4|\quad -|8|$
57. $-|-31|\quad |-5|$
58. $6\quad |-12|$
59. $|25|\quad |-23|$
60. $|19|\quad |-25|$
61. $-|-3|\quad |-9|$
62. $|24|\quad -18$
63. $-25\quad -|34|$

List the values from smallest to largest.

64. $6, 2, -1, |3|, |-5|$
65. $0, -4, -8, -|12|, |-10|$
66. $4, -2, 8, |-6|, -|3|$
67. $|9|, |4|, |-12|, |3|, |-5|$
68. $4, -2, -8, 0, 9$
69. $-|5|, -|6|, -|7|, |-8|, |-9|$
70. $-3, |0|, |-5|, |7|, |-12|$
71. $5, -|7|, -9, |15|, |-1|$
72. $12, 24, |36|, |-9|, |-45|$
73. $-8, -12, -|9|, -|20|, -|-18|$
74. $6, -4, -|-6|, |-8|, -2$
75. $3, 3.1, 3.4, |3.9|, |-3.6|$
76. $-2.1, -2, -2.4, |-2.8|, -|2.9|$
77. $-7, -7.1, -7.8, -|7.3|, |-7.4|$

Find the unknown number(s) in each of the following.

78. The absolute value of the number 6.
79. The absolute value of the number -8
80. All numbers whose absolute value is 4
81. All numbers whose absolute value is 10
82. All numbers such that if 6 is added to the number, the absolute value of the result is 11.
83. All numbers such that if 2 is added to the number, the absolute value of the result is 5,
84. All numbers a such that $|a| = |-a|$
85. All numbers a such that $|a| = a$
86. All numbers a such that $|a| = -a$
87. All numbers a such that $|a| = -3$
88. All numbers a such that $|a| = 5$
89. All numbers x such that $|x - 3| = |3 - x|$

1.4 Addition, Subtraction, Multiplication, and Division of Real Numbers

1. *Add real numbers.*
2. *Subtract real numbers.*
3. *Multiply real numbers.*
4. *Divide real numbers.*
5. *Write fractions with positive denominators.*

Addition of Real Numbers

1 To be successful in algebra you must have an understanding of how to add and subtract real numbers. We will use absolute value in explaining how real numbers may be added or subtracted. We first discuss how to add two numbers with the same sign, either both positive or both negative, and then, how to add two numbers with different signs, one positive and the other negative.

To Add Two Numbers with the Same Sign (Both Positive or Both Negative)

Add their absolute values and place the common sign before the sum.

The sum of two positive numbers will be a positive number and the sum of two negative numbers will be a negative number.

EXAMPLE 1 Add $-2 + (-5)$.

Solution: Since both numbers being added are negative, the sum will be negative. To find the sum, add the absolute values of these numbers and place a negative sign before the answer.

$$-2 + (-5)$$

$$\begin{aligned} |-2| &= 2 \\ |-5| &= \underline{5} \\ & 7 \end{aligned}$$

Since both numbers are negative, the sum must be negative. Thus

$$-2 + (-5) = -7 \quad ■$$

EXAMPLE 2 Add $-16 + (-25)$.

Solution:

$$-16 + (-25)$$

$$\begin{aligned} |-16| &= 16 \\ |-25| &= \underline{25} \\ & 41 \end{aligned}$$

Since both numbers are negative, the sum must be negative.

$$-16 + (-25) = -41 \quad ■$$

To Add Two Numbers with Different Signs (One Positive and the Other Negative)

Take the difference between the absolute value. The answer is positive if the positive number has the larger absolute value. The answer is negative if the negative number has the larger absolute value.

The sum of a positive number and a negative number may be either positive or negative. The sign of the answer will be the same as the sign of the number with the larger absolute value.

EXAMPLE 3 Add $5 + (-2)$.

Solution: Since the numbers being added are of opposite signs, we take the difference between the smaller absolute value and the larger. First evaluate each absolute value.

$$\begin{aligned} |5| &= 5 \\ |-2| &= 2 \end{aligned}$$

Now find the difference

$$\begin{array}{r} 5 \\ \underline{-2} \\ 3 \end{array}$$

The number 5 has a larger absolute value than the number -2, so the answer is positive.

$$5 + (-2) = 3 \quad ■$$

EXAMPLE 4 Add $-5 + 2$.

Solution:

$$|-5| = 5 \qquad 5$$
$$|2| = 2 \qquad \underline{-2}$$
$$3$$

The number -5 has a larger absolute value than the number 2, so the answer is negative.

$$-5 + 2 = -3 \quad ■$$

EXAMPLE 5 Evaluate $6 + (-8)$.

Solution:

$$|6| = 6 \qquad 8$$
$$|-8| = 8 \qquad \underline{-6}$$
$$2$$

The number -8 has a larger absolute value than the number 6, so the answer is negative.

$$6 + (-8) = -2 \quad ■$$

EXAMPLE 6 A submarine descended 400 feet below sea level. A short while later it descended another 250 feet. What was the submarine's depth with respect to sea level?

Solution: Consider motion in the downward direction to be negative and motion in the upward direction to be positive.

$$\text{distance} = -400 + (-250) = -650 \text{ feet}$$

The submarine is 650 feet below sea level. ■

Subtraction of Real Numbers

2 Consider the subtraction problem $5 - 2$. To evaluate this problem, we must subtract a positive 2 from a positive 5.

positive 2 (arrow pointing to $+2$)

$$5 - 2 \text{ means } 5 - (+2)$$

subtract (arrow pointing to each minus sign) subtract

Every subtraction problem can be expressed as an addition problem using the following rule:

> **Subtraction of Real Numbers**
>
> $$a - b = a + (-b)$$

The rule above says that **to subtract b from a, add the opposite (or additive inverse) of b to a.**

In the problem $5 - 2$, which means $5 - (+2)$, the opposite of $+2$ is -2. Thus

$$5 - 2 = 5 + (-2)$$

subtract positive 2 add negative 2

We can now find the sum of $5 + (-2)$ to be 3 using the method for adding real numbers presented earlier in this section. Therefore, $5 - 2 = 3$ ■

EXAMPLE 7 Evaluate $6 - 10$.

Solution: $6 - 10 = 6 + (-10) = -4$ ■

EXAMPLE 8 Evaluate $-8 - 4$.

Solution: $-8 - 4 = -8 + (-4) = -12$ ■

EXAMPLE 9 Evaluate $8 - (-10)$.

Solution: This problem is somewhat different since we are subtracting a negative number. The procedure to evaluate remains the same.

$$8 - (-10) = 8 + 10 = 18$$

subtract negative 10 add positive 10

Thus $8 - (-10) = 18$. ■

In Section 1.2 we showed that for any real number a,

$$\boldsymbol{-(-a) = a}$$

We can use this principle to evaluate such problems as $8 - (-10)$. Note that $-(-10) = 10$.

$$8 - (-10) = 8 + 10 = 18$$

EXAMPLE 10 Evaluate $-4 - (-12)$.

Solution: $-4 - (-12) = -4 + 12 = 8$ ■

EXAMPLE 11 Subtract 35 from -42.

Solution: $-42 - 35 = -77$ ■

EXAMPLE 12 Subtract -8 from -30.

Solution: $-30 - (-8) = -30 + 8 = -22$ ■

EXAMPLE 13 The highest point in the United States, Mt. McKinley in Alaska, is 20,320 feet above sea level. The lowest point in the United States, the Verdigris River in Kansas, is 680 feet below sea level. Find the vertical height difference between the two locations.

Solution: $20{,}320 - (-680) = 20{,}320 + 680 = 21{,}000$ feet ■

Addition and subtraction are often combined in the same problems, as the following examples illustrate. Unless parentheses are present, we evaluate the expression from left to right. When parentheses are used, we evaluate the expression within the parentheses first, then we evaluate from left to right.

EXAMPLE 14 Simplify $-2-(4-8)-3$.

Solution:

$$\begin{aligned} -2-(4-8)-3 &= -2-(-4)-3 \\ &= -2+4-3 \\ &= 2-3 \\ &= -1 \quad \blacksquare \end{aligned}$$

EXAMPLE 15 Simplify $6-(-2)+(7-13)-9$.

Solution:

$$\begin{aligned} 6-(-2)+(7-13)-9 &= 6+2+(-6)-9 \\ &= 8+(-6)-9 \\ &= 2-9 \\ &= -7 \quad \blacksquare \end{aligned}$$

EXAMPLE 16 Simplify $2-|-3|+4-(6-|-3|)$.

Solution: Begin by replacing the numbers in absolute value signs with their numerical equivalent, then evaluate.

$$\begin{aligned} 2-|-3|+4-(6-|-3|) &= 2-3+4-(6-3) \\ &= 2-3+4-(3) \\ &= -1+4-3 \\ &= 3-3 \\ &= 0 \quad \blacksquare \end{aligned}$$

Multiplication of Real Numbers

3 The following rules are used in determining the sign of the product when two numbers are multiplied.

Multiplication of Real Numbers

1. The product of two numbers with **like** signs is a **positive** number.
2. The product of two numbers with **unlike** signs is a **negative** number.

EXAMPLE 17 Evaluate $(4)(-3)$.

Solution: Since the numbers being multiplied have unlike signs (the first positive and the second negative) the product is negative.

$$(4)(-3) = -12 \quad \blacksquare$$

EXAMPLE 18 Evaluate $(-4)(-8)$.

Solution: Since both numbers being multiplied have the same sign (both negative) the product is positive.

$$(-4)(-8) = 32 \quad ■$$

EXAMPLE 19 Evaluate $\left(\frac{-3}{5}\right)\left|\frac{-6}{7}\right|$.

Solution: $\left|\frac{-6}{7}\right| = \frac{6}{7}$; therefore

$$\left(\frac{-3}{5}\right)\left|\frac{-6}{7}\right| = \left(\frac{-3}{5}\right)\left(\frac{6}{7}\right) = \frac{-3 \cdot 6}{5 \cdot 7} = \frac{-18}{35} \quad ■$$

EXAMPLE 20 Evaluate $4(-2)(-3)(1)$.

Solution:

$$\begin{aligned} 4(-2)(-3)(1) &= (-8)(-3)(1) \\ &= 24(1) \\ &= 24 \quad ■ \end{aligned}$$

When multiplying more than two numbers, the product will be negative when there are an odd number of negative numbers. The product will be positive when there are an even number of negative numbers.

Division of Real Numbers

4 The rules for division of real numbers are very similar to those for multiplication of real numbers.

Division of Real Numbers

1. The quotient of two numbers with **like** signs is a **positive** number.
2. The quotient of two numbers with **unlike** signs is a **negative number.**

EXAMPLE 21 Evaluate $-24 \div 6$.

Solution: Since the numbers have unlike signs, the quotient will be negative.

$$\frac{-24}{6} = -4 \quad ■$$

EXAMPLE 22 Evaluate $-64 \div (-4)$.

Solution: Since the numbers have like signs, both negative, the quotient is positive.

$$\frac{-64}{-4} = 16 \quad ■$$

EXAMPLE 23 Evaluate $\dfrac{-3}{8} \div \left|\dfrac{-2}{5}\right|$.

Solution: $\left|\dfrac{-2}{5}\right| = \dfrac{2}{5}$.

Thus

$$\frac{-3}{8} \div \left|\frac{-2}{5}\right| = \frac{-3}{8} \div \frac{2}{5}$$

Now invert the divisor and proceed as in multiplication.

$$\begin{aligned}\frac{-3}{8} \div \frac{2}{5} &= \frac{-3}{8} \cdot \frac{5}{2}\\ &= \frac{-3 \cdot 5}{8 \cdot 2}\\ &= -\frac{15}{16}\end{aligned}$$ ■

5 When the denominator of a fraction is a negative number, we generally rewrite the fraction with a positive denominator. To do this we make use of the following fact.

> If a and b represent any real numbers, $b \neq 0$, then
>
> $$\frac{a}{-b} = \frac{-a}{b} = -\frac{a}{b}$$

Thus when we have a quotient of $\dfrac{1}{-2}$, we rewrite it as either $\dfrac{-1}{2}$ or $-\dfrac{1}{2}$.

Exercise Set 1.4

Evaluate.

1. $4 + (-3)$
2. $9 + (-8)$
3. $12 + (-2)$
4. $14 + (-7)$
5. $-3 + 8$
6. $-4 + 12$
7. $-9 + 17$
8. $-36 + 19$
9. $-16 - (-5)$
10. $-32 - (-14)$
11. $35 - (-4)$
12. $-60 - 45$
13. $-53 - (-19)$
14. $19 + (-23)$
15. $4 - 8$
16. $7 + (-9)$
17. $5 - 7$
18. $-14 + 8$
19. $-36 + 5$
20. $40 - 24$
21. $24 - 40$
22. $-24 + 40$

Evaluate.

23. $4 + 6 - 3$
24. $7 - 4 - 8$
25. $2 - 3 - (-2)$
26. $5 + (-3) + (-4)$
27. $-6 - 4 - 2$
28. $(4 - 3) + (6 - 8)$
29. $-3 + (4 - 9) + 3$
30. $-2 + (4 - 6) - (3 - 8)$
31. $9 - (4 - 3) - (-2 - 1)$
32. $8 + (-4) + (-3 + 1)$
33. $-6 - 6 - (6 + 6) - 3$
34. $3 - (-4) + 6 - 3$

35. $-(-4+2)+(-6+3)+2$
36. $4-(8-9)+(-6+8)$
37. $|4|-|3|+|1|$
38. $6-|3|+4$
39. $3-|-8|-5$
40. $|9-4|-6$
41. $|6-9|-5$
42. $|12-5|-|5-12|$
43. $|6|-|-2|+3-8$
44. $-|-3|-|7|+(6+|-2|)$
45. $(-|3|+|5|)-(6-|-9|)$
46. $5-|-2|+3-|-5|$
47. $4-|8|+(4-6)-|12|$
48. $-(-9-4)-|-3|+2$

Evaluate.

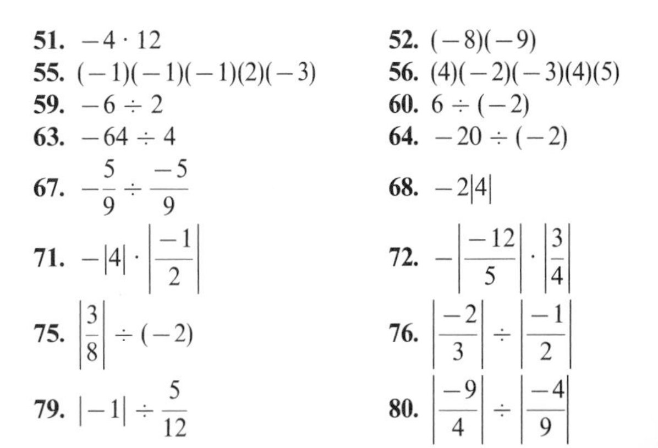

49. $4(-6)$
50. $-2(-8)$
51. $-4\cdot 12$
52. $(-8)(-9)$
53. $4(-12)$
54. $-1\cdot 80$
55. $(-1)(-1)(-1)(2)(-3)$
56. $(4)(-2)(-3)(4)(5)$
57. $(4)(-1)(6)(-2)(-2)$
58. $(1)(-3)(-4)(2)$
59. $-6\div 2$
60. $6\div(-2)$
61. $-3\div(-3)$
62. $-16\div 8$
63. $-64\div 4$
64. $-20\div(-2)$
65. $36\div(-4)$
66. $-80\div(-10)$
67. $-\frac{5}{9}\div\frac{-5}{9}$
68. $-2|4|$
69. $-3|8|$
70. $\left|-\frac{1}{2}\right|\cdot\left|\frac{-3}{4}\right|$
71. $-|4|\cdot\left|\frac{-1}{2}\right|$
72. $-\left|\frac{-12}{5}\right|\cdot\left|\frac{3}{4}\right|$
73. $\left|\frac{3}{5}\right|\cdot\left|\frac{-10}{6}\right|$
74. $\left|\frac{-4}{7}\right|\div\frac{1}{14}$
75. $\left|\frac{3}{8}\right|\div(-2)$
76. $\left|\frac{-2}{3}\right|\div\left|\frac{-1}{2}\right|$
77. $\frac{-5}{9}\div|-5|$
78. $\frac{-3}{8}\div\left|\frac{-5}{4}\right|$
79. $|-1|\div\frac{5}{12}$
80. $\left|\frac{-9}{4}\right|\div\left|\frac{-4}{9}\right|$

81. A submarine dives 350 feet. A short time later the submarine comes up 180 feet. Find the submarine's distance with respect to its starting point (consider distance in a downward direction as negative).

82. In New York City, the temperature during a 24-hour period dropped from 46°F to -12°F. Find the change in temperature.

83. Sue made a profit of \$3225 in the stock market over a given period. During the same period Nelson lost \$1088. Find the difference in their performances in the stock market.

84. On their first play, the East High football team lost 32 yards. On their second play, they gained 25 yards. What is the gain or loss for the two plays?

See Exercise 82.

Answer true if the statement is always true. If the statement is not always true, answer false.

85. The product of an odd number of negative numbers is a negative number.

86. The sum of two positive numbers is always a positive number.

87. The difference of two positive numbers is always a positive number.

88. The product of two positive numbers is always a positive number.

89. The quotient of two positive numbers is always a positive number.

90. The product of two negative numbers is always a negative number.

91. The product of two negative numbers is always a positive number.

92. The sum of two negative numbers is always a positive number.

93. The sum of a positive number and a negative number is always a negative number.

94. The product of a positive number and a negative number is always a negative number.

95. The difference of two negative numbers is always a negative number.

96. The difference of a positive number and a negative number is always a negative number.

97. The quotient of a positive number and a negative number is always a negative number.

98. The product of an even number of negative numbers is a negative number.

JUST FOR FUN

1. Evaluate $1 - 2 + 3 - 4 + \cdots + 99 - 100$. (*Hint:* Group in pairs of two numbers.)
2. Evaluate
$1 + 2 - 3 + 4 + 5 - 6 + 7 + 8 - 9 + 10 + 11 - 12 + \cdots + 22 + 23 - 24$.
(*Hint:* Examine in groups of three numbers.)
3. Evaluate $\dfrac{(1) \cdot |-2| \cdot (-3) \cdot |4| \cdot (-5)}{|-1| \cdot (-2) \cdot |-3| \cdot (4) \cdot |-5|}$.
4. Evaluate $\dfrac{(1)(-2)(3)(-4)(5) \cdots (97)(-98)}{(-1)(2)(-3)(4)(-5) \cdots (-97)(98)}$.
5. Evaluate $\dfrac{|-3| \cdot |6| \cdot |-1|(-|-2|)}{|2| \cdot (-|3|)}$.

1.5 Exponents and Roots

1 *Use exponents.*

2 *Use roots.*

To understand certain topics in algebra, you must know exponents and roots. We discuss exponents in more depth in Sections 5.1 and 5.2 and roots (or radicals) in Chapter 8. We introduce them here so that we can discuss two important topics, the use of parentheses and priority of operations.

1 If a, b, and c are integers, and $a \cdot b = c$, then a and b are said to be **factors** of c. If a and b are factors of c, then both a and b will divide c without remainder. For example,

$$\underset{\text{factor}}{\underset{\uparrow}{2}} \cdot \underset{\text{factor}}{\underset{\uparrow}{3}} = 6$$

Both 2 and 3 are factors of 6 since each divides 6 without a remainder.

$$\frac{\text{product} \rightarrow 6}{\text{factor} \rightarrow 2} = \underset{\text{factor}}{\underset{\uparrow}{3}} \qquad \frac{\text{product} \rightarrow 6}{\text{factor} \rightarrow 3} = \underset{\text{factor}}{\underset{\uparrow}{2}}$$

In the expression 3^2, the 3 is called the **base** and the 2 is called the **exponent.** 3^2 is read "three squared" or "three to the second power" and means

$$3^2 = \underbrace{3 \cdot 3}_{\text{2 factors of 3}}$$

The number 5^3 is read "five cubed" or "five to the third power" and means

$$5^3 = \underbrace{5 \cdot 5 \cdot 5}_{\text{3 factors of 5}}$$

In general, the number b to the nth power, written b^n, means

$$b^n = \underbrace{b \cdot b \cdot b \cdot b \cdot \cdots \cdot b}_{n \text{ factors of } b}$$

EXAMPLE 1 Evaluate each of the following.

(a) 4^2 (b) 5^3 (c) $(-2)^5$ (d) 1^{10} (e) $\left(\frac{-3}{4}\right)^3$

Solution:

(a) $4^2 = 4 \cdot 4 = 16$

(b) $5^3 = 5 \cdot 5 \cdot 5 = 125$

(c) $(-2)^5 = (-2)(-2)(-2)(-2)(-2) = -32$

(d) $1^{10} = 1$; 1 raised to any power will equal 1. Why?

(e) $\left(\frac{-3}{4}\right)^3 = \left(\frac{-3}{4}\right)\left(\frac{-3}{4}\right)\left(\frac{-3}{4}\right) = \frac{-27}{64}$ ■

COMMON STUDENT ERROR

Students should realize that $a^b \neq a \cdot b$.

Correct	Wrong
$2^4 = 2 \cdot 2 \cdot 2 \cdot 2 = 16$	~~$2^4 = 2 \cdot 4 = 8$~~

It is not necessary to write exponents of 1. Whenever we encounter a numerical value or a variable without an exponent, we assume that it has an exponent of 1. Thus 3 means 3^1, x means x^1, x^3y means x^3y^1, and $-xy$ means $-x^1y^1$.

A rule that we will be using in Chapter 5 is that any nonzero number, or letter, raised to the 0 power has a value of 1.

Zero Exponent Property

$$a^0 = 1 \qquad \text{for } a \neq 0$$

Examples of zero exponent property

$$3^0 = 1, \qquad x^0 = 1.$$

When a term contains an exponent, that exponent acts only on the one variable or number that directly precedes it, unless parentheses are used. For example, in the expression $2x^0$ the exponent acts only on the variable x: $2x^0 = 2(x^0) = 2(1) = 2$. In the expression $(2x)^0$ since the 2 and the x are in parentheses, the exponent acts on both the 2 and the x: $(2x)^0 = 1$.

EXAMPLE 2 Evaluate.

(a) 5^0 (b) $(2x)^0$ (c) $3x^0$ (d) $-4x^0$ (e) -2^0

Solution:
(a) $5^0 = 1$
(b) $(2x)^0 = 1$ Notice that the entire expression is raised to the 0 power because of the parentheses.
(c) $3x^0 = 3(1) = 3$ Notice that only the variable x is raised to the 0 power.
(d) $-4x^0 = -4(1) = -4$
(e) $-2^0 = -(2^0) = -1$ Notice that the exponent refers only to the 2 and not the negative sign preceding it ■

EXAMPLE 3 Evaluate $-x^2$ for the following values of x.

(a) 3 (b) -3

Solution: (a) $-x^2 = -(3^2) = -9$ (b) $-x^2 = -(-3)^2 = -(9) = -9$ ■

COMMON STUDENT ERROR

A negative sign directly preceding an expression that is raised to a power has the effect of negating that expression.

$$-3^2 \quad \text{means} \quad -(3^2) \quad \text{and not} \quad (-3)^2.$$
$$-x^2 \quad \text{means} \quad -(x^2) \quad \text{and not} \quad (-x)^2.$$

Example: Evaluate.

(a) -5^2 (b) $(-5)^2$

Solution:
(a) $-5^2 = -(5^2) = -25$
(b) $(-5)^2 = (-5)(-5) = 25$

Note that $-5^2 \neq (-5)^2$ since $-25 \neq 25$. Note also that $-x^2$ will always be a negative number for any nonzero value of x and that $(-x)^2$ will always be a positive number for any nonzero value of x. ■

EXAMPLE 4 Evaluate $-3^2 + (-1)^3 - 4^3 + (3 - 4)^0$.

Solution:
$$\begin{aligned} -3^2 + (-1)^3 - 4^3 + (3-4)^0 &= -(3^2) + (-1)^3 - (4^3) + (3-4)^0 \\ &= -9 + (-1) - 64 + 1 \\ &= -9 - 1 - 64 + 1 \\ &= -73 \end{aligned}$$ ■

Roots

2 The **principal or positive square root** of a number n, written $\sqrt{n}$, is the positive number that when multiplied by itself gives n.

EXAMPLE 5 Evaluate.

(a) $\sqrt{4}$ (b) $\sqrt{25}$ (c) $\sqrt{\dfrac{9}{4}}$

Solution: (a) $\sqrt{4} = 2$, since $2 \cdot 2 = 4$.
(b) $\sqrt{25} = 5$, since $5 \cdot 5 = 25$.
(c) $\sqrt{\frac{9}{4}} = \frac{3}{2}$, since $\frac{3}{2} \cdot \frac{3}{2} = \frac{9}{4}$. ■

In Section 1.1 we stated that the square root of 4, $\sqrt{4}$, is a rational number since it is equal to 2. The square roots of certain other numbers such as $\sqrt{2}$, $\sqrt{3}$, and $\sqrt{5}$ are irrational numbers. The decimal values of such numbers can never be given exactly since irrational numbers are non-terminating non-repeating decimal numbers. The approximate value of $\sqrt{2}$, and other irrational numbers, can be found in Appendix B or with a calculator.

$$\sqrt{2} \approx 1.41 \qquad \text{from Appendix B}$$
$$\sqrt{2} \approx 1.414213562 \qquad \text{from a calculator}$$

The concept used to explain square root can be expanded to explain cube roots and higher roots. The cube root of a number n is written $\sqrt[3]{n}$.

$$\sqrt[3]{n} = b \quad \text{if} \quad \underbrace{b \cdot b \cdot b}_{3 \text{ factors of } b} = n$$

For example, $\sqrt[3]{8} = 2$, because $2 \cdot 2 \cdot 2 = 8$. The expression $\sqrt[m]{n}$ is read the mth root of n.

$$\sqrt[m]{n} = b \quad \text{if} \quad \underbrace{b \cdot b \cdot b \cdot \cdots \cdot b}_{m \text{ factors of } b} = n$$

EXAMPLE 6 Evaluate.

(a) $\sqrt[3]{27}$ (b) $\sqrt[3]{64}$ (c) $\sqrt[4]{1}$ (d) $\sqrt[4]{16}$

Solution: (a) $\sqrt[3]{27} = 3$, since $3 \cdot 3 \cdot 3 = 27$.
(b) $\sqrt[3]{64} = 4$, since $4 \cdot 4 \cdot 4 = 64$.
(c) $\sqrt[4]{1} = 1$, since $1 \cdot 1 \cdot 1 \cdot 1 = 1$.
(d) $\sqrt[4]{16} = 2$, since $2 \cdot 2 \cdot 2 \cdot 2 = 16$. ■

EXAMPLE 7 Evaluate.

(a) $\sqrt[4]{81}$ (b) $\sqrt[3]{\frac{1}{27}}$ (c) $\sqrt[3]{-8}$ (d) $\sqrt[3]{-1}$

Solution: (a) $\sqrt[4]{81} = 3$, since $3 \cdot 3 \cdot 3 \cdot 3 = 81$.
(b) $\sqrt[3]{\frac{1}{27}} = \frac{1}{3}$, since $\left(\frac{1}{3}\right)\left(\frac{1}{3}\right)\left(\frac{1}{3}\right) = \frac{1}{27}$.
(c) $\sqrt[3]{-8} = -2$, since $(-2)(-2)(-2) = -8$.
(d) $\sqrt[3]{-1} = -1$, since $(-1)(-1)(-1) = -1$. ■

Note that in Example 7 (c) and (d) the cube root of a negative number is negative. Why is this so?

Calculator Corner

Exponents and Roots on a Calculator

To square a number on a calculator, you can multiply the number by itself. For example, to find 5^2, key in

[c] 5 [×] 5 [=] 25

A similar procedure can be used to evaluate other exponential expressions. Thus 6^4 can be found by pressing the following keys:

Many calculators have keys that simplify finding the value of an exponential expression. The [x^2] key can be used to square a number. The expression 5^2 can be evaluated on these calculators by pressing

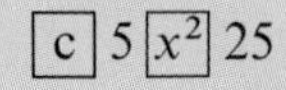

Other calculators have a [y^x] key. This key can be used to evaluate exponential expressions. When using the [y^x] key, first enter the base, y, then press the [y^x] key, then enter the exponent x, then press the [=] key. For example, to evaluate 6^4, follow the the following sequence.

[c] 6 [y^x] 4 [=] 1296

The square roots of numbers can be found on calculators with a square-root key, [$\sqrt{x}$]. To evaluate $\sqrt{25}$ on calculators that have this key, press

[c] 25 [$\sqrt{x}$] 5

Higher roots can be found on calculators that contain either one of two keys, [y^x] or [$\sqrt[x]{y}$]. To evaluate $\sqrt[4]{625}$ on a calculator having a [$\sqrt[x]{y}$] key, do the following:

[c] 625 [$\sqrt[x]{y}$] 4 [=] 5

Note that the radicand 625 is entered, then the [$\sqrt[x]{y}$] key is pressed, and then the root (or index) 4 is entered. When the [=] key is pressed, the answer 5 is displayed.

To evaluate $\sqrt[4]{625}$ on a calculator with a [y^x] key, follow this procedure:

[c] 625 [INV] [y^x] 4 [=] 5

Note that the radicand 625 is entered, then the inverse key [INV] is pressed, then the [y^x] key is pressed, and then the root 4 is entered. After the [=] key is pressed, the answer 5 is displayed.

Exercise Set 1.5

Evaluate.

1. 3^2 **2.** 4^2 **3.** 2^3 **4.** 1^5

5. 3^3 **6.** 5^2 **7.** 6^3 **8.** $(-2)^2$

9. $(-2)^3$
10. $(-3)^4$
11. $(-1)^3$
12. $(-5)^3$
13. $(-2)^5$
14. $\left(\frac{1}{3}\right)^4$
15. $\left(\frac{2}{3}\right)^3$
16. $\left(\frac{-3}{5}\right)^3$

Evaluate.

17. 6^0
18. x^0
19. $4x^0$
20. $(2x)^0$
21. $-3y^0$
22. $\left(\frac{1}{2}x\right)^0$
23. -7^0
24. $8x^0$

Evaluate.

25. $\sqrt{1}$
26. $\sqrt{9}$
27. $\sqrt{16}$
28. $\sqrt{25}$
29. $\sqrt{64}$
30. $\sqrt{169}$
31. $\sqrt{\frac{25}{36}}$
32. $\sqrt{\frac{81}{4}}$
33. $\sqrt{\frac{1}{4}}$
34. $\sqrt{\frac{4}{9}}$
35. $\sqrt{\frac{225}{81}}$
36. $\sqrt{\frac{9}{100}}$

Evaluate.

37. $\sqrt[3]{64}$
38. $\sqrt[3]{1}$
39. $\sqrt[3]{-8}$
40. $\sqrt[3]{-27}$
41. $\sqrt[3]{-64}$
42. $\sqrt[4]{16}$
43. $\sqrt[4]{1}$
44. $\sqrt[3]{8}$
45. $\sqrt[3]{125}$
46. $\sqrt[3]{-125}$
47. $\sqrt[3]{-216}$
48. $\sqrt[3]{\frac{1}{8}}$
49. $\sqrt[3]{\frac{1}{64}}$
50. $\sqrt[3]{\frac{1}{27}}$

Evaluate (a) x^2 and (b) $-x^2$ for the given value of x.

51. 3
52. 4
53. 1
54. -2
55. -1
56. -5
57. -3
58. 6
59. -4
60. 5

Evaluate (a) x^3 and (b) $-x^3$, for the given value of x.

61. 3
62. 2
63. 1
64. -5
65. -3
66. -1
67. -2
68. -4
69. 4
70. 5

Evaluate.

71. $4^2 + 3^2 - 2^2$
72. $(4-1)^2 + 2^3 - (-1)^2$
73. $2^2 + 3^2 + (-4)^2$
74. $(-1)^3 + 1^3 + 1^{10} + (-1)^{12}$
75. $(3-2)^3 + (2-3)^3$
76. $(-3)^3 - 2^2 - (-2)^2 + (4-4)^2$
77. $-2^2 - (2)^3 + (4-3)^0 + (2-4)^3$
78. $(-2)^2 + (-3)^2 + (-3)^3 - 4^2$
79. $(5-2)^2 + (2-5)^2 - 6^2 + 2^4$
80. $-2^4 - 3^2 - 1^6 + (4-2)^3$
81. $-(5-3)^2 - (2-4)^3 + 3^3$
82. $(-5-1)^2 + (3-2)^0 + (-4)^3$
83. $-(4-6)^2 + |3-4| + 2^2$
84. $|5-2| - \sqrt{4} - 3^2$
85. $5 - 2^0 + \sqrt{16} - (2-1)^3$
86. $|6-3| + (6-3)^2 - \sqrt{25}$
87. $(-3-1)^2 + (4-3)^5 + \sqrt{36} - |-19|$
88. $(4-2)^3 - (2-4)^2 + |7-11| - 3$
89. Explain why $\sqrt{-4}$ cannot be a real number.
90. Explain why an odd root of a positive number will be positive.
91. Explain why an odd root of a negative number will be negative.

JUST FOR FUN

1. Judy offers to give Karl \$1000 a day for a month (30 days) if he will give her 1 cent the first day and double the amount each day for the month.
(a) Write the amount that Karl will give Judy (in pennies) in exponential form.
(b) Determine the amount found in part (a) in dollars.
(c) After 30 days, what will be Judy's net profit or loss?

1.6 Priority of Operations

1 *Know the order of operations.*

2 *Use parentheses or brackets correctly.*

1 What is $2 + 3 \cdot 4$ equal to? Is it 20? Is it 14? To be able to answer questions of this type, we must know the priority of operations when evaluating a mathematical expression. You will often have to evaluate expressions containing multiple operations. To do so, follow the order of operations indicated below.

To Evaluate Mathematical Expressions

Use the following order:

1. First, evaluate the information within parentheses, ( ), or brackets, [ ]. If the expression contains nested parentheses (one pair of parentheses within another pair), evaluate the information in the innermost parentheses first.
2. Next, evaluate all terms containing exponents and roots.
3. Next, evaluate all multiplications and divisions moving from left to right.
4. Finally, evaluate all additions and subtractions moving from left to right.

Note that when evaluating expressions containing a fraction bar, we work separately above and below the fraction bar.

We can now answer the question posed above. Since multiplications are performed before additions:

$$2 + 3 \cdot 4 \quad \text{means} \quad 2 + (3 \cdot 4) = 2 + 12 = 14$$

2 Parentheses or brackets may be used (1) to change the order of operations to be followed in evaluating an algebraic expression or (2) to help clarify the understanding of an expression.

In the example above, $2 + 3 \cdot 4$, if we wished to have the addition performed before the multiplication, we could indicate this by placing parentheses about the $2 + 3$.

$$(2 + 3) \cdot 4 = 5 \cdot 4 = 20$$

Consider the expression $1 \cdot 3 + 2 \cdot 4$. According to the order of operations, multiplications are to be performed before additions. We can rewrite this expression as $(1 \cdot 3) + (2 \cdot 4)$. Notice that we did not change the priority of operations. The parentheses only help clarify the order to be followed.

Brackets are sometimes used in place of parentheses to help avoid confusion. For example, the expression $7((5 \cdot 3) + 6)$ may be easier to follow when written $7[(5 \cdot 3) + 6]$.

EXAMPLE 1 Evaluate $8 + 3 \cdot 5^2 - 7$.

Solution: Since there are no parentheses, we first evaluate 5^2.

$$8 + 3 \cdot 5^2 - 7 = 8 + 3 \cdot 25 - 7$$

Next, perform multiplications and divisions from left to right.

$$= 8 + 75 - 7$$

Next, perform additions and subtractions from left to right.

$$= 83 - 7$$
$$= 76 \quad ■$$

EXAMPLE 2 Evaluate $36 + 3[(12 - 4) \div 2]$.

Solution: First, evaluate the information in the innermost parentheses.

$$36 + 3[(12 - 4) \div 2] = 36 + 3[8 \div 2]$$

Next, evaluate the information within the brackets.

$$= 36 + 3(4)$$

Next, multiply; then add.

$$= 36 + 12$$
$$= 48 \quad ■$$

EXAMPLE 3 Evaluate $(4 \div 2) + [4(5 - 2)]^2$.

Solution: First, evaluate the information within parentheses.

$$\begin{aligned}(4 \div 2) + [4(5 - 2)]^2 &= 2 + [4(3)]^2 \\ &= 2 + (12)^2 \\ &= 2 + 144 \\ &= 146 \quad ■\end{aligned}$$

EXAMPLE 4 Evaluate $16 \div 8 \cdot 4 - 6^2 \div 2^2$.

Solution: Begin by squaring the 6 and the 2. Then perform the multiplications and divisions, moving from left to right.

$$\begin{aligned}16 \div 8 \cdot 4 - 6^2 \div 2^2 &= 16 \div 8 \cdot 4 - 36 \div 4 \\ &= 2 \cdot 4 - 36 \div 4 \\ &= 8 - 36 \div 4 \\ &= 8 - 9 \\ &= -1 \quad ■\end{aligned}$$

EXAMPLE 5 Evaluate $\dfrac{6 \div 2 + 5|7 - 3|}{2 + (3 - 5) \div 2}$.

Solution: Work separately above the fraction bar and below the fraction bar.

$$\begin{aligned} \frac{6 \div 2 + 5|7 - 3|}{2 + (3 - 5) \div 2} &= \frac{6 \div 2 + 5|4|}{2 + (-2) \div 2} \\ &= \frac{3 + 20}{2 + (-1)} \\ &= \frac{23}{1} = 23 \quad \blacksquare \end{aligned}$$

EXAMPLE 6 Evaluate $4x^2 - 2$ when $x = 3$.

Solution: Substitute 3 for each x in the expression, then evaluate.

$$\begin{aligned} 4x^2 - 2 &= 4(3)^2 - 2 \\ &= 4(9) - 2 \\ &= 36 - 2 \\ &= 34 \quad \blacksquare \end{aligned}$$

EXAMPLE 7 Evaluate $6 - (3x + 1) + 2x^2$ when $x = 4$.

Solution: Substitute 4 for each x in the expression, then evaluate.

$$\begin{aligned} 6 - (3x + 1) + 2x^2 &= 6 - [3(4) + 1] + 2(4)^2 \\ &= 6 - [12 + 1] + 2(16) \\ &= 6 - (13) + 32 \\ &= -7 + 32 \\ &= 25 \quad \blacksquare \end{aligned}$$

EXAMPLE 8 Evaluate $-x^3 - xy - y^2$ when $x = -2$ and $y = 5$.

Solution: Substitute -2 for each x and 5 for each y in the expression, then evaluate.

$$\begin{aligned} -x^3 - xy - y^2 &= -(-2)^3 - (-2)(5) - (5)^2 \\ &= -(-8) - (-10) - 25 \\ &= 8 + 10 - 25 \\ &= -7 \quad \blacksquare \end{aligned}$$

☐ *Calculator Corner*

We now know that $2 + 3 \cdot 4$ means $2 + (3 \cdot 4)$ and has a value of 14. What will a calculator display if you key in the following?

$$\boxed{c}\ 2\ \boxed{+}\ 3\ \boxed{\times}\ 4\ \boxed{=}$$

The answer depends on your calculator. Many calculators evaluate problems using the priority of operations discussed in this section.

Calculators using priority of operations:

$$\boxed{c}\ 2\ \boxed{+}\ 3\ \boxed{\times}\ 4\ \boxed{=}\ 14$$

Other calculators, generally the less expensive ones, will perform operations in the order they are entered.

Calculators that do not follow priority of operations:

$$\boxed{c}\ 2\ \boxed{+}\ 3\ \boxed{\times}\ 4\ \boxed{=}\ 20$$

Remember that in algebra, unless otherwise instructed by parentheses, we always perform multiplications and divisions before additions and subtractions.

Does your calculator use the priority of operations discussed in this section?

Exercise Set 1.6

Evaluate.

1. $6 + 4 \cdot 5$

2. $2 - 2^2 + 4$

3. $4^2 - 2 + 5$

4. $(6^2 - 2) \div (\sqrt{36} - 4)$

5. $6 \div 2 + 5 \cdot 4$

6. $4 \cdot 3 - 4 \cdot 5$

7. $24 \cdot 2 \div 4 \div 6$

8. $6 \div 2 + 36 \cdot 3$

9. $(\sqrt{4} - 3) \cdot (5 - 1)^3$

10. $20 - 6 \div 3 - 4$

11. $3 \cdot 7 + 4 \cdot 2$

12. $6 + \sqrt{9}(3 + 4)$

13. $2[1 - (4 \cdot 5)] + 6$

14. $[12 \div (4 \div 2)] - 5$

15. $(3^2 - 1) \div (3 + 1)$

16. $-4(5 - 2)^3 + 5$

17. $3[(4 + 6)^2 - 2]$

18. $2[3(8 - 2) \div 6]^3$

19. $\dfrac{6 - 4 \div 2}{8 - 3 + 6}$

20. $\dfrac{15 \div 3 + 2 \cdot 2}{\sqrt{25} \div 5 + 8 \div 2}$

21. $\dfrac{4 \cdot 6 \div 8 + 5}{3^2 - 4 \cdot 2 + 3}$

22. $\dfrac{4 \div 2 \cdot 3^2 - 1}{5 - (3 + 4)^2}$

23. $\dfrac{4 - (2 + 3)^2 - 6}{4(3 - 2) - 3^2}$

24. $\dfrac{6 \div 2 + 5^2 + 3}{4 - (-3 + 5) - 4^2}$

25. $\dfrac{2(-3) + 4 \cdot 5 - 3^2}{5 + \sqrt{4}(2^2 - 1)}$

26. $\dfrac{-(-3) + (-2) - (-4) - 3}{6 - (-2) + 3(5) - 4}$

27. $\dfrac{8 - 4 \div 2 \cdot 3 - 4}{5^2 - 3^2 \cdot 2 - 6}$

28. $\dfrac{8 - [4 - (3 - 1)^2]}{5 - (-3)^2 + 4 \div 2}$

29. $-2|-3 - 5| - 4$

30. $-3|-4 + 2| - 6 \cdot 3$

31. $3|4 - 6| - 2|-4 - 2| + 3^2$

32. $-2|-3|-6 \div |2| + 3^2$

33. $12 - 15 \div |5| + 2(|4| - 2)^2$

34. $\dfrac{4 - |-12| \div |3|}{2(4 - |5|) + 9}$

35. $\dfrac{6 - 2|9 - 4| + 8}{4 - |-4| + 4^2 \div 2^2}$

36. $\dfrac{6 - |-3| - 4|6 - 2|}{5 - 6 \cdot 2 \div |-4|}$

Evaluate.

37. $-3x^2 - 4$ when $x = 1$

38. $2x^2 + x$ when $x = 3$

39. $5x^2 - 2x + 5$ when $x = 3$

40. $-3x^2 + 6x + 5$ when $x = 5$

41. $3(x - 2)^2$ when $x = 7$

42. $4(x + 1)^2 - 6x$ when $x = 5$

43. $4(x-3)(x+4)$ when $x=1$

44. $3x(x-1)+5$ when $x=-4$

45. $-6x+3y$ when $x=2, y=4$

46. $6x+3y-5$ when $x=1, y=-3$

47. $4(x+y)^2+4x-3y$ when $x=2, y=-3$

48. $(4x^2-3y)-5$ when $x=4, y=-2$

49. $3(a+b)^2+4(a+b)-6$ when $a=4, b=-1$

50. $4xy-6x^2+3$ when $x=5, y=2$

51. $x^3y^2-6xy+3x$ when $x=2, y=3$

52. $\frac{6x^2}{3}+\frac{2x}{2}$ when $x=2$

53. $\frac{1}{2}(x^2+y^2-2xy)$ when $x=2, y=-3$

54. $\frac{5x}{3}-\frac{6y}{4}+3$ when $x=6, y=3$

55. $x^2y^4-y^3+3(x+y)$ when $x=2, y=-1$

56. $(x-3)^2+(y-5)^2$ when $x=-2, y=3$

57. $\frac{x^2}{25}+\frac{y^2}{9}$ when $x=0, y=2$

58. $\frac{(x-3)^2}{9}+\frac{(y+5)^2}{16}$ when $x=4, y=3$

Write an algebraic expression for each of the following. Then evaluate the expression for the given value of the variable.

59. Multiply the variable x by 3. To this product add 6. Now square this sum. Find the value of this expression when $x=3$.

60. Subtract 3 from x. Square this difference. Subtract 5 from this value. Now square this value. Find the value of this expression when $x=-1$.

61. 6 is added to the product of 3 and x. This expression is then multiplied by 6. 9 is then subtracted from this product. Find the value of the expression when $x=3$.

62. The sum of x and y is multiplied by 2. 5 is subtracted from this product. This expression is then squared. Find the value of the expression when $x=2$ and $y=-3$.

63. 3 is added to x. This sum is divided by twice y. This quotient is then squared. Finally, 3 is subtracted from this expression. Find the value of the expression when $x=5$ and $y=2$.

JUST FOR FUN

1. Evaluate $[(3 \div 6)^2+4]^2+3 \cdot 4 \div 12 \div 3$.

2. Evaluate $[4[3(x^2-2)]^2+4]^2$ when $x=2$.

3. Evaluate $[-2(3x^2+4)^2]^2 \div (3x^2-2)$ when $x=-2$.

4. Evaluate $\dfrac{2x+4-y\left(2+\frac{3}{x}\right)}{\frac{y-2}{6}+\frac{3x^2}{4}}$ when $x=2, y=3$.

5. Evaluate $\dfrac{3\left(\left|\frac{3}{5}\right|-\left|-\frac{2}{3}\right|\right)}{6\left|-\frac{1}{2}\right|-2\left|\frac{1}{3}\right|}$.

Summary

Glossary

Absolute value: The distance of a number from 0 on the number line. The absolute value of any real number will be greater than or equal to 0.

Additive identity element: 0.

Additive inverse: For any number a its additive inverse is $-a$.

Algebraic expression (or expression): Any combination of numbers, variables, exponents, mathematical symbols and operations.

Empty or null set: A set containing no elements, symbolized $\varnothing$ or $\{ \}$.

Finite set: A set that contains a finite number of elements.

Inequality symbols: $<, \leq, >, \geq, \neq$.

Infinite set: A set that has an infinite number of elements.

Intersection of sets: The intersection of set A and set B, $A \cap B$, is the set of elements which belong to both set A and set B.

Multiplicative identity element: 1.

Multiplicative inverse: For any number a, $a \neq 0$, its multiplication inverse is $1/a$.

Principal (or positive) square root: The principal square root of a number n, written $\sqrt{n}$, is the positive number that when multiplied by itself gives n.

Set: A collection of objects or elements.

Subset: Set A is a subset of set B, $A \subseteq B$, if every element of set A is also an element of set B.

Union of sets: The union of sets A and B, $A \cup B$, is the set of elements which belong to either set A or set B.

Variable: A letter used to represent a number.

Important Facts

Sets of Numbers

Real Numbers	$\{x \mid x \text{ is a point on the number line}\}$
Natural or counting numbers	$\{1, 2, 3, 4, 5, \ldots\}$
Whole numbers	$\{0, 1, 2, 3, 4, \ldots\}$
Integers	$\{\ldots, -3, -2, -1, 0, 1, 2, 3, \ldots\}$
Rational numbers	$\left\{\frac{p}{q} \,\middle|\, p \text{ and } q \text{ are integers, } q \neq 0\right\}$
Irrational numbers	$\{x \mid x \text{ is a real number that is not rational}\}$

Properties of the Real Number System

Commutative properties	$a + b = b + a, \quad ab = ba$
Associative properties	$(a + b) + c = a + (b + c), \quad (ab)c = a(bc)$
Identity properties	$a + 0 = 0 + a = a, \quad a \cdot 1 = 1 \cdot a = a$
Inverse properties	$a + (-a) = (-a) + a = 0, \quad a \cdot \frac{1}{a} = \frac{1}{a} \cdot a = 1$
Distributive property	$a(b + c) = ab + ac$
Multiplication property of 0	$a \cdot 0 = 0 \cdot a = 0$
Double-negative property	$-(-a) = a$

Absolute Value

$$|a| = \begin{cases} a, & a \geq 0 \\ -a, & a < 0 \end{cases}$$

Zero Exponent Property

$$a^0 = 1, \quad a \neq 0$$

Review Exercises

[1.1] List each set in roster form.

1. $A = \{x | x \text{ is a natural number between 2 and 7}\}$
2. $B = \{x | x \text{ is a whole number multiple of 3}\}$

Place either $\in$ or $\notin$ in the space provided to make the statement true.

3. $0 \quad \{0, 1, 2, 3\}$
4. $\{3\} \quad \{0, 1, 2, 3\}$
5. $\{5\} \quad \{4, 5, 6\}$
6. $8 \quad \{1, 2, 3, 4, \ldots\}$

Place either $\subseteq$ or $\not\subseteq$ in the space provided to make the statement true.

7. $\{3\} \quad \{1, 2, 3\}$
8. $0 \quad \{0, 1, 2, 3, \ldots\}$
9. $5 \quad \{3, 4, 5, 6\}$
10. $\{8\} \quad \{1, 2, 3, 4, \ldots\}$

Let N = set of natural numbers, W = set of whole numbers, I = set of integers, Q = set of rational numbers, H = set of irrational numbers, $\mathbb{R}$ = set of real numbers. Insert either $\subseteq$ or $\not\subseteq$ to make the statement true.

11. $N \quad W$
12. $Q \quad \mathbb{R}$
13. $I \quad Q$
14. $N \quad I$
15. $H \quad \mathbb{R}$
16. $Q \quad H$

Consider the set of numbers $\{-3, 4, 6, \frac{1}{2}, \sqrt{5}, \sqrt{3}, 0, \frac{15}{27}, -\frac{1}{5}, 1.47\}$. List the elements of the set that are:

17. Natural numbers
18. Whole numbers
19. Integers
20. Rational numbers
21. Irrational numbers
22. Real numbers

Answer true or false.

23. $\sqrt{3}$ is an irrational number.
24. A real number cannot be divided by 0.
25. $\dfrac{0}{1}$ is not a real number.
26. Every rational number and every irrational number is a real number.
27. $0, \frac{3}{5}, -2$, and 4 are all rational numbers.

Find $A \cup B$ and $A \cap B$ for each set A and B.

28. $A = \{1, 2, 3, 4, 5\}, \quad B = \{2, 3, 4, 5\}$
29. $A = \{3, 5, 7, 9\}, \quad B = \{2, 4, 6, 8\}$
30. $A = \{1, 2, 3, 4, \ldots\}, \quad B = \{2, 4, 6, \ldots\}$
31. $A = \{4, 6, 9, 10, 11\}, \quad B = \{3, 5, 9, 10, 12\}$

[1.2] Name the given property.

32. $3 + 4 = 4 + 3$
33. $x + 4 = 4 + x$
34. $3(x + 2) = 3x + 6$
35. $xy = yx$
36. $(x + 3) + 2 = x + (3 + 2)$
37. $a + 0 = a$
38. $(3x)y = 3(xy)$
39. $a \cdot 1 = a$
40. $-(-5) = 5$
41. $3(0) = 0$
42. $4 + 0 = 4$
43. $5 \cdot 1 = 5$
44. $x + (-x) = 0$
45. $x \cdot \dfrac{1}{x} = 1$
46. $(-2)\left(-\dfrac{1}{2}\right) = 1$
47. $(x + y) = 1(x + y)$

Fill in the statement on the right side of the equal sign using the property given.

48. $x + 3 =$ commutative property
49. $3(x + 5) =$ distributive property
50. $(x + 6) + (-4) =$ associative property
51. $3 \cdot x =$ commutative property

52. $(9 \cdot x) \cdot y =$ associative property

53. $4(x - y + 5) =$ distributive property

54. $a + 0 =$ identity property

55. $1 \cdot a =$ identity property

56. $a + (-a) =$ inverse property

57. $a \cdot \frac{1}{a} =$ inverse property

58. $-(-a) =$ double-negative property

[1.3] Insert either <, >, or = between the two numbers to make the statement true.

59. $3 \quad 2$

60. $1 \quad 4$

61. $-2 \quad 3$

62. $-4 \quad -6$

63. $-8 \quad 0$

64. $-4 \quad -3.9$

65. $1.06 \quad 1.6$

66. $-1.06 \quad -1.6$

67. $|3| \quad 3$

68. $|-3| \quad 3$

69. $|4| \quad |6|$

70. $|-4| \quad |-6|$

71. $13 \quad |-5|$

72. $|-12| \quad 4$

73. $|8| \quad -9$

74. $-|-2| \quad -5$

Write the numbers from smallest to largest.

75. $4, -2, -5, |7|$

76. $0, \frac{3}{5}, 2.3, |-3|$

77. $|-7|, |-5|, 3, -2$

78. $-4, -2, -2.1, -|3|$

79. $-4, 6, -|-3|, 5$

80. $|1.6|, |-2.3|, -3, 0$

[1.4–1.6] Evaluate.

81. $4 - 2 + 3 - 4$

82. $(3)(-4) - 6(8) - 3$

83. $-4|6| - 3(-4)$

84. $(4 - 6) - (-3 + 5) + 12$

85. $3|-2| - (4 - 3) + 2(-3)$

86. $|-16| \div (-4) + 2$

87. $(6 - 9) \div (9 - 6)$

88. $(4^2 - 6) - \sqrt{4} + 8$

89. $|6 - 3| \div 3 + 4 \cdot 8 - 12$

90. $\sqrt{36} \div 2 + |4 - 2| + 4^2$

91. $3^2 - 6 \cdot 9 + 4 \div 2^2 - 3$

92. $4 - (2 - 9)^0 + 3^2 \div 1 + 3$

93. $4^2 - (2 - 3^2)^2 + 4^3$

94. $-3^2 + 14 \div 2 \cdot 3 - 6$

95. $(-3)^2 - (-2)^2 + 20 + 4 \div 2$

96. $\dfrac{8 - 4 \div 2 + 3 \cdot 2}{\sqrt{36} \div 2^2 - 9}$

97. $\dfrac{-(4 - 6)^2 - 3(-2) + |-6|}{18 - 9 \div 3 \cdot 5}$

98. $\dfrac{8 - [5 - (-3 + 2)] \div 2}{|5 - 3| - |5 - 8| \div 3}$

99. $\dfrac{9|3 - 5| - 5|4| \div 10}{-3(5) - 2 \cdot 4 \div 2}$

Evaluate each expression for the values given.

100. $2x^2 + 3x + 1$ when $x = 2$

101. $(x - 2)^2 + 3x$ when $x = -2$

102. $4x^2 - 3y^2 + 5$ when $x = 1$ and $y = -2$

103. $-3x^2y + 6xy^2 - 2xy$ when $x = 1$ and $y = 3$

104. $(x - 2)^2 + (y - 4)^2 + 3$ when $x = 2$ and $y = 2$

105. $4(x - 3) + 5(y - 3) - 4$ when $x = -1$ and $y = -3$

106. $-x^2 + 3xy^2 - 6y^3$ when $x = 3$ and $y = 4$

107. $-x^2y - 6xy^2 + 4y^3$ when $x = -2$ and $y = 3$

Practice Test

1. List $A = \{x | x \text{ is a natural number greater than } 5\}$ in roster form.

Insert either $\subseteq$ or $\not\subseteq$ in the space provided to make the statement true.

2. $3 \quad \{1, 2, 3, 4\}$

3. $\{5\} \quad \{1, 2, 3, 4, 5\}$

Answer true or false.

4. Every rational number is a real number.

5. Every whole number is a natural number.

6. The union of the set of rational numbers and the set of irrational numbers is the set of real numbers.

Consider the set of numbers $\{-\frac{3}{5}, 2, -4, 0, \frac{19}{12}, 2.57, \sqrt{8}, \sqrt{2}, -1.92.\}$. List the elements of the set that are:

7. Rational numbers

8. Real numbers

Find $A \cup B$ and $A \cap B$ for sets A and B.

9. $A = \{8, 10, 11, 14\}, \quad B = \{5, 7, 8, 9, 10\}$

10. $A = \{1, 3, 5, 7, \ldots\}, \quad B = \{3, 5, 7, 9, 11\}$

Insert either $>$, $<$, or $=$ to make the statement true.

11. $-4 \quad |-9|$

12. $|-3| \quad -|5|$

13. List from smallest to largest: $|3|$, $-|4|$, -2, 6.

Name the property.

14. $3(x + 4) = 3x + 12$

15. $(x + y) + 3 = x + (y + 3)$

16. $3x + 4y = 4y + 3x$

17. $-4\left(-\frac{1}{4}\right) = 1$

18. $a + 0 = a$

Evaluate each of the following.

19. $4 - 6(-2) + 3^2$

20. $5^2 + 16 \div 4 - 3 \cdot 2$

21. $\dfrac{-3|4 - 8| \div 2 + 4}{\sqrt{36} + 18 \div 3^2}$

22. $\dfrac{-6^2 + 3(4 - |6|) \div 6}{4 - (-3) + 12 \div 4 \cdot 5}$

23. $\dfrac{[4 - (2 - 5)]^2 + 6 \div 2 \cdot 5}{|4 - 6| + |-6| \div 2}$

Evaluate each expression for the given values of x and y.

24. $-x^2 + 2xy + y^2$ when $x = 2$ and $y = 3$

25. $(x - 5)^2 + 2xy^2 - 6$ when $x = 2$ and $y = -3$

2 Linear Equations and Inequalities

See Section 2.3, Just For Fun Exercise 6

2.1 Solving Linear Equations

1. *Identify the reflexive, symmetric, and transitive properties.*
2. *Combine like terms.*
3. *Solve equations.*
4. *Identify conditional equations, inconsistent equations, and identities.*

1 In elementary algebra you learned how to solve linear equations. We briefly review these procedures in this section. Before we do so, we need to introduce three useful properties of equalities: the reflexive property, the symmetric property, and the transitive property.

Properties of Equalities

For all real numbers a, b and c:

1. $a = a$	*reflexive property*
2. If $a = b$, then $b = a$	*symmetric property*
3. If $a = b$ and $b = c$, then $a = c$	*transitive property*

Examples of reflexive property

$$3 = 3$$
$$x + 5 = x + 5$$
$$x^2 + 2x - 3 = x^2 + 2x - 3$$

Examples of symmetric property

If $x = 3$, then $3 = x$

If $y = x + 4$, then $x + 4 = y$

If $y = x^2 + 2x - 3$, then $x^2 + 2x - 3 = y$

Examples of transitive property

If $x = a$ and $a = 4y$, then $x = 4y$

If $a + b = c$ and $c = 4r$, then $a + b = 4r$

If $4k + 3r = 2m$ and $2m = 5w + 3$, then $4k + 3r = 5w + 3$

You will find that these three properties will be used often in the text.

Combining Terms

2 When an algebraic expression consists of several parts, the parts that are added or subtracted are called the **terms** of the expression. The expression

$$6x^2 - 3(x + y) - 4 + \frac{x + 2}{5}$$

has four terms: $6x^2$, $-3(x + y)$, -4, and $\frac{x + 2}{5}$.

The + and − signs that break the expression up into terms are a part of a term. However, when listing the terms of an expression it is not necessary to list the + sign at the beginning of a term.

Expression	*Terms*
$4x^2 - 3x - 7$	$4x^2, \quad -3x, \quad -7$
$-5x^3 + 3x^2y - 2$	$-5x^3, \quad 3x^2y, \quad -2$
$4(x+3) + 2x + 5(x-2) + 1$	$4(x+3), \quad 2x, \quad 5(x-2), \quad 1$
$\frac{x+3}{2} + 4x^2y^3 - x + 6$	$\frac{x+3}{2}, \quad 4x^2y^3, \quad -x, \quad 6$

The numerical part (constant) of a term is called its **numerical coefficient** or simply its **coefficient.** In the term $6x^2$ the 6 is the numerical coefficient. When the numerical coefficient is 1 or -1, we generally do not write the numeral 1. For example, x means $1x$, $-x^2y$ means $-1x^2y$, and $(x + y)$ means $1(x + y)$.

Term	*Numerical coefficient*
$\frac{2x}{3}$	$\frac{2}{3}$
$-4(x+2)$	-4
$\frac{x-2}{3}$	$\frac{1}{3}$
$-(x+y)$	-1

Note that $\frac{x-2}{3}$ means $\frac{1}{3}(x-2)$ and $-(x+y)$ means $-1(x+y)$.

The **degree of a term** is the sum of the exponents on the variables in the term. For example, $3x^2$ is a second-degree term, and $-4x$ is a first-degree term ($-4x$ means $-4x^1$). The number 3 can be written as $3x^0$, so the number 3 (and every other constant) is a zero-degree term. The term $6x^2y^3$ is a fifth-degree term since $2 + 3 = 5$. The term $4xy^5$ is a sixth degree term since the sum of the exponents is $1 + 5$ or 6.

Like terms are terms that have the same variables with the same exponents. For example, $3x$ and $5x$ are like terms, $2x^2$ and $-3x^2$ are like terms, and $3x^2y$ and $-2x^2y$ are like terms. Terms that are not like terms are said to be **unlike terms.**

To **simplify an expression** means to combine all like terms in the expression. To combine like terms, we can use the distributive property.

Examples of combining like terms

$$5x - 2x = (5-2)x = 3x$$

$$3x^2 - 5x^2 = (3-5)x^2 = -2x^2$$

$$-7x^2y + 3x^2y = (-7+3)x^2y = -4x^2y$$

$$4(x-y) - (x-y) = 4(x-y) - 1(x-y) = (4-1)(x-y) = 3(x-y)$$

EXAMPLE 1 Simplify each of the following. If an expression cannot be simplified, so state.

(a) $-2x + 5 + 3x - 7$
(b) $7x^2 - 2x^2 + 3x + 4$
(c) $2x - 3y + 5x - 6y + 3$

Solution: (a) $-2x + 5 + 3x - 7 = \underbrace{-2x + 3x}_{x} \underbrace{+ 5 - 7}_{-2}$ place like terms together

This expression simplifies to $x - 2$.

(b) $7x^2 - 2x^2 + 3x + 4 = 5x^2 + 3x + 4$

(c) $2x - 3y + 5x - 6y + 3 = 2x + 5x - 3y - 6y + 3$ place like terms together
$= 7x - 9y + 3$ ■

We are permitted to rearrange the terms in an expression because of the commutative and associative properties discussed earlier.

EXAMPLE 2 Simplify $7 - (2x - 5) - 3(2x + 4)$.

Solution:

$7 - (2x - 5) - 3(2x + 4) = 7 - 1(2x - 5) - 3(2x + 4)$
$= 7 - 2x + 5 - 6x - 12$ distributive property
$= -2x - 6x + 7 + 5 - 12$ rearrange terms
$= -8x$ combine like terms

Thus $7 - (2x - 5) - 3(2x + 4) = -8x$. ■

Solving Equations

3 An **equation** is a mathematical statement of equality. An equation must contain an equal sign and a mathematical expression on each side of the equal sign.

Examples of equations

$$x + 4 = -7$$
$$2x^2 - 4 = -3x + 5$$

The **solution** of an equation is the number or numbers that make the equation a true statement. The solution to the equation $x + 2 = 5$ is 3. The **solution set** of an equation is the set of real numbers that make the equation true. The solution set for the equation $x + 2 = 5$ is $\{3\}$.

Two or more equations with the same solution set are called **equivalent equations.** The equations $2x + 3 = 9$, $x + 2 = 5$, and $x = 3$ are all equivalent equations since the solution set for each is $\{3\}$. Equations are generally solved by starting with a given equation and producing a series of simpler equivalent equations.

In this chapter we will discuss how to solve **linear equations in one variable.** A linear equation is an equation that can be written in the form $ax + b = c, a \neq 0$. Notice that the degree of the highest-powered term in a linear equation is 1; for this reason linear equations are also called **first-degree equations.**

To solve equations we use the addition and multiplication properties to isolate the variable on one side of the equal sign.

Addition Property

If $a = b$, then $a + c = b + c$ for any a, b, and c.

The addition property states that the same number can be added to both sides of an equation without changing the solution to the original equation. Since subtraction is defined in terms of addition, the addition property also allows us to subtract the same number from both sides of an equation.

Multiplication Property

If $a = b$, then $a \cdot c = b \cdot c$ for any a, b, and c.

The multiplication property states that both sides of an equation can be multiplied by the same number without changing the solution. Since division is defined in terms of multiplication, the multiplication property also allows us to divide both sides of an equation by the same nonzero number.

To solve an equation we will often have to use a combination of properties to isolate the variable. Our goal is to get the variable all by itself on one side of the equation. A general procedure to solve linear equations follows.

To Solve Linear Equations

1. Eliminate all fractions by multiplying both sides of the equation by the least common denominator.
2. Use the distributive property to remove any parentheses.
3. Combine like terms on the same side of the equal sign.
4. Use the addition property to rewrite the equation with all terms containing the variable on one side of the equal sign and all terms not containing the variable on the other side of the equal sign. It may be necessary to use the addition property a number of times to accomplish this. Repeated use of the addition property will eventually result in an equation of the form $ax = b$.
5. Use the multiplication property to isolate the variable. This will give an answer of the form $x =$ some number.
6. Check the solution in the original equation.

The **least common denominator of a set of numbers,** LCD, (also called the **least common multiple,** LCM) is the smallest number that is divisible (without remainder) by each of the numbers. If you forgot how to find the LCD, review Appendix A or ask your teacher.

Now let us work some problems.

EXAMPLE 3 Solve the equation $2x + 4 = 9$.

Solution:

$$2x + 4 = 9$$

$$2x + 4 - 4 = 9 - 4 \qquad \text{subtract 4 from both sides of the equation}$$

$$2x = 5$$

$$\frac{\overset{1}{\cancel{2}}x}{\underset{1}{\cancel{2}}} = \frac{5}{2} \qquad \text{divide both sides of the equation by 2}$$

$$x = \frac{5}{2}$$

Check:

$$2x + 4 = 9$$

$$\cancel{2}\left(\frac{5}{\cancel{2}}\right) + 4 = 9$$

$$5 + 4 = 9$$

$$9 = 9 \qquad \text{true}$$

Since the answer checks the solution is $\frac{5}{2}$. ■

Whenever an equation contains like terms on the same side of the equal sign, combine the like terms before using the addition or multiplication properties.

EXAMPLE 4 Solve the equation $-4 = 3(x - 5) + 2x - 6$.

Solution:

$$-4 = 3(x - 5) + 2x - 6 \qquad \text{use distributive property}$$

$$-4 = 3x - 15 + 2x - 6 \qquad \text{combine like terms}$$

$$-4 = 5x - 21$$

$$21 - 4 = 5x - 21 + 21 \qquad \text{add 21 to both sides of the equation}$$

$$17 = 5x$$

$$\frac{17}{5} = \frac{5x}{5} \qquad \text{divide both sides of the equation by 5}$$

$$\frac{17}{5} = x$$

The solution is $\frac{17}{5}$. ■

To save space we will not always check our answers. You should, however, check all your answers.

EXAMPLE 5 Solve the equation $2x + 8 - 3x = -2(3x - 5) - 12$.

Solution: First use the distributive property and then combine like terms.

$$2x + 8 - 3x = -2(3x - 5) - 12$$
$$2x + 8 - 3x = -6x + 10 - 12$$
$$-x + 8 = -6x - 2$$
$$6x - x + 8 = -6x + 6x - 2 \quad \text{add } 6x \text{ to both sides of equation}$$
$$5x + 8 = -2$$
$$5x + 8 - 8 = -2 - 8 \quad \text{subtract 8 from both sides of equation}$$
$$5x = -10$$
$$\frac{5x}{5} = \frac{-10}{5} \quad \text{divide both sides of the equation by 5}$$
$$x = -2 \quad ■$$

In some of the following examples, we will omit some of the intermediate steps in determining the solution to an equation. Below we illustrate how the solution may be shortened.

Solution	*Shortened form of solution*
$x + 4 = 6$	$x + 4 = 6$
$x + 4 - 4 = 6 - 4$ ⟵ do this step mentally	$x = 2$
$x = 2$	
$3x = 6$	$3x = 6$
$\frac{3x}{3} = \frac{6}{3}$ ⟵ do this step mentally	$x = 2$
$x = 2$	

When an equation contains fractions, we often begin by multiplying *both* sides of the equation by the least common denominator. **This will result in each and every term** in the equation being multiplied by the least common denominator.

EXAMPLE 6 Solve the equation $5 - \frac{2x}{3} = -9$.

Solution: The least common denominator is 3. Multiply both sides of the equation by 3 to eliminate all fractions.

$$5 - \frac{2x}{3} = -9$$
$$3\left(5 - \frac{2x}{3}\right) = 3(-9)$$
$$3(5) - \overset{1}{\cancel{3}}\left(\frac{2x}{\underset{1}{\cancel{3}}}\right) = -27$$
$$15 - 2x = -27$$
$$-2x = -42$$
$$x = 21 \quad ■$$

EXAMPLE 7 Solve the equation $\frac{x}{4} + 3 = 2x - \frac{5}{3}$.

Solution: Multiply both sides of the equation by the least common denominator 12.

$$12\left(\frac{x}{4} + 3\right) = 12\left(2x - \frac{5}{3}\right)$$

$$\overset{3}{\cancel{12}}\left(\frac{x}{\underset{1}{\cancel{4}}}\right) + 12(3) = 12(2x) - \overset{4}{\cancel{12}}\left(\frac{5}{\underset{1}{\cancel{3}}}\right)$$

$$\begin{aligned} 3x + 36 &= 24x - 20 \\ 36 &= 21x - 20 \\ 56 &= 21x \\ \frac{56}{21} &= x \end{aligned}$$

■

EXAMPLE 8 Solve the equation $-\frac{2}{5}(x + 3) + 4 = \frac{1}{3}(x - 4)$.

Solution: Multiply both sides of the equation by the least common denominator 15 to remove fractions.

$$-\frac{2}{5}(x + 3) + 4 = \frac{1}{3}(x - 4)$$

$$15\left[-\frac{2}{5}(x + 3) + 4\right] = 15\left[\frac{1}{3}(x - 4)\right]$$

$$\overset{3}{\cancel{15}}\left(-\frac{2}{\underset{1}{\cancel{5}}}\right)(x + 3) + 15(4) = \overset{5}{\cancel{15}}\left(\frac{1}{\underset{1}{\cancel{3}}}\right)(x - 4)$$

$$\begin{aligned} -6(x + 3) + 60 &= 5(x - 4) \\ -6x - 18 + 60 &= 5x - 20 \\ -6x + 42 &= 5x - 20 \\ 42 &= 11x - 20 \\ 62 &= 11x \\ \frac{62}{11} &= x \end{aligned}$$

■

Identities and Inconsistent Equations

4 All the equations discussed thus far have been **conditional equations.** They are true only under specific conditions. For example, in Example 8 the equation was true only when $x = \frac{62}{11}$.

Consider the following equation $2x + 1 = 5x + 1 - 3x$. Solving the equation we obtain

$$
\begin{aligned}
2x + 1 &= 5x + 1 - 3x \\
2x + 1 &= 2x + 1 \\
2x + 1 - 1 &= 2x + 1 - 1 && \text{subtract 1 from both sides of the equation} \\
2x &= 2x \\
2x - 2x &= 2x - 2x && \text{subtract } 2x \text{ from both sides of the equation} \\
0 &= 0 && \text{a true statement}
\end{aligned}
$$

This equation, $2x + 1 = 5x + 1 - 3x$, is an example of an identity. An **identity** is an equation that is true for all real numbers. If at any point while solving an equation you realize that both sides of the equations are identical, as in

$$2x + 1 = 2x + 1,$$

the equation is an identity. The solution to $2x + 1 = 5x + 1 - 3x$ is all real numbers. **The solution to any identity is all real numbers.** If you continue to solve an equation that is an identity, you will end up with $0 = 0$, a true statement.

Now consider the equation $2(3x + 1) = 9x + 3 - 3x$.

$$
\begin{aligned}
2(3x + 1) &= 9x + 3 - 3x \\
6x + 2 &= 6x + 3 \\
6x + 2 - 2 &= 6x + 3 - 2 && \text{subtract 2 from both sides of the equation} \\
6x &= 6x + 1 \\
6x - 6x &= 6x - 6x + 1 && \text{subtract } 6x \text{ from both sides of the equation} \\
0 &= 1 && \text{a false statement}
\end{aligned}
$$

Since $0 = 1$ is never a true statement, this equation has no solution. An equation that has no solution is said to be an **inconsistent equation.** When solving an equation that turns out to be inconsistent, do not leave the answer blank. Write "no solution" as the answer. **An inconsistent equation has no solution.**

Every linear equation will be a conditional equation with exactly one solution, an identity with an infinite number of solutions, or inconsistent with no solution. Table 1 summarizes this information.

Table 1

Type of Linear Equation	Solution
Conditional equation	Has exactly one real solution
Identity	Is true for every real number; has an infinite number of solutions
Inconsistent equation	Has no solution

Exercise Set 2.1

Name the property indicated.

1. $2 = 2$
2. If $x = 5$, then $5 = x$
3. If $x + 2 = 3$, then $3 = x + 2$
4. If $x = 3$, and $3 = y$, then $x = y$
5. If $x + 1 = a$ and $a = 2y$, then $x + 1 = 2y$
6. $-3 = -3$
7. $x + 2 = x + 2$
8. If $x = 4$, then $x + 3 = 4 + 3$
9. If $x = 2$, then $x - 2 = 2 - 2$
10. If $x = 3$, then $3 = x$
11. If $2x = 4$, then $3(2x) = 3(4)$
12. If $5x = 4$, then $\frac{1}{5}(5x) = \frac{1}{5}(4)$
13. If $2x = 7$, then $\frac{1}{2}(2x) = \frac{1}{2}(7)$
14. If $x = 5$, then $x - 2 = 5 - 2$
15. If $x = 3$, then $x - 3 = 3 - 3$
16. If $x + 2 = 4$, then $x + 2 - 2 = 4 - 2$
17. If $x - 5 = 3$, then $x - 5 + 5 = 3 + 5$
18. If $5x = 3$, then $\frac{5x}{2} = \frac{3}{2}$
19. If $2x = 6$, then $\frac{2x}{2} = \frac{6}{2}$
20. If $x = 2$, then $2 = x$
21. If $x + 2 = x + y$, then $x + y = x + 2$
22. If $x - 3 = x + y$, and $x + y = z$, then $x - 3 = z$

Give the degree of the term.

23. $4x$
24. $-6x^2$
25. $3xy$
26. $18x^2y^3$
27. $\frac{1}{2}x^4y$
28. 7
29. -3
30. $-5x$
31. $3x^4y^6z^3$
32. x^4y^6
33. $3x^5y^6z$
34. $-2x^4y^7z^8$

Simplify. If an expression cannot be simplified, so state.

35. $8x + 7 + 7x - 12$
36. $3x^2 + 4x + 5$
37. $5x^2 - 3x + 2x - 5$
38. $6x^2 - 9x + 3 - 4x - 7$
39. $-4x^2 - 3x - 5x + 7$
40. $-3x + 4x - 7 - 2x + 1$
41. $2y - 5y + 3 - 6$
42. $7y + 3x - 7 + 4x - 2y$
43. $6y^2 + 6xy + 3$
44. $4x^2 - x^2 - 3x - 3x^2 + 4$
45. $xy + 3xy + y^2 - 2$
46. $3x^2y + 4xy^2 - 2x^2$
47. $6(x + 5) + 2(x + 5)$
48. $4(x - 3) + 2(x - 3) + 4$
49. $5 + 2(2x + 4) + 3$
50. $6 - 3(2x + 4) - 4x$
51. $4(x - 3) + 6(x - 3) + 4x - 2$
52. $5(x - 1) - 3(x - 1) + 2$
53. $4 - 6(3x + 2) - x + 4$
54. $3(x + y) - 4(x + y) - 3$
55. $4(x - 2y) + 3(x - 2y) - 6x$
56. $2x - 3(6y - 5) + 6y - 5x$

Solve each equation. If an equation has no solution, so state.

57. $2x + 3 = 5$
58. $4x - 9 = 3$
59. $4 = 6 - 5x$
60. $4x + 3 = -12$
61. $\frac{3}{5}x = 9$
62. $-\frac{2}{3}x = -12$
63. $-3 = 6 - x$
64. $-\frac{x}{4} = 8$
65. $6 = 4 - 5x$
66. $\frac{10}{3} = x + 6$
67. $-2 = \frac{-4x}{5} + 9$
68. $3(x - 2) = 12$

69. $12 = -2(2x - 6)$

70. $6 - 4x = -15$

71. $2x + 3 - 4x = 9$

72. $3(2x - 4) + 3(x + 1) = 9$

73. $-(x - 4) + 3x = -12$

74. $6 - (x + 2) = -3$

75. $\frac{1}{2}(x - 4) = 8$

76. $\frac{2}{3}(2x - 6) + 4 = 8$

77. $6 - 2(x - 3) + 5 = 15$

78. $\frac{x + 3}{4} = 6$

79. $\frac{2x - 5}{3} = -5$

80. $\frac{3x - 7}{5} + 2 = 9$

81. $\frac{4 - 3x}{2} + x = 4$

82. $\frac{1}{2}(3x - 5) = 6$

83. $3(2x - 3) + 9 = 0$

84. $-\frac{3}{5}(15 - 2x) = -3$

85. $x + 3 = 2x - 5$

86. $3x - 4 = 6 - 2x$

87. $3(x + 1) = 4x$

88. $6(x + 2) = -3(2 - x)$

89. $4x - 5(x + 3) = 2x - 3$

90. $2(x - 3) - (2x + 4) = -6x$

91. $4x - 6x + 2 = -4x + 12 - 2x$

92. $\frac{x}{3} + 2 = x + 4$

93. $2(x - 3) + 2x = 4x - 5$

94. $\frac{11 - 3x}{4} = 2x$

95. $\frac{x - 25}{3} = 2x$

96. $\frac{4}{3} + 4x = 2 - x$

97. $4(2 - 3x) = -6x - 2(3x - 4)$

98. $\frac{2}{3}x = \frac{9}{4}x$

99. $\frac{1}{2}(2x + 1) = \frac{1}{4}(x - 4)$

100. $\frac{4}{3} + \frac{x}{4} = x$

101. $\frac{7x}{5} = 2x + 3$

102. $4x - 3 = -5x - (x - 1)$

103. $\frac{2}{3}(x - 4) = \frac{2}{3}(4 - x)$

104. $\frac{2(x + 2)}{5} = \frac{x}{3}$

105. $\frac{3x}{4} - 2 = 6 + \frac{x}{3}$

106. $\frac{x - 8}{5} + \frac{x}{3} = \frac{-8}{5}$

107. $\frac{x + 1}{4} = \frac{x - 4}{2} - \frac{2x - 3}{4}$

108. What is a conditional equation?

109. What is an identity?

110. What is an inconsistent equation?

111. What are equivalent equations?

112. Consider the equation $x = 4$. Give three equivalent equations.

113. Consider the equation $2x = 5$. Give three equivalent equations.

JUST FOR FUN

Solve each of the following equations.

1. $-\frac{3}{5}(x + 2) - \frac{4}{3}(2x - 3) + 4 = \frac{1}{2}(x + 4) - 6x + 3(5 - x)$

2. $2[-3[4(x + 3) + 2] + 4] = 3[2(x + 3) + 5] + x + 6$

3. $\frac{x}{3} + \frac{x - 2}{4} + \frac{2x - 3}{5} = \frac{x - 3}{6} + 4(x - 7) - x + 2$

2.2 Word Problems

1 *Translate a statement into an algebraic expression.*

2 *Solve word problems.*

1 The next few sections will give you an idea of some of the many uses of algebra in real-life situations. Whenever possible, we include other relevant applications throughout the text.

Perhaps the most difficult part of solving a word problem is transforming the problem into an equation. Here are some examples of phrases represented as algebraic expressions.

Phrase	*Algebraic expression*
a number increased by 4	$x + 4$
twice a number	$2x$
5 less than a number	$x - 5$
a number subtracted from 9	$9 - x$
6 subtracted from a number	$x - 6$
one-eighth of a number	$\frac{1}{8}x$ or $\frac{x}{8}$
2 more than 3 times a number	$3x + 2$
4 less than 6 times a number	$6x - 4$
3 times the sum of a number and 5	$3(x + 5)$

The variable x was used in the algebraic expressions, but any variable could have been used to represent the unknown quantity.

EXAMPLE 1 Express each phrase as an algebraic expression.

(a) The distance, d, increased by 15 miles.
(b) 3 less than 4 times the area.
(c) 4 times a number decreased by 5.

Solution: (a) $d + 15$
(b) $4A - 3$
(c) $4n - 5$ ■

EXAMPLE 2 Write each of the following as an algebraic expression.

(a) The cost of purchasing x shirts at $4 each.
(b) The distance traveled in t hours at 55 miles per hour.
(c) The number of calories in x potato chips if each potato chip has 10 calories.
(d) The number of cents in n nickels.
(e) Eight percent commission on x dollars.

Solution: (a) We can reason like this: one shirt would cost 1(4) dollars; two shirts, 2(4) dollars; three shirts, 3(4) dollars; four shirts, 4(4) dollars; and so on. Continuing this reasoning process we can see that x shirts would cost $x(4)$ or $4x$ dollars. We can use the same reasoning process to complete each of the following.

(b) $55t$
(c) $10x$
(d) $5n$
(e) $0.08x$

Note in part (e) that 8% is written as 0.08 in decimal form. ■

Sometimes in a problem two numbers are related to each other in a certain way. We often represent one of the numbers as a variable and the other as an expression containing that variable. We generally let the less complicated description be represented by the variable and write the second (more complex expression) in terms of the variable. Some examples are illustrated below.

Phrase	*One number*	*Second number*
Peter's age now and Peter's age in 5 years	x	$x + 5$
one number is 3 times the other	x	$3x$
one number is 7 less than the other	x	$x - 7$
two consecutive integers	x	$x + 1$
two consecutive odd (or even) integers	x	$x + 2$
the sum of two numbers is 10	x	$10 - x$
a 6-foot piece of lumber cut in two pieces	x	$6 - x$
a number and the number increased by 7%	x	$x + 0.07x$
a number and the number decreased by 10%	x	$x - 0.10x$

EXAMPLE 3 For each of the following relationships, select a variable to represent one quantity and express the second quantity in terms of the first.

(a) Jan is 12 years older than her sister.
(b) The speed of the second train is 1.2 times the speed of the first.
(c) \$90 is split between David and his brother.
(d) It takes Tom 3 hours longer than Roberta to complete the task.
(e) Hilda has \$4 more than twice the amount of money Hector has.
(f) The length of a rectangle is 2 units less than 3 times its width.

Solution: (a) Sister x; Jan, $x + 12$
(b) First train, x; second train, $1.2x$
(c) Amount David has, x; amount brother has, $90 - x$
(d) Roberta, x; Tom, $x + 3$
(e) Hector, x; Hilda, $2x + 4$
(f) Width, x; length, $3x - 2$ ■

2 The word **"is"** in a word problem often means **"is equal to"** and is represented by an equal sign, =.

Verbal statement	*Algebraic equation*
3 more than a number *is* 9	$x + 3 = 9$
4 less than 3 times a number *is* 5	$3x - 4 = 5$
a number decreased by 4 *is* 3 more than twice the number	$x - 4 = 2x + 3$
the product of two consecutive integers *is* 20	$x(x + 1) = 20$
one number is 2 more than 5 times the other number; the sum of the two numbers *is* 62	$x + (5x + 2) = 62$
a number increased by 15% *is* 90	$x + 0.15x = 90$
a number decreased by 12% *is* 38	$x - 0.12x = 38$
the sum of a number and the number increased by 4% *is* 204	$x + (x + 0.04x) = 204$
the cost of renting a VCR for x days at $6 per day *is* $72	$6x = 72$
the distance traveled in n days when 600 miles *is* traveled per day is 1500 miles	$600n = 1500$

Although there are many types of word problems, the general procedure used to solve all word problems is basically the same.

To Solve a Word Problem

1. Read the problem carefully.
2. If possible, draw a sketch to illustrate the problem.
3. Identify the quantity you are being asked to find.
4. Choose a variable to represent the quantity, *and write down exactly what it represents.*
5. Write the word problem as an equation.
6. Solve the equation for the unknown quantity.
7. Answer the question asked.
8. Check the solution in the *original word problem.*

Many students solve the equation for the variable but do not answer the question or questions asked.

EXAMPLE 4 Four subtracted from 3 times a number is 29. Find the number.

Solution: *Step 3:* We are asked to find the unknown number.

Step 4: Let x = unknown number.

Step 5: Write the equation.

$$3x - 4 = 29$$

Step 6: Solve the equation.

$$3x - 4 = 29$$
$$3x = 33$$
$$x = 11$$

Step 7: Answer the question.

The number is 11.

Step 8: Check the solution in the original word problem.

Four subtracted from 3 times a number is 29.

$$3(11) - 4 = 29$$
$$33 - 4 = 29$$
$$29 = 29 \quad \text{true}$$ ■

EXAMPLE 5 The sum of two numbers is 47. Find the two numbers if one number is 2 more than 4 times the other number.

Solution:

Let x = smaller number

then $4x + 2$ = larger number

smaller number + larger number = sum of two numbers

$$x + 4x + 2 = 47$$
$$5x + 2 = 47$$
$$5x = 45$$
$$x = 9$$

The smaller number = 9; the larger number is $4x + 2 = 4(9) + 2 = 38$.

Check: The sum of two numbers is 47.

$$9 + 38 = 47$$
$$47 = 47 \quad \text{true}$$ ■

In Example 5, if you gave the answer 9, it would not be correct. The question asked you to find both numbers. The two numbers are 9 and 38.

EXAMPLE 6 The population of Springfield is 96,000. If the population is increasing by 5000 per year, how many years will it take to reach 125,500?

Solution:

Let n = number of years

then $5000n$ = population increase in n years

present population	+	increase in population over n years	=	future population
96,000	+	$5000n$	=	125,000

$$96{,}000 + 5000n = 125{,}500$$
$$5000n = 29{,}500$$
$$n = 5.9 \text{ years}$$

The population of Springfield will increase to 125,500 in 5.9 years. ■

EXAMPLE 7 For two consecutive numbers, one less than twice the smaller subtracted from 4 times the larger is 19. Find the two numbers.

Solution:

$$\text{Let } x = \text{smaller of two consecutive number}$$
$$\text{then } x + 1 = \text{larger of two consecutive numbers}$$

$$\text{one less than twice the smaller} = 2x - 1$$
$$\text{four times the larger} = 4(x + 1)$$

One less than twice the smaller *subtracted from* four times the larger *is* 19.

$$4(x + 1) - (2x - 1) = 19$$
$$4x + 4 - 2x + 1 = 19$$
$$2x + 5 = 19$$
$$2x = 14$$
$$x = 7$$
$$\text{smaller number} = x = 7$$
$$\text{larger number} = x + 1 = 8$$

Check: four times larger − one less than twice smaller = 19

$$4(8) - [2(7) - 1] = 19$$
$$32 - (14 - 1) = 19$$
$$32 - 13 = 19$$
$$19 = 19 \quad \text{true}$$

■

COMMON STUDENT ERROR

When an expression containing more than one term is to be subtracted from another, the entire expression must be subtracted, not just the first term. In Example 7, had we written

$$4(x + 1) - 2x - 1 = 19$$

we would have obtained the wrong answer. The parentheses are needed around the entire expression $2x - 1$ to show that the entire expression is to be subtracted.

Example: Subtract $3x + 4$ from $2x$.

Correct	*Wrong*
$2x - (3x + 4)$	~~$2x - 3x + 4$~~

Example: Subtract $2x - 5$ from $x - 3$.

Correct	*Wrong*
$x - 3 - (2x - 5)$	~~$x - 3 - 2x - 5$~~

EXAMPLE 8 A chemical company's bonus plan states that employees will receive a bonus of 6% of their yearly salary at the end of the calendar year. If Mr. Riordan received a total of \$23,320 for the year, including the bonus, what was his yearly salary?

Solution:

$$\text{Let } x = \text{yearly salary}$$
$$\text{then } 0.06x = \text{bonus}$$

$$\begin{aligned} \text{yearly salary} + \text{bonus} &= \text{total received} \\ x + 0.06x &= 23{,}320 \\ 1.06x &= 23{,}320 \\ x &= \frac{23{,}320}{1.06} = \$22{,}000 \end{aligned}$$

Mr. Riordan's yearly salary is \$22,000. ■

EXAMPLE 9 Budget Rent-a-Car offers cars at \$30 a day plus 24 cents a mile. Hertz rents the same car for \$44 a day plus 16 cents a mile. How many miles would you have to drive in 1 day so that the cost of renting from Hertz would equal the cost of renting from Budget?

Solution:

$$\begin{aligned} \text{Let } x &= \text{number of miles} \\ \text{then } 0.24x &= \text{cost of traveling } x \text{ miles in Budget's car} \\ \text{and } 0.16x &= \text{cost of traveling } x \text{ miles in Hertz's car} \end{aligned}$$

$$\begin{aligned} \text{Budget's cost} &= \text{Hertz's cost} \\ \text{daily fee} + \text{mileage cost} &= \text{daily fee} + \text{mileage cost} \\ 30 + 0.24x &= 44 + 0.16x \\ 0.24x &= 14 + 0.16x \\ 0.08x &= 14 \\ x &= \frac{14}{0.08} = 175 \text{ miles} \end{aligned}$$

If you plan to drive less than 175 miles, the Budget car will be less expensive. ■

EXAMPLE 10 Mr. Lopez, a farmer, wishes to put up a fence in a rectangular shape to contain his cows. He wishes the length of the rectangle to be 150 feet more than its width. If Mr. Lopez has 2700 feet of fencing, what should be the dimensions of the rectangle?

Solution: The formula for the perimeter of a rectangle is $P = 2l + 2w$. We are told that Mr. Lopez has 2700 feet to construct the fence; therefore, the perimeter, P, must be 2700. This gives $2700 = 2l + 2w$. Note that this equation contains two variables, l and w. To solve the equation, the equation must be expressed in terms of a single variable. We must therefore try to eliminate one of the variables by expressing the length in terms of the width or the width in terms of the length.

Let x = width of rectangle,

then $x + 150$ = length of rectangle

Note that the length of the rectangle is expressed in terms of its width (see Fig. 2.1).

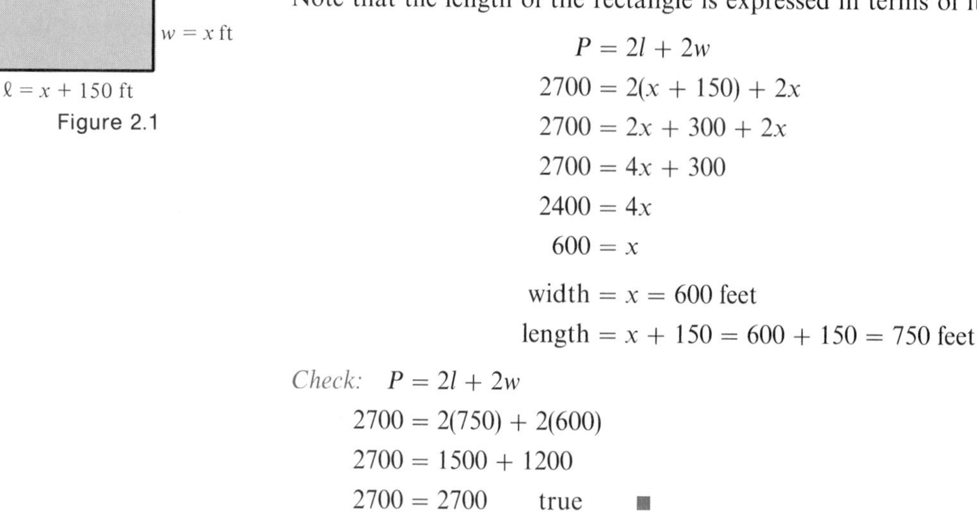

Figure 2.1

$$P = 2l + 2w$$
$$2700 = 2(x + 150) + 2x$$
$$2700 = 2x + 300 + 2x$$
$$2700 = 4x + 300$$
$$2400 = 4x$$
$$600 = x$$

$$\text{width} = x = 600 \text{ feet}$$
$$\text{length} = x + 150 = 600 + 150 = 750 \text{ feet}$$

Check: $P = 2l + 2w$

$$2700 = 2(750) + 2(600)$$
$$2700 = 1500 + 1200$$
$$2700 = 2700 \qquad \text{true}$$ ■

Exercise Set 2.2

(a) Write an equation that can be used to solve the problem and (b) find the solution to the problem.

1. One number is 5 times another number. The sum of the two numbers is 24. Find the numbers.
2. Kathy is 15 years older than Dawn. The sum of their ages is 41. Find Kathy's and Dawn's ages.
3. The sum of two consecutive integers is 51. Find the two integers.
4. The sum of two consecutive even integers is 78. Find the two integers.
5. Twice a number decreased by 8 is 38. Find the number.
6. For two consecutive integers, the smaller plus 3 times the larger is 39. Find the integers.
7. One train travels 3 times as far as another. The sum of their distances is 48 miles. Find the distance traveled by each train.
8. The sum of three consecutive integers is 48. Find the three integers.
9. The sum of three consecutive even integers is 66. Find the three integers.
10. When $\frac{1}{3}$ of a number is added to 10 the sum is 15. Find the number.
11. When $\frac{3}{5}$ of a number is subtracted from 10, the difference is 4. Find the number.
12. The larger of two numbers is 5 times the smaller. Find the two numbers if twice the smaller equals 3 less than $\frac{1}{2}$ the larger.
13. The larger of two integers is two more than 4 times the smaller. Find the two numbers if 5 times the smaller number is 2 more than $\frac{1}{2}$ the larger number.
14. The larger of two integers is $\frac{5}{2}$ the smaller. If twice the smaller is subtracted from twice the larger, the difference is 12. Find the two numbers.
15. The sum of the angles of a triangle is 180°. Find the three angles if the two base angles are the same and the third angle is twice as large as the other two angles.
16. Find the three angles of a triangle if one angle is 20° greater than the smallest angle and the third angle is twice the smallest angle.
17. An isosceles triangle is a triangle with two sides of equal length. Find the three sides of the isosceles triangle if the larger side is 15 inches greater than the two smaller sides and the sum of the lengths of the three sides is 45 inches.
18. A number increased by 8% is 54. Find the number.
19. The monthly rental fee for a telephone from the Rochester

Telephone Company is $2.89. The telephone store is selling telephones for $37.99. How long would it take for the monthly rental fee to equal the cost of a new phone?

20. It cost the Ranieri's $10.50 a week to wash and dry their clothes at the corner laundry. If a washer and drier cost a total of $798, how many weeks will it take for the laundry cost to equal the cost of a washer and drier?
21. The cost of renting a log splitter is $16 per day. If a new log splitter costs $236, how long would it take for the rental fee to equal the purchase price of the log splitter?
22. Bill Wicker buys a monthly bus pass, which entitles the owner to unlimited bus travel for $30 per month. Without the pass each bus ride costs $1.25. How many rides per month would Bill have to make so that it is less expensive to purchase the pass?
23. The cost of renting an automobile is $20 plus 18 cents a mile. Determine the maximum distance that Tony can drive if he has only $38.
24. A small city is growing by 300 per year. If the present population is 5200, how long will it take for the population to reach 8800?
25. A city with a population of 15,000 is decreasing by 800 per year. In how many years will the population drop to 8000?
26. The cost of renting an automobile is $25 a day plus 12 cents per mile. How far can Kendra drive in 1 day if she has only $40?
27. Mrs. Scilla used her own car for a 1-day 350-mile business trip. When an employee uses their own car for business the company reimburses the employee a fixed dollar amount plus 18 cents a mile. Mrs. Scilla cannot remember the fixed amount but knows she was reimbursed a total of $83.00. Find the fixed amount the company reimburses their employees.
28. Emil receives a flat weekly salary of $240 plus a 12% commission on the total dollar volume of all sales he makes. What must his dollar volume be in a week if he is to make a total weekly salary of $540?
29. The Computer Store has reduced the price of a computer by 15%. What is the original price of the computer if the sale price is $1275?
30. Monroe County has a 7% sales tax. What is the maximum price of a car if the total cost, including tax, is to be $8500?
31. Douglas is on a diet and can have only 400 calories for lunch. He orders a hot dog and French fries. If the hot dog and roll contain a total of 250 calories and each French fry contains 20 calories, how many French fries can he eat?
32. After Mrs. Englers is seated in a restaurant, she realizes that she only has $9.25. If she must pay 7% sales tax and wishes to leave a 15% tip, what is the maximum price of a lunch she can order?
33. United Airlines wants the cost of a one-way ticket between Jacksonville, Florida, and Houston, Texas, to be exactly $156, including tax. What should be the price of the one–way ticket if there is a 7% sales tax on the cost of the ticket?
34. A hot dog vendor wants the total cost of a hot dog plus a $7\frac{1}{2}\%$ sales tax to be exactly $1. Find the cost of the hot dog.
35. Melissa Bryant wishes to sell her paintings for $500, which includes a $6\frac{3}{4}\%$ sales tax. How much should she charge for each painting?
36. The Penfield Racquet Club has two payment plans. One plan is a flat $40 per month. The second is to pay $8.50 per hour for court time. If court time can be rented only in 1-hour intervals, how many hours would one have to play per month to make it advantageous to select the flat fee?
37. The American Health Center offers two membership plans. One plan calls for a flat yearly payment of $510. A second plan calls for a $150 yearly fee plus a $6 payment for each day the center is used. How often, over a period of a year, would Ms. McKane have to use the center so that the cost of the second plan equals the cost of the first plan?

See Exercise 29.

38. The Midtown Tennis Club offers two payment plans for its members. Plan 1 is a monthly fee of \$20 plus \$8 per hour of court rental time. Plan 2 is no monthly fee but court time costs \$16.25 per hour. How many hours would Mr. Burnham have to play per month so that plan 1 becomes advantageous?

39. The length of a rectangle is to be 2 feet greater than twice its width. Find the dimensions of the rectangle if the perimeter is to be 40 feet. Use $P = 2l + 2w$.

40. Mrs. Levin is planning to build a sandbox for her children. She wants its length to be 3 feet more than its width. Find the length and width of the sandbox if only 22 feet of lumber is available to form the frame.

41. The width of a rectangle is 1 meter more than $\frac{1}{2}$ its length. Find the length and width of the rectangle if its perimeter is 20 meters.

42. Ken Smith, a landscape architect, wishes to fence in two equal areas as illustrated in the figure. If both areas are squares and the total amount of fencing used is 91 meters, find the dimension of each square.

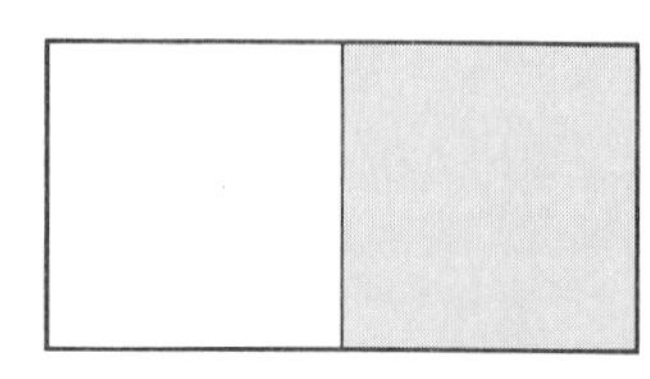

43. Taryn wishes to build a bookcase with four shelves. The height of the bookcase should be 3 feet greater than the width. If only 30 feet of wood is available to build the bookcase, what should the dimensions of the bookcase be?

44. Shun wishes to fence in three rectangular areas along a river bank as illustrated in the following figure. Each rectangle is to have the same dimensions and the length of each rectangle is to be 1 meter greater than its width. Find the length and width of each rectangle if the total amount of fencing used is 81 meters.

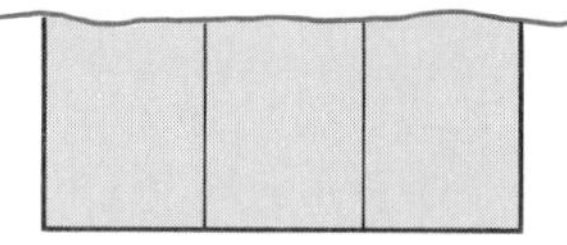

45. A Volkswagen Rabbit with a diesel engine costs \$7525 and gets 47 miles per gallon. The Rabbit with its standard engine costs \$6069 and gets 29 miles per gallon. How many miles will Jack Robinson have to drive in the diesel before he makes up the extra cost of the car? Assume that diesel fuel costs \$0.87 per gallon and gasoline costs \$0.90 per gallon.

JUST FOR FUN

1. Pick any number, multiply it by 2, add 33, subtract 13, divide by 2, and subtract the number you started with. You should end with the number 10. Show that this procedure will result in the answer 10 for any number, n, selected.

2.3 Motion and Mixture Problems

1. *Solve motion (or rate) word problems.*
2. *Solve mixture word problems.*

1 A formula with many useful applications is

$$\text{amount} = \text{rate} \cdot \text{time}$$

The "amount" in this formula can be a measure of many different quantities, including distance or length, area, or volume.

When applying this formula, we must make sure that the units are consistent. For example, if the rate is given in inches per minute, the time must be given in minutes.

If the rate is given as gallons per hour, the time must be given in hours. Problems that can be solved using this formula are called motion (or rate) problems because "rate" indicates motion of some type.

A nurse giving a patient an intravenous injection may use this formula to determine the drip rate of the fluid being injected. A company drilling for oil or water may use this formula to determine the amount of time needed to reach the goal. This formula may also be used when determining how fast a train or plane must travel to be at a certain point at a certain time.

EXAMPLE 1 A patient is to receive 800 cubic centimeters (cc) of glucose intravenously over an 8-hour period. What should be the average intravenous flow rate?

Solution: We are given the volume and the time and are asked to find the rate.

$$\text{volume} = \text{rate} \cdot \text{time}$$
$$800 = r \cdot 8$$
$$800 = 8r$$
$$r = 100 \text{ cc/hr}$$

Therefore, the intravenous flow rate should be set at 100 cc/hr. ■

EXAMPLE 2 A conveyer belt is transporting uncut lasagna noodles at a rate of 1.5 feet per second. A cutting blade is activated at regular intervals to cut the lasagna into 10.8-inch lengths. At what time intervals should the cutting blade be activated?

Solution: Since the rate is given in feet per second and the lasagna length is given in inches, one of these quantities must be changed. Since 1 foot equals 12 inches:

$$1.5 \text{ feet per second} = (1.5)(12) = 18 \text{ inches per second}$$
$$\text{length} = \text{rate} \cdot \text{time}$$
$$10.8 = 18 \cdot t$$
$$t = \frac{10.8}{18} = 0.6 \text{ second}$$

The blade should cut at 0.6-second intervals. ■

Sometimes when a problem has two different rates it is helpful to put the information in tabular form to help analyze the problem.

EXAMPLE 3 Two trains leave San Jose at the same time, traveling in the same direction on parallel tracks. One train travels at 80 miles per hour, the other travels at 60 miles per hour. In how many hours will they be 144 miles apart? (See Fig. 2.2)

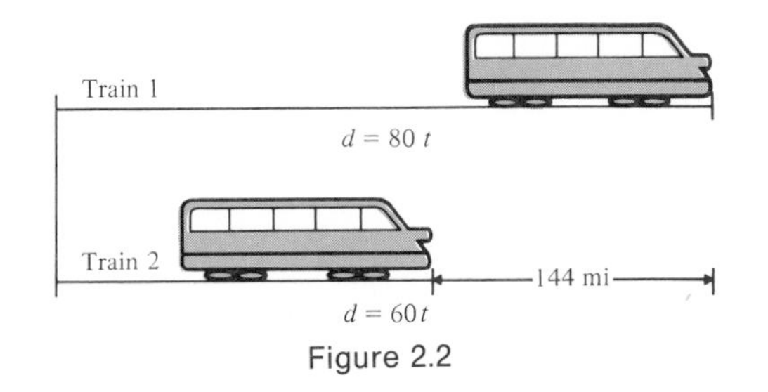

Figure 2.2

Solution: To solve this problem we will use the formula distance = rate · time.

Let t = time when they are 144 miles apart

Train	Rate	Time	Distance
1	80	t	$80t$
2	60	t	$60t$

The difference in their distances is 144 miles. Thus

$$\begin{aligned} \text{distance train 1} - \text{distance train 2} &= 144 \\ 80t - 60t &= 144 \\ 20t &= 144 \\ t &= 7.2 \end{aligned}$$

In 7.2 hours the trains will be 144 miles apart. ■

EXAMPLE 4 Mrs. Sanders and her daughter Teresa jog regularly. Mrs. Sanders jogs at 5 miles per hour, Teresa at 4 miles per hour. Teresa begins jogging at noon. Mrs. Sanders begins at 12:30 P.M. and travels the same straight path.

(a) Determine at what time the mother and daughter will meet.
(b) How far from the starting point will they be when they meet?

Solution: (a) Since Mrs. Sanders is the faster jogger, she will cover the same distance in a shorter period. When they meet they will both have traveled the same distance.

Let t = time Teresa is jogging

then $t - \frac{1}{2}$ = time Mrs. Sanders is jogging

Jogger	Rate	Time	Distance
Teresa	4	t	$4t$
Mrs. Sanders	5	$t - \frac{1}{2}$	$5(t - \frac{1}{2})$

When they meet they will both have covered the same distance from the starting point.

$$\begin{aligned} 4t &= 5\left(t - \frac{1}{2}\right) \\ 4t &= 5t - \frac{5}{2} \\ -t &= \frac{-5}{2} \\ t &= \frac{5}{2} \end{aligned}$$

They will meet in $2\frac{1}{2}$ hours after Teresa begins jogging or at 2:30 P.M.

(b) The distance can be found using either Mrs. Sanders' or Teresa's rate. We will use Teresa's.

$$d = rt$$
$$= 4\left(\frac{5}{2}\right) = \frac{20}{2} = 10 \text{ miles}$$

They will meet 10 miles from the starting point. ■

In Example 4, would the answer have changed if we let t represent the time Mrs. Sanders is jogging rather than the time Teresa is jogging? Try it and see.

Mixture Problems

2 Any problem where two or more quantities are combined to produce a different quantity, or where a single quantity is separated into two or more different quantities may be considered a mixture problem.

EXAMPLE 5 Liz Kendall has $10,000 that she wishes to invest in two stocks, Kaneb Services and Allied Chemical Corporation. Kaneb Stock is selling at $18 a share and Allied Chemical is selling at $42 a share.

(a) How many shares of each stock will she purchase if she wishes to purchase five times as many shares of Kaneb as Allied (only whole shares of stock may be purchased).
(b) How much money will be left?

Solution: (a) Let x = number of shares of Allied Chemical
then $5x$ = number of shares of Kaneb

The amount of money to purchase a stock is found by multiplying the cost of a stock by the number of shares purchased. Let us again construct a table.

Stock	Price	Number of Shares	Cost of Stock
Allied	42	x	$42x$
Kaneb	18	$5x$	$18(5x)$

$$\text{cost of Allied} + \text{cost of Kaneb} = \text{total cost}$$
$$42(x) + 18(5x) = 10{,}000$$
$$42x + 90x = 10{,}000$$
$$132x = 10{,}000$$
$$x = \frac{10{,}000}{132} = 75.76$$

Since Liz can purchase only whole shares, 75 shares of Allied and $5(75) = 375$ shares of Kaneb will be purchased.

(b) Total spent = (no. of shares Allied · cost) + (no. of shares Kaneb · cost)

$$\begin{aligned} &= (75 \cdot 42) + (375 \cdot 18) \\ &= 3150 + 6750 \\ &= 9900 \end{aligned}$$

The amount of money remaining is \$10,000 − \$9900 = \$100 ■

EXAMPLE 6 Ali, a pharmacist, has both 6% and 15% phenobarbitol solutions. He receives a prescription for 0.5 liter of an 8% phenobarbitol solution. How much of each solution must he mix to fill the prescription?

Solution:

$$\begin{aligned} \text{Let } x &= \text{number of liters of 6\% solution} \\ \text{then } 0.5 - x &= \text{number of liters of 15\% solution} \end{aligned}$$

The amount of phenobarbitol in a solution is found by multiplying the percent strength of phenobarbitol in the solution by the volume of the solution (see Fig. 2.3).

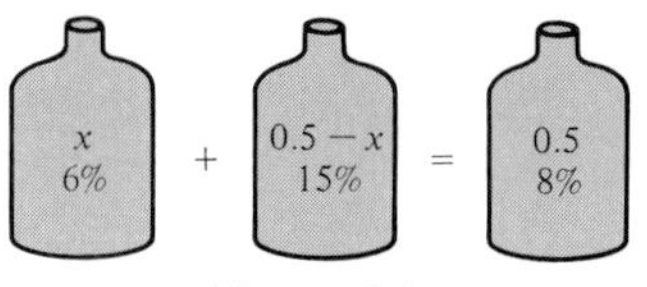

Figure 2.3

Strength of Solution	Number of Liters	Amount of Phenobarbitol
0.06	x	$0.06x$
0.15	$0.5 - x$	$0.15(0.5 - x)$
0.08	0.5	$0.08(0.5)$

amount of phenobarbitol in 6% solution + amount of phenobarbitol in 15% solution = amount of phenobarbitol in mixture

$$\begin{aligned} 0.06x + 0.15(0.5 - x) &= 0.08(0.5) \\ 0.06x + 0.075 - 0.15x &= 0.04 \\ 0.075 - 0.09x &= 0.04 \\ -0.09x &= 0.04 - 0.075 \\ -0.09x &= -0.035 \\ x &= \frac{-0.035}{-0.09} = 0.39 \text{ (to nearest hundredth)} \end{aligned}$$

Ali must mix 0.39 liter of the 6% solution and $0.5 - x$ or $0.5 - 0.39 = 0.11$ liter of the 15% solution to make 0.5 liter of an 8% solution. ■

EXAMPLE 7 A hot dog stand in Chicago sells hot dogs for \$1.00 each and potato knishes for \$1.25 each. If the total sales for the day equal \$307.50, and a total of 278 items were sold, how many of each item were sold.

Solution:

Let $x =$ number of hot dogs sold
then $278 - x =$ number of knishes sold

Item	Cost of Item	Number of Items	Total Value
hot dogs	1.00	x	$1.00x$
knishes	1.25	$278 - x$	$1.25(278 - x)$

total sales of hot dogs + total sales of knishes = total sales

$$1.00x + 1.25(278 - x) = 307.50$$
$$1.00x + 347.5 - 1.25x = 307.50$$
$$-.25x + 347.5 = 307.50$$
$$-.25x = -40$$
$$x = \frac{-40}{-.25} = 160$$

Therefore 160 hot dogs and 278 − 160 or 118 knishes were sold. ■

Exercise Set 2.3

For each exercise, write an equation that can be used to solve the problem. Solve the equation and answer the question asked.

1. Holley is on a cross-country ski team. If she averages 8 miles per hour, how far will she ski in 1.4 hours?
2. How fast must a car travel to cover 140 miles in $3\frac{1}{2}$ hours?
3. An airplane travels 580 miles per hour. How long will it take the plane to travel 3800 miles?
4. Mr. Donovan can mow 12,000 square feet per hour. How long will it take him to mow an area of 26,000 square feet?
5. The Apollo II took approximately 87 hours to reach the moon, a distance of about 238,000 miles. Find the Apollo's average speed.
6. A drain valve will drain a swimming pool at a rate of 600 gallons of water per hour. How long will it take to drain a pool containing 13,500 gallons of water?
7. A tree grows an average of 3.7 feet per year. How many feet will the tree grow in $10\frac{1}{2}$ years?
8. A hose can fill a 12,000-gallon swimming pool in $5\frac{2}{3}$ hours. Find the rate of flow of water through the hose.
9. At what rate must a photocopying machine make copies if it is to make 125 copies in 2.3 minutes?
10. Sandra can lay 52 bricks per hour. How long will it take her to lay 412 bricks?
11. A laser can cut through a steel door at the rate of $\frac{1}{5}$ centimeter per minute. How thick is the door if it requires 32 minutes to cut through the steel door?
12. If a submarine travels 135 knots in $6\frac{3}{5}$ hours, find the average speed of the submarine.
13. Under certain conditions, sound travels at 1080 feet per second. If a loud explosion occurs $2\frac{1}{2}$ miles away from you, how long will it take before you hear the noise? There are 5280 feet in a mile.
14. A patient is to receive 1200 cubic centimeters of intravenous fluid over a period of 3 hours. What should be the intravenous flow rate?
15. Two planes leave Cleveland at the same time. One plane flies east at 550 miles per hour. The other flies west at 650 miles per hour. In how many hours will they be 3000 miles apart?
16. Two trains leave Jacksonville going in the same direction along parallel tracks. Train 1 travels at 75 miles per hour, train 2 at 60 miles per hour. In how many hours will they be 105 miles apart?
17. Two families on vacation leave Pittsburgh and travel in opposite directions. After 3 hours the two cars are 330 miles apart. If one car travels at 60 miles per hour, find the speed of the other car.
18. Two rockets are launched from Cape Canaveral. The first rocket, launched at noon, will travel at 8000 miles per hour. The second rocket will be launched some time later and travel at 9500 miles per hour. When should the second

rocket be launched if the rockets are to meet at a distance of 38,000 miles from Earth?

19. Under certain conditions, sound travels at approximately 1080 feet per second. If you yell down a canyon and hear your echo in 1.7 seconds, approximately how deep is the canyon?

20. Two trains 520 miles apart travel toward each other on different tracks. One travels at 60 miles per hour, the other at 70 miles per hour. When will they pass each other?

For each exercise, write an equation that can be used to solve the problem. Solve the equation and answer the question asked.

21. Prime Computer's stock is selling at \$45 per share and American Motors' stock is selling at \$14 per share. Bob Davis has a maximum of \$9000 to invest. He wishes to purchase five times as many shares of American Motors as of Prime Computer.
   (a) How many shares of each will he purchase? Stocks can be purchased only in whole shares.
   (b) How much money will be left?

22. Mr. Templeton invests \$11,000 for one year, part at 9% and part at 10%. How much money was invested at each interest rate if the total interest earned from both investments is \$1050? Use interest = principal · rate · time.

23. Ms. McKane invested \$8000 for one year, part at 7% and part at $5\frac{1}{4}\%$. If she earned a total interest of \$458.50, how much was invested at each rate?

24. The Bernhams decide to invest \$6000 in two stocks, Sears Department Stores and U.S. Steel. They wish to purchase three times as many shares of Sears as of U.S. Steel.
   (a) If Sears is selling at \$34 a share and U.S. Steel at \$23, how many shares of each will be purchased?
   (b) How much money will be left?

25. Barbara Anders invested \$8000 for one year, part at 6% and part at 10%. How much was invested in each account if the same amount of interest was received from each account?

26. Hal invested \$4000 for one year in two savings accounts. He invested part of the money at 9% interest and the rest at 14%. If the total interest received is \$485, how much was invested in each account?

27. Connie invested \$6500 for one year in two savings accounts. She invested part of the money at 6% and the rest at 9%. If the total interest received is \$516, how much was invested in each account?

28. The admission at an ice hockey game is \$5.50 for adults and \$3.50 for children. A total of 650 tickets were sold, how many tickets were sold to children and how many to adults if a total of \$3095 was collected?

29. A movie ticket for an adult costs \$4.00 and a child's ticket costs \$2.25. A total of 172 tickets are sold for a given show. If \$597 is collected for the show, how many adults and how many children attended the show?

30. Victor has a total of 33 dimes and quarters. The total value of the coins is \$4.50. How many dimes and how many quarters does he have?

31. Jackie has a total of 83 nickels and dimes. The total value of the coins is \$5.35. How many nickels and how many dimes does Jackie have?

32. Dan has a total of 12 bills in his wallet. Some are \$5 and the rest are \$10 bills. The total value of the bills is \$115. How many \$5 bills and how many \$10 bills does he have?

33. Diedre holds two part-time jobs. One job pays \$4.00 per hour and the other pays \$4.50 per hour. Last week she earned a total of \$78 and worked for a total of 18 hours. How many hours did she work on each job?

34. Jim Kelly sells almonds for \$4.00 per pound, and walnuts for \$3.20 per pound. How many pounds of each should he mix to produce a 30-pound mixture that costs \$3.50 per pound?

35. How many pounds of coffee costing \$3.20 per pound must Larry mix with 18 pounds of coffee costing \$2.80 per pound to produce a mixture that costs \$3.10 per pound?

36. Milton mixes cookies costing \$1.60 per pound with cookies costing \$1.90 per pound. How many pounds of each should he use to make 60 pounds that sell for \$1.70 per pound?

37. How many ounces of water should the chemist add to 16 ounces of a 25% sulfuric acid solution to reduce it to a 10% solution?

38. How many ounces of pure vinegar should the cook add to 40 ounces of a 10% vinegar solution to make it a 25% vinegar solution?

39. How many quarts of pure antifreeze should Mr. Alberts add to 10 quarts of a 20% antifreeze solution to make a 50% antifreeze solution?

40. Lionel, a chemist, has 1 liter of a 20% sulfuric acid solution. How much of a 12% sulfuric acid solution must be mixed to make a 15% sulfuric acid solution?

41. Six quarts of punch contain 12% alcohol. How much pure soda water must the Joneses add to reduce the alcohol content to 10%?

42. A certain type of engine uses a fuel mixture of 15 parts of gasoline to 1 part of oil. How much pure gasoline must be mixed with a gasoline–oil mixture, which is 75% gasoline, to make 8 quarts of the mixture to run the engine?

43. Fifty pounds of a cement–sand mixture is 40% sand. How many pounds of sand must be added for the resulting mixture to be 60% sand?

44. Two brothers, Roland and Alexander, organize a business such that Roland receives 12% more of the total profits than Alexander. At the end of their first year the business's total profit is $52,000. How much should each brother receive?

45. Treetop Juice Company sells 8-ounce cans of apple juice for 40 cents and 8-ounce cans of apple drink for 16 cents each. They wish to market and sell cans of a juice drink for 25 cents that is part juice and part drink. How many ounces of each will be used if the juice drink is to be sold in 8-ounce cans?

46. Sundance dairy has 400 quarts of whole milk containing 5% butterfat. How many quarts of low-fat milk containing 1.5% butterfat should be added to produce milk containing 2% butterfat?

47. Some states allow a husband and wife to file individual state tax returns (on a single form) even though they file a joint federal return. It is usually to the taxpayer's advantage to do this when both the husband and wife work. The smallest amount of tax owed (or the largest refund) will occur when the husband's and wife's taxable incomes are the same.

Mr. Clar's 1988 taxable income was $18,200 and Mrs. Clar's income for that year was $22,450. The Clars' total tax deduction for the year was $6400. This deduction can be divided between Mr. and Mrs. Clar in any way they wish. How should the $6400 be divided between them to result in each person's having the same taxable income and therefore the greatest tax return?

JUST FOR FUN

1. (a) A videocassette tape is 246 meters in length. Some videocassette recorders have three tape speeds: SP for standard play; LP for long play, and SLP for super long play (some recorders refer to SLP as EP for extended play). If the tape runs for 2 hours at SP or 4 hours at LP or 6 hours at SLP, find the rate of speed of the tape at all three speeds.

(b) If a new 8-hour tape is to be developed for the SLP rate, find the length of the new tape.

2. Two cars labeled *A* and *B* are in a 500-lap race. Each lap is 1 mile. The lead car, *A*, is averaging 125 miles per hour when it reaches the halfway point. Car *B* is exactly 6.2 laps behind.

(a) Find the average speed of car *B*.

(b) At that instant how far behind, in seconds, is car *B* from car *A*?

3. Radar and sonar determine the distance of an object by emitting radio waves that travel at the speed of light, approximately 1000 feet per microsecond (a millionth of a second) in air. The device also determines the time it takes for its signal to travel to the object and return from the object.

Object ⇄ Radar

If radar determines that it takes 0.6 second to receive the echo of its signal, how far is the object from the radar device?

4. Lester knits at a rate of eight stitches per minute. He is planning to knit an afghan 4 feet by 6 feet. How long will it take Lester to knit the afghan if the instructions on the skein of wool indicate that four stitches equal 1 inch and six rows equal 1 inch?

5. The radiator of an automobile has a capacity of 16 quarts. It is presently filled with a 20% antifreeze solution. How many quarts must Jorge drain and replace with pure antifreeze to make the radiator contain a 50% antifreeze solution?

6. Before a bicycle rally, Seymour calculates that if he races at 15 miles per hour, he will pass the noon check-point 1 hour too soon; if he travels at 10 miles per hour, he will arrive an hour late. How far away is the check-point? What is the starting time?

2.4 Formulas

1 *Use subscripts and Greek letters in formulas.*

2 *Solve for a variable in an equation or formula.*

1 **Literal equations** are equations that have more than one letter. **Formulas** are literal equations that are used to represent a scientific or real-life principal in mathematical terms.

Examples of literal equations	*Examples of formulas*
$5y = 2x + 3$	$A = P(1 + rt)$ from Business
$x + 2y + 3z = 5$	$V = \frac{1}{2}at^2$ from Physics

Often a formula will contain subscripts. **Subscripts** are numbers (or other variables) placed below and to the right of variables. They are often used to help clarify a formula. For example, if a formula contains two velocities, the original velocity and the final velocity, these velocities might be symbolized as V_0 and V_f, respectively. Subscripts are read using the word "sub"; for example, V_f is read "V sub f" and x_2 is read "x sub 2."

Many mathematical and scientific formulas use *Greek letters*. Examples of the use of Greek letters are given in Table 2.

Table 2 **USE OF GREEK LETTERS IN SELECTED FORMULAS**

Formula	Greek Letter	Greek Letter Represents:
$A = \pi r^2$	π (pi)	A constant
$S = r\theta$	θ (theta)	Angle measurement
$v = \omega r$	v (nu)	Linear velocity
	ω (omega)	Angular velocity
$z = \dfrac{x - \mu}{\sigma}$	μ (mu)	Mean of set of data
	σ (sigma)	Standard deviation of a set of data
$\mu = f\lambda$	λ (lambda)	Length of wavelength
$R = e\sigma T^4$	σ (sigma)	A constant

Other Greek letters commonly used are: γ (gamma), α (alpha), ε (epsilon), ρ (rho), δ (delta), ϕ (phi), β (beta), and χ (chi). The Greek alphabet, like our own, has both upper and lower case letters. For example, σ represents the lower case sigma and Σ represents the upper case sigma.

2 Often in science or mathematics courses we are given a formula or equation solved for one variable and asked to solve it for a different variable. To do this, treat each variable in the equations, except the one you are solving for, as if it were a constant. Then solve for the desired variable using the properties discussed previously. To solve for a given variable it is necessary to get that variable all by itself on one side of the equal sign.

In Chapter 3 we will graph equations. To graph an equation it is sometimes necessary to solve the equation for the variable y. Example 1 illustrates one procedure for doing this.

EXAMPLE 1 Solve the equation $2x - 3y = 6$ for y.

Solution: We must isolate the term containing y.

$$2x - 3y = 6$$

$$2x - 2x - 3y = -2x + 6 \quad \text{subtract } 2x \text{ from both sides of equation}$$

$$-3y = -2x + 6$$

$$\frac{-3y}{-3} = \frac{-2x + 6}{-3} \quad \text{divide both sides of equation by } -3$$

$$y = \frac{-2x + 6}{-3}$$

Since we do not want to leave the answer with a negative number in the denominator we multiply both the numerator and the denominator by -1 to get

$$y = \frac{2x - 6}{3} \quad \text{or} \quad y = \frac{2}{3}x - 2 \quad \blacksquare$$

EXAMPLE 2 The formula for the area, A, of a triangle is $A = \frac{1}{2}bh$, where b is the length of the base and h is the height. Solve this formula for the height, h.

Solution: We are asked to express the height, h, of the triangle in terms of the triangle's area, A, and base, b. Since we are solving for h, we must isolate the h on one side of the equation. We must use the appropriate properties to remove the $\frac{1}{2}$ and the b from the right side of the equation.

$$A = \frac{1}{2}bh$$

$$2A = 2\left(\frac{1}{2}\right)bh \quad \text{multiply both sides of the equation by 2}$$

$$2A = bh$$

$$\frac{2A}{b} = \frac{bh}{b} \quad \text{divide both sides of the equation by } b$$

$$\frac{2A}{b} = h \quad \text{or} \quad h = \frac{2A}{b} \quad \blacksquare$$

EXAMPLE 3 A formula used in the study of statistics is $z = (x - \mu)/\sigma$. Solve this equation for x.

Solution: Note that this formula contains the Greek letters μ (mu) and σ (sigma). Treat them the same as you would any other letter.

$$z = \frac{x - \mu}{\sigma}$$

$$\sigma \cdot z = \frac{x - \mu}{\sigma} \cdot \sigma \quad \text{multiply both sides of equation by } \sigma$$

$$\sigma z = x - \mu$$

$$\sigma z + \mu = x - \mu + \mu \quad \text{add } \mu \text{ to both sides of equation}$$

$$\sigma z + \mu = x$$

The answer may be written in a number of other forms. Other acceptable answers include $x = \mu + \sigma z$ and $x = \mu + z\sigma$. $\blacksquare$

EXAMPLE 4 A formula that is of importance to most of us is the *tax free yield formula*, $T_f = T_a(1 - F)$. This formula can be used to convert a taxable yield, T_a, into its equivalent tax free yield, T_f, where F is the federal income tax bracket of the individual.

(a) Mary is in a 28% income tax bracket. Find the equivalent tax-free yield of a 12% taxable investment.

(b) Solve this equation for T_a, that is, write an equation for taxable yield in terms of tax-free yield.

Solution: (a)

$$\begin{aligned} T_f &= .12(1 - .28) \\ &= .12(.72) \\ &= .0864 \text{ or } 8.64\% \end{aligned}$$

Thus, for Mary, or anyone else in a 28% income tax bracket, a 12% taxable yield is equivalent to an 8.64% tax-free yield.

(b)

$$T_f = T_a(1 - f)$$

$$\frac{T_f}{1-f} = \frac{T_a(1-f)}{1-f} \qquad \text{divide both sides of equation by } 1 - f$$

$$\frac{T_f}{1-f} = T_a \quad \text{or} \quad T_a = \frac{T_f}{1-f} \qquad ■$$

EXAMPLE 5 A formula used in banking is $A = P(1 + rt)$. A represents the amount that must be repaid to the bank when P dollars are borrowed at simple interest rate, r, for time, t. Solve this equation for time, t.

Solution:

$$A = P(1 + rt)$$

$$A = P + Prt \qquad \text{distributive property}$$

$$A - P = P - P + Prt \qquad \text{subtract } P \text{ from both sides of equation to isolate term containing the variable } t$$

$$A - P = Prt$$

$$\frac{A-P}{Pr} = \frac{Prt}{Pr} \qquad \text{divide both sides of equation by } Pr \text{ to isolate } t$$

$$\frac{A-P}{Pr} = t \qquad ■$$

Exercise Set 2.4

Solve each equation for y.

1. $2x + y = 3$

2. $3x + y = 5$

3. $2x + 3y = 6$

4. $3x + 2y = 8$

5. $x - y = 8$

6. $2x - y = -5$

7. $2x - 4y = 6$
8. $5x - 3y = -4$
9. $2y = 8x - 3$
10. $-4y = -6x - 4$
11. $4x = 2y - 6$
12. $5x = -y + 4$
13. $6 - 3y = 3x$
14. $5x + 2y = 9$
15. $-3y + 8 = 9x$
16. $3x - 4y = 12$

Solve for the variable indicated.

17. $A = lw$, for l
18. $d = rt$, for t
19. $d = rt$, for r
20. $C = \pi d$, for d
21. $A = \frac{1}{2}bh$, for b
22. $C = 2\pi r$, for r
23. $i = prt$, for t
24. $P = 2l + 2w$, for w
25. $P = 2l + 2w$, for l
26. $V = \frac{1}{3}Bh$, for B
27. $V = \frac{1}{3}Bh$, for h
28. $V = lwh$, for h
29. $V = \pi r^2 h$, for h
30. $V = \frac{1}{3}lwh$, for l
31. $z = \frac{x - \mu}{\sigma}$, for σ
32. $z = \frac{x - \mu}{\sigma}$, for μ
33. $P = I^2 R$, for R
34. $I = P + Prt$, for r
35. $A = \frac{1}{2}h(b_1 + b_2)$, for h
36. $A = \frac{1}{2}h(b_1 + b_2)$, for b_1
37. $y = mx + b$, for m
38. $IR + Ir = E$, for R
39. $y - y_1 = m(x - x_1)$, for m
40. $R_T = \frac{R_1 + R_2}{2}$, for R_1
41. $S = \frac{n}{2}(f + l)$, for n
42. $S = \frac{n}{2}(f + l)$, for l
43. $\mathrm{C} = \frac{5}{9}(\mathrm{F} - 32)$, for F
44. $\mathrm{F} = \frac{9}{5}\mathrm{C} + 32$, for C
45. $y = \frac{kx}{z}$, for z
46. $\frac{P_1}{T_1} = \frac{P_2}{T_2}$, for T_2
47. $F = \frac{km_1 m_2}{d^2}$, for m_1
48. $A = \frac{r^2\theta}{2}$, for θ
49. $z = \frac{\bar{x} - \mu}{\frac{\sigma}{\sqrt{n}}}$, for $\bar{x}$
50. $z = \frac{\bar{x} - \mu}{\frac{\sigma}{\sqrt{n}}}$, for σ

JUST FOR FUN

1. A formula used in the study of electronics is $R_T = \frac{R_1 R_2}{R_1 + R_2}$. Solve this equation for R_1. Hint – factoring is required.
2. A formula useful in finding the center of gravity (or weighted mean) is

$$x = \frac{m_1 x_1 + m_2 x_2 + m_3 x_3}{m_1 + m_2 + m_3}$$

(a) Solve the equation above for x_1.
(b) Solve for m_1.

2.5 Solving Linear Inequalities

1. *Solve inequalities.*
2. *Graph solutions on the number line, and express solutions in interval notation and as solution sets.*
3. *Solve "and" type compound inequalities.*
4. *Solve continued inequalities.*
5. *Solve "or" type compound inequalities.*

---

1 The inequality symbols are as follows:*

Inequality Symbols

$>$	greater than
$\geq$	greater than or equal to
$<$	less than
$\leq$	less than or equal to

A mathematical expression containing one or more of the symbols above is called an **inequality.** The direction of the inequality symbol is sometimes called the **sense** of the inequality.

Examples of inequalities in one variable are

$$2x + 3 \leq 5, \qquad 4x > 3x - 5, \qquad -3 \leq -x + 5, \qquad 2x + 3 \geq 0$$

To solve an inequality, we must isolate the variable on one side of the inequality symbol. To isolate the variable, we use the same basic techniques used in solving equations.

Properties Used to Solve Inequalities

1. If $a > b$, then $a + c > b + c$
2. If $a > b$, then $a - c > b - c$
3. If $a > b$, *and* $c > 0$, then $ac > bc$
4. If $a > b$ *and* $c > 0$, then $\frac{a}{c} > \frac{b}{c}$
5. If $a > b$ *and* $c < 0$, then $ac < bc$
6. If $a > b$ *and* $c < 0$, then $\frac{a}{c} < \frac{b}{c}$

The first two properties state that the same number can be added to or subtracted from both sides of an inequality. The third and fourth properties state that both sides of an inequality can be multiplied or divided by any positive real number.

* $\neq$, not equal to, is also an inequality. $\neq$ means $<$ or $>$. Thus $2 \neq 3$ means $2 < 3$ or $2 > 3$.

The last two properties indicate that when both sides of an inequality are multiplied or divided by a negative number, the sense of the inequality changes.

Example of multiplication by a negative number		*Example of division by a negative number*	
Multiply both sides of the inequality by -1 and change the sense of the inequality.	$4 > 2$ $-1(4) < -1(2)$ $-4 < -2$	$10 \geq 4$ $\dfrac{10}{-2} \leq \dfrac{4}{-2}$ $-5 \leq -2$	Divide both sides of the inequality by -2 and change the sense of the inequality.

EXAMPLE 1 Solve the inequality $2x + 6 < 12$.

Solution:

$$2x + 6 < 12$$
$$2x + 6 - 6 < 12 - 6$$
$$2x < 6$$
$$\frac{2x}{2} < \frac{6}{2}$$
$$x < 3$$

Note that the solution set is $\{x | x < 3\}$. Any number less than 3 satisfies the inequality. ■

2 The solution set to an inequality in one variable can be graphed on the number line or written in interval notation.

Solution of inequality	*Solution indicated on number line*	*Solution represented in interval notation*
$x > a$		(a, ∞)
$x \geq a$		$[a, \infty)$
$x < a$		$(-\infty, a)$
$x \leq a$		$(-\infty, a]$
$a < x < b$		(a, b)
$a \leq x \leq b$		$[a, b]$
$a < x \leq b$		$(a, b]$
$a \leq x < b$		$[a, b)$

Note that a shaded circle on the number line indicates that the end point is part of the solution, and an unshaded circle indicates that the end point is not part of the solution. In interval notation brackets, [ ], are used to indicate that the end points are part of the solution; parentheses, ( ), indicate that the end points are not part of the solution. The symbol ∞ is read "infinity"; it indicates that the solution set continues indefinitely.

Solution of inequality	*Solution illustrated on number line*	*Solution represented in interval notation*
$x \geq 5$	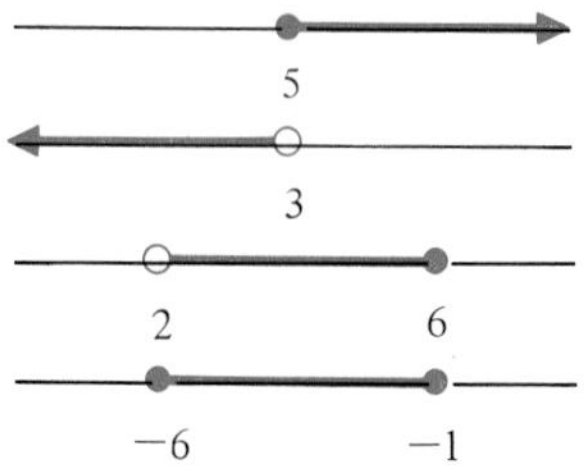	$[5, \infty)$
$x < 3$		$(-\infty, 3)$
$2 < x \leq 6$		$(2, 6]$
$-6 \leq x \leq -1$		$[-6, -1]$

EXAMPLE 2 Solve the following inequality and give the answer both on the number line and in interval notation.

$$3(x - 2) \leq 5x + 8$$

Solution:

$$3(x - 2) \leq 5x + 8$$
$$3x - 6 \leq 5x + 8$$
$$3x - 5x - 6 \leq 5x - 5x + 8$$
$$-2x - 6 \leq 8$$
$$-2x - 6 + 6 \leq 8 + 6$$
$$-2x \leq 14$$
$$\frac{-2x}{-2} \geq \frac{14}{-2}$$
$$x \geq -7$$

Number line 

Interval notation $[-7, \infty)$

Note that the solution set is $\{x | x \geq -7\}$. ■

In Example 2 we illustrated the answer on the number line, in interval notation, and as a solution set. Your instructor may indicate which way he or she prefers the answer be given.

EXAMPLE 3 Solve the inequality $\frac{1}{2}(4x + 14) \geq 5x + 4 - 3x - 10$.

Solution:

$$\frac{1}{2}(4x + 14) \geq 5x + 4 - 3x - 10$$
$$\frac{1}{2}(4x + 14) \geq 2x - 6$$
$$\left(\frac{1}{2}\right)(4x) + \left(\frac{1}{2}\right)(14) \geq 2x - 6$$
$$2x + 7 \geq 2x - 6$$
$$2x - 2x + 7 \geq 2x - 2x - 6$$
$$7 \geq -6$$

Since 7 is always greater than or equal to -6, the solution set is the set of all real numbers, $\mathbb{R}$. The solution set can also be indicated on the number line or given in interval notation.

or $(-\infty, \infty)$ ■

If the solution to Example 3 were $7 \leq -6$, the answer would have been no solution, since 7 is never less than or equal to -6. When an inequality has no solution, its solution set is the empty or null set, $\varnothing$ or $\{\ \}$.

EXAMPLE 4 Solve the inequality $\frac{4-2y}{3} \geq \frac{2y}{4} - 3$.

Solution: Multiply both sides of the inequality by the least common denominator, 12.

$$\overset{4}{\cancel{12}}\left(\frac{4-2y}{\underset{1}{\cancel{3}}}\right) \geq 12\left(\frac{2y}{4} - 3\right)$$

$$4(4) + 4(-2y) \geq \overset{3}{\cancel{12}}\left(\frac{2y}{\underset{1}{\cancel{4}}}\right) + 12(-3)$$

$$16 - 8y \geq 6y - 36$$

$$16 \geq 14y - 36$$

$$52 \geq 14y$$

$$\frac{52}{14} \geq y$$

$$\frac{26}{7} \geq y \quad \text{or} \quad y \leq \frac{26}{7}$$

$\frac{26}{7}$

In interval notation the answer is $(-\infty, \frac{26}{7}]$. The solution set is $\{y \mid y \leq \frac{26}{7}\}$. ■

HELPFUL HINT

> Note that in Example 4 we indicated that $\frac{26}{7} \geq y$ can be written $y \leq \frac{26}{7}$. Generally, when writing a solution to an inequality we write the variable on the left.
>
> For example,
>
> $a < x$ means $x > a$ (inequality symbol points to a in both cases)
>
> $a > x$ means $x < a$ (inequality symbol points to x in both cases)
>
> $-6 < x$ means $x > -6$ (inequality symbol points to -6 in both cases)
>
> and $-3 > x$ means $x < -3$. (inequality symbol points to x in both cases)

EXAMPLE 5 A small single-engine airplane can carry a maximum weight of 1500 pounds when its gas tank is full. Nancy Johnson, the pilot, has to transport boxes weighing 80 pounds.

(a) Write an inequality that can be used to determine the maximum number of boxes that Nancy can safely place on her plane if she weighs 125 pounds.

(b) Find the number of boxes that Nancy can transport.

Solution: (a) Let n = number of boxes.

$$\text{Nancy's weight} + \text{weight of } n \text{ boxes} \leq 1500$$
$$125 + 80n \leq 1500$$

(b)
$$125 + 80n \leq 1500$$
$$80n \leq 1375$$
$$n \leq 17.2$$

Therefore, Nancy can transport up to 17 boxes per trip. ■

EXAMPLE 6 A taxi's fare is \$1.75 for the first half-mile and \$1.10 for each additional half-mile. Any additional part of a half-mile will be rounded up to the next half-mile.

(a) Write an inequality that can be used to determine the maximum distance that Karen can travel if she has only \$12.35.
(b) Find the maximum distance that Karen can travel.

Solution: (a) Let x = number of half-miles after the first
then $1.10x$ = cost of traveling x additional half-miles

$$\text{cost of first half-mile} + \text{cost of additional half-miles} \leq \text{total cost}$$
$$1.75 + 1.10x \leq 12.35$$

(b)
$$1.75 + 1.10x \leq 12.35$$
$$1.10x \leq 10.60$$
$$x \leq \frac{10.60}{1.10}$$
$$x \leq 9.64$$

Karen can travel a distance less than or equal to 9 half-miles after the first half-mile, for a total of 10 half-miles, or 5 miles. If Karen travels for 10 half-miles after the first, she will owe $1.75 + 1.10(10) = 12.75$, which is more than she has. ■

EXAMPLE 7 For a business to realize a profit, the revenue (or income), R, must be greater than the cost, C. That is, a profit will be obtained only when $R > C$ (the company breaks even when $R = C$). A company that produces playing cards has a weekly cost equation of $C = 1525 + 1.7x$ and a weekly revenue equation of $R = 4.2x$, where x is the number of decks of playing cards produced and sold in a week. How many decks of cards must be produced and sold for the company to make a profit?

Solution: The company will make a profit when $R > C$ or

$$4.2x > 1525 + 1.7x$$
$$2.5x > 1525$$
$$x > \frac{1525}{2.5}$$
$$x > 610$$

The company will make a profit when more than 610 decks are produced and sold in a week. ■

Compound Inequalities

3 A **compound inequality** is formed by joining two inequalities with the word *and* or *or*.

Examples of compound inequalities

$$3 < x \quad \text{and} \quad x < 5$$
$$x + 4 > 3 \quad \text{or} \quad 2x - 3 < 6$$
$$4x - 6 \geq -3 \quad \text{and} \quad x - 6 < 5$$

The solution of a compound inequality using the word *and* is all the numbers that make *both* parts of the inequality true. Consider

$$3 < x \quad \text{and} \quad x < 5$$

What are the numbers that satisfy both inequalities? The numbers that satisfy both may be easier to see if we graph the solution to each inequality on a number line (see Fig. 2.4). Note that the numbers that satisfy both inequalities are the numbers between 3 and 5. The solution set is $\{x | 3 < x < 5\}$.

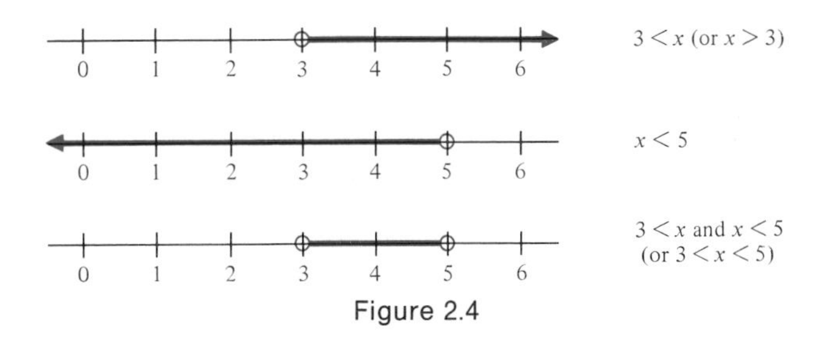

Figure 2.4

Recall from Chapter 1 that the intersection of two sets is the set of elements common to both sets. To find the solution set to an inequality containing the word *and*, take the *intersection* of the solution sets of the two inequalities.

EXAMPLE 8 Solve $x + 2 \leq 5$ and $2x - 4 > -2$.

Solution: Begin by solving each inequality separately.

$$\begin{aligned} x + 2 &\leq 5 \qquad & 2x - 4 &> -2 \\ x &\leq 3 & 2x &> 2 \\ & & x &> 1 \end{aligned}$$

Now take the intersection of the sets $\{x | x \leq 3\}$ and $\{x | x > 1\}$. Figure 2.5 illustrates the solution set is $\{x | 1 < x \leq 3\}$.

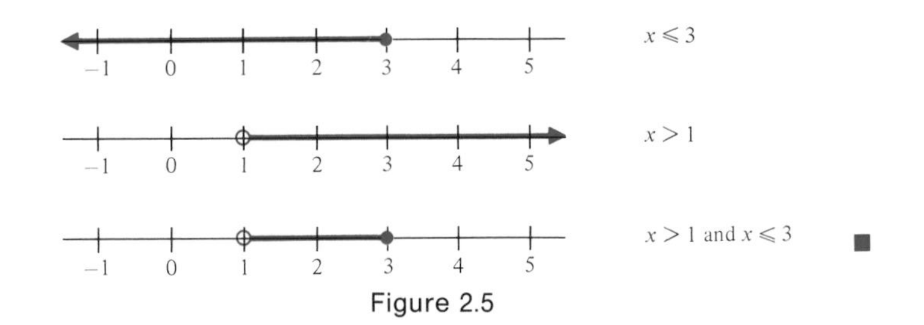

Figure 2.5 ■

4 Sometimes a compound inequality using the word *and* as the connecting word can be written in a shorter form. For example, $3 < x$ and $x < 5$ can be written as $3 < x < 5$. The word *and* does not appear when the inequality is written in this form, but it is implied. Inequalities written in the form $a < x < b$ are called **continued inequalities.** The compound inequality $1 < x + 5$ and $x + 5 \leq 7$ can be written $1 < x + 5 \leq 7$.

EXAMPLE 9 Solve $1 < x + 5 \leq 7$.

Solution: $1 < x + 5 \leq 7$ means $1 < x + 5$ and $x + 5 \leq 7$. Solve each inequality separately.

$$1 < x + 5 \qquad x + 5 \leq 7$$
$$-4 < x \qquad x \leq 2$$

Remember that $-4 < x$ means $x > -4$. Figure 2.6 illustrates that the solution set is $\{x \mid -4 < x \leq 2\}$. In interval notation the answer is $(-4, 2]$.

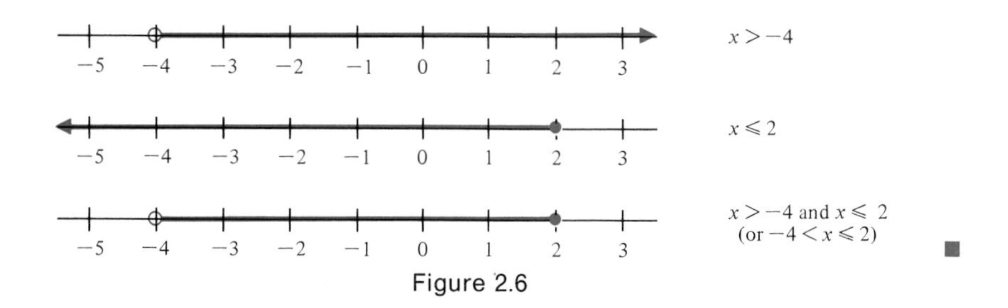

Figure 2.6

The inequality in Example 9, $1 < x + 5 \leq 7$, could have been solved in another way. We can still use the properties discussed earlier to solve continued inequalities. However, when working with such inequalities, whatever we do to one part we must do to all three parts. In Example 9, we could have subtracted 5 from all three parts to isolate the variable and obtain the answer.

$$1 < x + 5 \leq 7$$
$$1 - 5 < x + 5 - 5 \leq 7 - 5$$
$$-4 < x \leq 2$$

Note that this answer is the same as the answer obtained in example 9.

EXAMPLE 10 Solve the inequality.

$$-3 \leq 2x - 7 < 8$$

Solution: We wish to isolate the variable x. We begin by adding 7 to all three parts of the inequality.

$$-3 + 7 \leq 2x - 7 + 7 < 8 + 7$$
$$4 \leq 2x < 15$$

Now divide all three parts of the inequality by 2.

$$\frac{4}{2} \leq \frac{2x}{2} < \frac{15}{2}$$
$$2 \leq x < \frac{15}{2}$$

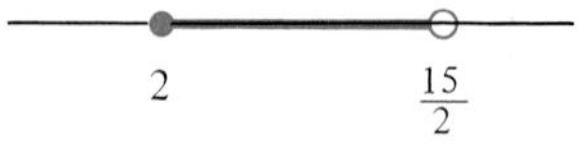

or $\left[2, \frac{15}{2}\right)$ ■

EXAMPLE 11 Solve the inequality.

$$-2 < \frac{4 - 3x}{5} < 8$$

Solution: Multiply all three parts by 5 to eliminate the denominator.

$$-2(5) < \cancel{5}\left(\frac{4 - 3x}{\cancel{5}}\right) < 8(5)$$
$$-10 < 4 - 3x < 40$$
$$-10 - 4 < 4 - 4 - 3x < 40 - 4$$
$$-14 < -3x < 36$$

At this point we divide all three parts of the inequality by -3. Remember that when we multiply or divide an inequality by a negative number, the sense of the inequality changes.

$$\frac{-14}{-3} > \frac{-3x}{-3} > \frac{36}{-3}$$
$$\frac{14}{3} > x > -12$$

Although $\frac{14}{3} > x > -12$ is correct, we generally write continued inequalities with the lesser value on the left. We will therefore rewrite the answer as

$$-12 < x < \frac{14}{3}$$

The answer may also be illustrated on the number line or written in interval notation.

or $(-12, \frac{14}{3})$ ■

HELPFUL HINT

You must be very careful when writing the solution to a continued inequality. In Example 11 we can change the answer from

$$\frac{14}{3} > x > -12 \quad \text{to} \quad -12 < x < \frac{14}{3}$$

This is correct since both say the x is greater than -12 and less than $\frac{14}{3}$. Notice that the inequality symbol in both cases is pointing to the smaller number.

In Example 11, had we written the answer $\frac{14}{3} < x < -12$, we would have given the **wrong** solution. Remember that the inequality $\frac{14}{3} < x < -12$ means that $\frac{14}{3} < x$ and $x < -12$. There is no number that is both greater than $\frac{14}{3}$ and less than -12. Also, by examining the inequality $\frac{14}{3} < x < -12$, it appears as if we are saying that -12 is a greater number than $\frac{14}{3}$, which is obviously *wrong*.

It would also be **wrong** to write the answer

$$\cancel{-12 < x > \tfrac{14}{3}} \quad \text{or} \quad \cancel{\tfrac{14}{3} < x > -12}.$$

EXAMPLE 12 An average greater than or equal to 80 and less than 90 will result in a final grade of B in a course. Steven received grades of 85, 90, 68, and 70 on his first four exams in the course. Within what range of grades will Steven's fifth and last exam result in his receiving a final grade of B in the course?

Solution: Let $x =$ Steven's last exam grade.

$$80 \le \text{average of five exams} < 90$$

$$80 \le \frac{85 + 90 + 68 + 70 + x}{5} < 90$$

$$80 \le \frac{313 + x}{5} < 90$$

$$400 \le 313 + x < 450$$

$$400 - 313 \le x < 450 - 313$$

$$87 \le x < 137$$

Steven would need a minimum grade of 87 to obtain a final grade of B. If the highest grade he could receive on the test is 100, it is impossible for him to obtain a final grade of A (90 average or higher). ■

5 The solution to a compound inequality using the word *or* is all the numbers that make *either* of the inequalities a true statement. Consider the compound inequality

$$x > 3 \qquad \text{or} \qquad x < 5$$

What are the numbers that satisfy the inequality? Let us graph the solution to each inequality on the number line (see Fig. 2.7). Note that every real number satisfies at least one of the two inequalities. The solution set to the compound inequality is all real numbers, $\mathbb{R}$.

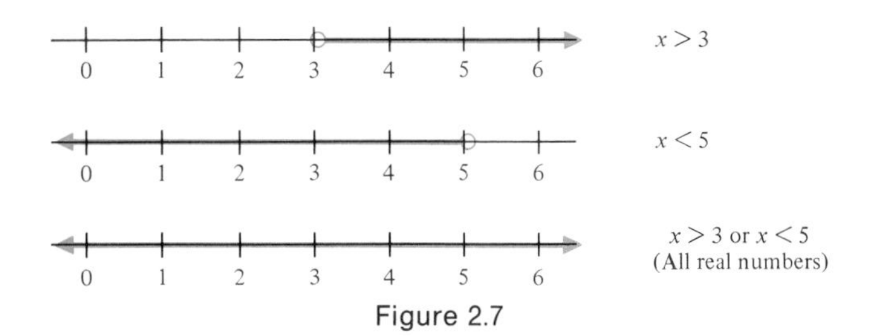

Figure 2.7

Recall from Chapter 1 that the *union* of two sets is the set of elements that belong to *either* of the sets. To find the solution set of an inequality containing the word *or*, take the union of the solution sets of the two inequalities that comprise the compound inequality.

EXAMPLE 13 Solve $x + 3 \le -1$ or $-4x + 3 < -5$.

Solution: Solve each inequality separately.

$$x + 3 \le -1 \qquad -4x + 3 < -5$$
$$x \le -4 \qquad -4x < -8$$
$$x > 2$$

Now graph each solution on number lines and then find the union (Fig. 2.8).

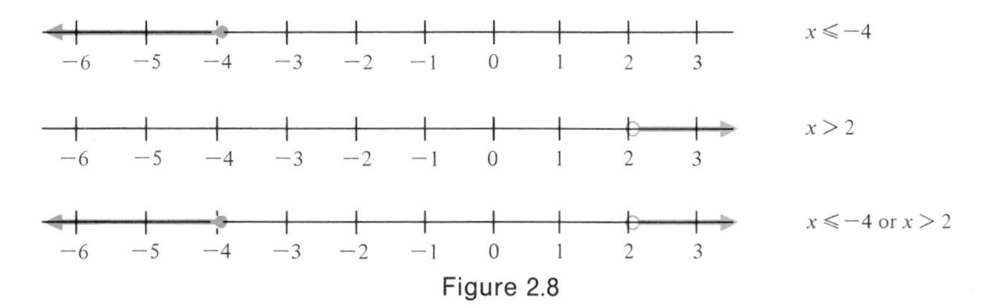

Figure 2.8

The solution set is $\{x | x \le -4\} \cup \{x | x > 2\}$. The union of these two sets can be written as $\{x | x \le -4 \text{ or } x > 2\}$. Thus the solution set is $\{x | x \le -4 \text{ or } x > 2\}$. In interval notation the answer is $(-\infty, 4] \cup (2, \infty)$. ■

We often encounter inequalities in our daily lives. For example, on a highway there may be a minimum speed of 30 mph and a maximum speed of 55 mph. A restaurant may have a sign stating that the maximum capacity is 300 people; and the minimum takeoff speed of an airplane may be 125 miles per hour.

Exercise Set 2.5

Solve the inequality and graph the solution on the number line.

1. $x + 3 < 8$
2. $x + 5 \ge -5$
3. $2x > 12$
4. $3x \le -5$
5. $2x + 3 > 4$
6. $3 - x < -4$
7. $4x + 5 > 3x$
8. $x + 4 < 2x - 3$
9. $4x + 3 \le -2x + 9$

10. $4(x - 2) \leq 4x - 8$

11. $14 > 3a - 6$

12. $4b - 6 \geq 2b + 12$

13. $-(x - 3) + 4 \leq -2x + 5$

14. $\frac{y}{3} + \frac{2}{5} \leq 4$

15. $2y - 6y + 10 \leq 2(-2y + 3)$

16. $\frac{c + 3}{2} + 5 > c + 2$

17. $(w - 5) \leq \frac{3}{4}(2w + 6)$

18. $4 - 3x < 7 + 2x + 4$

19. $4 + \frac{3x}{2} < 6$

20. $\frac{3y - 6}{2} > \frac{2y + 5}{6}$

21. $\frac{5 - 4y}{3} \leq 5 - 4y$

22. $\frac{3(x - 2)}{5} > \frac{5(2 - x)}{3}$

23. $x + 1 < 3(x + 2) - 2x$

24. $\frac{1}{2}\left(\frac{3}{5}y + 4\right) \leq \frac{1}{3}(y - 6)$

Solve the inequality by the method illustrated in Examples 10 and 11. Give the solution in interval notation.

25. $4 < x + 3 < 9$

26. $-2 \leq x - 5 < 7$

27. $-3 < 5x \leq 8$

28. $-2 < -4x < 8$

29. $4 \leq 2x - 3 < 7$

30. $-12 < 3x - 5 \leq -4$

31. $\frac{1}{2} < 3x + 4 < 6$

32. $-7 < 4 - 3x < 9$

33. $-6 < 2(x - 3) < 8$

34. $4 < \frac{4x - 3}{2} \leq 12$

35. $-12 \leq \frac{4 - 3x}{-5} < 2$

36. $\frac{3}{5} < \frac{-x - 5}{3} < 6$

37. $6 \leq -3(2x - 4) < 12$

38. $-7 < \frac{4 - 2x}{3} < \frac{1}{3}$

39. $-15 < \frac{3(x - 2)}{5} \leq 0$

40. $0 < \frac{2(x - 3)}{5} \leq 12$

41. $1 < \frac{4 - 6x}{2} < 5$

42. $\frac{1}{8} \leq 4 - 2(x + 3) \leq 5$

Solve the inequality and indicate the solution set.

43. $x < 4$ and $x > 2$

44. $x < 4$ or $x > 2$

45. $x < 2$ and $x > 4$

46. $x < 2$ or $x > 4$

47. $x + 2 < 3$ and $x + 1 > -2$

48. $2x - 3 \leq 5$ or $2x - 8 \geq 4$

49. $5x - 3 \leq 7$ or $-x + 3 < -5$

50. $-2x - 3 < 2$ and $x + 6 > 4$

51. $3x - 6 \leq 4$ or $2x - 3 < 5$

52. $-x + 6 > -3$ or $4x - 2 < 12$

53. $4x + 5 \geq 5$ and $3x - 4 \leq 2$

54. $x - 3 > -5$ and $-2x - 4 > -2$

55. $5x - 3 > 10$ and $4 - 3x < -2$

56. $x - 4 > 4$ or $3x - 5 \geq 1$

57. $4 - x < -2$ or $3x - 1 < -1$

58. $-x + 3 < 0$ or $2x - 5 \geq 3$

For each exercise, set up an inequality that can be used to solve the problem. Solve the problem and find the desired value.

59. Cal, a janitor, must move a large shipment of books from the first floor to the fifth floor. The sign on the elevator reads "maximum weight 900 pounds." If each box of books weigh 80 pounds, find the maximum number of boxes that Cal can place on the elevator.

60. If the janitor in Exercise 59, weighing 170 pounds, must ride up with the boxes, find the number of boxes of books that can be placed in the elevator.

61. A twin-engine plane must have a takeoff load of no more than 1800 pounds. If the passengers have a total weight of 725 pounds, find the maximum weight of luggage that the plane can carry.

62. A telephone operator informs a customer in a phone booth that the charge for calling Denver, Colorado, is $4.25 for the first 3 minutes and 48 cents for each additional minute. Any additional part of a minute will be rounded up to the nearest minute. Find the maximum time the customer can talk if he has only $9.50.

63. A downtown parking garage in Austin charges $0.75 for the first hour and $0.50 for each additional hour. How long can you park in the garage if you wish to pay no more than $3.75?

64. A truck can carry a maximum load of 3400 pounds. If the passengers weigh 585 pounds, how many bags of crushed stones weighing 120 pounds each can the truck carry?

65. Miriam is considering writing and publishing her own book. She estimates her revenue equation to be $R = 6.42x$ and her cost equation to be $C = 10{,}025 + 1.09x$, where x is the number of books she sells. Find the number of books she must sell to make a profit. See Example 7.

66. Mark Nottingham is considering opening a dry-cleaning store. He estimates his cost equation to be $C = 8000 + 0.08x$ and his revenue equation to be $R = 1.85x$, where x is the number of garments dry cleaned in a year. How many garments must be dry-cleaned in a year for Mark to make a profit?

67. A nonprofit organization can purchase a \$40 bulk-mailing permit and then send bulk mail at a rate of 5.2 cents per piece. Without the permit each piece of bulk mail costs 11 cents. How many pieces of bulk mail would have to be mailed for it to be financially worthwhile for an organization to purchase the bulk-mailing permit?

68. The cost for mailing a package first class is 22 cents for the first ounce and 17 cents for each additional ounce. What is the maximum weight of a package that can be mailed first class for \$4.95?

69. To receive an A in a course you must obtain an average of 90 or higher on five exams. If Ray's first four exam grades are 90, 87, 96, and 95, what is the minimum grade Ray can receive on the fifth exam to get an A in the course?

70. To pass a course you need an average grade of 60 or more. If Maria's grades are 65, 72, 90, 47, and 62, find the minimum grade Maria can get on her sixth and last exam and pass the course.

71. For air to be considered "clean," the average of three pollutants must be less than 3.2 parts per million. If the first two pollutants are 2.7 and 3.42 ppm, what values of the third pollutant will result in clean air?

72. Ms. Mahoney's grades on her first four exams are 87, 92, 70, and 75. An average greater than or equal to 80 and less than 90 will result in a final grade of B. What range of grades on Ms. Mahoney's fifth and last exam will result in a final grade of B? Assume a maximum grade of 100.

73. The water acidity in a pool is considered normal when the average pH reading of three daily measurements is between 7.2 and 7.8. If the first two pH readings are 7.48 and 7.85, find the range of pH values for the third reading that will result in the acidity level being normal.

JUST FOR FUN

1. Russell's first five exams were 82, 90, 74, 76, and 68. The final exam for the course is to count one-third in computing the final average. A final average greater than or equal to 80 and less than 90 will result in a final grade of B. What range of final-exam grades will result in Russell receiving a final grade of B in the course? Assume a maximum grade of 100 is possible.

2.6 Solving Equations and Inequalities Containing Absolute Value

1. *Solve equations containing absolute value.*
2. *Solve inequalities of the form $|x| < a$.*
3. *Solve inequalities of the form $|x| > a$.*

Equations Containing Absolute Value

1 In Section 1.3 we introduced the concept of absolute value. We stated that the absolute value of a number may be considered the distance (without sign) from the number 0 on the number line. The absolute value of 3, written $|3|$, is 3 since it is 3 units from 0 on the number line. Similarly, the absolute value of -3, written $|-3|$, is also 3 since it is 3 units from 0 on the number line (see Fig. 2.9).

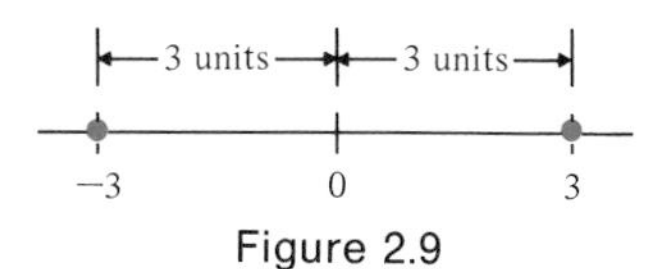

Figure 2.9

Consider the equation $|x| = 3$; what values of x make this equation true? We know that $|3| = 3$ and $|-3| = 3$. The solutions to $|x| = 3$ are 3 and -3. When solving the equation $|x| = 3$ we are finding the values that are 3 units from 0 on the number line. When solving an equation of the form $|x| = a, a \geq 0$, we are finding the values that are a units from 0 on the number line.

Equations of the Form $|x| = a$

If $|x| = a$ and $a > 0$, then $x = a$ or $x = -a$

EXAMPLE 1 Solve the equation $|x| = 4$.

Solution: Using the given rule, we get $x = 4$ or $x = -4$. The solution set is $\{-4, 4\}$. ■

EXAMPLE 2 Solve the equation $|x| = 0$.

Solution: The only real number whose absolute value equals 0 is 0. Thus $|x| = 0$ has the solution set $\{0\}$. ■

EXAMPLE 3 Solve the equation $|x| = -2$.

Solution: The absolute value of a number is never negative, so there are no solutions to this equation. The solution set is $\varnothing$. ■

EXAMPLE 4 Solve the equation $|2w - 1| = 5$.

Solution: If we consider $2w - 1$ to be x, then $2w - 1$ must be 5 units from 0 on the number line. Thus the quantity $2w - 1$ must be equal to 5 or -5.

$$\begin{aligned} 2w - 1 &= 5 \quad \text{or} \quad & 2w - 1 &= -5 \\ 2w &= 6 \quad \text{or} \quad & 2w &= -4 \\ w &= 3 \quad \text{or} \quad & w &= -2 \end{aligned}$$

Check:

$$\begin{aligned} w = 3,\quad |2(3) - 1| &= 5 & w = -2,\quad |2(-2) - 1| &= 5 \\ |6 - 1| &= 5 & |-4 - 1| &= 5 \\ |5| &= 5 & |-5| &= 5 \\ 5 &= 5 \quad \text{true} & 5 &= 5 \quad \text{true} \end{aligned}$$

The solution set is $\{-2, 3\}$. ■

EXAMPLE 5 Solve the equation $|\frac{2}{3}z - 6| + 4 = 6$.

Solution: First subtract 4 from both sides of the equation to get the absolute value alone on one side of the equation.

$$\left|\frac{2}{3}z - 6\right| + 4 = 6$$

$$\left|\frac{2}{3}z - 6\right| = 2$$

Now we proceed as before.

$$\begin{aligned} \frac{2}{3}z - 6 &= 2 \quad \text{or} \quad & \frac{2}{3}z - 6 &= -2 \\ \frac{2}{3}z &= 8 \quad \text{or} \quad & \frac{2}{3}z &= 4 \\ 2z &= 24 \quad \text{or} \quad & 2z &= 12 \\ z &= 12 \quad \text{or} \quad & z &= 6 \end{aligned}$$

The solution set is $\{6, 12\}$. ■

Inequalities Containing Absolute Value

2 We just showed how to solve equations of the form $|x| = a$. Now you will learn how to solve inequalities containing absolute values. We will work first with inequalities of the form $|x| < a$.

Earlier we stated that the solution set to $|x| = 3$ were the values that were exactly 3 units from 0 on the number line. The solution set to $|x| = 3$ is $\{-3, 3\}$. Similarly, we can state that $|x| < 3$ is the set of values that are less than 3 units from the number 0 on the number line. This includes all real numbers between -3 and 3 (see Figure 2.10). The solution set of $|x| < 3$ is $\{x \mid -3 < x < 3\}$.

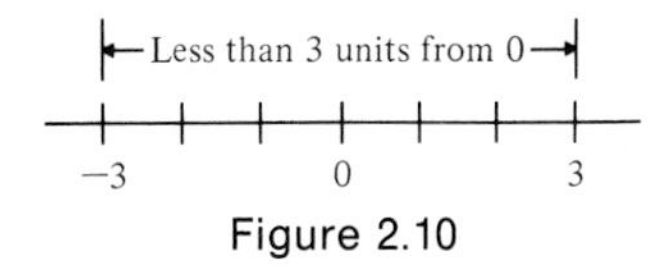

Figure 2.10

When we are asked to find the solution set to an inequality of the form $|x| < a$, we are finding the set of values that are less than a units from 0 on the number line. The solution set to $|x| \leq a$ is the set of values that are less than *or equal to a* units from 0 on the number line.

We can use the same reasoning process to solve more complicated problems, as shown in Example 6.

EXAMPLE 6 Solve the inequality $|2x - 3| < 5$.

Solution: The solution to this inequality will be the set of values such that the distance between $2x - 3$ and 0 on the number line will be less than 5 units (see Fig. 2.11). Using Fig. 2.11, we can see that $-5 < 2x - 3 < 5$.

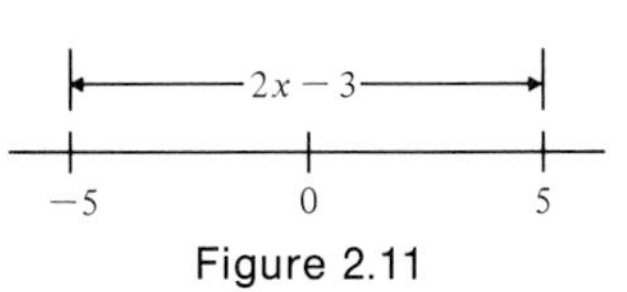

Figure 2.11

Solving we get

$$-5 < 2x - 3 < 5$$
$$-2 < 2x < 8$$
$$-1 < x < 4$$

The solution set is $\{x \mid -1 < x < 4\}$. When x is any number between -1 and 4, $2x - 3$ will be a number less than 5 units from 0 on the number line (or a number between -5 and 5). ■

Using the same reasoning process, we can see that to solve inequalities of the form $|x| < a$, we use the following procedure:

> **Inequalities of the Form |x| < a**
>
> If $|x| < a$ and $a > 0$, then $-a < x < a$

EXAMPLE 7 Solve the inequality $|3x - 4| \leq 5$ and graph the solution on the number line.

Solution: Since this inequality is of the form $|x| \leq a$, we write

$$-5 \leq 3x - 4 \leq 5$$
$$-1 \leq 3x \leq 9$$
$$-\frac{1}{3} \leq x \leq 3$$

$-\frac{1}{3}$ 3

■

EXAMPLE 8 Solve the inequality $|4 - x| + 1 < 3$ and graph the solution on the number line.

Solution: First isolate the absolute value by subtracting 1 from both sides of the inequality. Then solve as in the previous examples.

$$|4 - x| + 1 < 3$$
$$|4 - x| < 2$$
$$-2 < 4 - x < 2$$
$$-6 < -x < -2$$
$$-1(-6) > -1(-x) > -1(-2)$$
$$6 > x > 2$$
$$\text{or} \quad 2 < x < 6$$

■

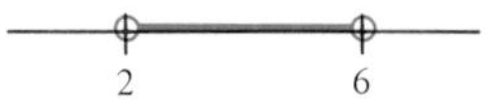

3 Now we look at inequalities of the form $|x| > a$. Consider $|x| > 3$. This inequality represents the set of values that are greater than 3 units from 0 on the number line. The solution to $|x| > 3$ is $x > 3$ or $x < -3$ (see Fig. 2.12).

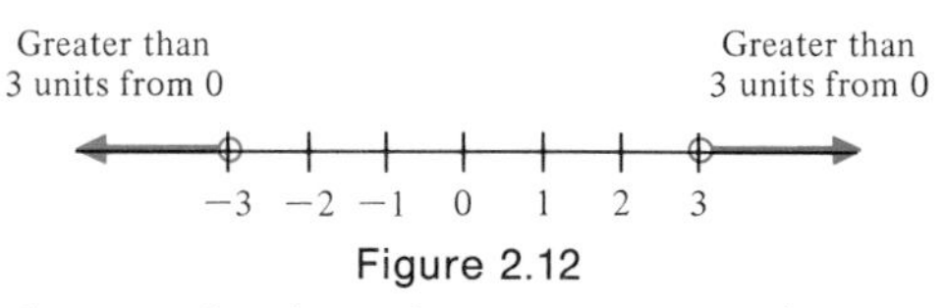

Figure 2.12

Similarly, $|x| > a$ is the set of values that are greater than a units from 0 on the number line.

EXAMPLE 9 Solve the inequality $|2x - 3| > 5$ and graph the solution on the number line.

Solution: The solution to $|2x - 3| > 5$ is the set of values such that the distance between $2x - 3$ and 0 on the number line will be greater than 5. The quantity $2x - 3$ must either be less than -5 or greater than $+5$ (see Fig. 2.13). Since $2x - 3$ must be either less than -5 or greater than 5 we set up and solve the following compound inequality

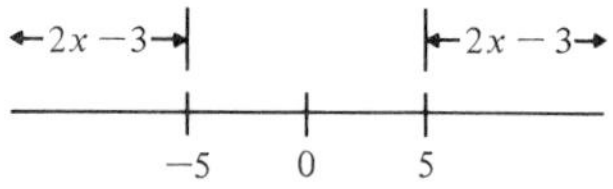

Figure 2.13

$$2x - 3 < -5 \quad \text{or} \quad 2x - 3 > 5$$
$$2x < -2 \qquad\qquad 2x > 8$$
$$x < -1 \qquad\qquad x > 4$$

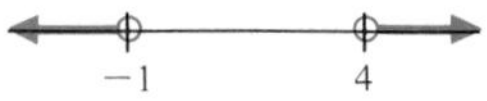

The solution set to $|2x - 3| > 5$ is $\{x | x < -1 \text{ or } x > 4\}$.

When x is any number less than -1 or greater than 4, $2x - 3$ will be greater than 5 units from 0 on the number line (or a number less than -5 or greater than 5). ■

Using the same reasoning process we can see that to solve inequalities of the form $|x| > a$ we use the following procedure.

> **Inequalities of the Form |x| > a**
>
> If $|x| > a$ and $a > 0$, then $x < -a$ or $x > a$

EXAMPLE 10 Solve the inequality $|2x - 5| \geq 3$ and graph the solution on the number line.

Solution: Since this inequality is of the form $|x| \geq a$ we use the procedure given above.

$$2x - 5 \leq -3 \quad \text{or} \quad 2x - 5 \geq 3$$
$$2x \leq 2 \qquad\qquad 2x \geq 8$$
$$x \leq 1 \qquad\qquad x \geq 4$$

The solution set is $\{x | x \leq 1 \text{ or } x \geq 4\}$.

−1 0 1 2 3 4 5 ■

EXAMPLE 11 Solve the inequality $\left|\dfrac{3x - 4}{2}\right| \geq \dfrac{5}{12}$.

Solution: Since this inequality is of the form $|x| \geq a$ we write,

$$\frac{3x - 4}{2} \leq -\frac{5}{12} \quad \text{or} \quad \frac{3x - 4}{2} \geq \frac{5}{12}$$

Now multiply both sides of the inequality by the least common denominator, 12.

$$\overset{6}{\cancel{12}}\left(\frac{3x - 4}{\underset{1}{\cancel{2}}}\right) \leq \frac{-5}{\cancel{12}} \cdot \cancel{12} \qquad \overset{6}{\cancel{12}}\left(\frac{3x - 4}{\underset{1}{\cancel{2}}}\right) \geq \frac{5}{\cancel{12}} \cdot \cancel{12}$$
$$6(3x - 4) \leq -5 \qquad 6(3x - 4) \geq 5$$
$$18x - 24 \leq -5 \qquad 18x - 24 \geq 5$$
$$18x \leq 19 \qquad 18x \geq 29$$
$$x \leq \frac{19}{18} \quad \text{or} \quad x \geq \frac{29}{18}$$

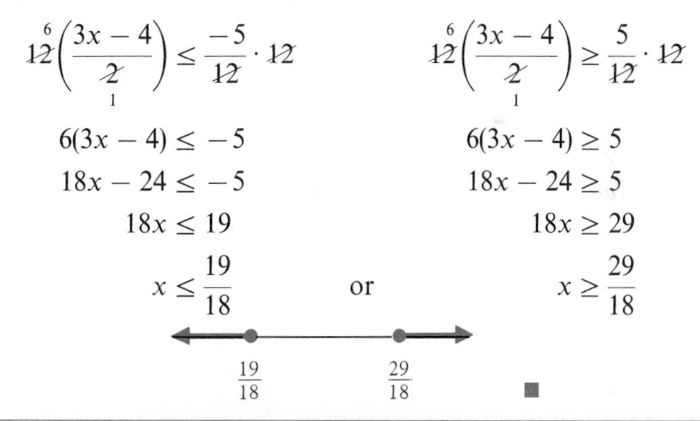

■

> **Summary of Procedures for Solving Equations and Inequalities Containing Absolute Value**
>
> For $a > 0$
>
> If $|x| = a$, then $x = a$ or $x = -a$
>
> If $|x| < a$, then $-a < x < a$
>
> If $|x| > a$, then $x < -a$ or $x > a$

Note that inequalities of the form $|x| > a$, where a is a negative value (such as $|2x - 3| > -2$) are true for all values of x. The solution set for such inequalities will be all real numbers. Inequalities of the form $|x| < a$, where a is a negative value (such as $|2x - 3| < -2$) are never true. The solution set for such inequalities will be the empty set.

Exercise Set 2.6

Find the solution set for each equation.

1. $|x| = 5$
2. $|y| = 7$
3. $|x| = 12$
4. $|y| = 3$
5. $|x| = 0$
6. $|w| = 4$
7. $|x| = -2$
8. $|x + 1| = 5$
9. $|x - 3| = 5$
10. $|x + 5| = 7$
11. $|3 + y| = \frac{3}{5}$
12. $|4 - x| = 5$
13. $|2w + 4| = 6$
14. $|3x - 4| = 0$
15. $|5 - 3x| = \frac{1}{2}$
16. $|4x + 2| = 9$
17. $|3(y + 4)| = 12$
18. $|4(x - 2)| = 18$
19. $\left|\frac{x - 3}{4}\right| = 5$
20. $\left|\frac{2x - 4}{5}\right| = 12$
21. $\left|\frac{3z + 5}{6}\right| = 9$
22. $|2x - 4| + 2 = 10$
23. $|4 - 2x| - 5 = 5$
24. $\left|\frac{3x - 2}{4}\right| - 5 = 1$
25. $\left|\frac{5x - 3}{2}\right| + 2 = 6$
26. $\left|\frac{2x + 3}{2}\right| + 1 = 4$

Find the solution set for each inequality.

27. $|y| \le 5$
28. $|x| \le 9$
29. $|x - 3| < 5$
30. $|x - 7| \le 9$
31. $|7 - x| < 5$
32. $|2y + 4| < 1$
33. $|3z - 5| \le 5$
34. $|x - 3| - 2 < 3$
35. $|2x - 5| + 3 \le 10$
36. $|2x + 3| - 5 \le 10$
37. $|4 - 3x| - 4 < 11$
38. $|4 + 3x| \le 9$
39. $|x - 5| \le \frac{1}{2}$
40. $\left|\frac{w + 4}{3}\right| < 4$
41. $\left|\frac{2x - 1}{3}\right| \le \frac{5}{3}$
42. $\left|5 - \frac{3x}{4}\right| < 8$
43. $\left|\frac{4c - 4}{5}\right| \le 8$
44. $\left|2\left(\frac{3 - x}{5}\right)\right| < \frac{9}{5}$

Find the solution set for each inequality.

45. $|x| > 3$
46. $|y| \ge 5$
47. $|z| \ge 2$
48. $|x + 4| > 5$
49. $|5 - x| \ge 3$
50. $|x + 5| > 9$
51. $|3x + 1| > 4$
52. $|4 - 3y| \ge 8$
53. $|5 + 2x| \ge 3$
54. $\left|\frac{6 + 2z}{3}\right| > 2$
55. $\left|\frac{5 - 3w}{4}\right| \ge 10$
56. $\left|\frac{3x + 4}{5}\right| > \frac{7}{5}$
57. $|4x - 3| + 2 > 7$
58. $|2x - 1| - 4 \ge 8$
59. $|2 - 3x| - 4 \ge -2$
60. $\left|\frac{2x - 3}{4}\right| - 1 > 3$
61. $\left|\frac{x}{2} + 4\right| \ge 5$
62. $\left|3\left(x + \frac{1}{2}\right)\right| > 5$
63. $\left|4 - \frac{3x}{5}\right| \ge 9$

JUST FOR FUN

1. Find all values of x such that $|x - 3| = |3 - x|$
2. Find all values of x and y such that $|x - y| = |y - x|$
3. Solve $|x + 6| = |2x - 3|$
4. Solve $|x - 3| = |x + 5|$

Summary

Glossary

Coefficient (or numerical coefficient): The numerical part of a term.
Compound inequality: Two inequalities joined with the word *and* or *or*.
Conditional equation: An equation true only under specific conditions.
Continued inequality: An inequality of the form $a < x < b$.
Degree of a term: The sum of the exponents in a term.
Equation: A mathematical statement of equality.
Equivalent equations: Equations with the same solution set.
Formula: An equation used to represent a scientific or real-life principle in mathematical terms.
Identity: An equation true for all real numbers.
Inconsistent equation: An equation that has no solution.
Inequality: A mathematical expression containing one or more inequality symbols.
Least common denominator: The smallest number divisible by a given set of numbers.
Like terms: Terms that have the same variables with the same exponents.
Linear equation: The standard form of a linear equation in one variable is $ax + b = c$, $a \neq 0$. A linear equation is also called a first-degree equation.
Sense of an inequality: The direction of the inequality symbol.
Solution of an equation: The number or numbers that make the equation true.
Solution set of an equation: The set of real numbers which make the equation true.
Subscript: Numbers or letters to the right of and below a variable.
Terms: The parts added or subtracted in an algebraic expression.
Unlike terms: Terms that are not "like" terms.

Important Facts

Properties of Equality

Reflexive property: $a = a$.
Symmetric property: If $a = b$, then $b = a$.
Transitive property: If $a = b$ and $b = c$, then $a = c$.
Addition property: If $a = b$, then $a + c = b + c$.
Multiplication property: If $a = b$, then $ac = bc$.

Absolute Value

If $|x| = a$, then $x = a$ or $x = -a$.
If $|x| < a$, then $-a < x < a$.
If $|x| > a$, then $x < -a$ or $x > a$.

Review Exercises

[2.1] State the degree of the term.

1. $15x^4y^6$
2. $6x$
3. $-4xyz^5$

Simplify the expression. If an expression cannot be simplified, so state.

4. $x^2 + 3x + 6$
5. $x^2 + 2xy + 6x^2 - 4$
6. $3(x + 4) - 3x - 4$
7. $-(x + y) + 3x - 5y + 6$

Solve each equation. If an equation has no solution, so state.

8. $\dfrac{x - 4}{5} = 9 - x$
9. $3(x + 2) - 6 = 4(x - 5)$
10. $3 + \dfrac{x}{2} = \dfrac{5}{6}$
11. $-6 - 2x = \dfrac{1}{2}(4x + 12)$
12. $2\left(\dfrac{x}{2} - 4\right) = 3\left(x + \dfrac{1}{3}\right)$
13. $3x - 4 = 6x + 4 - 3x$
14. $4 - \dfrac{3}{x} = 5$

[2.2] Write an equation that can be used to solve the problem. Solve the problem and check your answer.

15. The sum of a number and 4 times the number is 80. Find the number.
16. Hassan is 4 years older than his sister. The sum of their ages is 36. Find the age of Hassan and his sister.
17. Four times a number increased by 12 is 32. Find the number.
18. One-fourth of a number plus 6 is 11. Find the number.
19. A number decreased by 60% is 20. Find the number.
20. The sum of two consecutive integers is 49. Find the integers.
21. The sum of two consecutive odd integers is 28. Find the integers.
22. A number decreased by 10% is 180. Find the number.
23. The larger of two numbers is 1 less than twice the smaller. When the smaller is subtracted from the larger the difference is 9. Find the two numbers.
24. The cost of renting an automobile is $20 a day plus 15 cents a mile. How far can a person drive in 1 day on $25.40?
25. A blouse has been reduced by 12%. The sale price is $22. Find the original price.
26. Find the three angles of a triangle if one angle is 25° greater than the smallest angle and the other angle is 5° less than twice the smallest angle.
27. The West Ridge Fitness Center has two membership plans. The first plan is a flat $60 per month fee with no other costs. The second plan is $25 per month plus a $4.00 per visit charge. How many visits would Mike have to make per month to make it advantageous for him to select the first plan?

[2.3] Solve the following motion and mixture problems.

28. Two trains leave Portland at the same time traveling in opposite directions. One train travels at 60 miles per hour and the other at 90 miles per hour. In how many hours will they be 400 miles apart?
29. Two trains leave Tucson at the same time, along parallel tracks, traveling in the same direction. The faster train travels at 80 miles per hour. Find the speed of the other train if the trains are 90 miles apart after 3 hours.
30. Kenji wishes to put part of his $15,000 in a money-market account earning 12% interest and the balance in a savings account earning 5.5% interest. How much money should he invest in each account if he wishes the total interest for the year to be $900? Use $i = prt$.
31. A gasoline distribution company needs 1200 gallons of 89%-octane gasoline. The distributor only has gasoline rated at 86% octane and gasoline rated at 91% octane. How much of each gasoline should be mixed to obtain the desired gasoline?
32. Mr. Tomlins, a grocer, has two coffees, one selling for $4.00 per pound and the other for $4.80 per pound. How many pounds of each type of coffee should he mix to make 40 pounds of coffee to sell for $4.50 per pound?

[2.4] Solve for the variable indicated.

33. $A = lw$ for l

34. $A = \pi r^2 h$ for h

35. $P = 2l + 2w$ for w

36. $d = rt$ for r

37. $y = mx + b$ for m

38. $2x - 3y = 5$, for y

39. $P_1V_1 = P_2V_2$ for V_2

40. $S = \dfrac{3a + b}{2}$ for a

41. $K = 2(d + l)$ for l

42. $I = p + prt$ for t

43. $A = \dfrac{1}{2}h(b_1 + b_2)$, for b_1

44. $w = V_o t - 2l$ for t

[2.5] Solve the inequality. Graph the solution on the real number line.

45. $x - 3 \geq 4$

46. $2 - x \leq 5$

47. $2x + 4 > 9$

48. $16 \leq 4x - 5$

49. $\dfrac{4x + 3}{5} > -3$

50. $2(x - 3) > 3x + 4$

51. $-4(x - 2) \leq 6x + 4$

52. $\dfrac{x}{4} \geq 5 - 2x$

Write an inequality that can be used to solve the problem. Solve the inequality and answer the question.

53. A small airplane has a maximum load of 1525 pounds if it is to take off safely. If the passengers weigh 468 pounds, how many 80-pound boxes can be safely transported on the plane?

54. Jack, a telephone operator, informs a customer in a phone booth that the charge for calling Tallahassee, Florida, is \$4.50 for the first 3 minutes and 95 cents each additional minute and any part thereof. How long can the customer talk if he has \$8.65?

55. A fitness center guarantees that you will lose a minimum of 3 pounds the first week and $1\frac{1}{2}$ pounds each additional week. Find the minimum amount of time needed to lose 27 pounds.

Solve the inequality. Indicate the solution in interval notation.

56. $1 < x - 4 < 7$

57. $2 \leq x + 5 < 8$

58. $3 < 2x - 4 < 8$

59. $-12 < 6 - 3x < -2$

60. $-1 \leq \dfrac{2x - 3}{4} < 5$

61. $-8 < \dfrac{4 - 2x}{3} < 0$

62. Manuel's first four exam grades are 94, 73, 72, and 80. If a final average greater than or equal to 80 and less than 90 is needed to receive a final grade of B in the course, what range of grades on the fifth and last exam will result in Manuel receiving a B in the course? Assume a maximum grade of 100.

Find the solution set to each compound inequality.

63. $x < 3$ and $2x - 4 > -10$

64. $2x - 1 > 5$ or $3x - 2 \leq 7$

65. $3x + 5 > 2$ or $6 - x < 1$

66. $4x - 3 \leq 7$ and $2x - 1 \geq 3$

67. $4x - 5 < 11$ and $-3x - 4 \geq 8$

68. $\dfrac{5x - 3}{2} > 7$ or $\dfrac{2x - 1}{3} \leq -3$

[2.6] Find the solution set to each of the following.

69. $|x| = 4$

70. $|x| < 3$

71. $|x| \geq 4$

72. $|x - 4| = 9$

73. $|x - 2| \geq 5$

74. $|x + 6| < 5$

75. $|4 - 2x| = 5$

76. $|3 - 2x| < 7$

77. $|4 - 3x| \geq 5$

78. $\left|\dfrac{x - 3}{4}\right| = 5$

79. $\left|\dfrac{2x - 3}{5}\right| = 1$

80. $\left|\dfrac{x - 4}{3}\right| < 6$

81. $\left|\dfrac{4 - x}{3}\right| \geq \dfrac{1}{2}$

82. $|3(x + 2)| \leq \dfrac{9}{2}$

83. $|4(2 - x)| > 5$

84. $|x + 3| - 2 < 7$

85. $|2x - 3| + 4 \geq 10$

86. $\left|\dfrac{x - 4}{2}\right| - 3 > 5$

Practice Test

1. State the degree of the term $-6xy^2z^3$.
2. Solve the equation $3(x - 2) = 4(4 - x) + 5$.
3. Solve the equation $\frac{3}{5} - \frac{x}{2} = 4$.
4. Solve the equation $\frac{3x}{4} - 1 = 5 + \frac{2x - 1}{3}$.

For each problem, write an equation that can be used to solve the problem. Solve the equation and answer the question asked.

5. The sum of two consecutive integers is 47. Find the two integers.
6. The population of Urbandale is decreasing at a rate of 600 per year. If the current population is 12,000, how long will it take for the population to drop to 9520?
7. The cost of renting an automobile is $18 a day and 15 cents a mile. How far can Valerie drive in 1 day on $35?
8. Two joggers start at the same point at the same time and jog in opposite directions. Homer jogs at 4 miles per hour, while Frances jogs at $5\frac{1}{4}$ miles per hour. How far apart will they be in $1\frac{1}{4}$ hours?
9. How many liters of 12% salt solution must be added to 10 liters of 25% salt solution to get a 20% salt solution?
10. Solve for b in the equation $c = \frac{a - 3b}{2}$.
11. Solve $A = \frac{1}{2}h(b_1 + b_2)$ for b_2
12. Solve the following inequality and graph the solution on the number line.

$$\frac{6 - 2x}{5} \geq -12$$

13. Solve the inequality and write the solution in interval notation.

$$-4 < \frac{x + 4}{2} < 8$$

14. Find the solution set to the equation.

$$|x - 4| = 5$$

Find the solution set to the inequalities.

15. $|2x - 3| + 1 > 6$
16. $\left|\frac{2x - 3}{4}\right| \leq \frac{1}{2}$

3 Graphing Linear Equations

See Section 3.5, Table 1

3.1 The Cartesian Coordinate System, Distance, and Midpoint Formulas

1. *Plot points in the Cartesian coordinate system.*
2. *Find the distance between two points.*
3. *Find the midpoint of a line segment.*

In this chapter we discuss procedures for drawing graphs. A graph is a picture that shows the relationship between two or more variables in an equation. Many algebraic relationships are easier to understand if we can see a visual picture of them.

1 Before learning how to construct a graph, you must know the **Cartesian (or rectangular) coordinate system.**

The Cartesian coordinate system, named after the French mathematician and philosopher René Descartes (1596–1650), consists of two axes (or number lines) in a plane drawn perpendicular to each other (see Fig. 3.1). Note how the two axes yield four **quadrants,** labeled I, II, III, and IV.

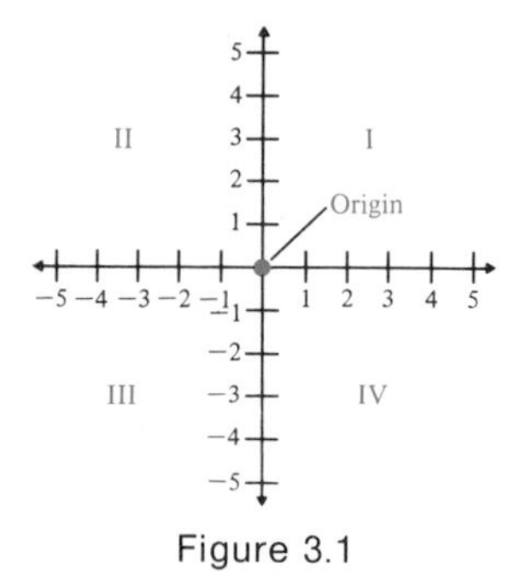

Figure 3.1

The horizontal axis is called the ***x* axis.** The vertical axis is called the ***y* axis.** The point of intersection of the two axes is called the **origin.** Starting from the origin and moving to the right the numbers increase; moving to the left the numbers decrease. Starting from the origin and moving up the numbers increase; moving down the numbers decrease.

To graph a point it is necessary to know both its x coordinate and y coordinate. An **ordered pair** is used to give the two coordinates of a point. If, for example, the x coordinate of a point is 3 and the y coordinate is 5, the ordered pair representing the point is (3, 5). Note that the x coordinate is always the first coordinate listed in the ordered pair. The point representing the ordered pair (3, 5) is plotted in Fig. 3.2.

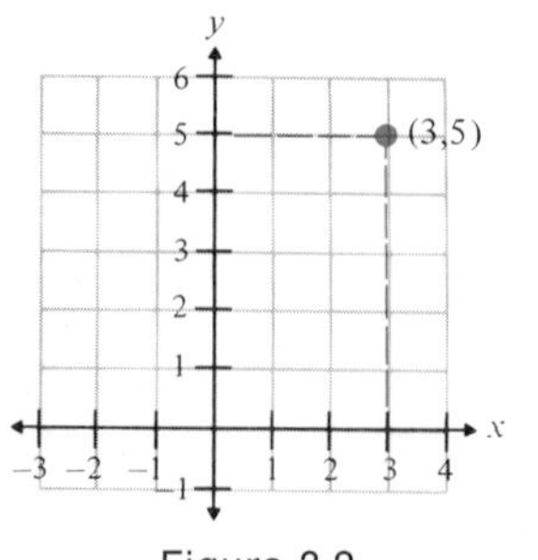

Figure 3.2

EXAMPLE 1 Plot each of the following points on the same set of axes.

(a) $A(4, 2)$ (b) $B(0, -3)$ (c) $C(-3, 1)$ (d) $D(4, 0)$

Solution: See Fig. 3.3.

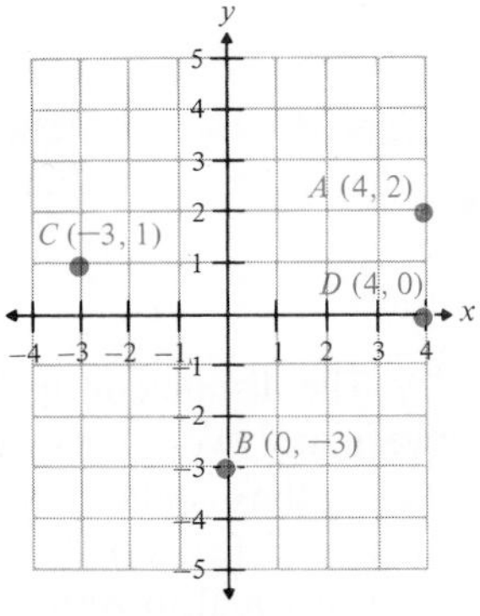

Figure 3.3

Notice that when the x coordinate is 0, as in part (b), the point is on the y axis. When the y coordinate is 0, as in part (d), the point is on the x axis. ■

EXAMPLE 2 List the ordered pair for each of the points shown in Figure 3.4.

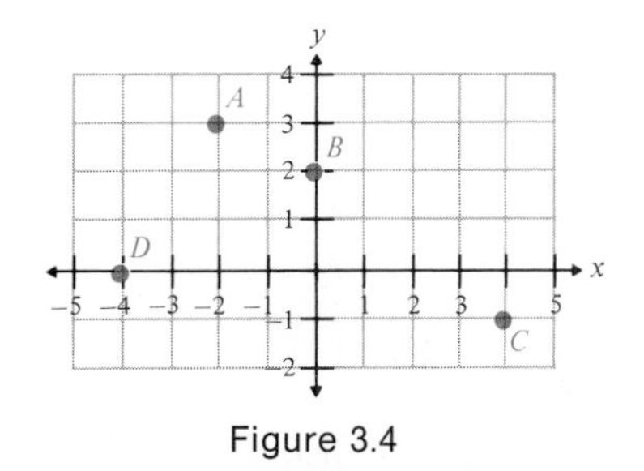

Figure 3.4

Solution: Remember to give the x value first in the ordered pair.

Point	Ordered Pair
A	$(-2, 3)$
B	$(0, 2)$
C	$(4, -1)$
D	$(-4, 0)$

■

2 Now we will see how to find the distance between any two points in a plane. After this we will show how to find the midpoint of a given line segment. You need these two concepts to understand conic sections (Chapter 10).

Distance Between Two Points

To find the distance, d, between two points we use the distance formula.

> **Distance Formula**
>
> The distance, d, between any two points (x_1, y_1) and (x_2, y_2), can be found by the distance formula
>
> $$d = \sqrt{(x_2 - x_1)^2 + (y_2 - y_1)^2}$$

The distance formula will be derived in Section 8.6. Note that the distance between any two points will always be a positive number. Can you explain why? When finding the distance it makes no difference which point we designate as point 1 (x_1, y_1) or point 2 (x_2, y_2). Note that the square of any real number will always be greater than or equal to zero. For example, $(5 - 2)^2 = (2 - 5)^2 = 9$.

EXAMPLE 3 Determine the distance between the points $(-1, 7)$ and $(-4, 3)$.

Solution: First plot the points (Fig. 3.5). Call $(-1, 7)$ point 2 and $(-4, 3)$ point 1. Thus (x_2, y_2) represents $(-1, 7)$ and (x_1, y_1) represents $(-4, 3)$. Now use the distance formula to find the distance, d.

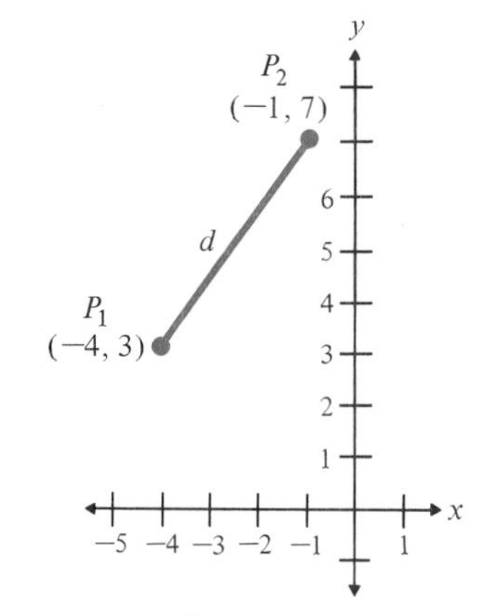

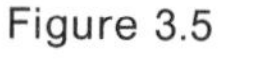
Figure 3.5

$$\begin{aligned} d &= \sqrt{(x_2 - x_1)^2 + (y_2 - y_1)^2} \\ &= \sqrt{[-1 - (-4)]^2 + (7 - 3)^2} \\ &= \sqrt{(-1 + 4)^2 + 4^2} \\ &= \sqrt{3^2 + 4^2} \\ &= \sqrt{9 + 16} \\ &= \sqrt{25} \\ &= 5 \end{aligned}$$

Thus the distance between the points $(-1, 7)$ and $(-4, 3)$ is 5 units. ■

If in Example 3 we had selected $(-4, 3)$ for point 2 and $(-1, 7)$ for point 1, our results would not have changed.

$$\begin{aligned} d &= \sqrt{(x_2 - x_1)^2 + (y_2 - y_1)^2} \\ &= \sqrt{[-4 - (-1)]^2 + (3 - 7)^2} \\ &= \sqrt{(-4 + 1)^2 + (-4)^2} \\ &= \sqrt{(-3)^2 + (-4)^2} \\ &= \sqrt{9 + 16} \\ &= \sqrt{25} \\ &= 5 \end{aligned}$$

When using the distance formula, do not expect your distance to always come out as a rational number. If your answer is an irrational number such as $\sqrt{187}$, leave your answer in terms of the square root.

HELPFUL HINT

> Students will sometimes begin finding the distance correctly using the distance formula but will forget to take the square root of the sum $(x_2 - x_1)^2 + (y_2 - y_1)^2$ to obtain the correct answer.

Midpoint of a Line Segment

3 It is often necessary to find the midpoint of a line segment between two given points. To do this, we use the midpoint formula.

> **Midpoint Formula**
>
> Given any two points (x_1, y_1) and (x_2, y_2), the point halfway between the given points can be found by the midpoint formula:
>
> $$\text{midpoint} = \left(\frac{x_1 + x_2}{2}, \frac{y_1 + y_2}{2}\right)$$

EXAMPLE 4 Determine the midpoint of the line segment between the points $(-3, 7)$ and $(4, 2)$.

Solution: It makes no difference which points we label (x_1, y_1) and (x_2, y_2). Let us replace (x_1, y_1) with $(-3, 7)$ and (x_2, y_2) with $(4, 2)$ (see Figure 3.6).

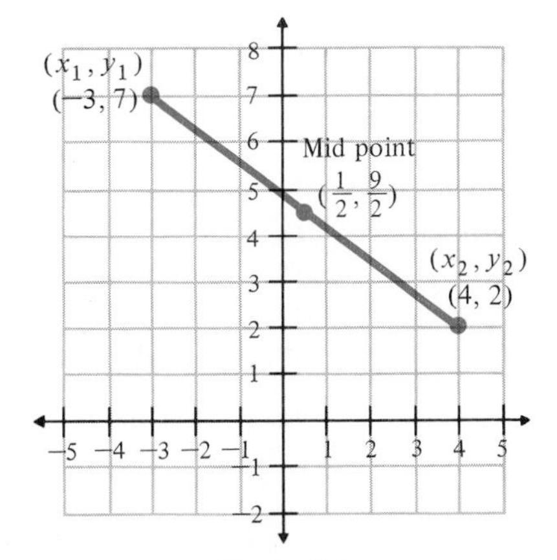

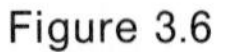
Figure 3.6

$$\begin{aligned}\text{midpoint} &= \left(\frac{x_1 + x_2}{2}, \frac{y_1 + y_2}{2}\right)\\ &= \left(\frac{-3 + 4}{2}, \frac{7 + 2}{2}\right)\\ &= \left(\frac{1}{2}, \frac{9}{2}\right)\end{aligned}$$

The point $(\frac{1}{2}, \frac{9}{2})$ is halfway between the points $(-3, 7)$ and $(4, 2)$. ■

Exercise Set 3.1

1. List the ordered pairs corresponding to the following ing points.

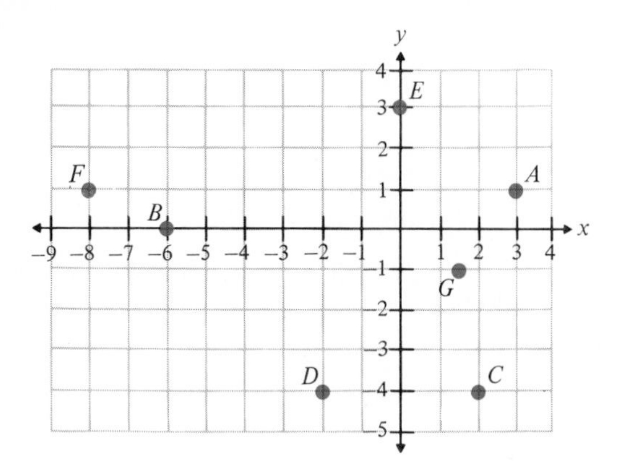

2. List the ordered pairs corresponding to the following points.

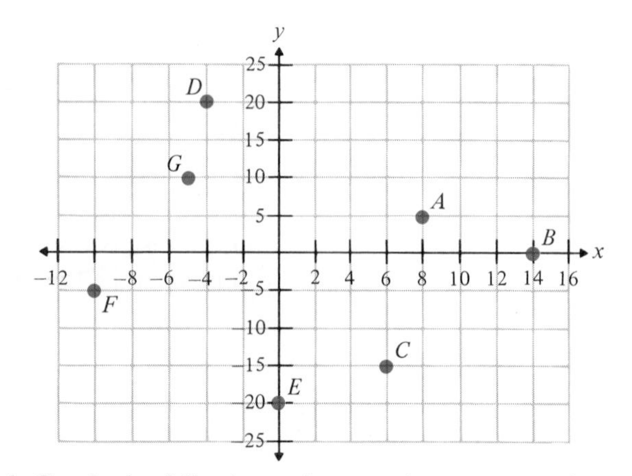

3. Graph the following points on the same set of axes.
(a) $(4, 2)$ (b) $(-6, 2)$
(c) $(0, -5)$ (d) $(-2, 0)$

4. Graph the following points on the same set of axes.
(a) $(-4, -2)$ (b) $(3, 2)$
(c) $(2, -6)$ (d) $(-3, 5)$

Determine the distance between the points.

5. $(2, -2)$ and $(2, -5)$
6. $(-5, 5)$ and $(-5, 1)$
7. $(-4, 3)$ and $(5, 3)$
8. $(-8, 7)$ and $(-3, 7)$
9. $(-1, -1)$ and $(3, 2)$
10. $(1, 4)$ and $(-3, 1)$
11. $(-1, -4)$ and $(4, 8)$
12. $(-1, 3)$ and $(2, 6)$
13. $(-3, -5)$ and $(6, -2)$
14. $(5, 3)$ and $(-5, -3)$
15. $(0, 6)$ and $(5, -1)$
16. $(9, 12)$ and $(5, 11)$
17. $(0, 0)$ and $(-4, -2)$
18. $(3, -1)$ and $\left(\frac{1}{2}, 4\right)$
19. $\left(\frac{3}{4}, 2\right)$ and $\left(-\frac{1}{2}, 6\right)$
20. $(4, 0)$ and $\left(-\frac{3}{5}, -4\right)$

Determine the midpoint of the line segment between the points.

21. $(5, 2)$ and $(-1, 4)$
22. $(1, 4)$ and $(2, 6)$
23. $(-5, 3)$ and $(5, -3)$
24. $(0, 8)$ and $(4, -6)$
25. $(-2, -8)$ and $(-6, -2)$
26. $(4, 7)$ and $(1, -3)$
27. $(1, -6)$ and $(-8, -4)$
28. $(12, 4)$ and $(-12, 4)$
29. $(0, 0)$ and $(4, -8)$
30. $\left(3, \frac{1}{2}\right)$ and $(2, -4)$
31. $\left(\frac{5}{2}, 3\right)$ and $\left(2, \frac{9}{2}\right)$
32. $\left(-\frac{5}{2}, -\frac{11}{2}\right)$ and $\left(-\frac{7}{2}, \frac{3}{2}\right)$

33. When the distance between two different points is found using the distance formula, why must the distance always be a positive number?

3.2 Graphing Linear Equations

1. *Write a linear equation in standard form.*
2. *Know what a graph represents.*
3. *Graph linear equations by plotting points.*
4. *Graph linear equations using intercepts.*
5. *Apply graphing to practical problems.*

1 All the equations that we graph in this chapter will be straight lines. These equations are called linear equations. A **linear equation** is an equation whose graph will be a straight line. Linear equations are also called **first-degree equations** since the degree of their highest-powered term is the first degree.

A linear equation may be written in a number of different forms. One is standard form.

Standard Form of a Linear Equation

$$ax + by = c$$

where a, b, and c are real numbers, and a and b are not both 0.

Examples of linear equations in standard form

$$2x + 3y = 4$$
$$-x + 5y = -2$$

2 Consider the linear equation in two variables, $y = x + 1$. What is the solution? Since the equation contains two variables, its solutions must contain two numbers, one for each variable. One set of numbers that satisfies this equation is $x = 1$ and $y = 2$. To see that this is true, we substitute both values into the equation at the same time and see that the equation checks.

$$y = x + 1$$
$$2 = 1 + 1$$
$$2 = 2 \quad \text{true}$$

One solution to the equation $y = x + 1$ is the ordered pair $(1, 2)$. However, the equation $y = x + 1$ has many other solutions. If you check the ordered pairs $(2, 3)$, $(3, 4)$, $(-1, 0)$, $(\frac{1}{2}, \frac{3}{2})$, you will see that they are all solutions to the equation $y = x + 1$. How many possible solutions does the equation $y = x + 1$ have? The equation $y = x + 1$ has an unlimited or *infinite number* of possible solutions. Since it is not possible to list all the specific solutions to the equation, we illustrate them with a graph. **A graph of an equation is an illustration of the set of points that satisfy the equation.**

Graphing Equations by Plotting Points

3 All linear equations will be straight lines when graphed. Since only two points are needed to draw a straight line, when graphing linear equations we only need to find, and plot, two ordered pairs that satisfy the equation. But it is always a good idea to use a third ordered pair as a check. If the three points are not in a straight line, you have made a mistake. A set of points in a straight line is said to be **collinear.**

One method of finding ordered pairs that satisfy an equation is to solve the equation for y. Then substitute values for x and find the corresponding values of y.

EXAMPLE 1 Graph the equation $y = 3x + 6$.

Solution: This equation is already solved for y. We will find three ordered pairs that satisfy the equation by arbitrarily selecting three values for x, substituting them in the equation, and finding the corresponding values for y. In this equation we let x have values of 0, 2 and -3.

$$y = 3x + 6$$

x value	*y value*
$x = 0$	$y = 3(0) + 6 = 6$
$x = 2$	$y = 3(2) + 6 = 12$
$x = -3$	$y = 3(-3) + 6 = -3$

x	y
0	6
2	12
−3	−3

Now plot the three ordered pairs on the same set of axes (Fig. 3.7). Since the three points are collinear, everything appears correct. Connect the three points with a straight line. Place arrows at the ends of the line to show that the line continues infinitely in both directions.

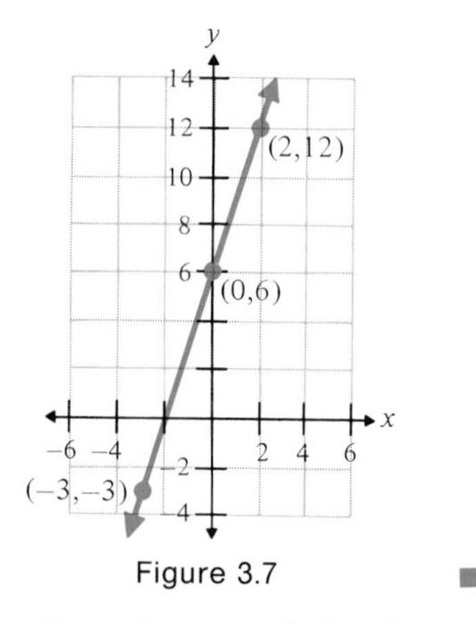

Figure 3.7 ■

To plot the equation $y = 3x + 6$, we used the three values $x = 0$, $x = 2$, and $x = -3$. We could have picked three entirely different values and obtained exactly the same graph. When selecting values to substitute for x, use whatever values make the equation easy to evaluate.

The graph in Example 1 represents the set of all ordered pairs that satisfy the equation $y = 3x + 6$. If we select any point on this line, the ordered pair representing that point will be a solution to the equation $y = 3x + 6$. Similarly, any solution to the equation will be represented by a point on the line.

EXAMPLE 2 Graph the equation $-2x + 3y = -6$.

Solution: We will first solve the equation for y. This will make it easier to select values to substitute for x that can be quickly evaluated.

$$-2x + 3y = -6$$
$$3y = 2x - 6$$
$$y = \frac{2x - 6}{3}$$
$$y = \frac{2x}{3} - \frac{6}{3}$$
$$y = \frac{2x}{3} - 2$$

Now we will select values for x that make $2x/3$ integral values. $2x/3$ will have integral values when x is a multiple of 3. We will therefore select $x = 0$, 3, and 6.

$$y = \frac{2x}{3} - 2$$

x value	*y value*
$x = 0$	$y = \frac{2(0)}{3} - 2 = -2$
$x = 3$	$y = \frac{2(3)}{3} - 2 = 0$
$x = 6$	$y = \frac{2(6)}{3} - 2 = 2$

x	y
0	−2
3	0
6	2

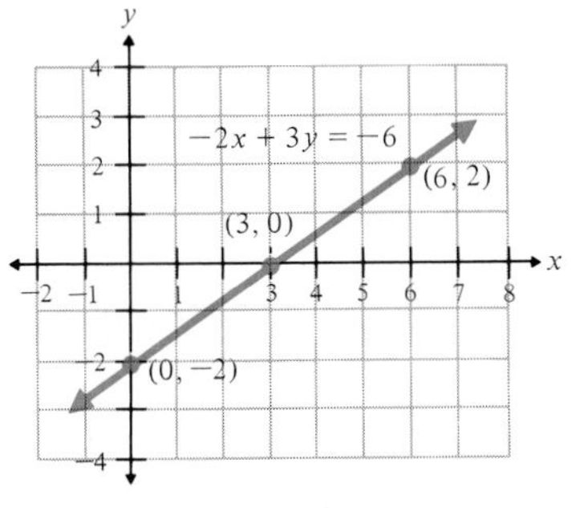

Figure 3.8

Now plot the points and draw the graph (Fig. 3.8). ■

EXAMPLE 3 Graph the equation $2y = 4x - 12$.

Solution: By first solving the equation for y it will be easier to determine ordered pairs that satisfy the equation.

$$2y = 4x - 12$$
$$y = \frac{4x - 12}{2} = \frac{4x}{2} - \frac{12}{2} = 2x - 6$$

Now select values for x and solve for y in the equation $y = 2x - 6$.

$$y = 2x - 6$$

x value	*y value*
0	$y = 2(0) - 6 = -6$
2	$y = 2(2) - 6 = -2$
3	$y = 2(3) - 6 = 0$

x	y
0	−6
2	−2
3	0

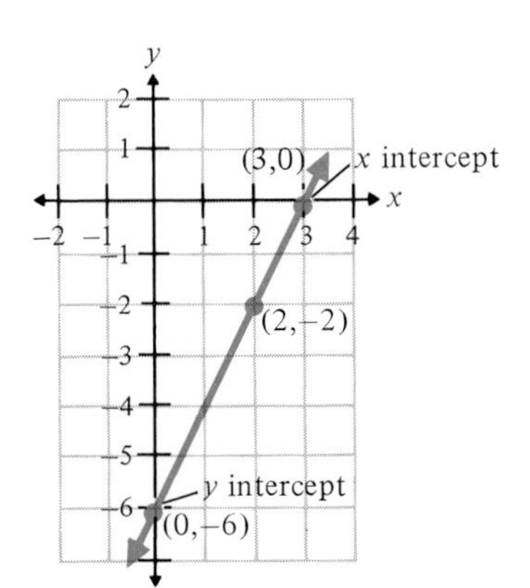

Figure 3.9

Plot the points and draw the straight line. (Fig. 3.9). ■

Graphing Equations Using Intercepts

4 Let's examine two points on the graph shown in Fig. 3.9. Note that the graph crosses the x axis at the point (3, 0). Therefore, 3 is called the **x intercept.** The graph crosses the y axis at the point (0, −6). Therefore, −6 is called the **y intercept.** It is often convenient to graph linear equations by finding their x and y intercepts.

> **x and y Intercepts**
>
> To find the y intercept, set $x = 0$ and solve for y.
>
> To find the x intercept, set $y = 0$ and solve for x.

EXAMPLE 4 Graph the equation $3y = 6x + 12$ by plotting the x and y intercepts.

Solution: To find the y intercept (where the graph crosses the y axis), set $x = 0$ and solve for y.

$$3y = 6x + 12$$
$$3y = 6(0) + 12$$
$$3y = 0 + 12$$
$$3y = 12$$
$$y = \frac{12}{3} = 4$$

The graph crosses the y axis at $y = 4$. The ordered pair representing the y intercept is (0, 4).

To find the x intercept (where the graph crosses the x axis), set $y = 0$ and solve for x.

$$3y = 6x + 12$$
$$3(0) = 6x + 12$$
$$0 = 6x + 12$$
$$-12 = 6x$$
$$-\frac{12}{6} = x$$
$$-2 = x$$

The graph crosses the x axis at $x = -2$. The ordered pair representing the x intercept is (−2, 0).

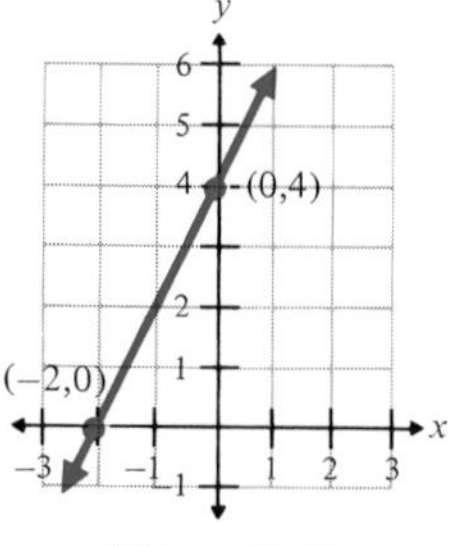

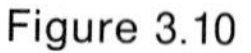
Figure 3.10

Now plot the intercepts and draw the graph (Fig. 3.10). ■

When graphing equations using intercepts you must be particularly careful. Since you are plotting only two points, you have no check point. If one of your intercepts is wrong, your graph will be wrong. When graphing by plotting intercepts, you may wish to plot a third point as a checkpoint.

EXAMPLE 5 Graph the equation $y = 3$.

Solution: This equation can be written as $y = 3 + 0x$. Thus for any value of x selected, y will be 3. The graph of $y = 3$ is illustrated in Fig. 3.11.

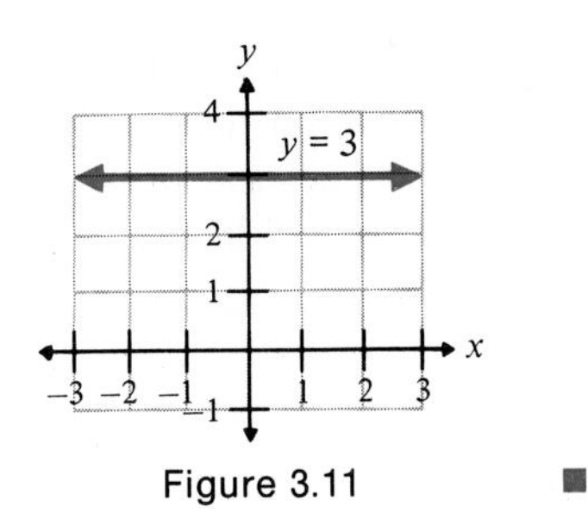

Figure 3.11 ■

The graph of any equation of the form $y = a$ will always be a horizontal line for any real number a.

EXAMPLE 6 Graph the equation $x = -2$.

Solution: This equation can be written as $x = -2 + 0y$. Thus for every value of y selected, x will have a value of -2 (see Fig. 3.12).

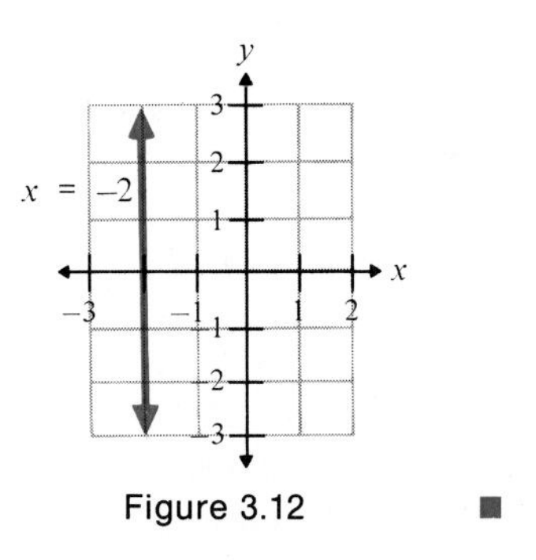

Figure 3.12 ■

The graph of any equation of the form $x = a$ will always be a vertical line for any real number a.

Applications of Graphing

5 Graphs are often used to show the relationship between variables. The axes of a graph do not have to be labeled x and y; they can be any designated variables. Consider the following example.

EXAMPLE 7 The yearly profit, p, of a tire store can be estimated by the formula $p = 20x - 30{,}000$ where x is the number of tires sold per year.

(a) Draw a graph of profits versus tires sold for up to 6,000 tires.
(b) Estimate the number of tires that must be sold for the company to break even.
(c) Estimate the number of tires sold if the company has a $40,000 profit.

Solution: (a) The minimum number of tires that can be sold is 0. Therefore negative values do not have to be indicated on the horizontal axis. We will arbitrarily select 3 values for x and find the corresponding values of p. The graph is illustrated in Fig. 3.13.

x	p
0	−30,000
2,000	10,000
5,000	70,000

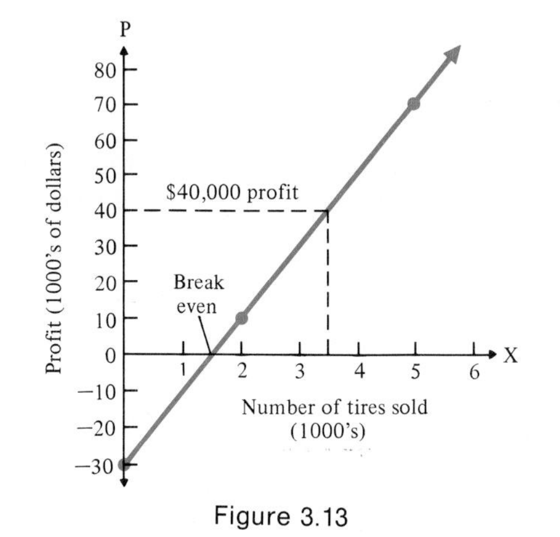

Figure 3.13

(b) To break even approximately 1500 tires must be sold.
(c) To make a $40,000 profit, approximately 3500 tires must be sold.

Sometimes it is difficult to read an exact answer from a graph. To determine the exact number of tires needed to break even substitute 0 for p in the equation $p = 20x - 30{,}000$ and solve for x. To determine the exact number of tires needed to obtain a $40,000 profit, substitute 40,000 for p and solve the equation for x. ■

Exercise Set 3.2

Graph each equation by solving the equation for y, selecting three arbitrary values for x, and finding the corresponding values of y.

1. $y = 4$

2. $x = 6$

3. $x = -2$

4. $y = 5$

5. $y = 4x - 2$

6. $y = -x + 3$

7. $y = x + 2$

8. $y = x - 4$

9. $y = -\frac{1}{2}x + 5$

10. $2y = 2x + 4$

11. $6x - 2y = 4$

12. $4x - y = 5$

13. $5x - 2y = 8$

14. $-2x + 4y = 8$

15. $6x + 5y = 30$

16. $-2x - 3y = 6$

17. $-6x - y = -7$

18. $8y - 16x = 24$

19. $y = 20x + 40$

20. $2y - 50 = 100x$

21. $y = \frac{2}{3}x$

22. $y = -\frac{3}{5}x$

23. $y = \frac{1}{2}x + 4$

24. $y = -\frac{2}{5}x + 2$

25. $2y = 3x + 6$

26. $4x - 6y = 10$

27. $4x + 3y = 9$

28. $6x + 3y = 5$

29. $-2x + 5y = 15$

30. $5x - 2y = -7$

31. $-4x - 3y = -12$

32. $4x = 5y + 2$

Graph each equation using x and y intercepts.

33. $y = 8x + 4$

34. $y = -2x + 6$

35. $y = 2x + 3$

36. $y = -3x + 8$

37. $y = -6x + 5$

38. $y = 4x - 8$

39. $4y + 3x = 12$

40. $-2x + 3y = 10$

41. $4x = 3y - 9$

42. $7x + 14y = 21$

43. $\frac{1}{2}x + 2y = 4$

44. $30x + 25y = 50$

45. $6x - 12y = 24$

46. $25x + 50y = 100$

47. $8y = 6x - 12$

48. $-3y - 2x = -6$

49. $\frac{1}{3}x + \frac{1}{4}y = 12$

50. $12x + 36y = -72$

51. $-16y = 4x + 96$

52. $\frac{1}{3}x - 2y = 6$

53. $30y + x = 45$

54. $120x - 360y = 720$

55. $40x + 6y = 40$

56. $20x - 240 = -60y$

57. Using the formula distance = rate · time, $d = rt$, draw a graph of distance versus time for a constant rate of 50 miles per hour.

58. Using the simple interest formula interest = principal · rate · time, $i = prt$, draw a graph of interest versus time for a principal of \$1000 and a rate of 8%.

59. The profit of a company that produces bicycles can be approximated by the formula $P = 60x - 80{,}000$, where x is the number of bicycles produced and sold.

(a) Draw a graph of profit versus the number of bicycles sold (for up to 5000 bicycles).

(b) Estimate the number of bicycles that must be sold for the company to break even.

(c) Estimate the number of bicycles that must be sold for the company to make a \$150,000 profit.

60. The auto rental fee from an auto rental agency is \$15 a day plus 12 cents a mile.

(a) Write an equation expressing rental fee, F, in terms of miles, m.

(b) Draw a graph illustrating the rental fee versus the mileage for up to 200 miles.

(c) Estimate the rental fee for 1 day if Mary drives 60 miles

(d) Estimate the number of miles Mary has driven if the rental fee is \$32.

61. The weekly cost of operating a taxi is \$50 plus 12 cents per mile.

(a) Write an equation expressing weekly cost, c, in terms of miles, m.

(b) Draw a graph illustrating weekly cost versus the number of miles, up to 200, driven per week.

(c) How many miles would Jack have to drive for the weekly cost to be \$70?

(d) If the weekly cost is \$60, how many miles did Jack drive?

62. Ellen Branston's weekly salary is \$200 plus 15% commission on her weekly sales.

(a) Write an equation expressing weekly salary, s, in terms of weekly sales, x.

(b) Draw a graph of weekly salary versus weekly sales, for up to \$5000 in sales.

(c) What is Ellen's weekly salary if her sales were \$4000?

(d) If her salary for the week is \$400, what are her weekly sales?

63. Ms. Tocci, a real estate agent, makes \$150 per week plus a 1% sales commission on each property she sells.

(a) Write an equation expressing her weekly salary, s, in terms of sales commission, x.

(b) Draw a graph of her salary versus her weekly sales, for sales up to \$100,000.

(c) If she sells one house per week for \$80,000, what will be her weekly salary?

See Exercise 61

JUST FOR FUN

1. Graph $y = |x|$.
2. Graph $y = |x + 1|$.
3. Graph $y = |x - 2|$.
4. Graph $y = |x| - 2$.
5. Graph $y = \begin{cases} x + 3, & x > 4 \\ 3x - 5, & x \le 4 \end{cases}$.
6. Graph $y = \begin{cases} 2x - 3, & x \ge 2 \\ -3x + 7, & x < 2 \end{cases}$.

3.3 Slope of a Line

1 *Find the slope of a line.*

2 *Determine when two lines are parallel or perpendicular.*

1 The **slope of a line** is the ratio of the vertical change to the horizontal change between any selected points on the line. As an example, consider the two points (3, 6) and (1, 2) on the line in Figure 3.14(a). If we draw a line parallel to the x axis through the point (1, 2) and a line parallel to the y axis through the point (3, 6), the two lines intersect at (3, 2) (see Fig. 3.14b). From Fig. 3.14b we can determine the slope of the line. The vertical change (along the y axis) is $6 - 2$, or 4 units. The horizontal change (along the x axis) is $3 - 1$, or 2 units. Thus the slope of the line through these two points is $4 \div 2$ or 2.

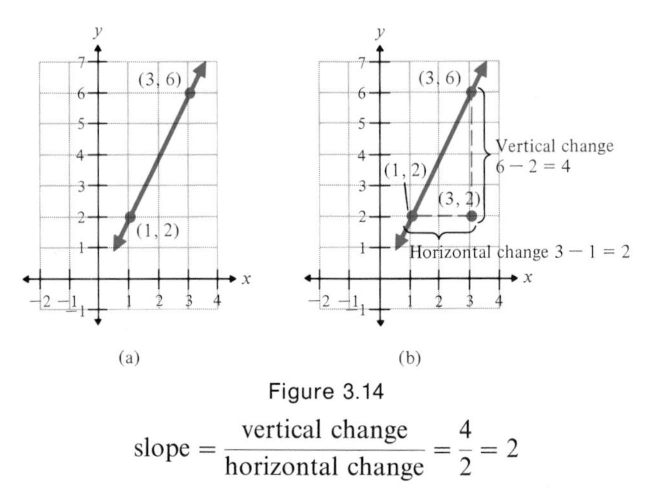

Figure 3.14

$$\text{slope} = \frac{\text{vertical change}}{\text{horizontal change}} = \frac{4}{2} = 2$$

By examining the line connecting these two points, we can see that for each two units the graph moves up the y axis it moves 1 unit to the right on the x axis (see Fig. 3.15).

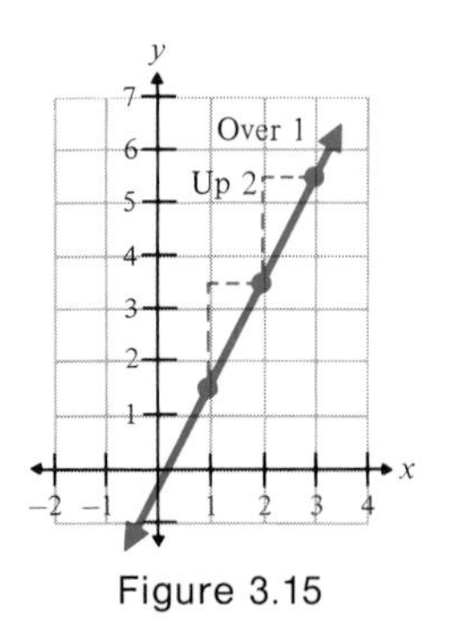

Figure 3.15

Let's now determine the procedure to find the slope of a line passing through the two points (x_1, y_1) and (x_2, y_2). Consider Fig. 3.16. The vertical change can be found by subtracting y_1 from y_2. The horizontal change can be found by subtracting x_1 from x_2.

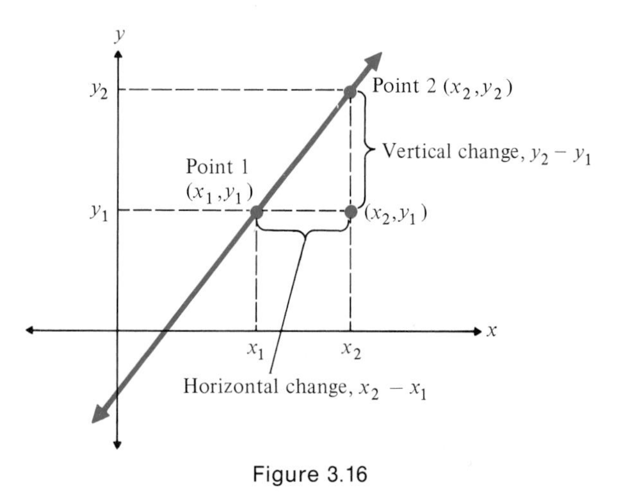

Figure 3.16

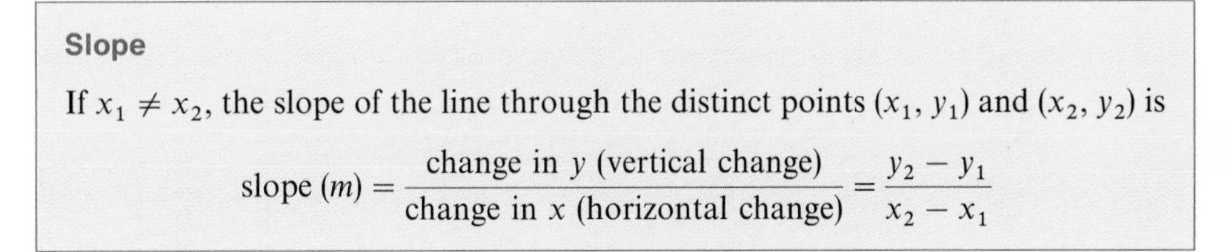

Slope

If $x_1 \neq x_2$, the slope of the line through the distinct points (x_1, y_1) and (x_2, y_2) is

$$\text{slope } (m) = \frac{\text{change in } y \text{ (vertical change)}}{\text{change in } x \text{ (horizontal change)}} = \frac{y_2 - y_1}{x_2 - x_1}$$

It makes no difference which two points on the line are selected when finding the slope of a line. It also makes no difference which point you label (x_1, y_1) or (x_2, y_2). The Greek capital letter delta, Δ, is often used to represent the words "the change in." Thus the slope is sometimes indicated as

$$m = \frac{\Delta y}{\Delta x} = \frac{y_2 - y_1}{x_2 - x_1}$$

EXAMPLE 1 Find the slope of the line in Fig. 3.17.

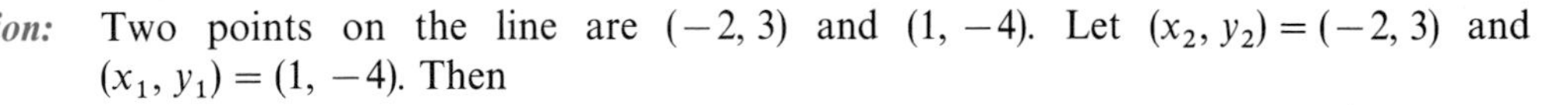

Solution: Two points on the line are $(-2, 3)$ and $(1, -4)$. Let $(x_2, y_2) = (-2, 3)$ and $(x_1, y_1) = (1, -4)$. Then

$$m = \frac{y_2 - y_1}{x_2 - x_1} = \frac{3 - (-4)}{-2 - 1}$$

$$= \frac{3 + 4}{-3}$$

$$= -\frac{7}{3}$$

Figure 3.17

The slope of the line is $-\frac{7}{3}$. Note that if we had let $(x_1, y_1) = (-2, 3)$ and $(x_2, y_2) = (1, -4)$, the slope would remain the same. Try and see. ■

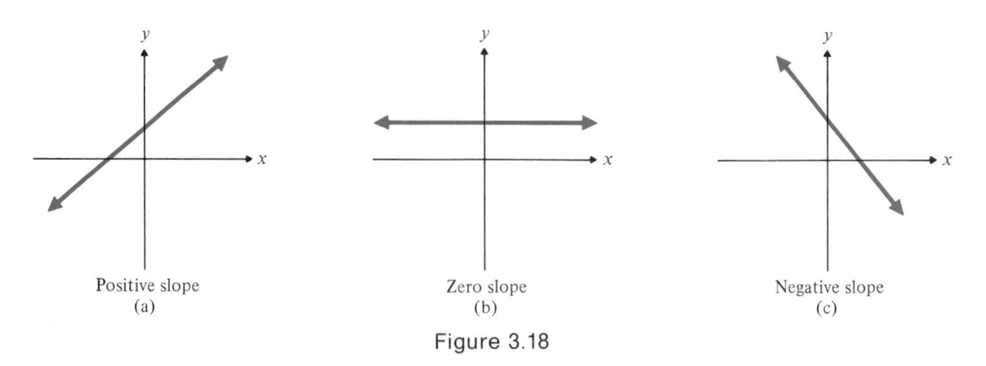

Figure 3.18

A line that rises going from left to right, Fig. 3.18a, has a **positive slope.** A line that neither rises nor falls going from right to left, Fig. 3.18b, has **zero slope.** And a line that falls going left to right, Fig. 3.18c, has a **negative slope.**

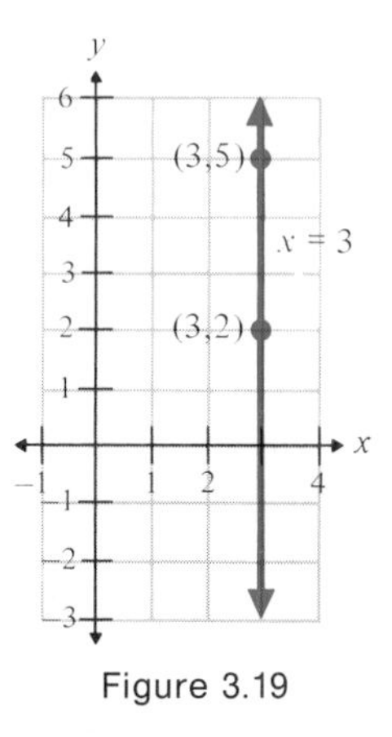

Figure 3.19

Consider the graph of $x = 3$ (Fig. 3.19). What is its slope? The graph is a vertical line and goes through the points (3, 2) and (3, 5). Let the point (3, 5) represent (x_2, y_2) and let (3, 2) represent (x_1, y_1). Then the slope of the line is

$$m = \frac{y_2 - y_1}{x_2 - x_1} = \frac{5 - 2}{3 - 3} = \frac{3}{0}$$

Since it is meaningless to divide by 0, we say that the slope of this line does not exist. **The slope of any vertical line does not exist.**

Parallel and Perpendicular Lines

2 Two lines in the same plane are **parallel** when they do not intersect no matter how far they are extended. Figure 3.20 illustrates two parallel lines, l_1 and l_2. For two lines not to intersect they must rise or fall at the same rate. That is, their slopes must be the same. **Two lines are parallel if their slopes are the same, and two lines with the same slope are parallel lines.** If line l_1 has slope m_1 and line l_2 has slope m_2, and if $m_1 = m_2$, then lines l_1 and l_2 must be parallel lines.

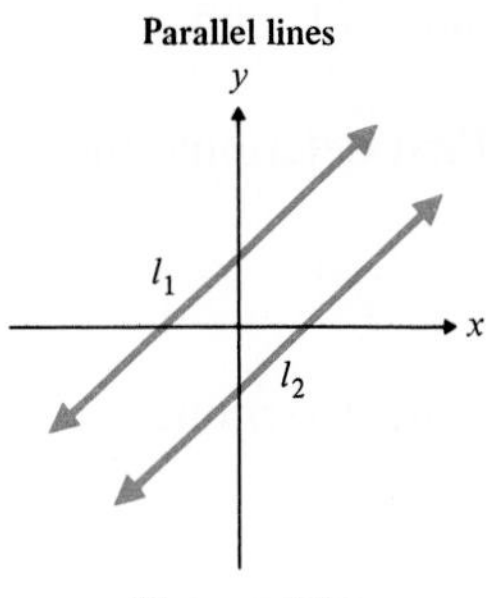

Figure 3.20

EXAMPLE 2 Two points on l_1 are (1, 6) and $(-1, 2)$. Two points on l_2 are (2, 3) and $(-1, -3)$. Determine if l_1 and l_2 are parallel lines.

Solution: First determine the slope of l_1.

$$m_1 = \frac{6-2}{1-(-1)} = \frac{4}{2} = 2$$

Now determine the slope of l_2.

$$m_2 = \frac{3-(-3)}{2-(-1)} = \frac{6}{3} = 2$$

Since l_1 and l_2 have the same slope, 2, the two lines are parallel. ■

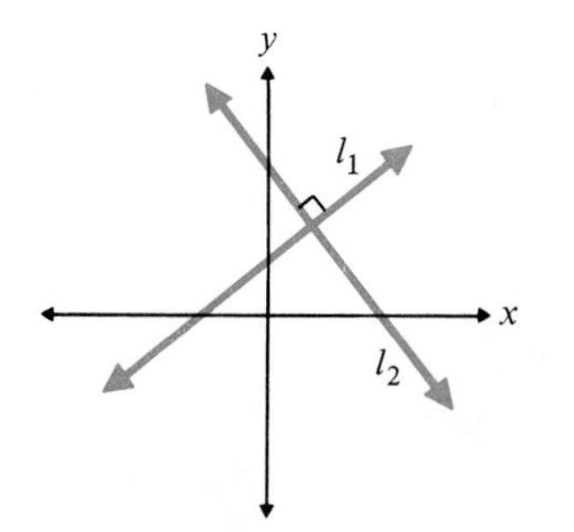

Figure 3.21

Two lines that cross at right angles (90° angles) are said to be **perpendicular lines.** Two perpendicular lines are illustrated in Fig. 3.21. The square where the two lines meet is used to indicate that the two lines meet at a right angle.

Two lines will be perpendicular to each other when their slopes are negative reciprocals. For any slope a its negative reciprocal is $-1/a$.

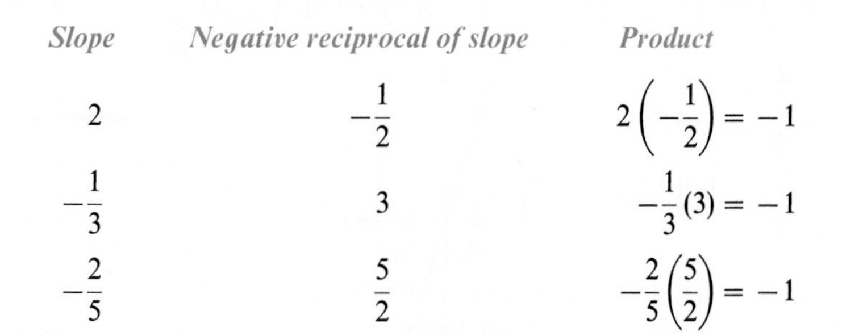

Slope	*Negative reciprocal of slope*	*Product*
2	$-\frac{1}{2}$	$2\left(-\frac{1}{2}\right) = -1$
$-\frac{1}{3}$	3	$-\frac{1}{3}(3) = -1$
$-\frac{2}{5}$	$\frac{5}{2}$	$-\frac{2}{5}\left(\frac{5}{2}\right) = -1$

Notice that the product of a number and its negative reciprocal equals -1. If l_1 has slope m_1 and l_2 has slope m_2, and if $m_1 m_2 = -1$, then l_1 and l_2 must be perpendicular lines.

EXAMPLE 3 Two points on l_1 are (6, 3) and (2, −3). Two points on l_2 are (0, 2) and (6, −2). Determine if l_1 and l_2 are perpendicular lines.

Solution: First determine the slope of l_1.

$$m_1 = \frac{3-(-3)}{6-2} = \frac{6}{4} = \frac{3}{2}$$

Now determine the slope of l_2.

$$m_2 = \frac{2-(-2)}{0-6} = \frac{4}{-6} = \frac{-2}{3}$$

Now determine if $m_1 m_2 = -1$. If so, the lines are perpendicular.

$$m_1 m_2 = \frac{3}{2}\left(-\frac{2}{3}\right) = -1$$

Since the product of the slopes equals −1, the lines are perpendicular. Note that each slope is the negative reciprocal of the other. ■

Note: Any horizontal line is perpendicular to any vertical line, although the negative reciprocal test cannot be applied.

Exercise Set 3.3

Find the slope of the line through the given points. If the slope of the line does not exist, so state.

1. (1, 5) and (2, −3)
2. (3, 1) and (5, 4)
3. (5, 2) and (1, 4)
4. (5, 1) and (2, 4)
5. (−1, 4) and (0, 3)
6. (2, 3) and (−2, 3)
7. (4, 2) and (4, −1)
8. (6, −2) and (−1, −2)
9. (−3, 4) and (−1, −6)
10. (4, −3) and (3, −4)
11. (2, 5) and (−1, 5)
12. (−2, 3) and (7, −3)
13. (2, −4) and (−5, −3)
14. (−4, 0) and (0, −6)
15. (2, 0) and (−4, −2)
16. (−6, 2) and (4, −3)

Find the slope of the line in each of the given figures. If the slope of the line does not exist, so state.

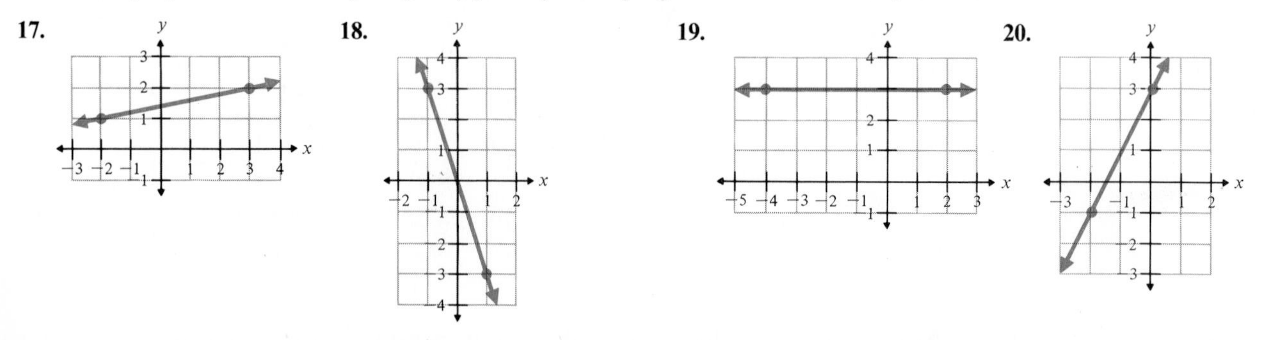

21. 22. 23. 24.

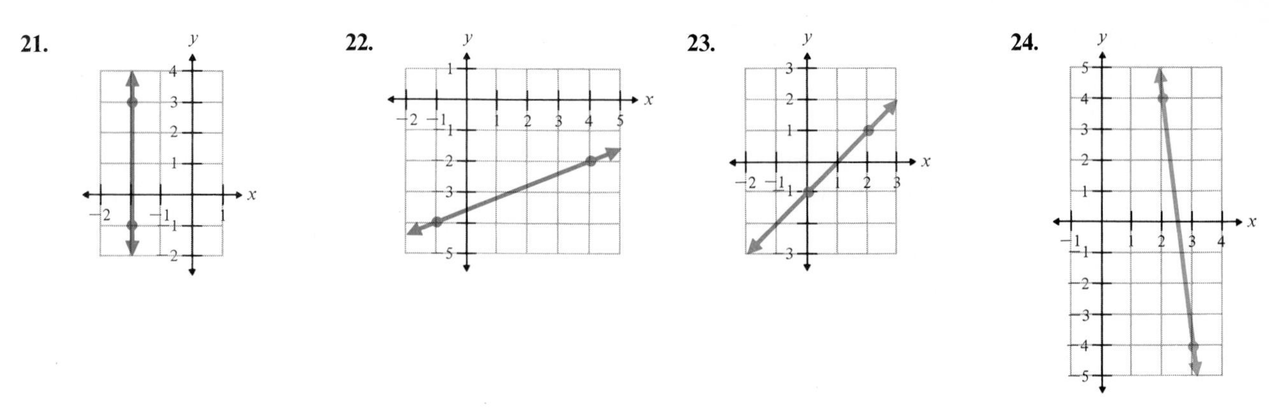

Two points on l_1 and two points on l_2 are given. Determine if l_1 is parallel to l_2, l_1 is perpendicular to l_2, or neither.

25. l_1: (0, 4) and (2, 8); l_2: (0, −1) and (3, 5)
26. l_1: (−1, 0) and (2, 3); l_2: (3, 2) and (4, −1)
27. l_1: (3, 2) and (−1, −2), l_2: (2, 0) and (3, −1)
28. l_1: (2, 3) and (3, 2); l_2: (1, 4) and (4, 1)
29. l_1: (3, 4) and (−2, 3); l_2: (0, −3) and (2, −1)
30. l_1: (6, 2) and (3, 1); l_2: (−6, −2) and (−3, −1)
31. l_1: (0, 2) and (6, −2); l_2: (4, 0) and (6, 3)
32. l_1: (−1, 3) and (4, 2); l_2: (1, −3) and (4, 2)
33. l_1: (1, 5) and (−2, −1); l_2: (1, −2) and (3, 2)
34. l_1: (1, −6) and (0, −4); l_2: (0, −6) and (2, −2)

Solve for the given variable if the line through the two given points is to have the given slope.

35. (6, a) and (3, 4), $m = 1$
36. (1, 0) and (4, x), $m = 3$
37. (5, b) and (2, −4), $m = 2$
38. (6, 1) and (4, d), $m = 3$
39. (2, −3) and (3, c), $m = -1$
40. (y, −1) and (3, 2), $m = -3$
41. (x, 2) and (3, −4), $m = 2$
42. (−2, −3) and (x, 4), $m = \frac{1}{2}$
43. (3, 5) and (x, 3), $m = \frac{2}{3}$
44. (−4, −1) and (y, 2), $m = \frac{-3}{5}$

45. Explain how to find the slope of a given line.
46. Explain what it means when the slope of a line is positive.
47. Explain what it means when the slope of a line is negative.
48. What is the slope of a horizontal line? Explain why this is so.
49. Explain why the slope of a vertical line does not exist.
50. When finding the slope of a line, how does the slope change if we interchange (x_1, y_1) and (x_2, y_2)?

3.4 Slope-Intercept and Point-Slope Forms of a Line

1 *Write linear equations in slope-intercept form.*
2 *Write linear equations in point-slope form.*

1 A linear equation written in the form $y = mx + b$ is said to be in **slope-intercept form.**

Slope-Intercept Form of a Line

$$y = mx + b$$

where ***m* is the slope** of the line and ***b* is the *y* intercept** of the line.

Examples of equations in slope-intercept form

$$y = 3x - 6 \qquad y = \frac{1}{2}x + \frac{3}{2}$$

This form is called the slope-intercept form because the m represents the slope of the graph and the b represents the y intercept.

$$\overset{\text{slope}}{\downarrow}\qquad\overset{y\text{ intercept}}{\downarrow}$$

$$y = mx + b$$

Equation	*Slope*	*y intercept*
$y = 3x - 6$	3	-6
$y = \frac{1}{2}x + \frac{3}{2}$	$\frac{1}{2}$	$\frac{3}{2}$

> To write an equation in slope-intercept form, solve the equation for y.

EXAMPLE 1 Write the equation $-3x + 4y = 8$ in slope-intercept form. State the slope and y intercept.

Solution: Solve for y.

$$-3x + 4y = 8$$

$$4y = 3x + 8$$

$$y = \frac{3x + 8}{4}$$

$$y = \frac{3}{4}x + \frac{8}{4}$$

$$y = \frac{3}{4}x + 2$$

The slope is $\frac{3}{4}$; the y intercept is 2. ■

EXAMPLE 2 Write the equation of the line illustrated in Fig. 3.22.

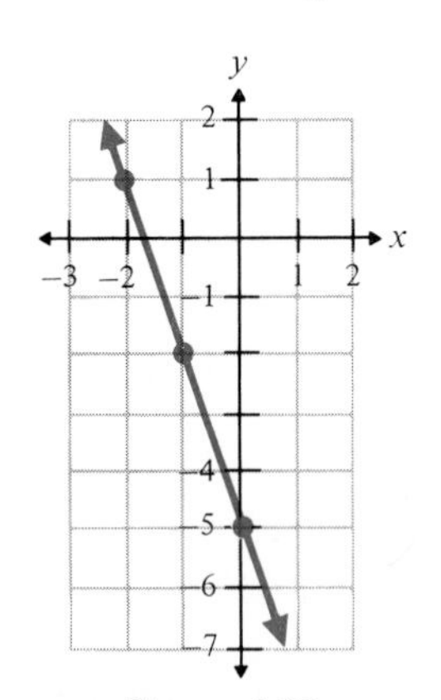

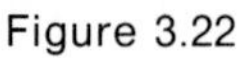
Figure 3.22

Solution: If we can determine the line's slope and its y intercept, we can write the equation in slope-intercept form. By looking at the figure we can determine the y intercept is -5.

Notice that y changes 3 units for each unit change of x. Also note that the slope is negative since the line falls as it moves to the right. Therefore, the slope of the line is -3. We could also find the slope by taking two points on the line and finding $\Delta y/\Delta x$ for the two points selected.

Since the slope is -3 and the y intercept is -5, the equation of the line is $y = -3x - 5$. ■

EXAMPLE 3 (a) Determine if the following lines are parallel.

$$2x - y = -4$$
$$2y = 4x - 2$$

(b) Graph both equations on the same set of axes.

Solution: (a) Recall from Section 3.3 that two lines are parallel when they have the same slope. To compare the slopes of the two lines, write each in slope-intercept form by solving each equation for y.

$$\begin{aligned} 2x - y &= -4 \\ -y &= -2x - 4 \\ y &= 2x + 4 \end{aligned} \qquad \begin{aligned} 2y &= 4x - 2 \\ y &= \frac{4x - 2}{2} \\ y &= 2x - 1 \end{aligned}$$

Since both lines have the same slope, 2, they are parallel.

(b) Both lines are graphed in Fig. 3.23.

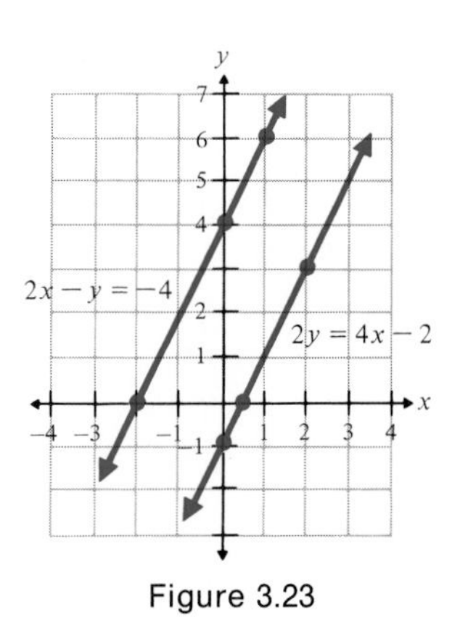

Figure 3.23 ■

EXAMPLE 4 (a) Determine if the equations $2x + 3y = 6$ and $3x - 2y = 12$ will be perpendicular lines when graphed.
(b) Graph both equations on the same set of axes.

Solution: (a) To determine if the equations will be perpendicular lines when graphed, we must find and compare their individual slopes. Recall that two lines are perpendicular when their slopes are negative reciprocals of each other. If the product of their slopes equal -1, the lines will be perpendicular. To find the slopes, we will write the equations in slope intercept form by solving the equations for y.

$$2x + 3y = 6 \qquad\qquad 3x - 2y = 12$$

$$3y = -2x + 6 \qquad\qquad -2y = -3x + 12$$

$$y = \frac{-2x + 6}{3} \qquad\qquad y = \frac{-3x + 12}{-2}$$

$$y = \frac{-2}{3}x + 2 \qquad\qquad y = \frac{3}{2}x - 6$$

Since $\frac{3}{2}$ is the negative reciprocal of $-\frac{2}{3}$, $\frac{3}{2}(-\frac{2}{3}) = -1$, the lines will be perpendicular when graphed.
(b) Both lines are graphed in Fig. 3.24

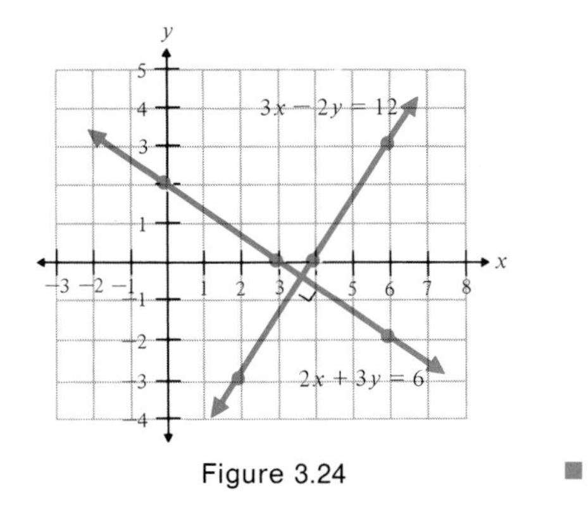

Figure 3.24 ■

EXAMPLE 5 Give the equation of the line that is parallel to the line $2x + 4y = 8$ and has a y intercept of 5.

Solution: If we know the slope of a line and its y intercept, we can use the slope-intercept form, $y = mx + b$, to write the equation. The slope of the given equation can be found by solving for y.

$$2x + 4y = 8$$

$$4y = -2x + 8$$

$$y = \frac{-2x + 8}{4}$$

$$y = -\frac{1}{2}x + 2$$

Two lines are parallel when they have the same slope. Therefore, the slope of the line parallel to the given line must be $-\frac{1}{2}$. Since its slope is $-\frac{1}{2}$ and its y intercept is 5, its equation must be

$$y = -\frac{1}{2}x + 5 \quad ■$$

EXAMPLE 6 Give the equation of the line that is perpendicular to the line $2x + 4y = 8$ and has a y intercept of 5.

Solution: Two lines are perpendicular when their slopes are negative reciprocals of each other. From Example 5 we know that the slope of the given line is $-\frac{1}{2}$. Therefore, the slope of a line perpendicular to the given line must be 2. The line perpendicular to the given line has a y intercept of 5. Thus the equation we are seeking is

$$y = 2x + 5 \quad ■$$

Point-Slope Form of a Line

2 When the slope of a line and a point on the line are known, we can use the **point-slope form** to determine the equation of the line. The point-slope form can be developed by beginning with the slope between any two points (x, y) and (x_1, y_1) on a line.

$$m = \frac{y - y_1}{x - x_1} \qquad \text{or} \qquad \frac{m}{1} = \frac{y - y_1}{x - x_1}$$

now cross-multiply to obtain

$$m(x - x_1) = y - y_1 \qquad \text{or} \qquad y - y_1 = m(x - x_1)$$

Point-Slope Form of a Line

$$y - y_1 = m(x - x_1)$$

where ***m* is the slope** of the line and **(x_1, y_1) is a point on the line.**

EXAMPLE 7 Write an equation of the line that goes through the point (2, 3) and has a slope of 4.

Solution: The slope, m, is 4. The point on the line is (2, 3); call this (x_1, y_1). Substitute 4 for m, 2 for x_1, and 3 for y_1, in the point-slope form of a line.

$$\begin{aligned} y - y_1 &= m(x - x_1) \\ y - 3 &= 4(x - 2) \\ y - 3 &= 4x - 8 \\ y &= 4x - 5 \end{aligned}$$

The graph of $y = 4x - 5$ has a slope of 4 and passes through the point (2, 3). ■

EXAMPLE 8 Find an equation of the line through the points $(-1, -3)$ and $(4, 2)$.

Solution: To use the point-slope form, we must first find the slope between the two points. To determine the slope, let's designate $(-1, -3)$ as (x_1, y_1) and $(4, 2)$ as (x_2, y_2).

$$m = \frac{y_2 - y_1}{x_2 - x_1} = \frac{2 - (-3)}{4 - (-1)} = \frac{2 + 3}{4 + 1} = \frac{5}{5} = 1$$

We can use either point (one at a time) in determining the equation of the line. This example will be worked out using both points to show that the solutions obtained are identical.

Use point $(-1, -3)$ as (x_1, y_1):

$$\begin{aligned} y - y_1 &= m(x - x_1) \\ y - (-3) &= 1[x - (-1)] \\ y + 3 &= x + 1 \\ y &= x - 2 \end{aligned}$$

Use point $(4, 2)$ as (x_1, y_1):

$$\begin{aligned} y - y_1 &= m(x - x_1) \\ y - 2 &= 1(x - 4) \\ y - 2 &= x - 4 \\ y &= x - 2 \end{aligned}$$

The solutions are identical. ■

EXAMPLE 9 Determine, in standard form, the equation of the line that is parallel to $5x - 3y = 12$ and passes through the point $(-4, 6)$.

Solution: First find the slope of the given line.

$$\begin{aligned} 5x - 3y &= 12 \\ -3y &= -5x + 12 \\ y &= \frac{-5x + 12}{-3} \\ y &= \frac{5}{3}x - 4 \end{aligned}$$

Since the slope of the given line is $\frac{5}{3}$, the slope of any line parallel to it must also be $\frac{5}{3}$. We now know the slope of the line, $\frac{5}{3}$, and a point on the line, $(-4, 6)$. We can therefore use the point-slope form to determine the equation.

$$\begin{aligned} y - y_1 &= m(x - x_1) \\ y - 6 &= \frac{5}{3}[x - (-4)] \end{aligned}$$

We now proceed to write the equation in standard form.

$$y - 6 = \frac{5}{3}(x + 4)$$

$$3(y - 6) = 5(x + 4)$$

$$3y - 18 = 5x + 20$$

$$-5x + 3y - 18 = 20$$

$$-5x + 3y = 38$$

Note that $5x - 3y = -38$ is also an acceptable answer. ■

EXAMPLE 10 Determine, in slope-intercept form, the equation of the line that is perpendicular to $5y = -10x + 7$ and passes through the point $(4, \frac{1}{3})$.

Solution:

$$5y = -10x + 7$$

$$y = \frac{-10x + 7}{5}$$

$$y = -2x + \frac{7}{5}$$

Since the slope of the given line is -2, the slope of the line perpendicular to it must be $\frac{1}{2}$. The line we are seeking passes through the point $(4, \frac{1}{3})$. Using the point-slope form, we obtain

$$y - y_1 = m(x - x_1)$$

$$y - \tfrac{1}{3} = \tfrac{1}{2}(x - 4)$$

Multiply both sides of the equation by the least common denominator, 6, to eliminate fractions.

$$6\left(y - \frac{1}{3}\right) = 6\left[\frac{1}{2}(x - 4)\right]$$

$$6y - 2 = 3(x - 4)$$

$$6y - 2 = 3x - 12$$

Proceed to write the equation in slope-intercept form.

$$6y = 3x - 10$$

$$y = \frac{3x - 10}{6}$$

$$y = \frac{3x}{6} - \frac{10}{6}$$

$$y = \frac{1}{2}x - \frac{5}{3}$$ ■

HELPFUL HINT

We have discussed 3 forms of a linear equation, they are

Standard form: $ax + by = c$

Slope-intercept form: $y = mx + b$

Point-slope form: $y - y_1 = m(x - x_1)$

Consider the equation $2y = 3x + 4$, we can write this equation in all 3 forms as follows:

Standard form: $-3x + 2y = 4$ (or $3x - 2y = -4$)

Slope-intercept form: $y = \frac{3}{2}x + 2$
(solve for y)

Point-slope form: $y - 2 = \frac{3}{2}x$ or $y - 2 = \frac{3}{2}(x - 0)$

Note that to go from slope-intercept form to point-slope form we subtracted 2 from both sides of the equation. Also note that the point represented in point-slope form is (0, 2), or the y intercept.

Exercise Set 3.4

Write an equation of the given line.

1.

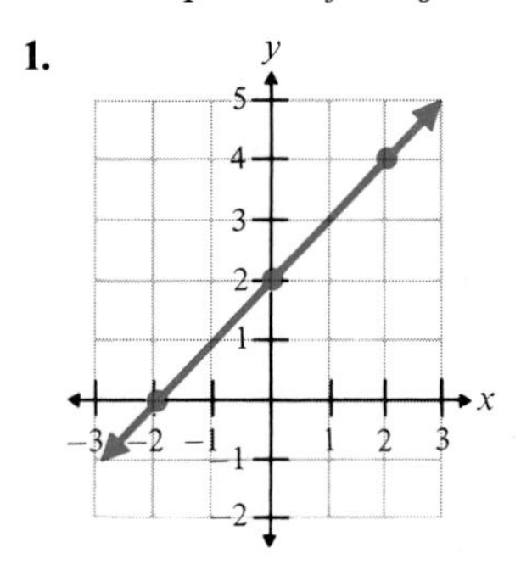

2.

4.

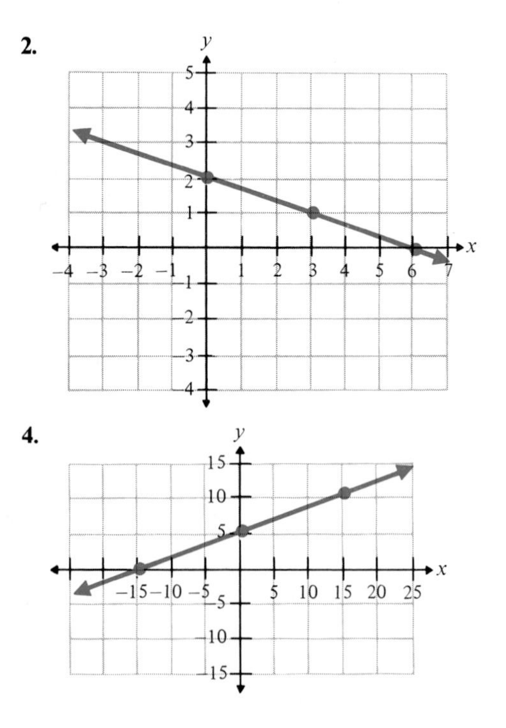

3.

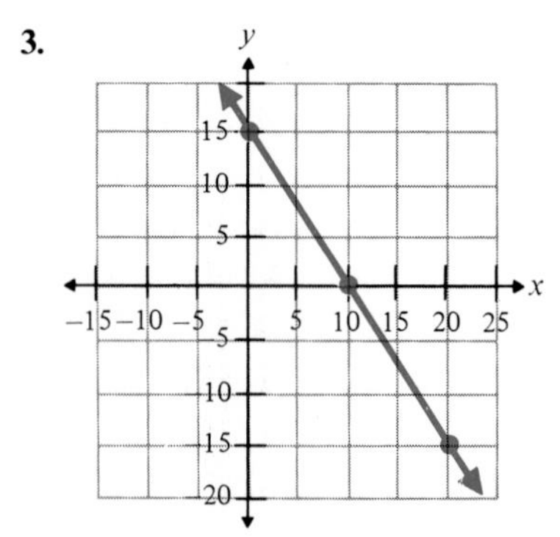

Determine if the two given lines are parallel, perpendicular, or neither.

5. $y = 2x - 4$
$y = 2x + 3$

6. $2x + 3y = 6$
$y = -\frac{2}{3}x + 5$

7. $4x + 2y = 8$
$8x = 4 - 4y$

8. $3x - 5y = 10$
$3y + 5x = 5$

9. $2x + 5y = 10$
$-x + 3y = 9$

10. $6x + 2y = 8$
$4x - 9 = -y$

11. $y = \frac{1}{2}x - 6$
$-3y = 6x + 9$

12. $2y - 6 = -5x$
$y = -\frac{5}{2}x - 2$

13. $y = 2x - 6$
$x = -2y - 4$

14. $3x + 4y = 12$
$4x + 3y = 12$

15. $2x + y - 6 = 0$
$6x + 3y = 12$

16. $x - 3y = -9$
$y = 3x + 6$

17. $-6x + 4y = 6$
$3y = -2x + 12$

18. $-4x + 6y = 12$
$2x - 3y = 6$

Find the equation of a line with the properties given. Write the equation in slope-intercept form.

19. Slope $= 4$, through $(2, 3)$
20. Slope $= -2$, through $(-4, 5)$
21. Slope $= -1$, through $(6, 0)$
22. Slope $= \frac{1}{2}$, through $(-1, -5)$
23. Slope $= -\frac{2}{3}$, through $(-1, -2)$
24. Slope $= \frac{3}{5}$, through $(4, -2)$
25. Through $(4, 6)$ and $(-2, 1)$
26. Through $(-4, -2)$ and $(-2, 1)$
27. Through $(6, 3)$ and $(5, 2)$
28. Through $(-4, 6)$ and $(4, -6)$
29. Through $(1, 0)$ and $(-2, 4)$
30. Through $(10, 3)$ and $(0, -2)$

Find the equation of a line with the properties given. Write the equation in the form indicated.

31. Through $(1, 4)$ parallel to $y = 2x + 4$ (slope-intercept form)
32. Through $(3, -2)$ parallel to $y = -3x + 6$ (standard form)
33. Through $(\frac{1}{2}, 3)$ parallel to $2x + 3y - 9 = 0$ (standard form)
34. Through $(\frac{1}{5}, -\frac{2}{3})$ parallel to $-3x = 2y + 6$ (slope-intercept form)
35. Through $(-4, \frac{3}{4})$ parallel to $y = \frac{2}{3}x - 5$ (slope-intercept form)
36. Through $(2, 3)$ perpendicular to $y = 2x - 3$ (slope-intercept form)
37. Through $(-2, 4)$ perpendicular to $4x - 2y = 8$ (standard form)
38. Through $(\frac{1}{2}, -3)$ perpendicular to $5x = -2y + 3$ (standard form)
39. Through $(-\frac{2}{3}, -4)$ perpendicular to $\frac{1}{2}x = y - 6$ (slope-intercept form)
40. With x intercept 2 and y intercept 3 (standard form)
41. With x intercept $\frac{1}{2}$ and y intercept $-\frac{1}{4}$ (standard form)
42. Through $(2, 5)$ and parallel to the line with x intercept 1 and y intercept 3 (slope-intercept form)
43. Through $(4, -2)$ and parallel to the line with x intercept -3 and y intercept 2 (slope-intercept form)
44. Through $(-3, 4)$ and perpendicular to the line with x intercept 2 and y intercept 2 (standard form)
45. Through $(6, 2)$ and perpendicular to the line with x intercept 2 and y intercept -3 (slope-intercept form)
46. Through the point $(2, 1)$ parallel to the line through the points $(3, 5)$ and $(-2, 3)$ (slope-intercept form)
47. Through the point $(6, -2)$ perpendicular to the line through the points $(-2, \frac{1}{2})$ and $(4, 3)$ (standard form)
48. In this chapter we have discussed three forms of an equation. Name the three forms, and give one equation illustrating each of the forms.
49. Consider the equation $y - x = 2$. Write the equation in:
(a) Standard form
(b) Slope-intercept form
(c) Point-slope form
50. Consider the equation $4 = 2y + 4x$. Write the equation in:
(a) Standard form
(b) Slope-intercept form
(c) Point-slope form

3.5 Relations and Functions

1. *Identify relations.*
2. *Find the domain and range of a relation.*
3. *Identify functions.*
4. *Use function notation.*
5. *Graph a linear function.*

1 If you plan to take additional mathematics courses, an understanding of relations and functions will be very helpful. In this section you are introduced to these important concepts. The function concept is discussed and expanded further in later sections of the text.

A **relation** is any set of ordered pairs. A relation may be indicated by (1) a set of ordered pairs, (2) a table of values, (3) a graph, (4) a rule, or (5) an equation. For example, each of the following indicates a relation.

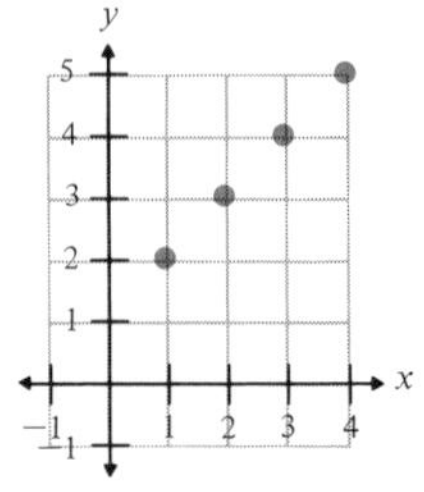

1. Set of ordered pairs $\{(1, 2), (2, 3), (3, 4), (4, 5)\}$

2. Table of values

x	1	2	3	4
y	2	3	4	5

3. Graph (see figure on left)

4. Rule: For each integer from 1 to 4 inclusive, add 1 to obtain its corresponding value.

5. Equation: $y = x + 1$, for $1 \leq x \leq 4$, $x \in N$

Note that the five examples listed above all indicate the same relation.

Consider the following. An apple cost 20 cents. What will be your cost if you purchase 0 apples, 1 apple, 2 apples, and so on? We can indicate this using Table 1.

Table 1

Number of Apples, n	Cost, c
0	0
1	0.20
2	0.40
3	0.60
⋮	⋮
10	2.00
⋮	⋮

In general, we can see that the cost for purchasing n apples will be 20 cents times the number of apples, or $0.20n$. We can represent the cost of purchasing n apples, where n is a whole number, by the equation $c = 0.20n$. In the equation $c = 0.20n$ the cost, c, depends on the number of apples, n; thus we call c the *dependent variable* and n the *independent variable.*

2 In any relation the set of values that can be used for the independent variable is called its **domain.** The set of values that represent the dependent variable is called its **range.**

Consider the equation for the cost of apples, $c = 0.20n$, what is its domain and what is its range? The domain of this relation is the set of "input values" that can be used to represent n, the number of apples. Since we cannot purchase a fractional part of an apple or a negative amount of apples, the domain is the set of numbers $\{0, 1, 2, 3, \ldots\}$. Note that the values on the left side of Table 1 are the elements that make up the domain. When the values in the domain, 0, 1, 2, 3, . . . , are substituted for n in the formula $c = 0.20n$, the values we get out are 0.00, 0.20, 0.40, 0.60, ... These values appear on the right side of Table 1. The range is the set of these "output values." The range is $\{0.00, 0.20, 0.40, 0.60, \ldots\}$.

If we list the table of values in Table 1 as a set of ordered pairs, we get $\{(0, 0.00), (1, 0.20), (2, 0.40), (3, 0.60), \ldots\}$. Note that the *domain is the set of first coordinates in the set of ordered pairs,* and the *range is the set of second coordinates in the set of ordered pairs.*

When a graph is given, its domain and range can be determined by observation as illustrated in Example 1.

EXAMPLE 1 State the domain and range of the relation shown in Fig. 3.25.

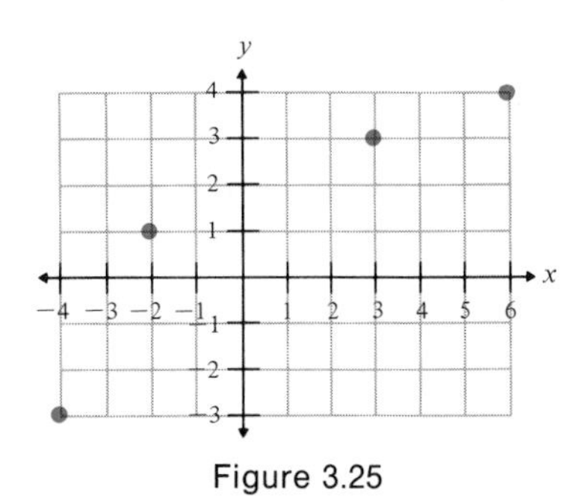

Figure 3.25

Solution: The domain is the set of x values (first coordinate in the set of ordered pairs).

$$\text{Domain:} \quad \{-4, -2, 3, 6\}$$

The range is the set of y values (second coordinate in the set of ordered pairs).

$$\text{Range:} \quad \{-3, 1, 3, 4\}$$

The numbers in the domain and range were listed from smallest to largest; however, you may list the numbers in any order. ■

EXAMPLE 2 State the domain and range of the relation shown in Fig. 3.26.

Solution: The domain is the set of x values. All values of x between -2 and 6 inclusive are indicated on the graph. We can indicate this using set builder notation, as discussed in Section 1.1.

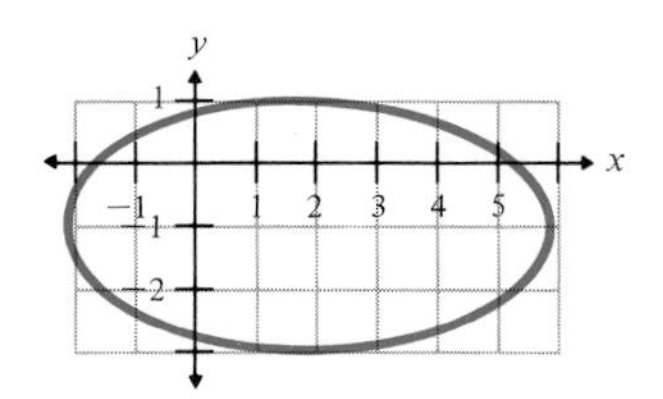

Figure 3.26

$$\text{Domain:} \quad \{x \mid -2 \le x \le 6\}$$

The range is the set of y values. All values of y between -3 and 1 inclusive are indicated on the graph.

$$\text{Range:} \quad \{y \mid -3 \le y \le 1\} \quad ■$$

EXAMPLE 3 Determine the domain and range of the relation shown in Fig. 3.27.

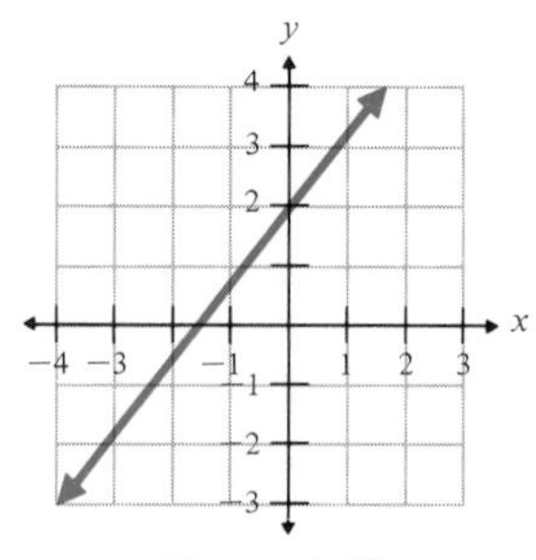

Figure 3.27

Solution: Since the line is extended indefinitely, every value of x will be included in the domain. The domain is the set of real numbers.

$$\text{Domain: } \mathbb{R}$$

The range will also be the set of real numbers since all values of y are included on the graph.

$$\text{Range: } \mathbb{R} \quad \blacksquare$$

EXAMPLE 4 Determine the domain and range in each of the following relations.

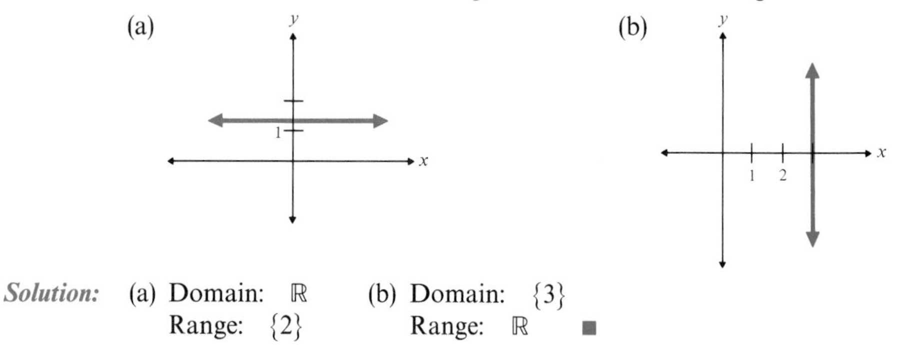

Solution: (a) Domain: $\mathbb{R}$ (b) Domain: $\{3\}$
Range: $\{2\}$ Range: $\mathbb{R}$ ■

> **Functions**
>
> A **function** is a relation in which no two ordered pairs have the same first coordinate and a different second coordinate.

3 A function is a special type of relation. For a relation to be a function each first coordinate in the set of ordered pairs must have a unique second coordinate. Is the set of ordered pairs $\{(1, 4), (2, 3), (3, 5), (-1, 3), (0, 6)\}$ a function? Do any of the ordered pairs have the same first coordinate and a different second coordinate? *Since no two ordered pairs have the same first coordinate the set of ordered pairs is a function.* Note that the second coordinate in the ordered pair may repeat.

Now consider a second set of ordered pairs: $\{(-1, 3), (4, 2), (3, 1), (2, 6), (3, 5)\}$. Is this set of ordered pairs a function? Since two ordered pairs, namely (3, 1) and (3, 5), have the same first coordinate, this set of ordered pairs is not a function.

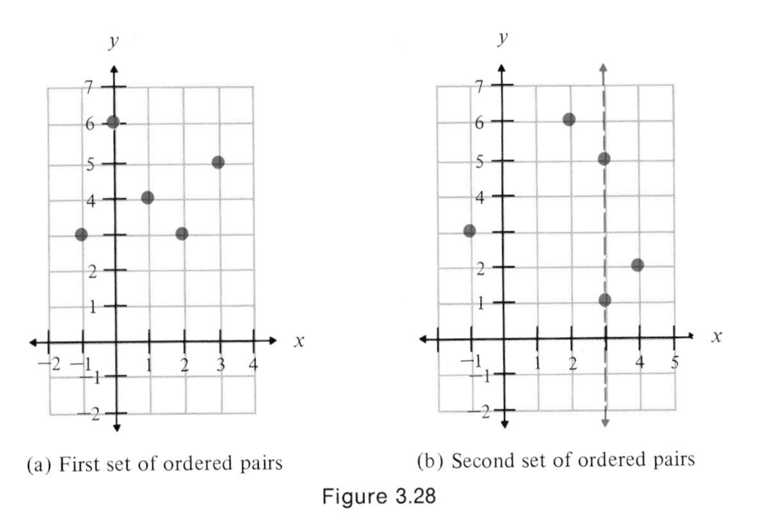

(a) First set of ordered pairs (b) Second set of ordered pairs

Figure 3.28

Let us graph each set of ordered pairs and observe them visually (Fig. 3.28). Note in part (a) that each x value has a unique y value. If a vertical line is drawn through any point, no other points are intersected. This relation is therefore a function. In part (b), if we draw a vertical line through the point (3, 1), it will intersect the point (3, 5). Thus each x value does not have a unique y value and the set of points is not a function.

To determine if a graph is a function, we can use the **vertical line test.** If a vertical line can be drawn through any part of the graph and the line intersects another part of the graph, the graph is not a function. If a vertical line cannot be drawn to intersect the graph at more than one point, the graph is a function.

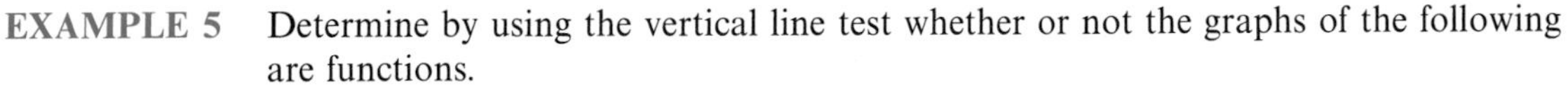

EXAMPLE 5 Determine by using the vertical line test whether or not the graphs of the following are functions.

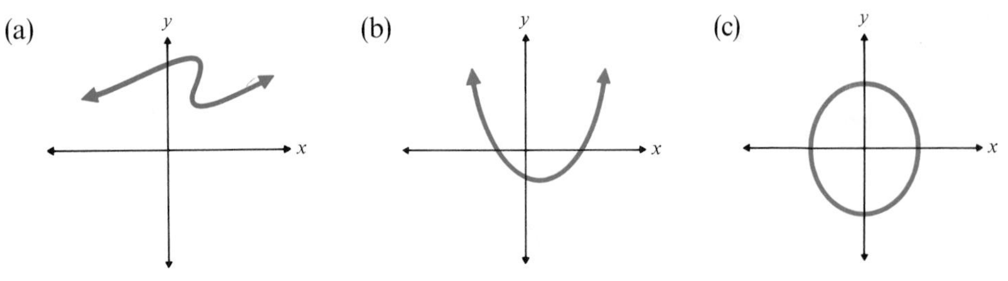

Solution: (a) Not a function (b) Is a function (c) Not a function

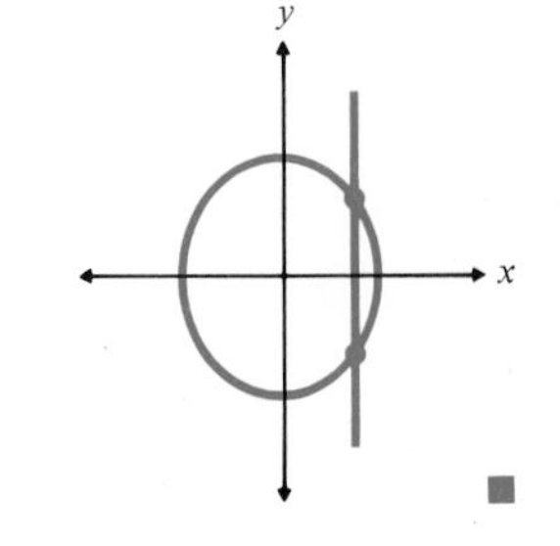

■

EXAMPLE 6 State the range and domain of the function illustrated in Fig. 3.29.

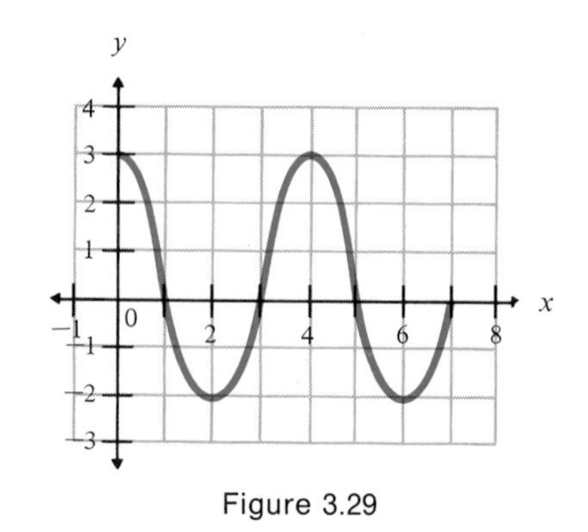

Figure 3.29

Solution: The domain is the set of x values, 0 through 7. Thus the domain is $\{x|0 \leq x \leq 7\}$. The range is the set of y values, -2 through 3. The range is therefore $\{y|-2 \leq y \leq 3\}$. ■

Function Notation

4 Consider the equation $y = 3x + 2$. By examining its graph (Fig. 3.30) we can see that it is a function.

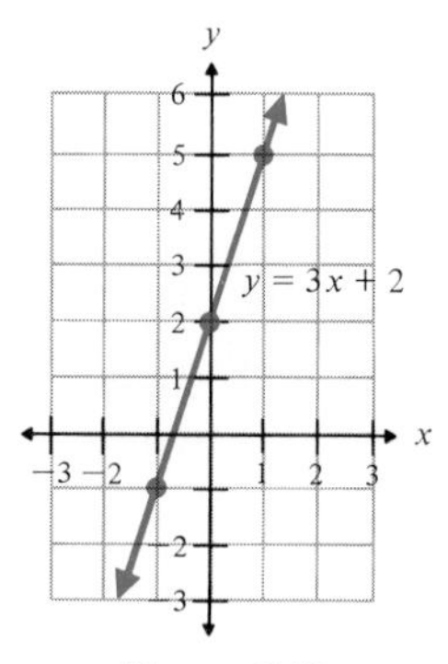

Figure 3.30

Since the value of y depends on the value of x and the equation is a function, we say that **y is a function of x.** The notation $y = f(x)$ is used to show that y is a function of the variable x. For this example we can write

$$y = f(x) = 3x + 2$$

The notation $f(x)$ is read "f of x" and *does not mean* f times x. Other letters may be used to indicate functions. For example $g(x)$ and $h(x)$ also represent functions of x.

To evaluate a function for a specific value of x substitute that value for x. For example

$$f(x) = 3x + 2$$
$$f(1) = 3(1) + 2 = 5$$
$$f(-4) = 3(-4) + 2 = -10$$
$$f(0) = 3(0) + 2 = 2$$

The notation $f(1)$ is read "f at 1," $f(-4)$ is read "f at -4," and so on.

EXAMPLE 7 Graph $f(x) = 2x - 3$.

Solution: Remember that $f(x)$ is the same as y. We can make a chart of values by substituting values of x and finding the corresponding value of y [or of $f(x)$].

$$y = f(x) = 2x - 3$$

$x = 0 \quad y = f(0) = 2(0) - 3 = -3$

$x = 2 \quad y = f(2) = 2(2) - 3 = 1$

$x = 3 \quad y = f(3) = 2(3) - 3 = 3$

x	y
0	−3
2	1
3	3

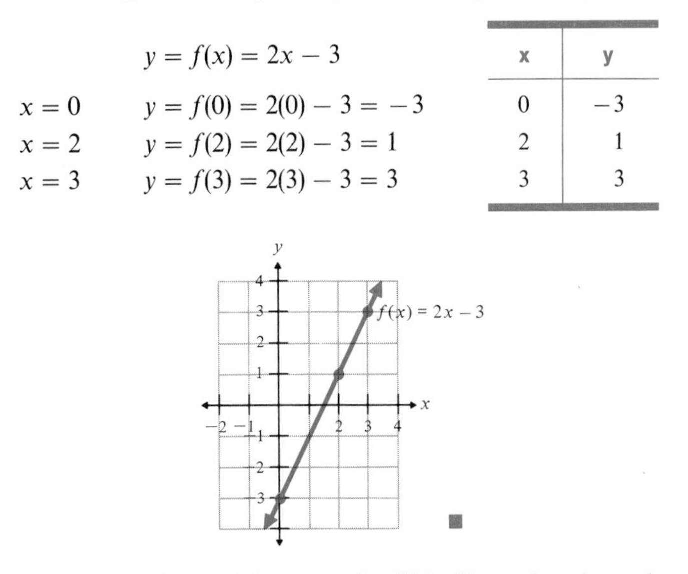

All equations of the form $f(x) = ax + b$ will be **linear functions,** that is, straight lines that are also functions. All linear equations, other than equations of the form $x = a$, will be functions.

The graph of $x = -1$ is given in Fig. 3.31. Note that it is not a function since it does not pass the vertical line test. However, equations of the form $y = b$ are functions since they do pass the vertical line test. The graph of $y = 1$ is given in Fig. 3.32.

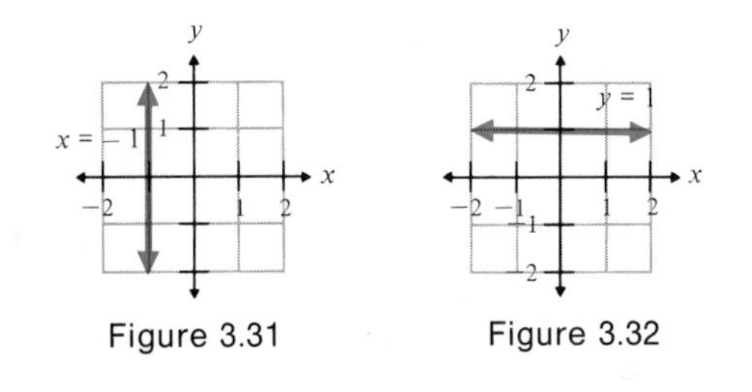

Figure 3.31

Figure 3.32

Exercise Set 3.5

Determine which of the relations are also functions. Give the range and domain of each relation or function.

1. $\{(1, 4), (2, 2), (3, 5), (4, 3), (5, 1)\}$
2. $\{(1, 1), (4, 4), (3, 3), (2, 2), (4, 1)\}$
3. $\{(3, -1), (5, 0), (1, 2), (4, 4), (2, 2), (7, 5)\}$
4. $\left\{(-1, 1), (0, -3), (3, 4), (4, 5), \left(-2, \frac{1}{2}\right)\right\}$
5. $\{(5, 0), (3, -4), (2, -1), (5, 2), (1, 1)\}$
6. $\{(6, 3), (-3, 4), (0, 3), (5, 2), (3, 5), (2, 5)\}$
7. $\left\{\left(\frac{1}{2}, \frac{2}{3}\right), (3, 0), (2, -1), (5, -3), (-2, 2), (0, 5)\right\}$
8. $\left\{\left(\frac{1}{5}, 2\right), \left(2, \frac{1}{2}\right), \left(\frac{2}{3}, 0\right), (-3, 2), (-3, -3), (5, 1)\right\}$
9. $\{(6, 0), (2, -3), (1, 5), (1, 0), (1, 2)\}$
10. $\{(3, -3), (3, -7), (3, -9), (3, 5)\}$
11. $\{(0, 3), (1, 3), (2, 2), (1, -1), (2, -7)\}$
12. $\{(3, 5), (2, 5), (1, 5), (0, 5), (-1, 5)\}$

13.

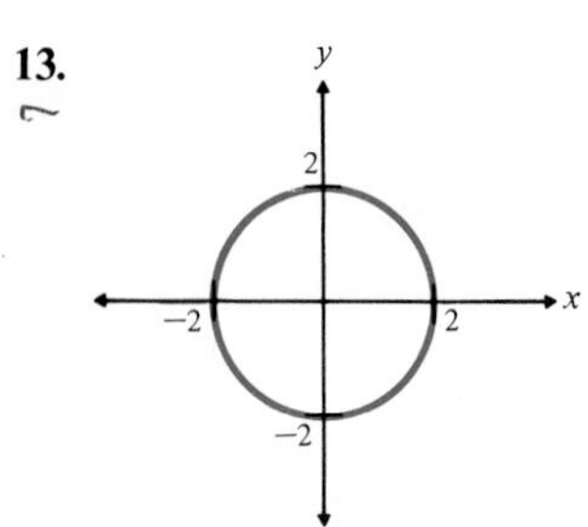

14.

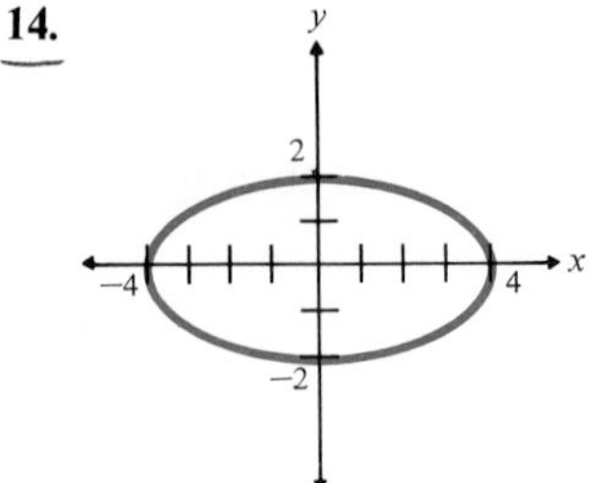

15.

16.

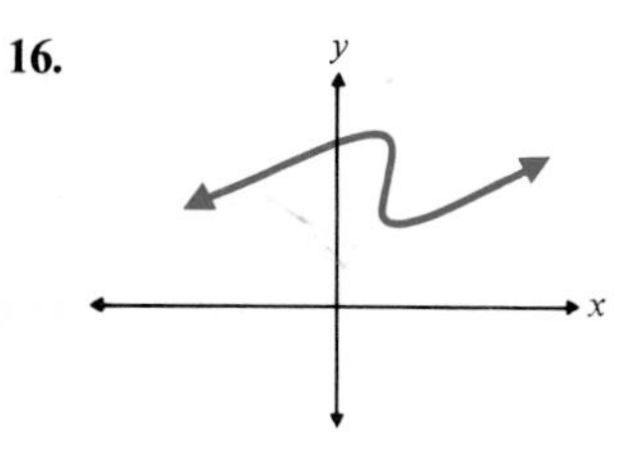

17.

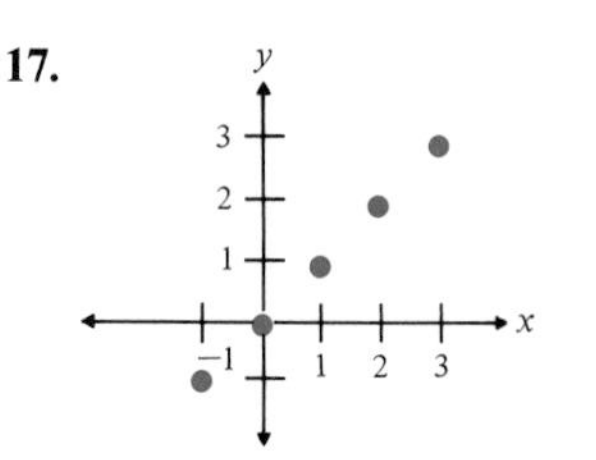

18.

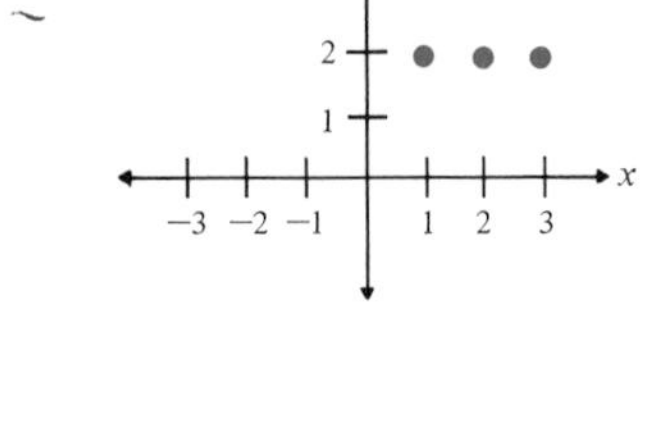

19.

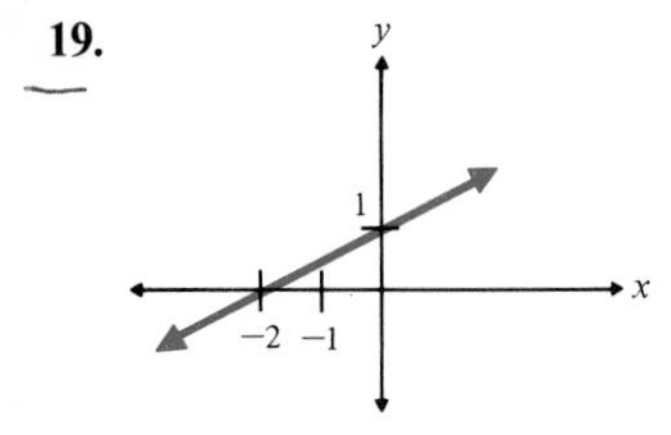

20.

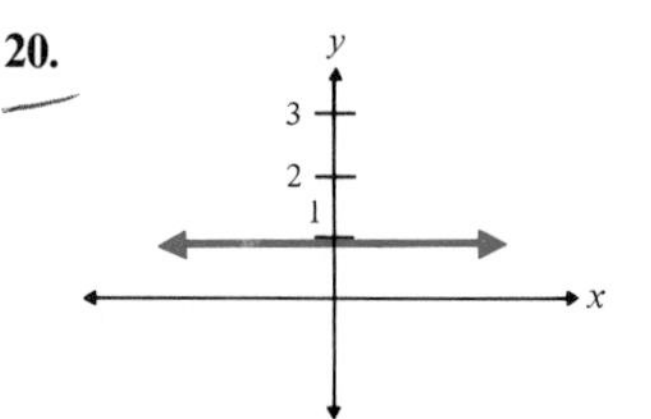

21.

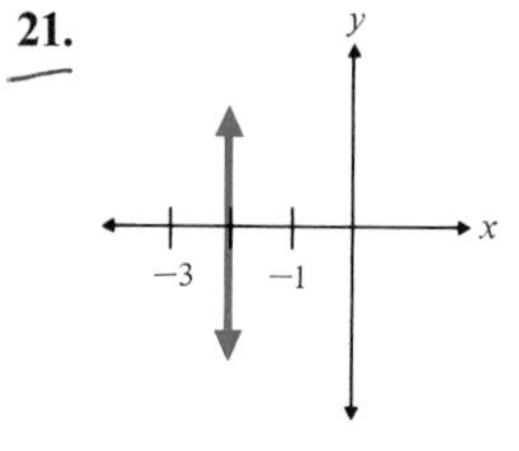

22.

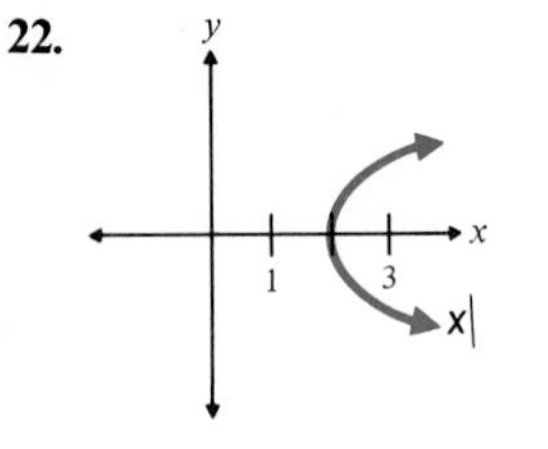

23.

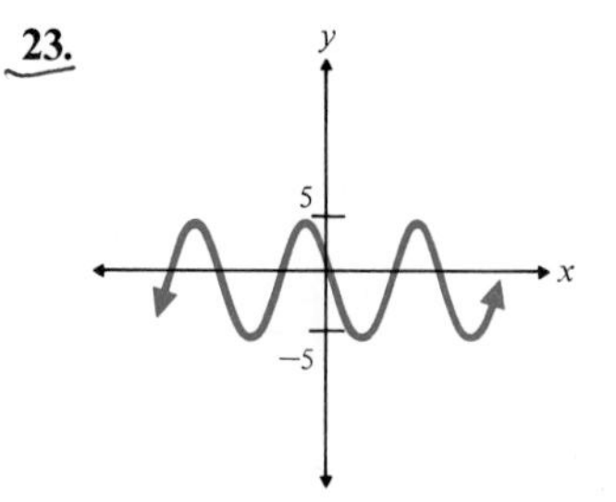

24.

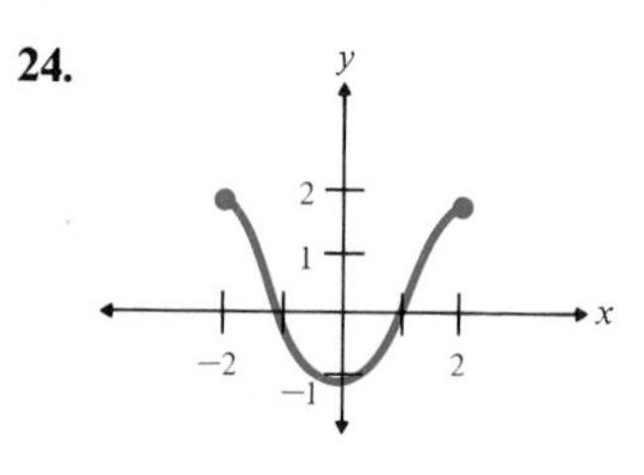

Evaluate the functions at the values indicated.

25. $f(x) = 2x + 7$; find **(a)** $f(3)$ **(b)** $f(-2)$

26. $f(x) = -3x + 4$; find **(a)** $f(0)$ **(b)** $f(1)$

27. $f(x) = 5x - 6$; find **(a)** $f(2)$ **(b)** $f(3)$

28. $f(x) = 2x + 6$; find **(a)** $f(-3)$ **(b)** $f(\frac{1}{2})$

29. $f(x) = 3 - 2x$; find **(a)** $f(2)$ **(b)** $f(\frac{1}{2})$

30. $f(x) = \frac{1}{2}x - 4$; find **(a)** $f(10)$ **(b)** $f(-4)$

31. $f(x) = \frac{2}{3}x - 3$; find **(a)** $f(3)$ **(b)** $f(-12)$

32. $f(x) = -\frac{3}{4}x + \frac{1}{2}$; find **(a)** $f(2)$ **(b)** $f(0)$

Graph each of the following functions.

33. $f(x) = 3x + 1$

34. $f(x) = -x + 3$

35. $f(x) = 2x - 1$

36. $f(x) = 4x - 2$

37. $f(x) = -x - 2$

38. $f(x) = 3x - 5$

39. $f(x) = \frac{1}{2}x + 3$

40. $f(x) = \frac{1}{3}x + 1$

41. $g(x) = 2x - 6$

42. $g(x) = 6x - 2$

Answer the following questions.

43. What is a relation?
44. What is a function?
45. Are all relations also functions? Explain.
46. Are all functions also relations? Explain.
47. Explain how the vertical line test is used to determine if a relation is a function.
48. What is the domain of a relation?
49. What is the range of a relation?

3.6 Graphing Linear Inequalities

1 *Graph linear inequalities.*

A linear inequality results when the equal sign in a linear equation is replaced with an inequality sign. Examples of linear inequalities in two variables are:

$$2x + 3y > 2 \qquad 3y < 4x - 6$$
$$-x - 2y \leq 3 \qquad 5x \geq 2y - 3$$

To Graph a Linear Inequality

1. Replace the inequality symbol with an equal sign.
2. Draw the graph of the equation in step 1. If the original inequality contains a $\geq$ or $\leq$ symbol, draw the graph using a solid line. If the original inequality contains a $>$ or $<$ symbol, draw the graph using a dashed line.
3. Select any point not on the line and determine if this point is a solution to the original inequality. If the point selected is a solution, shade the region on the side of the line containing this point. If the selected point does not satisfy the inequality, shade the region on the side of the line not containing this point.

EXAMPLE 1 Graph the inequality $y < 2x - 4$.

Solution: Graph the equation $y = 2x - 4$. Since the original inequality contains a less-than sign, $<$, use a dashed line when drawing the graph (see Fig. 3.33). The dashed line indicates that the points on this line are not solutions to the inequality $y < 2x - 4$. Select a point not on the line and determine if this point satisfies the inequality. Often the easiest point to use is the origin, (0, 0).

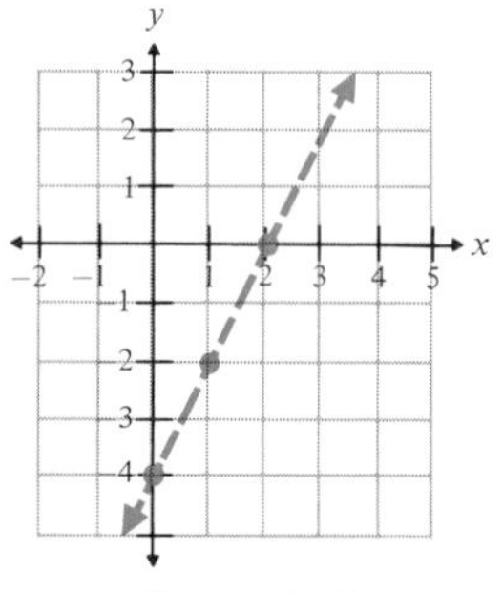

Figure 3.33

$$y < 2x - 4$$
$$0 < 2(0) - 4$$
$$0 < 0 - 4$$
$$0 < -4 \qquad \text{false}$$

Since 0 is not less than -4, $0 \nless -4$, the point (0, 0) does not satisfy the inequality. The solution will be all the points on the opposite side of the line from the point (0, 0). Shade in this region (Fig. 3.34). Every point in the shaded area satisfies the given inequality. Let's check a few selected points *A*, *B*, and *C*.

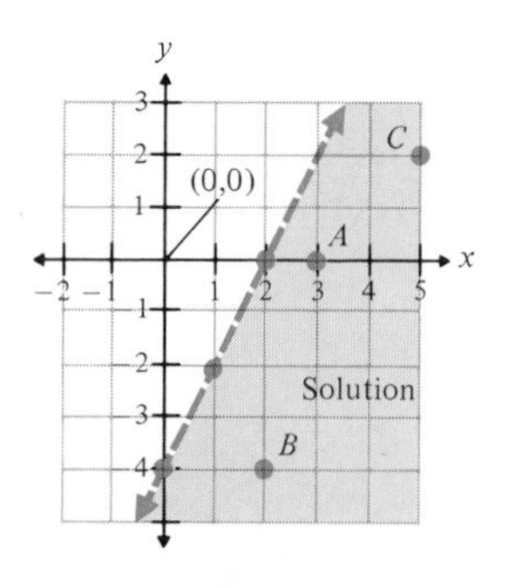

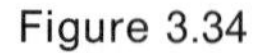
Figure 3.34

Point A	*Point B*	*Point C*
(3, 0)	(2, −4)	(5, 2)
$y < 2x - 4$	$y < 2x - 4$	$y < 2x - 4$
$0 < 2(3) - 4$	$-4 < 2(2) - 4$	$2 < 2(5) - 4$
$0 < 2$ true	$-4 < 0$ true	$2 < 6$ true

EXAMPLE 2 Graph the inequality $y \geq -\frac{1}{2}x$.

Solution: Graph the equation $y = -\frac{1}{2}x$. Since the inequality is $\geq$, we use a solid line to indicate that the points on the line are solutions to the inequality (Fig. 3.35). Since the point (0, 0) is on the line, we cannot select that point to find the solution. Let's arbitrarily select the point (3, 1).

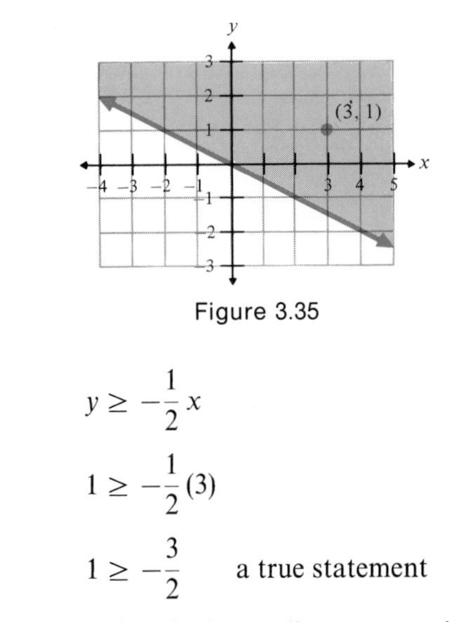

Figure 3.35

$$y \geq -\frac{1}{2}x$$

$$1 \geq -\frac{1}{2}(3)$$

$$1 \geq -\frac{3}{2} \quad \text{a true statement}$$

Since the point (3, 1) satisfies the inequality, every point on the same side of the line as (3, 1) will also satisfy the inequality $y \geq -\frac{1}{2}x$. Shade this region as indicated. Every point in the shaded region as well as every point on the line satisfies the inequality. ■

Exercise Set 3.6

Graph each inequality.

1. $x > 3$

2. $y < -2$

3. $x \geq \frac{5}{2}$

4. $y < x$

5. $y \geq 2x$

6. $y > -2x$

7. $y < 2x + 1$

8. $y \geq 3x - 1$

9. $y < -3x + 4$

10. $y \geq 2x + 4$

11. $y \geq \frac{1}{2}x - 4$

12. $y < 3x + 5$

13. $y \leq \frac{1}{3}x + 6$

14. $y > 6x + 1$

15. $y \leq -3x + 5$

16. $y \leq \frac{2}{3}x + 3$

17. $y > 5x - 9$

18. $y > \frac{2}{3}x - 1$

19. $2x + y < 4$

20. $3x - 4y \leq 12$

21. $2x \leq 5y + 10$

22. $5x - y \leq 4$

23. $-2x - 4y > 8$

24. $-5x + 3y \geq 9$

25. When graphing an inequality containing $\geq$ or $\leq$, why will points on the line be solutions to the inequality?

26. When graphing an inequality containing $>$ or $<$, why will points on the line not be solutions to the inequality?

JUST FOR FUN

1. Graph $y < |x|$.
2. Graph $y \geq |x + 4|$,

1.

2.

Summary

Glossary

Cartesian Coordinate System (or Rectangular Coordinate System): Two number lines drawn perpendicular to each other, creating four quadrants.

Collinear Points: Points in a straight line.

Domain: The set of first coordinates in a set of ordered pairs.

Function: A relation in which no two ordered pairs have the same first coordinate and a different second coordinate.

Graph: An illustration of the set of points that satisfy an equation.

Linear Function: Equations of the form $f(x) = ax + b$ are linear functions.
Negative Reciprocals: Two real numbers whose product is -1.
Negative Slope: A line has a negative slope when it falls as it moves from left to right.
Ordered Pair: The x and y coordinates of a point listed in parentheses, x first.
Origin: The point of intersection of the x and y axes.
Parallel Lines: Two lines in the same plane that do not intersect no matter how far they are extended. Two lines are parallel when they have the same slope.
Perpendicular Lines: Two lines that cross at right angles. Two lines are perpendicular when their slopes are negative reciprocals.
Positive Slope: A line has a positive slope when it rises as it moves from left to right.
Range: The set of second coordinates in a set of ordered pairs.
Relation: Any set of ordered pairs.
Slope of a Line: The ratio of the vertical change to the horizontal change between any two points on a line.
***x* Axis:** The horizontal axis in the Cartesian coordinate system.
***x* Intercept:** The value of x where a graph crosses the x axis.
***y* Axis:** The vertical axis in the Cartesian coordinate system.
***y* Intercept:** The value of y where a graph crosses the y axis.

Important Facts

Distance Formula $d = \sqrt{(x_2 - x_1)^2 + (y^2 - y_1)^2}$

Midpoint Formula $\left(\dfrac{x_1 + x_2}{2}, \dfrac{y_1 + y_2}{2}\right)$

Slope of a Line, *m* $m = \dfrac{\Delta y}{\Delta x} = \dfrac{y_2 - y_1}{x_2 - x_1}$

Standard Form of a Line $ax + by = c$

Slope-Intercept Form of a Line $y = mx + b$

Point-Slope Form of a Line $y - y_1 = m(x - x_1)$

To find the *x* intercept, set $y = 0$ and solve the equation for x.
To find the *y* intercept, set $x = 0$ and solve the equation for y.
To write an equation in slope-intercept form, solve the equation for y.

Review Exercises

[3.1]

1. Plot the ordered pairs on the same set of axes.

(a) $A(5, 3)$ **(b)** $B(0, 6)$ **(c)** $C\left(5, \frac{1}{2}\right)$
(d) $D(-4, 3)$ **(e)** $E(-6, -1)$ **(f)** $F(-2, 0)$

Find the distance between the two given points.

2. $(0, 0), (3, -4)$
3. $(6, 2), (2, -1)$
4. $(-2, -3), (3, 9)$
5. $(-4, 3), (-2, 5)$
6. $(3, 4), (5, 4)$
7. $(-3, 5), (-3, -8)$

Find the midpoint of the line segment between the two given points.

8. $(6, 5), (-1, 0)$
9. $(-6, -1), (5, 3)$
10. $(5, 5), (-2, -2)$
11. $(0, -8), (-2, 7),$
12. $(-6, 4), (-2, -5)$
13. $(-6, 4), (4, 8)$

[3.2] Graph the equation by the method of your choice.

14. $y = 4$

15. $x = -2$

16. $y = 4x$

17. $y = 2x - 1$

18. $y = -3x + 4$

19. $y = -\frac{1}{2}x + 2$

20. $6x + 3y = 6$

21. $2x - 3y = 12$

22. $2y = 3x - 6$

23. $4x - y = 8$

24. $5x - 2y = 10$

25. $3x = 6y + 9$

26. $25x - 50y = 200$

27. $3x - 2y = 150$

28. $\frac{2}{3}x = \frac{1}{4}y + 20$

[3.3–3.4] Determine the slope and y intercept of the equation.

29. $y = -x + 5$

30. $y = 3x + 5$

31. $y = -4x + \frac{1}{2}$

32. $2x + 3y = 9$

33. $3x + 6y = 9$

34. $4y = 6x + 12$

35. $3x + 5y = 12$

36. $9x + 7y = 15$

37. $36x - 72y = 144$

38. $x = -2$

39. $y = 6$

40. $y = -4x$

Determine the slope of the line through the two given points.

41. $(4, 6), (5, -1)$

42. $(-2, 3), (4, 1)$

43. $(-3, 5), (0, 6)$

44. $(-4, -2), (-3, 5)$

Two points on l_1 and two points on l_2 are given. Determine if l_1 is parallel to l_2, l_1 is perpendicular to l_2, or neither.

45. l_1: $(4, 3)$ and $(0, -3)$; l_2: $(1, -1)$ and $(2, -2)$

46. l_1: $(3, 2)$ and $(2, 3)$; l_2: $(4, 1)$ and $(1, 4)$

47. l_1: $(4, 0)$ and $(1, 3)$; l_2: $(5, 2)$ and $(6, 3)$

48. l_1: $(-3, 5)$ and $(2, 3)$; l_2: $(-4, -2)$ and $(-1, 2)$

Solve for the given variable if the line through the two given points is to have the given slope.

49. $(5, a)$ and $(4, 2)$; $m = 1$

50. $(3, 0)$ and $(5, x)$; $m = 3$

51. $(-2, -1)$ and $(4, y)$; $m = -6$

52. $(x, 2)$ and $(5, -2)$; $m = 2$

Write the equation of the line.

53.

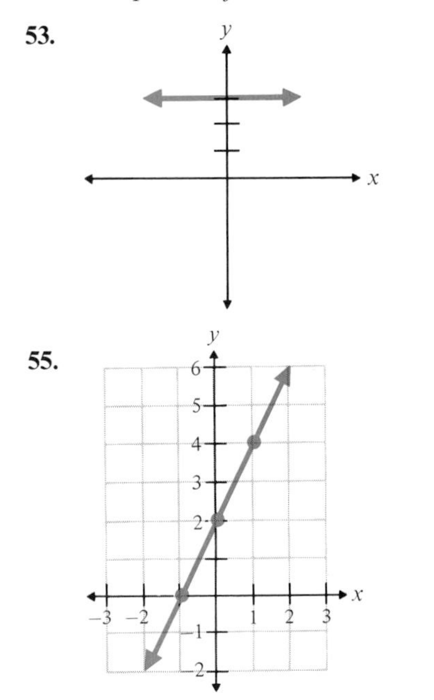

54.

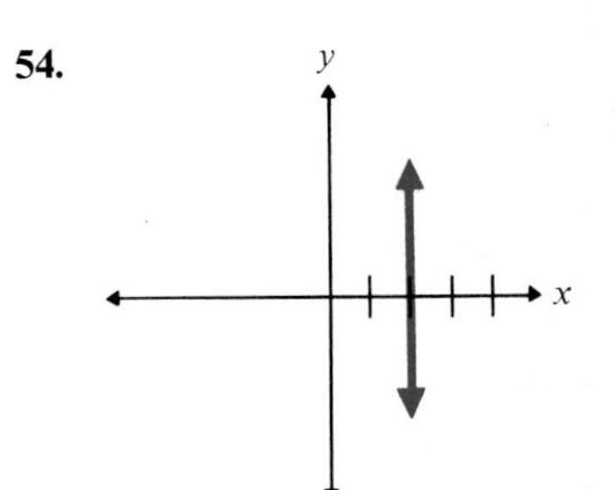

55.

56.

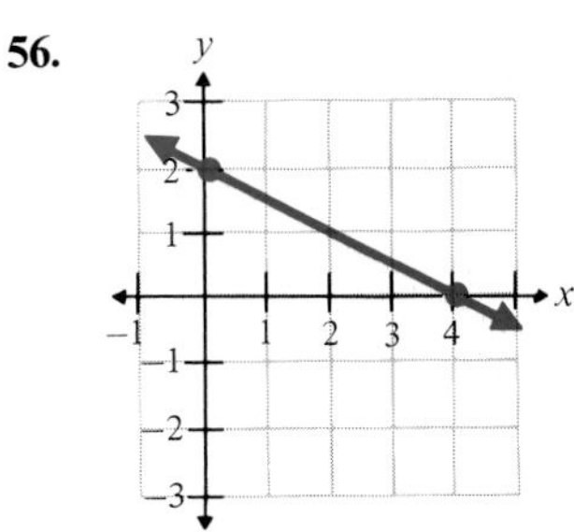

Determine if the two lines are parallel, perpendicular, or neither.

57. $y = 3x - 6$
$6y = 18x + 6$

58. $2x - 3y = 9$
$-3x - 2y = 6$

59. $y = \frac{4}{9}x + 5$
$4x = 9y + 9$

60. $4x = 6y + 3$
$-2x = -3y + 10$

61. $y = \frac{2}{5}x - 5$
$5x + 2y + 10 = 0$

62. $4x - 2y = 10$
$-2x + 4y = -8$

Find the equation of the line with the properties given. Write the answer in slope-intercept form.

63. Slope = 2, through (3, 4)

64. Slope = −3, through (−1, 5)

65. Slope = $-\frac{2}{3}$, through (3, 2)

66. Through (4, 3) and (2, 1)

67. Through (−2, 3) and (0, −4)

68. Through (−6, 2) parallel to $y = 3x - 4$

69. Through (4, −2) parallel to $2x - 5y = 6$

70. Through (−3, 1) perpendicular to $y = \frac{3}{5}x + 5$

71. Through (4, 2) perpendicular to $4x - 2y = 8$

72. The yearly profit of a bagel company can be estimated by the formula $p = 0.1x - 5000$, where x is the number of bagels sold per year.

(a) Draw a graph of profits versus bagels sold for up to 250,000 bagels.

(b) Estimate the number of bagels that must be sold for the company to break even.

(c) Estimate the number of bagels sold if the company has a $20,000 profit.

73. Draw a graph illustrating the interest on a $12,000 loan for a 1-year period for various interest rates up to 20%.

[3.5] Give the range and domain of each relation.

74. {(3, 4), (−2, 5), (0, −1), (6, 9)}

75. {($\frac{1}{2}$, 2), (4, −6), (5, 3), (2, −1)}

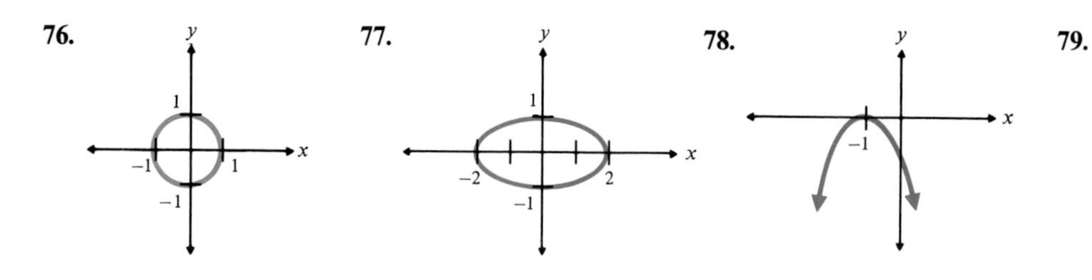

79.

Determine which of the following relations are functions.

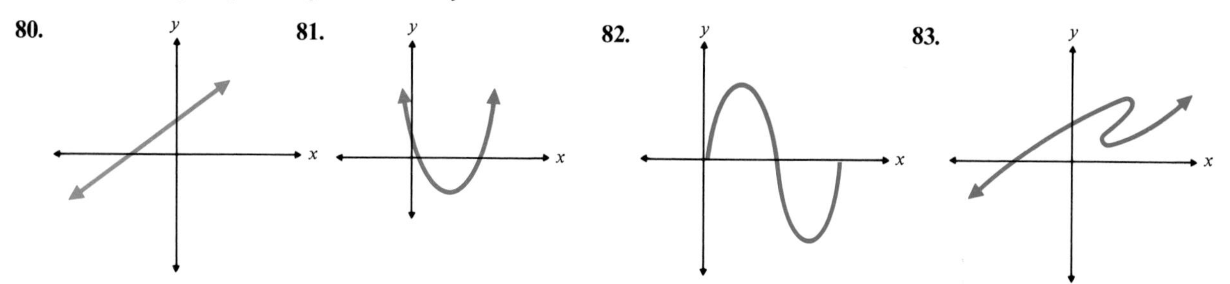

84. $\{(0, 4), (5, 6), (-2, 4), (1, 3)\}$

85. $\{(6, -2), (-3, 5), (5, 2), (-1, 4)\}$

86. $\{(3, 2), (4, 2), (5, 4), (6, 6)\}$

87. $\{(1, 4), (2, 4), (3, 4), (4, 4)\}$

88. $\{(-3, 2), (4, 3), (3, 2), (4, 6)\}$

89. $\{(0, 4), (1, 3), (5, 2), (1, -1)\}$

Evaluate the functions at the values indicated.

90. $f(x) = 3x + 4$; find **(a)** $f(2)$ **(b)** $f(-3)$

91. $f(x) = -2x + 5$; find **(a)** $f(-1)$ **(b)** $f(\frac{1}{2})$

92. $f(x) = -\frac{1}{2}x + 4$; find **(a)** $f(4)$ **(b)** $f(3)$

93. $f(x) = 6 - 2x$; find **(a)** $f(0)$ **(b)** $f(-5)$

Graph each of the following functions.

94. $f(x) = 2x + 4$

95. $f(x) = 3x - 5$

96. $f(x) = -4x + 2$

97. $g(x) = \dfrac{1}{2}x + 2$

[3.6] Graph the given inequality.

98. $y \geq -3$

99. $x < 4$

100. $y < 3x$

101. $y > 2x + 1$

102. $y \le 4x - 3$

103. $y \ge 6x + 5$

104. $y < -x + 4$

105. $y \le \frac{1}{3}x - 2$

Practice Test

1. Find the distance between the points (1, 3) and (−2, −1).
2. Find the slope and y intercept of $4x - 9y = 15$.
3. Write the equation of the following graph.

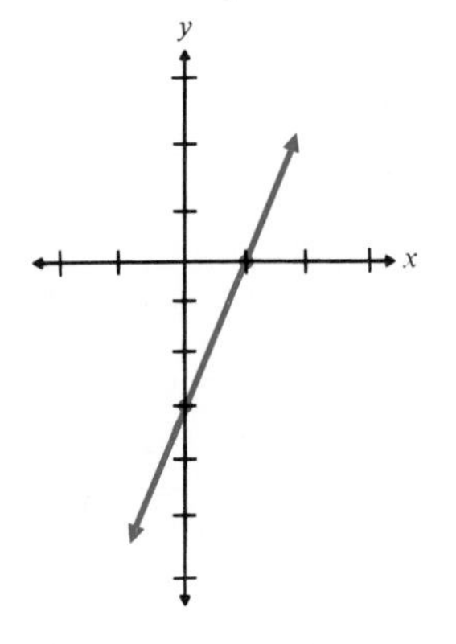

4. Write the equation of the line with a slope of 4 passing through the point (−1, 3). Write the equation in slope intercept form.
5. Write the equation of the line (in slope-intercept form) passing through the points (3, −1) and (−4, 2).
6. Write the equation of the line (in slope-intercept form) passing through the point (−1, 4) perpendicular to $2x + 3y = 6$.
7. Graph $y = 2x - 2$.
8. Graph $2x + 3y = 10$.
9. Graph $4x = -y + 10$.
10. State the domain and range of the following relation.

$$\{(4, 0), (2, -3), (\tfrac{1}{2}, 2), (6, 9)\}.$$

11. Determine which, if any, of the following are functions.

(a) **(b)**

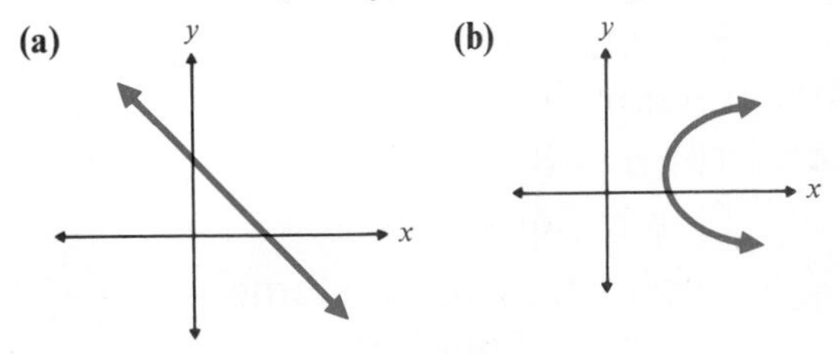

(c) $\{(1, 2), (3, 4), (-1, 2), (5, 0)\}$

12. $f(x) = 3x - 5$; find **(a)** $f(2)$ **(b)** $f(-4)$
13. Graph $f(x) = \frac{2}{3}x + 1$.
14. Graph $y \ge -3x + 5$.
15. Graph $y \le 4x - 2$.

4 Systems of Linear Equations and Inequalities

See Section 4.3, Exercise 7.

4.1 Solving Systems of Linear Equations

1. *Solve a system of linear equations graphically.*
2. *Solve a system of linear equations by substitution.*
3. *Solve a system of linear equations by the addition method.*

1 It is often necessary to find a common solution to two or more linear equations. We refer to the equations in this type of problem as **simultaneous linear equations** or as a **system of linear equations.**

$$\left.\begin{array}{l}(1)\ y = x + 5\\(2)\ y = 2x + 4\end{array}\right\}\ \text{system of linear equations}$$

A solution to a system of equations is an ordered pair or pairs that satisfy *all* equations in the system. The only solution to the system above is (1, 6).

Check in equation (1)	*Check in equation (2)*
(1, 6)	(1, 6)
$y = x + 5$	$y = 2x + 4$
$6 = 1 + 5$	$6 = 2(1) + 4$
$6 = 6$ true	$6 = 6$ true

The ordered pair (1, 6) satisfies *both* equations and is a solution to the system of equations.

A system of equations may consist of more than two equations. If a system consists of three equations in three variables, say x, y, and z, the solution will be an *ordered triple* of the form (x, y, z). If the ordered triple (x, y, z) is a solution to the system, it must satisfy all three equations in the system. A system with three equations and three unknowns is referred to as a *third-order system.* A system of equations may have more than three equations and three variables but we will not discuss them in this book.

Graphing Method

1 To solve a system of linear equations in two variables graphically, graph all equations in the system on the same set of axes. The solution to the system will be the ordered pair (or pairs) common to all the lines, or the point of intersection of all lines in the system.

When two lines are graphed, three situations are possible, as illustrated in Fig. 4.1.

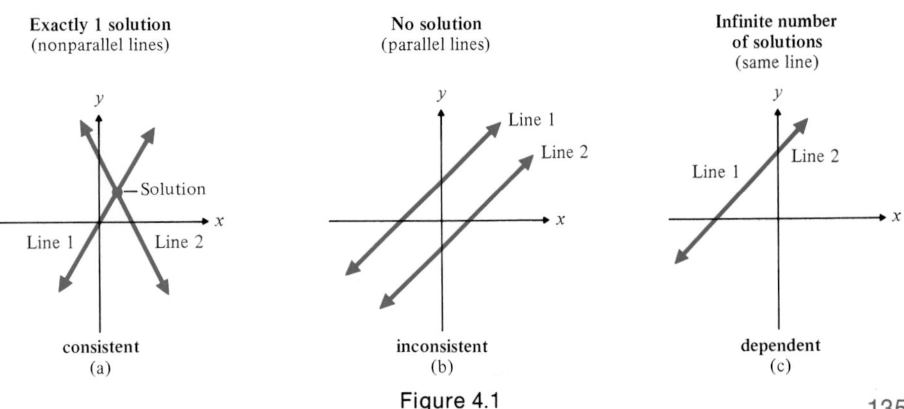

Figure 4.1

In Fig. 4.1a, line 1 and line 2 are nonparallel lines and intersect at exactly one point. This system of equations has *exactly one solution.* This is an example of a **consistent** system of equations. A consistent system of equations is a system of equations that has a solution.

Line 1 and line 2 of Fig. 4.1b are different but parallel lines. The lines do not intersect, and this system of equations has *no solution.* This is an example of an **inconsistent** system of equations. An inconsistent system of equations is a system of equations that has no solution.

In Fig. 4.1c, line 1 and line 2 are actually the same line. In this case every point on the line satisfies both equations and is a solution to the system of equations. This system has *an infinite number of solutions.* This is an example of a **dependent** system of equations. A dependent system of linear equations is a system of equations where both equations represent the same line. *Note that a dependent system is also a consistent system since it has a solution.*

We can determine if a system of linear equations is consistent, inconsistent, or dependent by writing each equation in slope-intercept form and comparing the slopes and y intercepts. Note that if the slopes of the lines are different, Fig. 4.1a, the system is consistent. If the slopes are the same but the y intercepts different, Fig. 4.1b, the system is inconsistent, and if both the slopes and the y intercepts are the same, Fig. 4.1c, the system is dependent.

EXAMPLE 1 Without graphing the equations determine if the following system of equations is consistent, inconsistent, or dependent.

$$2x + y = 3$$
$$4x + 2y = 12$$

Solution: Write each equation in slope intercept form.

$$2x + y = 3 \qquad\qquad 4x + 2y = 12$$
$$y = -2x + 3 \qquad\qquad 2y = -4x + 12$$
$$y = -2x + 6$$

Since both equations have the same slope, -2, and different y intercepts the lines are parallel lines. Therefore the system is inconsistent and has no solution. ■

EXAMPLE 2 Solve the following system of equations graphically.

$$y = x + 2$$
$$y = -x + 4$$

Solution: Graph both equations on the same set of axes, Fig. 4.2.

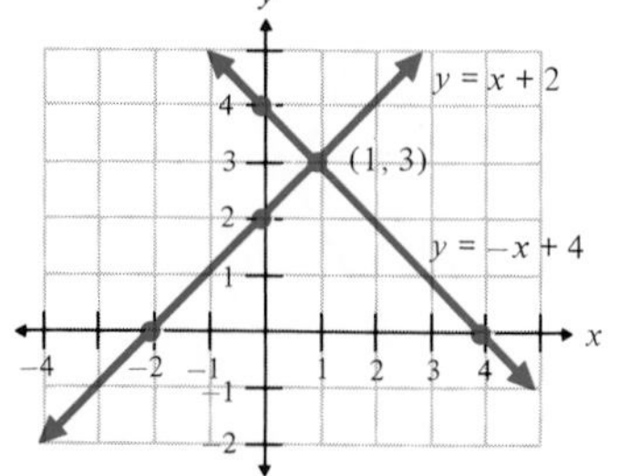

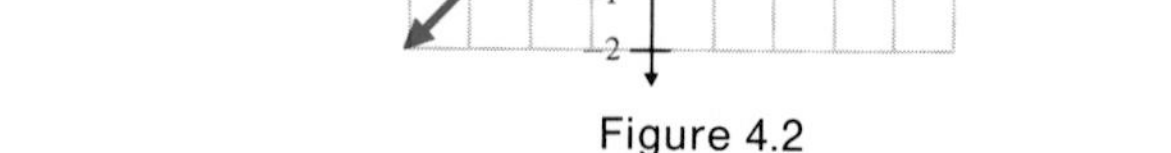

Figure 4.2

The solution is the point of intersection of the two lines, (1, 3). ■

Substitution Method

Often, an exact solution to system of equations may be difficult to find on a graph. When an exact answer is necessary, the system should be solved algebraically, either by substitution or by addition (elimination) of equations.

To Solve a System of Equations by Substitution

1. Solve for a variable in either equation. (If possible solve for a variable with a numerical coefficient of 1 to avoid working with fractions.)
2. Substitute the expression found for the variable in step 1 into the other equation.
3. Solve the equation determined in step 2 to find the value of one variable.
4. Substitute the value found in step 3 into the equation from step 1. Solve the equation to find the remaining variable.

EXAMPLE 3 Solve the system of equations by substitution.

$$y = 2x + 5$$
$$y = -4x + 2$$

Solution: Since both equations are already solved for y we can substitute $2x + 5$ for y in the second equation, and then solve for the remaining variable x.

$$\begin{aligned} 2x + 5 &= -4x + 2 \\ 6x + 5 &= 2 \\ 6x &= -3 \\ x &= -\frac{1}{2} \end{aligned}$$

Now find y be substituting $-\frac{1}{2}$ for x in either of the original equations. We will use the first equation.

$$\begin{aligned} y &= 2x + 5 \\ &= 2\left(-\frac{1}{2}\right) + 5 \\ &= -1 + 5 \\ &= 4 \end{aligned}$$

The solution is $(-\frac{1}{2}, 4)$. ■

EXAMPLE 4 Solve the system of equations by substitution.

$$2x + y = 11$$
$$x + 3y = 18$$

Solution: Begin by solving for one of the variables in either of the equations. You may solve for either of the variables; however, if you solve for a variable with a numerical coefficient of 1, you may avoid working with fractions. In this equation the y term in $2x + y = 11$ and the x term in $x + 3y = 18$ both have a numerical coefficient of 1.

Let's solve for y in $2x + y = 11$.

$$2x + y = 11$$
$$y = -2x + 11$$

Next, substitute $-2x + 11$ for y in the *other equation,* $x + 3y = 18$, and solve for the remaining variable, x.

$$x + 3y = 18$$
$$x + 3\overbrace{(-2x + 11)} = 18$$
$$x - 6x + 33 = 18$$
$$-5x + 33 = 18$$
$$-5x = -15$$
$$x = 3$$

Finally substitute $x = 3$ in the equation $y = -2x + 11$ and solve for y.

$$y = -2x + 11$$
$$y = -2(3) + 11 = 5$$

The solution is the ordered pair (3, 5) ■

If, when solving a system of equations by either substitution or the addition method, you arrive at an equation that is false, such as $5 = 6$ or $0 = 3$, the system is inconsistent and has no solution. If you obtain an equation that is true, such as $6 = 6$ or $0 = 0$, the system is dependent and has an infinite number of solutions.

HELPFUL HINT

> You may successfully solve for one of the variables and forget to solve for the other. Remember that a solution must contain a numerical value for each variable in the system.

Addition Method

3 A third, and often the easiest method of solving a system of equations is the addition or elimination method. The object of this process is to obtain two equations whose sum will be an equation containing only one variable. Keep in mind that your immediate goal is to obtain one equation containing only one unknown.

EXAMPLE 5 Solve the following system of equations using the addition method.

$$x + y = 6$$
$$2x - y = 3$$

Solution: Note that one equation contains a $+y$ and the other contains a $-y$. By adding the equations, we can eliminate the variable y and obtain one equation containing only one unknown, x.

$$\begin{array}{r} x + y = 6 \\ 2x - y = 3 \\ \hline 3x \quad\; = 9 \end{array}$$

Now solve for the remaining variable, x.

$$\frac{3x}{3} = \frac{9}{3}$$
$$x = 3$$

Finally, solve for y by inserting $x = 3$ in either of the original equations.

$$x + y = 6$$
$$3 + y = 6$$
$$y = 3$$

The solution is (3, 3). ■

To Solve a System of Equations by the Addition (or Elimination) Method

1. If necessary, rewrite each equation so that the terms containing variables appear on the left side of the equal sign and any constants appear on the right side of the equal sign.
2. If necessary, multiply one or both equations by a constant(s) so that when the equations are added, the resulting sum will contain only one variable.
3. Add the respective sides of the equations. This will result in a single equation containing only one variable.
4. Solve for the variable in the equation in step 3.
5. Substitute the value found in step 4 into either of the original equations. Solve that equation to find the value of the remaining variable.

In Step 2 of the procedure we indicate that it may be necessary to multiply both sides of an equation by a constant. In this text we will use brackets, [ ], to indicate multiplication of *an entire equation* by a real number. Thus 4[ ] means to multiply the entire equation within the brackets by 4, and in general a[ ] means to multiply the entire equation within the brackets by the real number a.

EXAMPLE 6 Solve the following system of equations using the addition method.

$$\begin{aligned} 2x + y &= 11 \\ x + 3y &= 18 \end{aligned}$$

Solution: The object of the addition process is to obtain two equations whose sum will be an equation containing only one variable. To eliminate the variable x we multiply the second equation by -2 and add the two equations.

$$\begin{aligned} 2x + y &= 11 \qquad &\text{gives} \qquad 2x + y &= 11 \\ -2[x + 3y &= 18] & -2x - 6y &= -36 \end{aligned}$$

Now add:

$$\begin{array}{r} 2x + y = 11 \\ -2x - 6y = -36 \\ \hline -5y = -25 \\ y = 5 \end{array}$$

Solve for x.

$$\begin{aligned} 2x + y &= 11 \\ 2x + 5 &= 11 \\ 2x &= 6 \\ x &= 3 \end{aligned}$$

The solution is (3, 5). Note that we could have first eliminated the variable y by multiplying the first equation by -3 and then adding. ■

EXAMPLE 7 Solve the following system of equations using the addition method.

$$\begin{aligned} 2x + 3y &= 6 \\ 5x - 4y &= -8 \end{aligned}$$

Solution: The x variable can be eliminated by multiplying the first equation by -5 and the second by 2 and then adding the equations.

$$\begin{aligned} -5[2x + 3y &= 6] \qquad &\text{gives} \qquad -10x - 15y &= -30 \\ 2[5x - 4y &= -8] & 10x - 8y &= -16 \end{aligned}$$

$$\begin{array}{r} -10x - 15y = -30 \\ 10x - 8y = -16 \\ \hline -23y = -46 \\ y = 2 \end{array}$$

The same value could be obtained for y by multiplying the first equation by 5 and the second by -2 and then adding. Try it now and see.

Solve for x.

$$\begin{aligned} 2x + 3y &= 6 \\ 2x + 3(2) &= 6 \\ 2x + 6 &= 6 \\ 2x &= 0 \\ x &= 0 \end{aligned}$$

The solution to this system is (0, 2). ■

EXAMPLE 8 Solve the following system of equations using the addition method.

$$\begin{aligned} 2x + 3y &= 7 \\ 5x - 7y &= -3 \end{aligned}$$

Solution: We can eliminate the variable x by multiplying the first equation by -5 and the second by 2.

$$\begin{aligned} -5[2x + 3y &= 7] \quad \text{gives} \quad & -10x - 15y &= -35 \\ 2[5x - 7y &= -3] & 10x - 14y &= -6 \end{aligned}$$

$$\begin{aligned} -10x - 15y &= -35 \\ 10x - 14y &= -6 \\ \hline -29y &= -41 \\ y &= \frac{41}{29} \end{aligned}$$

We can now find x by substituting $y = \frac{41}{29}$ into one of the original equations and solving for x. If you try this, you will see that although it can be done, it gets pretty messy. An easier method that can be used to solve for x is to go back to the original equations and eliminate the variable y.

$$\begin{aligned} 7[2x + 3y &= 7] \quad \text{gives} \quad & 14x + 21y &= 49 \\ 3[5x - 7y &= -3] & 15x - 21y &= -9 \end{aligned}$$

$$\begin{aligned} 14x + 21y &= 49 \\ 15x - 21y &= -9 \\ \hline 29x &= 40 \\ x &= \frac{40}{29} \end{aligned}$$

The solution is $(\frac{40}{29}, \frac{41}{29})$. ■

EXAMPLE 9 Solve the system of equations.

$$x + \frac{4}{3}y = 2$$

$$y = -\frac{2}{3}x + \frac{5}{2}$$

Solution: First eliminate the fractions in the equations by multiplying each equation by its least common denominator. To eliminate fractions we multiply the first equation by 3 and the second by 6.

$$x + \frac{4}{3}y = 2 \qquad y = -\frac{2}{3}x + \frac{5}{2}$$

$$3\left(x + \frac{4}{3}y\right) = 3 \cdot 2 \qquad 6 \cdot y = 6\left(-\frac{2}{3}x + \frac{5}{2}\right)$$

$$3x + 4y = 6 \qquad 6y = -4x + 15$$

The new system of equations is

$$\begin{aligned} 3x + 4y &= 6 \\ 6y &= -4x + 15 \end{aligned} \qquad \text{or} \qquad \begin{aligned} 3x + 4y &= 6 \\ 4x + 6y &= 15 \end{aligned}$$

Now solve the system. We will solve for y by eliminating the terms containing x.

$$\begin{aligned} 4[3x + 4y &= 6] \\ -3[4x + 6y &= 15] \end{aligned} \qquad \text{gives} \qquad \begin{aligned} 12x + 16y &= 24 \\ -12x - 18y &= -45 \\ \hline -2y &= -21 \\ y &= \frac{21}{2} \end{aligned}$$

Now solve for x, by eliminating the y terms.

$$\begin{aligned} 3[3x + 4y &= 6] \\ -2[4x + 6y &= 15] \end{aligned} \qquad \text{gives} \qquad \begin{aligned} 9x + 12y &= 18 \\ -8x - 12y &= -30 \\ \hline x \qquad &= -12 \end{aligned}$$

We could also have found x by substituting $y = \frac{21}{2}$ in either of the original equations and then solving for x. The solution to the system is $(-12, \frac{21}{2})$. ■

EXAMPLE 10 Solve the following system of equations using the addition method.

$$2x + y = 3$$

$$4x + 2y = 12$$

Solution:

$$\begin{aligned} -2[2x + y &= 3] \\ 4x + 2y &= 12 \end{aligned} \qquad \text{gives} \qquad \begin{aligned} -4x - 2y &= -6 \\ 4x + 2y &= 12 \end{aligned}$$

$$\begin{aligned} -4x - 2y &= -6 \\ 4x + 2y &= 12 \\ \hline 0 &= 6 \qquad \text{false} \end{aligned}$$

Since $0 = 6$ is a false statement, this system has no solution. The system is inconsistent and the lines will be parallel when graphed. ■

EXAMPLE 11 Solve the following system of equations using the addition method.

$$x - \frac{1}{2}y = 2$$
$$y = 2x - 4$$

Solution: First align the x and y terms on the left side of the equation.

$$x - \frac{1}{2}y = 2$$
$$2x - \quad y = 4$$

Now proceed as in previous examples.

$$-2\left[x - \frac{1}{2}y = 2\right] \quad \text{gives} \quad -2x + y = -4$$
$$2x - \quad y = 4 \qquad\qquad\qquad 2x - y = \quad 4$$

$$\begin{aligned} -2x + y &= -4 \\ 2x - y &= \quad 4 \\ \hline 0 &= \quad 0 \quad \text{true} \end{aligned}$$

Since $0 = 0$ is a true statement, the system is dependent and has an infinite number of solutions. Both equations represent the same line. ■

We have illustrated three methods that can be used to solve a system of linear equations: graphing, substitution, and the addition method. When you are given a system of equations, which method should you use to solve the system? When you need an exact solution, graphing should not be used. Of the two algebraic methods, the addition method may be the easiest to use if there are no numerical coefficients of 1 in the system. If one or more of the equations has a coefficient of 1, you may wish to use either method. We will present a fourth method, using determinants, in the optional Section 4.4.

Exercise Set 4.1

Determine which, if any, of the ordered pairs or ordered triples satisfy the system of linear equations.

1. $y = -6x$
$y = -2x + 8$
(a) $(0, 0)$ **(b)** $(-4, 16)$ **(c)** $(-2, 12)$

2. $x + 2y = 4$
$y = 3x + 3$
(a) $(0, 2)$ **(b)** $(-2, 3)$ **(c)** $(4, 15)$

3. $y = 2x + 4$
$y = 2x - 1$
(a) $(0, 4)$ **(b)** $(3, 10)$ **(c)** $(-2, 0)$

4. $2x - 3y = 6$
$y = \frac{2}{3}x - 2$
(a) $(3, 0)$ **(b)** $(3, -2)$ **(c)** $(1, -\frac{4}{3})$

5. $y = -x + 4$
$2y = -2x + 8$
(a) $(2, 5)$ **(b)** $(0, 4)$ **(c)** $(5, -1)$

6. $3x - 4y = 8$
$2y = \frac{3}{2}x - 4$
(a) $(1, -6)$ **(b)** $(-\frac{1}{3}, -\frac{9}{4})$ **(c)** $(0, -2)$

7. $2x + 3y = 6$
$-2x + 5 = y$
(a) $(\frac{1}{2}, \frac{5}{3})$ **(b)** $(2, 1)$ **(c)** $(\frac{9}{4}, \frac{1}{2})$

8. $x + 2y - z = -5$
$2x - y + 2z = 8$
$3x + 3y + 4z = 5$
(a) $(1, 3, -2)$ **(b)** $(1, -2, 2)$ **(c)** $(0, 8, -2)$

9. $4x + y - 3z = 1$
$2x - 2y + 6z = 11$
$-6x + 3y + 12z = -4$
(a) $(2, -1, -2)$ **(b)** $(\frac{1}{2}, 2, 1)$ **(c)** $(\frac{1}{2}, -3, \frac{2}{3})$

10. $2x - 3y + z = 1$
$x + 2y + z = -1$
$3x - y + 3z = 4$
(a) $(1, 1, 4)$ **(b)** $(-3, -1, 4)$ **(c)** $(0, 3, 10)$

Write each equation in slope-intercept form. Without graphing the equations, state whether the system of equations is consistent, inconsistent, or dependent. Also indicate whether the system has exactly one solution, no solution, or an infinite number of solutions.

11. $2y = -x + 5$
$x - 2y = 1$

12. $2x + y = 6$
$2x - y = 6$

13. $3y = 2x + 3$
$y = \frac{2}{3}x - 2$

14. $y = \frac{1}{2}x + 4$
$2y = x + 8$

15. $2x - 3y = 4$
$3x - 2y = -2$

16. $x + 2y = 6$
$2x + y = 4$

17. $2x = 3y + 4$
$6x - 9y = 12$

18. $x - y = 3$
$2x - 2y = -2$

19. $y = \frac{3}{2}x + \frac{1}{2}$
$3x - 2y = -\frac{1}{2}$

20. $x - y = 3$
$\frac{1}{2}x - 2y = -6$

In Exercises 21 through 30, determine the solution to the system of equations graphically.

21. $y = x + 4$
$y = -x + 2$

22. $y = 2x + 4$
$y = -3x - 6$

23. $y = 2x - 1$
$2y = 4x + 6$

24. $y = -2x - 1$
$x + 2y = 4$

25. $2x + 3y = 6$
$4x = -6y + 12$

26. $x + y = 1$
$3x - y = -5$

27. $x + 3y = 4$
$x = 1$

28. $2x - 5y = 10$
$y = \frac{2}{5}x - 2$

29. $y = -5x + 5$
$y = 2x - 2$

30. $2x - y = -4$
$2y = 4x - 6$

Find the solution to each system of equations by substitution.

31. $x + 2y = 9$
$x = 2y + 1$

32. $y = x + 2$
$2y = -x - 2$

33. $x + y = 6$
$x = y$

34. $2x + y = 3$
$2y = 6 - 4x$

35. $2x + y = 3$
$2x + y + 5 = 0$

36. $y = 2x + 4$
$y = -2$

37. $x = -1$
$x + y + 5 = 2$

38. $y = \frac{1}{3}x - 2$
$x - 3y = 6$

39. $x - \frac{1}{2}y = 2$
$y = 2x - 4$

40. $2x + 3y = 7$
$6x - y = 1$

41. $3x + y = -1$
$y = 3x + 5$

42. $y = -2x + 5$
$x + 3y = 0$

43. $y = 2x - 13$
$-4x - 7 = 9y$

44. $x = y + 4$
$3x + 7y = -18$

45. $5x - 2y = -7$
$5 = y - 3x$

46. $5x - 4y = -7$
$x - \frac{3}{5}y = -2$

47. $x = 3y + 5$
$y = \frac{2}{3}x$

48. $x + 2y = 4$
$x + \frac{1}{2}y = 4$

49. $\frac{1}{2}x - \frac{1}{3}y = 2$
$\frac{1}{4}x + \frac{2}{3}y = 6$

50. $\frac{1}{2}x + \frac{1}{3}y = 13$
$\frac{1}{5}x + \frac{1}{8}y = 5$

Solve each system of equations using the addition method.

51. $x + y = -2$
$x - y = 4$

52. $x - y = 12$
$x + y = 2$

53. $-x + y = 5$
$x + 2y = 1$

54. $x + y = 0$
$-x + y = -2$

55. $3x + 2y = 15$
$x - 2y = -7$

56. $3x + 3y = 18$
$4x - y = 4$

57. $3x + y = 6$
$-6x - 2y = 10$

58. $2x + y = 14$
$-3x + y = -2$

59. $2x + y = 6$
$3x - 2y = 16$

60. $4x - 3y = 8$
$-2x + 5y = 14$

61. $2x - 5y = 13$
$5x + 3y = 17$

62. $4x = 2y + 6$
$y = 2x - 3$

63. $3y = 2x + 4$
$3y = 2x + 4$

64. $5x + 4y = 10$
$-3x - 5y = 7$

65. $4x - 3y = 8$
$-3x + 4y = 9$

66. $2x - y = 8$
$3x + y = 6$

67. $3x + 4y = 2$
$2x = -5y - 1$

68. $2x = 5y + 13$
$5x = -3y + 17$

69. $2y = -5x - 3$
$4x - 7y = 3$

70. $3x + 4y = 2$
$2x = -5y - 1$

71. $4x + 5y = 3$
$2x - 3y = 4$

72. $2x + 3y = 5$
$-3x - 4y = -2$

73. $2x - \frac{1}{3}y = 6$
$5x - y = 4$

74. $\frac{x}{2} - \frac{y}{3} = 1$
$\frac{x}{4} - \frac{y}{9} = \frac{2}{3}$

75. $\frac{x}{3} = 4 - \frac{y}{4}$
$3x = 4y$

76. $\frac{x}{4} - 3 = \frac{y}{6}$
$y = \frac{x}{2} + 2$

77. $\frac{x}{5} + \frac{y}{2} = 4$
$\frac{2x}{3} - y = \frac{8}{3}$

78. $\frac{2x}{3} - 4 = \frac{y}{2}$
$x - 3y = \frac{1}{3}$

79. Explain how you can determine, without graphing or solving, whether a system of two linear equations is consistent, inconsistent, or dependent.

80. When solving a system of equations by addition or substitution, how will you know if the system is inconsistent?

81. When solving a system of equations by addition or substitution, how will you know if the system is dependent?

JUST FOR FUN

Solve each of the following using the addition method.

1. $\frac{x+2}{2} - \frac{y+4}{3} = 4$
$\frac{x+y}{2} = \frac{1}{2} + \frac{x-y}{3}$

2. $\frac{5x}{2} + 3y = \frac{9}{2} + y$
$\frac{1}{4}x - \frac{1}{2}y = 6x + 12$

Solve the following system of equations, where *a* and *b* represent any nonzero constants by the (a) substitution and (b) addition methods.

3. $4ax + 3y = 19$
$-ax + y = 4$

4. $ax = 2 - by$
$-ax + 2by - 1 = 0$

4.2 Third-Order Systems of Linear Equations

1 *Solve third-order systems of equations.*

A third-order system consists of three equations with three unknowns. The solution of a third-order system will be an ordered triple. A fourth-order system is one that consists of four equations and four unknowns. The procedures discussed in this section can be expanded to solve fourth- and higher-order systems.

Three methods that can be used to solve a third-order system are: substitution, the addition method, and determinants. Determinants will be discussed in Section 4.4.

EXAMPLE 1 Solve the system of equations given below by substitution.

$$x = 4$$
$$2x + y = 20$$
$$-x + 4y + 2z = 24$$

Solution: Substitute 4 for x in the equation $2x + y = 20$, and solve for y.

$$2x + y = 20$$
$$2(4) + y = 20$$
$$8 + y = 20$$
$$y = 12$$

Now substitute $x = 4$ and $y = 12$ in the last equation and solve for z.

$$-x + 4y + 2z = 24$$
$$-(4) + 4(12) + 2z = 24$$
$$-4 + 48 + 2z = 24$$
$$44 + 2z = 24$$
$$2z = -20$$
$$z = -10$$

Check: The solution must be checked in all three original equations.

$x = 4$	$2x + y = 20$	$-x + 4y + 2z = 24$
$4 = 4$ true	$2(4) + 12 = 20$	$-(4) + 4(12) + 2(-10) = 24$
	$20 = 20$ true	$24 = 24$ true

The solution is the ordered triple $(4, 12, -10)$. Remember that the ordered triple lists the x value first, the y value second, and the z value third. ■

Not every third-order system can be solved by substitution. When a third-order system cannot be solved using substitution, we can find the solution by the addition method as illustrated below.

EXAMPLE 2 Solve the system of equations.

(a) $3x + 2y + z = 4$
(b) $2x - 3y + 2z = -7$
(c) $x + 4y - z = 10$

Solution: For the sake of clarity the three equations have been labeled (a), (b), and (c). To solve this system of equations, we must first obtain two equations containing the same two variables. This is done by selecting two equations and using the addition method to eliminate one of the variables. For example, by adding equations (a) and (c) the variable z will be eliminated. Next we use a different pair of equations [either (a) and (b) or (b) and (c)] and use the addition method to eliminate the same variable that was eliminated previously. If we multiply equation (a) by -2 and add it to equation (b), the variable z will again be eliminated. We will then have two equations containing only two unknowns. Let us now work this example and then discuss it further.

$$\begin{array}{lrcl} \text{(a)} & 3x + 2y + z &=& 4 \\ \text{(c)} & x + 4y - z &=& 10 \\ \hline & 4x + 6y \quad &=& 14 \quad \text{(d)} \end{array}$$

Now select a different set of equations and eliminate the variable z.

$$\begin{array}{llcl} \text{(a)} & -2[3x + 2y + \ z = \ \ 4] & \text{gives} & -6x - 4y - 2z = \ -8 \\ \text{(b)} & \quad 2x - 3y + 2z = -7 & & \quad 2x - 3y + 2z = \ -7 \\ \cline{4-4} & & & -4x - 7y \qquad = -15 \quad \text{(e)} \end{array}$$

We now have a system consisting of two equations with two unknowns.

$$\begin{array}{lrcr} \text{(d)} & 4x + 6y &=& 14 \\ \text{(e)} & -4x - 7y &=& -15 \end{array}$$

Next, we solve for one of the variables using a method presented earlier. If we add the two equations, the variable x will be eliminated.

$$\begin{array}{rcr} 4x + 6y &=& 14 \\ -4x - 7y &=& -15 \\ \hline -y &=& -1 \\ y &=& 1 \end{array}$$

Next, we substitute $y = 1$ in either one of the two equations containing only two variables [(d) or (e)] and solve for x.

$$\begin{array}{lrcl} \text{(d)} & 4x + 6y &=& 14 \\ & 4x + 6(1) &=& 14 \\ & 4x + 6 &=& 14 \\ & 4x &=& 8 \\ & x &=& 2 \end{array}$$

Finally, substitute $x = 2$, $y = 1$ in any of the original equations and solve for z.

$$\begin{array}{lrcl} \text{(a)} & 3x + 2y + z &=& 4 \\ & 3(2) + 2(1) + z &=& 4 \\ & 6 + 2 + z &=& 4 \\ & 8 + z &=& 4 \\ & z &=& -4 \end{array}$$

The solution is the ordered triple $(2, 1, -4)$. ■

We selected first to eliminate the variable z by using equations (a) and (c) and then equations (a) and (b). We could have elected to eliminate either variable x or variable y first. For example, we could have eliminated variable x by multiplying equation (c) by -2 and then adding it to equation (b). We could also eliminate variable x by multiplying equation (c) by -3 and then adding it to equation (a). Solve the system above by first eliminating the variable x.

EXAMPLE 3 Solve the system of equations.

$$\begin{aligned} 2x - 3y + 2z &= -1 \\ x + 2y &= 14 \\ x - 3z &= -5 \end{aligned}$$

Solution: We will select to eliminate variable y from the first two equations.

$$\begin{array}{ll} 2[2x - 3y + 2z = -1] & \text{gives} \quad 4x - 6y + 4z = -2 \\ 3[\ x + 2y = 14] & \phantom{\text{gives} \quad} 3x + 6y = 42 \\ \hline & \phantom{\text{gives} \quad} 7x + 4z = 40 \end{array}$$

We now have two equations containing only the variables x and z.

$$\begin{aligned} 7x + 4z &= 40 \\ x - 3z &= -5 \end{aligned}$$

Let's now eliminate the variable x.

$$\begin{array}{lr} 7x + 4z = 40 & \text{gives} \quad 7x + 4z = 40 \\ -7[x - 3z = -5] & -7x + 21z = 35 \\ \hline & 25z = 75 \\ & z = 3 \end{array}$$

Now we solve for x by using one of the equations containing the variables x and z.

$$\begin{aligned} x - 3z &= -5 \\ x - 3(3) &= -5 \\ x - 9 &= -5 \\ x &= 4 \end{aligned}$$

Finally, solve for y using any one of the original equations that contains a y.

$$\begin{aligned} x + 2y &= 14 \\ 4 + 2y &= 14 \\ 2y &= 10 \\ y &= 5 \end{aligned}$$

The solution is (4, 5, 3).

Check:

$$\begin{array}{rrr} 2x - 3y + 2z = -1 & x + 2y = 14 & x - 3z = -5 \\ 2(4) - 3(5) + 2(3) = -1 & 4 + 2(5) = 14 & 4 - 3(3) = -5 \\ 8 - 15 + 6 = -1 & 4 + 10 = 14 & 4 - 9 = -5 \\ -1 = -1 & 14 = 14 & -5 = -5 \\ \text{true} & \text{true} & \text{true} \end{array}$$

■

Geometrical Interpretation of Three Variables and Three Unknowns

When we have a system of linear equations in two variables we can find its solution graphically using the Cartesian coordinate system. A linear equation in three variables, x, y, and z, can be graphed on a coordinate system with three axes drawn perpendicular to each other (see Fig. 4.3).

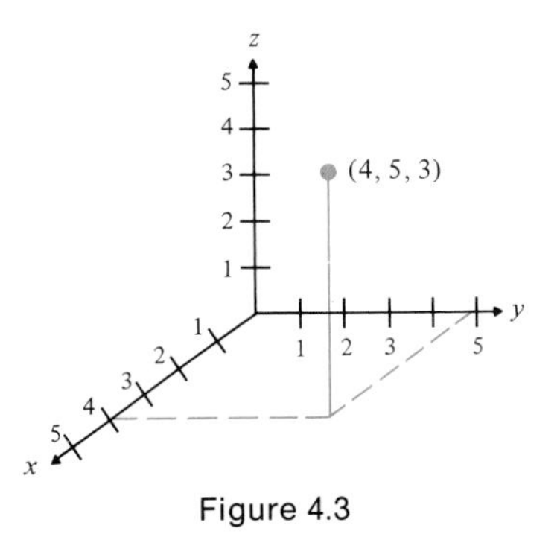

Figure 4.3

A point plotted in this type of three-dimensional system would appear to be a point in space. If we were to graph an equation such as $x + 2y + 3z = 4$, we would find that its graph would be a plane, not a line. In Example 3 we indicated the solution to be the ordered triple (4, 5, 3). This means that the three planes, one from each of the three given equations, all intersect at the point (4, 5, 3) in this three-dimensional system.

Exercise Set 4.2

Solve by substitution.

1. $x = 1$
$2x + y = 4$
$-3x - y + 4z = 15$

2. $2x + 3y = 9$
$4x - 6z = 12$
$y = 5$

3. $5x - 6z = -17$
$3x - 4y + 5z = -1$
$2z = -6$

4. $2x - 5y = 12$
$-3y = -9$
$2x - 3y + 4z = 8$

5. $x + 2y = 6$
$3y = 9$
$x + 2z = 12$

6. $x - y + 5z = -4$
$3x - 2z = 6$
$4z = 2$

Solve using the addition method.

7. $x + y - z = -3$
$x + z = 2$
$2x - y + 2z = 3$

8. $x - 3y = 13$
$2y + z = 1$
$y - 2z = 11$

9. $x + y + z = 4$
$x - 2y - z = 1$
$2x - y - 2z = -1$

10. $2x - 2y + 3z = 5$
$2x + y - 2z = -1$
$4x - y - 3z = 0$

11. $2x - y - z = 4$
$4x - 3y - 2z = -2$
$8x - 2y - 3z = 3$

12. $x + 2y - 3z = 5$
$x + y + z = 0$
$3x + 4y + 2z = -1$

13. $x + 2y + 2z = 1$
$2x - y + z = 3$
$4x + y + 2z = 0$

14. $2x + 2y - z = 2$
$3x + 4y + z = -4$
$5x - 2y - 3z = 5$

15. What does the graph of an equation in two variables, such as $x + 2y = 3$ represent? What does a graph of an equation in three variables, such as $x + 2y + 3z = 6$ represent?

An equation in three variables, x, y and z, represents a plane. Consider a system of equations consisting of three variables in three equations. Answer the following questions.

16. If the three planes are parallel to one another, the system will be inconsistent. Can you explain why?

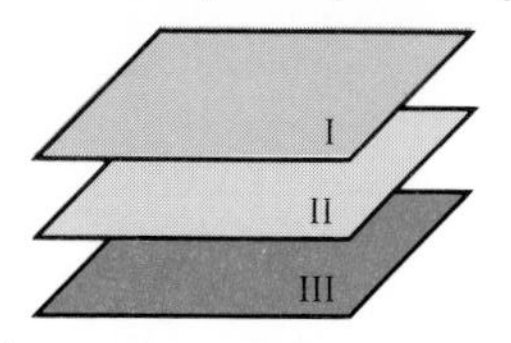

17. If two of the planes are parallel to each other and the third plane intersects each of the other two planes, how many points will be common to all three planes?

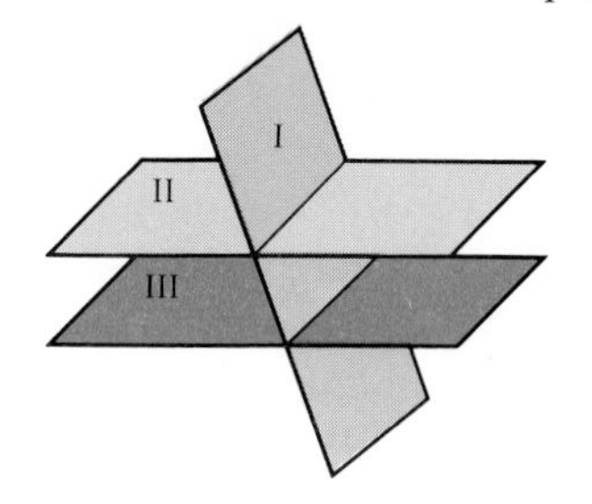

18. If the three planes are as illustrated in the figure below, how many points will be common to all three planes?

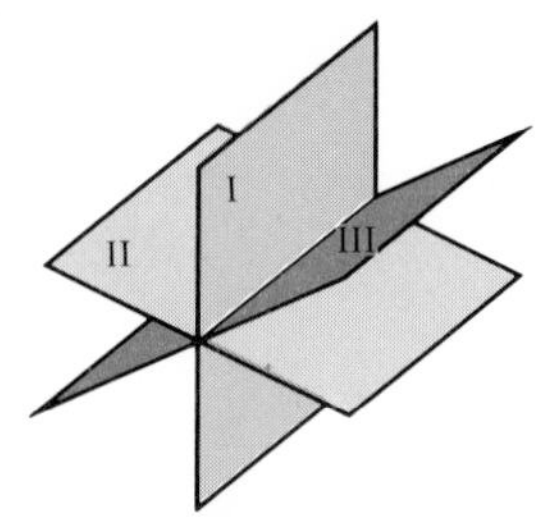

19. If the three planes are as illustrated as in the figure below, how many points will be common to all three planes?

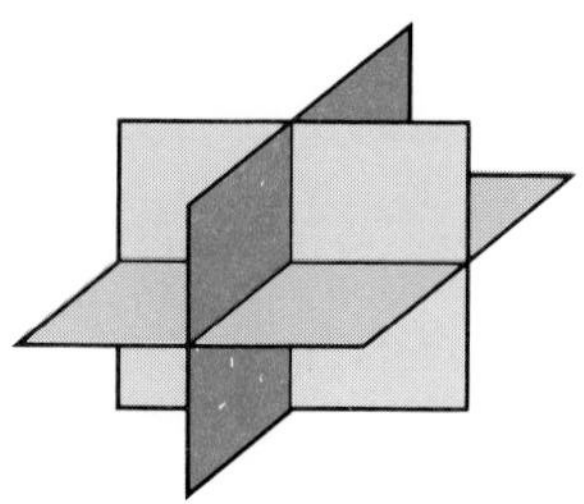

JUST FOR FUN

1. Solve the system.

$$x - \frac{2}{3}y - \frac{2z}{3} = -2$$

$$\frac{2x}{3} + y - \frac{2z}{3} = \frac{1}{3}$$

$$-\frac{x}{4} + y - \frac{z}{4} = \frac{3}{4}$$

2. Find the solution to the fourth-order system.

$$3a + 2b - c = 0$$

$$2a + 2c + d = 5$$

$$a + 2b - d = -2$$

$$2a - b + c + d = 2$$

4.3 Applications of Systems of Linear Equations

1 *Use systems of equations to solve practical application problems.*

Many of the applications solved in earlier chapters using only one variable can now be solved using two variables. The following example illustrates how Example 5 of Section 2.2 can be solved using two variables.

EXAMPLE 1 The sum of two numbers is 47. Find the two numbers if one number is 2 more than 4 times the other number.

Solution: Let x = smaller number
y = larger number

Statement	*Equation*
The sum of two numbers is 47	$x + y = 47$
One number is 2 more than 4 times the other number	$y = 4x + 2$

Systems of equations

$$x + y = 47$$
$$y = 4x + 2$$

We will solve this system using substitution. Substitute $4x + 2$ in place of y in the first equation.

$$x + y = 47$$
$$x + (4x + 2) = 47$$
$$5x + 2 = 47$$
$$5x = 45$$

Smaller number: $x = 9$

Larger number: $y = 4x + 2$
$= 4(9) + 2 = 38$

The two numbers are 9 and 38. This answer checks with the solution obtained in Section 2.2. ■

EXAMPLE 2 A plane travels 600 miles per hour with the wind and 450 miles per hour against the wind. Find the speed of the wind and the speed of the plane in still air.

Solution: Let x = speed of plane in still air
y = speed of wind

Speed of plane going with wind: $x + y = 600$
Speed of plane going against wind: $x - y = 450$

$$\begin{aligned} x + y &= 600 \\ x - y &= 450 \\ \hline 2x \quad &= 1050 \\ x &= 525 \end{aligned}$$

The plane's speed is 525 miles per hour in still air.

$$x + y = 600$$
$$525 + y = 600$$
$$y = 75$$

The wind's speed is 75 miles per hour. ■

EXAMPLE 3 A motorboat takes 3 hours to make a downstream trip when the water current is 2 miles per hour. The motorboat requires 4 hours to make the return trip against the current.

(a) Find the speed of the motorboat in still water

(b) Find the one-way distance.

Solution: (a) The following diagrams may be helpful. Recall from earlier sections that distance = speed · time.

Let $x =$ the speed of the boat in still water

then $x + 2 =$ speed of boat with current

and $x - 2 =$ speed of boat against current

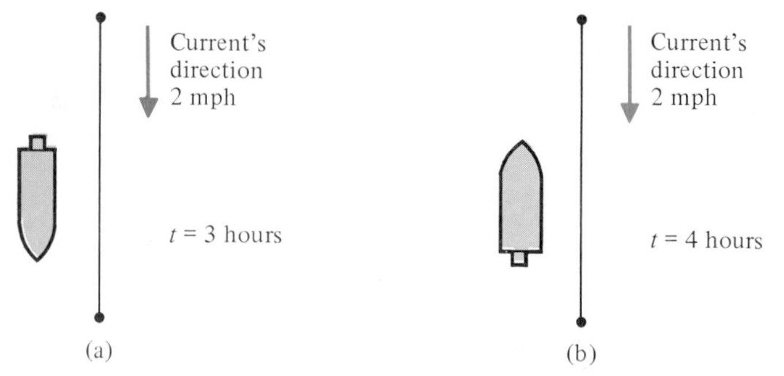

Downstream: moving with current ***Upstream: moving against current***

$$\text{distance} = \text{speed} \cdot \text{time} \qquad \text{distance} = \text{speed} \cdot \text{time}$$

$$d = (x + 2)(3) \qquad d = (x - 2)(4)$$

$$= 3x + 6 \qquad = 4x - 8$$

The system of equations is

$$d = 3x + 6$$
$$d = 4x - 8$$

Since the distance traveled downstream equals the distance traveled upstream, we set the two distances equal to each other.

$$\text{distance downstream} = \text{distance upstream}$$
$$3x + 6 = 4x - 8$$
$$-x + 6 = -8$$
$$-x = -14$$
$$x = 14$$

The speed of the motorboat in still water is 14 miles per hour.

(b) The distance can now be found by substituting $x = 14$ in either of the distance equations.

$$d = 3x + 6 \quad \text{or} \quad d = 4x - 8$$
$$= 3(14) + 6 \qquad = 4(14) - 8$$
$$= 42 + 6 \qquad = 56 - 8$$
$$= 48 \qquad = 48$$

Thus the one-way distance is 48 miles. ■

EXAMPLE 4 Each year, Ms. Martinez, a real estate salesperson, has the option of selecting the method by which her salary will be determined. Option 1 is a straight 6% of her total sales, and option 2 is a weekly salary of $100 plus 2% of her total sales. Determine the weekly sales (in dollars) needed for option 1 to result in the same weekly income as option 2.

Solution: We will set up a formula for each option, expressing income (I) in terms of weekly sales (s).

$$\text{Option 1:} \quad I = 0.06s$$
$$\text{Option 2:} \quad I = 100 + 0.02s$$

To solve this problem by substitution, set the two I's equal to each other.

$$0.06s = 100 + 0.02s$$
$$0.06s - 0.02s = 100 + 0.02s - 0.02s$$
$$0.04s = 100$$
$$s = \frac{100}{0.04} = \$2500$$

Thus if Ms. Martinez's weekly sales equal $2500, she would receive the same salary under each option. If her weekly sales are higher than $2500, she would receive a higher income with option 1. ■

EXAMPLE 5 A major university is doing research to determine the most cost-efficient method to set up a number of computer terminals throughout the university. The university is considering two options. Option 1 is a $60,000 minicomputer whose terminals cost $1000 each. Option 2 is a $20,000 network system whose terminals cost $3000 each. How many terminals would the university have to install to make the total cost of the network system equal to the total cost of the minicomputer?

Solution: The network system has a much smaller initial cost ($20,000 versus $60,000); however, its terminals cost more ($3000 versus $1000).

Let n = number of terminals.

$$\text{Option 1:} \quad \text{cost} = \text{cost of minicomputer} + \text{cost of } n \text{ terminals}$$
$$c = 60{,}000 + 1000n$$
$$\text{Option 2:} \quad \text{cost} = \text{cost of network} + \text{cost of } n \text{ terminals}$$
$$c = 20{,}000 + 3000n$$

The system of equations is

$$c = 60{,}000 + 1000n$$
$$c = 20{,}000 + 3000n$$

Now, solve by substitution

$$60{,}000 + 1000n = 20{,}000 + 3000n$$
$$40{,}000 + 1000n = 3000n$$
$$40{,}000 = 2000n$$
$$\frac{40{,}000}{2000} = n$$
$$20 = n$$

The total cost of the network system equals the total cost of the minicomputer when 20 terminals are used. ■

EXAMPLE 6 Martin mixes a 50% saltwater solution with a 75% saltwater solution to get 60 liters of a 60% saltwater solution. How many liters of the 50% solution and the 75% solution does he mix?

Solution: Let x = number of liters of 50% solution
y = number of liters of 75% solution

Statement	*Equation*
Total volume of combination is 60 liters	$x + y = 60$
Salt content of 50% solution + salt content of 75% solution = salt content of mixture	$0.5x + 0.75y = 0.6(60)$

System of equations

$$x + y = 60$$
$$0.5x + 0.75x = 0.6(60)$$

Solve for y in first equation.

$$x + y = 60$$
$$y = 60 - x$$

Substitute $60 - x$ for y in the second equation.

$$0.5x + 0.75y = 0.6(60)$$
$$0.5x + 0.75(60 - x) = 36$$
$$0.5x + 45 - 0.75x = 36$$
$$-0.25x + 45 = 36$$
$$-0.25x = -9$$
$$x = \frac{-9}{-0.25} = 36$$

Now solve for y.

$$x + y = 60$$
$$36 + y = 60$$
$$y = 24$$

Thus Martin mixed 36 liters of 50% solution with 24 liters of 75% solution to obtain 60 liters of 60% solution. ■

The following example illustrates a practical application of a third-order system.

EXAMPLE 7 Hobson, Inc. has a small manufacturing plant that makes three types of inflatable boats: one-person, two-person, and four-person models. Each boat requires the service of three departments: cutting, assembly, and packaging. The cutting, assembly, and packaging departments are allowed to use a total of 380, 330, and 120 person-hours per week, respectively. The time requirements for each boat and department are specified in the table below. Determine how many of each type of boat Hobson's must produce each week for its plant to operate at full capacity.

	Time (hr)		
Department	One-Person Boat	Two-Person Boat	Four-Person Boat
Cutting	0.6	1.0	1.5
Assembly	0.6	0.9	1.2
Packaging	0.2	0.3	0.5

Let x = number of one-person boats
y = number of two-person boats
z = number of four-person boats

Then the total number of cutting hours for the three types of boats must equal 380 person-hours.

$$0.6x + 1.0y + 1.5z = 380$$

The total number of assembly hours must equal 330 person-hours.

$$0.6x + 0.9y + 1.2z = 330$$

The total number of packaging hours must equal 120 person-hours.

$$0.2x + 0.3y + 0.5z = 120$$

System of equations

$$\begin{aligned} 0.6x + 1.0y + 1.5z &= 380 \\ 0.6x + 0.9y + 1.2z &= 330 \\ 0.2x + 0.3y + 0.5z &= 120 \end{aligned}$$

Let's first eliminate the variable x.

$$\begin{aligned} 0.6x + 1.0y + 1.5z &= 380 \\ -1[0.6x + 0.9y + 1.2z &= 330] \end{aligned} \quad \text{gives} \quad \begin{aligned} 0.6x + 1.0y + 1.5z &= 380 \\ -0.6x - 0.9y - 1.2z &= -330 \\ \hline 0.1y + 0.3z &= 50 \end{aligned}$$

$$\begin{aligned} 0.6x + 1.0y + 1.5z &= 380 \\ -3[0.2x + 0.3y + 0.5z &= 120] \end{aligned} \quad \text{gives} \quad \begin{aligned} 0.6x + 1.0y + 1.5z &= 380 \\ -0.6x - 0.9y - 1.5z &= -360 \\ \hline 0.1y \qquad\qquad &= 20 \end{aligned}$$

Note that when we added the last two equations, both variables x and z were eliminated at the same time. We can therefore find y.

$$0.1y = 20$$

$$y = \frac{20}{0.1} = 200$$

Now find z using the equation $0.1y + 0.3z = 50$.

$$0.1(200) + 0.3z = 50$$

$$20 + 0.3z = 50$$

$$0.3z = 30$$

$$z = \frac{30}{0.3} = 100$$

Finally, find x.

$$0.6x + 1.0y + 1.5z = 380$$

$$0.6x + 1.0(200) + 1.5(100) = 380$$

$$0.6x + 200 + 150 = 380$$

$$0.6x + 350 = 380$$

$$0.6x = 30$$

$$x = \frac{30}{0.6} = \frac{300}{6} = 50$$

Thus Hobson's should produce 50 one-person boats, 200 two-person boats, and 100 four-person boats. ■

Exercise Set 4.3

In each exercise, (a) express the problem as a system of linear equations and (b) use the method of your choice to find the solution to the problem.

1. The sum of two consecutive even integers is 54. Find the two numbers.
2. The sum of two consecutive integers is 54. Find the two numbers.
3. The sum of two consecutive odd integers is 76. Find the two numbers.
4. The difference of two numbers is 25. Find the two numbers if the larger is 1 less than 3 times the smaller.
5. The difference of two numbers is 28. Find the two numbers if the larger is 3 times the smaller.
6. A plane can travel 540 miles per hour with the wind and 490 miles per hour against the wind. Find the speed of the plane in still air and the speed of the wind.
7. Carlos can paddle a canoe 4.5 miles in 1 hour with the current and 3.2 miles in 1 hour against the current. Find the speed of the canoe in still water and the speed of the current.
8. Steve Trinter, an electronics salesman, can make a weekly salary of \$200 plus 5% commission on sales, or a weekly salary consisting of a straight 15% commission of sales. Determine the amount of sales necessary for the 15% straight commission salary to equal the \$200 plus 5% commission salary.
9. Hertz Automobile Rental Agency charges a daily fee of \$30 plus 14 cents a mile. National Automobile Rental Agency charges a daily fee of \$16 plus 24 cents a mile for the same car. What distance would you have to drive in 1 day to make the cost of renting from Hertz equal to the cost of renting from National?
10. Mr. Lee, a druggist, needs 1000 milliliters of a 10% phenobarbital solution. He has only 5% and 25% phenobarbital solution available. How many milliliters of each solution should he mix to obtain the desired solution?
11. The ABC Printing Company charges a set up fee of \$1600,

See Exercise 11.

plus \$6 per book it prints. The XYZ Printing Company has a fixed fee of \$1200, plus an \$8 fee per book it prints. How many books would have to be ordered to make the ABC Company's total charges equal the XYZ Company's total charges?

12. The total cost of printing a political leaflet consists of a fixed charge and an additional charge for each leaflet. If the total cost for 1000 leaflets is \$550 and the total cost for 2000 leaflets is \$800, find the fixed charge and the charge for each leaflet.

13. Mrs. Spinelli runs a grocery store. She wishes to mix 30 pounds of coffee to sell for a total cost of \$100. To obtain the mixture, Mrs. Spinelli will mix coffee that sells for \$3 per pound with coffee that sells for \$5 per pound. How many pounds of each type of coffee should she use?

14. In chemistry class, Gina has an 80% acid solution and a 50% acid solution. How much of each solution must she mix to get 100 liters of a 75% acid solution?

15. Mario owns a dairy. He has milk that is 5% butterfat and skim milk, without butterfat. How much 5% milk and how much skim milk should he mix to make 100 gallons of milk that is 3.5% butterfat?

16. Pierre's recipe for Quiche Lorraine calls for 2 cups (16 ounces) of light cream that is 20% milk fat. It is often difficult to find light cream with 20% milk fat at the supermarket. What is commonly found is heavy cream, which is 36% milk fat, and half-and-half, which is 10.5% milk fat. How much of the heavy cream and how much of the half-and-half should Pierre mix to obtain the mixture necessary for the recipe?

17. Mr. and Mrs. McAdams invest a total of \$8000 in two savings accounts. One account gives 10% interest and the other 8%. Find the amount placed in each account if they receive a total of \$750 in interest after 1 year. Use interest = principal · rate · time.

18. The Webers wish to invest a total of \$12,500 in two savings accounts. One account is giving 10% interest and the other $5\frac{1}{4}$%. The Webers wish their interest from the two accounts to be at least \$1200 at the end of the year. Find the minimum amount that can be placed in the account giving 10% interest.

19. If the minicomputer system in Example 5 costs \$60,000 plus \$1500 per terminal, and the network system costs \$20,000 plus \$3000 per terminal, how many terminals would have to be ordered for the cost of the minicomputer system to equal the cost of the network system?

20. Two cars start at the same point and travel in opposite directions. One car travels at 80 kilometers per hour and the other at 65 kilometers per hour. In how many hours will they be 435 kilometers apart?

21. Two trains start at the same point going in the same direction on parallel tracks. One train travels at 70 miles per hour and the other at 42 miles per hour. In how many hours will they be 154 miles apart?

22. An airplane takes 4 hours to fly a distance of 2850 miles with the wind. The return trip against the wind takes $4\frac{3}{4}$ hours. Find the speed of the plane in still air and the speed of the wind.

23. The Tree Sweet Juice Company sells 8-ounce cans of apple juice for 40 cents and 8-ounce cans of apple drink for 16 cents each. The company wishes to market and sell cans of juice drink for 25 cents which is part juice and part drink. How many ounces of each should be used if the juice drink is to be sold in 8-ounce cans?

24. Animals in an experiment are to be kept on a strict diet. Each animal is to receive, among other things, 20 grams of protein and 6 grams of carbohydrates. The scientist only has two food mixes available of the following compositions.

	Protein (%)	Carbohydrate (%)
Mix *A*	10	6
Mix *B*	20	2

How many grams of each mix should be used to obtain the right diet for a single animal?

25. Two brothers jog to school daily. The older jogs at 9 miles per hour, the younger at 5 miles per hour. When the older brother reaches the school, the younger brother is $\frac{1}{2}$ mile from the school. How far is the school from the boys' house?

26. By weight, one alloy of brass is 70% copper and 30% zinc. Another alloy of brass is 40% copper and 60% zinc. How many grams of each of these alloys need to be melted and combined to obtain 300 grams of a brass alloy that is 60% copper and 40% zinc?

27. In the study of electronics it is necessary to analyze current flow through certain paths of a circuit. The study of three branches (A, B, and C) of a circuit yields the following results:

$$\begin{aligned} I_A + I_B + I_C &= 0 \\ -8I_B + 10I_C &= 0 \\ 4I_A - 8I_B &= 6 \end{aligned}$$

where I_A, I_B, and I_C represent the current in branches A, B, and C, respectively. Determine the current in each branch of the circuit.

28. In the study of physics we often study the forces acting on an object. For three forces, F_1, F_2, and F_3, acting on a beam, the following equations were obtained.

$$\begin{aligned} 3F_1 + F_2 - F_3 &= 2 \\ F_1 - 2F_2 + F_3 &= 0 \\ 4F_1 - F_2 + F_3 &= 3 \end{aligned}$$

Find the three forces.

JUST FOR FUN

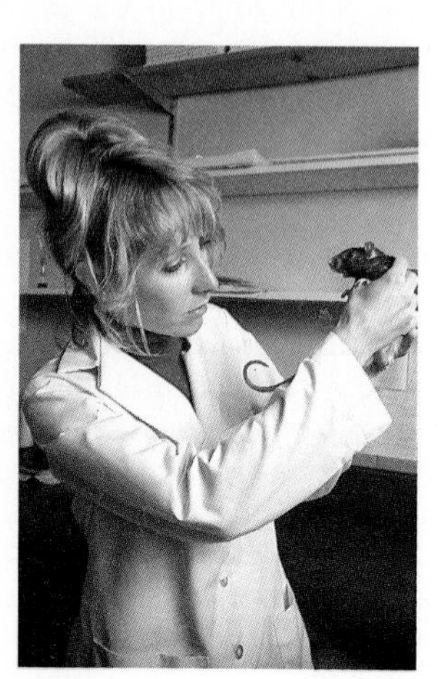

See Just for Fun 1.

1. In an article published in the *Journal of Comparative Physiology and Psychology* J. S. Brown discusses how we often approach a situation with mixed emotions. For example, when a person is asked to give a speech, he may be a little apprehensive about his ability to do a good job. At the same time, he would like the recognition that goes along with making the speech. J. S. Brown performed an experiment on trained rats. He placed their food in a metal box. He used that same box to administer small electrical shocks to the mice. Therefore, the rats "wished" to go into the box to receive food, yet did not "wish" to go into the box for fear of receiving a small shock. Using the appropriate apparatus, Brown arrived at the following relationships:

$$\text{pull (in grams) toward food} = -\frac{1}{5}d + 70 \qquad 30 < d < 172.5$$

$$\text{pull (in grams) away from shock} = -\frac{4}{3}d + 230 \qquad 30 < d < 172.5$$

where d is the distance in centimeters from the box (and food).

(a) Find, using the substitution method, the distance at which the pull toward the food equals the pull away from the shock.

(b) If the rat is placed 100 cm from the box (or food), what will the rat do?

2. An 8% solution, a 10% solution, and a 20% solution of sulfuric acid are to be mixed to get 100 milliliters of a 12% solution. If the volume of acid from the 8% solution is to equal half the volume of acid from the other two solutions, how much of each solution is needed?

3. By volume, one alloy is 60% copper, 30% zinc, and 10% nickel. A second alloy has percentages 50, 30, and 20, respectively, of the three metals. A third alloy is 30% copper and 70% nickel. How much of each alloy must be mixed so that 100 pounds of the resulting alloy is 40% copper, 15% zinc, and 45% nickel?

4.4 Solving Systems of Equations by Determinants (Optional)

1. *Find the value of a second-order determinant.*
2. *Use Cramer's Rule to solve second-order systems of equations.*
3. *Find the value of a third-order determinant.*
4. *Use Cramer's Rule to solve third-order systems of equations.*

Systems of linear equations can also be solved using determinants. Determinants are particularly useful when solving third- and higher-order systems of equations.

A **determinant** is a square array of numbers enclosed between two vertical bars. Examples of determinants are:

$$\begin{vmatrix} 4 & -3 \\ 0 & 5 \end{vmatrix} \qquad \begin{vmatrix} 3 & 0 & 5 \\ 4 & -2 & 3 \\ 2 & \frac{1}{2} & -1 \end{vmatrix}$$

(a) (b)

The numbers that make up the array are called the **elements** of the determinant. The elements of determinant (a) are 4, -3, 0, and 5.

Determinant (a) is a **second-order determinant** since it has two rows and two columns of elements. Determinant (b) is a **third-order determinant.** Determinants can be of order greater than 3.

The **principal diagonal** of a determinant is the line of elements from the upper left corner to the lower right corner. The **secondary diagonal** of a determinant is the line of elements from the lower left-hand corner to the upper right-hand corner.

$$\begin{vmatrix} a_1 & b_1 \\ a_2 & b_2 \end{vmatrix} \qquad \begin{vmatrix} a_1 & b_1 \\ a_2 & b_2 \end{vmatrix}$$

principal diagonal secondary diagonal

1 Every determinant represents a number. The **value of a second-order determinant** is the product of the elements in its principal diagonal minus the product of the elements in its secondary diagonal.

Value of a Second-Order Determinant

$$\begin{vmatrix} a_1 & b_1 \\ a_2 & b_2 \end{vmatrix} = a_1b_2 - a_2b_1$$

EXAMPLE 1 Find the value of the determinant $\begin{vmatrix} 4 & 6 \\ -3 & 2 \end{vmatrix}$.

Solution: Here $a_1 = 4$, $a_2 = -3$, $b_1 = 6$, $b_2 = 2$.

$$\begin{vmatrix} 4 & 6 \\ -3 & 2 \end{vmatrix} = 4(2) - (-3)(6)$$
$$= 8 + 18$$
$$= 26$$

The determinant has a value of 26. ■

EXAMPLE 2 Find the value of the determinant $\begin{vmatrix} -3 & 4 \\ 1 & 5 \end{vmatrix}$.

Solution:

$$\begin{vmatrix} -3 & 4 \\ 1 & 5 \end{vmatrix} = (-3)(5) - (1)(4)$$
$$= -15 - 4$$
$$= -19 \quad ■$$

2 Consider the system of equations

$$(1) \quad a_1x + b_1y = c_1$$
$$(2) \quad a_2x + b_2y = c_2$$

To eliminate the variable y, we can multiply both sides of equation (1) by b_2 and both sides of equation (2) by $-b_1$ and then add.

$$\begin{array}{rr} b_2[a_1x + b_1y = c_1] & a_1b_2x + b_1b_2y = c_1b_2 \\ -b_1[a_2x + b_2y = c_2] & -a_2b_1x - b_1b_2y = -c_2b_1 \\ \hline & (a_1b_2 - a_2b_1)x \qquad = c_1b_2 - c_2b_1 \\ & x = \dfrac{c_1b_2 - c_2b_1}{a_1b_2 - a_2b_1} \end{array}$$

We can solve the system for y in a similar manner.

$$\begin{array}{rr} -a_2[a_1x + b_1y = c_1] & -a_1a_2x - a_2b_1y = -a_2c_1 \\ a_1[a_2x + b_2y = c_2] & a_1a_2x + a_1b_2y = a_1c_2 \\ \hline & (a_1b_2 - a_2b_1)y = a_1c_2 - a_2c_1 \\ & y = \dfrac{a_1c_2 - a_2c_1}{a_1b_2 - a_2b_1} \end{array}$$

Both x and y have the same denominator, namely $a_1b_2 - a_2b_1$. Note that

$$\begin{vmatrix} a_1 & b_1 \\ a_2 & b_2 \end{vmatrix} = a_1b_2 - a_2b_1$$

The numerator of the expression used to find x is $c_1b_2 - c_2b_1$. Note that

$$\begin{vmatrix} c_1 & b_1 \\ c_2 & b_2 \end{vmatrix} = c_1b_2 - c_2b_1$$

The numerator of the expression used to find y is $a_1c_2 - a_2c_1$. Note that

$$\begin{vmatrix} a_1 & c_1 \\ a_2 & c_2 \end{vmatrix} = a_1c_2 - a_2c_1$$

Using the information above and making the appropriate substitutions, we can express both x and y as a quotient of two determinants.

$$x = \frac{c_1b_2 - c_2b_1}{a_1b_2 - a_2b_1} = \frac{\begin{vmatrix} c_1 & b_1 \\ c_2 & b_2 \end{vmatrix}}{\begin{vmatrix} a_1 & b_1 \\ a_2 & b_2 \end{vmatrix}} \qquad y = \frac{a_1c_2 - a_2c_1}{a_1b_2 - a_2b_1} = \frac{\begin{vmatrix} a_1 & c_1 \\ a_2 & c_2 \end{vmatrix}}{\begin{vmatrix} a_1 & b_1 \\ a_2 & b_2 \end{vmatrix}}$$

A second order system of equations can be solved by Cramer's Rule.

Cramer's Rule

For a system of equations of the form

$$a_1x + b_1y = c_1$$
$$a_2x + b_2y = c_2$$

$$x = \frac{\begin{vmatrix} c_1 & b_1 \\ c_2 & b_2 \end{vmatrix}}{\begin{vmatrix} a_1 & b_1 \\ a_2 & b_2 \end{vmatrix}} = \frac{D_x}{D} \quad \text{and} \quad y = \frac{\begin{vmatrix} a_1 & c_1 \\ a_2 & c_2 \end{vmatrix}}{\begin{vmatrix} a_1 & b_1 \\ a_2 & b_2 \end{vmatrix}} = \frac{D_y}{D}$$

For discussion purposes, we will denote the three determinants as:

$$D = \begin{vmatrix} a_1 & b_1 \\ a_2 & b_2 \end{vmatrix} \qquad D_x = \begin{vmatrix} c_1 & b_1 \\ c_2 & b_2 \end{vmatrix} \qquad D_y = \begin{vmatrix} a_1 & c_1 \\ a_2 & c_2 \end{vmatrix}$$

determinant of numerical coefficients	a's in determinant D replaced by constants, c's	b's in determinant D replaced by constants, c's

Note that

$$x = \frac{D_x}{D} \quad \text{and} \quad y = \frac{D_y}{D}.$$

EXAMPLE 3 Use determinants to evaluate the following system.

$$2x + y = 6$$
$$3x + y = 5$$

Solution: Both equations are given in the desired form, $ax + by = c$. We will refer to $2x + y = 6$ as equation 1 and $3x + y = 5$ as equation 2.

$$\begin{array}{ccc} a_1 & b_1 & c_1 \\ \downarrow & \downarrow & \downarrow \\ 2x + & 1y = & 6 \\ 3x + & 1y = & 5 \\ \uparrow & \uparrow & \uparrow \\ a_2 & b_2 & c_2 \end{array}$$

We now find D, D_x, and D_y.

$$D = \begin{vmatrix} a_1 & b_1 \\ a_2 & b_2 \end{vmatrix} = \begin{vmatrix} 2 & 1 \\ 3 & 1 \end{vmatrix} = 2(1) - 3(1) = -1$$

$$D_x = \begin{vmatrix} c_1 & b_1 \\ c_2 & b_2 \end{vmatrix} = \begin{vmatrix} 6 & 1 \\ 5 & 1 \end{vmatrix} = 6(1) - 5(1) = 1$$

$$D_y = \begin{vmatrix} a_1 & c_1 \\ a_2 & c_2 \end{vmatrix} = \begin{vmatrix} 2 & 6 \\ 3 & 5 \end{vmatrix} = 2(5) - 3(6) = -8$$

$$x = \frac{D_x}{D} = \frac{1}{-1} = -1$$

$$y = \frac{D_y}{D} = \frac{-8}{-1} = 8$$

Thus the solution is $x = -1$, $y = 8$ or the ordered pair $(-1, 8)$.

Check:

$$\begin{aligned} 2x + y &= 6 \\ 2(-1) + 8 &= 6 \\ -2 + 8 &= 6 \\ 6 &= 6 \quad \text{true} \end{aligned} \qquad \begin{aligned} 3x + y &= 5 \\ 3(-1) + 8 &= 5 \\ -3 + 8 &= 5 \\ 5 &= 5 \quad \text{true} \end{aligned}$$

■

It makes no difference which equations you label 1 and 2 as long as you remain consistent. Rework the example letting $3x + y = 5$ represent equation 1 and $2x + y = 6$, equation 2.

EXAMPLE 4 Solve the following system using determinants.

$$\begin{aligned} 2x - 4y &= 8 \\ 3x + 5y &= -10 \end{aligned}$$

Solution:

$$\begin{aligned} a_1 &= 2, & b_1 &= -4, & c_1 &= 8 \\ a_2 &= 3, & b_2 &= 5, & c_2 &= -10 \end{aligned}$$

$$D = \begin{vmatrix} a_1 & b_1 \\ a_2 & b_2 \end{vmatrix} = \begin{vmatrix} 2 & -4 \\ 3 & 5 \end{vmatrix} = 2(5) - 3(-4) = 22$$

$$D_x = \begin{vmatrix} c_1 & b_1 \\ c_2 & b_2 \end{vmatrix} = \begin{vmatrix} 8 & -4 \\ -10 & 5 \end{vmatrix} = 8(5) - (-10)(-4) = 0$$

$$D_y = \begin{vmatrix} a_1 & c_1 \\ a_2 & c_2 \end{vmatrix} = \begin{vmatrix} 2 & 8 \\ 3 & -10 \end{vmatrix} = 2(-10) - (3)(8) = -44$$

$$x = \frac{D_x}{D} = \frac{0}{22} = 0$$

$$y = \frac{D_y}{D} = \frac{-44}{22} = -2$$

Thus the solution is $(0, -2)$. ■

When solving a system using determinants, if you obtain the quotient 0/0 for any of the variables, the two equations represent the same line and there are an infinite number of solutions. If you obtain a quotient of the form $a/0$, $a \neq 0$, the two lines are parallel and there is no solution.

3 A third-order determinant is evaluated as follows:

$$\begin{vmatrix} a_1 & b_1 & c_1 \\ a_2 & b_2 & c_2 \\ a_3 & b_3 & c_3 \end{vmatrix} = a_1 \underbrace{\begin{vmatrix} b_2 & c_2 \\ b_3 & c_3 \end{vmatrix}}_{\text{minor determinant of } a_1} - a_2 \underbrace{\begin{vmatrix} b_1 & c_1 \\ b_3 & c_3 \end{vmatrix}}_{\text{minor determinant of } a_2} + a_3 \underbrace{\begin{vmatrix} b_1 & c_1 \\ b_2 & c_2 \end{vmatrix}}_{\text{minor determinant of } a_3}$$

This method of evaluating the determinant is called **expansion of the determinant by the minors of the first column.**

The determinant $\begin{vmatrix} b_2 & c_2 \\ b_3 & c_3 \end{vmatrix}$ is called the minor determinant of a_1.

The minor determinant of a_1 is found by crossing out the elements in the same row and column in which the element a_1 appears.

$$\begin{matrix} \not{a_1} & \not{b_1} & \not{c_1} \\ \not{a_2} & b_2 & c_2 \\ \not{a_3} & b_3 & c_3 \end{matrix}$$

The remaining elements form the minor determinant of a_1.

$$\begin{vmatrix} b_2 & c_2 \\ b_3 & c_3 \end{vmatrix}$$

The minor determinant of a_2 is found similarly.

$$\begin{matrix} a_1 & b_1 & c_1 \\ a_2 & b_2 & c_2 \\ a_3 & b_3 & c_3 \end{matrix} \qquad \text{minor determinant of } a_2 \quad \begin{vmatrix} b_1 & c_1 \\ b_3 & c_3 \end{vmatrix}$$

The minor determinant of a_3 is found below.

$$\begin{matrix} a_1 & b_1 & c_1 \\ a_2 & b_2 & c_2 \\ a_3 & b_3 & c_3 \end{matrix} \qquad \text{minor determinant of } a_3 \quad \begin{vmatrix} b_1 & c_1 \\ b_2 & c_2 \end{vmatrix}$$

EXAMPLE 5 Evaluate $\begin{vmatrix} 4 & -2 & 6 \\ 3 & 5 & 0 \\ 1 & -3 & -1 \end{vmatrix}$.

Solution:

$$\begin{vmatrix} 4 & -2 & 6 \\ 3 & 5 & 0 \\ 1 & -3 & -1 \end{vmatrix} = 4\begin{vmatrix} 5 & 0 \\ -3 & -1 \end{vmatrix} - 3\begin{vmatrix} -2 & 6 \\ -3 & -1 \end{vmatrix} + 1\begin{vmatrix} -2 & 6 \\ 5 & 0 \end{vmatrix}$$

$$\begin{aligned} &= 4[5(-1) - (-3)0] - 3[(-2)(-1) - (-3)6] + 1[(-2)0 - 5(6)] \\ &= 4(-5 + 0) - 3(2 + 18) + 1(0 - 30) \\ &= 4(-5) - 3(20) + 1(-30) \\ &= -20 - 60 - 30 \\ &= -110 \end{aligned}$$

The determinant has a value of -110. ■

In Example 5 we found the solution using expansion of the determinants by the minor determinants of the first column. To evaluate a determinant, we can use expansion of the determinant by the minor determinants of any row or any column. To determine if the product of the element and its minor determinant is to be added or subtracted in obtaining the result, we use the following chart.

$$\begin{matrix} + & - & + \\ - & + & - \\ + & - & + \end{matrix}$$

If the element is in a position marked with a $+$, the product is to be added. If the element is in a position marked with a $-$, the product is to be subtracted.

EXAMPLE 6 Evaluate $\begin{vmatrix} 4 & -2 & 6 \\ 3 & 5 & 0 \\ 1 & -3 & -1 \end{vmatrix}$ using expansion of the determinant by the minors of the second row.

Solution: The second row of the chart is −, +, −. Thus the first product is to be subtracted, the second product added, and the third product subtracted.

$$\begin{vmatrix} 4 & -2 & 6 \\ 3 & 5 & 0 \\ 1 & -3 & -1 \end{vmatrix} = \underset{\text{subtract}}{\underset{\uparrow}{-3}}\begin{vmatrix} -2 & 6 \\ -3 & -1 \end{vmatrix} \underset{\text{add}}{\underset{\uparrow}{+5}}\begin{vmatrix} 4 & 6 \\ 1 & -1 \end{vmatrix} \underset{\text{subtract}}{\underset{\uparrow}{-0}}\begin{vmatrix} 4 & -2 \\ 1 & -3 \end{vmatrix}$$

$$= -3[(-2)(-1) - (-3)6] + 5[4(-1) - 1(6)] - 0[4(-3) - 1(-2)]$$
$$= -3(2 + 18) + 5(-4 - 6) - 0(-12 + 2)$$
$$= -3(20) + 5(-10) - 0$$
$$= -60 - 50$$
$$= -110$$ ■

Note that the same answer was obtained by evaluating the determinant using the first column or the second row. When evaluating a determinant containing one or more 0's in a particular row or column, you may wish to evaluate the determinant by expansion of the minor determinants of that row or column.

4 Cramer's rule can be extended to third-order systems of equations as follows:

To evaluate the system

$$a_1x + b_1y + c_1z = d_1$$
$$a_2x + b_2y + c_2z = d_2$$
$$a_3x + b_3y + c_3z = d_3$$

with

$$D = \begin{vmatrix} a_1 & b_1 & c_1 \\ a_2 & b_2 & c_2 \\ a_3 & b_3 & c_3 \end{vmatrix} \qquad D_x = \begin{vmatrix} d_1 & b_1 & c_1 \\ d_2 & b_2 & c_2 \\ d_3 & b_3 & c_3 \end{vmatrix}$$

$$D_y = \begin{vmatrix} a_1 & d_1 & c_1 \\ a_2 & d_2 & c_2 \\ a_3 & d_3 & c_3 \end{vmatrix} \qquad D_z = \begin{vmatrix} a_1 & b_1 & d_1 \\ a_2 & b_2 & d_2 \\ a_3 & b_3 & d_3 \end{vmatrix}$$

then

$$x = \frac{D_x}{D}, \qquad y = \frac{D_y}{D}, \qquad z = \frac{D_z}{D}$$

Note that the denominators of the expressions for x, y, and z are all the same determinant, D. Note that constants, the d's, replace the a's, the numerical coefficients of the x terms, in D_x. The d's replace the b's, the numerical coefficients of the y terms in D_y. And the d's replace the c's, the numerical coefficients of the z terms in D_z.

EXAMPLE 7 Solve the following system of equations using determinants.

$$\begin{aligned} 3x - 2y - z &= -6 \\ 2x + 3y - 2z &= 1 \\ x - 4y + z &= -3 \end{aligned}$$

Solution:

$$\begin{array}{llll} a_1 = 3 & b_1 = -2 & c_1 = -1 & d_1 = -6 \\ a_2 = 2 & b_2 = 3 & c_2 = -2 & d_2 = 1 \\ a_3 = 1 & b_3 = -4 & c_3 = 1 & d_3 = -3 \end{array}$$

We will use expansion of the minor determinants by the first row to evaluate D.

$$D = \begin{vmatrix} 3 & -2 & -1 \\ 2 & 3 & -2 \\ 1 & -4 & 1 \end{vmatrix} = 3\begin{vmatrix} 3 & -2 \\ -4 & 1 \end{vmatrix} - (-2)\begin{vmatrix} 2 & -2 \\ 1 & 1 \end{vmatrix} + (-1)\begin{vmatrix} 2 & 3 \\ 1 & -4 \end{vmatrix}$$

$$= 3(-5) + 2(4) - 1(-11)$$

$$= -15 + 8 + 11 = 4$$

We will evaluate D_x using expansion of the determinant by the minor determinant of the first column.

$$D_x = \begin{vmatrix} -6 & -2 & -1 \\ 1 & 3 & -2 \\ -3 & -4 & 1 \end{vmatrix} = (-6)\begin{vmatrix} 3 & -2 \\ -4 & 1 \end{vmatrix} - (1)\begin{vmatrix} -2 & -1 \\ -4 & 1 \end{vmatrix} + (-3)\begin{vmatrix} -2 & -1 \\ 3 & -2 \end{vmatrix}$$

$$= -6(-5) - 1(-6) - 3(7)$$

$$= 30 + 6 - 21 = 15$$

We will use expansion of the determinant by the minor determinants of the first row to evaluate D_y.

$$D_y = \begin{vmatrix} 3 & -6 & -1 \\ 2 & 1 & -2 \\ 1 & -3 & 1 \end{vmatrix} = 3\begin{vmatrix} 1 & -2 \\ -3 & 1 \end{vmatrix} - (-6)\begin{vmatrix} 2 & -2 \\ 1 & 1 \end{vmatrix} + (-1)\begin{vmatrix} 2 & 1 \\ 1 & -3 \end{vmatrix}$$

$$= 3(-5) + 6(4) - 1(-7)$$

$$= -15 + 24 + 7 = 16$$

We will evaluate D_z using expansion of the determinant by the minor determinants of the first row.

$$D_z = \begin{vmatrix} 3 & -2 & -6 \\ 2 & 3 & 1 \\ 1 & -4 & -3 \end{vmatrix} = 3\begin{vmatrix} 3 & 1 \\ -4 & -3 \end{vmatrix} - (-2)\begin{vmatrix} 2 & 1 \\ 1 & -3 \end{vmatrix} + (-6)\begin{vmatrix} 2 & 3 \\ 1 & -4 \end{vmatrix}$$

$$= 3(-5) + 2(-7) - 6(-11)$$

$$= -15 - 14 + 66 = 37$$

We found that $D = 4$, $D_x = 15$, $D_y = 16$, and $D_z = 37$.

$$x = \frac{D_x}{D} = \frac{15}{4} \qquad y = \frac{D_y}{D} = \frac{16}{4} = 4 \qquad z = \frac{D_z}{D} = \frac{37}{4}$$

The solution to the system is $(\frac{15}{4}, 4, \frac{37}{4})$. Note the ordered triple lists x, y, and z, in this order. ■

When we are given a third-order system of equations in which one or more equations are missing a variable, we insert the variable with a coefficient of 0. This helps in aligning like terms. For example:

$$\begin{aligned} 2x - 3y + 2z &= -1 \\ x + 2y \qquad &= 14 \\ x \qquad - 3z &= -5 \end{aligned} \quad \text{is written} \quad \begin{aligned} 2x - 3y + 2z &= -1 \\ x + 2y + 0z &= 14 \\ x + 0y - 3z &= -5 \end{aligned}$$

when solving the system using determinants. Furthermore, it is very important to place the numbers in the correct column. In this example:

$$D = \begin{vmatrix} 2 & -3 & 2 \\ 1 & 2 & 0 \\ 1 & 0 & -3 \end{vmatrix} \qquad D_x = \begin{vmatrix} -1 & -3 & 2 \\ 14 & 2 & 0 \\ -5 & 0 & -3 \end{vmatrix}$$

$$D_y = \begin{matrix} 2 & -1 & 2 \\ 1 & 14 & 0 \\ 1 & -5 & -3 \end{matrix} \qquad D_z = \begin{vmatrix} 2 & -3 & -1 \\ 1 & 2 & 14 \\ 1 & 0 & -5 \end{vmatrix}$$

When you are solving for a variable in a third order system, if the denominator, D, has a value of 0 and any numerator (D_x, D_y, and D_z) does not have a value of zero, then the system is inconsistent and has no solution. If D, D_x, D_y, and D_z are all 0, then the system is dependent and there are infinitely many solutions.

Exercise Set 4.4

Solve the system of equations using determinants.

1. $x + 2y = 5$
$x - 2y = 1$

2. $3x - 2y = 4$
$3x + y = -2$

3. $x - 2y = -1$
$x + 3y = 9$

4. $3x - y = 3$
$4x - 3y = 14$

5. $3x + 4y = 8$
$2x - 3y = 9$

6. $6x + 3y = -4$
$9x + 5y = -6$

7. $2x = y + 5$
$6x + 2y = -5$

8. $x + 5y = 3$
$2x + 10y = 6$

9. $3x = -4y - 6$
$3y = -5x + 1$

10. $5x - 5y = 3$
$x - y = -2$

Solve the system of equations using determinants.

11. $x + y - z = -3$
$x + z = 2$
$2x - y + 2z = 3$

12. $2x - y + 3z = 0$
$x + 2y - z = 5$
$2y + z = 1$

13. $-x + y = 1$
$y - z = 2$
$x + z = -2$

14. $$\begin{aligned} -x + 2y + 3z &= -1 \\ -3x - 3y + z &= 0 \\ 2x + 3y + z &= 2 \end{aligned}$$

15. $$\begin{aligned} 2x + 2y + 2z &= 0 \\ -x - 3y + 7z &= 15 \\ 3x + y + 4z &= 21 \end{aligned}$$

16. $$\begin{aligned} x - 2y + 3z &= 4 \\ 2x - y + z &= -5 \\ x + y - z &= -2 \end{aligned}$$

17. $$\begin{aligned} x - y + 2z &= 3 \\ x - y + z &= 1 \\ 2x + y + 2z &= 2 \end{aligned}$$

18. $$\begin{aligned} 2x + y - 2 &= 0 \\ 3x + 2y + z &= 3 \\ x - 3y - 5z &= 5 \end{aligned}$$

19. $$\begin{aligned} x + 2y + z &= 1 \\ x - y + z &= 1 \\ 2x + y + 2z &= 2 \end{aligned}$$

20. $$\begin{aligned} 2x + y - 2z &= -4 \\ x + y + z &= 1 \\ x + y + 2z &= 3 \end{aligned}$$

21. Describe a determinant, a second-order determinant, and a third-order determinant.

22. Given a second-order determinant of the form $\begin{vmatrix} a_1 & b_1 \\ a_2 & b_2 \end{vmatrix}$, how will the value of the determinant change if the a's are switched with the b's, $\begin{vmatrix} b_1 & a_1 \\ b_2 & a_2 \end{vmatrix}$?

23. Given a second-order determinant of the form $\begin{vmatrix} a_1 & b_1 \\ a_2 & b_2 \end{vmatrix}$, how will the value of the determinant change if the a's are switched with each other and the b's are switched with each other, $\begin{vmatrix} a_2 & b_2 \\ a_1 & b_1 \end{vmatrix}$?

4.5 Solving Systems of Linear Inequalities

1. *Solve systems of linear inequalities.*
2. *Solve systems of linear inequalities containing absolute value.*

1 In Section 3.6 we showed how to graph linear inequalities in two variables. In Section 4.1 we learned how to solve systems of equations graphically. In this section we show how to solve systems of linear inequalities graphically. The solution to a system of linear inequalities is the set of points that satisfies all inequalities in the system. Although a system of linear inequalities may contain more than two inequalities, in this book we consider only systems with two inequalities.

To Solve a System of Linear Inequalities

Graph each inequality on the same set of axes. The solution is the set of points that satisfies all the inequalities in the system.

EXAMPLE 1 Determine the solution to the system of inequalities.

$$\begin{aligned} x + y &\leq 6 \\ y &> 2x - 3 \end{aligned}$$

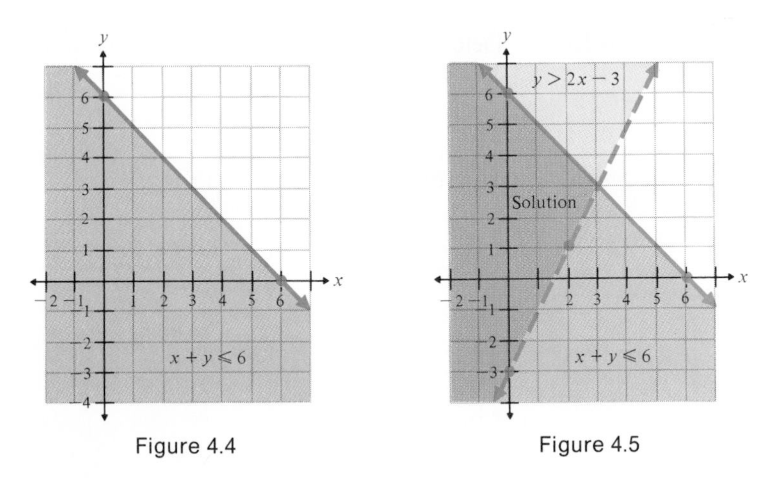

Figure 4.4

Figure 4.5

Solution: First graph the inequality $x + y \leq 6$ (see Fig. 4.4). Now on the same set of axes graph the inequality $y > 2x - 3$ (Fig. 4.5). The solution is the set of points common to both inequalities. It is the part of the graph that contains both shadings. The dashed line is not part of the solution, but the part of the solid line that satisfies both inequalities is. ■

EXAMPLE 2 Determine the solution to the system of inequalities.

$$2x + 3y \geq 4$$
$$2x - y > -6$$

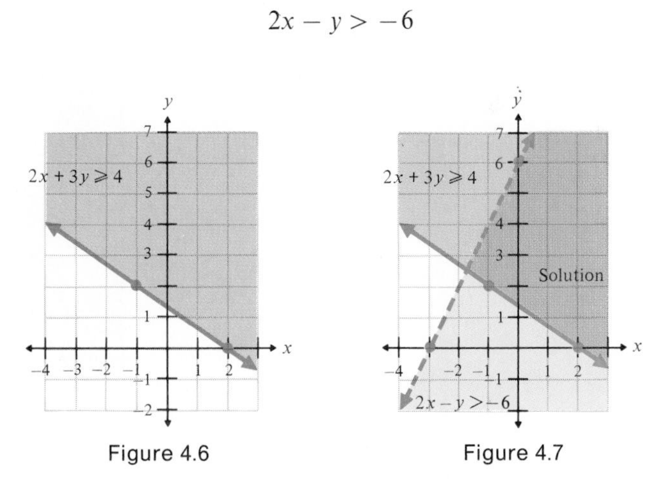

Figure 4.6

Figure 4.7

Solution: Graph $2x + 3y \geq 4$ (see Fig. 4.6). Graph $2x - y > -6$ on the same set of axes (Fig. 4.7). The solution is the part of the graph with both shadings and the part of the solid line that satisfies both inequalities. ■

EXAMPLE 3 Determine the solution to the system of inequalities.

$$y < 4$$
$$x > -2$$

Solution: The solution is illustrated in Fig. 4.8.

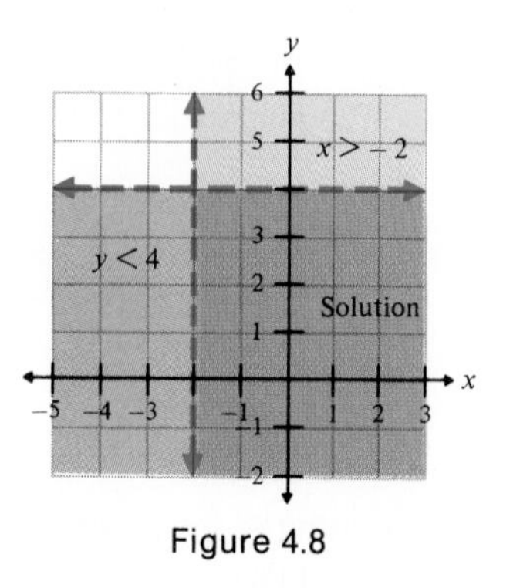

Figure 4.8 ■

Graphing Inequalities with Absolute Value (Optional)

2 Now we will graph on the Cartesian coordinate system a few inequalities containing absolute value.

EXAMPLE 4 Graph $|x| < 3$ in the Cartesian coordinate system.

Solution: From Section 2.7 we know that $|x| < 3$ means $-3 < x < 3$. Now draw dashed vertical lines through -3 and 3 and shade the area between the two (Fig. 4.9).

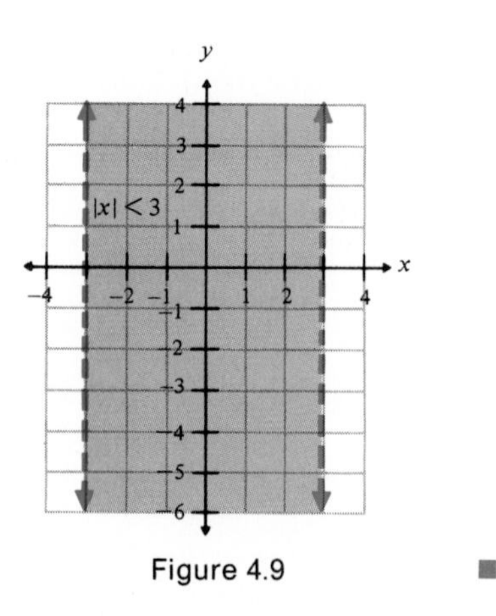

Figure 4.9 ■

EXAMPLE 5 Graph $|y + 1| > 3$ in the Cartesian coordinate system.

Solution: From Section 2.7 we know that $|y + 1| > 3$ means $y + 1 > 3$ or $y + 1 < -3$. Solve each inequality.

$$y + 1 > 3 \quad \text{or} \quad y + 1 < -3$$
$$y > 2 \qquad\qquad y < -4$$

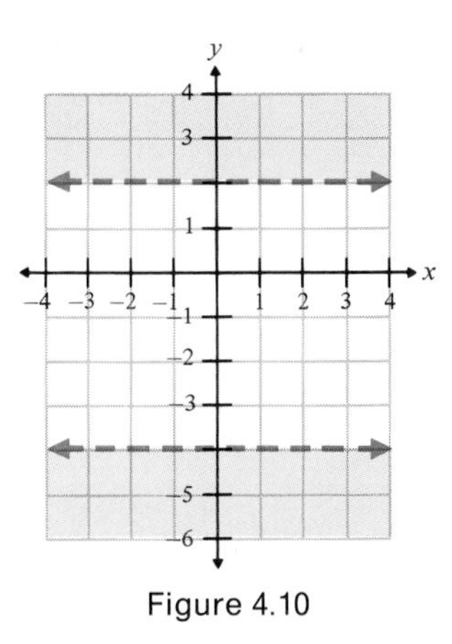

Figure 4.10

Now graph both inequalities and take the *union* of the two graphs. The solution is the shaded area in Fig. 4.10. ■

EXAMPLE 6 Graph the system of inequalities.

$$|x| < 3$$
$$|y + 1| > 3$$

Solution: Draw both inequalities on the same set of axes. Therefore, we combine the graph drawn in Example 4 with the graph drawn in Example 5 (see Fig. 4.11). The points common to both inequalities form the solution to the system.

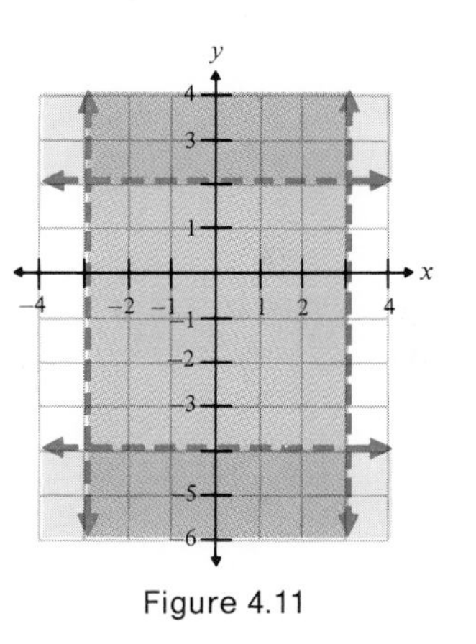

Figure 4.11 ■

Exercise Set 4.5

Determine the solution to each system of inequalities.

1. $x - y > 2$
$y < -2x + 3$

2. $y \geq 3x - 2$
$y > -4x$

3. $y \leq x - 4$
$y < -2x + 4$

4. $2x + 3y < 6$
$4x - 2y \geq 8$

5. $y < x$
$y \geq 3x + 2$

6. $-x + 3y \geq 6$
$-2x - y > 4$

7. $4x - 2y < 6$
$y \leq -x + 4$

8. $y \leq 3x + 4$
$y > 2$

9. $-4x + 5y < 20$
$x \geq -3$

10. $3x - 4y \leq 6$
$y > -x + 4$

11. $x \leq 4$
$y \geq -2$

12. $x \geq 0$
$x - 3y < 6$

13. $5x + 2y > 10$
$3x - y > 3$

14. $3x + 2y > 8$
$x - 5y < 5$

15. $-2x < y + 4$
$3x \geq y$

16. $y \leq 4x - 6$
$2x + 4y < 6$

17. $\frac{1}{2}x + 3y > 6$
$y < 3x - 4$

18. $\frac{1}{2}x + \frac{1}{3}y \geq 2$
$2x - 3y \leq -6$

19. $4x - y < 6$
$y \geq \frac{2}{3}x + 1$

20. $y \leq -5x + 3$
$y < \frac{1}{3}x + 3$

Determine the solution to each of the following systems.

21. $|x| < 3$
$y > x$

22. $|y| > 2$
$y \leq x + 3$

23. $|x| > 1$
$y \leq 3x + 2$

24. $2x + y > -1$
$|x| \leq 2$

25. $|y| < 4$
$y \geq -2x + 2$

26. $|x| \leq 3$
$|y| > 2$

27. $|x| \geq 1$
$|y| \geq 2$

28. $|x| < 2$
$|y| \geq 3$

29. $|x + 2| < 3$
$|y| > 4$

30. $|x - 3| \geq 2$
$x + y < 5$

31. $|x - 2| > 1$
$y > -2$

32. $|x - 3| \leq 4$
$|y + 2| \leq 1$

33. $|x - 3| > 4$
$|y + 1| \leq 3$

34. $|x - 2| \leq 3$
$|y + 3| \geq 2$

35. Is it possible for a system of linear inequalities to have no solution? If you answered *yes*, explain how this can occur.

JUST FOR FUN

Determine the solution to each of the following systems.

1. $|2x - 3| - 1 > 3$
$|y - 2| < 3$

2. $|x + \frac{3}{2}| > \frac{5}{2}$
$|2y - \frac{1}{2}| \leq \frac{3}{2}$

3. $y < |x|$
$y < 4$

4. $y \geq |x - 2|$
$y \leq -|x - 2|$

Summary

Glossary

Consistent system of equations: A system of equations that has a solution.

Dependent system of equations: A system of equations that has more than one point as a solution.

Determinant: A square array of numbers enclosed between two vertical bars. A determinant represents a number.

Elements of a determinant: The numbers that make up the array of numbers in a determinant are called its elements.

Inconsistent system of equations: A system of equations that has no solution.

Second-order determinant: A determinant that has two rows and two columns of elements.

Solution to a system of linear equations: The ordered pair or pairs that satisfy all equations in the system.

System of linear equations: Two or more linear equations taken as a system.

System of linear inequalities: Two or more linear inequalities taken as a system.

Third-order determinant: A determinant that has three rows and three columns of elements.

Third-order system of linear equations: A system of linear equations consisting of three equations with three unknowns.

Important Facts

Value of a Second-Order Determinant

$$\begin{vmatrix} a_1 & b_1 \\ a_2 & b_2 \end{vmatrix} = a_1b_2 - a_2b_1$$

Cramer's Rule For a system of equations of the form

$$a_1x + b_1y = c_1$$
$$a_2x + b_2y = c_2$$

$$x = \frac{\begin{vmatrix} c_1 & b_1 \\ c_2 & b_2 \end{vmatrix}}{\begin{vmatrix} a_1 & b_1 \\ a_2 & b_2 \end{vmatrix}} = \frac{D_x}{D} \quad \text{and} \quad y = \frac{\begin{vmatrix} a_1 & c_1 \\ a_2 & c_2 \end{vmatrix}}{\begin{vmatrix} a_1 & b_1 \\ a_2 & b_2 \end{vmatrix}} = \frac{D_y}{D}$$

Value of a Third-Order Determinant

minor determinant of a_1 ↓ minor determinant of a_2 ↓ minor determinant of a_3 ↓

$$\begin{vmatrix} a_1 & b_1 & c_1 \\ a_2 & b_2 & c_2 \\ a_3 & b_3 & c_3 \end{vmatrix} = a_1 \begin{vmatrix} b_2 & c_2 \\ b_3 & c_3 \end{vmatrix} - a_2 \begin{vmatrix} b_1 & c_1 \\ b_3 & c_3 \end{vmatrix} + a_3 \begin{vmatrix} b_1 & c_1 \\ b_2 & c_2 \end{vmatrix}$$

To evaluate the system

$$a_1x + b_1y + c_1z = d_1$$
$$a_2x + b_2y + c_2z = d_2$$
$$a_3x + b_3y + c_3z = d_3$$

with

$$D = \begin{vmatrix} a_1 & b_1 & c_1 \\ a_2 & b_2 & c_2 \\ a_3 & b_3 & c_3 \end{vmatrix} \qquad D_x = \begin{vmatrix} d_1 & b_1 & c_1 \\ d_2 & b_2 & c_2 \\ d_3 & b_3 & c_3 \end{vmatrix}$$

$$D_y = \begin{vmatrix} a_1 & d_1 & c_1 \\ a_2 & d_2 & c_2 \\ a_3 & d_3 & c_3 \end{vmatrix} \qquad D_z = \begin{vmatrix} a_1 & b_1 & d_1 \\ a_2 & b_2 & d_2 \\ a_3 & b_3 & d_3 \end{vmatrix}$$

Then $x = \frac{D_x}{D}$, $y = \frac{D_y}{D}$, and $z = \frac{D_z}{D}$.

Review Exercises

[4.1] Write each equation in slope-intercept form. Without graphing or solving the system of equations, state whether the system of linear equations is consistent, inconsistent, or dependent. Also indicate whether the system has exactly one solution, no solution, or an infinite number of solutions.

1. $x + 2y = 8$
$3x + 6y = 12$

2 $y = -3x - 6$
$2x + 3y = 8$

3. $y = \frac{1}{2}x + 4$
$x + 2y = 8$

4. $6x = 4y - 8$
$4x = 6y + 8$

Determine the solution to the system of equations graphically.

5. $y = x + 3$
$y = 2x + 5$

6. $x = -2$
$y = 3$

7. $2x + 2y = 8$
$2x - y = -4$

8. $y = x - 3$
$2x - 2y = 6$

Find the solution to the system of equations by substitution.

9. $y = 2x + 1$
$y = 3x - 2$

10. $y = -x + 5$
$y = 2x - 1$

11. $y = 2x - 8$
$2x - 5y = 0$

12. $-x + 3y = 9$
$x = -2y + 1$

13. $2x + y = 5$
$3x + 2y = 8$

14. $2x - y = 6$
$x + 2y = 13$

15. $3x + y = 17$
$2x - 3y = 4$

16. $x = -3y$
$x + 4y = 6$

Find the solution to the system of equations using the addition method.

17. $x + y = 6$
$x - y = 10$

18. $x + 2y = -3$
$2x - 2y = 6$

19. $2x + 3y = 4$
$x + 2y = -6$

20. $x + y = 12$
$2x + y = 5$

21. $4x - 3y = 8$
$2x + 5y = 8$

22. $-2x + 3y = 15$
$3x + 3y = 10$

23. $x + \frac{2}{5}y = \frac{9}{5}$
$x - \frac{3}{2}y = -2$

24. $2x + 2y = 8$
$y = 4x - 3$

25. $y = -\frac{3}{4}x + \frac{10}{4}$
$x + \frac{5}{4}y = \frac{7}{2}$

26. $2x - 5y = 12$
$x - \frac{4}{3}y = -2$

Determine the solution to the third-order system using substitution or the addition method.

27. $x + 2y = 12$
$4x = 8$
$3x - 4y + 5z = 20$

28. $3x + 4y - 5z = 10$
$4x + 2z = 16$
$2z = -4$

29. $x - 5y + 5z = -6$
$3x + 3y - z = 10$
$x + 3y + 2z = 5$

30. $-x - y - z = -6$
$2x + 3y - z = 7$
$-3x + y + z = -6$

31. $3y - 2z = -4$
$3x - 5z = -7$
$2x + y = 6$

32. $3x + 2y - 5z = 19$
$2x - 3y + 3z = -15$
$5x - 4y - 2z = -2$

(a) Express the problem as a system of linear equations and (b) use the method of your choice to find the solution to the problem.

33. The sum of two numbers is 48. Find the two numbers if the larger is 3 less than twice the smaller

34. The difference of two numbers is 18. Find the two numbers if the larger is 4 times the smaller.

35. Each year Ellen has the option of selecting one of two retirement plans. Option 1 is a yearly rate of 8% of the employee's salary. Option 2 is $500 plus 3% of the salary. At what salary will the retirement benefit from option 1 equal that from option 2?

36. Curtis has a 30% acid solution and a 50% acid solution. How much of each must he mix to get 6 liters of a 40% acid solution?

37. A plane can travel 600 miles per hour with the wind and 530 miles per hour against the wind. Find the speed of the wind and the speed of the plane in still air.

38. It takes Anna in her motorboat 2 hours to travel downstream to a store when the current is 2 miles per hour. The return trip against the current takes her $2\frac{1}{2}$ hours. Find (a) the speed of the boat in still water and (b) the one-way distance.

Determine the solution to the system of equations using determinants.

39. $5x + 6y = 14$
$x - 3y = 7$

40. $3x + 5y = -2$
$5x + 3y = 2$

41. $4x + 3y = 2$
$7x - 2y = -11$

42. $x + y + z = 8$
$x - y - z = 0$
$x + 2y + z = 9$

43. $4x - 2y - 3z = 5$
$-8x - y + z = -5$
$2x + y + 2z = 5$

44. $x + 2y - 4z = 17$
$2x - y + z = -9$
$2x - y - 3z = -1$

45. $y + 3z = 4$
$-x - y + 2z = 0$
$x + 2y + z = 1$

46. $2x - 3y + 4z = 3$
$5x + 4y - 2z = 7$
$-3x + 2y - 5z = -6$

[4.5] Determine the solution to the system of inequalities.

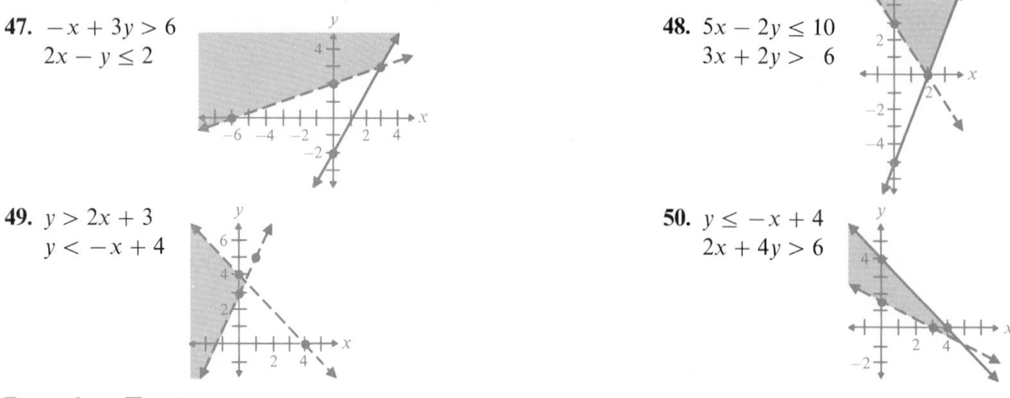

47. $-x + 3y > 6$
$2x - y \leq 2$

48. $5x - 2y \leq 10$
$3x + 2y > 6$

49. $y > 2x + 3$
$y < -x + 4$

50. $y \leq -x + 4$
$2x + 4y > 6$

Practice Test

Write each equation in slope-intercept form. Then determine, without solving the system, whether the system of equations is consistent inconsistent, or dependent. State whether the system has exactly one solution, no solution, or an infinite number of solutions.

1. $4x + 3y = -6$
$6y = 8x + 4$

2. $5x + 3y = 9$
$5x - 3y = 9$

Solve the system of equations graphically.

3. $y = 3x - 2$
$y = -2x + 8$

Solve the system of equations by substitution.

4. $y = 4x - 5$
$y = 2x + 7$

5. $3x + y = 8$
$x - y = 6$

Solve the system of equations using the addition method.

6. $2x + y = 5$
$x + 3y = -10$

7. $\frac{3}{2}x + y = 6$
$x - \frac{5}{2}y = -4$

(a) In Exercise 8, express the problem as a system of linear equations and (b) use the method of your choice to find the solution to the problem.

8. Max has cashews that sell for \$4 a pound and peanuts that sell for \$2.50 a pound. How much of each must he mix to get 20 pounds of a mixture that sells for \$3.00 per pound?

9. Solve the system of equations

$$x = 2$$
$$2x + 3y = 10$$
$$-x + 3y - 2z = 10$$

10. Solve the following system of equations

$$x + y + z = 2$$
$$-2x - y + z = 1$$
$$x - 2y - z = 1$$

Graph the system of inequalities and indicate its solution.

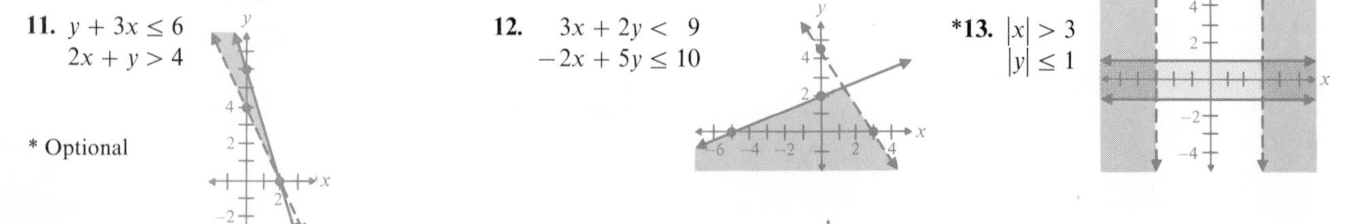

11. $y + 3x \leq 6$
$2x + y > 4$

12. $3x + 2y < 9$
$-2x + 5y \leq 10$

***13.** $|x| > 3$
$|y| \leq 1$

* Optional

5 Polynomials

See Section 5.1, Just For Fun 2b.

5.1 Exponents and Scientific Notation

1. *Learn the meaning of exponents.*
2. *Learn the product rule for exponents.*
3. *Learn the quotient rule for exponents.*
4. *Learn the negative exponent rule.*
5. *Use two or more rules to simplify expressions.*
6. *Write numbers in scientific notation.*
7. *Change numbers in scientific notation to numbers without exponents.*
8. *Use scientific notation in calculations.*

---

1 Before we can discuss polynomials, we need to know exponents. In this section we will review and then expand our knowledge of exponents. Recall from Section 1.5 that

$$2^3 = \underbrace{2 \cdot 2 \cdot 2}_{\text{3 factors of 2}}$$

and

$$x^m = \underbrace{x \cdot x \cdot x \cdot x \cdots x}_{m \text{ factors of } x}$$

The quantity 2^3 is called an **exponential expression.** The 2 is the **base** and the 3 is the **exponent** of the expression.

2 Consider the multiplication $x^3 \cdot x^5$.

$$x^3 \cdot x^5 = (x \cdot x \cdot x) \cdot (x \cdot x \cdot x \cdot x \cdot x) = x^8$$

This problem could also be evaluated using the product rule for exponents.

> **Product Rule for Exponents**
>
> If m and n are natural numbers and a is any real number, then
>
> $$a^m \cdot a^n = a^{m+n}$$

To multiply expressions in exponential form, maintain the common base and add the exponents.

$$x^3 \cdot x^5 = x^{3+5} = x^8$$

EXAMPLE 1 Simplify each of the following.

(a) $3^2 \cdot 3^3$ (b) $x^3 \cdot x^9$ (c) $x \cdot x^6$

Solution:

(a) $3^2 \cdot 3^3 = 3^{2+3} = 3^5 = 243$

(b) $x^3 \cdot x^9 = x^{3+9} = x^{12}$

(c) $x \cdot x^6 = x^1 \cdot x^6 = x^{1+6} = x^7$ ■

COMMON STUDENT ERROR

When multiplying expressions in exponential form, *do not multiply the common base.*

Correct	*Wrong*
$2^2 \cdot 2^3 = 2^5$	~~$2^2 \cdot 2^3 = 4^5$~~

When multiplying two expressions written in exponential form, *do not multiply the exponents, add the exponents.*

Correct	*Wrong*
$2^2 \cdot 2^3 = 2^5$	~~$2^2 \cdot 2^3 = 2^6$~~

3 Consider the division $x^7 \div x^4$.

$$\frac{x^7}{x^4} = \frac{\overset{1}{\cancel{x}} \cdot \overset{1}{\cancel{x}} \cdot \overset{1}{\cancel{x}} \cdot \overset{1}{\cancel{x}} \cdot x \cdot x \cdot x}{\underset{1}{\cancel{x}} \cdot \underset{1}{\cancel{x}} \cdot \underset{1}{\cancel{x}} \cdot \underset{1}{\cancel{x}}} = x \cdot x \cdot x = x^3$$

The problem above could also be solved using the quotient rule for exponents.

Quotient Rule For Exponents

If a is any nonzero real number and m and n are nonzero integers, then

$$\frac{a^m}{a^n} = a^{m-n}$$

To divide expressions in exponential form, maintain the common base and subtract the exponents.

$$\frac{x^7}{x^4} = x^{7-4} = x^3$$

EXAMPLE 2 Simplify each of the following.

(a) $\frac{5^4}{5^2}$ (b) $\frac{x^5}{x^2}$ (c) $\frac{y^2}{y^5}$

Solution: (a) $\frac{5^4}{5^2} = 5^{4-2} = 5^2 = 25$ (b) $\frac{x^5}{x^2} = x^{5-2} = x^3$ (c) $\frac{y^2}{y^5} = y^{2-5} = y^{-3}$ ■

Notice in Example 2(c) that the answer contains a negative exponent. Let's do part (c) again by dividing out common factors.

$$\frac{y^2}{y^5} = \frac{\overset{1}{\cancel{y}} \cdot \overset{1}{\cancel{y}}}{\underset{1}{\cancel{y}} \cdot \underset{1}{\cancel{y}} \cdot y \cdot y \cdot y} = \frac{1}{y^3}$$

From the process above and from Example 2(c), we can reason that $y^{-3} = 1/y^3$.

4 This problem could also be done using the negative exponent rule.

Negative Exponent Rule

For any nonzero real number a and any whole number m,

$$a^{-m} = \frac{1}{a^m}, \qquad a \neq 0$$

EXAMPLE 3 Write each of the following without negative exponents.

(a) 2^{-3} (b) x^{-2} (c) $\frac{1}{x^{-3}}$ (d) $\frac{2}{y^{-5}}$

Solution: (a) $2^{-3} = \frac{1}{2^3} = \frac{1}{8}$ (b) $x^{-2} = \frac{1}{x^2}$

(c) $\frac{1}{x^{-3}} = \frac{1}{\frac{1}{x^3}} = \frac{1}{1} \cdot \frac{x^3}{1} = x^3$ (d) $\frac{2}{y^{-5}} = \frac{2}{\frac{1}{y^5}} = \frac{2}{1} \cdot \frac{y^5}{1} = 2y^5$ ■

HELPFUL HINT

Notice that **a factor can be moved from a numerator to a denominator or from a denominator to a numerator simply by changing the sign of the *exponent*.**

Other examples

$$2 = \frac{1}{2^{-1}} \qquad x^{-3} = \frac{1}{x^3}$$

$$6^{-1} = \frac{1}{6} \qquad x^5 = \frac{1}{x^{-5}}$$

COMMON STUDENT ERROR

Correct	*Wrong*	*Wrong*
$\frac{1}{3^2} = 3^{-2}$	$\frac{1}{3^2} = 3^2$	$\frac{1}{3^2} = -3^2$
$\frac{1}{x^5} = x^{-5}$	$\frac{1}{x^5} = x^5$	$\frac{1}{x^5} = -x^5$

*An important concept that you must understand is that **an exponent applies only to the number or variable immediately preceding it unless parentheses are used.*** For example, when considering xy^2, only the y is squared. Similarly, in the expression xy^{-1} only the y is raised to the negative 1 power.

$$xy^{-1} = x \cdot y^{-1} = x \cdot \frac{1}{y} = \frac{x}{y}$$

Other examples

$$2x^{-1} = 2 \cdot \frac{1}{x} = \frac{2}{x}$$

$$-x^{-2} = -1 \cdot x^{-2} = -1 \cdot \frac{1}{x^2} = \frac{-1}{x^2}$$

$$-5x^{-4} = -5 \cdot \frac{1}{x^4} = \frac{-5}{x^4}$$

COMMON STUDENT ERROR

Remember that the exponent refers only to the symbol or number directly preceding it. The correct procedure for evaluating the expression, -6^2, is

$$-6^2 = -(6)^2 = -(6)(6) = -36 \qquad \textbf{correct}$$

If the number -6 was to be squared, it would be written $(-6)^2$.

$$(-6)^2 = (-6)(-6) = 36 \qquad \textbf{correct}$$

Students often make the following error.

$$\cancel{-6^2 = (-6)(-6) = 36} \qquad \textbf{wrong}$$

EXAMPLE 4 Write each of the following without negative exponents.

(a) $-3^2x^2y^{-3}$ (b) $(-4)^2x^{-2}yz^{-3}$ (c) $4^{-2}x^{-1}y^2$ (d) $\dfrac{3xz^2}{y^{-4}}$

Solution: (a) $-3^2x^2y^{-3} = -\dfrac{9x^2}{y^3}$ (b) $(-4)^2x^{-2}yz^{-3} = \dfrac{16y}{x^2z^3}$

(c) $4^{-2}x^{-1}y^2 = \dfrac{y^2}{4^2x^1} = \dfrac{y^2}{16x}$ (d) $\dfrac{3xz^2}{y^{-4}} = 3xy^4z^2$ ■

5 Now look at some examples that combine a number of these properties.

EXAMPLE 5 Use the rules for exponents to simplify the expression. Write the answer without negative exponents.

(a) $4^2 \cdot 4^{-4}$ (b) $y^4 \cdot y^{-6}$ (c) $\dfrac{2^3}{2^{-2}}$ (d) $\dfrac{x^{-5}}{x^{-2}}$

Solution: (a) $4^2 \cdot 4^{-4} = 4^{2+(-4)} = 4^{-2} = \dfrac{1}{4^2} = \dfrac{1}{16}$

(b) $y^4 \cdot y^{-6} = y^{4+(-6)} = y^{-2} = \dfrac{1}{y^2}$

(c) $\dfrac{2^3}{2^{-2}} = 2^{3-(-2)} = 2^{3+2} = 2^5 = 32$

(d) $\dfrac{x^{-5}}{x^{-2}} = x^{-5-(-2)} = x^{-5+2} = x^{-3} = \dfrac{1}{x^3}$ ■

EXAMPLE 6 Simplify and write the answer without negative exponents.

$$\frac{(x^{-3})(2x^6)}{x^{-5}}$$

Solution:

$$\frac{(x^{-3})(2x^6)}{x^{-5}} = \frac{2x^{-3+6}}{x^{-5}}$$
$$= \frac{2x^3}{x^{-5}}$$
$$= 2x^8 \quad ■$$

EXAMPLE 7 Simplify and write the answer without negative exponents.

$$\left(\frac{4x^3y^{-2}}{3xy^5}\right)\left(\frac{6x^4y^3}{4x^{-2}y^2}\right)$$

Solution: Begin by simplifying each factor and writing each factor without negative exponents.

$$\left(\frac{4x^3y^{-2}}{3xy^5}\right)\left(\frac{6x^4y^3}{4x^{-2}y^2}\right) = \left(\frac{4x^2}{3y^7}\right)\left(\frac{3x^6y}{2}\right)$$

Now simplify further.

$$= \frac{\overset{2}{\cancel{4}}x^2}{\underset{2}{\cancel{3}}y^7} \cdot \frac{\overset{1}{\cancel{3}}x^6y}{\underset{1}{\cancel{2}}}$$
$$= \frac{2x^8}{y^6} \quad ■$$

EXAMPLE 8 Simplify (assume that all variables used as exponents are integers).

(a) $x^{3b} \cdot x^{4b+5}$ (b) $\dfrac{y^{3r-2}}{y^{2r+4}}$ (c) $\dfrac{(x^{2p+4})(x^{p-2})}{x^{3p+1}}$

Solution: (a) By the product rule for exponents,

$$x^{3b} \cdot x^{4b+5} = x^{3b+(4b+5)} = x^{7b+5}$$

(b) By the quotient rule for exponents,

$$\frac{y^{3r-2}}{y^{2r+4}} = y^{3r-2-(2r+4)} = y^{r-6}$$

(c) Using both the product and quotient rule, we get

$$\frac{(x^{2p+4})(x^{p-2})}{x^{3p+1}} = \frac{x^{(2p+4)+(p-2)}}{x^{3p+1}}$$
$$= \frac{x^{3p+2}}{x^{3p+1}}$$
$$= x^{3p+2-(3p+1)}$$
$$= x^{3p+2-3p-1}$$
$$= x^1 \text{ or } x \quad ■$$

Scientific Notation

6 When working with scientific problems, we often deal with very large and very small numbers. For example, the distance from the earth to the sun is about 93,000,000 miles. The wavelength of a yellow color of light is about 0.0000006 meter. Because it is difficult to work with many zeros, scientists often express such numbers with exponents. For example, the number 93,000,000 might be written 9.3×10^7 and the number 0.0000006 might be written 6.0×10^{-7}. Numbers such as 9.3×10^7 and 6.0×10^{-7} are in a form called **scientific notation.** Each number written in scientific notation is written with a number greater than or equal to 1 and less than 10 $(1 \leq a < 10)$ multiplied by some power of 10.

Examples of numbers in scientific notation

$$3.2 \times 10^6$$
$$4.762 \times 10^3$$
$$8.07 \times 10^{-2}$$
$$1 \times 10^{-5}$$

Consider the number 32,400:

$$\begin{aligned} 32{,}400 &= 3.2 \times 10{,}000 \\ &= 3.2 \times 10^4 \end{aligned}$$

Note that $10{,}000 = 10^4$. Also note that there are four zeros in 10,000, the same number as the exponent in 10^4. The procedure for writing a number in scientific notation follows.

To Write a Number in Scientific Notation

1. Move the decimal in the original number to the right or left until you obtain a number greater than or equal to 1 and less than 10.
2. Count the number of places you have moved the decimal to obtain the number in step 1. If the decimal was moved to the left, the count is to be considered positive. If the decimal was moved to the right, the count is to be considered negative.
3. Multiply the number obtained in step 1 by 10 raised to the count (power) found in step 2.

EXAMPLE 9 Write the following numbers using scientific notation.

(a) 10,700 (b) 0.000386 (c) 972,000 (d) 0.0083

Solution: (a) 10,700 means 10,700.

$10{,}700. = 1.07 \times 10^4$

4 places to left

(b) $0.000386 = 3.86 \times 10^{-4}$

4 places to right

(c) $972{,}000. = 9.72 \times 10^5$

5 places to left

(d) $0.0083 = 8.3 \times 10^{-3}$

3 places to right ■

7

To Convert from a Number Given in Scientific Notation

1. Observe the exponent on the power of 10.
2. (a) If the exponent is positive, move the decimal in the number to the right the same number of places as the exponent. It may be necessary to add zeros to the number.
   (b) If the exponent is negative, move the decimal in the number to the left the same number of places as the exponent. It may be necessary to add zeros.

EXAMPLE 10 Write each number without exponents.

(a) 2.1×10^4 (b) 6.28×10^{-3} (c) 7.95×10^8

Solution: (a) Moving the decimal four places to the right gives

$$2.1 \times 10^4 = 2.1 \times 10{,}000 = 21000$$

(b) $6.28 \times 10^{-3} = 0.00628$ Move the decimal three places to the left.

(c) $7.95 \times 10^8 = 795{,}000{,}000$ Move the decimal eight places to the right. ■

8 We can use the rules of exponents discussed in this section when working with numbers written in scientific notation.

EXAMPLE 11 Simplify $(3.6 \times 10^6)(2 \times 10^{-4})$.

Solution:
$$(3.6 \times 10^6)(2 \times 10^{-4}) = (3.6 \times 2)(10^6 \times 10^{-4})$$
$$= 7.2 \times 10^2$$
$$= 720 \quad ■$$

EXAMPLE 12 Simplify $\dfrac{6.2 \times 10^{-5}}{2 \times 10^{-3}}$.

Solution:
$$\frac{6.2 \times 10^{-5}}{2 \times 10^{-3}} = \left(\frac{6.2}{2}\right)\left(\frac{10^{-5}}{10^{-3}}\right)$$
$$= 3.1 \times 10^{-5-(-3)}$$
$$= 3.1 \times 10^{-5+3}$$
$$= 3.1 \times 10^{-2}$$
$$= 0.031 \quad ■$$

EXAMPLE 13 Multiply (42,100,000)(0.008).

Solution: Change each number to scientific notation form.

$$(42{,}100{,}000)(0.008) = (4.21 \times 10^7)(8 \times 10^{-3})$$
$$= (4.21 \times 8)(10^7 \times 10^{-3})$$
$$= 33.68 \times 10^4$$
$$= 336{,}800 \quad ■$$

Calculator Corner

What will your calculator show when you multiply very large or very small numbers? The answer depends on whether your calculator has the ability to display an answer in scientific notation form. On calculators without this ability, you will probably get an error message since the answer will be too large or too small for the display.

Example: On a calculator without scientific notation

C 8000000 × 600000 = Error

If your calculator has the ability to give an answer in scientific notation form, you would get an answer given in scientific notation, as follows:

Example:

C 8000000 × 600000 = 4.8 12

The 4.8 12 means 4.8×10^{12}.

Example:

C .0000003 × .004 = 1.2 −9

The 1.2 −9 means 1.2×10^{-9}.

Exercise Set 5.1

Simplify and write the answer without any negative exponents.

1. 3^{-2}
2. 5^{-2}
3. 1^{-2}
4. x^{-4}
5. $3x^{-2}$
6. $5y^{-3}$
7. $\frac{1}{x^{-1}}$
8. $\frac{1}{y^{-3}}$
9. $\frac{1}{x^{-4}}$
10. $\frac{1}{2x^{-3}}$
11. $\frac{3}{5y^{-2}}$
12. $\frac{2x}{y^{-3}}$
13. $\frac{6x^4}{y^{-1}}$
14. $\frac{3xy^{-3}}{4}$
15. $\frac{5x^{-2}y^{-3}}{2z^{-1}}$
16. $\frac{4x^{-3}y}{z^4}$
17. $\frac{5x^{-2}y^{-3}}{z^{-4}}$
18. $\frac{10xy^5}{2z^{-3}}$

Simplify and write the answer without any negative exponents.

19. $3^2 \cdot 3$
20. $5 \cdot 5^3$
21. $2^4 \cdot 2^{-3}$
22. $6^3 \cdot 6^{-4}$
23. $7 \cdot 7^{-3}$
24. $x^2 \cdot x$
25. $x^2 \cdot x^4$
26. $x^6 \cdot x^{-2}$
27. $x^{-4} \cdot x^3$
28. $\frac{3^4}{3^2}$
29. $\frac{5^3}{5^5}$
30. $\frac{5^2}{5^{-2}}$
31. $\frac{7^{-5}}{7^{-3}}$
32. $\frac{x^5}{x^4}$
33. $\frac{x^9}{x^2}$
34. $\frac{x^3}{x^{-2}}$
35. $\frac{x^{-2}}{x}$
36. $\frac{x^0}{x^{-3}}$
37. $\frac{y^3}{2y^5}$
38. $\frac{3y^{-2}}{y^3}$
39. $\frac{6x^4}{x^{-3}}$
40. $\frac{4x^7}{x^7}$
41. $\frac{x^{-3}}{x^{-5}}$
42. $2x^{-4} \cdot 6x^{-3}$

43. $(-3y^{-2})(-y^3)$
44. $(3x^4y^{-2})(2xy^{-3})$
45. $(2x^{-3}y^{-4})(6x^{-4}y^7)$
46. $(-2x^3y^4)(-x^{-3}y^5)$
47. $(5x^2y^{-2}z^4)(-2x^5y^2z)$
48. $(-3x^{-4}y^6z^{-4})(2x^3yz^3)$
49. $(2x^4y^7z^9)(4x^3y^{-5}z^{-12})$
50. $\dfrac{24x^3y^2}{8xy}$
51. $\dfrac{27x^5y^{-4}}{9x^3y^2}$
52. $\dfrac{6x^{-2}y^3}{2x^4y}$
53. $\dfrac{9xy^{-4}}{3x^{-2}y}$
54. $\dfrac{(x^{-2})(4x^2)}{x^3}$
55. $\dfrac{(2x^4)(6xy^3)}{4y^3}$
56. $\dfrac{(4xy^5)(5x^4y^3)}{2x^5y^9}$
57. $\dfrac{(-3x^{-1}y^{-2})(2x^4y^{-3})}{6xy^4}$
58. $\dfrac{(x^4y)(3x^4y^{-3}z)}{6x^8y^2z^4}$
59. $\left(\dfrac{3x^2y^5}{2z^{-1}}\right)\left(\dfrac{4z^3}{9x^4y}\right)$
60. $\left(\dfrac{5x^2y^3}{4z^3}\right)\left(\dfrac{8xy^6}{2z^3}\right)$
61. $\left(\dfrac{2x^5}{y^{-3}}\right)\left(\dfrac{3xy^{-2}}{z^{-3}}\right)$
62. $\left(\dfrac{2x^{-2}y^{-3}}{z^3}\right)\left(\dfrac{x^3y^5}{z^{-2}}\right)$

Simplify each of the following. Assume that all variables represent integers.

63. $x^{4a} \cdot x^{3a+4}$
64. $y^{4r-2} \cdot y^{-2r+3}$
65. $w^{5b-2} \cdot w^{2b+3}$
66. $d^{5x+3} \cdot d^{-2x-3}$
67. $\dfrac{x^{2w+3}}{x^{w-4}}$
68. $\dfrac{y^{5m-1}}{y^{7m-1}}$
69. $\dfrac{(x^{3p+5})(x^{2p-3})}{x^{4p-1}}$
70. $\dfrac{(s^{2t-3})(s^{-t+5})}{s^{2t+4}}$

Express each number in scientific notation.

71. 3700
72. 3,610,000
73. 900
74. 0.00062
75. 0.047
76. 0.0000462
77. 19,000
78. 5,260,000,000
79. 0.00000186
80. 0.0003
81. 0.00000914
82. 37,000

Express each number without exponents.

83. 5.2×10^3
84. 1.63×10^{-4}
85. 4×10^7
86. 6.15×10^5
87. 2.13×10^{-5}
88. 9.64×10^{-7}
89. 3.12×10^{-1}
90. 4.6×10^1
91. 9×10^6
92. 7.3×10^4
93. 5.35×10^2
94. 1.04×10^{-2}

Express each number without exponents.

95. $(5 \times 10^3)(3 \times 10^4)$
96. $(2 \times 10^{-3})(3 \times 10^2)$
97. $(2.1 \times 10^1)(3 \times 10^{-4})$
98. $(1.6 \times 10^{-2})(4 \times 10^{-3})$
99. $\dfrac{6.4 \times 10^5}{2 \times 10^3}$
100. $\dfrac{8 \times 10^{-3}}{2 \times 10^1}$
101. $\dfrac{8.4 \times 10^{-6}}{4 \times 10^{-4}}$
102. $\dfrac{25 \times 10^3}{5 \times 10^{-2}}$
103. $\dfrac{4 \times 10^5}{2 \times 10^4}$
104. $\dfrac{16 \times 10^3}{8 \times 10^{-3}}$
105. (700,000)(6,000,000)
106. (0.0006)(5,000,000)
107. (0.003)(0.00015)
108. (230,000)(3000)
109. $\dfrac{1{,}400{,}000}{700}$
110. $\dfrac{20{,}000}{0.0005}$
111. $\dfrac{0.00004}{200}$
112. $\dfrac{(0.0012)(400{,}000)}{0.000006}$
113. $\dfrac{(150{,}000)(.000003)}{0.0005}$

114. The distance to the sun is 93,000,000 miles. If a spacecraft travels at a speed of 3100 miles per hour, how long will it take for it to reach the sun?

115. A computer can do one calculation in 0.0000004 second. How long would it take a computer to do a trillion (10^{12}) calculations?

JUST FOR FUN

1. The Richter scale is used to measure the intensity of earthquakes. An earthquake that measures 1 on the Richter scale is one that is barely detected by instruments. An earthquake that measures 2 on the Richter scale is 10 times as intense as an earthquake that measures 1 on the Richter scale. An earthquake that measures 3 on the Richter scale is 10 times as intense as one that measures 2, and $10 \cdot 10$ or 100 times as intense as one that measures 1, and so on.

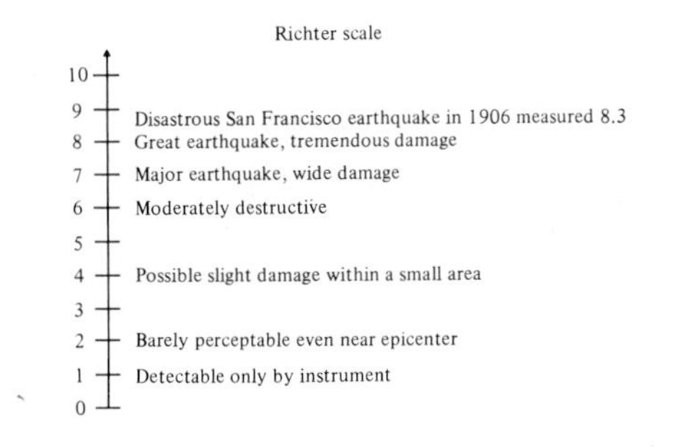

(a) Use each Richter scale number as an exponent of a power of 10 and give its equivalent value. For example, 0 gives $10^0 = 1$, 1 gives $10^1 = 10$, 2 gives $10^2 = 100$, and so on.

(b) How many times more intense is an earthquake that measures 6 than an earthquake that measures 2 on the Richter scale?

(c) On May 2, 1983, an earthquake measuring 6.5 on the Richter scale did major damage to Coalinga, California. How many times more intense was the great San Francisco earthquake than the earthquake that damaged Coalinga?

2. A *light year* is the distance that light travels in one year.

(a) Find the number of miles in a light year if light travels at 1.86×10^5 miles *per second.*

(b) If the earth is 93,000,000 miles from the sun, how long does it take for light from the sun to reach the earth?

(c) Our Milky Way galaxy is about 6.25×10^{16} miles across. If a spaceship could travel at half the speed of light, how long would it take for the craft to travel from one end of the galaxy to the other?

5.2 More on Exponents

1. *Learn the zero exponent rule.*
2. *Learn the three power rules.*
3. *Use two or more rules to simplify expressions.*

1 In Section 5.1 we introduced a number of properties of exponents. In this section we discuss the zero exponent rule and the power rules of exponents. For the sake of clarity, we will refer to the three power rules we discuss as power rules 1, 2, and 3. First, we discuss the zero exponent rule.

Consider the following: Any nonzero number divided by itself is 1. Therefore

$$\frac{x^5}{x^5} = 1$$

By the quotient rule for exponents,

$$\frac{x^5}{x^5} = x^{5-5} = x^0$$

Since $x^0 = \frac{x^5}{x^5}$ and $\frac{x^5}{x^5} = 1$, by the transitive property of equalities,

$$x^0 = 1$$

Here is the zero exponent rule.

Zero Exponent Rule

If a is any nonzero real number, then

$$a^0 = 1$$

We must specify that $a \neq 0$ because 0^0 is not a real number.

EXAMPLE 1 Simplify (assume that $x \neq 0$).

(a) x^0 (b) $3x^0$ (c) $(5x)^0$ (d) $-(a+b)^0$

Solution: (a) $x^0 = 1$ (b) $3x^0 = 3(1) = 3$ (c) $(5x)^0 = 1$ (d) $-(a+b)^0 = -(1) = -1$ ■

2 Consider the problem $(x^3)^2$.

$$(x^3)^2 = x^3 \cdot x^3 = x^{3+3} = x^6$$

The problem above could also be evaluated using power rule 1.

Power Rule for Exponents

If a is a real number and m and n are integers, then

$$(a^m)^n = a^{m \cdot n} \qquad \text{Power Rule 1}$$

To raise an expression in exponential form to a power, maintain the base and multiply the exponents.

$$(x^3)^2 = x^{3 \cdot 2} = x^6$$

EXAMPLE 2 Simplify each of the following, and write the answers without negative exponents.

(a) $(2^3)^2$ (b) $(x^3)^5$ (c) $(y^3)^{-5}$ (d) $(3^{-2})^3$

Solution: (a) $(2^3)^2 = 2^{3 \cdot 2} = 2^6 = 64$ (b) $(x^3)^5 = x^{3 \cdot 5} = x^{15}$

(c) $(y^3)^{-5} = y^{3(-5)} = y^{-15} = \frac{1}{y^{15}}$ (d) $(3^{-2})^3 = 3^{-2(3)} = 3^{-6} = \frac{1}{3^6}$ or $\frac{1}{729}$ ■

HELPFUL HINT

> Students often confuse the product rule
>
> $$a^m \cdot a^n = a^{m+n}$$
>
> with the power rule
>
> $$(a^m)^n = a^{m \cdot n}$$
>
> Note that, for example, $(x^3)^2 = x^6$, not x^5.

Two additional forms of the Power Rule for exponents are given below.

> **Power Rules for Exponents**
>
> If a and b are real numbers and m is an integer, then
>
> $$(ab)^m = a^m b^m \qquad \text{Power Rule 2}$$
>
> $$\left(\frac{a}{b}\right)^m = \frac{a^m}{b^m},\ b \neq 0 \qquad \text{Power Rule 3}$$

Note that when an expression within parentheses is raised to a power, each factor in the parentheses is raised to that power.

EXAMPLE 3 Simplify each of the following. Write answers without negative exponents.

(a) $(4x^3y^{-2})^3$ (b) $(3xy^{-3})^{-2}$ (c) $\left(\dfrac{3}{x^{-2}}\right)^4$

Solution: (a) $(4x^3y^{-2})^3 = 4^3x^9y^{-6}$

$$= 4^3x^9 \cdot \frac{1}{y^6}$$

$$= \frac{64x^9}{y^6}$$

(b) $(3xy^{-3})^{-2} = 3^{-2}x^{-2}y^6$

$$= \frac{1}{3^2} \cdot \frac{1}{x^2} \cdot y^6$$

$$= \frac{y^6}{9x^2}$$

(c) $\left(\dfrac{3}{x^{-2}}\right)^4 = \dfrac{3^4}{x^{-8}}$

$$= \frac{81}{x^{-8}}$$

$$= 81x^8 \quad ■$$

Two or more rules can be used together to simplify problems, as illustrated in the following examples.

EXAMPLE 4 Simplify and write the answer without negative exponents.

$$\left(\frac{6x^2y^{-2}}{5}\right)^3\left(\frac{2xy^4}{3}\right)^2$$

Solution:

$$\left(\frac{6x^2y^{-2}}{5}\right)^3\left(\frac{2xy^4}{3}\right)^2 = \frac{6^3x^6y^{-6}}{5^3} \cdot \frac{2^2x^2y^8}{3^2}$$

$$= \frac{\overset{24}{\cancel{216}}x^6}{125y^6} \cdot \frac{4x^2y^8}{\underset{1}{\cancel{9}}}$$

$$= \frac{96x^8y^2}{125} \quad ■$$

EXAMPLE 5 Simplify and write the answer without negative exponents.

(a) $\left(\dfrac{6x^2y^4}{2x^2y}\right)^2$ (b) $\left(\dfrac{3x^4y^{-2}}{6xy^3}\right)^3$

Solution: Problems involving exponents can often be solved in more than one way. In general it will be easier to divide out common factors before peforming the other operations.

(a) $\left(\dfrac{6x^2y^4}{2x^2y}\right)^2 = (3y^3)^2$

$= 9y^6$

(b) $\left(\dfrac{3x^4y^{-2}}{6xy^3}\right)^3 = \left(\dfrac{x^3}{2y^5}\right)^3$

$= \dfrac{x^9}{2^3y^{15}}$

$= \dfrac{x^9}{8y^{15}}$ ■

Summary of Rules of Exponents

For all real numbers a and b and all integers m and n.

Product rule	$a^m \cdot a^n = a^{m+n}$	
Quotient rule	$\dfrac{a^m}{a^n} = a^{m-n}$,	$a \neq 0$
Negative exponent rule	$a^{-m} = \dfrac{1}{a^m}$,	$a \neq 0$
Zero exponent rule	$a^0 = 1$,	$a \neq 0$
Power rules	$(a^m)^n = a^{mn}$	
	$(ab)^m = a^m b^m$	
	$\left(\dfrac{a}{b}\right)^m = \dfrac{a^m}{b^m}$,	$b \neq 0$

Exercise Set 5.2

Evaluate. Assume all bases represented by variables are non-zero.

1. x^0
2. 3^0
3. $4x^0$
4. $5y^0$
5. $-(7x)^0$
6. $-2x^0$
7. $-3x^0$
8. $(x + y)^0$
9. $-(a + b)^0$
10. $3(a + b)^0$
11. $-2(x - y)^0$
12. $3x^0 + 4y^0$
13. $2(3x^0 + 3y^0)$
14. $-4(x^0 - 3y^0)$

Simplify and write the answers without negative exponents.

15. $x^0 \cdot x^4$
16. $(2^2)^3$
17. $(3^2)^2$
18. $(3^2)^{-1}$
19. $(2^3)^{-2}$
20. $(x^2)^3$
21. $(x^4)^3$
22. $(x^3)^{-5}$
23. $(y^5)^0$
24. $(3y^2)^0$
25. $(y^0)^3$
26. $(x^{-3})^{-2}$
27. $(-x)^2$
28. $(-x)^3$
29. $(-x)^{-3}$
30. $(-2x^{-2})^3$
31. $3(x^4)^{-2}$
32. $(2x^{-3})^2$
33. $(3x^2)^3$
34. $(4y^2)^2$

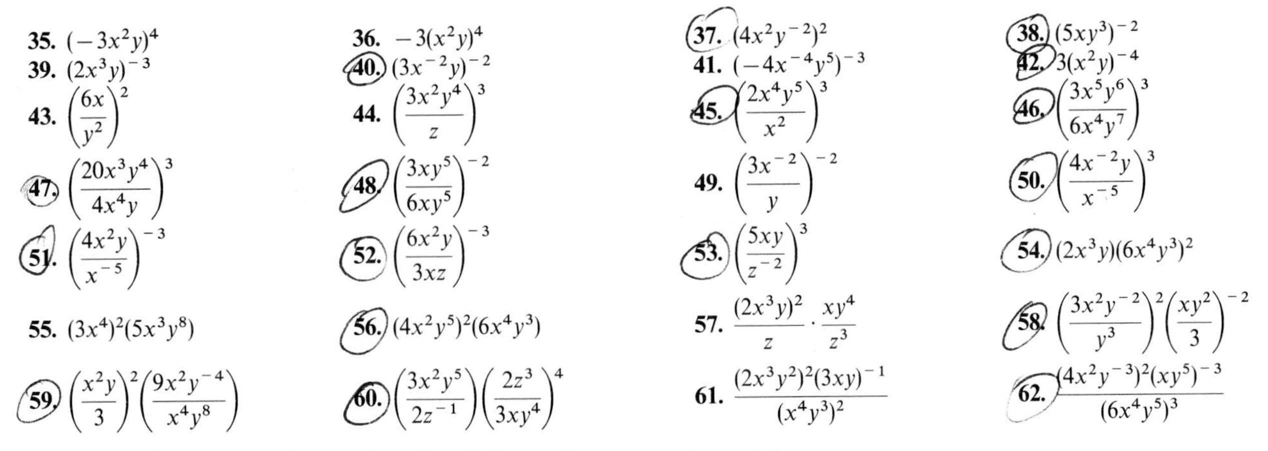

35. $(-3x^2y)^4$ 36. $-3(x^2y)^4$ 37. $(4x^2y^{-2})^2$ 38. $(5xy^3)^{-2}$

39. $(2x^3y)^{-3}$ 40. $(3x^{-2}y)^{-2}$ 41. $(-4x^{-4}y^5)^{-3}$ 42. $3(x^2y)^{-4}$

43. $\left(\frac{6x}{y^2}\right)^2$ 44. $\left(\frac{3x^2y^4}{z}\right)^3$ 45. $\left(\frac{2x^4y^5}{x^2}\right)^3$ 46. $\left(\frac{3x^5y^6}{6x^4y^7}\right)^3$

47. $\left(\frac{20x^3y^4}{4x^4y}\right)^3$ 48. $\left(\frac{3xy^5}{6xy^5}\right)^{-2}$ 49. $\left(\frac{3x^{-2}}{y}\right)^{-2}$ 50. $\left(\frac{4x^{-2}y}{x^{-5}}\right)^3$

51. $\left(\frac{4x^2y}{x^{-5}}\right)^{-3}$ 52. $\left(\frac{6x^2y}{3xz}\right)^{-3}$ 53. $\left(\frac{5xy}{z^{-2}}\right)^3$ 54. $(2x^3y)(6x^4y^3)^2$

55. $(3x^4)^2(5x^3y^8)$ 56. $(4x^2y^5)^2(6x^4y^3)$ 57. $\frac{(2x^3y)^2}{z}\cdot\frac{xy^4}{z^3}$ 58. $\left(\frac{3x^2y^{-2}}{y^3}\right)^2\left(\frac{xy^2}{3}\right)^{-2}$

59. $\left(\frac{x^2y}{3}\right)^2\left(\frac{9x^2y^{-4}}{x^4y^8}\right)$ 60. $\left(\frac{3x^2y^5}{2z^{-1}}\right)\left(\frac{2z^3}{3xy^4}\right)^4$ 61. $\frac{(2x^3y^2)^2(3xy)^{-1}}{(x^4y^3)^2}$ 62. $\frac{(4x^2y^{-3})^2(xy^5)^{-3}}{(6x^4y^5)^3}$

Simplify each expression. Assume that all variables used as exponents are integers.

63. $x^{-m}(x^{3m+2})^2$ 64. $y^{3b+2}\cdot(y^{2b+4})^2$ 65. $(b^{5y-2})^y\cdot b^{5y}$

66. $\frac{x^{3a}\cdot(x^{2a})^3}{x^{4a-2}}$ 67. $\frac{(m^{-5y+2})(m^{2y+3})}{m(m^{4y+1})}$ 68. $\frac{(w^{2p+4})^2(w^{-p-8})}{w^3(w^{4p-3})}$

JUST FOR FUN

1. Consider the expression $y = x^{10}$, where x is positive. Determine (using a calculator with a y^x key if available) the value of y when:
   (a) $x = 1$, $x = 2$, $x = 3$, $x = 4$
   (b) $x = 0.3$, $x = 0.2$, $x = 0.1$
   (c) Describe what happens to y if values greater than 1 are substituted for x, if 1 is substituted for x, and if values less than 1 are substituted for x.

5.3 Addition and Subtraction of Polynomials

1. *Identify polynomials.*
2. *Find the degree of a polynomial.*
3. *Add polynomials.*
4. *Subtract polynomials.*

1 A **polynomial** is a finite sum of terms in which all variables have whole-number exponents and no variables appear in a denominator.

A polynomial with terms of the form ax^n is called a **polynomial in x.**

Examples of polynomials in x	*Not polynomials*
$5x$	$x^{1/2}$ (fractional exponent)
$6x^2 - \frac{1}{2}x + 4$	$2x^{-1}$ (negative exponent)

Polynomials can be in more than one variable as illustrated below.

Examples of polynomials in x and y

$$3xy - 6x^2y$$
$$4x^2y - 3xy^2 + 5$$

A polynomial of only one term is called a **monomial.** A **binomial** is a two-term polynomial, and a **trinomial** is a three-term polynomial. Polynomials containing more than three terms are not given special names. The term "poly" is a prefix meaning many.

Examples of monomials	*Examples of binomials*	*Examples of trinomials*
4	$x + 4$	$x^2 - 2x + 1$
$6x$	$x^2 - 6x$	$6x^2 + 3xy - 2y^2$
$\frac{1}{5}xyz^3$	$x^2y - y^2$	$\frac{1}{2}x + 3y - 6x^2y^2$

2 In Section 2.1 we stated that the **degree of a term** is the sum of the exponents on the variables in the term. Thus $3x^2y^3z$ is of degree 6 $(2 + 3 + 1 = 6)$.

The **degree of a polynomial** is the same as that of its highest-degree term.

Polynomial	*Degree of polynomial*
$8x^3 + 2x^2 + 3x + 4$	third (x^3 is highest-degree term)
4	zero (4 or $4x^0$ is highest-degree term)
$6x^2 + 4x^2y^5 - 6$	seventh ($4x^2y^5$ is highest-degree term)

The polynomials $2x^3 + 4x^2 - 6x + 3$ and $4x^2 - 3xy + 5y^2$ are examples of polynomials in **descending order** of the variable x because the exponents on the variable x descend (or get lower) as the terms go from left to right. Polynomials are often written in descending order of a given variable.

EXAMPLE 1 Write each of the following polynomials in descending order of the variable x.

(a) $3x + 4x^2 - 6$ (b) $xy - 6x^2 + 3y^2$

Solution: (a) $3x + 4x^2 - 6 = 4x^2 + 3x - 6$

(b) $xy - 6x^2 + 3y^2 = -6x^2 + xy + 3y^2$ ■

Addition of Polynomials

3

Adding Polynomials

To add polynomials, combine the like terms of the polynomials.

EXAMPLE 2 Simplify $(4x^2 - 6x + 3) + (2x^2 + 5x - 1)$.

Solution:

$$(4x^2 - 6x + 3) + (2x^2 + 5x - 1)$$
$$= 4x^2 - 6x + 3 + 2x^2 + 5x - 1 \quad \text{remove parentheses}$$
$$= \underbrace{4x^2 + 2x^2}_{6x^2} \underbrace{- 6x + 5x}_{-x} + \underbrace{3 - 1}_{+2} \quad \text{rearrange terms}$$
$$= 6x^2 - x + 2 \quad \text{combine like terms}$$
■

EXAMPLE 3 Simplify $(3x^2y - 4xy + y) + (x^2y + 2xy + 3y - 2)$.

Solution:

$$(3x^2y - 4xy + y) + (x^2y + 2xy + 3y - 2)$$
$$= 3x^2y - 4xy + y + x^2y + 2xy + 3y - 2 \quad \text{remove parentheses}$$
$$= \underbrace{3x^2y + x^2y}_{4x^2y} \underbrace{- 4xy + 2xy}_{-2xy} \underbrace{+ y + 3y}_{+4y} - 2 \quad \text{rearrange terms}$$
$$= 4x^2y - 2xy + 4y - 2 \quad \text{combine like terms}$$
■

To Add Polynomials in Columns

1. Arrange polynomials in descending order, one under the other, with like terms in the same columns.
2. Find the sum of the terms in each column.

EXAMPLE 4 Add $4x^2 - 2x + 2$ and $-2x^2 - x + 4$ using columns.

Solution:

$$\begin{array}{r} 4x^2 - 2x + 2 \\ -2x^2 - x + 4 \\ \hline 2x^2 - 3x + 6 \end{array}$$
■

EXAMPLE 5 Add $(3x^3 + 2x - 4)$ and $(2x^2 - 6x - 3)$ using columns.

Solution: Since the polynomial $3x^3 + 2x - 4$ does not have an x^2 term, we will add the term $0x^2$ to the polynomial. This procedure sometimes helps in aligning like terms.

$$\begin{array}{r} 3x^3 + 0x^2 + 2x - 4 \\ 2x^2 - 6x - 3 \\ \hline 3x^3 + 2x^2 - 4x - 7 \end{array}$$
■

Subtraction of Polynomials

4

To Subtract Polynomials

1. Remove parentheses from the polynomial being subtracted, and change the sign of every term of the polynomial being subtracted.
2. Combine like terms.

EXAMPLE 6 Simplify $(3x^2 - 2x + 5) - (x^2 - 3x + 4)$.

Solution:

$(3x^2 - 2x + 5) - (x^2 - 3x + 4)$

$= 3x^2 - 2x + 5 - x^2 + 3x - 4$ remove parentheses (change sign of each term being subtracted)

$= 3x^2 - x^2 - 2x + 3x + 5 - 4$ rearrange terms

$= 2x^2 + x + 1$ combine like terms ■

EXAMPLE 7 Subtract $(-x^2 - 2x + 3)$ from $(x^3 + 4x + 6)$.

Solution:

$(x^3 + 4x + 6) - (-x^2 - 2x + 3)$

$= x^3 + 4x + 6 + x^2 + 2x - 3$ remove parentheses

$= x^3 + x^2 + 4x + 2x + 6 - 3$ rearrange terms

$= x^3 + x^2 + 6x + 3$ combine like terms ■

EXAMPLE 8 Simplify $x^2y - 4xy^2 + 5 - (2x^2y - 3y^2 + 4)$.

Solution:

$x^2y - 4xy^2 + 5 - (2x^2y - 3y^2 + 4)$

$= x^2y - 4xy^2 + 5 - 2x^2y + 3y^2 - 4$ remove parentheses

$= x^2y - 2x^2y - 4xy^2 + 3y^2 + 5 - 4$ rearrange terms

$= -x^2y - 4xy^2 + 3y^2 + 1$ combine like terms

Note that $-x^2y$ and $-4xy^2$ are not like terms since the variables have different exponents. Also, $-4xy^2$ and $3y^2$ are not like terms since the $3y^2$ does not contain the variable x. ■

To Subtract Polynomials in Columns

1. Write the polynomial being subtracted below the polynomial from which it is being subtracted. List like terms in the same column.
2. Change the sign of each term in the polynomial being subtracted. (This step can be done mentally if you like.)
3. Find the sum of the terms in each column.

EXAMPLE 9 Subtract $4x^2 - 2x + 4$ from $2x^2 + 3x + 6$ using columns.

Solution: First align like terms.

$$\begin{array}{r} 2x^2 + 3x + 6 \\ -\,(4x^2 - 2x + 4) \end{array}$$

Change the signs of the terms of the polynomial being subtracted and add.

$$\begin{array}{r} 2x^2 + 3x + 6 \\ -4x^2 + 2x - 4 \\ \hline -2x^2 + 5x + 2 \end{array}$$

■

Exercise Set 5.3

Indicate those expressions that are polynomials. If the polynomial has a specific name, for example, "monomial," "binomial," and so on, give the name. If the expression is not a polynomial, state so.

1. $5y$
2. $5x^2 - 6x + 9$
3. -10
4. $5x^{-3}$
5. $4 - 5z$
6. $6xy + 3z$
7. $8x^2 - 2x + 8y^2$
8. $x - 3$
9. $3x^{1/2} + 2xy$
10. $-2x^2 + 5x^{-1}$
11. $2xy + 5y^2$
12. $4 - 3x$

Write the polynomial in descending order of the variable x. If the polynomial is already in descending order, state so. Give the degree of each polynomial.

13. $-8 - 4x - x^2$
14. $2x + 4 - x^2$
15. $6y^2 + 3xy + 10x^2$
16. $3x^3 - x + 4$
17. $5 + 2x^3 + x^2 - 3x$
18. $-4 + x - 3x^2 + 4x^3$
19. $-2x^4 + 5x^2 - 4$
20. $6x^3 + 4xy^2 - 5x^2y + 7y^3$
21. $5xy^2 + 3x^2y - 6 - 2x^3$
22. $4y^2 - 2xy - 3x^2 + 5x^4$

Simplify.

23. $(6x + 3) + (x - 5)$
24. $(2x - 3) + (3x - 4)$
25. $(3x - 4) - (2x + 2)$
26. $(6x + 3) - (4x - 2)$
27. $(-3x + 4) + (-2x + 5)$
28. $(-7x + 4) + (-2x - 3)$
29. $(-12x - 3) - (-5x - 7)$
30. $(12x - 3) - (-2x + 7)$
31. $(3y - 8) + (4y - 3)$
32. $(2x + 3y) + (4y - 3x)$
33. $(x^2 - 6x + 3) - (2x + 5)$
34. $(x - 4) - (3x^2 - 4x + 6)$
35. $(x + y - z) + (2x - y + 3)$
36. $(4y^2 + 6y - 3) - (2y^2 + 6)$
37. $(-2x^2 + 3x - 9) + (-2x - 3)$
38. $(4x^2 - 6x + 3) - (3x + 7)$
39. $(5x - 7) + (2x^2 + 3x + 12)$
40. $(-3x + 8) + (-2x^2 - 3x - 5)$
41. $(6y^2 - 6y + 4) - (-2y^2 - y + 7)$
42. $(5z^2 - 2z + 4) - (3z + 5)$
43. $(3x^2 - 4x + 8) + (2x^2 + 5x + 12)$
44. $(x^2 - 6x + 7) + (-x^2 + 3x + 5)$
45. $(-2x^2 + 4x - 5) - (5x^2 + 3x + 7)$
46. $(5x^2 - x - 1) - (-3x^2 - 2x - 5)$
47. $(-3x^2 - 4x + 8) + (2x^2 + 5x + 12)$
48. $(3x^2y + 2xy^2) + (-3xy^2 - x^2y + 1)$
49. $(5x^2 - x + 12) - (x + 5)$
50. $(6x^2y - 3xy) - (4x^2y + 2xy)$
51. $(-3x^3 + 4x^2y + 3xy^2) + (2x^3 - x^2y + xy^2)$
52. $(9x^3 - 2x^2 + 4x - 7) + (2x^3 - 6x^2 - 4x + 3)$
53. $(-2xy^2 + 4) - (-7xy^2 + 12)$
54. $(x^2 + 6xy + y^2) - (3x^2 + 6xy + 3y^2)$
55. $(x^2 + xy - y^2) + (2x^2 - 3xy + y^2)$
56. $(x^2y + 6x^2 - 3xy^2) + (-x^2y - 12x^2 + 4xy^2)$
57. $(4x^3 - 6x^2 + 5x - 7) - (2x^2 + 6x - 3)$
58. $(4x^3 - 3x^2y + 4) - (2y^2 + 8)$
59. $(4x^2y + 2x - 3) + (3x^2y - 5x + 5)$
60. $(x^2y + x - y) + (2x^2y + 2x - 6y + 3)$
61. $(9x^3 - 4) - (x^2 + 5)$
62. $(9x^3 + 6x^2y + 3xy^2) - (-5x^2y + 5)$
63. Add $x^2 - 2x + 4$ and $3x + 12$
64. Add $4x^2 - 6x + 5$ and $-2x - 8$
65. Subtract $(4x - 6)$ from $(3x + 5)$
66. Subtract $(-x^2 + 3x + 5)$ from $(4x^2 - 6x + 2)$
67. Add $-2x^2 + 4x - 12$ and $-x^2 - 2x$
68. Add $5x^2 + x + 9$ and $2x^2 - 12$
69. Subtract $(5x^2 - 6)$ from $(2x^2 - 4x + 8)$
70. Subtract $(x^2 - 6)$ from $(6x^2 - 5x + 3)$
71. Add $3x^2 + 4x - 5$ and $4x^2 + 3x - 8$
72. Add $-5x^2 - 3$ and $x^2 + 2x - 9$
73. Subtract $(-6y^2 + 3y - 4)$ from $(9y^2 - 3y)$
74. Subtract $(4x^2 + 7x - 9)$ from $(x^3 - 6x + 3)$
75. Add $6x^2 + 3xy$ and $-2x^2 + 4xy + 3y$
76. Add $4x^2 + 3y^2 + 4$ and $-3x^2 - 7 + y^2$
77. Subtract $(-4x^2 + 6x)$ from $(x^3 - 6)$
78. Subtract $(2x^2 - 6x + 4)$ from $(6x + 8)$
79. Add $4x^2 + 3x + y^2$ and $4x - 3y - 5y^2$
80. Add $x^2 + 3y$ and $y^2 + 5y + 7$
81. Subtract $(5x^2y + 8)$ from $(-2x^2y + 6xy^2 + 8)$
82. Subtract $(6x^2y + 3xy)$ from $(2x^2y + 12xy)$

5.4 Multiplication of Polynomials

1. *Multiply a monomial by a polynomial.*
2. *Multiply a binomial by a binomial.*
3. *Square a binomial.*
4. *Find the product of the sum and difference of the same two terms (difference of two squares).*
5. *Multiply a polynomial by a polynomial.*

1 When multiplying two polynomials, each term of one polynomial must be multiplied by each term of the other polynomial. When one of the polynomials is a monomial and the other is not, we can use the expanded form of the distributive property to perform the multiplication.

Distributive Property

$$a(b + c + d + \cdots + n) = ab + ac + ad + \cdots + an$$

EXAMPLE 1 Multiply $3x(4x + 5)$.

Solution:
$$3x(4x + 5) = (3x)(4x) + (3x)(5)$$
$$= 12x^2 + 15x \quad ■$$

EXAMPLE 2 Multiply $-4x^2(5x^3 - 3x + 5)$.

Solution:
$$-4x^2(5x^3 - 3x + 5) = (-4x^2)(5x^3) + (-4x^2)(-3x) + (-4x^2)(5)$$
$$= -20x^5 + 12x^3 - 20x^2 \quad ■$$

EXAMPLE 3 Multiply $2xy(3x^2y + 6xy^2 + 4)$.

Solution:
$$2xy(3x^2y + 6xy^2 + 4) = (2xy)(3x^2y) + (2xy)(6xy^2) + (2xy)(4)$$
$$= 6x^3y^2 + 12x^2y^3 + 8xy \quad ■$$

Multiplying a Binomial by a Binomial

2 Consider multiplying $(a + b)(c + d)$. Treating $(a + b)$ as a single term and using the distributive property, we get

$$(a + b)(c + d) = (a + b)c + (a + b)d$$
$$= ac + bc + ad + bd$$

When multiplying a binomial by a binomial, each term of the first binomial must be multiplied by each term of the second binomial and all the results added together.

FOIL Method

A convenient method for finding the product of two binomials is the FOIL method. Consider

$$(a + b)(c + d)$$

F stands for **first:** Multiply the first terms of each binomial together.

$$\overset{\text{F}}{(a + b)(c + d)} \qquad \text{product } ac$$

O stands for **outer:** Multiply the two outer terms together.

$$\overset{\text{O}}{(a + b)(c + d)} \qquad \text{product } ad$$

I stands for **inner:** Multiply the two inner terms together.

$$\overset{\text{I}}{(a + b)(c + d)} \qquad \text{product } bc$$

L stands for **last:** Multiply the last terms together.

$$\overset{\text{L}}{(a + b)(c + d)} \qquad \text{product } bd$$

The answer will be the sum of the products.

$$(a + b)(c + d) = ac + ad + bc + bd$$

EXAMPLE 4 Multiply $(3x + 2)(x - 5)$ using the FOIL method.

Solution:

L, F, I, O

$$(3x + 2)(x - 5)$$

$$\begin{aligned} &\quad\;\; \text{F} \qquad\quad\; \text{O} \qquad\quad\;\; \text{I} \qquad\quad \text{L} \\ &(3x)(x) + (3x)(-5) + (2)(x) + (2)(-5) \\ &= 3x^2 - 15x + 2x - 10 = 3x^2 - 13x - 10 \end{aligned}$$

■

EXAMPLE 5 Multiply $(3x^2 + 6)(x - 2y)$.

Solution: $(3x^2 + 6)(x - 2y)$

$$\begin{aligned} &\quad\;\; \text{F} \qquad\qquad \text{O} \qquad\qquad \text{I} \qquad\quad \text{L} \\ &(3x^2)(x) + (3x^2)(-2y) + (6)(x) + (6)(-2y) \\ &= 3x^3 - 6x^2y + 6x - 12y \end{aligned}$$

■

EXAMPLE 6 Multiply $(x + 4)^2$.

Solution:
$$\begin{aligned}(x + 4)^2 &= (x + 4)(x + 4)\\ &= x^2 + 4x + 4x + 16\\ &= x^2 + 8x + 16\end{aligned}$$ ■

Square of a Binomial

3 Example 6 is the square of a binomial. We must often square a binomial, so we have special formulas for doing so.

> **Square of a Binomial**
>
> $$(a + b)^2 = a^2 + 2ab + b^2$$
> $$(a - b)^2 = a^2 - 2ab + b^2$$

The square of a binomial is the sum of the square of the first term, twice the product of the two terms, and the square of the last term.

EXAMPLE 7 Expand $(3x + 5)^2$.

Solution:
$$\begin{aligned}(3x + 5)^2 &= (3x)^2 + 2(3x)(5) + (5)^2\\ &= 9x^2 + 30x + 25\end{aligned}$$ ■

EXAMPLE 8 Expand $(2x - 3y)^2$.

Solution:
$$\begin{aligned}(2x - 3y)^2 &= (2x)^2 - 2(2x)(3y) + (3y)^2\\ &= 4x^2 - 12xy + 9y^2\end{aligned}$$ ■

EXAMPLE 9 Expand $(4x^2 - 3y)^2$.

Solution:
$$\begin{aligned}(4x^2 - 3y)^2 &= (4x^2)^2 - 2(4x^2)(3y) + (3y)^2\\ &= 16x^4 - 24x^2y + 9y^2\end{aligned}$$ ■

Examples 7 through 9 could also be done using the FOIL method.

COMMON STUDENT ERROR

Do not forget the middle term when squaring a binomial.

Correct	*Wrong*
$(x + 2)^2 = (x + 2)(x + 2)$	~~$(x + 2)^2 = x^2 + 4$~~
$= x^2 + 4x + 4$	
$(x - 3)^2 = (x - 3)(x - 3)$	~~$(x - 3)^2 = x^2 + 9$~~
$= x^2 - 6x + 9$	

EXAMPLE 10 Expand $[x + (y - 1)]^2$.

Solution: This looks more complicated than the previous examples, but it is done the same way. Treat x as the first term and $(y - 1)$ as the second term.

$$\begin{aligned}[x + (y - 1)]^2 &= (x)^2 + 2(x)(y - 1) + (y - 1)^2 \\ &= x^2 + (2x)(y - 1) + y^2 - 2y + 1 \\ &= x^2 + 2xy - 2x + y^2 - 2y + 1\end{aligned}$$

None of the six terms above are like terms, therefore no terms can be combined. Note that $(y - 1)^2$ is also the square of a binomial and was expanded as such. ■

EXAMPLE 11 Use the FOIL method to multiply $(x + 6)(x - 6)$.

Solution:

$$\begin{aligned}(x + 6)(x - 6) &= x^2 - 6x + 6x - 36 \\ &= x^2 - 36\end{aligned}$$ ■

Difference of Two Squares

4 Note that in Example 11 the other and inner terms add to 0. By examining Example 11, we see that the product of the sum and difference of the same two terms is the difference of the squares of the terms.

> **Product of Sum and Difference of the Same Two Terms**
>
> $$(a + b)(a - b) = a^2 - b^2$$

To multiply two binomials that differ only in the sign between their two terms, subtract the square of the second term from the square of the first term. Note that $a^2 - b^2$ represents a **difference of two squares.**

EXAMPLE 12 Multiply $(3x + 4)(3x - 4)$.

Solution: $(3x + 4)(3x - 4) = (3x)^2 - (4)^2 = 9x^2 - 16$ ■

EXAMPLE 13 Multiply $(3x + 5y)(3x - 5y)$.

Solution: $(3x + 5y)(3x - 5y) = (3x)^2 - (5y)^2 = 9x^2 - 25y^2$ ■

EXAMPLE 14 Multiply $(5x + y^3)(5x - y^3)$.

Solution: $(5x + y^3)(5x - y^3) = (5x)^2 - (y^3)^2 = 25x^2 - y^6$ ■

EXAMPLE 15 Multiply $[4x + (3y + 2)][(4x - (3y + 2)]$.

Solution: Treat $4x$ as the first term and $3y + 2$ as the second term. Then we have the sum and difference of the same two terms.

$$\begin{aligned}[4x + (3y + 2)][4x - (3y + 2)] &= (4x)^2 - (3y + 2)^2 \\ &= 16x^2 - (9y^2 + 12y + 4) \\ &= 16x^2 - 9y^2 - 12y - 4\end{aligned}$$ ■

Multiplying a Polynomial by a Polynomial

5 It is often easier to multiply polynomials vertically. The method to use is very similar to the procedure used for multiplying whole numbers. When multiplying two polynomials, every term of the first polynomial must be multiplied by every term of the second polynomial. The procedure is illustrated in Example 16.

EXAMPLE 16 Multiply $x^2 - 3x + 2$ by $2x^2 - 3$.

Solution: Place the longer polynomial on top, then multiply. Make sure you align like terms as you multiply so that the terms can be added

$$\begin{array}{lrrrrrl}
 & & & x^2 & -3x & +2 & \\
 & & & & 2x^2 & -3 & \\
\hline
-3(x^2-3x+2) \longrightarrow & & & -3x^2 & +9x & -6 & \text{multiply top expression by } -3 \\
2x^2(x^2-3x+2) \longrightarrow & 2x^4 & -6x^3 & +4x^2 & & & \text{multiply top expression by } 2x^2 \\
\hline
 & 2x^4 & -6x^3 & +x^2 & +9x & -6 & \text{add like terms in columns}
\end{array}$$

■

EXAMPLE 17 Multiply $3x^2 + 6xy - 5y^2$ by $x + 3y$.

Solution:

$$\begin{array}{lrrrrl}
 & & 3x^2 & +6xy & -5y^2 & \\
 & & & x & +3y & \\
\hline
3y(3x^2+6xy-5y^2) \longrightarrow & & 9x^2y & +18xy^2 & -15y^3 & \text{multiply top expression by } 3y \\
x(3x^2+6xy-5y^2) \longrightarrow & 3x^3 & +6x^2y & -5xy^2 & & \text{multiply top expression by } x \\
\hline
 & 3x^3 & +15x^2y & +13xy^2 & -15y^3 & \text{add like terms in columns}
\end{array}$$

■

Exercise Set 5.4

Multiply as indicated.

1. $2(x + 3)$
2. $3x(x - 4)$
3. $4(x - 3)$
4. $-5x(x + 2)$
5. $-4x(-2x + 6)$
6. $-x(x + 2)$
7. $3x(x^2 + 3x - 1)$
8. $-x(2x^2 - 6x + 5)$
9. $-2x(x^2 - 2x + 5)$
10. $-3x(-2x^2 + 5x - 6)$
11. $5x^2(-4x^2 + 6x - 4)$
12. $2y^3(3y^2 + 2y - 6)$
13. $6xy(x^3 - 3x^2y + 4y^2)$
14. $-3x^2y(-2x^4y^2 + 3xy^3 + 4)$
15. $3x^4(2xy^2 + 5x^7 - 6y)$
16. $\frac{1}{2}x^2y(4x^5y^2 + 3x - 6y^2)$
17. $\frac{2}{3}yz(3x + 4y - 9y^2)$
18. $-\frac{3}{5}x^2y\left(-\frac{2}{3}xy^4 + \frac{1}{9}xy + 3\right)$

Multiply using the FOIL method.

19. $(x - 4)(x + 5)$
20. $(4x - 6)(3x - 5)$
21. $(3x + 1)(x - 3)$
22. $(6 - 2x)(5 + 3x)$
23. $(2x - 3)(4x + 2)$
24. $(6z + 3)(-4z + 5)$
25. $(3y + 4)(2y - 3)$
26. $(3y + 4)(y + 1)$
27. $(x - y)(x + y)$
28. $(4 - 6z)(4 - 6z)$
29. $(2x^2 - 3)(2x - 3)$
30. $(6x - y)(2x + y)$
31. $(4 - x)(3 + 2x^2)$
32. $(6y - 2x)(5x - 3)$
33. $(2x + 3)(4 - 2x)$
34. $(3x + 4)(4y - 2x)$
35. $(x + y)(y + z)$
36. $(x + 2y)(2x - 3)$
37. $(2x^2 - 3y)(3x^2 + 2y)$
38. $(x + 3)(2y^2 - 5)$
39. $(4x^2 - 3y)(2y^2 - 3x)$
40. $(3xy^2 + y)(4x - 3xy)$

Multiply using either the square of a binomial procedure or the difference of two squares procedure.

41. $(x + 4)(x - 4)$
42. $(x + 3)^2$
43. $(2x - 1)(2x + 1)$
44. $(x + 2)(x + 2)$
45. $(2x - 3y)^2$
46. $(4x - 2y)(4x + 2y)$
47. $(4x - 5y)^2$
48. $(3m + 2n)^2$
49. $(4a - 3b)^2$
50. $(a - 3b)^2$
51. $(2r + 4s)^2$
52. $(3w^2 + 4)(3w^2 - 4)$
53. $(2x + 5y)^2$
54. $(4x^2 + 3)^2$
55. $(2y^2 - 5w)^2$
56. $(3x^2 - 4y)(3x^2 + 4y)$
57. $(5m^2 + 2n)(5m^2 - 2n)$
58. $[a + (b + 2)][a - (b + 2)]$
59. $[(3m + 2) + n][(3m + 2) - n]$
60. $[3 + (x - y)][3 - (x - y)]$
61. $[(5x + 1) + 6y][(5x + 1) - 6y]$
62. $[x + (y + 3)]^2$
63. $[4 - (x - 3y)]^2$
64. $[5x + (2y + 3)]^2$
65. $[y + (4 - 2x)]^2$
66. $[w - (y + 2)]^2$
67. $[w + (3x - 4)][w - (3x + 4)]$
68. $[3p + (2w - 3)][3p - (2w - 3)]$

Multiply vertically.

69. $(x^2 - 3x + 2)(x - 4)$
70. $(2x^2 - 3x + 4)(-3x + 4)$
71. $(y^2 - 3y + 4)(3 - 2y)$
72. $(5x^3 + 4x^2 - 6x + 2)(x + 5)$
73. $(x + 3)(2x^2 + 4x - 3)$
74. $(2x + 3)(4x^2 - 5x + 2)$
75. $(5x + 4)(x^2 - x + 4)$
76. $(2x - 5)(3x^2 - 4x + 7)$
77. $(7x - 3)(-2x^2 - 4x + 1)$
78. $(x - 2)(4x^2 + 9x - 2)$
79. $(x^2y - 3xy^2)(x + 2y)$
80. $(x^2 + y^2 + 2)(x + 2)$
81. $(a + b)(a^2 - ab + b^2)$
82. $(x - 2y)(x^2 + 2xy + 3y^2)$
83. $(a + 2b)(a^2 - 2ab + 4b^2)$
84. $(2m + n)(3m^2 - mn + 2n^2)$
85. $(a - 3b)(2a^2 - ab + 2b^2)$
86. $(3p + n)(p^2 - 3pn + n^2)$
87. $(2x - 3y)(3x^2 + 4xy - 2y^2)$
88. $(3x^2 + 6xy - 5y^2)(x - 3y)$
89. $(x^3 - x + 3)(x^2 - 2x - 4)$
90. $(x^3 - 2x^2 + 5x - 6)(2x^2 - 3x + 4)$
91. $(x + 3)^3$
92. $(x - 2)^3$
93. $(2x + 3)^3$

JUST FOR FUN

Perform each of the following polynomial multiplications.

1. $[(y + 1) - (x + 2)]^2$.
2. Multiply $(x - 3y)^4$.

5.5 Division of Polynomials

1. *Divide a polynomial by a monomial.*
2. *Divide a polynomial by a binomial.*

Dividing a Polynomial by a Monomial

1 In division of polynomials, division by zero is not permitted. When the denominator (or divisor) contains a variable, the variable cannot be a value that will result in the denominator being zero. (We discuss this concept further in Chapter 7.)

> **To divide a polynomial by a monomial,** divide each term of the polynomial by the monomial.

EXAMPLE 1 Divide $\dfrac{4x^2 - 8x}{2x}$.

Solution: $\dfrac{4x^2 - 8x}{2x} = \dfrac{4x^2}{2x} - \dfrac{8x}{2x} = 2x - 4$ ■

EXAMPLE 2 Divide $\dfrac{4x^3 - 6x^2 + 8x - 3}{2x}$.

Solution:

$$\frac{4x^3 - 6x^2 + 8x - 3}{2x} = \frac{4x^3}{2x} - \frac{6x^2}{2x} + \frac{8x}{2x} - \frac{3}{2x}$$
$$= 2x^2 - 3x + 4 - \frac{3}{2x}$$ ■

EXAMPLE 3 Divide $\dfrac{4x^2y - 6x^4y^3 - 3x^5y^2 + 5x}{2xy^2}$

Solution:

$$\frac{4x^2y - 6x^4y^3 - 3x^5y^2 + 5x}{2xy^2} = \frac{4x^2y}{2xy^2} - \frac{6x^4y^3}{2xy^2} - \frac{3x^5y^2}{2xy^2} + \frac{5x}{2xy^2}$$
$$= \frac{2x}{y} - 3x^3y - \frac{3x^4}{2} + \frac{5}{2y^2}$$ ■

Dividing a Polynomial by a Binomial

2 We divide a polynomial by a binomial in much the same way as we perform long division.

EXAMPLE 4 Divide $\dfrac{x^2 + 7x + 10}{x + 2}$.

Solution: Rewrite the division problem as

$$x + 2 \overline{\big) x^2 + 7x + 10}$$

Divide x^2 (the first term in $x^2 + 7x + 10$) by x (the first term in $x + 2$).

$$\frac{x^2}{x} = x$$

Place the quotient, x, above the term containing x in the divisor

$$\begin{array}{r} x \\ x + 2 \overline{\big) x^2 + 7x + 10} \end{array}$$

Next, multiply the x by $x + 2$ as you would do in long division and place the product under their like terms.

times

$$\begin{array}{rl} x & \\ x + 2 \overline{\big) x^2 + 7x + 10} & \\ x^2 + 2x & \leftarrow x(x + 2) \end{array}$$

equals

Now subtract $x^2 + 2x$ from $x^2 + 7x$ by changing the signs of $x^2 + 2x$ and adding.

$$\begin{array}{r|l} & x \\ \hline x+2 & x^2 + 7x + 10 \\ & -x^2 - 2x \\ \hline & 5x \end{array}$$

Next, bring down the 10, the next term.

$$\begin{array}{r|l} & x \\ \hline x+2 & x^2 + 7x + 10 \\ & x^2 + 2x \\ \hline & 5x + 10 \end{array}$$

Determine the quotient of $5x$ divided by x.

$$\frac{5x}{x} = +5$$

Place the $+5$ above the constant in the dividend and multiply 5 by $x + 2$. Finally, finish the problem by subtracting.

$$\begin{array}{r|l} & x + 5 \\ \hline x+2 & x^2 + 7x + 10 \\ & x^2 + 2x \\ \hline & 5x + 10 \\ & 5x + 10 \leftarrow 5(x+2) \\ \hline & 0 \end{array}$$

(times; equals)

Thus $\dfrac{x^2 + 7x + 10}{x + 2} = x + 5$. There is no remainder. ■

EXAMPLE 5 Divide $\dfrac{6x^2 - 5x + 5}{2x + 3}$.

Solution: In this problem we will mentally change the signs of the terms being subtracted, then add.

$$\begin{array}{r|l} & 3x - 7 \\ \hline 2x+3 & 6x^2 - 5x + 5 \\ & 6x^2 + 9x \leftarrow 3x(2x+3) \\ \hline & -14x + 5 \\ & -14x - 21 \leftarrow -7(2x+3) \\ \hline & 26 \leftarrow \text{remainder} \end{array}$$

Thus $\dfrac{6x^2 - 5x + 5}{2x + 3} = 3x - 7 + \dfrac{26}{2x + 3}$. ■

When you are dividing a polynomial by a binomial, you should list both the polynomial and binomial in descending order. If a given powered term is missing, it is often helpful to include that term with a numerical coefficient of 0. For example, when dividing $(6x^2 + x^3 - 4) \div (x - 2)$, we rewrite the problem as $(x^3 + 6x^2 + 0x - 4) \div (x - 2)$ before beginning the division.

EXAMPLE 6 Divide $(3x^5 + 4x^2 - 12x - 17) \div (x^2 - 2)$.

Solution: Whenever a power of x is missing, we will add that power of x with a coefficient of 0 to help align like terms.

$$\begin{array}{r|l}
 & 3x^3 + \qquad\quad 6x + 4 \\
x^2 + 0x - 2 & 3x^5 + 0x^4 + 0x^3 + 4x^2 - 12x - 17 \\
 & \underline{3x^5 + 0x^4 - 6x^3} \leftarrow 3x^3(x^2 + 0x - 2) \\
 & 6x^3 + 4x^2 - 12x \\
 & \underline{6x^3 + 0x^2 - 12x} \leftarrow 6x(x^2 + 0x - 2) \\
 & 4x^2 + 0x - 17 \\
 & \underline{4x^2 + 0x - 8} \leftarrow 4(x^2 + 0x - 2) \\
 & -9
\end{array}$$

To find the answer, we performed the divisions

$$\frac{3x^5}{x^2} = 3x^3, \qquad \frac{6x^3}{x^2} = 6x, \qquad \frac{4x^2}{x^2} = 4.$$

The quotients $3x^3$, $6x$, and 4 were placed above their like terms in the dividend. The answer is $3x^3 + 6x + 4 - \dfrac{9}{x^2 - 2}$ ■

Exercise Set 5.5

Divide as indicated.

1. $\dfrac{6x + 8}{2}$
2. $\dfrac{3x + 6}{2}$
3. $\dfrac{4x^2 + 2x}{2x}$
4. $\dfrac{5y^3 + 6y^2 + 3y}{3y}$
5. $\dfrac{12x^2 - 4x - 8}{4}$
6. $\dfrac{6x^2 + 3x + 12}{2x}$
7. $\dfrac{15y^6 + 5y^2}{5y^4}$
8. $\dfrac{4x^5 - 6x^4 + 12x^3}{4x^2}$
9. $\dfrac{12x^3 + 6x^2 + 3x + 9}{6x^2}$
10. $\dfrac{6x^2y - 9xy^2}{3xy}$
11. $\dfrac{4x^2y^2 - 8xy^3 + 3y^4}{2y^2}$
12. $\dfrac{-5x^3y^2 + 10xy - 6}{10x}$
13. $\dfrac{15x^{12} - 5x^9 + 30x^6}{5x^6}$
14. $\dfrac{6x^2y - 12x^3y^2 + 9y^3}{2xy^2}$
15. $\dfrac{3xyz + 6xyz^2 - 9x^3y^5z^7}{6xy}$
16. $\dfrac{a^2b^2c - 6abc^2 + 5a^3b^5}{2abc^2}$

Divide as indicated.

17. $\dfrac{x^2 + 4x + 3}{x + 1}$
18. $\dfrac{x^2 + 7x + 10}{x + 5}$
19. $\dfrac{2x^2 + 13x + 15}{x + 5}$

20. $\dfrac{2x^2 + 13x + 15}{2x + 3}$

21. $\dfrac{6x^2 + 16x + 8}{3x + 2}$

22. $\dfrac{2x^2 + x - 10}{x - 2}$

23. $\dfrac{2x^2 + x - 10}{2x + 5}$

24. $\dfrac{2x^2 + 7x - 15}{2x - 3}$

25. $\dfrac{x^2 - 25}{x - 5}$

26. $\dfrac{4x^2 - 9}{2x - 3}$

27. $\dfrac{8x^2 + 6x - 27}{4x + 9}$

28. $\dfrac{x^3 + 3x^2 + 5x + 4}{x + 1}$

29. $\dfrac{4x^3 + 12x^2 + 7x - 3}{2x + 3}$

30. $\dfrac{9x^3 - 3x^2 - 3x + 4}{3x + 2}$

31. $\dfrac{2x^3 - 3x^2 - 3x + 4}{x - 1}$

32. $\dfrac{2x^3 + 6x - 4}{x + 4}$

33. $\dfrac{4x^3 - 5x}{2x - 1}$

34. $\dfrac{9x^3 - x + 3}{3x - 2}$

35. $\dfrac{-x^3 - 6x^2 + 2x - 3}{x - 1}$

36. $\dfrac{3x^5 + 4x^2 - 12x - 8}{x^2 - 2}$

37. $\dfrac{2x^4 - 8x^3 + 19x^2 - 33x + 15}{x^2 - x + 5}$

38. $\dfrac{3x^4 + 4x^3 - 32x^2 - 5x - 20}{x + 4}$

39. $\dfrac{3x^4 + 4x^3 - 32x^2 - 5x - 20}{3x^3 - 8x^2 - 5}$

JUST FOR FUN

1. Divide $\dfrac{2x^3 - x^2y - 7xy^2 + 2y^3}{x - 2y}$

2. Divide $\dfrac{3x^3 - 5}{3x - 2}$ *Hint:* Answer contains fractions.

5.6 Synthetic Division (Optional)

1 *Divide polynomials by binomials using synthetic division.*

1 When a polynomial is divided by a binomial of the form $x - a$, the division process can be greatly shortened by a process called **synthetic division.** Consider the following examples.

$$
\begin{array}{r|l}
 & 2x^2 + 5x - 4 \\
\hline
x - 3 & 2x^3 - x^2 - 19x + 15 \\
 & 2x^3 - 6x^2 \\
\cline{2-2}
 & 5x^2 - 19x \\
 & 5x^2 - 15x \\
\cline{2-2}
 & -4x + 15 \\
 & -4x + 12 \\
\cline{2-2}
 & 3
\end{array}
\qquad
\begin{array}{r|rrrr}
 & 2 & +5 & -4 & \\
\hline
1 - 3 & 2 & -1 & -19 & 15 \\
 & 2 & -6 & & \\
 & & 5 & -19 & \\
 & & 5 & -15 & \\
 & & & -4 & 15 \\
 & & & -4 & 12 \\
 & & & & 3
\end{array}
$$

Note that the variables do not play a role in determining the numerical coefficients of the quotient. This division problem can be done more quickly and easily using synthetic division as outlined below.

1. Write the dividend in descending powers of x. Then list the numerical coefficients of each term in the dividend. If a given powered term is missing, place a 0 in the appropriate position to serve as a placeholder. In the preceding problem the numerical coefficients of the dividend are

$$2 \quad -1 \quad -19 \quad 15$$

2. When dividing by a binomial of the form $x - a$, place $\underline{|a}$ to the right of the line of numbers from part 1. In this problem we are dividing by $x - 3$; thus $a = 3$. We write

$$2 \quad -1 \quad -19 \quad 15 \quad \underline{|3}$$

3. Bring down the left-hand number as follows:

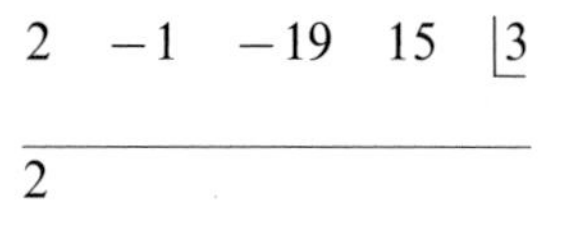

4. Multiply $2 \cdot 3$ to get 6. Place 6 under the -1. Then add $-1 + 6$ to get 5.

$$\begin{array}{rrrrl} 2 & -1 & -19 & 15 & \underline{|3} \\ & 6 & & & \\ \hline 2 & 5 & & & \end{array}$$

5. Multiply $5 \cdot 3$ to get 15. Place 15 under -19. Then add to get -4. Repeat this procedure as illustrated.

$$\begin{array}{rrrrl} 2 & -1 & -19 & 15 & \underline{|3} \\ & 6 & 15 & -12 & \\ \hline 2 & 5 & -4 & 3 & \end{array}$$

Note that the first three numerical values are identical to the numerical values obtained in the quotient when worked out by long division. The last digit, the 3, is identical to the remainder obtained by long division. The quotient must be one degree less than the dividend since we are dividing by $x - 3$. Since the original dividend was a third-degree polynomial, the quotient must be a second-degree polynomial.

The quotient is therefore $2x^2 + 5x - 4 + \dfrac{3}{x - 3}$.

EXAMPLE 1 Divide using synthetic division.

$$(6 - x^2 + x^3) \div (x + 2)$$

Solution: First list the dividend in descending order of x.

$$(x^3 - x^2 + 6) \div (x + 2)$$

Since there is no x term, insert a 0 as a placeholder when listing the numerical coefficients.

$$\begin{array}{rrrr|l} 1 & -1 & 0 & 6 & \underline{-2} \\ & -2 & 6 & -12 & \\ \hline 1 & -3 & 6 & -6 & \end{array}$$

note that $x + 2 = x - (-2)$ and therefore $a = -2$

Since the dividend is a third-degree equation, the quotient must be second degree. The quotient is $x^2 - 3x + 6 - \dfrac{6}{x+2}$. ■

EXAMPLE 2 Use synthetic division to divide.

$$(3x^4 + 11x^3 - 20x^2 + 7x + 35) \div (x + 5)$$

Solution:

$$\begin{array}{rrrrr|l} 3 & 11 & -20 & 7 & 35 & \underline{-5} \\ & -15 & 20 & 0 & -35 & \\ \hline 3 & -4 & 0 & 7 & 0 & \end{array}$$

Since the dividend is of the fourth degree, the quotient must be of the third degree. The quotient is $3x^3 - 4x^2 + 0x - 7$ with no remainder. This can be simplified to $3x^3 - 4x^2 - 7$. ■

EXAMPLE 3 Use synthetic division to divide

$$(3x^3 - 6x^2 + 4x + 5) \div \left(x - \frac{1}{2}\right)$$

Solution:

$$\begin{array}{rrrr|l} 3 & -6 & 4 & 5 & \underline{\frac{1}{2}} \\ & \frac{3}{2} & -\frac{9}{4} & \frac{7}{8} & \\ \hline 3 & -\frac{9}{2} & \frac{7}{4} & \frac{47}{8} & \end{array}$$

The solution is

$$3x^2 - \frac{9}{2}x + \frac{7}{4} + \frac{47}{8(x - \frac{1}{2})} \quad \text{or} \quad 3x^2 - 4.5x + 1.75 + \frac{5.875}{x - \frac{1}{2}}$$ ■

Exercise Set 5.6

Divide using synthetic division.

1. $(x^2 + x - 6) \div (x - 2)$
2. $(x^2 - 4x - 32) \div (x + 4)$
3. $(x^2 + 5x - 6) \div (x + 6)$
4. $(x^2 + 12x + 32) \div (x + 4)$
5. $(x^2 + 5x - 12) \div (x - 3)$
6. $(2x^2 - 9x + 15) \div (x - 6)$
7. $(3x^2 - 7x - 10) \div (x - 4)$
8. $(x^3 + 6x^2 + 4x - 7) \div (x + 5)$
9. $(4x^3 - 3x^2 + 2x) \div (x - 1)$
10. $(x^3 - 7x^2 - 13x + 5) \div (x - 2)$

11. $(3x^3 + 7x^2 - 4x + 12) \div (x + 3)$

12. $(3x^4 - 25x^2 - 20) \div (x - 3)$

13. $(5x^3 - 6x^2 + 3x - 6) \div (x + 1)$

14. $(y^4 - 1) \div (y - 1)$

15. $(x^4 + 4x^3 - x^2 - 16x - 4) \div (x - 2)$

16. $(2x^4 - x^2 + 5x - 12) \div (x - 3)$

17. $(y^5 + y^4 - 10) \div (y + 1)$

18. $(z^5 + 4z^4 - 10) \div (z + 1)$

19. $(3x^3 + 2x^2 - 4x + 1) \div \left(x - \frac{1}{3}\right)$

20. $(8x^3 - 6x^2 - 5x + 3) \div \left(x + \frac{3}{4}\right)$

21. $(2x^4 - x^3 + 2x^2 - 3x + 1) \div \left(x - \frac{1}{2}\right)$

22. $(9y^3 + 9y^2 - y + 2) \div \left(y + \frac{2}{3}\right)$

JUST FOR FUN

1. Synthetic division can be used to divide polynomials by binomials of the form $ax - b$, $a \neq 1$. To perform this division, divide $ax - b$ by a to obtain $x - \frac{b}{a}$. Then place b/a to the right of the numerical coefficients of the polynomial. Work the problem as explained previously. After summing the numerical values below the line, divide all of them, except the remainder, by a. Write the quotient of the problem using these numbers.

(a) Use this procedure to divide $(9x^3 + 9x^2 + 5x + 12)$ by $(3x + 5)$.
(b) Explain why we do not divide the remainder by a.

5.7 Polynomial Functions

1. *Identify polynomial functions.*
2. *Evaluate polynomial functions.*
3. *Use polynomial functions in practical applications.*

1 The concept of function was first introduced in Section 3.5. In this section we expand on this concept. Since functions are so important to mathematics, and are a unifying concept, we also discuss them further in later chapters.

In Section 3.5 we discussed linear functions of the form $f(x) = ax + b$. Linear functions are a specific type of **polynomial function.** There are many other types of polynomial functions. The general form of a polynomial function is given below.

Polynomial Function

$$f(x) = a_n x^n + a_{n-1}x^{n-1} + a_{n-2}x^{n-2} + a_{n-3}x^{n-3} + \cdots + a_1 x + a_0$$

where all exponents on x are whole numbers and $a_n, a_{n-1}, a_{n-2}, \ldots, a_1, a_0$ are all real numbers with $a_n \neq 0$.

Examples of polynomial functions

$f(x) = 3x + 4$ (linear function—first degree)

$f(x) = 5x^2 - \frac{1}{2}x + 3$ (quadratic function—second degree)

$f(x) = 6x^3 - 4x$ (cubic function—third degree)

$f(x) = \sqrt{2}x^4 - 6x$ (fourth-degree function)

Note that the right side of each of these functions is a polynomial since all exponents on x are whole numbers.

The graph of every equation of the form $y = a_n x^n + a_{n-1}x^{n-1} + a_{n-2}x^{n-2} + \cdots + a_1 x + a_0$, where all exponents on x are whole numbers, will pass the vertical line test. Every equation of this form is therefore a function. We will graph some polynomial functions in Section 5.8.

2 To evaluate a polynomial function for a specific value of the variable, substitute the value in the function whenever the variable appears.

EXAMPLE 1 $f(x) = 3x^3 - 6x^2 + 2x - 1$

Evaluate (a) $f(2)$ (b) $f(-2)$

Solution: (a)
$$\begin{aligned} f(x) &= 3x^3 - 6x^2 + 2x - 1 \\ f(2) &= 3(2)^3 - 6(2)^2 + 2(2) - 1 \\ &= 3(8) - 6(4) + 4 - 1 \\ &= 24 - 24 + 4 - 1 \\ &= 3 \end{aligned}$$

(b)
$$\begin{aligned} f(x) &= 3x^3 - 6x^2 + 2x - 1 \\ f(-2) &= 3(-2)^3 - 6(-2)^2 + 2(-2) - 1 \\ &= 3(-8) - 6(4) - 4 - 1 \\ &= -24 - 24 - 4 - 1 \\ &= -53 \end{aligned}$$ ■

EXAMPLE 2 $f(x) = 3x + 2$

Find (a) $f(1)$ (b) $f(a)$ (c) $f(a + b)$

Solution: (a)
$$\begin{aligned} f(x) &= 3x + 2 \\ f(1) &= 3(1) + 2 \\ &= 3 + 2 = 5 \end{aligned}$$

(b) How do we find $f(a)$? To find $f(1)$ we substituted 1 for each x in the function. Similarly, to find $f(a)$ we substitute a for each x in the function.

$$\begin{aligned} f(x) &= 3x + 2 \\ f(a) &= 3a + 2 \end{aligned}$$

(c) To find $f(a + b)$ we substitute $(a + b)$ for each x in the function.

$$\begin{aligned} f(x) &= 3x + 2 \\ f(a + b) &= 3(a + b) + 2 \\ &= 3a + 3b + 2 \end{aligned}$$ ■

EXAMPLE 3 $f(x) = x^2 + 2x - 3$

Find (a) $f(3)$ (b) $f(a)$ (c) $f(a + b)$

Solution: (a)
$$f(x) = x^2 + 2x - 3$$
$$f(3) = 3^2 + 2(3) - 3$$
$$= 9 + 6 - 3 = 12$$

(b)
$$f(x) = x^2 + 2x - 3$$
$$f(a) = a^2 + 2a - 3$$

(c)
$$f(x) = x^2 + 2x - 3$$
$$f(a + b) = (a + b)^2 + 2(a + b) - 3$$
$$= (a + b)(a + b) + 2(a + b) - 3$$
$$= a^2 + 2ab + b^2 + 2a + 2b - 3$$ ■

COMMON STUDENT ERROR

Students often make the incorrect assumptions that

$$\cancel{f(a + b) = f(a) + f(b)} \quad \textbf{wrong}$$
$$\cancel{f(a - b) = f(a) - f(b)} \quad \textbf{wrong}$$

To see why this is wrong, consider the function

$$f(x) = x^2 - 2x + 3$$
$$f(1) = 1^2 - 2(1) + 3 = 2$$
$$f(2) = 2^2 - 2(2) + 3 = 3$$
$$f(3) = 3^2 - 2(3) + 3 = 6$$

Note:
$$f(1 + 2) \neq f(1) + f(2)$$
$$f(3) \neq f(1) + f(2)$$
$$6 \neq 2 + 3$$
$$6 \neq 5$$

Applications of Polynomial Functions

3 Some applications of polynomials were discussed in Chapter 2. The examples discussed in that chapter were generally of the first degree. Now we examine additional applications of polynomial functions.

EXAMPLE 4 A polygon is a closed figure with straight line segments as sides. The number of different diagonals, d, in a polygon is a function of the number of sides n in the polygon.

$$d = f(n) = \frac{1}{2}n^2 - \frac{3}{2}n$$

(a) How many diagonals has a quadrilateral (four sides)?
(b) How many diagonals has an octagon (eight sides)?

Solution: (a) $n = 4$, $\quad d = f(4) = \frac{1}{2}(4)^2 - \frac{3}{2}(4)$

$$= \frac{1}{2}(16) - 6$$

$$= 8 - 6 = 2$$

A quadrilateral has 2 diagonals.

(b) $n = 8$, $\quad d = f(8) = \frac{1}{2}(8)^2 - \frac{3}{2}(8)$

$$= \frac{1}{2}(64) - 12$$

$$= 32 - 12 = 20 \quad ■$$

EXAMPLE 5 Neil Armstrong became the first person to walk on the moon on July 20, 1969. The velocity, v, of his spacecraft (the Eagle), in meters per second, was a function of time before touchdown, t.

$$v = f(t) = 3.2t + 0.45$$

The height, h, of the spacecraft above the moon's surface, in meters, was also a function of time before touchdown.

$$h = f(t) = 1.6t^2 + 0.45t$$

What was the velocity of the spacecraft and distance from the surface of the moon at:

(a) 5 seconds from touchdown?
(b) 2 seconds from touchdown?
(c) touchdown?

Solution: (a) To find the velocity and height at 5 seconds from touchdown, substitute $t = 5$ into the appropriate formulas.

$$v = f(t) = 3.2t + 0.45$$
$$v = f(5) = 3.2(5) + 0.45$$
$$= 16.0 + 0.45 = 16.45 \text{ meters per second}$$

$$h = f(t) = 1.6t^2 + 0.45t$$
$$h = f(5) = 1.6(5)^2 + 0.45(5)$$
$$= 1.6(25) + 2.25$$
$$= 40 + 2.25 = 42.25 \text{ meters}$$

(b) At 2 seconds from touchdown

$$v = 3.2(2) + 0.45 = 6.4 + 0.45 = 6.85 \text{ meters per second}$$

$$h = 1.6(2)^2 + 0.45(2) = 1.6(4) + 0.45(2)$$
$$= 6.4 + 0.9 = 7.3 \text{ meters}$$

(c) At touchdown, $t = 0$

$$v = 3.2(0) + 0.45 = 0 + 0.45 = 0.45 \text{ m/s}$$

Thus touchdown velocity was 0.45 meter per second.

$$h = 1.6(0^2) + 0.45(0) = 1.6(0) + 0 = 0 + 0 = 0$$

Thus at touchdown the Eagle was on the moon and the distance from the moon was 0. ■

Exercise Set 5.7

1. $f(x) = 3x - 1$; find
(a) $f(1)$ **(b)** $f(-3)$ **(c)** $f(a)$

2. $f(x) = 6x - 5$; find
(a) $f(3)$ **(b)** $f(-4)$ **(c)** $f(b)$

3. $f(x) = 2x + 3$; find
(a) $f(4)$ **(b)** $f(-2)$ **(c)** $f(a + b)$

4. $f(x) = -x + 4$; find
(a) $f(-1)$ **(b)** $f(h)$ **(c)** $f(a + b)$

5. $f(x) = x^2 - x + 4$; find
(a) $f(2)$ **(b)** $f(3)$ **(c)** $f(-1)$

6. $f(x) = x^2 + 5x - 6$; find
(a) $f(4)$ **(b)** $f(-2)$ **(c)** $f(0)$

7. $f(x) = 2x^3 - x + 3$; find
(a) $f(-1)$ **(b)** $f(2)$ **(c)** $f(-2)$

8. $f(x) = 3x^2 - x$; find
(a) $f(4)$ **(b)** $f(2)$ **(c)** $f(-3)$

9. $f(x) = -x^2 + 4x - 3$; find
(a) $f(2)$ **(b)** $f(-4)$ **(c)** $f(c)$

10. $f(x) = x^2 + 2x + 5$; find
(a) $f(4)$ **(b)** $f(b)$ **(c)** $f(b + 1)$

11. $f(x) = 3x^2 - 2x + 7$; find
(a) $f(-2)$ **(b)** $f(2)$ **(c)** $f(a + 3)$

12. $f(x) = x^2 - x + 1$; find
(a) $f(4)$ **(b)** $f(a)$ **(c)** $f(a + 2)$

13. $f(x) = -x^2 - 2x + 5$; find
(a) $f(-2)$ **(b)** $f(c)$ **(c)** $f(c + 2)$

14. $f(x) = -3x^2 + 4x - 6$; find
(a) $f(-1)$ **(b)** $f(-x)$ **(c)** $f(x + 1)$

15. $f(x) = 2x^2 - 3x + 5$; find
(a) $f(5)$ **(b)** $f(-x)$ **(c)** $f(2x)$

16. $f(x) = 4x^2 - 3x$; find
(a) $f(-2)$ **(b)** $f(4)$ **(c)** $f(h + 2)$

17. $f(x) = x^2 + 3x - 4$; find
(a) $f(h)$ **(b)** $f(h + 4)$ **(c)** $f(a + h)$

18. $f(x) = 2x^2 - x + 4$; find
(a) $f(h)$ **(b)** $f(x - 1)$ **(c)** $f(a + h)$

19. $f(x) = 2x^2 - 3x + 1$; find
(a) $f(h)$ **(b)** $f(x + 3)$ **(c)** $f(x + h)$

20. $f(x) = 3x^2 - 2x + 5$; find
(a) $f(b)$ **(b)** $f(x + 2)$ **(c)** $f(a + b)$

Solve each word problem.

21. The sum s of the first n even counting numbers is given by the function $s = f(n) = n^2 + n$. Find the sum of:
(a) The first 10 even numbers
(b) The first 15 even numbers

22. Use the function in Example 4 to find the number of diagonals in a figure with:
(a) 10 sides
(b) 6 sides

23. Use the function in Example 5 to determine the speed and height above the surface of the moon at:
(a) 6 seconds from touchdown
(b) 2.5 seconds from touchdown

24. The temperature, T, in degrees Celsius, in a sauna, n minutes after being turned on is given by the function $T = f(n) = -0.03n^2 + 1.5n + 14$. Find the sauna's temperature after:
(a) 3 minutes
(b) 12 minutes

25. The stopping distance, d, in meters, for a car traveling v kilometers per hour is given by the function $d = f(v) = 0.18v + 0.01v^2$. Find the stopping distance for speeds of:
(a) 50 km/hr
(b) 25 km/hr

26. The approximate number of accidents in one month, n, involving drivers x years of age can be approximated by the function $n = f(x) = 2x^2 - 150x + 4000$. Find the approximate number of accidents in one month involving:
(a) 18-year-olds
(b) 25-year-olds

27. The number of centimeters that a specific spring will stretch, s, when a mass, m, in kilograms is attached to it is found by the function $s = f(m) = 3.4m - 0.3m^2$ (for $m \leq 6$ kg). How much will the spring stretch if the following masses are attached to it?
(a) 2 kg
(b) 4.2 kg

28. The profits earned, P, in millions of dollars, from constructing an office building having x stories can be approximated by the function $P = f(x) = 0.02x^2 + 0.1x - 0.3$. Find the approximate profit earned from construction of an office building of:
(a) 3 stories
(b) 5 stories

29. The total number of oranges, N, in a square pyramid whose base is n by n oranges (see the figure) is given by the function

$$N = f(n) = \frac{1}{3}n^3 + \frac{1}{2}n^2 + \frac{1}{6}n$$

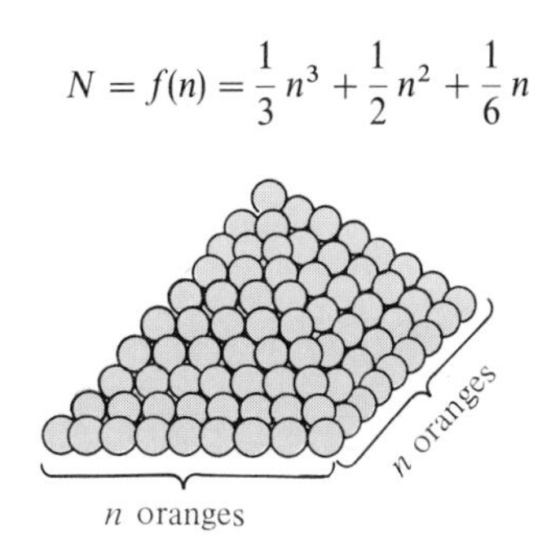

Find the number of oranges if the base is:
(a) 6 by 6 oranges
(b) 8 by 8 oranges

30. If the cost of a ticket to a rock concert is increased by x dollars, the estimated increase in revenue, R, in thousands of dollars is given by the function $R = f(x) = 24 + 5x - x^2$, $x < 8$. Find the increase in revenue if the cost of the ticket is increased by:
(a) \$1
(b) \$4

JUST FOR FUN

1. $f(x) = x^3 - 2x^2 + 6x + 3$; find $f(x + 3)$.

2. $f(x) = 2x^2 + 3x - 4$; find:
(a) $f(x + h)$
(b) $f(x + h) - f(x)$
(c) $\dfrac{f(x + h) - f(x)}{h}$

3. (a) Write a function in d that can be used to find the shaded area in the figure. Use 3.14 for π.
(b) Find the shaded area when $d = 4$ feet.
(c) Find the shaded area when $d = 6$ feet.

5.8 Graphing Polynomial Functions

1. *Understand how to graph polynomial functions.*
2. *Graph quadratic functions.*
3. *Find the axis of symmetry and vertex of a parabola.*
4. *Graph cubic functions.*

1 In Chapter 3 we graphed linear functions. Other polynomial functions can be graphed in much the same way. To graph polynomial functions we can substitute

values for x, find the corresponding values for y, and plot the points on the Cartesian coordinate system. When graphing linear functions, we had to plot only two points to draw the line. When plotting other polynomial functions, we must be sure to plot a sufficient number of points to get a true picture of the graph. The graphs of polynomial functions will be smooth curves and will pass the vertical line test for functions.

After plotting points, connect the points to get a smooth curve. When drawing the smooth curve, start with the point on the graph with the smallest x value and draw to the point having the next larger value of x. Continue this way through all of the points. Draw arrow tips on the ends of the graph to show that the graph continues in the same direction.

2 The first function we will graph in this section is a quadratic function.

Quadratic Function

Any function of the form

$$f(x) = ax^2 + bx + c, \qquad a \neq 0$$

where a, b, and c are real numbers, is a **quadratic function.**

In Section 3.5 we learned that $y = f(x)$. Therefore any quadratic equation of the form $y = ax^2 + bx + c$, $a \neq 0$, will also be a quadratic function.

Every quadratic function will have the shape of a **parabola** (see Fig. 5.1) when graphed. When graphing a function of the form

$$f(x) = ax^2 + bx + c \text{ (or } y = ax^2 + bx + c)$$

the sign of the numerical coefficient of the squared term, a, will determine whether the parabola will open upward or downward. When the squared term is positive, the parabola will open upward, as in Fig. 5.1a. When the squared term is negative the parabola will open downward, as shown in Fig. 5.1b. The **vertex** is the lowest point on a parabola that opens upward and the highest point on a parabola that opens downward.

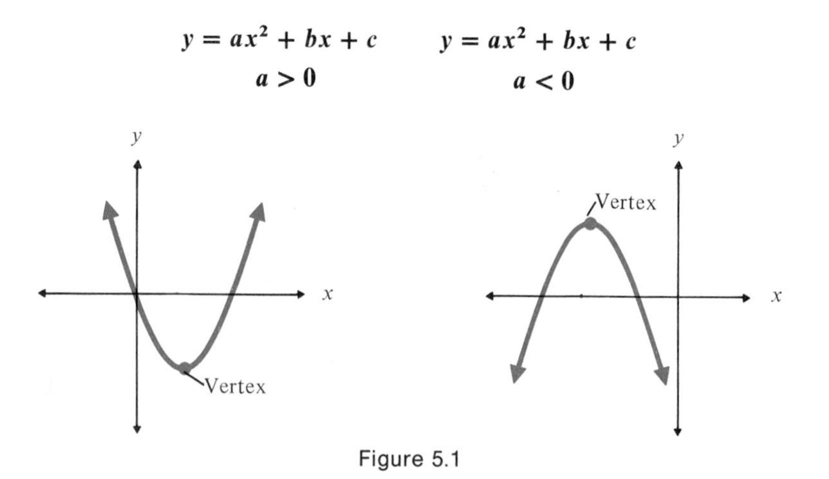

Figure 5.1

When graphing a quadratic equation, make sure you plot a sufficient number of points to show whether the parabola is opening upward or downward. We will graph a quadratic equation in Example 1.

EXAMPLE 1 Graph $f(x) = x^2 - 4x + 3$.

Solution: Since $f(x)$ is the same as y, graphing $f(x) = x^2 - 4x + 3$ is the same as graphing $y = x^2 - 4x + 3$. Make a table of values by substituting values for x and solving for $f(x)$ or y.

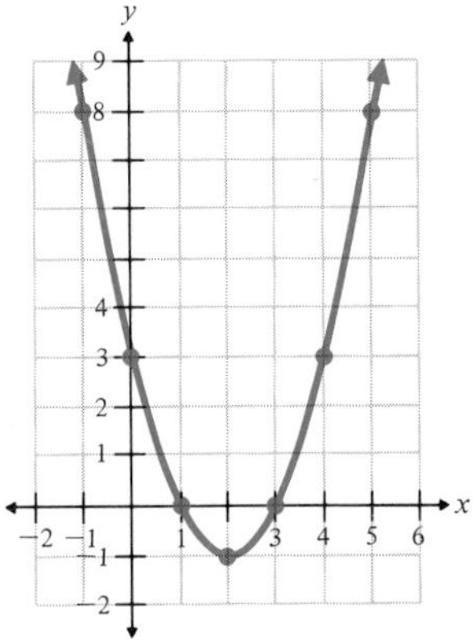

$$f(x) = x^2 - 4x + 3$$
$$f(-1) = (-1)^2 - 4(-1) + 3 = 1 + 4 + 3 = 8$$
$$f(0) = 0^2 - 4(0) + 3 = 0 - 0 + 3 = 3$$
$$f(1) = 1^2 - 4(1) + 3 = 1 - 4 + 3 = 0$$
$$f(2) = 2^2 - 4(2) + 3 = 4 - 8 + 3 = -1$$
$$f(3) = 3^2 - 4(3) + 3 = 9 - 12 + 3 = 0$$
$$f(4) = 4^2 - 4(4) + 3 = 16 - 16 + 3 = 3$$
$$f(5) = 5^2 - 4(5) + 3 = 25 - 20 + 3 = 8$$

x	y
−1	8
0	3
1	0
2	−1
3	0
4	3
5	8

Figure 5.2

Now plot the points and connect them with a smooth curve (see Fig. 5.2). ■

Notice that the graph in Figure 5.2 is a function since it passes the vertical line test discussed in Section 3.5. The *domain* of this function, the set of values that can be used for x, is the set of real numbers, $\mathbb{R}$. The *range*, the corresponding set of values of y, is the set of real numbers greater than or equal to -1.

$$\text{Domain: } \mathbb{R}$$
$$\text{Range: } \{y \mid y \geq -1\}$$

3 When graphing quadratic functions how do we decide what values to use for x? When the location of the vertex is unknown, this is a difficult question to answer. When the location of the vertex is known, it becomes more obvious which values to use.

Let us examine the parabola in Example 1 more closely, Figure 5.3.

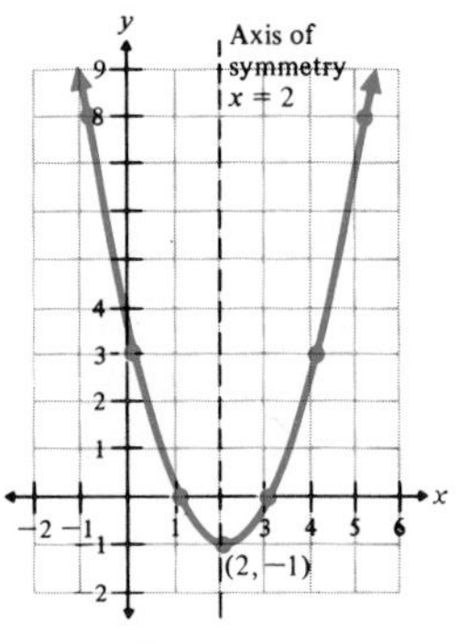

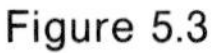
Figure 5.3

Notice the parabola is *symmetric* about a vertical line through the vertex. This means that if we folded the page along this imaginary line, called the *axis of symmetry*, the right and left sides would concide. Every parabola that opens upward or downward will have an axis of symmetry that will be a vertical line through its vertex. If we can determine the location of the axis of symmetry we can use it as a guide in selecting values for x. When quadratic functions of the form $y = ax^2 + bx + c$ are graphed, the axis of symmetry of the parabola will be $x = -b/2a$. We will derive this formula in Section 9.6.

Axis of Symmetry

Given an equation of the form $y = ax^2 + bx + c$, its graph will be a parabola with axis of symmetry

$$x = \frac{-b}{2a}$$

Note that in Figure 5.3 the axis of symmetry is $x = 2$. We can find this by the axis of symmetry formula as follows:

$$f(x) = x^2 - 4x + 3$$
$$a = 1, \quad b = -4, \quad c = 3$$
$$x = \frac{-b}{2a} = \frac{-(-4)}{2(1)} = \frac{4}{2} = 2$$

The equation of the axis of symmetry is $x = 2$. Note that the x coordinate of the vertex of the parabola is also at 2. The y coordinate of the vertex can now be found by substituting 2 for x in the function.

$$\begin{aligned} f(x) &= x^2 - 4x + 3 \\ f(2) &= 2^2 - 4(2) + 3 \\ &= 4 - 8 + 3 \\ &= -1 \end{aligned}$$

Therefore when $x = 2$, $f(x)$ or $y = -1$. The coordinates of the vertex of the parabola are $(2, -1)$.

Now we will graph another quadratic function, this time making use of the axis of symmetry in selecting values for x.

EXAMPLE 2 Graph $f(x) = -x^2 + 2x + 3$.

Solution: Since the coefficient of the squared term, -1, is less than 0, this parabola will open downward. Now find the axis of symmetry

$$x = \frac{-b}{2a} = \frac{-(2)}{2(-1)} = \frac{-2}{-2} = 1$$

The parabola will be symmetric about the line $x = 1$. Since the axis of symmetry is $x = 1$, we will choose values for x of $-2, -1, 0, 1, 2, 3, 4$. Note that we selected three values less than 1 and three values greater than 1. Now we find the corresponding values of y and draw the graph as shown in Fig. 5.4.

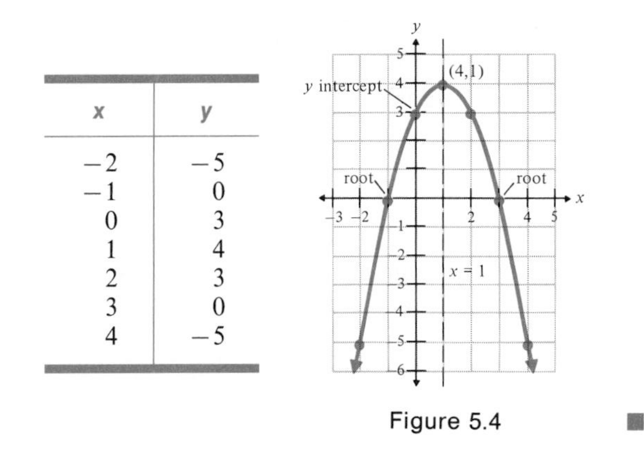

x	y
-2	-5
-1	0
0	3
1	4
2	3
3	0
4	-5

Figure 5.4 ■

The parabola in Fig. 5.4 is a smooth curve. When graphing a parabola, or any polynomial function, it is often helpful to plot each point as it is determined. If the point does not appear to be part of the curve, check your calculations. Note that in Fig. 5.4 the domain is all real numbers, $\mathbb{R}$, and the range is the set of values less than or equal to 4, $\{y|y \leq 4\}$.

In Fig. 5.4 we see that the vertex is at the point (4, 1), and the y intercept, where the graph crosses the y axis, is 3. The y intercept of any graph can be determined by substituting $x = 0$ into the equation and evaluating the equation. For the function in Example 2

$$f(x) = -x^2 + 2x + 3$$
$$f(0) = -0^2 + 2(0) + 3 = 3$$

Can you explain why this procedure always gives the y intercept?

Also indicated in Fig. 5.4 are the x intercepts of the graphs. The x intercepts, where the graph crosses the x axis, are also called the **roots** of the equation. Therefore the roots of $f(x) = -x^2 + 2x + 3$ are -1 and 3. In Chapter 9 we will discuss finding the roots of quadratic equations by using procedures other than graphing.

In this section we selected quadratic functions such that their parabolas will have vertices whose coordinates are integer values. In Section 9.6, we will discuss quadratic functions in more depth and consider parabolas whose vertices do not have x coordinates that are integer values.

4 Now we will graph a cubic, or third degree, function by plotting points. Graphing third degree or higher functions by this method may not result in a totally accurate graph. To graph third degree or higher functions accurately requires a knowledge of calculus. However this method will be sufficient for our needs.

EXAMPLE 3 Graph $f(x) = x^3 - 3x + 1$.

Solution: Select values for x and find the corresponding values of $f(x)$, or y. Plot the points and draw a smooth curve from point to point, (see Fig. 5.5).

$$f(x) = x^3 - 3x + 1$$

$$f(-3) = (-3)^3 - 3(-3) + 1 = -27 + 9 + 1 = -17$$
$$f(-2) = (-2)^3 - 3(-2) + 1 = -8 + 6 + 1 = -1$$
$$f(-1) = (-1)^3 - 3(-1) + 1 = -1 + 3 + 1 = 3$$
$$f(0) = 0^3 - 3(0) + 1 = 0 - 0 + 1 = 1$$
$$f(1) = 1^3 - 3(1) + 1 = 1 - 3 + 1 = -1$$
$$f(2) = 2^3 - 3(2) + 1 = 8 - 6 + 1 = 3$$
$$f(3) = 3^3 - 3(3) + 1 = 27 - 9 + 1 = 19$$

x	y
−3	−17
−2	−1
−1	3
0	1
1	−1
2	3
3	19

Figure 5.5

Notice that the points $(-3, -17)$ and $(3, 19)$ listed in the table were not plotted on the graph in Fig. 5.5. Their y values are too small and too large respectively. The arrows on the graph indicate the graph continues in the same direction. The graph would pass through these two points if the vertical axis was extended. Unfortunately, there is no easy method to determine the vertices of the graph of a cubic equation. The graph in Fig. 5.5 is a function whose range and domain are both all real numbers, $\mathbb{R}$.

In Example 3 we graphed $f(x) = x^3 - 3x + 1$. If we multiply each term on the right side of the function by -1 we obtain $f(x) = -x^3 + 3x - 1$. What will the graph of $f(x) = -x^3 + 3x - 1$ look like? We will graph this function in Example 4.

EXAMPLE 4 Graph $f(x) = -x^3 + 3x - 1$.

Solution:

$$f(x) = -x^3 + 3x - 1$$

$$f(-3) = -(-3)^3 + 3(-3) - 1 = 17$$
$$f(-2) = -(-2)^3 + 3(-2) - 1 = 1$$
$$f(-1) = -(-1)^3 + 3(-1) - 1 = -3$$
$$f(0) = -(0)^3 + 3(0) - 1 = -1$$
$$f(1) = -(1)^3 + 3(1) - 1 = 1$$
$$f(2) = -(1)^3 + 3(2) - 1 = -3$$
$$f(3) = -(3)^3 + 3(3) - 1 = -19$$

x	y
−3	17
−2	1
−1	−3
0	−1
1	1
2	−3
3	−19

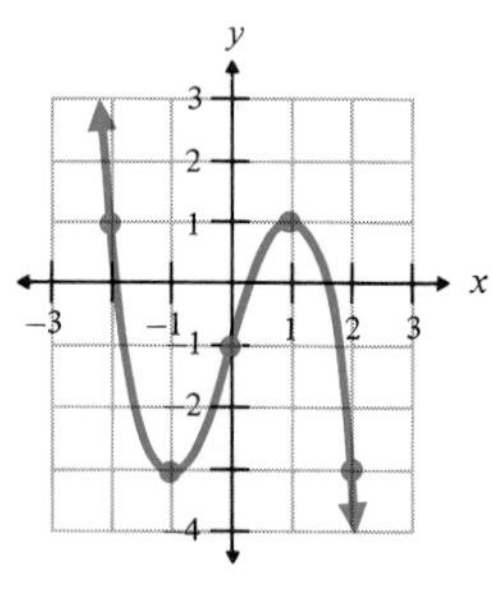

Figure 5.6

The graph is given in Fig. 5.6. ■

Notice for each point (x, y) on the graph of $f(x) = x^3 - 3x + 1$ in Fig. 5.5, the corresponding point on the graph of $f(x) = -x^3 + 3x - 1$ in Fig. 5.6 is $(x, -y)$. The graph in Fig. 5.5 is inverted to obtain the graph in Fig. 5.6.

When graphing quadratic functions we stated the sign of a, the coefficient of the squared term, determines whether the parabola opens upward or downward. When a is positive, the parabola opens upward and when a is negative, the parabola opens downward. Similarily when graphing a cubic function, the sign of the coefficient of the cubed term determines whether the graph will eventually continue to increase or decrease as x increases. If the coefficient of the cubed term is positive, as in Example 3, the graph will eventually continue to increase, or rise, as x increases. If the coefficient of the cubed term is negative, as in Example 4, the graph will eventually continue to decrease, or fall, as x increases. Can you explain why this must happen?

Exercise Set 5.8

Indicate the axis of symmetry, the coordinates of the vertex, and whether the parabola opens up or down.

1. $y = x^2 + 2x - 7$
2. $y = x^2 + 4x - 9$
3. $y = -x^2 + 4x - 6$
4. $y = 3x^2 + 6x - 9$
5. $y = -3x^2 + 6x + 8$
6. $y = x^2 + 8x - 6$
7. $y = -4x^2 - 8x - 12$
8. $y = 2x^2 + 4x + 6$
9. $y = x^2 - x + 2$
10. $y = -x^2 + x + 8$
11. $y = 4x^2 + 12x - 5$
12. $y = -2x^2 - 6x - 5$

Graph each quadratic function and give its domain and range.

13. $y = x^2 - 1$
14. $y = x^2 + 4$
15. $y = -x^2 + 3$

16. $y = -x^2 - 2$
17. $y = x^2 + 2x + 3$
18. $y = x^2 + 4x + 3$

19. $y = x^2 + 2x - 15$
20. $y = -x^2 + 10x - 21$
21. $y = -x^2 + 4x - 5$

22. $y = x^2 + 8x + 15$
23. $y = x^2 + 6x - 2$
24. $y = x^2 - 6x + 4$

25. $y = x^2 - 6x + 9$
26. $y = x^2 - 6x$
27. $f(x) = -x^2 - 4x + 4$

28. $f(x) = x^2 - 4x + 4$

29. $f(x) = 2x^2 - 4x - 6$

30. $f(x) = x^2 - 2x + 1$

31. $f(x) = -x^2 + 4x - 8$

32. $f(x) = 2x^2 - 4x$

33. $f(x) = x^2 - 2x - 15$

34. $f(x) = -2x^2 - 8x + 4$

35. $f(x) = -2x^2 + 8x - 1$

36. $f(x) = 3x^2 - 6x + 1$

37. $f(x) = 3x^2 + 6x - 9$

38. $f(x) = 4x^2 + 8x + 4$

39. $f(x) = 2x^2 - 4x + 1$

40. $f(x) = -3x^2 - 6x + 5$

Graph each cubic function.

41. $y = x^3$

42. $y = x^3 + 1$

43. $y = x^3 + x$

44. $y = x^3 + 2x - 1$

45. $y = x^3 + 2x^2 - 3x - 1$

46. $f(x) = x^3 - x^2 + 2x$

47. $f(x) = 2x^3 + x - 8$

48. $f(x) = -x^3 + 3x$

49. $f(x) = -x^3 + x - 6$

50. $f(x) = -2x^3 + 6x^2 + 2x - 6$

51. Explain how to determine if the graph of a quadratic function opens up or down.

52. What is the name given to the graph of a quadratic function?

53. Consider the function $f(x) = x^3$. What happens to y as x increases? As x decreases?

54. Consider the function $f(x) = -x^3$. What happens to y as x increases? As x decreases?

55. Consider the function $f(x) = x^4$. What happens to y as x increases from -3 to 3?

56. Consider the function $f(x) = -x^4$. What happens to y as x increases from -3 to 3.

JUST FOR FUN

1. Graph $f(x) = x^4 - 3x^2 + 6$.
2. Graph $f(x) = x^4 - 2x^2 + 3x - 4$.

Summary

Glossary

Binomial: A two-term polynomial.

Degree of a polynomial: The same as the highest-degree term in the polynomial.

Descending order of the variable: Polynomial written so that the exponents on the variable decrease as terms go from left to right.

Monomial: A one-term polynomial.

Polynomial: A finite sum of terms in which all variables have whole-number exponents and no variables appear in a denominator.

Polynomial function: A function of the form $f(x) = a_nx^n + a_{n-1}x^{n-1} + a_{n-2}x^{n-2} + \cdots + a_1x + a_0$.

Quadratic function: A function of the form $f(x) = ax^2 + bx + c$, $a \neq 0$.

Scientific notation: Writing large and small numbers as a number greater than or equal to 1 and less than 10 multiplied by some power of 10.

Synthetic division: A shortened process of dividing a polynomial by a binomial of the form $x - a$.

Trinomial: A three-term polynomial.

Important Facts

Rules for Exponents

1. $a^m \cdot a^n = a^{m+n}$ — product rule
2. $\dfrac{a^m}{a^n} = a^{m-n}, \quad a \neq 0$ — quotient rule
3. $a^{-m} = \dfrac{1}{a^m}, \quad a \neq 0$ — negative exponent rule
4. $a^0 = 1, \quad a \neq 0$ — zero exponent rule
5. $(a^m)^n = a^{mn}$
   $(ab)^m = a^m b^m$
   $\left(\dfrac{a}{b}\right)^m = \dfrac{a^m}{b^m}, \quad b \neq 0$ — power rules

FOIL Method to Multiply Two Binomials

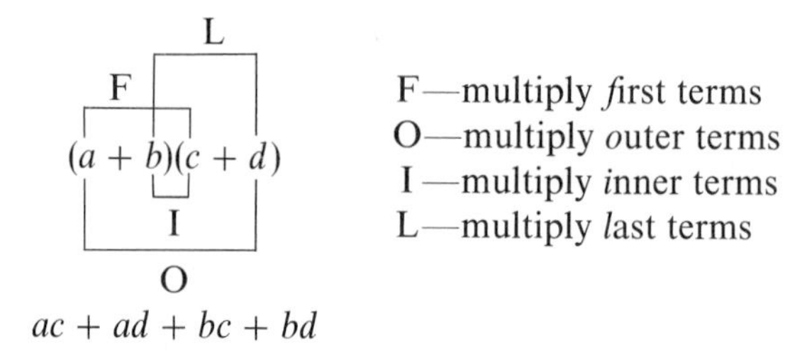

$$ac + ad + bc + bd$$

Special Product Formula

$$(a + b)^2 = a^2 + 2ab + b^2$$
$$(a - b)^2 = a^2 - 2ab + b^2$$
square of a binomial

$$(a + b)(a - b) = a^2 - b^2$$
product of sum and difference of the same two terms (or difference of two squares)

Review Exercises

[5.1–5.2] Simplify and write the answer without negative exponents.

1. $4^2 \cdot 4^1$
2. $x^3 \cdot x^5$
3. $y^5 \cdot y^2$
4. $\dfrac{3^4}{3^1}$
5. $\dfrac{x^6}{x^2}$
6. $\dfrac{y^{12}}{y^3}$
7. $\dfrac{y^5}{y^6}$
8. $\dfrac{x^4}{x^{-3}}$
9. $x^4 \cdot x^{-7}$
10. $3^{-2} \cdot 3^{-1}$
11. $2^{-3} \cdot 2^{-2}$
12. $3x^0$
13. $(3x^2)^2$
14. $(6xy^2)(-2xy^4)$
15. $(7x^2y^5)(-3xy^4)$
16. $(4x^2y^{-3})(2x^{-4}y^2)$
17. $(2x^5y^{-4})(3x^2y)(5x^{-2}y^{-3})$
18. $\dfrac{6x^{-3}y^5}{2x^2y^{-2}}$
19. $\dfrac{12x^{-3}y^{-4}}{4x^{-2}y^5}$
20. $\dfrac{(5x^3y^2)(2xy^4z)}{20x^4y^{-2}z}$
21. $\left(\dfrac{3x^4y^2}{2x^{-2}y^3}\right)\left(\dfrac{4x^2y^{-3}}{6xy^4}\right)$
22. $\left(\dfrac{5x^2y}{x}\right)^3$
23. $\left(\dfrac{x^5y}{-3y^2}\right)^2$
24. $\left(\dfrac{2xy^3}{y^4}\right)^2$
25. $\left(\dfrac{-5x^{-2}y}{z^3}\right)^3$
26. $\left(\dfrac{6xy^3}{z^2}\right)^{-2}$
27. $\left(\dfrac{9x^{-2}y}{3xy}\right)^{-3}$
28. $(-2x^{-3}y^2)^{-4}$
29. $\left(\dfrac{5x^{-2}y^3}{xy^4}\right)^3$
30. $\left(\dfrac{2x^4y^6}{z^3}\right)^2(2x^2y^4)$
31. $\left(\dfrac{16x^4y^3z^{-2}}{4x^5y^2z^3}\right)^3$
32. $\left(\dfrac{2x^4y}{z^3}\right)^2\left(\dfrac{3x^{-2}y^4}{3xy^6}\right)$

Express each number in scientific notation.

33. 0.0000742
34. 260,000
35. 183,000
36. 0.000001

Express each answer without exponents.

37. $(6.2 \times 10^3)(4 \times 10^4)$
38. $(25 \times 10^{-3})(1.2 \times 10^6)$
39. $\dfrac{18 \times 10^3}{9 \times 10^5}$
40. $\dfrac{4000000}{0.02}$
41. $\dfrac{0.00016}{4000}$
42. $(0.004)(500000)$

[5.3] Indicate if the expression is a polynomial. If the expression is a polynomial, (a) give the special name of the polynomial if it has one, (b) write the polynomial in descending order of the variable x, and (c) give the degree of the polynomial.

43. $5 - x$
44. $x^2 - 3 + 5x$
45. 6
46. $x^2 - y^2 + xy$
47. $x^5y^3 - 6xy^3 + x^4y$
48. $-3 - 9x^2y + 6xy^3 + 2x^4$
49. $3x^2 + 6x^{-1} + 4$
50. $3x^2 + 2x^{1/2} - 6$

Add or subtract as indicated.

51. $(4x + 3) + (6x - 8)$
52. $(5x^2 + 3x - 6) + (2x^2 - 7x - 9)$
53. $(3x^2 + 6x + 4) - (x^2 + 2x)$
54. $(2x^3 - 4x^2 - 3x) - (4x^2 - 3x + 9)$
55. $(3x^2y + 6xy - 5y^2) - (4y^2 + 3xy)$
56. $(-6xy + 6y^2 - 3x) - (y^2 + 3xy + 6x)$

[5.4] Multiply as indicated.

57. $4x(x^2 + 2x + 3)$
58. $-2xy^2(x^3 + x^2y^5 - 6y)$
59. $(3x + 4y)(4x - 5y)$
60. $(3xy + 1)(2x + 3y)$
61. $(x + 5)^2$
62. $(x - 3)^2$
63. $(2x + 5)^2$
64. $(3x - 2y)^2$
65. $(x + 6)(x - 6)$
66. $(x + y)(x - y)$
67. $(2x + 3)(2x - 3)$
68. $(5xy - 6)(5xy + 6)$
69. $(2x - 5y^2)(2x + 5y^2)$
70. $[(x + 3y) + 2]^2$
71. $[(x + 3y) + 2][(x + 3y) - 2]$
72. $(3x^2 + 4x - 6)(2x - 3)$
73. $(4x^3 + 6x - 5)(x + 3)$
74. $(x^2y + 6xy + y^2)(x + y)$

[5.5] Divide as indicated.

75. $\dfrac{15y^3 + 6y}{3y}$
76. $\dfrac{9xy - 6y^2 + 3y}{3y}$
77. $\dfrac{4x^3y^2 + 8x^2y^3 + 12xy^4}{8xy^3}$
78. $(x^2 + x - 12) \div (x - 3)$
79. $(6x^2 - 11x + 3) \div (3x - 1)$
80. $(4x^3 + 12x^2 + x - 12) \div (2x + 3)$
81. $(4x^4 - 7x^2 - 5x + 4) \div (2x - 1)$

[5.6] Use synthetic division to obtain the quotient.

82. $(3x^3 - 2x^2 + 10) \div (x - 3)$
83. $(2y^5 - 10y^3 + y - 1) \div (y + 1)$
84. $(x^5 - 20) \div (x - 2)$
85. $(2x^3 + x^2 + 5x - 3) \div \left(x - \dfrac{1}{2}\right)$

[5.7] Evaluate each of the following functions at the indicated values.

86. $f(x) = 3x^2 - 4x - 1$; find **(a)** $f(2)$ **(b)** $f(-5)$
87. $f(x) = x^3 - 2x + 3$; find **(a)** $f(0)$ **(b)** $f(2)$
88. $f(x) = (x - 2)^2 + 5$; find **(a)** $f(2)$ **(b)** $f(-3)$
89. $f(x) = x^2 + 2x - 1$; find **(a)** $f(a)$ **(b)** $f(a + 2)$
90. $f(x) = x^2 - x + 3$; find **(a)** $f(a)$ **(b)** $f(a + b)$

[5.8] Graph each of the following functions and give the domain and range.

91. $f(x) = x^2 - 4x + 4$

92. $y = x^2 - 1$

93. $f(x) = 2x^2 - 4x + 3$

Graph each of the following functions.

94. $f(x) = x^3$

95. $y = x^3 + 1$

96. $y = x^3 + 2x - 3$

97. The number of baskets of apples, N, that are produced by x trees in a small orchard ($x \leq 100$) is given by the function $N = f(x) = 40x - 0.2x^2$. How many baskets of apples are produced by:
(a) 20 trees?
(b) 50 trees?

98. If ball is dropped from the top of a 100-foot building, its height above the ground, h, at any time, t, can be found by the function $h = f(t) = -16t^2 + 100$, $-t \leq 2.5$. Find the height of the ball at:
(a) 1 second
(b) 2 seconds

Practice Test

Simplify and write the answer without negative exponents.

1. $\left(\frac{3x^2y^3}{9x^5y^{-2}}\right)^2$

2. $\left(\frac{-3x^3y^{-2}}{y^5}\right)^2$

3. (a) Give the specific name of the following polynomial, (b) write the polynomial in descending powers of the variable x, and (c) state the degree of the polynomial.

$$-4x^2y^3 + 2x - 6x^4$$

Perform the operation indicated.

4. $(4x^3 - 3x - 4) - (2x^2 - 5x - 12)$

5. $(12x^6 - 6xy^2 + 15) \div 3x$

6. $(3x + y)(y - 2x)$

7. $(2x^2 + 3xy - 6y^2)(2x + y)$

8. $(2x^2 - 7x + 10) \div (2x + 3)$

9. $(6x^2y + 3y^2 + 5x) - (4x^2y + 2x - 4y^2)$

10. $(2x^3 - x^2 + 5x - 7) \div (x - 3)$

11. $3x^2y^4(-2x^5y^2 + 6x^2y^3 - 3x)$

12. $(2x + 3y)^2$

13. Use synthetic division to obtain the quotient.

$$(3x^4 - 12x^3 - 60x + 4) \div (x - 5)$$

14. $f(x) = x^3 - 2x + 3$; find $f(-3)$.

15. Graph $f(x) = x^2 - 4x + 2$ and give its domain and range.

16. A town presently has a population of 6000. The chamber of commerce estimates that the town's population, P, in years can be approximated by the function

$$P = f(n) = 12n^2 + 10n + 6000, \qquad n \leq 10$$

Find the town's expected population in 8 years.

6 Factoring

See Section 6.5, Exercise 65.

6.1 Factoring a Monomial from a Polynomial and Factoring by Grouping

1. *Find the greatest common factor.*
2. *Factor a monomial from a polynomial.*
3. *Factor by grouping.*

Factoring is the opposite of the multiplication process. For example, in Chapter 5 we learned that

$$3x^2(6x + 3xy + 5x^3) = 18x^3 + 9x^3y + 15x^5$$

and

$$(6x + 3y)(2x - 5y) = 12x^2 - 24xy - 15y^2$$

In this chapter we learn how to determine the factors of a given expression. For example, we may show that

$$18x^3 + 9x^3y + 15x^5 = 3x^2(6x + 3xy + 5x^3)$$

and

$$12x^2 - 24xy - 15y^2 = (6x + 3y)(2x - 5y)$$

1 In earlier chapters we stated that if $a \cdot b = c$, then a and b are said to be **factors** of c. A given expression may have many factors. What are the integer factors of the number 12?

$$\begin{aligned} 1 \cdot 12 &= 12 \qquad & (-1)(-12) &= 12 \\ 2 \cdot 6 &= 12 & (-2)(-6) &= 12 \\ 3 \cdot 4 &= 12 & (-3)(-4) &= 12 \end{aligned}$$

Note that the factors of 12 are ± 1 (read "plus or minus 1"), ± 2, ± 3, ± 4, ± 6, ± 12. Generally when asked to list the factors of a positive number, we list only the positive factors, although it should be understood the negatives of these factors are also factors. We would say the factors of 12 are 1, 2, 3, 4, 6, and 12.

What are the factors of $6x^3$?

Factors	*Factors*
$1 \cdot 6x^3 = 6x^3$	$x \cdot 6x^2 = 6x^3$
$2 \cdot 3x^3 = 6x^3$	$2x \cdot 3x^2 = 6x^3$
$3 \cdot 2x^3 = 6x^3$	$3x \cdot 2x^2 = 6x^3$
$6 \cdot x^3 = 6x^3$	$6x \cdot x^2 = 6x^3$

Some factors of $6x^3$ are 1, 2, 3, 6, x, $2x$, $3x$, $6x$, x^2, $2x^2$, $3x^2$, $6x^2$, x^3, $2x^3$, $3x^3$, and $6x^3$. The opposite or negative of each of these factors is also a factor.

To factor a monomial from a polynomial, we must determine the greatest common factor (GCF) of each term in the polynomial. After the GCF is determined, we use the distributive property (in reverse) to factor the expression. The **greatest common factor** of two or more expressions is the largest factor that divides (without remainder) each expression. Consider the three numbers 12, 18, and 24. The GCF of these three numbers is 6 since 6 is the largest number that divides (is a factor of) each of these numbers. If you forgot how to find the GCF of a set of numbers, review an arithmetic or elementary algebra text before going any further.

The GCF of a collection of terms containing variables is easily found. Consider the terms x^3, x^4, x^5, and x^6. The GCF of these terms is x^3, since x^3 is the highest power of x that divides all four terms. Note that the GCF of a collection of terms will be the *lowest power of the common variable.*

EXAMPLE 1 Find the GCF of the following terms.

$$y^{12}, \quad y^4, \quad y^9, \quad y^7$$

Solution: Note that y^4 is the lowest power of y that appears in any of the four given terms. The GCF is therefore y^4. ■

EXAMPLE 2 Find the GCF of the following terms.

$$x^3y^2, \quad xy^4, \quad x^4y^5$$

Solution: The lowest power of x that appears in any of three terms is x (or x^1). The lowest power of y that appears in any of the three terms is y^2. Thus the GCF of the three terms is xy^2. ■

EXAMPLE 3 Find the GCF of the following terms.

$$6x^2y^3, \quad 9x^3y^4, \quad 24x^4$$

Solution: The GCF is $3x^2$. Since y does not appear in $24x^4$, it is not part of the GCF. ■

EXAMPLE 4 Find the GCF of the following terms.

$$6(x-3)^2, \quad 5(x-3), \quad 18(x-3)^4$$

Solution: The three numerical values have no common factor other than 1. The lowest power of $(x-3)$ in any of the three terms is $(x-3)$. Thus the GCF of the three terms is $(x-3)$. ■

Factoring a Monomial from a Polynomial

2

To Factor a Monomial from a Polynomial

1. Determine the greatest common factor of all terms in the polynomial.
2. Write each term as the product of the GCF and its other factor.
3. Use the distributive property to *factor out* the GCF.

EXAMPLE 5 Factor $2x + 6$.

Solution: The GCF is 2.

$$2x + 6 = 2 \cdot x + 2 \cdot 3 \qquad \text{write each term as a product of the GCF and some other factor}$$

$$= 2(x+3) \qquad \text{distributive property}$$

Note that the GCF, 2, was *factored out* from each term in the polynomial. The $2x + 6$ in factored form is $2(x + 3)$. The factors of $2x + 6$ are 2 and $(x + 3)$. ■

To check the factoring process, multiply the factors using the distributive property. The product should be the expression with which you began. For instance in Example 5,

Check: $2(x + 3) = 2 \cdot x + 2 \cdot 3 = 2x + 6$

EXAMPLE 6 Factor $15x^4 - 5x^3 + 20x^2$.

Solution: The GCF is $5x^2$.

$$\begin{aligned} 15x^4 - 5x^3 + 20x^2 &= 5x^2 \cdot 3x^2 - 5x^2 \cdot x + 5x^2 \cdot 4 \\ &= 5x^2(3x^2 - x + 4) \end{aligned}$$ ■

EXAMPLE 7 Factor $20x^2y^3 + 6xy^4 - 12x^3y^5$.

Solution: The GCF is $2xy^3$.

$$\begin{aligned} 20x^2y^3 + 6xy^4 - 12x^3y^5 &= 2xy^3 \cdot 10x + 2xy^3 \cdot 3y - 2xy^3 \cdot 6x^2y^2 \\ &= 2xy^3(10x + 3y - 6x^2y^2) \end{aligned}$$

Check: $2xy^3(10x + 3y - 6x^2y^2) = 20x^2y^3 + 6xy^4 - 12x^3y^5$ ■

EXAMPLE 8 Factor $3x(5x - 2) + 4(5x - 2)$.

Solution: The GCF is $(5x - 2)$. Factoring out the GCF gives

$$3x(5x - 2) + 4(5x - 2) = (3x + 4)(5x - 2)$$

We could have also used the commutative property of multiplication to rewrite the expression as

$$(5x - 2)3x + (5x - 2)4 = (5x - 2)(3x + 4)$$

The answers $(3x + 4)(5x - 2)$ and $(5x - 2)(3x + 4)$ are equivalent answers and either may be given. ■

EXAMPLE 9 Factor $9(2x - 5) + 6(2x - 5)^2$.

Solution: The GCF is $3(2x - 5)$.

$$\begin{aligned} 9(2x - 5) + 6(2x - 5)^2 &= 3(2x - 5) \cdot 3 + 3(2x - 5) \cdot 2(2x - 5) \\ &= 3(2x - 5)[3 + 2(2x - 5)] \end{aligned}$$

Now combine like terms:

$$\begin{aligned} &= 3(2x - 5)[3 + 4x - 10] \\ &= 3(2x - 5)(4x - 7) \end{aligned}$$ ■

Factoring by Grouping

3 When a polynomial contains 4 terms it may be possible to factor the polynomial by grouping. To factor by grouping, remove common factors from groups of terms. This procedure is illustrated in the following example. Factoring by grouping is important because we use it to factor trinomials in Section 6.2.

EXAMPLE 10 Factor $ax + ay + bx + by$.

Solution: There is no factor (other than 1) common to all four terms. However, a is common to the first two terms and b is common to the last two terms. Factor a from the first two terms and b from the last two terms.

$$ax + ay + bx + by = a(x + y) + b(x + y)$$

Now $(x + y)$ is common to both terms. Factor out $(x + y)$.

$$a(x + y) + b(x + y) = (a + b)(x + y)$$

Thus $ax + ay + bx + by = (a + b)(x + y)$. ■

To Factor by Grouping

1. Arrange the four terms into two groups of two terms each. Each group of two terms must have a GCF.
2. Factor the GCF from each group of two terms.
3. If the two terms formed in Step 2 have a GCF, factor out that GCF.

EXAMPLE 11 Factor $6x^2 + 9x + 8x + 12$ by grouping.

Solution: Factor a $3x$ from the first two terms and a 4 from the last two terms. Then factor the GCF, $2x + 3$, from the resulting two terms.

$$\begin{aligned} 6x^2 + 9x + 8x + 12 &= 3x(2x + 3) + 4(2x + 3) \\ &= (3x + 4)(2x + 3) \end{aligned}$$ ■

Factoring by grouping problems may be checked by multiplying the factors using the FOIL method. Check the answer to Example 11 now.

EXAMPLE 12 Factor $ax - x + a - 1$ by grouping.

Solution:

$$\begin{aligned} ax - x + a - 1 &= x(a - 1) + 1(a - 1) \\ &= (x + 1)(a - 1) \end{aligned}$$

Note that $a - 1$ was expressed as $1(a - 1)$. ■

EXAMPLE 13 Factor $4x^2 - 2x - 2x + 1$ by grouping.

Solution: When $2x$ is factored from the first two terms, we get

$$4x^2 - 2x - 2x + 1 = 2x(2x - 1) - 2x + 1$$

Now factor -1 from the last two terms to get a common factor of $2x - 1$.

$$\begin{aligned} &= 2x(2x - 1) - 1(2x - 1) \\ &= (2x - 1)(2x - 1) \quad \text{or} \quad (2x - 1)^2 \end{aligned}$$ ■

EXAMPLE 14 Factor $2x^2 + 4xy + 3xy + 6y^2$.

Solution: This problem contains two variables, x and y. The procedure to factor is basically the same. We will factor out a $2x$ from the first two terms and a $3y$ from the last two terms.

$$\begin{aligned} 2x^2 + 4xy + 3xy + 6y^2 &= 2x(x + 2y) + 3y(x + 2y) \\ &= (2x + 3y)(x + 2y) \quad \blacksquare \end{aligned}$$

EXAMPLE 15 Factor $6r^4 - 9r^3s + 8rs - 12s^2$.

Solution: Factor a $3r^3$ from the first two terms and a $4s$ from the last two terms.

$$\begin{aligned} 6r^4 - 9r^3s + 8rs - 12s^2 &= 3r^3(2r - 3s) + 4s(2r - 3s) \\ &= (3r^3 + 4s)(2r - 3s) \quad \blacksquare \end{aligned}$$

Exercise Set 6.1

Factor out the greatest common factor. If an expression cannot be factored, so state.

1. $8n + 8$
2. $12x + 15$
3. $13x + 5$
4. $6x^2 + 3x - 9$
5. $16x^2 - 12x - 6$
6. $27y^3 - 9y^2 + 18y$
7. $20p^3 - 18p^2 + 12p$
8. $6x^3 - 8x^2 - x$
9. $7x^5 - 9x^4 + 3x^3$
10. $45y^{12} + 30y^{10}$
11. $24y^{15} - 9y^3 + 3y$
12. $38x^4 - 16x^5 - 9x^3$
13. $x + 3xy^2$
14. $2x^2y - 6x + 6x^3$
15. $6x + 5y + 5xy$
16. $3x^2y + 6x^2y^2 + 3xy$
17. $16xy^2z + 4x^3y - 8$
18. $80x^5y^3z^4 - 36x^2yz^3$
19. $40x^2y^2 + 16xy^4 + 64xy^3$
20. $40x^2y^4z + 8x^6y^2z^2 - 4x^3y$
21. $36xy^2z^3 + 36x^3y^2z + 9x^2yz$
22. $19x^4y^{12}z^{13} - 8x^5y^3z^9$
23. $14y^3z^5 - 28y^3z^6 - 9xy^2z^2$
24. $7x^4y^9 - 21x^3y^7z^5 - 35y^8z^9$
25. $24x^6 + 8x^4 - 4x^3y$
26. $44x^5y + 11x^3y + 22x^2$
27. $48x^2y + 16xy^2 + 33xy$
28. $52x^2y^2 + 16xy^3 + 26z$
29. $x(x + 2) + 3(x + 2)$
30. $5x(2x - 5) + 3(2x - 5)$
31. $7x(4x - 3) - 1(4x - 3)$
32. $3x(7x + 1) - 2(7x + 1)$
33. $4x(2x + 1)^2 + 1(2x + 1)$
34. $3x(4x - 5)^3 + 1(4x - 5)^2$
35. $4x(2x + 1) + 2x + 1$
36. $3x(4x - 5) + 4x - 5$
37. $5x(x + 3)^2 - 3(x + 3)$
38. $3x(2x + 5) - 6(2x + 5)^2$
39. $6x^5(2x + 7) + 4x^3(2x + 7) - 2x^2(2x + 7)$
40. $3x^2(b - 4) - 2x(b - 4) + 5(b - 4)$
41. $6x^3(2x + 5) - 2x^2(2x + 5) - (2x + 5)$
42. $5a(3x - 2)^5 + 4(3x - 2)^4$
43. $4p(2r - 3)^7 - 3(2r - 3)^6$
44. $4a^2(5a - 3) + 2a(5a - 3) - 3(5a - 3)$

Factor by grouping.

45. $x^2 + 3x - 5x - 15$
46. $x^2 + 3x - 2x - 6$
47. $4x^2 + 6x - 6x - 9$
48. $5x^2 - 10x + 3x - 6$
49. $3x^2 + 9x + x + 3$
50. $x^2 + 4x + x + 4$
51. $4x^2 - 2x - 2x + 1$
52. $2x^2 + 6x - x - 3$
53. $8x^2 - 20x - 4x + 10$
54. $8x^2 - 4x - 20x + 10$
55. $ax + ay + bx + by$
56. $2b + 2c + ab + ac$
57. $3ac + 3ad + 2bc + 2bd$
58. $35x^2 - 40xy + 21xy - 24y^2$
59. $15a^2 - 18ab - 20ab + 24b^2$

60. $15b^2 - 20bc - 18bc + 24c^2$
61. $x^2 + 2xy - 3xy - 6y^2$
62. $x^2 - 3xy + 2xy - 6y^2$
63. $6x^2 - 9xy + 2xy - 3y^2$
64. $3x^2 - 18xy + 4xy - 24y^2$
65. $10x^2 - 12xy - 25xy + 30y^2$
66. $12x^2 - 9xy + 4xy - 3y^2$
67. $x^3 + xy - x^2y - y^2$
68. $18a^2 + 3ax^2 - 6ax - x^3$
69. $2x^3y^3 + 6xyz^2 - 3x^2y^4 - 9y^2z^2$
70. $2a^4b - 2ac^2 - 3a^3bc + 3c^3$.
71. $3x^5 - 15x^3 + 2x^3 - 10x$
72. $8r^2 + 6rs - 12rs - 9s^2$
73. $3p^3 + 3pq^2 + 2p^2q + 2q^3$
74. $16r^3 - 4r^2s^2 - 4rs + s^3$

JUST FOR FUN

1. Factor $4x^2(x-3)^3 - 6x(x-3)^2 + 4(x-3)$
2. Factor
  (a) $(x+1)^2 + (x+1)$
  (b) $(x+1)^3 + (x+1)^2$
  (c) $(x+1)^n + (x+1)^{n-1}$
  (d) $(x+1)^{n+1} + (x+1)^n$
3. Factor $4x(x+5)^{-2} + 2x(x+5)^{-1}$.
4. Factor $x(2x-3)^{-1/2} + 6(2x-3)^{1/2}$.

6.2 Factoring Trinomials

1. *Factor trinomials of the form $x^2 + bx + c$.*
2. *Factor trinomials of the form $ax^2 + bx + c$, $a \neq 1$ using grouping.*
3. *Factor trinomials of the form $ax^2 + bx + c$, $a \neq 1$, using trial and error.*
4. *Factor using substitution.*

1 In this section we learn how to factor trinomials of the form $ax^2 + bx + c$, $a \neq 0$.

Trinomials	*Coefficients*
$3x^2 + 2x - 5$	$a = 3, \quad b = 2, \quad c = -5$
$-\frac{1}{2}x^2 - 4x + 3$	$a = -\frac{1}{2}, \quad b = -4, \quad c = 3$

To Factor Trinomials of the Form $x^2 + bx + c$ (note $a = 1$)

1. Find two numbers (or factors) whose product is c, and whose sum is b.
2. The factors of the trinomial will be of the form

$$(x \quad)(x \quad)$$

↑ one factor determined in Step 1 ↑ other factor determined in Step 1

This procedure is illustrated in the following examples.

EXAMPLE 1 Factor $x^2 - x - 12$.

Solution: $a = 1, b = -1, c = -12$. We must determine two numbers whose product is c, -12, and whose sum is b, -1.

Factors of -12	*Sum of factors*
$(1)(-12)$	$1 + (-12) = -11$
$(2)(-6)$	$2 + (-6) = -4$
$(3)(-4)$	$3 + (-4) = -1$
$(4)(-3)$	$4 + (-3) = 1$
$(6)(-2)$	$6 + (-2) = 4$
$(12)(-1)$	$12 + (-1) = 11$

The numbers we are seeking are 3 and -4.

$$x^2 - x - 12 = (x \underbrace{+ 3}_{\text{one factor of } -12})(x \underbrace{- 4}_{\text{other factor of } -12})$$

■

Notice in Example 1 we listed all the factors of -12. After the two factors are found whose product is c and whose sum is b, there is no need to go further in listing the factors. The factors were listed here to illustrate the fact that, for example, $(2)(-6)$ is a different set of factors than $(-2)(6)$. Note that as the positive factor increases the sum of the factors increases.

EXAMPLE 2 Factor $x^2 - 5x - 6$.

Solution: We must find two numbers whose product is -6 and whose sum is -5. The numbers are 1 and -6. Note: $(1)(-6) = -6$ and $1 + (-6) = -5$.

$$x^2 - 5x - 6 = (x + 1)(x - 6)$$

Since the factors may be placed in any order $(x - 6)(x + 1)$ is also an acceptable answer. ■

HELPFUL HINT

> Trinomial factoring problems can be checked using the FOIL method. For example, to check Example 2,
>
> $$(x + 1)(x - 6) = x^2 - 6x + x - 6 = x^2 - 5x - 6$$
>
> Notice that you obtain the trinomial you started with.

The procedures used to factor trinomials of the form $x^2 + bx + c$ can be used on other types of trinomials, as illustrated in the following example.

EXAMPLE 3 Factor $x^2 + 2xy - 15y^2$.

Solution: We must find two numbers whose product is -15 and whose sum is 2. The two numbers are 5 and -3. Note $(5)(-3) = -15$ and $5 + (-3) = 2$. Since the last term of the trinomial contains a y^2, the second term of each factor must contain a y.

$$x^2 + 2xy - 15y^2 = (x + 5y)(x - 3y)$$

Check: $(x + 5y)(x - 3y) = x^2 - 3xy + 5xy - 15y^2$
$= x^2 + 2xy - 15y^2$ ■

If each term of a trinomial has a common factor, use the distributive property to remove the common factor before following the procedure outlined earlier.

EXAMPLE 4 Factor $3x^2 - 6x - 72$.

Solution: The factor 3 is common to all three terms of the trinomial. Factoring out the 3 gives

$$3x^2 - 6x - 72 = 3(x^2 - 2x - 24)$$

The 3 that was factored out is a part of the answer but plays no further part in the factoring process. Now continue to factor $x^2 - 2x - 24$ in the usual manner. We must find two numbers whose product is -24 and whose sum is -2. The numbers are -6 and 4.

$$3(x^2 - 2x - 24)$$
$$= 3(x - 6)(x + 4)$$

Therefore $3x^2 - 6x - 72 = 3(x - 6)(x + 4)$. ■

2 Now we will look at some examples of factoring trinomials of the form

$$ax^2 + bx + c = 0,$$

where $a \neq 1$. Two methods of solving this type of trinomial will be illustrated. The first method makes use of factoring by grouping a procedure that was presented in Section 5.1. The second method, trial and error, involves trying various combinations until the correct combination is found.

Method 1 (Using Grouping)

To Factor Trinomials of the Form $ax^2 + bx + c$, $a \neq 1$, Using Grouping

1. Find two numbers whose product is $a \cdot c$ and whose sum is b.
2. Rewrite the bx term using the numbers found in Step 1.
3. Factor by grouping.

EXAMPLE 5 Factor $2x^2 - 5x - 12$.

Solution: $a = 2$, $b = -5$, and $c = -12$. We must find two numbers whose product is $a \cdot c$, or $2(-12) = -24$, and whose sum is b, -5. The two numbers are -8 and 3. Note: $(-8)(3) = -24$ and $-8 + 3 = -5$. Now rewrite the bx term, $-5x$, using the -8 and 3.

$$2x^2 - 5x - 12 = 2x^2 \overbrace{-\,8x + 3x}^{-5x} - 12$$

Now factor by grouping as explained in Section 5.1.

$$= 2x(x - 4) + 3(x - 4)$$
$$= (2x + 3)(x - 4) \quad ■$$

Note that in Example 5 we wrote $-5x$ as $-8x + 3x$. The same answer would be obtained if we wrote $-5x$ as $3x - 8x$. Therefore, it makes no difference which factor is listed first when factoring by grouping.

EXAMPLE 6 Factor $12x^2 - 19x + 5$.

Solution: We must find two numbers whose product is $(12)(5) = 60$ and whose sum is -19. Since the product of the numbers is positive and their sum is negative, the two numbers must both be negative. Why?

The two numbers are -15 and -4. Note: $(-15)(-4) = 60$ and $-15 + (-4) = -19$.

$$12x^2 - 19x + 5 = 12x^2 \overbrace{-\,15x - 4x}^{-19x} + 5$$
$$= 3x(4x - 5) - 1(4x - 5)$$
$$= (3x - 1)(4x - 5) \quad ■$$

Try Example 6 again, this time writing $-19x$ as $-4x - 15x$. If you do it correctly, you should get the same answer.

Method 2 (Trial and Error)

3 Now we will look at the trial and error method of factoring trinomials. As an aid in our explanation we will multiply $(2x + 3)(x + 1)$ using the FOIL method.

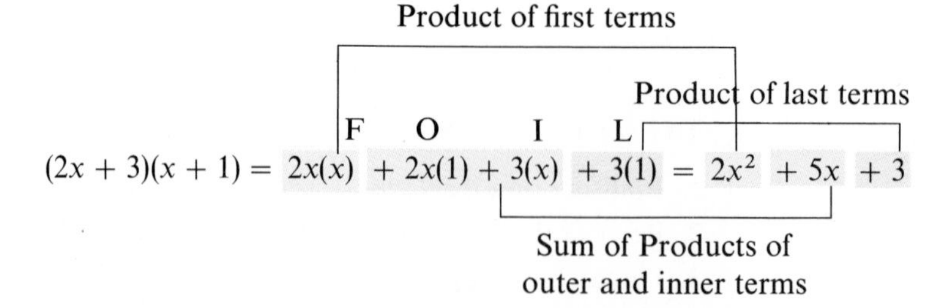

When given the trinomial $2x^2 + 5x + 3$ to factor, you should realize that the product of the first terms of the factors must be $2x^2$, the product of the last terms must be 3, and the sum of the products of the outer and inner terms must be $5x$.

To factor $2x^2 + 5x + 3$ we begin as shown here.

$$2x^2 + 5x + 3 = (2x \quad)(x \quad) \qquad \text{product of first terms is } 2x^2$$

Now fill in the second terms using positive integers whose product is 3. Only positive integers will be considered since the product of the last terms is positive, and the sum of the products of the outer and inner terms is also positive. The two possibilities are as follows:

$$\left.\begin{array}{l}(2x + 1)(x + 3)\\(2x + 3)(x + 1)\end{array}\right\} \text{Product of last terms is 3}$$

To determine which is the correct factoring process find the sum of the products of the outer terms and inner terms. If either one has a sum of $5x$, the middle term of the trinomial, then that factoring process is the correct one.

$$(2x + 1)(x + 3) = 2x^2 + 6x + x + 3 = 2x^2 + 7x + 3 \qquad \text{wrong middle term}$$

$$(2x + 3)(x + 1) = 2x^2 + 2x + 3x + 3 = 2x^2 + 5x + 3 \qquad \text{correct middle term}$$

Therefore, the factors of $2x^2 + 5x + 3$ are $2x + 3$ and $x + 1$. Thus

$$2x^2 + 5x + 3 = (2x + 3)(x + 1)$$

Note that if we had begun our factoring process by writing

$$2x^2 + 5x + 3 = (x \quad)(2x \quad)$$

we would have obtained the correct answer if we had continued with the procedure.

To Factor Trinomials of the Form $ax^2 + bx + c$, $a \neq 1$, Using Trial and Error

1. Write all pairs of factors of the coefficient of the squared term, a.
2. Write all pairs of factors of the constant, c.
3. Try various combinations of these factors until the correct middle term, bx, is found.

EXAMPLE 7 Factor $3x^2 - 13x + 10$.

Solution: The only factors of 3 are 1 and 3. Therefore we write

$$3x^2 - 13x + 10 = (3x \quad)(x \quad)$$

The number 10 has both positive and negative factors. However, since the product of the last terms must be positive $(+10)$, and the sum of the products of the outer and inner terms must be negative (-13), the two factors of 10 must both be negative. Why? The negative factors of 10 are $(-1)(-10)$ and $(-2)(-5)$. Below is a listing of the possible factors.

Possible factors	*Sum of Products of outer and inner terms*
$(3x - 1)(x - 10)$	$-31x$
$(3x - 10)(x - 1)$	$-13x$ ← correct middle term
$(3x - 2)(x - 5)$	$-17x$
$(3x - 5)(x - 2)$	$-11x$

Thus $3x^2 - 13x + 10 = (3x - 10)(x - 1)$. ■

HELPFUL HINT

When factoring a trinomial of the form $ax^2 + bx + c$, the sign of the constant, c, is very helpful in finding the solution. If $a > 0$, then:

1. When the constant, c, is positive, and the numerical coefficient of the x term, b, is positive, both numerical factors will be positive.

Example: $x^2 + 7x + 12 = (x + 3)(x + 4)$ — positive positive positive positive

2. When c is positive and b is negative, both numerical factors will be negative.

Example: $x^2 - 5x + 6 = (x - 2)(x - 3)$ — negative positive negative negative

3. When c is negative, one of the numerical factors will be positive and the other will be negative.

Example: $x^2 + x - 6 = (x + 3)(x - 2)$ — positive negative positive negative

EXAMPLE 8 Factor $6x^2 - 11x - 10$.

Solution: The factors of 6 will be either $6 \cdot 1$ or $2 \cdot 3$. Therefore the factors may be of the form $(6x \quad)(x \quad)$ or $(2x \quad)(3x \quad)$. When there is more than one set of factors for the first term, we generally try the medium-sized factors first. If this does not work, try other factors.

Thus we write

$$6x^2 - 11x - 10 = (2x \quad)(3x \quad)$$

The factors of -10 are $(-1)(10)$, $(1)(-10)$, $(-2)(5)$ and $(2)(-5)$. Since there are eight different factors of -10, there will be eight different pairs of possible factors to try. Can you list them? The correct factoring is

$$6x^2 - 11x - 10 = (2x - 5)(3x + 2) \quad ■$$

Note that in Example 8 we were fortunate to find that the factors were of the form $(2x \quad)(3x \quad)$. If we had not been able to find the correct factors using these, we would have tried $(6x \quad)(x \quad)$.

EXAMPLE 9 Factor $4x^2 + 10x - 6$.

Solution: Whenever the coefficient of the squared term is not 1, your first step should be to determine if there is a common factor to all of the terms. If so, factor it out. 2 is common to each term in this trinomial, so factor out a 2.

$$4x^2 + 10x - 6 = 2(2x^2 + 5x - 3)$$

Now continue to factor $2x^2 + 5x - 6$ using either Method 1 or Method 2.

$$4x^2 + 10x - 12 = 2(2x - 1)(x + 3) \quad ■$$

It is important for you to realize that not every trinomial can be factored by the methods presented in this section. Consider the following example.

EXAMPLE 10 Factor $2x^2 + 6x + 5$.

Solution: If you try to factor this using either Method 1 or Method 2 you will see it cannot be factored. Polynomials of this type are called **prime polynomials.** ■

Factoring Using Substitution

4 Sometimes a more complicated trinomial can be factored by substituting one variable for another. The next two examples illustrate **factoring using substitution.**

EXAMPLE 11 Factor $y^4 - y^2 - 6$.

Solution: If we can get this expression in the form $ax^2 + bx + c$, it will be easier to factor. Note that $(y^2)^2 = y^4$. If we substitute x for y^2 the trinomial becomes

$$y^4 - y^2 - 6 = (y^2)^2 - y^2 - 6$$
$$= x^2 - x - 6$$

Now proceed to factor $x^2 - x - 6$.

$$= (x + 2)(x - 3)$$

Finally, substitute y^2 in place of x to obtain

$$= (y^2 + 2)(y^2 - 3)$$

Thus $y^4 - y^2 - 6 = (y^2 + 2)(y^2 - 3)$. Note that x was substituted for y^2, and then y^2 was substituted back for x. ■

EXAMPLE 12 Factor $(x + 5)^2 + 3(x + 5) - 4$.

Solution: We will use basically the same procedure to factor this example as we used in Example 11. By substituting $a = x + 5$ in the equation, we obtain

$$(x + 5)^2 + 3(x + 5) - 4 = a^2 + 3a - 4$$

Now factor $a^2 + 3a - 4$.

$$= (a + 4)(a - 1)$$

Finally, replace a with $x + 5$ to obtain

$$= [(x + 5) + 4][(x + 5) - 1]$$
$$= (x + 9)(x + 4)$$

Note that a was substituted for $x + 5$, and then $x + 5$ was substituted back for a. ■

In Example 11 we used x in our substitution, whereas in Example 12 we used a. The letter selected is immaterial to the final answer.

Exercise Set 6.2

Factor each trinomial completely. If the trinomial cannot be factored, so state.

1. $x^2 + 7x + 6$
2. $x^2 - 6x + 10$
3. $p^2 - 3p - 10$
4. $y^2 - 12y + 11$
5. $x^2 - x - 6$
6. $w^2 - 7w + 9$
7. $x^2 + 16x + 64$
8. $x^2 - 16x + 64$
9. $x^2 - 34x + 64$
10. $x^2 - 11x - 30$
11. $a^2 - 18a + 45$
12. $x^2 - 11x + 10$
13. $y^2 - 9y + 15$
14. $p^2 - 17p - 60$
15. $x^2 - 4xy + 3y^2$
16. $x^2 - 6xy + 8y^2$
17. $a^2 - 11ab + 18b^2$
18. $z^2 - 6yz + 10y^2$
19. $x^2 - 12xy - 45y^2$
20. $x^2 + 11xy + 24y^2$
21. $5x^2 + 20x + 15$
22. $2x^2 - 14x + 12$
23. $4x^2 + 12x - 16$
24. $3y^2 - 33y + 54$
25. $x^3 - 3x^2 - 18x$
26. $x^3 + 11x^2 - 42x$
27. $x^3 - 5x^2 - 24x$
28. $x^3 - 10x^2 + 24x$
29. $2x^2 + 5x + 3$
30. $5p^2 - 8p + 3$
31. $4w^2 + 13w + 3$
32. $3x^2 + 14x - 5$
33. $5x^2 + 2x + 3$
34. $5y^2 - 16y + 3$
35. $3x^2 - 10x + 7$
36. $3x^2 - 11x - 6$
37. $2x^2 + 11x + 15$
38. $3w^2 - 2w - 8$
39. $5y^2 - 16y + 3$
40. $7x^2 + 43x + 6$
41. $3y^2 - 2y - 5$
42. $3x^2 - 22xy + 7y^2$
43. $4x^2 + 4xy - 3y^2$
44. $6x^3 + 5x^2 - 4x$
45. $8x^2 + 2x - 20$
46. $9x^2 - 15x - 36$
47. $12x^3 - 12x^2 - 45x$
48. $8x^2 - 8xy - 6y^2$
49. $12a^2 - 34ab + 24b^2$
50. $18w^2 + 18wz - 8z^2$
51. $x^3y - 3x^2y - 18xy$
52. $8x^3y + 24x^2y - 32xy$
53. $a^3b - a^2b - 12ab$
54. $a^3b + 2a^2b - 35ab$
55. $5x^2 + 25xy + 20y^2$
56. $3b^4c - 18b^3c^2 + 27b^2c^3$
57. $6p^3q^2 - 24p^2q^3 - 30pq^4$
58. $5a^3b^2 - 8a^2b^3 + 3ab^4$

Factor each trinomial completely.

59. $x^4 + x^2 - 6$
60. $x^4 - 3x^2 - 10$
61. $x^4 + 5x^2 + 6$
62. $x^4 - 2x^2 - 15$
63. $y^4 - 7y^2 - 30$
64. $3z^4 - 14z^2 - 5$
65. $6a^4 + 5a^2 - 25$
66. $(2x + 1)^2 + 2(2x + 1) - 15$
67. $4(x + 1)^2 + 8(x + 1) + 3$
68. $(2x + 3)^2 - (2x + 3) - 6$
69. $6(r + 3)^2 + 13(r + 3) + 5$
70. $6(a + 2)^2 - 7(a + 2) - 5$
71. $6(p - 5)^2 + 11(p - 5) + 3$
72. $2x^2y^2 + 3xy - 9$
73. $a^2b^2 - 8ab + 15$
74. $x^2y^2 + 10xy + 24$
75. $3x^2y^2 - 2xy - 5$
76. $3p^2q^2 + 11pq + 6$
77. $x^2(x + 3) + 3x(x + 3) + 2(x + 3)$
78. $x^2(x - 1) - x(x - 1) - 30(x - 1)$
79. $2a^2(5 - a) - 7a(5 - a) + 5(5 - a)$
80. $2y^2(y + 2) + 13y(y + 2) + 15(y + 2)$
81. $2x^2(x - 3) + 7x(x - 3) + 6(x - 3)$

82. When factoring any trinomial what should the first step always be?

83. If the factors of a polynomial are $(2x + 3y)(x - 4y)$, find the polynomial.

84. If we know one factor of the polynomial $x^2 + 3x - 18$ is $x - 3$, how can we find the other factor? Find the other factor.

85. If we know one factor of the polynomial $x^2 - xy - 6y^2$ is $x - 3y$, how can we find the other factor? Find the other factor.

JUST FOR FUN

1. Have you ever seen the "proof" that 1 is equal to 2? Here it is.

Let $a = b$; then square both sides of the equation.

$$a^2 = b^2$$

$$a^2 = b \cdot b$$

$$a^2 = ab \qquad \text{substitute } a = b$$

$$a^2 - b^2 = ab - b^2 \qquad \text{subtract } b^2 \text{ from both sides of equation}$$

$$(a + b)(a - b) = b(a - b) \qquad \text{factor both sides of the equation}$$

$$\frac{(a + b)\cancel{(a - b)}}{\cancel{a - b}} = \frac{b\cancel{(a - b)}}{\cancel{a - b}} \qquad \text{divide both sides of the equation by } (a - b) \text{ and divide out common factors}$$

$$a + b = b \qquad \text{substitute } a = b$$

$$b + b = b$$

$$2b = b \qquad \text{now divide both sides of the equation by } b$$

$$\frac{\overset{1}{2\cancel{b}}}{\underset{1}{\cancel{b}}} = \frac{\overset{1}{\cancel{b}}}{\underset{1}{\cancel{b}}}$$

$$2 = 1$$

Obviously, $2 \neq 1$. Therefore, we must have made an error somewhere. Can you find it?

2. Factor completely $4a^{2n} - 4a^n - 15$.

3. Factor completely $12x^{2n}y^{2n} + 2x^ny^n - 2$.

6.3 Special Factoring Formulas

1. *Factor the difference of two squares.*
2. *Factor perfect square trinomials.*
3. *Factor the sum and difference of two cubes.*

1 In this section we present some special factoring formulas: factoring the difference of two squares, perfect square trinomials, and the sum and difference of two cubes. It will be to your advantage to memorize these formulas.

The expression $x^2 - 9$ is an example of the difference of two squares.

$$x^2 - 9 = (x)^2 - (3)^2$$

To factor the difference of two squares, it is convenient to use the difference-of-two-squares formula. This formula was first presented in Section 5.4.

Difference of Two Squares

$$a^2 - b^2 = (a + b)(a - b)$$

EXAMPLE 1 Factor each of the following differences of squares using the difference-of-two-squares formula.

(a) $x^2 - 16$ (b) $16x^2 - 9y^2$ (c) $9y^2 - 64$

Solution: (a) $x^2 - 16 = (x)^2 - (4)^2$
$= (x + 4)(x - 4)$

(b) $16x^2 - 9y^2 = (4x)^2 - (3y)^2$
$= (4x + 3y)(4x - 3y)$

(c) $9y^2 - 64 = (3y)^2 - (8)^2$
$= (3y + 8)(3y - 8)$ ■

EXAMPLE 2 Factor each of the following differences of squares.

(a) $x^6 - y^4$ (b) $z^4 - 81x^6$

Solution: Rewrite each expression as a difference of two squares and then use the difference-of-two-squares formula to factor the expression.

(a) $x^6 - y^4 = (x^3)^2 - (y^2)^2$
$= (x^3 + y^2)(x^3 - y^2)$

(b) $z^4 - 81x^6 = (z^2)^2 - (9x^3)^2$
$= (z^2 + 9x^3)(z^2 - 9x^3)$ ■

EXAMPLE 3 Factor $4x^2 - 16y^2$ using the difference of two squares.

Solution: First remove the common factor, 4.

$$4x^2 - 16y^2 = 4(x^2 - 4y^2)$$

Now use the formula for the difference of two squares.

$$\begin{aligned} 4(x^2 - 4y^2) &= 4[(x)^2 - (2y)^2] \\ &= 4(x + 2y)(x - 2y) \end{aligned}$$ ■

EXAMPLE 4 Factor $x^4 - 16y^4$.

Solution: $x^4 - 16y^4 = (x^2)^2 - (4y^2)^2$
$= (x^2 + 4y^2)(x^2 - 4y^2)$

Note that $(x^2 - 4y^2)$ is also a difference of two squares. We use the difference-of-squares formula a second time to obtain

$= (x^2 + 4y^2)[(x)^2 - (2y)^2]$
$= (x^2 + 4y^2)(x + 2y)(x - 2y)$ ■

EXAMPLE 5 Factor $(x - 5)^2 - 9$ using the difference of two squares.

Solution: We can express $(x - 5)^2 - 9$ as a difference of two squares.

$$\begin{aligned} (x - 5)^2 - 9 &= (x - 5)^2 - 3^2 \\ &= [(x - 5) + 3][(x - 5) - 3] \\ &= (x - 2)(x - 8) \end{aligned}$$

Thus $(x - 5)^2 - 9$ factors into $(x - 2)(x - 8)$. ■

Note: It is not possible to factor the sum of two squares of the form $a^2 + b^2$ over the set of real numbers.

For example, it is not possible to factor $x^2 + 4$ since $x^2 + 4 = x^2 + 2^2$ and this is a sum of two squares.

2 In Section 5.4 we saw that

$$(a + b)^2 = a^2 + 2ab + b^2$$
$$(a - b)^2 = a^2 - 2ab + b^2$$

If we reverse the left and right sides of the two formulas above, we obtain two special factoring formulas.

Perfect Square Trinomials

$$a^2 + 2ab + b^2 = (a + b)^2$$
$$a^2 - 2ab + b^2 = (a - b)^2$$

The two trinomials above are called **perfect square trinomials** since each is the square of a binomial. To be a perfect square trinomial the first and last terms must be the squares of some expression and the middle term must be twice the product of the first and last terms. When you are given a trinomial to factor, first determine if it is perfect square trinomial before you attempt to factor, as explained in the preceding section. If it is a perfect square trinomial, you can factor using the formula given above.

Perfect square trinomials

$y^2 + 6y + 9$	or	$y^2 + 2(y)(3) + 3^2$
$9a^2b^2 - 24ab + 16$	or	$(3ab)^2 - 2(3ab)(4) + 4^2$
$(r + s)^2 + 6(r + s) + 9$	or	$(r + s)^2 + 2(r + s)(3) + 3^2$

Now let us factor some perfect square trinomials.

EXAMPLE 6 Factor $x^2 - 8x + 16$,

Solution: Since the first and last terms are squares, x^2, and 4^2, respectively, this trinomial might be a perfect square trinomial. To determine if it is, take twice the product of x and 4 to see if you obtain $8x$.

$$2(x)(4) = 8x$$

Since $8x$ is the middle term, and since the sign of the middle term is $-$, we factor as follows:

$$x^2 - 8x + 16 = (x - 4)^2 \quad ■$$

EXAMPLE 7 Factor $9x^4 - 12x^2 + 4$.

Solution: The first term is a square $(3x^2)^2$, as is the last term 2^2. Since $2(3x^2)(2) = 12x^2$, we factor as follows:

$$9x^4 - 12x^2 + 4 = (3x^2 - 2)^2 \quad ■$$

EXAMPLE 8 Factor $(a + b)^2 + 6(a + b) + 9$.

Solution: The first term $(a + b)^2$ is a square. The last term 9 is a square, 3^2. The middle term is $2(a + b)(3) = 6(a + b)$. Therefore, this is a perfect square trinomial. Thus

$$(a + b)^2 + 6(a + b) + 9 = [(a + b) + 3]^2 = (a + b + 3)^2 \quad ■$$

EXAMPLE 9 Factor $x^2 - 6x + 9 - y^2$

Solution: Since $x^2 - 6x + 9$ is a perfect square trinomial, we write

$$(x - 3)^2 - y^2$$

Now $(x - 3)^2 - y^2$ is a difference of two squares, thus

$$\begin{aligned}(x - 3)^2 - y^2 &= [(x - 3) + y][(x - 3) - y] \\ &= (x - 3 + y)(x - 3 - y)\end{aligned}$$

Thus $x^2 - 6x + 9 - y^2 = (x - 3 + y)(x - 3 - y)$ ■

Sum and Difference of Two Cubes

3 Earlier in this section we discussed the difference of two squares. Now we will consider the sum and difference of two cubes. Consider the product of $(a + b)(a^2 - ab + b^2)$.

$$\begin{array}{r} a^2 - ab + b^2 \\ a + b \\ \hline a^2b - ab^2 + b^3 \\ a^3 - a^2b + ab^2 \quad\quad \\ \hline a^3 \quad\quad\quad\quad\quad + b^3 \end{array}$$

Thus $a^3 + b^3 = (a + b)(a^2 - ab + b^2)$. Using multiplication we can also show that $a^3 - b^3 = (a - b)(a^2 + ab + b^2)$. The sum and the difference of two cubes are summarized in the box below.

Sum of Two Cubes

$$a^3 + b^3 = (a + b)(a^2 - ab + b^2)$$

Difference of Two Cubes

$$a^3 - b^3 = (a - b)(a^2 + ab + b^2)$$

EXAMPLE 10 Factor $x^3 + 27$.

Solution: Rewrite $x^3 + 27$ as a sum of two cubes. Then factor using the sum of two cubes formula.

$$\begin{aligned}x^3 + 27 &= (x)^3 + (3)^3 \\ &= (x + 3)[x^2 - x(3) + 3^2] \\ &= (x + 3)(x^2 - 3x + 9)\end{aligned}$$

Thus $x^3 + 27 = (x + 3)(x^2 - 3x + 9)$. ■

EXAMPLE 11 Factor $27x^3 - 8y^6$.

Solution: We first determine that $27x^3$ and $8y^6$ have no common factors other than 1. Since we can express both $27x^3$ and $8y^6$ as cubes, we can factor the given expression using the difference of two cubes formula.

$$\begin{aligned} 27x^3 - 8y^6 &= (3x)^3 - (2y^2)^3 \\ &= (3x - 2y^2)[(3x)^2 + (3x)(2y^2) + (2y^2)^2] \\ &= (3x - 2y^2)(9x^2 + 6xy^2 + 4y^4) \end{aligned}$$

Thus $27x^3 - 8y^6 = (3x - 2y^2)(9x^2 + 6xy^2 + 4y^4)$. ■

EXAMPLE 12 Factor $8y^3 - 64x^6$.

Solution: First factor the 8 common to both terms.

$$8y^3 - 64x^6 = 8(y^3 - 8x^6)$$

Next factor $y^3 - 8x^6$ by writing it as a difference of two cubes.

$$\begin{aligned} 8(y^3 - 8x^6) &= 8[(y)^3 - (2x^2)^3] \\ &= 8(y - 2x^2)[y^2 + y(2x^2) + (2x^2)^2] \\ &= 8(y - 2x^2)(y^2 + 2x^2y + 4x^4) \end{aligned}$$

Thus $8y^3 - 64x^6 = 8(y - 2x^2)(y^2 + 2x^2y + 4x^4)$. ■

COMMON STUDENT ERROR

Note that

$$(a + b)^2 = a^2 + 2ab + b^2$$
$$(a - b)^2 = a^2 - 2ab + b^2$$

Correct	*Wrong*
$a^3 + b^3 = (a + b)(a^2 - ab + b^2)$	~~$a^3 + b^3 = (a + b)(a^2 - 2ab + b^2)$~~
$a^3 - b^3 = (a - b)(a^2 + \underbrace{ab}_{\text{not } 2ab} + b^2)$	~~$a^3 - b^3 = (a - b)(a^2 + 2ab + b^2)$~~

Exercise Set 6.3

Use a special factoring formula to factor each expression. If the polynomial cannot be factored, so state.

1. $x^2 - 81$
2. $x^2 - 9$
3. $x^2 + 9$
4. $x^2 - 16$
5. $1 - 4x^2$
6. $1 - 9a^2$
7. $16 - x^2$
8. $4x^2 - 25$
9. $x^2 - 36y^2$
10. $y^4 - 49x^2$
11. $25 - 16y^4$
12. $x^6 - 144y^4$
13. $a^4 - 4b^4$
14. $16y^2 - 81x^2$
15. $81a^4 - 16b^2$
16. $x^6 - 4$
17. $x^2y^2 - 1$
18. $a^2b^2 - 49c^2$
19. $49 - 64x^2y^2$

20. $4a^2c^2 - 16x^2y^2$
21. $9x^2y^2 - 4x^2$
22. $25x^4 - 81y^6$
23. $25 - (x + y)^2$
24. $36 - (x - 6)^2$
25. $(x + y)^2 - 16$
26. $(2x + 3)^2 - 9$
27. $a^2 - (3b + 2)^2$
28. $a^2 + 2ab + b^2$
29. $x^2 + 10x + 25$
30. $a^2 - 12a + 36$
31. $25 - 10t + t^2$
32. $4 + 4x + x^2$
33. $y^2 - 8y + 16$
34. $4x^2 - 20xy + 25y^2$
35. $9y^2 + 6yz + z^2$
36. $9a^2 + 12a + 4$
37. $25a^2b^2 - 20ab + 4$
38. $9x^2y^2 + 24xy + 16$
39. $a^4 + 12a^2 + 36$
40. $w^4 + 16w^2 + 64$
41. $x^2 - 5xy + 25y^2$
42. $a^4 + 2a^2b^2 + b^4$
43. $(x + y)^2 + 2(x + y) + 1$
44. $(x + 1)^2 + 6(x + 1) + 9$
45. $a^4 - 2a^2b^2 + b^4$
46. $(w - 3)^2 + 8(w - 3) + 16$
47. $x^2 + 6x + 9 - y^2$
48. $x^2 - 2x + 1 - y^2$
49. $9 - (x^2 - 8x + 16)$
50. $25 - (x^2 + 4x + 4)$
51. $x^3 - 27$
52. $y^3 + 125$
53. $x^3 + y^3$
54. $a^3 - 8$
55. $a^3 - 125$
56. $x^3 - 8a^3$
57. $x^3 - 64$
58. $64 - a^3$
59. $y^3 + 1$
60. $w^3 - 64$
61. $27 - 8y^3$
62. $27y^3 - 8x^3$
63. $5x^3 + 40y^3$
64. $24x^3 - 81y^3$
65. $y^6 + x^9$
66. $8y^3 - 125x^6$
67. $5x^3 - 625y^3$
68. $16y^6 - 250x^3$
69. $(x + 1)^3 + 1$
70. $(x - 3)^3 + 8$
71. $(x - y)^3 - 27$

72. Find two values of b that will make $4x^2 + bx + 9$ a perfect square trinomial.
73. Find two values of c that will make $16x^2 + cx + 4$ a perfect square trinomial.
74. Find the value of c that will make $25x^2 + 20x + c$ a perfect square trinomial.
75. Find the value of d that will make $49x^2 - 42x + d$ a perfect square trinomial.

JUST FOR FUN

Factor each of the following.

1. (a) $x^2 - 7$ **(b)** $2x^2 - 15$
2. $(x - 8)^2 - (x - 5)^2$
3. $a^{2n} - 16a^n + 64$

6.4 A General Review of Factoring

1 *Factor problems using a combination of factoring procedures.*

1 We have presented a number of different factoring methods. Now we will combine problems and techniques from the previous sections.

A general procedure to follow to factor any polynomial follows.

To Factor a Polynomial

1. Determine if the polynomial has a greatest common factor other than 1. If so, factor out the GCF from every term in the polynomial.
2. If the polynomial has two terms (or is a binomial), determine if it is a difference of two squares or a sum or difference of two cubes. If so, factor using the appropriate formula.
3. If the polynomial has three terms (or is a trinomial), determine if it is a perfect square trinomial. If so, factor accordingly. If it is not, then factor the trinomial using the method discussed in Section 6.2.
4. If the polynomial has more than three terms, try factoring by grouping. If that does not work, see if 3 of the terms are the square of a binomial.
5. As a final step examine your factored polynomial to see if any factors listed have a common factor and can be factored further. If you find a common factor, factor it out at this point.

The following examples illustrate how to use the procedure.

EXAMPLE 1 Factor $3x^4 - 27x^2$.

Solution: First determine if there is a greatest common factor other than 1. Since $3x^2$ is common to both terms factor it out.

$$\begin{aligned} 3x^4 - 27x^2 &= 3x^2(x^2 - 9) \\ &= 3x^2(x + 3)(x - 3) \end{aligned}$$

Note that $x^2 - 9$ is a difference of two squares ■

EXAMPLE 2 Factor $3x^2y^2 - 24xy^2 + 48y^2$.

Solution: Begin by factoring the GCF, $3y^2$, from each term.

$$\begin{aligned} 3x^2y^2 - 24xy^2 + 48y^2 &= 3y^2(x^2 - 8x + 16) \\ &= 3y^2(x - 4)^2 \end{aligned}$$

Note that $x^2 - 8x + 16$ is a perfect square trinomial. If you did not recognize this, you would still obtain the correct answer by factoring the trinomial into $(x-4)(x-4)$ ■

EXAMPLE 3 Factor $24x^2 - 6xy + 16xy - 4y^2$.

Solution: Always begin by determining if the polynomial has a common factor. In this example the number 2 is common to all terms. Remove the common factor 2, then factor the remaining four-term polynomial by grouping.

$$\begin{aligned} 24x^2 - 6xy + 16xy - 4y^2 &= 2[12x^2 - 3xy + 8xy - 2y^2] \\ &= 2[3x(4x - y) + 2y(4x - y)] \\ &= 2(3x + 2y)(4x - y) \end{aligned}$$ ■

EXAMPLE 4 Factor $10a^2b - 15ab + 20b$.

Solution: $10a^2b - 15ab + 20b = 5b(2a^2 - 3a + 4)$
Since $2a^2 - 3a + 4$ cannot be factored, we stop here. ■

EXAMPLE 5 Factor $2x^4y + 54xy$.

Solution:
$$2x^4y + 54xy = 2xy(x^3 + 27)$$
$$= 2xy(x + 3)(x^2 - 3x + 9)$$

Note that $x^3 + 27$ is a sum of two cubes. ■

Exercise Set 6.4

Factor each of the following completely.

1. $3x^2 + 3x - 36$
2. $2x^2 - 16x + 32$
3. $2x^2y - 8xy + 16xy^2$
4. $6x^3y^2 + 10x^2y^3 + 8x^2y^2$
5. $4x^2 + 24x + 36$
6. $3x^3 - 12x^2 - 36x$
7. $2x^2 - 72$
8. $4x^2 - 4y^2$
9. $5x^5 - 45x$
10. $6x^2y^2z^2 - 24x^2y^2$
11. $3x^3 - 3x^2 - 12x^2 + 12x$
12. $2x^2y^2 + 6xy^2 - 10xy^2 - 30y^2$
13. $5x^4y^2 + 20x^3y^2 - 15x^3y^2 - 60x^2y^2$
14. $6x^2 - 15x - 9$
15. $x^4 - x^2y^2$
16. $4x^3 + 108$
17. $x^7y^2 - x^4y^2$
18. $x^4 - 16$
19. $x^5 - 16x$
20. $12x^2y^2 + 33xy^2 - 9y^2$
21. $4x^6 + 32y^3$
22. $12x^4 - 6x^3 - 6x^3 + 3x^2$
23. $2(a + b)^2 - 18$
24. $12x^3y^2 + 4x^2y^2 - 40xy^2$
25. $x^2 + 6xy + 9y^2$
26. $3x^2 - 30x + 75$
27. $(x + 2)^2 - 4$
28. $4y^4 - 36x^6$
29. $50r^2 - 162$
30. $pq + 6q + pr + 6r$
31. $(y + 3)^2 + 4(y + 3) + 4$
32. $b^4 + 2b^2 + 1$
33. $45a^4 - 30a^3 + 5a^2$
34. $(x + 1)^2 - (x + 1) - 6$
35. $x^3 + \frac{1}{27}$
36. $8y^3 - \frac{1}{8}$
37. $3x^3 + 2x^2 - 27x - 18$
38. $6y^3 + 14y^2 + 4y$
39. $a^3b - 16ab^3$
40. $x^6 + y^6$
41. $9 - (x^2 + 2xy + y^2)$
42. $x^2 - 2xy + y^2 - 25$

6.5 Solving Equations Using Factoring

1. *Know the standard form of a quadratic equation.*
2. *Use the zero-factor property to solve equations.*
3. *Use factoring to solve equations.*
4. *Use factoring to solve application problems.*

1 Quadratic equations in two variables, x and y, were introduced and graphed in Section 5.8. In this section we explain how to solve quadratic equations in one variable using factoring. Every quadratic equation has a second-degree term as its highest term.

Examples of quadratic Equations

$$3x^2 + 6x - 4 = 0$$
$$5x = 2x^2 - 4$$
$$(x + 4)(x - 3) = 0$$

Any quadratic equation can be written in standard form.

Standard Form of a Quadratic Equation

$$ax^2 + bx + c = 0 \qquad a \neq 0$$

where a, b, and c are real numbers.

Before going any further convert each of the three quadratic equations given at the bottom of page 246 to standard form, with $a > 0$.

2 To solve equations using factoring we make use of the **zero-factor property**.

Zero-Factor Property

For all real numbers a and b, if $a \cdot b = 0$, then either $a = 0$ or $b = 0$, or both a and $b = 0$.

The zero-factor property indicates that if the product of two factors equals zero, one (or both) of the factors must be zero.

EXAMPLE 1 Solve the equation $(x + 5)(x - 2) = 0$.

Solution: Since the product of the factors equals 0, according to the rule above, one or both factors must equal zero. Set each factor equal to 0 and solve each equation.

$$x + 5 = 0 \qquad \text{or} \qquad x - 2 = 0$$
$$x = -5 \qquad\qquad\qquad x = 2$$

Thus if x is either -5 or 2, the product of the factors is 0.

Check:

$$x = -5 \qquad\qquad\qquad x = 2$$
$$(x + 5)(x - 2) = 0 \qquad\qquad (x + 5)(x - 2) = 0$$
$$(-5 + 5)(-5 - 2) = 0 \qquad\qquad (2 + 5)(2 - 2) = 0$$
$$0(-7) = 0 \qquad\qquad 7(0) = 0$$
$$0 = 0 \quad \text{true} \qquad\qquad 0 = 0 \quad \text{true}$$

■

3

To Solve an Equation Using Factoring

1. Use the addition property to remove all terms from one side of the equation. This will result in one side of the equation being equal to 0.
2. Combine like terms in the equation and then factor.
3. Set each factor containing a variable equal to zero, solve the equations, and find the solutions.
4. Check the solutions in the original equation.

EXAMPLE 2 Solve the equation $2x^2 - 12x = 0$.

Solution: Since all terms are already on the left of the equal sign and the right side equals 0 we factor the left side.

$$2x^2 - 12x = 0$$
$$2x(x - 6) = 0$$

Now set each factor equal to zero.

$$2x = 0 \quad \text{or} \quad x - 6 = 0$$
$$x = 0 \qquad\qquad x = 6$$

The numbers 0 and 6 both satisfy the equation $2x^2 - 12x = 0$. ■

COMMON STUDENT ERROR

The zero-factor property can be used only when one side of the equation is equal to 0.

Correct	*Wrong*
$(x - 4)(x + 3) = 0$	~~$(x - 4)(x + 3) = 2$~~
$x - 4 = 0 \quad \text{or} \quad x + 3 = 0$	~~$x - 4 = 2 \quad \text{or} \quad x + 3 = 2$~~

Note that in the wrong process illustrated on the right the zero-factor property cannot be used since the right side of the equation is not equal to 0. Example 3 shows how to solve such problems correctly.

EXAMPLE 3 Solve the equation $(x - 1)(3x + 2) = 4x$.

Solution: Since the right side of the equation is not 0, we cannot use the zero-factor property. Begin by multiplying the factors on the left side of the equation. Then subtract $4x$ from both sides of the equation to obtain a 0 on the right side.

$$(x - 1)(3x + 2) = 4x$$
$$3x^2 - x - 2 = 4x$$
$$3x^2 - 5x - 2 = 0$$
$$(3x + 1)(x - 2) = 0$$

$$3x + 1 = 0 \quad \text{or} \quad x - 2 = 0$$
$$3x = -1 \qquad\qquad x = 2$$
$$x = -\frac{1}{3}$$

The solutions are $-\frac{1}{3}$ and 2. ■

EXAMPLE 4 Solve the equation $3x^2 + 2x - 12 = -7x$.

Solution:

$$3x^2 + 2x - 12 = -7x \quad \text{add } 7x \text{ to both sides of equation}$$
$$3x^2 + 9x - 12 = 0$$
$$3(x^2 + 3x - 4) = 0 \quad \text{factor out the 3}$$
$$3(x + 4)(x - 1) = 0 \quad \text{factor the trinomial}$$
$$x + 4 = 0 \quad \text{or} \quad x - 1 = 0 \quad \text{solve}$$
$$x = -4 \quad \text{or} \quad x = 1$$

Since the 3 that was factored out is an expression not containing a variable, we do not have to set it equal to zero. Only the numbers -4 and 1 satisfy the equation $3x^2 + 2x - 12 = -7x$. ■

EXAMPLE 5 Solve the equation $2x(x + 2) = x(x - 3) - 12$.

Solution:

$$2x(x + 2) = x(x - 3) - 12$$
$$2x^2 + 4x = x^2 - 3x - 12$$
$$x^2 + 7x + 12 = 0$$
$$(x + 4)(x + 3) = 0$$
$$x + 4 = 0 \quad \text{or} \quad x + 3 = 0$$
$$x = -4 \quad \text{or} \quad x = -3$$ ■

The equations in Examples 1 through 5 were all quadratic equations which we solved by factoring. Other methods that can be used to solve quadratic equations include completing the square and the quadratic formula; we discuss these methods in Chapter 9.

The zero-factor property can be extended to three or more factors as illustrated in Example 6.

EXAMPLE 6 Solve the equation $2x^3 + 5x^2 - 3x = 0$.

Solution: First factor, then set each factor containing an x equal to 0.

$$2x^3 + 5x^2 - 3x = 0$$
$$x(2x^2 + 5x - 3) = 0$$
$$x(2x - 1)(x + 3) = 0$$
$$x = 0 \quad \text{or} \quad 2x - 1 = 0 \quad \text{or} \quad x + 3 = 0$$
$$2x = 1 \quad \text{or} \quad x = -3$$
$$x = \frac{1}{2}$$

The numbers 0, $\frac{1}{2}$, and -3 all satisfy the given equation and are solutions to the equation. ■

Note that the equation in Example 6 is not a quadratic equation for its highest-degree term is 3, not 2. This is an example of a *cubic*, or *third-degree, equation.*

HELPFUL HINT

When solving an equation whose highest powered term has a negative coefficient, we generally make it positive by multiplying both sides of the equation by -1. This makes the factoring process easier.

For example to solve the equation

$$-x^2 + 5x + 6 = 0$$
$$-1(-x^2 + 5x + 6) = -1 \cdot 0$$
$$x^2 - 5x - 6 = 0$$

Now solve the equation $x^2 - 5x - 6 = 0$ to obtain the solution.

$$(x - 6)(x + 1) = 0$$
$$x - 6 = 0 \quad \text{or} \quad x + 1 = 0$$
$$x = 6 \qquad\qquad x = -1$$

The numbers 6 and -1 both satisfy the original equation, $-x^2 + 5x + 6 = 0$.

4 Now let us look at some applications problems that use factoring in their solution.

EXAMPLE 7 The area of a triangle is 27 square inches. Find the base and height if its height is 3 less than twice its base. Use area $= \frac{1}{2}$(base)(height)

Solution: Let x = base, then $2x - 3$ = height (Fig. 6.1)

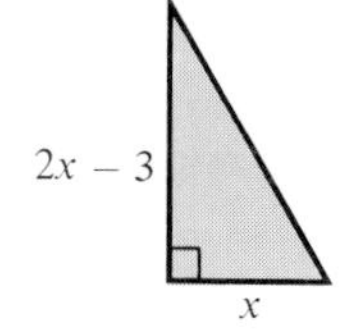

Figure 6.1

$$\text{area} = \frac{1}{2}(\text{base})(\text{height})$$

$$27 = \frac{1}{2}(x)(2x - 3) \qquad \text{multiply both sides of the equation by 2 to remove fractions}$$

$$2(27) = \not{2}\left[\frac{1}{\not{2}}(x)(2x - 3)\right]$$
$$54 = x(2x - 3)$$
$$54 = 2x^2 - 3x$$

or

$$2x^2 - 3x - 54 = 0$$
$$(2x + 9)(x - 6) = 0$$
$$2x + 9 = 0 \quad \text{or} \quad x - 6 = 0$$
$$2x = -9 \qquad\qquad x = 6$$
$$x = -\frac{9}{2}$$

Since the dimensions of a geometric figure cannot be negative, we can eliminate $x = -\frac{9}{2}$ as an answer to our problem.

$$\text{base} = x = 6 \text{ inches}$$
$$\text{height} = 2x - 3 = 2(6) - 3 = 9 \text{ inches}$$ ■

EXAMPLE 8 A projectile on top of a building 384 feet high is fired upward with a velocity of 32 feet per second. The projectile's distance, s, above the ground at any time t is given by the formula $s = -16t^2 + 32t + 384$. Find the time that it takes for the object to strike the ground.

Solution: When the object strikes the ground, its distance from the ground is 0. Substituting $s = 0$ into the equation gives

$$0 = -16t^2 + 32t + 384$$

or

$$-16t^2 + 32t + 384 = 0$$
$$-16(t^2 - 2t - 24) = 0$$
$$-16(t + 4)(t - 6) = 0$$

$$t + 4 = 0 \quad \text{or} \quad t - 6 = 0$$
$$t = -4 \qquad\qquad t = 6$$

Since it makes no sense to speak about a negative time, -4 is not a possible answer. The object will strike the ground in 6 seconds. ■

Exercise Set 6.5

Solve each equation.

1. $x(x + 5) = 0$
2. $3x(x - 5) = 0$
3. $5x(x + 9) = 0$
4. $2(x + 3)(x - 5) = 0$
5. $(2x + 5)(x - 3)(3x + 6) = 0$
6. $x(2x + 3)(x - 5) = 0$
7. $4x - 12 = 0$
8. $9x - 27 = 0$
9. $-x^2 + 12x = 0$
10. $x^2 + 4x = 0$
11. $9x^2 = -18x$
12. $x^2 + 6x + 5 = 0$
13. $x^2 + x - 12 = 0$
14. $x(x + 6) = -9$
15. $x(x - 12) = -20$
16. $3y^2 - 2 = -y$
17. $-z^2 - 3z = -18$
18. $3x^2 = -21x - 18$
19. $3x^2 - 6x - 72 = 0$
20. $x^3 = 3x^2 + 18x$
21. $x^3 + 19x^2 = 42x$
22. $3x^2 - 9x - 30 = 0$
23. $2y^2 + 22y + 60 = 0$
24. $w^2 + 45 + 18w = 0$
25. $-2x - 8 = -x^2$
26. $-9x + 20 = -x^2$
27. $-x^2 + 30x + 64 = 0$
28. $-y^2 + 12y - 11 = 0$
29. $x^3 - 3x^2 - 18x = 0$
30. $z^3 + 16z^2 = -64z$
31. $3p^2 = 22p - 7$
32. $5w^2 - 16w = -3$
33. $3r^2 + r = 2$
34. $3x^2 = 7x + 20$
35. $4x^3 + 4x^2 - 48x = 0$
36. $x^2 - 25 = 0$
37. $x^2 - 16x = 0$
38. $4x^2 - 9 = 0$
39. $x^3 - 16x = 0$
40. $2x^4 - 32x^2 = 0$
41. $(x + 4)^2 - 16 = 0$
42. $(2x + 5)^2 - 9 = 0$
43. $(x - 7)(x + 5) = -20$
44. $(x + 1)^2 = 3x + 7$
45. $2a^2 - 12 - 4a = a^2 - 3a$
46. $(x - 4)^2 - 4 = 0$
47. $(b - 1)(3b + 2) = 4b$
48. $2(a^2 - 3) - 3a = 2(a + 3)$
49. $2(x + 2)(x - 2) = (x - 2)(x + 3) - 2$
50. $2(a + 3)(a - 5) = 2(a - 1) + 8$
51. $2x^3 + 16x^2 + 30x = 0$
52. $18x^3 - 15x^2 = 12x$

For each exercise, write the problem as an equation. Solve the equation and answer the question.

53. The product of two consecutive positive integers is 72. Find the two integers.

54. The product of two consecutive positive even integers is 80. Find the two integers.

55. The product of two consecutive positive odd integers is 99. Find the two integers.

56. The product of two positive numbers is 108. Find the two numbers if one is 3 more than the other.

See Exercise 66.

57. The product of two positive numbers is 35. Find the two numbers if one number is 3 less than twice the other number.
58. The product of two positive integers is 36. Find the two integers if one number is 4 times the other.
59. The area of a rectangle is 36 square feet. Find the length and width if the length is 4 times the width.
60. The area of a rectangle is 54 square inches. Find the length and width if the length is 3 inches less than twice the width.
61. The base of a triangle is 6 centimeters greater than its height. Find the base and height if the area is 80 square centimeters.
62. The height of a triangle is 1 centimeter less than twice its base. Find the base and height if the triangle's area is 33 square centimeters.
63. If the sides of a square are increased by 4 meters, the area becomes 121 square meters. Find the length of a side of the original square.
64. If the sides of a square are increased by 6 meters, the area becomes 64 square meters. Find the length of a side of the original square.
65. A model rocket is to be launched from a hill 80 feet above sea level. The launch site is next to the ocean (sea level), and the rocket will fall into the ocean. The rocket's distance, s, above sea level at any time, t, is found by the equation $s = -16t^2 + 64t + 80$. Find the time it takes for the rocket to strike the ocean.
66. An object is dropped from a helicopter that is 256 feet above the ground. The distance of the object from the ground at any time t is given by the equation $s = -16t^2 + 256$. Find the time it takes for the object to strike the ground.
67. A television is thrown downward from the top of a 640-foot-tall building with a velocity of 48 feet per second. The television's distance, s, from the ground at any time, t, is found by the equation $s = -16t^2 - 48t + 640$. Find the time it takes for the television to strike the ground.

JUST FOR FUN

Solve the following equations.

1. $x^6 - 7x^3 - 60 = 0$
2. $(x + 3)^2 + 2(x + 3) = 24$
3. $x^4 - 5x^2 + 4 = 0$

Summary

Glossary

Factoring: The opposite of the multiplication process.

Greatest common factor: The greatest common factor of two or more expressions is the greatest factor that divides each expression.

Standard form of a quadratic equation: $ax^2 + bx + c = 0,\ a \neq 0$.

Zero-factor property: If $a \cdot b = 0$, then either $a = 0$ or $b = 0$ or both a and $b = 0$.

Important Facts

Special Factoring Formulas

$a^2 - b^2 = (a + b)(a - b)$ difference of two squares

$a^2 + 2ab + b^2 = (a + b)^2$, $a^2 - 2ab + b^2 = (a - b)^2$ perfect square trinomials

$a^3 + b^3 = (a + b)(a^2 - ab + b^2)$ sum of two cubes

$a^3 - b^3 = (a - b)(a^2 + ab + b^2)$ difference of two cubes

Note: The sum of two squares $a^2 + b^2$ cannot be factored over the set of real numbers.

Review Exercises

[6.1] *Find the greatest common factor for each set of terms.*

1. $40x^2, 36x^3, 16x^5$
2. $12xy, 36xy^2, 18x^2y$
3. $15x^3y^2z^5, -6x^2y^3, 30xy^4z$
4. $x(2x - 5), 3(2x - 5)^2, 5(2x - 5)^3$
5. $x(x + 5), x + 5, 2(x + 5)^2$
6. $2x, (x - 2), (x - 2)^2$

Factor. If an expression cannot be factored, so state.

7. $9x + 33$
8. $12x^2 + 4x - 8$
9. $60x^4 + 6x^9 - 18x^5y^2$
10. $24x^6 - 13y^5 + 6z$
11. $x(5x + 3) - 2(5x + 3)$
12. $2x(4x - 3) + 4x - 3$
13. $3x(x - 1)^2 - 2(x - 1)$
14. $4x(2x - 1) + 3(2x - 1)^2$
15. $6x^2y^2 - 12x^3y - 9xy^5$
16. $12xy^4z^3 + 6x^2y^3z^2 - 15x^3y^2z^3$

Factor by grouping.

17. $x^2 + 3x + 2x + 6$
18. $x^2 - 5x + 3x - 15$
19. $x^2 - 7x + 7x - 49$
20. $3x^2 + x + 9x + 3$
21. $5x^2 + 20x - x - 4$
22. $5x^2 - xy + 20xy - 4y^2$
23. $12x^2 - 8xy + 15xy - 10y^2$
24. $12x^2 + 15xy - 8xy - 10y^2$
25. $4x^2 + 24x - x - 6$
26. $12x^2 - 9xy - 4xy + 3y^2$

[6.2] *Factor each trinomial.*

27. $x^2 + 8x + 15$
28. $x^2 - 8x + 15$
29. $x^2 - 6x - 27$
30. $x^2 - 12x - 45$
31. $x^2 - 5xy - 50y^2$
32. $x^2 - 15xy - 54y^2$
33. $2x^2 + 16x + 32$
34. $3x^2 - 18x + 27$
35. $x^3 - 3x^2 - 18x$
36. $2x^3 - 30x^2 + 112x$
37. $3x^2 + 13x + 4$
38. $2x^2 - xy - 10y^2$
39. $4x^3 - 9x^2 + 5x$
40. $12x^3 + 61x^2 + 5x$
41. $x^4 - 3x^2 - 10$
42. $x^4 - x^2 - 20$
43. $(x + 5)^2 + 10(x + 5) + 24$

[6.3] *Use a special factoring formula to factor.*

44. $x^2 - 36$
45. $x^2 - 49$
46. $4x^2 - 16y^4$
47. $x^4 - 81$
48. $(x + 2)^2 - 9$
49. $(x - 3)^2 - 4$
50. $x^2 + 8x + 16$
51. $4x^2 - 12x + 9$
52. $9y^2 + 24y + 16$
53. $w^4 - 16w^2 + 64$
54. $x^3 + 8$
55. $x^3 - 8$
56. $8x^3 + 27$
57. $27x^3 - 8y^3$
58. $27x^3 - 8y^6$
59. $8y^6 - 125x^3$
60. $(x + 1)^3 - 8$

[6.4] Factor completely.

61. $x^2y^2-2xy^2-15y^2$
62. $3x^3-18x^2+24x$
63. $3x^3y^4+18x^2y^4-6x^2y^4-36xy^4$
64. $3y^5-27y$
65. $2x^3y+16y$
66. $5x^4y+20x^3y+20x^2y$
67. $6x^3-21x^2-12x$
68. $x^2+10x+25-y^2$
69. $3x^3+24y^3$
70. $x^2(x+4)+3x(x+4)-4(x+4)$
71. $4(2x+3)^2-12(2x+3)+5$
72. $4x^4+4x^2-3$
73. $(x-1)x^2-(x-1)x-2(x-1)$

[6.5] Solve the equation.

74. $(x-5)(3x+2)=0$
75. $2x^2-3x=0$
76. $15x^2+20x=0$
77. $x^2-2x-24=0$
78. $x^2+8x+15=0$
79. $x^2=-2x+8$
80. $3x^2+21x+30=0$
81. $x^3-6x^2+8x=0$
82. $6x^3+6x^2-12x=0$
83. $8x^2-3=-10x$
84. $4x^2-16=0$
85. $x(x+3)=2(x+4)-2$

Write the problem as an equation. Solve the equation and answer the question.

86. The product of two positive integers is 30. Find the integers if the larger is 4 less than twice the smaller.

87. The area of a rectangle is 63 square feet. Find the length and width of the rectangle if the length is 2 feet greater than the width.

88. The base of a triangle is 3 more than twice the height. Find the base and height if the area of the triangle is 22 square meters.

89. One square has a side 4 inches longer than the side of a second square. If the area of the larger square is 81 square inches, find the length of a side of each square.

90. A small rocket is projected upward from the top of a 144-foot-tall building with a velocity of 128 feet per second. The rocket's distance from the ground, s, at any time, t, is given by the formula $s=-16t^2+128t+144$. Find the time it takes for the rocket to strike the ground.

Practice Test

Factor completely.

1. $4x^2y-4x$
2. $3x^2+12x+2x+8$
3. $8x^2y^4-12xy^3-24x^5y^5$
4. $9x^3y^2+12x^2y^5-27xy^4$
5. $5(x-2)^2+15(x-2)$
6. $2x^2+4xy+3xy+6y^2$
7. $x^2-16x+63$
8. $x^2-7xy+12y^2$
9. $3x^3-6x^2-9x$
10. $6x^2-7x+2$
11. $5x^2+17x+6$
12. $81x^2-16y^4$
13. $27x^3y^6-8y^6$
14. $(x+3)^2+2(x+3)-3$
15. $2x^4+5x^2-18$

Solve the equation.

16. $2(x-5)(3x+2)=0$
17. $x^2-27=-6x$
18. $x^3+4x^2-5x=0$

See Exercise 19.

19. The area of a triangle is 28 square meters. If the base of the triangle is 2 meters greater than 3 times the height, find the base and height of the triangle.

20. An object is projected upward from the top of a 448-foot-tall building with an initial velocity of 48 feet per second. The distance of the object, s, from the ground at any time, t, is given by the equation $s=-16t^2+48t+448$. Find the time after which the object strikes the ground.

7 Rational Expressions and Equations

See Review Exercise 80.

7.1 Reducing Rational Expressions

1 *Find the domain of a rational expression.*

2 *Reduce a rational expression to its lowest terms.*

To be successful in understanding rational expressions, you must have a thorough understanding of the factoring techniques discussed in Chapter 6.

1 A **rational expression** (also called an **algebraic fraction**) is an algebraic expression of the form p/q, where p and q are polynomials and $q \neq 0$. Examples of rational expressions are

$$\frac{2}{3}, \quad \frac{x+3}{x}, \quad \frac{x^2+4x}{x-3}, \quad \frac{x}{x^2-4}$$

Note that the denominator of an algebraic expression cannot equal 0 since division by 0 is not permitted. In the expression $(x+3)/x$, x cannot have a value of 0 since the denominator would then equal 0. In $(x^2+4x)/(x-3)$, x cannot have a value of 3 for that would result in the denominator having a value of 0. What values of x cannot be used in the expression $x/(x^2-4)$? If you answered 2 and -2, you answered correctly.

Whenever we list a rational expression containing a variable in the denominator, we always assume that the value or values of the variable that make the denominator 0 are excluded.

In Section 3.5 we discussed the domain of relations and functions. When discussing rational expressions the **domain** will be the set of values that can be used to replace the variable. For example, in the expression $(x+2)/(x-3)$, the domain will be all real numbers except 3, $\{x | x \neq 3\}$. Note that 3 would make the denominator 0.

EXAMPLE 1 Find the domain of $\dfrac{x-2}{3x-8}$.

Solution: The domain will be all real numbers except those that make the denominator equal 0. What value will make the denominator equal 0? We can determine this by setting the denominator equal to 0 and solving the equation for x.

$$3x - 8 = 0$$
$$3x = 8$$
$$x = \frac{8}{3}$$

If x were $\frac{8}{3}$, the denominator would be 0. The domain is therefore all real numbers except $\frac{8}{3}$, $\{x | x \neq \frac{8}{3}\}$ ■

EXAMPLE 2 Find the domain of $\dfrac{3}{y^2 - 2y - 15}$.

Solution: We must determine which values of y will make the denominator equal 0. To do this, set the denominator equal to 0 and solve for y.

$$y^2 - 2y - 15 = 0$$
$$(y - 5)(y + 3) = 0$$
$$y - 5 = 0 \quad \text{or} \quad y + 3 = 0$$
$$y = 5 \qquad\qquad y = -3$$

Therefore, the domain will consist of all real numbers except -3 and 5.

$$\text{Domain: } \{y \mid y \neq -3,\ y \neq 5\}$$ ■

When we work problems containing rational expressions, we must make sure that we write the answer in the lowest terms. An algebraic fraction is **reduced to its lowest terms** when the numerator and denominator have no common factors other than 1. The fraction $\frac{6}{9}$ is not in reduced form because the 6 and 9 both contain the common factor of 3. When the 3 is factored out, the reduced fraction is $\frac{2}{3}$.

$$\frac{6}{9} = \frac{\overset{1}{\cancel{3}} \cdot 2}{\underset{1}{\cancel{3}} \cdot 3} = \frac{2}{3}$$

The rational expression $\dfrac{ab - b^2}{2b}$ is not in reduced form because both the numerator and denominator have a common factor of b. To reduce this expression, factor the b from each term in the numerator, then divide out the common factor b.

$$\frac{ab - b^2}{2b} = \frac{\cancel{b}(a - b)}{2\cancel{b}} = \frac{a - b}{2}$$

$\dfrac{ab - b^2}{2b}$ becomes $\dfrac{a - b}{2}$ when reduced to its lowest terms.

To Reduce Rational Expressions

1. Factor both numerator and denominator as completely as possible.
2. Divide both the numerator and the denominator by any common factors.

EXAMPLE 3 Reduce $\dfrac{x^2 + 2x - 3}{x + 3}$ to its lowest terms.

Solution: Factor the numerator, then divide out the common factor.

$$\frac{x^2 + 2x - 3}{x + 3} = \frac{\cancel{(x + 3)}(x - 1)}{\cancel{x + 3}} = x - 1$$

The rational expression reduces to $x - 1$. ■

When the terms in a numerator differ only in sign from the terms in a denominator, we can factor out -1 from either the numerator or denominator. **When -1 is factored from a polynomial, the sign of each term in the polynomial changes.** For example,

$$-2x + 3 = -1(2x - 3) = -(2x - 3)$$
$$6 - 5x = -1(-6 + 5x) = -(5x - 6)$$
$$-3x^2 + 5x - 6 = -1(3x^2 - 5x + 6) = -(3x^2 - 5x + 6)$$

EXAMPLE 4 Reduce $\dfrac{2x - 5}{5 - 2x}$.

Solution: Since each term in the numerator differs in sign from its corresponding term in the denominator we will factor -1 from each term in the denominator.

$$\frac{2x - 5}{5 - 2x} = \frac{2x - 5}{-1(-5 + 2x)} = \frac{\cancel{2x - 5}}{-\cancel{(2x - 5)}} = \frac{1}{-1} = -1$$

Note: Expressions that differ only in sign are said to be **opposites.** Thus $2x - 5$ and $5 - 2x$ are opposites. ■

EXAMPLE 5 Reduce $\dfrac{3x^2 + 19x - 14}{2 - 3x}$.

Solution:

$$\frac{3x^2 + 19x - 14}{2 - 3x} = \frac{(3x - 2)(x + 7)}{2 - 3x}$$
$$= \frac{\cancel{(3x - 2)}(x + 7)}{-\cancel{(3x - 2)}} = -(x + 7) \quad ■$$

EXAMPLE 6 Reduce $\dfrac{5x^2y + 10xy^2 - 25x^2y^3}{5x^2y}$.

Solution: Factor the numerator. The greatest common factor of each term in the numerator is $5xy$. Then divide out the common factors.

$$\frac{5x^2y + 10xy^2 - 25x^2y^3}{5x^2y} = \frac{5xy(x + 2y - 5xy^2)}{5x^2y}$$
$$= \frac{x + 2y - 5xy^2}{x} \quad ■$$

EXAMPLE 7 Reduce $\dfrac{x^2 - x - 12}{(x + 2)(x - 4) + x(x - 4)}$.

Solution: Factor the numerator and denominator. Note that each term in the denominator has a common factor of $x - 4$.

$$\frac{x^2 - x - 12}{(x + 2)(x - 4) + x(x - 4)} = \frac{(x + 3)(x - 4)}{(x - 4)[(x + 2) + x]}$$
$$= \frac{(x + 3)\cancel{(x - 4)}}{\cancel{(x - 4)}(2x + 2)}$$
$$= \frac{x + 3}{2x + 2}$$ ■

COMMON STUDENT ERROR

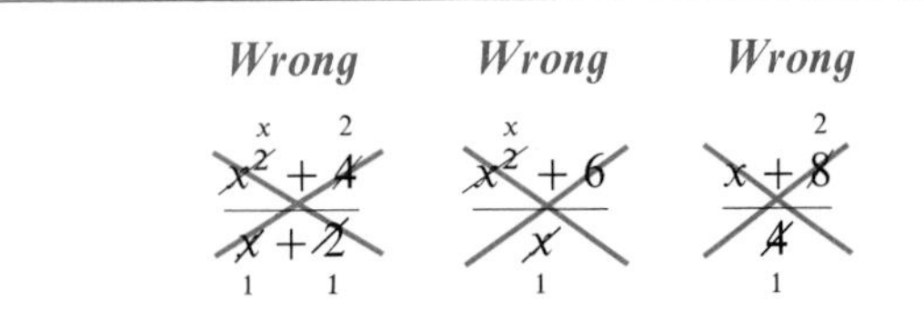

Remember that you can divide out only common **factors.** Only when expressions are **multiplied** can they be factors of the expression. None of the expressions above can be simplified from their original form.

Exercise Set 7.1

Determine the domain of each of the following.

1. $\dfrac{x}{x + 5}$

2. $\dfrac{2x}{2x - 6}$

3. $\dfrac{5}{x^2 - 9}$

4. $\dfrac{6}{x}$

5. $\dfrac{5 - x}{x^2}$

6. $\dfrac{0}{x - 2}$

7. $\dfrac{3}{x^2 + 4}$

8. $\dfrac{2}{(x - 2)^2}$

9. $\dfrac{3}{x^2 - 4}$

10. $\dfrac{-3}{x - \frac{1}{2}}$

11. $\dfrac{-2}{16 - r^2}$

12. $\dfrac{5}{(x + 3)(x - 2)}$

13. $\dfrac{3}{x^2 + 7x + 6}$

14. $\dfrac{x - 3}{x^2 + 4x - 21}$

15. $\dfrac{y - 1}{y - 1}$

16. $\dfrac{2x - 3}{3x - 7}$

17. $\dfrac{-2}{-8z + 15}$

18. $\dfrac{x - 3}{7 - 12x}$

19. $\dfrac{x - 3}{x^2 - 8x + 15}$

20. $\dfrac{4x - 6}{x^2 + 6x + 9}$

21. $\dfrac{4x - 3}{x^2 - 16}$

22. $\dfrac{3a^2 - 6a + 4}{2a^2 + 3a - 2}$

23. $\dfrac{4 - 2x}{x^3 + 9x}$

24. $\dfrac{w^2 - 3w + 4}{3w^2 + 7w + 4}$

Write each expression in reduced form.

25. $\dfrac{15}{25}$

26. $\dfrac{40}{220}$

27. $\dfrac{x - xy}{x}$

28. $\dfrac{x^2 - 2x}{x}$

29. $\dfrac{3x + xy}{y + 3}$

30. $\dfrac{x^2 - 4}{x + 2}$

31. $\dfrac{x^2 - 4x^4}{x}$

32. $\dfrac{5x^2 - 10xy}{25x}$

33. $\dfrac{6 - 4x^2}{18}$

34. $\dfrac{5x^2 - 25x}{10}$

35. $\dfrac{6x^2 + 12xy - 18y^2}{9}$

36. $\dfrac{4x^2y + 12xy + 18x^3y^3}{8xy^2}$

37. $\dfrac{4x - 2}{2 - 4x}$

38. $\dfrac{3x - 6}{2 - x}$

39. $\dfrac{3x - 5}{-9x + 15}$

40. $\dfrac{3 - 2x}{2x - 3}$

41. $\dfrac{x + 4}{x^2 + 9x + 20}$

42. $\dfrac{4x^2 - 9}{2x^2 - x - 3}$

43. $\dfrac{-2x - 6}{x^2 - x - 12}$

44. $\dfrac{x^2 - 2x - 24}{6 - x}$

45. $\dfrac{x^2 - 64}{x - 8}$

46. $\dfrac{4x^2 - 16x^4 + 6x^5y}{8x^3y}$

47. $\dfrac{x^2 + 5x + 6}{x^2 - 3x - 10}$

48. $\dfrac{x^2 - 10x + 24}{x^2 - 5x + 4}$

49. $\dfrac{x^2 + 8x + 15}{-x - 3}$

50. $\dfrac{-(x^2 - 4)}{x^2 - 7x + 10}$

51. $\dfrac{9 - x}{-(x^2 - 8x - 9)}$

52. $\dfrac{x^5 + 4x^4 - 6x^3}{x^2 + 4x - 6}$

53. $\dfrac{(x + 1)(x - 3) + (x + 1)(x - 2)}{2(x + 1)}$

54. $\dfrac{x(x - 1) + x(x - 4)}{2x - 5}$

55. $\dfrac{x^2 - 8x + 5x - 40}{x^2 - 2x + 5x - 10}$

56. $\dfrac{x^2 - 8x + 16}{x^2 + 3x - 4x - 12}$

57. $\dfrac{xy - yw + xz - zw}{xy + yw + xz + zw}$

58. $\dfrac{a^2 + 3a - ab - 3b}{a^2 - ab + 5a - 5b}$

59. $\dfrac{a^3 - b^3}{a^2 - b^2}$

60. $\dfrac{x^2 + 2x - 3}{x^3 + 27}$

61. Explain why $\dfrac{\sqrt{x}}{x + 1}$ is not a rational expression.

62. If a numerator and denominator are opposites of each other, what procedure can we follow to reduce the expression?

63. Consider the expression $\dfrac{1}{x}$.

(a) What is the value of the fraction when $x = 1$?

(b) What is the value of the fraction when $x = 10$?

(c) What is the value of the fraction when $x = 100$?

(d) What happens to the value of the fraction as the denominator increases?

64. Consider the fraction $\dfrac{1}{x}$.

(a) What is the value of the fraction when $x = 1$?

(b) What is the value of the fraction when $x = 0.5$?

(c) What is the value of the fraction when $x = 0.1$?

(d) What happens to the value of the fraction as the denominator approaches 0? (Assume $x > 0$.)

JUST FOR FUN

Polynomial expressions are of the form $a_nx^n + a_{n-1}x^{n-1} + \cdots + a_1x + a_0$ and polynomial functions are of the form $f(x) = a_nx^n + a_{n-1}x^{n-1} + \cdots + a_1x + a_0$. There are also rational expressions of the form p/q, p and q polynomials, $q \neq 0$ and rational functions of the form $f(x) = \dfrac{p}{q}$, p and q polynomials, $q \neq 0$. The graph of every rational function will pass the vertical line test for functions. A discussion of rational functions is beyond the scope of this course, but just for fun you may wish to try to graph one.

1. Consider the rational function $f(x) = \dfrac{x^2 - 4}{x - 2}$.

(a) Determine its domain.
(b) Graph the function. (*Hint:* See what happens if you factor the numerator.)

2. Consider the rational function $y = \dfrac{x^2 - 4x + 3}{x - 1}$.

(a) Determine its domain.
(b) Graph the function.

7.2 Multiplication and Division of Rational Expressions

1 *Multiply rational expressions.*
2 *Divide rational expressions.*

1 To multiply fractions, we often divide out their common factors, then multiply the numerators together, and multiply the denominators together.

Multiplication of Rational Expressions

Multiplication

$$\frac{a}{b} \cdot \frac{c}{d} = \frac{a \cdot c}{b \cdot d},\ b \neq 0,\ d \neq 0$$

We follow the same basic procedure to multiply rational expressions as illustrated in Example 1.

EXAMPLE 1 Multiply $\dfrac{3x^2}{z^2} \cdot \dfrac{2z^5}{9x}$.

Solution:

$$\frac{\overset{1}{\cancel{3}}\overset{x}{\cancel{x^2}}}{\underset{1}{\cancel{z^2}}} \cdot \frac{2\overset{z^3}{\cancel{z^5}}}{\underset{3}{\cancel{9}}\underset{1}{\cancel{x}}} = \frac{x \cdot 2z^3}{1 \cdot 3} = \frac{2xz^3}{3}$$ ■

To Multiply Rational Expressions

1. Factor all numerators and denominators as far as possible.
2. Divide out the common factors.
3. Multiply numerators together and multiply denominators together.
4. Reduce the answer when possible.

EXAMPLE 2 Multiply $\frac{x-5}{4x} \cdot \frac{x^2-2x}{x^2-7x+10}$.

Solution:
$$\frac{x-5}{4x} \cdot \frac{x^2-2x}{x^2-7x+10} = \frac{x-5}{4x} \cdot \frac{x(x-2)}{(x-2)(x-5)} = \frac{1}{4}$$ ■

EXAMPLE 3 Multiply $\frac{2x-5}{x-4} \cdot \frac{x^2-8x+16}{5-2x}$.

Solution:
$$\frac{2x-5}{x-4} \cdot \frac{x^2-8x+16}{5-2x} = \frac{2x-5}{x-4} \cdot \frac{(x-4)(x-4)}{5-2x}$$
$$= \frac{2x-5}{x-4} \cdot \frac{(x-4)(x-4)}{-1(2x-5)}$$
$$= \frac{x-4}{-1} = -(x-4) \quad \text{or} \quad -x+4$$ ■

EXAMPLE 4 Multiply $\frac{x^2-y^2}{x+y} \cdot \frac{x+2y}{2x^2-xy-y^2}$.

Solution:
$$\frac{(x^2-y^2)}{x+y} \cdot \frac{x+2y}{2x^2-xy-y^2} = \frac{(x+y)(x-y)}{x+y} \cdot \frac{x+2y}{(2x+y)(x-y)}$$
$$= \frac{x+2y}{2x+y}$$ ■

EXAMPLE 5 Multiply $\frac{ab-ac+bd-cd}{ab+ac+bd+cd} \cdot \frac{b^2+bc+bd+cd}{b^2+bd-bc-cd}$

Solution: Factor both numerators and denominators by grouping; then divide out common factors.

$$\frac{ab-ac+bd-cd}{ab+ac+bd+cd} \cdot \frac{b^2+bc+bd+cd}{b^2+bd-bc-cd}$$
$$= \frac{a(b-c)+d(b-c)}{a(b+c)+d(b+c)} \cdot \frac{b(b+c)+d(b+c)}{b(b+d)-c(b+d)}$$
$$= \frac{(a+d)(b-c)}{(a+d)(b+c)} \cdot \frac{(b+d)(b+c)}{(b-c)(b+d)} = 1$$ ■

Division of Rational Expressions

2 To divide numerical fractions, we invert the divisor and proceed as in multiplication.

$$\frac{a}{b} \div \frac{c}{d} = \frac{a}{b} \cdot \frac{d}{c} = \frac{ad}{bc}, \; b \neq 0, c \neq 0, d \neq 0$$

We follow the same basic procedure to divide rational expressions, as illustrated in Example 6.

EXAMPLE 6 Divide $\dfrac{12x^4}{5y^3} \div \dfrac{3x^5}{10y}$.

Solution:

$$\frac{12x^4}{5y^3} \div \frac{3x^5}{10y} = \frac{\overset{4}{\cancel{12}}\overset{1}{\cancel{x^4}}}{\underset{1}{\cancel{5}}\underset{y^2}{\cancel{y^3}}} \cdot \frac{\overset{2}{\cancel{10}}\overset{1}{\cancel{y}}}{\underset{1}{\cancel{3}}\underset{x}{\cancel{x^5}}}$$

$$= \frac{4 \cdot 2}{y^2x}$$

$$= \frac{8}{xy^2} \quad \blacksquare$$

To Divide Rational Expressions

Invert the divisor (the second or bottom fraction) and then multiply the resulting rational expressions.

EXAMPLE 7 Divide $\dfrac{x^2 - 9}{x + 4} \div \dfrac{x - 3}{x + 4}$.

Solution:

$$\frac{x^2 - 9}{x + 4} \cdot \frac{x + 4}{x - 3} = \frac{(x + 3)\cancel{(x - 3)}}{\cancel{x + 4}} \cdot \frac{\cancel{x + 4}}{\cancel{x - 3}}$$

$$= x + 3 \quad \blacksquare$$

EXAMPLE 8 Divide $\dfrac{12x^2 - 22x + 8}{3x} \div \dfrac{3x^2 + 2x - 8}{2x^2 + 4x}$.

Solution:

$$\frac{12x^2 - 22x + 8}{3x} \cdot \frac{2x^2 + 4x}{3x^2 + 2x - 8} = \frac{2(6x^2 - 11x + 4)}{3x} \cdot \frac{2x(x + 2)}{(3x - 4)(x + 2)}$$

$$= \frac{2\cancel{(3x - 4)}(2x - 1)}{3\cancel{x}} \cdot \frac{2\cancel{x}\cancel{(x + 2)}}{\cancel{(3x - 4)}\cancel{(x + 2)}}$$

$$= \frac{4(2x - 1)}{3} \quad \blacksquare$$

EXAMPLE 9 Divide $\dfrac{x^4 - y^4}{x - y} \div \dfrac{x^2 + xy}{x^2 - 2xy + y^2}$.

Solution:

$$\frac{x^4 - y^4}{x - y} \cdot \frac{x^2 - 2xy + y^2}{x^2 + xy} = \frac{(x^2 + y^2)(x^2 - y^2)}{x - y} \cdot \frac{(x - y)(x - y)}{x(x + y)}$$

$$= \frac{(x^2 + y^2)\cancel{(x + y)}\cancel{(x - y)}}{\cancel{x - y}} \cdot \frac{(x - y)(x - y)}{x\cancel{(x + y)}}$$

$$= \frac{(x^2 + y^2)(x - y)^2}{x} \quad \blacksquare$$

Exercise Set 7.2

Multiply as indicated.

1. $\dfrac{3x}{2y} \cdot \dfrac{y^3}{6}$
2. $\dfrac{15x^3y^2}{z} \cdot \dfrac{z}{5xy^3}$
3. $\dfrac{16x^2}{y^4} \cdot \dfrac{5x^2}{4y^2}$
4. $\dfrac{12x^2}{6y^2} \cdot \dfrac{36xy^5}{4}$
5. $\dfrac{3y^3}{8x} \cdot \dfrac{9x^2}{y^3}$
6. $\dfrac{45a^3b^3}{12c^3} \cdot \dfrac{4c}{9a^3b^5}$
7. $\dfrac{80m^4}{49x^5y^7} \cdot \dfrac{14x^{12}y^5}{25m^5}$
8. $\dfrac{32m}{5n^3} \cdot \dfrac{-15m^2n^3}{4}$
9. $\dfrac{6x^5y^3}{5z^3} \cdot \dfrac{6x^4}{5yz^4}$
10. $\dfrac{-18x^2y}{11z^2} \cdot \dfrac{22z^3}{x^2y^5}$
11. $(2x + 5) \cdot \dfrac{1}{4x + 10}$
12. $\dfrac{1}{4x - 3} \cdot (20x - 15)$
13. $\dfrac{x - 3}{x + 5} \cdot \dfrac{2x^2 + 10x}{2x - 6}$
14. $\dfrac{x^2 - 4}{x^2 - 9} \cdot \dfrac{x + 3}{x - 2}$
15. $\dfrac{3x - 2}{3x + 2} \cdot \dfrac{4x - 1}{1 - 4x}$
16. $\dfrac{x - 6}{2x + 5} \cdot \dfrac{2x}{-x + 6}$
17. $\dfrac{4 - x}{x - 4} \cdot \dfrac{x - 3}{3 - x}$
18. $\dfrac{5 - 2x}{x + 8} \cdot \dfrac{-x - 8}{2x - 5}$
19. $\dfrac{x^2 + 7x + 12}{x + 4} \cdot \dfrac{1}{x + 3}$
20. $\dfrac{x^2 + 3x - 10}{2x} \cdot \dfrac{x^2 - 3x}{x^2 - 5x + 6}$
21. $\dfrac{x^2 - 5x - 24}{2x^2 - 2x - 24} \cdot \dfrac{4x^2 + 4x - 24}{x^2 - 10x + 16}$
22. $\dfrac{4x + 4y}{xy^2} \cdot \dfrac{x^2y}{3x + 3y}$
23. $\dfrac{a^2 - b^2}{a} \cdot \dfrac{a^2 + ab}{a + b}$
24. $\dfrac{x^2 - 25}{x^2 - 7x + 10} \cdot \dfrac{2 - x}{x}$
25. $\dfrac{a^2 + 6a + 9}{a^2 - 4} \cdot \dfrac{2 - a}{a + 3}$
26. $\dfrac{5x^2 + 17x + 6}{x + 3} \cdot \dfrac{-x + 1}{x^2 - 3x + 2}$
27. $\dfrac{6x^2 - 14x - 12}{6x + 4} \cdot \dfrac{x + 3}{2x^2 - 2x - 12}$
28. $\dfrac{2x^2 - 9x + 9}{8x - 12} \cdot \dfrac{2x}{x^2 - 3x}$

29. $\dfrac{2x+4y}{x^2+4xy+4y^2} \cdot \dfrac{x+2y}{2}$

30. $\dfrac{x^2-y^2}{x^2+xy} \cdot \dfrac{3x^2+6x}{3x^2-2xy-y^2}$

31. $\dfrac{x^2-y^2}{8x^2-16xy+8y^2} \cdot \dfrac{4x-4y}{x+y}$

32. $\dfrac{2x^2-5x-3}{x(x+3)} \cdot \dfrac{x^3+8x^2+15x}{2x^2+11x+5}$

33. $\dfrac{6x^3-x^2-x}{2x^2+x-1} \cdot \dfrac{x^2-1}{x^3-2x^2+x}$

34. $\dfrac{x^3+27}{x^3} \cdot \dfrac{x^4}{x^3+x^2-6x}$

35. $\dfrac{x+2}{x^3-8} \cdot \dfrac{(x-2)^2}{x^2-4}$

36. $\dfrac{8x^2-8y^4}{x+y^2} \cdot \dfrac{2x}{8(x-y^2)}$

37. $\dfrac{2x^3-7x^2+3x}{x^2+2x-3} \cdot \dfrac{x^2+3x}{(x-3)^2}$

38. $\dfrac{x^2+2x-2x-4}{x^2+3x+4x+12} \cdot \dfrac{x^2+3x+3x+9}{x^2+3x-2x-6}$

39. $\dfrac{ac-ad+bc-bd}{ac+ad+bc+bd} \cdot \dfrac{pc+pd-qc-qd}{pc-pd+qc-qd}$

40. $\dfrac{a^2+2ab-ab-2b^2}{2a^2+2ad-2ab-2bd} \cdot \dfrac{a^2+ad-ac-cd}{a^2-ac+2ab-2bc}$

41. $\dfrac{x^2+5x+6}{x^2-x-20} \cdot \dfrac{2x^2+6x-8}{x^2-9} \cdot \dfrac{x^2-3x}{x-1}$

42. $\dfrac{x^2-1}{x^2+x} \cdot \dfrac{2x+2}{1-x^2} \cdot \dfrac{x^2+x-2}{x^2-x}$

43. $\dfrac{x^3+64}{x-2} \cdot \dfrac{x^2-4}{x+4} \cdot \dfrac{x}{x+2}$

Divide as indicated.

44. $\dfrac{5x^3}{4y} \div \dfrac{10x}{8y^4}$

45. $\dfrac{9x^3}{4} \div \dfrac{3}{16y^2}$

46. $\dfrac{25xy^2}{7z^3} \div \dfrac{5x^2y^2}{14z^2}$

47. $\dfrac{9y}{7z^2} \div \dfrac{3xy}{4z}$

48. $\dfrac{x^2y^5}{3z} \div \dfrac{3z}{2x}$

49. $\dfrac{12a^2}{4bc} \div \dfrac{3a^2}{bc}$

50. $\dfrac{-2xw}{y^5} \div \dfrac{6x^2}{y^6}$

51. $\dfrac{-xy}{a} \div \dfrac{-2ax}{6y}$

52. $\dfrac{7a^2b}{xy} \div \dfrac{14}{5xy^2}$

53. $2xz \div \dfrac{4xy}{z}$

54. $\dfrac{27x}{5y^2} \div 3x^2y^2$

55. $\dfrac{1}{7x^2y} \div \dfrac{1}{21x^3y}$

56. $\dfrac{6x+6y}{a} \div \dfrac{12x+12y}{a^2}$

57. $\dfrac{2a+2b}{3} \div \dfrac{a^2-b^2}{a-b}$

58. $\dfrac{3x^2+6x}{x} \div \dfrac{2x+4}{x^2}$

59. $\dfrac{1}{-x-4} \div \dfrac{x^2-7x}{x^2-3x-28}$

60. $\dfrac{x^2+10x+21}{x+7} \div x+3$

61. $\dfrac{x^2-9x+14}{x^2-5x+6} \div \dfrac{x^2-5x-14}{x+2}$

62. $(x-3) \div \dfrac{x^2+3x-18}{x}$

63. $\dfrac{1}{x^2-17x+30} \div \dfrac{1}{x^2+7x-18}$

64. $\dfrac{x^2-12x+32}{x^2-6x-16} \div \dfrac{x^2-x-12}{x^2-5x-24}$

65. $\dfrac{a-b}{9a+9b} \div \dfrac{a^2-b^2}{a^2+2a+1}$

66. $\dfrac{x^2-9x+8}{x+7} \div \dfrac{x-1}{x^2+11x+28}$

67. $\dfrac{(x+2)^2}{x-2} \div \dfrac{x^2-4}{2x-4}$

68. $\dfrac{x^2-4}{2y} \div \dfrac{2-x}{6xy}$

69. $\dfrac{x^2+7x+10}{1-x} \div \dfrac{x^2+2x-15}{x-1}$

70. $\dfrac{2x^2+9x+4}{x^2+7x+12} \div \dfrac{2x^2-x-1}{(x+3)^2}$

71. $\dfrac{a^2-b^2}{9} \div \dfrac{3a-3b}{27x^2}$

72. $\dfrac{x^2-y^2}{x^2-2xy+y^2} \div \dfrac{x+y}{x-y}$

73. $\dfrac{9x^2-9y^2}{6x^2y^2} \div \dfrac{3x+3y}{12x^2y^5}$

74. $\dfrac{x^4-y^8}{x^2+y^4} \div \dfrac{x^2-y^4}{3x^2}$

75. $\dfrac{(x^2-y^2)^2}{(x^2-y^2)^3} \div \dfrac{x^2+y^2}{x^4-y^4}$

76. $\dfrac{2x^4+4x^2}{6x^2+14x+4} \div \dfrac{x^2+2}{3x^2+x}$

77. $\dfrac{8a^3-1}{4a^2+2a+1} \div \dfrac{a-1}{(a-1)^2}$

78. $\dfrac{27x^3+64}{3x+4} \div \dfrac{18x^2-24x+32}{x}$

7.3 Addition and Subtraction of Rational Expressions

1. *Add or subtract rational expressions with a common denominator.*
2. *Find the least common denominator (LCD).*
3. *Add or subtract rational expressions with unlike denominators.*

1 Recall that when adding (or subtracting) two arithmetic fractions with a common denominator, we add (or subtract) the numerators while keeping the common denominator.

Addition

$$\frac{a}{c}+\frac{b}{c}=\frac{a+b}{c}, \qquad c \neq 0$$

Subtraction

$$\frac{a}{c}-\frac{b}{c}=\frac{a-b}{c}, \qquad c \neq 0$$

To add or subtract rational expressions with a common denominator, we use the same principle as illustrated in Example 1.

EXAMPLE 1 Add $\dfrac{3}{x+2}+\dfrac{x-4}{x+2}$.

Solution: Since the denominators are the same, we add the numerators and keep the common denominator.

$$\frac{3}{x+2}+\frac{x-4}{x+2}=\frac{3+(x-4)}{x+2}$$

$$=\frac{x-1}{x+2} \quad ■$$

To Add or Subtract Expressions with a Common Denominator

1. Add or subtract the numerators.
2. Place the sum or difference of the numerators found in part 1 over the common denominator.
3. Reduce the fraction if possible.

EXAMPLE 2 Add $\dfrac{x^2+3x-2}{(x+5)(x-2)}+\dfrac{4x+12}{(x+5)(x-2)}$.

Solution:

$$\frac{x^2+3x-2}{(x+5)(x-2)}+\frac{4x+12}{(x+5)(x-2)}=\frac{x^2+3x-2+(4x+12)}{(x+5)(x-2)}$$

$$=\frac{x^2+7x+10}{(x+5)(x-2)}$$

$$=\frac{\cancel{(x+5)}(x+2)}{\cancel{(x+5)}(x-2)}=\frac{x+2}{x-2} \quad ■$$

When subtracting rational expressions, be sure to subtract the entire numerator of the fraction being subtracted. Study the common student error that follows very carefully.

COMMON STUDENT ERROR

The error presented here is often made by students. Study the information presented, so you will not make this error.

How do you simplify this problem?

$$\frac{4x}{x-2} - \frac{2x+1}{x-2}$$

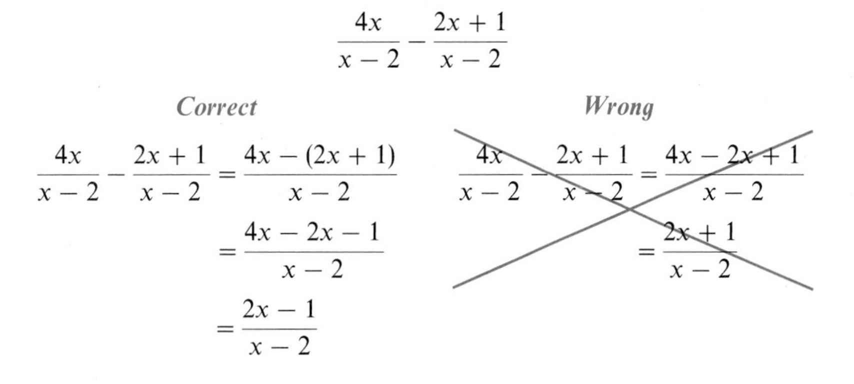

Correct
$\frac{4x}{x-2} - \frac{2x+1}{x-2} = \frac{4x-(2x+1)}{x-2}$
$= \frac{4x-2x-1}{x-2}$
$= \frac{2x-1}{x-2}$
Wrong
$\frac{4x}{x-2} - \frac{2x+1}{x-2} = \frac{4x-2x+1}{x-2}$
$= \frac{2x+1}{x-2}$

The procedure on the right is wrong because the *entire numerator*, $2x + 1$, must be subtracted from $4x$. In the procedure on the right only $2x$ was subtracted. Note that **the sign of each term** (not just the first term) **in the numerator of the fraction being subtracted must change.**

EXAMPLE 3 Subtract $\frac{3x}{x-6} - \frac{x^2-4x+6}{x-6}$.

Solution:

$$\frac{3x}{x-6} - \frac{x^2-4x+6}{x-6} = \frac{3x-(x^2-4x+6)}{x-6}$$

$$= \frac{3x-x^2+4x-6}{x-6}$$

$$= \frac{-x^2+7x-6}{x-6}$$

$$= \frac{-(x^2-7x+6)}{x-6}$$

$$= \frac{-\cancel{(x-6)}(x-1)}{\cancel{x-6}}$$

$$= -(x-1) \quad ■$$

2 To add two numerical fractions with *unlike denominators*, we must first obtain a common denominator.

EXAMPLE 4 Add $\frac{3}{5} + \frac{4}{7}$.

Solution: The least common denominator (LCD) of 5 and 7 is 35. Thirty-five is the smallest number divisible by both 5 and 7. Now proceed to rewrite each fraction so that it has a denominator equal to the LCD.

$$\frac{3}{5} + \frac{4}{7} = \frac{3}{5} \cdot \frac{7}{7} + \frac{4}{7} \cdot \frac{5}{5}$$

$$= \frac{21}{35} + \frac{20}{35} = \frac{41}{35} \quad \text{or} \quad 1\frac{6}{35} \qquad ■$$

To add or subtract rational expressions with unlike denominators, we must first write each expression with a common denominator.

To Find the Least Common Denominator of Rational Expressions

1. Factor each denominator completely. Factors in any given denominator that occur more than once should be expressed as powers [therefore, $(x + 5)(x + 5)$ should be expressed as $(x + 5)^2$].
2. List all different factors (other than 1) that appear in any of the denominators. When the same factor appears in more than one denominator, write the factor with the *highest* power that appears.
3. The least common denominator is the product of all the factors found in step 2.

EXAMPLE 5 Find the LCD.

$$\frac{3}{5x} - \frac{2}{x^2}$$

Solution: The factors that appear in the denominators are 5 and x. List each factor with its highest power. The LCD is the product of these factors.

$$\text{LCD} = 5 \cdot \underbrace{x^2}_{\text{highest power of } x} = 5x^2 \qquad ■$$

EXAMPLE 6 Find the LCD.

$$\frac{1}{18x^3y} + \frac{5}{27x^2y^3}$$

Solution: The LCD of 18 and 27 is 54. The variable factors are x and y.

$$\text{LCD} = 54x^3y^3 \qquad ■$$

EXAMPLE 7 Find the LCD.

$$\frac{3}{x} - \frac{2y}{x + 5}$$

Solution: The factors that appear are x and $(x + 5)$. Note that the x in the second denominator, $x + 5$, is not a factor of that denominator since the operation is addition rather than multiplication.

$$\text{LCD} = x(x + 5) \qquad ■$$

EXAMPLE 8 Find the LCD.

$$\frac{3}{2x^2 - 4x} + \frac{x^2}{x^2 - 4x + 4}$$

Solution: Factor both denominators.

$$\frac{3}{2x(x - 2)} + \frac{x^2}{(x - 2)^2}$$

The factors that appear are 2, x, and $x - 2$. List the highest powers of each of these factors that appear.

$$\text{LCD} = 2 \cdot x \cdot (x - 2)^2 = 2x(x - 2)^2$$ ■

EXAMPLE 9 Find the LCD.

$$\frac{5x}{x^2 - x - 12} - \frac{6x^2}{x^2 - 7x + 12}$$

Solution: Factor both denominators:

$$\frac{5x}{(x + 3)(x - 4)} - \frac{6x^2}{(x - 3)(x - 4)}$$

$$\text{LCD} = (x + 3)(x - 4)(x - 3)$$

Note that although $(x - 4)$ is a common factor of each denominator, the highest power of that factor that appears in either denominator is 1. ■

The method used to add or subtract rational expressions with unlike denominator is illustrated in Example 10.

EXAMPLE 10 Add $\frac{3}{x} + \frac{5}{y}$.

Solution: First determine the LCD.

$$\text{LCD} = xy$$

Now write each fraction with the LCD. We do this by multiplying **both** numerator and denominator of each fraction by any factors needed to obtain the LCD.

In this problem the fraction on the left must be multiplied by y/y and the fraction on the right must be multiplied by x/x.

$$\frac{y}{y} \cdot \frac{3}{x} + \frac{5}{y} \cdot \frac{x}{x} = \frac{3y}{xy} + \frac{5x}{xy}$$

By multiplying both the numerator and denominator by the same factor we are in effect multiplying by 1, which does not change the value of the fraction, only its appearance. Thus the new fraction is equivalent to the original fraction.

Now add the numerators while leaving the LCD alone.

$$\frac{3y}{xy} + \frac{5x}{xy} = \frac{3y + 5x}{xy} \quad \text{or} \quad \frac{5x + 3y}{xy}$$

Therefore, $\frac{3}{x} + \frac{5}{y} = \frac{5x + 3y}{xy}$. ■

To Add or Subtract Two Rational Expressions with Unlike Denominators

1. Determine the LCD.
2. Rewrite each fraction as an equivalent fraction with the LCD. This is done by multiplying both the numerator and denominator of each fraction by any factors needed to obtain the LCD.
3. Leave the denominator in factored form but multiply out the numerator.
4. Add or subtract the numerators while maintaining the LCD.
5. When possible, factor the remaining numerator and reduce fractions.

EXAMPLE 11 Add $\dfrac{5}{4x^2y} + \dfrac{3}{14xy^3}$.

Solution: The LCD is $28x^2y^3$. We must write each fraction with the denominator $28x^2y^3$. To do this, multiply the fraction on the left by $7y^2/7y^2$ and the fraction on the right by $2x/2x$.

$$\frac{7y^2}{7y^2} \cdot \frac{5}{4x^2y} + \frac{3}{14xy^3} \cdot \frac{2x}{2x} = \frac{35y^2}{28x^2y^3} + \frac{6x}{28x^2y^3}$$
$$= \frac{35y^2 + 6x}{28x^2y^3}$$

EXAMPLE 12 Add $\dfrac{3}{x+2} + \dfrac{4}{x}$.

Solution: We must write each fraction with the LCD $x(x + 2)$. To do this, multiply the fraction on the left by x/x and the fraction on the right by $(x + 2)/(x + 2)$.

$$\frac{x}{x} \cdot \frac{3}{(x+2)} + \frac{4}{x} \cdot \frac{(x+2)}{(x+2)} = \frac{3x}{x(x+2)} + \frac{4(x+2)}{x(x+2)}$$
$$= \frac{3x + 4(x+2)}{x(x+2)}$$
$$= \frac{3x + 4x + 8}{x(x+2)}$$
$$= \frac{7x + 8}{x(x+2)}$$

EXAMPLE 13 Subtract $\dfrac{x+2}{x-4} - \dfrac{x+3}{x+4}$.

Solution: The LCD is $(x - 4)(x + 4)$.

$$\frac{x+4}{x+4} \cdot \frac{x+2}{x-4} - \frac{x+3}{x+4} \cdot \frac{x-4}{x-4} = \frac{(x+4)(x+2)}{(x+4)(x-4)} - \frac{(x+3)(x-4)}{(x+4)(x-4)}$$

Use the FOIL method to multiply each numerator.

$$= \frac{x^2 + 6x + 8}{(x + 4)(x - 4)} - \frac{x^2 - x - 12}{(x + 4)(x - 4)}$$

$$= \frac{x^2 + 6x + 8 - (x^2 - x - 12)}{(x + 4)(x - 4)}$$

$$= \frac{x^2 + 6x + 8 - x^2 + x + 12}{(x + 4)(x - 4)}$$

$$= \frac{7x + 20}{(x + 4)(x - 4)} \quad ■$$

EXAMPLE 14 Add $\dfrac{4}{x - 3} + \dfrac{x + 5}{3 - x}$.

Solution: Note that each denominator is the opposite, or additive inverse, of the other. (The terms of one denominator differ only in sign from the terms of the other denominator.) When this special situation arises, we can multiply the numerator and denominator of either one of the fractions by -1 to obtain the LCD.

$$\frac{4}{x - 3} + \frac{x + 5}{3 - x} = \frac{4}{x - 3} + \frac{-1}{-1} \cdot \frac{(x + 5)}{(3 - x)}$$

$$= \frac{4}{x - 3} + \frac{-x - 5}{x - 3}$$

$$= \frac{-x - 1}{x - 3} \quad ■$$

EXAMPLE 15 Subtract $\dfrac{3x + 4}{2x^2 - 5x - 12} - \dfrac{2x - 3}{5x^2 - 18x - 8}$.

Solution: $\dfrac{3x + 4}{(2x + 3)(x - 4)} - \dfrac{2x - 3}{(5x + 2)(x - 4)}$

The LCD is $(2x + 3)(x - 4)(5x + 2)$.

$$\frac{3x + 4}{(2x + 3)(x - 4)} - \frac{2x - 3}{(5x + 2)(x - 4)}$$

$$= \frac{5x + 2}{5x + 2} \cdot \frac{3x + 4}{(2x + 3)(x - 4)} - \frac{2x - 3}{(5x + 2)(x - 4)} \cdot \frac{2x + 3}{2x + 3}$$

$$= \frac{15x^2 + 26x + 8}{(5x + 2)(2x + 3)(x - 4)} - \frac{4x^2 - 9}{(5x + 2)(2x + 3)(x - 4)}$$

$$= \frac{15x^2 + 26x + 8 - (4x^2 - 9)}{(5x + 2)(2x + 3)(x - 4)}$$

$$= \frac{15x^2 + 26x + 8 - 4x^2 + 9}{(5x + 2)(2x + 3)(x - 4)}$$

$$= \frac{11x^2 + 26x + 17}{(5x + 2)(2x + 3)(x - 4)} \quad ■$$

EXAMPLE 16 $\dfrac{x-1}{x-2} - \dfrac{x+1}{x+2} + \dfrac{x-6}{x^2-4}$.

Solution: $\dfrac{x-1}{x-2} - \dfrac{x+1}{x+2} + \dfrac{x-6}{x^2-4} = \dfrac{x-1}{x-2} - \dfrac{x+1}{x+2} + \dfrac{x-6}{(x+2)(x-2)}$

The LCD is $(x+2)(x-2)$.

$$\begin{aligned}
&\frac{x-1}{x-2} - \frac{x+1}{x+2} + \frac{x-6}{(x+2)(x-2)} \\
&= \frac{x+2}{x+2}\cdot\frac{x-1}{x-2} - \frac{x+1}{x+2}\cdot\frac{x-2}{x-2} + \frac{x-6}{(x+2)(x-2)} \\
&= \frac{x^2+x-2}{(x+2)(x-2)} - \frac{x^2-x-2}{(x+2)(x-2)} + \frac{x-6}{(x+2)(x-2)} \\
&= \frac{x^2+x-2-(x^2-x-2)+(x-6)}{(x+2)(x-2)} \\
&= \frac{x^2+x-2-x^2+x+2+x-6}{(x+2)(x-2)} \\
&= \frac{3x-6}{(x+2)(x-2)} \\
&= \frac{3\cancel{(x-2)}}{(x+2)\cancel{(x-2)}} \\
&= \frac{3}{x+2} \quad ■
\end{aligned}$$

Exercise Set 7.3

Add or subtract as indicated.

1. $\dfrac{2x-7}{3} - \dfrac{4}{3}$

2. $\dfrac{2x+3}{5} - \dfrac{x}{5}$

3. $\dfrac{x-4}{x} - \dfrac{x+4}{x}$

4. $\dfrac{-2x+6}{x^2+x-6} + \dfrac{3x-3}{x^2+x-6}$

5. $\dfrac{-x-4}{x^2-16} + \dfrac{2(x+4)}{x^2-16}$

6. $\dfrac{2x+4}{(x+2)(x-3)} - \dfrac{x+7}{(x+2)(x-3)}$

7. $\dfrac{x^2+2x}{3x} - \dfrac{x^2+5x+6}{3x}$

8. $\dfrac{x^2}{x+3} + \dfrac{9}{x+3}$

9. $\dfrac{4x+12}{3-x} - \dfrac{3x+15}{3-x}$

10. $\dfrac{x^2-2}{x^2+6x-7} - \dfrac{-4x+19}{x^2+6x-7}$

11. $\dfrac{-x^2}{x^2+5x-14} + \dfrac{x^2+x+7}{x^2+5x-14}$

12. $\dfrac{3x^2+15x}{x^3+2x^2-8x} + \dfrac{2x^2+5x}{x^3+2x^2-8x}$

13. $\dfrac{x^3-10x^2+35x}{x(x-6)} - \dfrac{x^2+5x}{x(x-6)}$

14. $\dfrac{3x^2+2x-16}{3x^2+4x} - \dfrac{4x-8}{3x^2+4x}$

Find the least common denominator.

15. $\dfrac{5x}{x+1}+\dfrac{6}{x+2}$

16. $\dfrac{-4}{8x^2y^2}+\dfrac{7}{5x^4y^5}$

17. $\dfrac{x+3}{16x^2y}-\dfrac{x^2}{3x^3}$

18. $\dfrac{9}{(x-4)(x+3)}-\dfrac{x+8}{x-4}$

19. $6x^2+\dfrac{9x}{x-3}$

20. $\dfrac{x^2+3}{18x}-\dfrac{x-7}{12(x+5)}$

21. $\dfrac{x-2}{x^2-5x-24}+\dfrac{3}{x^2+11x+24}$

22. $\dfrac{6x+5}{x^2-4}-\dfrac{3x}{x^2-5x-14}$

23. $\dfrac{6}{x+3}-\dfrac{x+5}{x^2-4x+3}$

24. $\dfrac{2x}{4x^2+7x+3}-\dfrac{3}{2x^2-x-3}$

25. $\dfrac{3x-5}{6x^2+13x+6}+\dfrac{3}{3x^2+5x+2}$

26. $\dfrac{6x}{(x-2)(x+3)}+\dfrac{2x+4}{x^2-4}-\dfrac{2x}{2x+4}$

27. $\dfrac{3}{x-2}+\dfrac{x}{x^2+x-6}+\dfrac{x+3}{x+4}$

28. $\dfrac{4}{x^2-9}+\dfrac{x}{(x+3)^2}-\dfrac{5}{x+3}$

Add or subtract as indicated.

29. $\dfrac{4}{3x}+\dfrac{2}{x}$

30. $\dfrac{4}{3y}-\dfrac{1}{4y}$

31. $\dfrac{6}{x^2}+\dfrac{3}{2x}$

32. $3+\dfrac{5}{x}$

33. $\dfrac{5}{6y}+\dfrac{3}{4y^2}$

34. $\dfrac{3x}{4y}+\dfrac{5}{6xy}$

35. $\dfrac{5}{12x^4y}-\dfrac{1}{5x^2y^3}$

36. $\dfrac{3}{4xy^3}+\dfrac{1}{6x^2y}$

37. $\dfrac{4x}{3xy}+2$

38. $\dfrac{1}{a-1}+\dfrac{1}{a}$

39. $\dfrac{4}{x-3}-\dfrac{2}{x}$

40. $\dfrac{5}{b-2}+\dfrac{3x}{2-b}$

41. $\dfrac{x}{x-y}-\dfrac{x}{y-x}$

42. $\dfrac{b}{a-b}+\dfrac{a+b}{b}$

43. $\dfrac{2}{x-3}+\dfrac{4}{x-1}$

44. $\dfrac{3}{x-2}+\dfrac{1}{2-x}$

45. $\dfrac{3}{x-2}-\dfrac{1}{2-x}$

46. $\dfrac{x+5}{x-5}-\dfrac{x-5}{x+5}$

47. $\dfrac{x+7}{x+3}-\dfrac{x-3}{x+7}$

48. $\dfrac{x}{x^2-9}-\dfrac{4(x-3)}{x+3}$

49. $\dfrac{4x}{x-4}+\dfrac{x+4}{x+1}$

50. $\dfrac{x-2}{x+5}-\dfrac{x}{x^2+4x-5}$

51. $\dfrac{2x+1}{x-5}-\dfrac{4}{x^2-3x-10}$

52. $\dfrac{x}{x+1}+\dfrac{1}{x^2+2x+1}$

53. $\dfrac{-x^2+5x}{(x-5)^2}+\dfrac{x+1}{x-5}$

54. $\dfrac{4}{(2x-3)(x+4)}-\dfrac{3}{(x+4)(x-4)}$

55. $\dfrac{x+2}{x^2+x-2}+\dfrac{3}{x^2-1}$

56. $\dfrac{3(x-3)}{x^2+2x-8}+\dfrac{2(x+1)}{x^2-3x+2}$

57. $\dfrac{x+y}{y}+\dfrac{y}{x-y}$

58. $\dfrac{5x}{x^2-9x+8}-\dfrac{3(x+2)}{x^2-6x-16}$

59. $\dfrac{4xy}{x^2-y^2}+\dfrac{x-y}{x+y}$

60. $\dfrac{2x+3}{x-2}+\dfrac{x-1}{x^2+3x-10}$

61. $\dfrac{3x}{2x-3}+\dfrac{3x+6}{2x^2+x-6}$

62. $\dfrac{7}{3x^2+x-4}+\dfrac{9x+2}{3x^2-2x-8}$

63. $\dfrac{3x-4}{4x+1}+\dfrac{3x+6}{4x^2+9x+2}$

64. $\dfrac{x+3}{3x^2+6x-9}+\dfrac{x-3}{6x^2-15x+9}$

65. $\dfrac{x-y}{x^2-4xy+4y^2}+\dfrac{x-3y}{x^2-4y^2}$

66. $\dfrac{x+2y}{x^2-xy-2y^2}-\dfrac{y}{x^2-3xy+2y^2}$

67. $\dfrac{2x}{x-3}-\dfrac{2x}{x+3}+\dfrac{36}{x^2-9}$

68. $\dfrac{3}{x-1}+\dfrac{4}{x+1}+\dfrac{x+2}{x^2-1}$

69. $\dfrac{x^2+2}{x^2-x-2}+\dfrac{1}{x+1}-\dfrac{x}{x-2}$

70. $\dfrac{2}{x^2-16}+\dfrac{x+1}{x^2+8x+16}+\dfrac{3}{x-4}$

71. $\dfrac{3x+2}{x-5}+\dfrac{x}{3x+4}-\dfrac{7x^2+24x+28}{3x^2-11x-20}$

JUST FOR FUN

Add or subtract as indicated.

1. $\dfrac{3}{x^2-x-6}-\dfrac{2}{x^2+x-6}$

2. $\dfrac{3}{x^3+27}+\dfrac{4}{x^2-9}$

7.4 Complex Fractions

1. *Simplify complex fractions by multiplying by a common denominator.*
2. *Simplify complex fractions by simplifying numerator and denominator.*

1 A **complex fraction** is one that has a fractional expression in its numerator or its denominator or both its numerator and denominator. Examples of complex fractions are:

$$\frac{\frac{2}{3}}{5},\quad \frac{\frac{x+1}{x}}{3x},\quad \frac{\frac{x}{y}}{x+1},\quad \frac{\frac{a+b}{a}}{\frac{a-b}{b}},\quad \frac{3+\frac{1}{x}}{\frac{1}{x^2}+\frac{3}{x}}$$

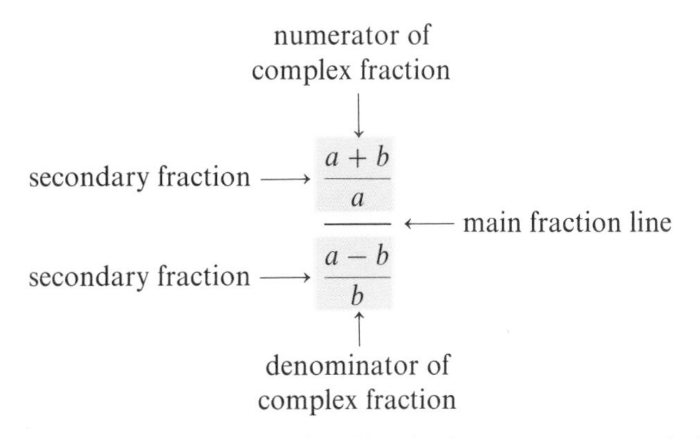

The expression above the main fraction line is the numerator, and the expression below the main fraction line is the denominator of the complex fraction. We will explain two methods that can be used to simplify complex fractions.

Method 1

1 The first method involves multiplying both the numerator and denominator of the complex fraction by a common denominator.

To Simplify a Complex Fraction (by multiplying by a common denominator)

1. Find the least common denominator of each of the two secondary fractions.
2. Next find the LCD of the complex fraction. The LCD of the complex fraction will be the LCD of the two expressions found in step 1.
3. Multiply both secondary fractions by the LCD of the complex fraction found in step 2.
4. Simplify when possible.

This procedure is illustrated in Example 1.

EXAMPLE 1 Simplify $\dfrac{\frac{2}{3}+\frac{3}{4}}{\frac{3}{4}-\frac{1}{2}}$.

Solution: *Step 1:* The LCD of the numerator of the complex fraction is 12. The LCD of the denominator is 4.

Step 2: The LCD of the complex fraction is the LCD of 12 and 4, which is 12.

Step 3: Multiply both secondary fractions by 12.

$$\frac{12\left(\frac{2}{3}+\frac{3}{4}\right)}{12\left(\frac{3}{4}-\frac{1}{2}\right)}=\frac{12\left(\frac{2}{3}\right)+12\left(\frac{3}{4}\right)}{12\left(\frac{3}{4}\right)-12\left(\frac{1}{2}\right)}$$

Step 4: Simplify.

$$= \frac{\overset{4}{\cancel{12}}\left(\frac{2}{\cancel{3}}\right) + \overset{3}{\cancel{12}}\left(\frac{3}{\cancel{4}}\right)}{\overset{3}{\cancel{12}}\left(\frac{3}{\cancel{4}}\right) - \overset{6}{\cancel{12}}\left(\frac{1}{\cancel{2}}\right)}$$

$$= \frac{8+9}{9-6} = \frac{17}{3} \qquad ■$$

EXAMPLE 2 Simplify $\dfrac{\dfrac{2}{x^2} - \dfrac{3}{x}}{\dfrac{x}{5}}$.

Solution: The LCD of the numerator of the complex fraction is x^2. The LCD of the denominator is 5. Therefore the LCD of the complex fraction is $5x^2$. Multiply the numerator and denominator by $5x^2$.

$$\frac{5x^2\left(\frac{2}{x^2} - \frac{3}{x}\right)}{5x^2\left(\frac{x}{5}\right)} = \frac{5x^2\left(\frac{2}{x^2}\right) - 5x^2\left(\frac{3}{x}\right)}{5x^2\left(\frac{x}{5}\right)}$$

$$= \frac{10 - 15x}{x^3}$$

$$= \frac{5(2-3x)}{x^3} \qquad ■$$

EXAMPLE 3 Simplify $\dfrac{a + \dfrac{1}{b}}{b + \dfrac{1}{a}}$.

Solution: Multiply both numerator and denominator of the complex fraction by its LCD, ab.

$$\frac{ab\left(a + \frac{1}{b}\right)}{ab\left(b + \frac{1}{a}\right)} = \frac{a^2b + a}{ab^2 + b} = \frac{a\cancel{(ab+1)}}{b\cancel{(ab+1)}} = \frac{a}{b} \qquad ■$$

EXAMPLE 4 Simplify $\dfrac{a^{-1} + ab^{-2}}{ab^{-2} - a^{-2}b^{-1}}$.

Solution: Rewrite each expression without negative exponents.

$$\frac{\frac{1}{a} + \frac{a}{b^2}}{\frac{a}{b^2} - \frac{1}{a^2b}}$$

The LCD of the numerator is ab^2. The LCD of the denominator is a^2b^2. The LCD of the complex fraction is a^2b^2.

$$\frac{a^2b^2\left(\frac{1}{a}+\frac{a}{b^2}\right)}{a^2b^2\left(\frac{a}{b^2}-\frac{1}{a^2b}\right)}=\frac{ab^2+a^3}{a^3-b}$$ ■

Method 2

2

To Simplify a Complex Fraction (by simplifying numerator and denominator)

Complex fractions can also be simplified as follows:

1. Add or subtract each secondary fraction as indicated.
2. Invert and multiply the denominator of the complex fraction by the numerator of the complex fraction.
3. Simplify when possible.

Example 5 will show how Example 4 can be completed by the alternative procedure given above.

EXAMPLE 5 Simplify $\dfrac{a^{-1}+ab^{-2}}{ab^{-2}-a^{-2}b^{-1}}$.

Solution:

$$\frac{a^{-1}+ab^{-2}}{ab^{-2}-a^{-2}b^{-1}}=\frac{\frac{1}{a}+\frac{a}{b^2}}{\frac{a}{b^2}-\frac{1}{a^2b}}$$

Now add the fractions in the numerator and subtract the fractions in the denominator. The lowest common denominator of the numerator of the complex fraction is ab^2. The LCD of the denominator of the complex fraction is a^2b^2.

$$\frac{\frac{b^2}{b^2}\cdot\frac{1}{a}+\frac{a}{b^2}\cdot\frac{a}{a}}{\frac{a^2}{a^2}\cdot\frac{a}{b^2}-\frac{1}{a^2b}\cdot\frac{b}{b}}=\frac{\frac{b^2}{ab^2}+\frac{a^2}{ab^2}}{\frac{a^3}{a^2b^2}-\frac{b}{a^2b^2}}=\frac{\frac{a^2+b^2}{ab^2}}{\frac{a^3-b}{a^2b^2}}$$

Now invert the denominator of the complex fraction and multiply it by the numerator.

$$=\frac{a^2+b^2}{\cancel{a}\cancel{b^2}}\cdot\frac{a^{\cancel{2}}\cancel{b^2}}{a^3-b}$$

$$=\frac{a(a^2+b^2)}{a^3-b}\quad\text{or}\quad\frac{a^3+ab^2}{a^3-b}$$

Note that we obtain the same answer by either method. ■

When doing the exercises, unless a particular method is specified, you may use either procedure.

Exercise Set 7.4

Simplify.

1. $\dfrac{1+\frac{3}{5}}{2+\frac{1}{5}}$

2. $\dfrac{1-\frac{9}{16}}{3+\frac{4}{5}}$

3. $\dfrac{2+\frac{3}{8}}{1+\frac{1}{3}}$

4. $\dfrac{\frac{3}{5}+\frac{2}{7}}{\frac{1}{5}+\frac{5}{6}}$

5. $\dfrac{\frac{4}{9}-\frac{3}{8}}{4-\frac{3}{5}}$

6. $\dfrac{1-\frac{x}{y}}{x}$

7. $\dfrac{\frac{x^2y}{4}}{\frac{2}{x}}$

8. $\dfrac{\frac{15a}{b^2}}{\frac{b^3}{5}}$

9. $\dfrac{\frac{8x^2y}{3z^3}}{\frac{4xy}{9z^5}}$

10. $\dfrac{\frac{36x^4}{5y^4z^5}}{\frac{9xy^2}{15z^5}}$

11. $\dfrac{x+\frac{1}{y}}{\frac{x}{y}}$

12. $\dfrac{x-\frac{x}{y}}{\frac{1+x}{y}}$

13. $\dfrac{\frac{9}{x}+\frac{3}{x^2}}{3+\frac{1}{x}}$

14. $\dfrac{\frac{2}{a}+\frac{1}{2a}}{a+\frac{a}{2}}$

15. $\dfrac{3-\frac{1}{y}}{2-\frac{1}{y}}$

16. $\dfrac{\frac{x}{x-y}}{\frac{x^2}{y}}$

17. $\dfrac{\frac{x}{y}-\frac{y}{x}}{\frac{x+y}{x}}$

18. $\dfrac{1}{\frac{1}{x}+y}$

19. $\dfrac{\frac{a^2}{b}-b}{\frac{b^2}{a}-a}$

20. $\dfrac{\frac{1}{x}+\frac{2}{x^2}}{2+\frac{1}{x^2}}$

21. $\dfrac{\frac{a}{b}-2}{\frac{-a}{b}+2}$

22. $\dfrac{\frac{x^2-y^2}{x}}{\frac{x+y}{x^3}}$

23. $\dfrac{\frac{4x+8}{3x^2}}{\frac{4x}{6}}$

24. $\dfrac{\frac{a}{a+1}+1}{\frac{2a+1}{a-1}}$

25. $\dfrac{\frac{x}{4}-\frac{1}{x}}{1+\frac{x+4}{x}}$

26. $\dfrac{1+\frac{x}{x+1}}{\frac{2x+1}{x-1}}$

27. $\dfrac{\frac{1}{x-1}+1}{\frac{1}{x+1}-1}$

28. $\dfrac{\frac{a+1}{a-1}+\frac{a-1}{a+1}}{\frac{a+1}{a-1}-\frac{a-1}{a+1}}$

29. $\dfrac{\frac{a-2}{a+2}-\frac{a+2}{a-2}}{\frac{a-2}{a+2}+\frac{a+2}{a-2}}$

30. $\dfrac{\frac{5}{5-x}+\frac{6}{x-5}}{\frac{3}{x}+\frac{2}{x-5}}$

Simplify.

31. $2a^{-2} + b$

32. $3a^{-2} + b^{-1}$

33. $(a^{-1} + b^{-1})^{-1}$

34. $\dfrac{a^{-1} + b^{-1}}{ab}$

35. $\dfrac{a^{-1} + b^{-1}}{\dfrac{1}{ab}}$

36. $\dfrac{a^{-1} + 1}{b^{-1} - 1}$

37. $\dfrac{\dfrac{a}{b} + a^{-1}}{\dfrac{b}{a} + a^{-1}}$

38. $\dfrac{a^{-1} + b^{-1}}{a^{-1}}$

39. $\dfrac{x^{-1} - y^{-1}}{x^{-1} + y^{-1}}$

40. $\dfrac{x^{-2} + \dfrac{1}{x}}{x^{-1} + x^{-2}}$

41. $\dfrac{a^{-1} + b^{-1}}{(a + b)^{-1}}$

42. $\dfrac{3a^{-1} - b^{-1}}{(a - b)^{-1}}$

43. $2x^{-1} - (3y)^{-1}$

44. $\dfrac{\dfrac{5}{x} + \dfrac{1}{y}}{(x - y)^{-1}}$

45. The efficiency of a jack, E, is given by the formula

$$E = \frac{\frac{1}{2}h}{h + \frac{1}{2}}$$

where h is determined by the pitch of the jack's thread.

Pitch

Determine the efficiency of a jack whose value of h is:

(a) $\frac{2}{3}$ **(b)** $\frac{4}{5}$

46. If two resistors with resistances R_1 and R_2 are connected in parallel, their combined resistance, R_T, can be found from the following formula:

$$R_T = \frac{1}{\dfrac{1}{R_1} + \dfrac{1}{R_2}}$$

Simplify the right side of the formula.

47. If three resistors with resistances R_1, R_2, and R_3 are connected in parallel, their combined resistance can be found by the following formula:

$$R_T = \frac{1}{\dfrac{1}{R_1} + \dfrac{1}{R_2} + \dfrac{1}{R_3}}$$

Simplify the right side of this formula.

48. A formula used in the study of optics is

$$f = (p^{-1} + q^{-1})^{-1}$$

where p is the object's distance from a lens, q is the image distance from the lens, and f is the focal length of the lens. Express the right side of the formula without any negative exponents.

JUST FOR FUN

Simplify.

1. $\dfrac{1}{2a + \dfrac{1}{2a + \dfrac{1}{2a}}}$

2. $\dfrac{1}{x + \dfrac{1}{x + \dfrac{1}{x + 1}}}$

7.5 Solving Equations Containing Rational Expressions

1. *Solve equations containing fractions by multiplying by the LCD.*
2. *Know when proposed solutions must be checked.*
3. *Solve equations containing fractions by cross-multiplication.*
4. *Solve application problems using rational equations.*

1 In Sections 7.1 through 7.4 we presented techniques to add, subtract, multiply, and divide rational expressions. In this section we present a method for solving equations containing fractions.

To Solve Equations Containing Fractions

1. Determine the LCD of all fractions in the equation.
2. Multiply **both** sides of the equation by the LCD. This will result in every term in the equation being multiplied by the LCD.
3. Remove any parentheses and combine like terms on each side of the equation.
4. Solve the equation using the properties discussed in earlier sections.
5. Check the solution in the original equation.

In Step 2 we multiply both sides of the equation by the LCD to eliminate fractions from the equation.

EXAMPLE 1 Solve $\frac{x}{3} + 2x = 7$.

Solution:

$$3\left(\frac{x}{3} + 2x\right) = 7 \cdot 3 \quad \text{multiply both sides of the equation by the LCD, 3}$$

$$\not{3}\left(\frac{x}{\not{3}}\right) + 3 \cdot 2x = 7 \cdot 3 \quad \text{this has the effect of multiplying each term in the equation by the LCD}$$

$$x + 6x = 21$$

$$7x = 21$$

$$x = 3$$

Check:

$$\frac{x}{3} + 2x = 7$$

$$\frac{3}{3} + 2(3) = 7$$

$$1 + 6 = 7$$

$$7 = 7 \quad \text{true}$$

■

EXAMPLE 2 Solve $\frac{3}{4} + \frac{5x}{9} = \frac{x}{6}$.

Solution: Multiply both sides of the equation by the LCD, 36.

$$36\left(\frac{3}{4} + \frac{5x}{9}\right) = \frac{x}{6} \cdot 36$$

$$\overset{9}{\cancel{36}}\left(\frac{3}{\cancel{4}}\right) + \overset{4}{\cancel{36}}\left(\frac{5x}{\cancel{9}}\right) = \frac{x}{\cancel{6}} \cdot \overset{6}{\cancel{36}}$$

$$27 + 20x = 6x$$

$$27 = -14x$$

$$x = \frac{27}{-14} \quad \text{or} \quad \frac{-27}{14}$$ ■

2 **Whenever a variable appears in any denominator you must check your proposed answer in the original equation. When checking if a proposed answer makes any denominator equal to zero, that value is not a solution to the equation.** Such values are called **extraneous roots** or **extraneous solutions.** An extraneous root is a number obtained when solving an equation that is not a solution to the original equation.

EXAMPLE 3 Solve $3 - \frac{4}{x} = \frac{5}{2}$.

Solution: Multiply both sides of the equation by the LCD, $2x$.

$$2x\left(3 - \frac{4}{x}\right) = \left(\frac{5}{2}\right) \cdot 2x$$

$$2x(3) - 2\cancel{x}\left(\frac{4}{\cancel{x}}\right) = \left(\frac{5}{\cancel{2}}\right)\cancel{2}x$$

$$6x - 8 = 5x$$

$$x - 8 = 0$$

$$x = 8$$

Check:

$$3 - \frac{4}{x} = \frac{5}{2}$$

$$3 - \frac{4}{8} = \frac{5}{2}$$

$$3 - \frac{1}{2} = \frac{5}{2}$$

$$\frac{5}{2} = \frac{5}{2} \quad \text{true}$$

Since 8 checks, it is the solution to the equation. ■

In this book some checks will be omitted to conserve space. You should however check all answers when a variable appears in any denominator.

3 A **proportion** is a rational equation of the form

$$\frac{a}{b} = \frac{c}{d}, \quad b \neq 0, \quad d \neq 0$$

Sometimes a proportion may be solved more quickly by **cross-multiplication** than by multiplying both sides of the equation by the LCD.

Cross-Multiplication

$$\text{If} \quad \frac{a}{b} = \frac{c}{d} \quad \text{then} \quad ad = bc, \quad b \neq 0, \quad d \neq 0$$

EXAMPLE 4 Solve the following proportion using cross-multiplication.

$$\frac{3}{x+4} = \frac{4}{x-1}$$

Solution:

$$3(x-1) = 4(x+4)$$
$$3x - 3 = 4x + 16$$
$$-x - 3 = 16$$
$$-x = 19$$
$$x = -19$$

If we check -19 in the original equation we will see that it is the solution to the equation. ■

The LCD of Example 4 is $(x+4)(x-1)$. If we multiply both sides of the equation in Example 4 by the LCD, we would obtain the same solution but it would involve a little more work. Try it and see.

Now let's examine some examples that involve quadratic equations.

EXAMPLE 5 Solve $x + \dfrac{12}{x} = -7$.

Solution:

$$x \cdot \left(x + \frac{12}{x}\right) = -7 \cdot x \qquad \text{multiply both sides of the equation by } x$$

$$x(x) + x\left(\frac{12}{x}\right) = -7x$$
$$x^2 + 12 = -7x$$
$$x^2 + 7x + 12 = 0$$
$$(x+3)(x+4) = 0$$

$$x + 3 = 0 \qquad \text{or} \qquad x + 4 = 0$$
$$x = -3 \qquad\qquad x = -4$$

Checks of -3 and -4 will show that they are solutions to the equation. ■

EXAMPLE 6 Solve $\dfrac{x^2}{x-4} = \dfrac{16}{x-4}$.

Solution:

$$\cancel{(x-4)} \cdot \frac{x^2}{\cancel{x-4}} = \frac{16}{\cancel{x-4}} \cdot \cancel{(x-4)}$$

$$x^2 = 16$$

$$x^2 - 16 = 0 \qquad \text{this is a difference of squares}$$

$$(x+4)(x-4) = 0$$

$$x + 4 = 0 \qquad \text{or} \qquad x - 4 = 0$$

$$x = -4 \qquad\qquad x = 4$$

Check:

$x = -4$		$x = 4$
$\dfrac{x^2}{x-4} = \dfrac{16}{x-4}$		$\dfrac{x^2}{x-4} = \dfrac{16}{x-4}$
$\dfrac{(-4)^2}{-4-4} = \dfrac{16}{-4-4}$		$\dfrac{(4)^2}{4-4} = \dfrac{16}{4-4}$
$\dfrac{16}{-8} = \dfrac{16}{-8}$		$\dfrac{16}{0} = \dfrac{16}{0}$
$-2 = -2$	true	no solution

Since 4 results in a denominator of 0, 4 is *not* a solution to the equation. The 4 is an extraneous root. The only solution to the equation is -4. ■

Notice that the equation in Example 6 is a proportion. Many proportions are difficult to solve by cross-multiplication. What would happen if you attempted to solve the proportion in Example 6 by cross-multiplication?

EXAMPLE 7 Solve $\dfrac{2x}{x^2-4} + \dfrac{1}{x-2} = \dfrac{2}{x+2}$.

Solution:

$$\frac{2x}{(x+2)(x-2)} + \frac{1}{x-2} = \frac{2}{x+2}$$

Multiply both sides of the equation by the LCD, $(x+2)(x-2)$.

$$(x+2)(x-2) \cdot \left[\frac{2x}{(x+2)(x-2)} + \frac{1}{x-2}\right] = \frac{2}{x+2} \cdot (x+2)(x-2)$$

$$\cancel{(x+2)}\cancel{(x-2)} \cdot \frac{2x}{\cancel{(x+2)}\cancel{(x-2)}} + (x+2)\cancel{(x-2)} \cdot \frac{1}{\cancel{x-2}} = \frac{2}{\cancel{x+2}} \cdot \cancel{(x+2)}(x-2)$$

$$2x + (x+2) = 2(x-2)$$

$$2x + x + 2 = 2x - 4$$

$$3x + 2 = 2x - 4$$

$$x + 2 = -4$$

$$x = -6$$

A check will show that -6 is the solution. ■

4 Now let us look at some applications of fractional equations.

EXAMPLE 8 A formula frequently used in optics is

$$\frac{1}{p} + \frac{1}{q} = \frac{1}{f}$$

where p represents the distance of the object from a mirror (or lens), q represents the distance of the image from the mirror (or lens), and f represents the focal length of the mirror (or lens). If a mirror has a focal length of 10 centimeters, how far from the mirror will the image appear when the object is 30 centimeters from the mirror?

Solution: The object distance, p, is 30 centimeters and the focal length, f, is 10 centimeters. We are asked to find the image distance, q.

$$\frac{1}{p} + \frac{1}{q} = \frac{1}{f}$$

$$\frac{1}{30} + \frac{1}{q} = \frac{1}{10}$$

Multiply both sides of the equation by the LCD, $30q$.

$$30q\left(\frac{1}{30} + \frac{1}{q}\right) = 30q\left(\frac{1}{10}\right)$$

$$\cancel{30}q\left(\frac{1}{\cancel{30}}\right) + 30\cancel{q}\left(\frac{1}{\cancel{q}}\right) = \overset{3}{\cancel{30}}q\left(\frac{1}{\cancel{10}}\right)$$

$$q + 30 = 3q$$

$$30 = 2q$$

$$15 = q$$

Thus the image will appear at a distance of 15 centimeters from the mirror. ■

EXAMPLE 9 In electronics the total resistance, R_T, of resistors wired in a parallel circuit is determined by the formula

$$\frac{1}{R_T} = \frac{1}{R_1} + \frac{1}{R_2} + \frac{1}{R_3} + \cdots + \frac{1}{R_n}$$

where $R_1, R_2, R_3, \ldots, R_n$ are the resistances of the individual resistors (measured in ohms) in the circuit.

Find the total resistance if two resistors, one of 200 ohms and the other of 300 ohms, are wired in a parallel circuit.

Solution: Since there are only two resistances, we use the formula

$$\frac{1}{R_T} = \frac{1}{R_1} + \frac{1}{R_2}$$

Let $R_1 = 200$ ohms and $R_2 = 300$ ohms; then

$$\frac{1}{R_T} = \frac{1}{200} + \frac{1}{300}$$

Multiply both sides of the equation by the LCD, $600R_T$.

$$600R_T \cdot \frac{1}{R_T} = 600R_T\left(\frac{1}{200} + \frac{1}{300}\right)$$

$$600\cancel{R_T} \cdot \frac{1}{\cancel{R_T}} = \overset{3}{\cancel{600}}R_T\left(\frac{1}{\cancel{200}}\right) + \overset{2}{\cancel{600}}R_T\left(\frac{1}{\cancel{300}}\right)$$

$$600 = 3R_T + 2R_T$$

$$600 = 5R_T$$

$$R_T = \frac{600}{5} = 120$$

Thus the total resistance of the parallel circuit is 120 ohms. ■

EXAMPLE 10 If three identical resistors are to be wired in parallel, what should be the resistance of each resistor if the total resistance of the circuit is to be 300 ohms?

Solution: Let x = resistance of each resistor.

$$\frac{1}{R_T} = \frac{1}{R_1} + \frac{1}{R_2} + \frac{1}{R_3}$$

Since R_1, R_2, and R_3 are all the same value,

$$\frac{1}{300} = \frac{1}{x} + \frac{1}{x} + \frac{1}{x}$$

$$\frac{1}{300} = \frac{3}{x}$$

$$x = 900$$

Each of the three resistors should be 900 ohms. ■

Exercise Set 7.5

Solve each equation and then check your solution.

1. $\frac{2}{5} = \frac{x}{10}$
2. $\frac{3}{k} = \frac{9}{6}$
3. $\frac{5}{12} = \frac{20}{x}$
4. $\frac{x}{8} = \frac{-15}{4}$
5. $\frac{a}{25} = \frac{12}{10}$
6. $\frac{9c}{10} = \frac{9}{5}$
7. $\frac{9}{3b} = \frac{-6}{2}$
8. $\frac{5}{8} = \frac{2b}{80}$
9. $\frac{x+4}{9} = \frac{5}{9}$
10. $\frac{1}{4} = \frac{z+1}{8}$
11. $\frac{4x+5}{6} = \frac{7}{2}$
12. $\frac{a}{5} = \frac{a-3}{2}$
13. $\frac{6x+7}{10} = \frac{2x+9}{6}$
14. $\frac{n}{10} = 9 - \frac{n}{5}$
15. $\frac{x}{3} - \frac{3x}{4} = \frac{1}{12}$
16. $\frac{2}{8} + \frac{3}{4} = \frac{w}{5}$
17. $\frac{3}{4} - x = 2x$
18. $\frac{2}{y} + \frac{1}{2} = \frac{5}{2y}$
19. $\frac{5}{3x} + \frac{3}{x} = 1$
20. $\frac{x}{4} - \frac{x}{6} = \frac{1}{4}$
21. $\frac{x-1}{x-5} = \frac{4}{x-5}$
22. $\frac{2x+3}{x+1} = \frac{3}{2}$
23. $\frac{5y-3}{7} = \frac{15y-2}{28}$
24. $\frac{2}{x+1} = \frac{1}{x-2}$

25. $\frac{5}{-x-6}=\frac{2}{x}$

26. $\frac{4}{y-3}=\frac{6}{y+3}$

27. $\frac{x-2}{x+4}=\frac{x+1}{x+10}$

28. $\frac{x-3}{x+1}=\frac{x-6}{x+5}$

29. $x-\frac{4}{3x}=-\frac{1}{3}$

30. $\frac{b}{2}-\frac{4}{b}=-\frac{7}{2}$ 1,

31. $\frac{2x-1}{3}-\frac{3x}{4}=\frac{5}{6}$

32. $x+\frac{3}{x}=\frac{12}{x}$

33. $x+\frac{6}{x}=-5$

34. $\frac{15}{x}+\frac{9x-7}{x+2}=9$

35. $\frac{3y-2}{y+1}=4-\frac{y+2}{y-1}$

36. $\frac{2b}{b+1}=2-\frac{5}{2b}$

37. $\frac{1}{x+3}+\frac{1}{x-3}=\frac{-5}{x^2-9}$

38. $c-\frac{c}{3}+\frac{c}{5}=26$

39. $\frac{2}{x-3}-\frac{4}{x+3}=\frac{8}{x^2-9}$

40. $\frac{x+1}{x+3}+\frac{x-3}{x-2}=\frac{2x^2-15}{x^2+x-6}$

41. $\frac{y}{2y+2}+\frac{2y-16}{4y+4}=\frac{2y-3}{y+1}$

42. $\frac{3}{x+3}+\frac{5}{x+4}=\frac{12x+19}{x^2+7x+12}$

43. $\frac{1}{x+2}+\frac{1}{x-2}=\frac{4}{x^2-4}$

44. $\frac{4r-1}{r^2+5r-14}=\frac{1}{r-2}-\frac{2}{r+7}$

45. $\frac{5}{x^2+4x+3}+\frac{2}{x^2+x-6}=\frac{3}{x^2-x-2}$

46. $\frac{2}{x^2+2x-8}-\frac{1}{x^2+9x+20}=\frac{4}{x^2+3x-10}$

47. Refer to Example 8. Find the distance of the image from the mirror if the object is 12 inches from the mirror and the focal length is 6 inches.
48. If the object distance is 12 inches and the image distance is 20 inches, find the focal length of the mirror.
49. If the focal length of the mirror is 8 centimeters and the image distance from the mirror is 15 centimeters, find the object's distance from the mirror.
50. If the object distance is 15 centimeters and the image distance is 9 centimeters, find the focal length of the mirror.
51. A mirror has a focal length of 12 centimeters. Find the object's distance and image distance if the image distance is 3 times the object distance.
52. A mirror has a focal length of 2 inches. Find the object's distance and image distance if the object's distance is 3 inches more than the image distance.
53. Refer to Examples 9 and 10. What is the total resistance in the circuit if resistors of 200 ohms and 700 ohms are connected in parallel?
54. What is the total resistance in the circuit if resistors of 500 ohms and 750 ohms are connected in parallel?
55. What is the total resistance in the circuit if resistors of 300 ohms, 500 ohms, and 3000 ohms are connected in parallel?
56. Three resistors of identical resistance are to be connected in parallel. What should be the resistance of each resistor if the circuit is to have a total resistance of 700 ohms?
57. What is an extraneous root?
58. Under what circumstances is it necessary to check your answers for extraneous roots?

JUST FOR FUN

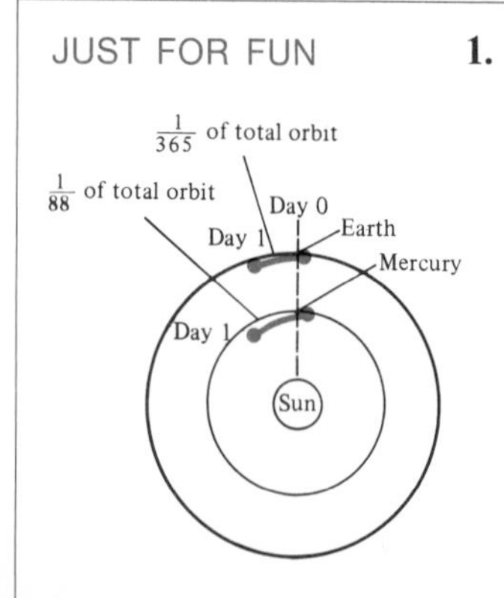

1. The synodic period of Mercury is the time required for swiftly moving Mercury to gain one lap on Earth in their orbits around the sun. If the orbital periods (in Earth days) of the two planets are designated P_m and P_e, Mercury will be seen on the average to move $1/P_m$ of a revolution per day, while Earth moves $1/P_e$ of a revolution per day in pursuit. Mercury's daily gain on Earth is $(1/P_m - 1/P_e)$ of a revolution, so that the time for Mercury to gain one complete revolution on Earth, the synodic period s, may be found by the formula

$$\frac{1}{s}=\frac{1}{P_m}-\frac{1}{P_e}$$

If $P_e = 365$ days and P_m is 88 days, find the synodic period in units of terrestrial (Earth) days.

7.6 Applications of Rational Equations

1. *Set up and solve work problems.*
2. *Set up and solve number problems.*
3. *Set up and solve motion problems.*

Some applications of rational equations were illustrated in Section 7.5. In this section we examine some additional applications. We study work problems first.

1 Problems where two or more machines or people work together to complete a certain task are sometimes referred to as work problems. Work problems often involve equations containing fractions. Generally, work problems are based on the fact that the work done by person 1 (or machine 1) plus the work done by person 2 (or machine 2) is equal to the total amount of work done by both people (or both machines).

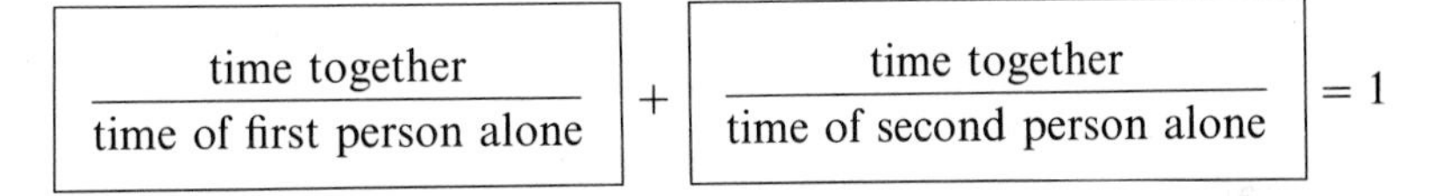

$$\frac{\text{time together}}{\text{time of first person alone}} + \frac{\text{time together}}{\text{time of second person alone}} = 1$$

or fractional part by first person + fractional part by second person = 1

EXAMPLE 1 Bob can mow Mr. Richard's lawn in 3 hours. Steve can mow Mr. Richard's lawn in 4 hours. How long will it take to mow the lawn if both Bob and Steve work together?

Solution: Let x = time, in hours, for both boys together to mow the

Bob can mow the entire lawn in 3 hours. Thus in 1 hour he can mow $\frac{1}{3}$ of the lawn by himself. In two hours he can mow $\frac{2}{3}$ of the lawn and in x hours he can mow $x/3$ of the lawn. Steve can mow the lawn in 4 hours. Thus in 1 hour he can mow $\frac{1}{4}$ of the lawn by himself. In 2 hours he can mow $\frac{2}{4}$ of the lawn and in x hours he can mow $x/4$ of the lawn. To solve this problem, we make use of the fact that:

$$\begin{matrix}\text{part of lawn mowed} \\ \text{by Bob in } x \text{ hours}\end{matrix} + \begin{matrix}\text{part of lawn mowed} \\ \text{by Steve in } x \text{ hours}\end{matrix} = 1 \text{ (whole lawn mowed)}$$

$$\frac{x}{3} + \frac{x}{4} = 1$$

Multiply both sides of the equation by the LCD, 12; then solve for x.

$$12\left(\frac{x}{3} + \frac{x}{4}\right) = 12 \cdot 1$$

$$12\left(\frac{x}{3}\right) + 12\left(\frac{x}{4}\right) = 12$$

$$4x + 3x = 12$$

$$7x = 12$$

$$x = \frac{12}{7} \quad \text{or} \quad 1\frac{5}{7} \text{ hours}$$

The two boys together can mow the lawn in $1\frac{5}{7}$ hours. Note that this is less time than it takes either boy to mow the lawn by himself (which is what we expect.) ■

EXAMPLE 2 Water pump A can drain an olympic-size pool in 20 hours of continuous operation. Water pump B can drain the same pool in 30 hours of continuous operation. How long will it take both water pumps working together to drain the pool?

Solution: Let t = total time for pumps operating together to drain pool.

$$\begin{matrix}\text{proportion of water} \\ \text{drained by pump } A \\ \text{in } t \text{ hours}\end{matrix} + \begin{matrix}\text{proportion of water} \\ \text{drained by pump } B \\ \text{in } t \text{ hours}\end{matrix} = 1 \text{ (whole pool drained)}$$

$$\frac{t}{20} + \frac{t}{30} = 1$$

$$60\left(\frac{t}{20} + \frac{t}{30}\right) = 60 \cdot 1$$

$$60\left(\frac{t}{20}\right) + 60\left(\frac{t}{30}\right) = 60$$

$$3t + 2t = 60$$

$$5t = 60$$

$$t = 12$$

Thus the two pumps working together can drain the pool in 12 hours. ■

EXAMPLE 3 A tank can be filled by one pipe in 4 hours and can be emptied by another pipe in 6 hours. If the valves to both pipes are open, how long will it take to fill the tank?

Solution: As one pipe is filling the tank, the other is emptying the tank. Thus the pipes are working against each other.

Let x = amount of time to fill the tank.

$$\begin{matrix}\text{proportion of water tank} \\ \text{filled in } x \text{ hours}\end{matrix} - \begin{matrix}\text{proportion of water tank} \\ \text{emptied in } x \text{ hours}\end{matrix} = 1 \text{ (total tank filled)}$$

$$\frac{x}{4} - \frac{x}{6} = 1$$

$$12\left(\frac{x}{4} - \frac{x}{6}\right) = 12 \cdot 1$$

$$12\left(\frac{x}{4}\right) - 12\left(\frac{x}{6}\right) = 12$$

$$3x - 2x = 12$$

$$x = 12$$

The tank will be filled in 12 hours. ■

2 Now let us look at a problem where we must find a given number.

EXAMPLE 4 What number multiplied by the numerator and added to the denominator of the fraction $\frac{4}{7}$ makes the resulting fraction equal $\frac{5}{3}$?

Solution: Let x = unknown number.

$$\frac{4x}{7+x} = \frac{5}{3} \qquad \text{now cross-multiply}$$

$$3(4x) = 5(7 + x)$$

$$12x = 35 + 5x$$

$$7x = 35$$

$$x = 5$$

The number is 5.

Check:

$$\frac{4 \cdot 5}{7+5} = \frac{20}{12} = \frac{5}{3} \quad ■$$

EXAMPLE 5 When the reciprocal of 3 times a number is subtracted from 1, the result is the reciprocal of twice the number. Find the number.

Solution: Let x = unknown number. Then $3x$ is 3 times the number and $\frac{1}{3x}$ is the reciprocal of 3 times the number. Twice the number is $2x$ and $\frac{1}{2x}$ is the reciprocal of twice the number.

$$1 - \frac{1}{3x} = \frac{1}{2x} \qquad \text{the LCD is } 6x$$

$$6x\left(1 - \frac{1}{3x}\right) = \frac{1}{2x} \cdot 6x$$

$$6x(1) - 6x\left(\frac{1}{3x}\right) = 6x\left(\frac{1}{2x}\right)$$

$$6x - 2 = 3$$

$$6x = 5$$

$$x = \frac{5}{6}$$

The number is $\frac{5}{6}$. ■

3 The last type of problem we will look at involves motion. Recall that we discussed motion problems earlier in Section 2.3. In that section we learned that distance = rate · time. Sometimes it is convenient to solve for the time when solving motion problems.

$$\text{time} = \frac{\text{distance}}{\text{rate}}$$

EXAMPLE 6 Amy Schumacher can fly her plane 300 miles against the wind in the same time it takes her to fly 400 miles with the wind. If the wind blows at 20 miles per hour, find the speed of the plane in still air.

Solution: Let $x =$ speed of plane in still air. Let us set up a table to help analyze the problem.

Plane	d	r	t
Against wind	300	$x - 20$	$\frac{300}{x - 20}$
With wind	400	$x + 20$	$\frac{400}{x + 20}$

Since the times are the same we set up and solve the following equation:

$$\frac{300}{x - 20} = \frac{400}{x + 20}$$
$$300(x + 20) = 400(x - 20)$$
$$300x + 6000 = 400x - 8000$$
$$6000 = 100x - 8000$$
$$14000 = 100x$$
$$140 = x$$

The speed of the plane in still air is 140 miles per hour. ■

EXAMPLE 7 Mary Kay rides her bike to and from her home to San Francisco City College. Going to school she rides mostly downhill and averages 15 miles per hour. Coming home, mostly uphill, she averages only 6 miles per hour. If it takes her $\frac{1}{2}$ hour longer for her to get home than to ride to school, how far is the college from her home?

Solution: Let $x =$ the distance from her home to the college. Note that in this problem the times are not equal. Her time returning is $\frac{1}{2}$ hour longer than going. Therefore, to make the times equal we must add $\frac{1}{2}$ hour to her time going (or subtract $\frac{1}{2}$ hour from her time returning).

	d	r	t
Going	x	15	$\frac{x}{15}$
Returning	x	6	$\frac{x}{6}$

$$\text{time going} + \frac{1}{2}\text{ hour} = \text{time returning}$$

$$\frac{x}{15} + \frac{1}{2} = \frac{x}{6}$$

$$30\left(\frac{x}{15}\right) + 30\left(\frac{1}{2}\right) = 30\left(\frac{x}{6}\right)$$

$$2x + 15 = 5x$$

$$15 = 3x$$

$$5 = x$$

Therefore, Mary Kay lives 5 miles from San Francisco City College. ■

EXAMPLE 8 The number ④ train in the New York City subway system goes from Woodlawn/Jerome Avenue in the Bronx to Flatbush Avenue/Brooklyn College in Brooklyn. The total one-way distance between these two stops is 24.2 miles. On this route two tracks run parallel to each other, one for the local train and the other for the express train. The local train stops at every station (48 stops), while the express stops at only certain selected stations (33 stations). The local and express trains leave Woodlawn/Jerome Avenue at the same time. When the express reaches Flatbush Avenue/Brooklyn College, the local is at Wall Street, 7.8 miles from Flatbush. If the express averages 5.2 miles per hour faster than the local, find the speed of the two trains.

Solution: Let x = speed of local
then $x + 5.2$ = speed of express

In the same time period that the express reaches the end of the line, 24.2 miles, the local will have traveled $24.2 - 7.8 = 16.4$ miles.

Train	d	r	t
Local	16.4	x	$\frac{16.4}{x}$
Express	24.2	$x + 5.2$	$\frac{24.2}{x + 5.2}$

$$\frac{16.4}{x} = \frac{24.2}{x + 5.2}$$

$$16.4(x + 5.2) = 24.2x$$

$$16.4x + 85.28 = 24.2x$$

$$85.28 = 7.8x$$

$$10.9 = x$$

The local averages 10.9 miles per hour and the express averages $10.9 + 5.2 = 16.1$ miles per hour. ■

Exercise Set 7.6

Solve each problem.

1. At the Community Savings Bank it takes one computer 4 hours to complete a certain job and a second computer 5 hours to complete the same job. How long will it take the two computers together to complete the job?
2. Mr. Dell can fertilize the farm in 6 hours. Mrs. Dell can fertilize the farm in 7 hours. How long will it take them to fertilize the farm if they work together?
3. Jason can paint a house in 8 days. Patti can paint a house in 6 days. How long will it take them to paint the house together?
4. A $\frac{1}{2}$-inch-diameter hose can fill a swimming pool in 8 hours. A $\frac{4}{5}$-inch-diameter hose can fill the same pool in 5 hours. How long will it take to fill the pool when both hoses are used?
5. Ramon can mow a lawn on a rider lawn mower in 4 hours. Donna can mow the lawn in 6 hours with a push lawn mower. How long will it take them to mow the lawn together?
6. A conveyer belt operating at full speed can fill a tank with top soil in 3 hours. When a valve at the bottom of the tank is opened, the tank will empty in 4 hours. If the conveyer belt is operating at full speed and the valve at the bottom of the tank is open, how long will it take to fill the tank?
7. When only the cold water valve is opened, a washtub will fill in 8 minutes. When only the hot water valve is opened, the washtub will fill in 12 minutes. When the drain of the washtub is open, it will drain completely in 7 minutes. If both the hot and cold water valves are open and the drain is open, how long will it take for the washtub to fill?
8. What number multiplied by the numerator and added to the denominator of the fraction $\frac{4}{3}$ makes the resulting fraction $\frac{5}{2}$?
9. What number added to the numerator and multiplied by the denominator of the fraction $\frac{3}{2}$ makes the resulting fraction $\frac{1}{8}$?
10. One number is twice another. The sum of their reciprocals is $\frac{3}{4}$. Find the numbers.
11. The sum of the reciprocals of two consecutive integers is $\frac{11}{30}$. Find the two integers.
12. The sum of the reciprocals of two consecutive even integers is $\frac{5}{12}$. Find the two integers.
13. When a number is added to both the numerator and denominator of the fraction $\frac{5}{7}$, the resulting fraction is $\frac{4}{5}$. Find the number added.
14. When 3 is added to twice the reciprocal of a number, the sum is $\frac{31}{10}$. Find the number.

See Exercise 19

15. The reciprocal of 3 less than a certain number is twice the reciprocal of 6 less than twice the number. Find the number(s).
16. If 3 times a number is added to twice the reciprocal of the number, the answer is 5. Find the number(s).
17. If 3 times the reciprocal of a number is subtracted from twice the reciprocal of the square of the number, the difference is -1. Find the number(s).
18. A Greyhound Bus can travel 400 kilometers in the same time that an Amtrak train can travel 600 kilometers. If the speed of the train is 40 kilometers per hour greater than that of the bus, find the speeds of the bus and train.
19. The speed of a boat in still water is 20 miles per hour. It takes the same amount of time for the boat to travel 3 miles downstream (with the current) as it does to travel 2 miles upstream (against the current). Find the speed of the current.
20. The speed of an airplane in no wind is 500 miles per hour. If the plane travels 300 miles with a tailwind (pushing the plane) in the same amount of time that it travels 280 miles against the headwind, find the speed of the wind.

21. The rate of a bicyclist is 8 miles per hour faster than that of a jogger. If the bicyclist travels 10 miles in the same amount of time that the jogger travels 5 miles, find the rate of the jogger.
22. Two rockets are launched. The first travels at 18,000 miles per hour. The second rocket is launched $\frac{1}{2}$ hour later in the same direction, and travels at 20,000 miles per hour. When the rockets are the same distance from the earth how far from the earth will they be?
23. Two trains leave at the same time and travel along parallel tracks. One train travels 10 miles per hour faster than the other. If the faster train travels 120 miles in the same time the slower train travels 100 miles, what are the speeds of the two trains?
24. Two cross-country skiers ski along the same path. One skier averages 8 miles per hour, while the other averages 6 miles per hour. If it takes the slower skier $\frac{1}{2}$ hour longer than the faster skier to reach the designated resting point, how far is the resting point from the starting point?
25. The current of a river is 3 miles per hour. It takes a motorboat a total of 3 hours to travel 12 miles upstream and return 12 miles downstream. What is the speed of the boat in still water?

7.7 Variation

1 *Write an equation expressing direct variation.*

2 *Write an equation expressing inverse variation.*

3 *Write an equation expressing joint variation.*

4 *Write an equation containing a combination of variations.*

In Sections 7.5 and 7.6 we saw many applications of equations containing fractions. In this section we see still more applications of rational equations.

1 Many scientific formulas are expressed in terms of variations. A **variation** is an equation that relates one variable to one or more other variables using the operations of multiplication or division (or both operations). There are essentially three types of variation problems: direct, inverse, and joint variation.

Direct Variation

In **direct variation** the two related variables will both increase together or both decrease together; that is, as one increases so does the other, and as one decreases so does the other.

Consider a car traveling at 30 miles an hour. The car travels 30 miles in 1 hour, 60 miles in 2 hours, and 90 miles in 3 hours. Notice that as the time increases, the distance traveled increases, and as the time decreases, the distance traveled decreases.

Also note that the ratio of distance to time is a constant (30) in each case:

$$\frac{\text{distance}}{\text{time}} = \frac{30}{1} = 30, \qquad \frac{60}{2} = 30, \qquad \frac{90}{3} = 30$$

The formula used to calculate distance traveled is

$$\text{distance} = \text{speed} \cdot \text{time}$$

Since the speed has been specified as a constant, 30 miles per hour, the formula can be written

$$d = 30t$$

We say that distance varies directly as time or that distance is directly proportional to time.

The above equation is an example of a direct variation.

Direct Variation

The general form of a direct variation is

$$x = ky$$

In this formula, k is called the **constant of proportionality** or the variation constant.

EXAMPLE 1 The circumference of a circle, C, is directly proportional (or varies directly) to its radius, r. Write the equation for the circumference of a circle if the constant of proportionality, k, is 2π.

Solution: $C = kr$ (C varies directly as r)
$C = 2\pi r$ (constant of proportionality is 2π) ■

EXAMPLE 2 The resistance (R) of a wire varies directly as its length (L).

(a) Write this variation as an equation.
(b) Find the resistance (measured in ohms) of a 20-foot length of wire assuming that the constant of proportionality for the wire is 0.007.

Solution: (a) $R = kL$
(b) $R = 0.007(20) = 0.14$
The resistance of the wire is 0.14 ohm. ■

In certain variation problems the constant of proportionality, k, may not be known. In such cases it can often be found by substituting given values in the variation and solving for k.

EXAMPLE 3 The gravitational force of attraction (F) between an object and the earth is directly proportional to the mass (m) of the object. If the force of attraction is 640 when the object's mass is 20, find the constant of proportionality.

Solution:

$$F = km$$

$$640 = k20$$

$$\frac{640}{20} = \frac{20k}{20}$$

$$32 = k$$

Thus the constant of proportionality is 32. ■

EXAMPLE 4 x varies directly as y. If x is 80 when y is 20, find x when y is 90.

Solution: Since the constant of proportionality is not given, we must first find k using the given information.

$$x = ky$$
$$80 = k20$$
$$\frac{80}{20} = \frac{20k}{20}$$
$$4 = k$$

We now use $k = 4$ to find x when y is 90.

$$x = ky$$
$$x = 4(90)$$
$$x = 360$$

Thus when y equals 90, x equals 360. ■

Inverse Variation

2 A second type of variation is **inverse variation.** When two quantities vary inversely it means that as one quantity increases, the other quantity decreases, and vice versa.

To explain inverse variation we will use the distance formula, distance = speed · time. If we solve for time we get time = distance/speed. Assume that the distance is fixed at 120 miles, then

$$\text{time} = \frac{120}{\text{speed}}$$

Note that at a speed of 120 miles per hour, it would take 1 hour to cover this distance. At a speed of 60 miles an hour, it would take 2 hours. At 30 miles an hour, it would take 4 hours. Note that as the speed decreases the time increases, and vice versa. Also note that the product of the speed and the time is a constant:

$$120 \cdot 1 = 120 \qquad 60 \cdot 2 = 120 \qquad 30 \cdot 4 = 120$$

The above equation can be written

$$t = \frac{120}{s}$$

This equation is an example of an inverse variation, where the time and speed are inversely proportional. The constant of proportionality is 120.

Inverse Variation

The general form of an inverse variation is

$$x = \frac{k}{y} \qquad \text{or} \qquad xy = k$$

Two quantities vary inversely, or are inversely proportional, when as one quantity increases the other quantity decreases. In the example just mentioned as the speed increases the time decreases, and vice versa. Thus the speed and time are inversely proportional to each other.

EXAMPLE 5 The illuminance (I) of a light source varies inversely as the square of the distance (d) from the source. Assuming that the illuminance is 75 units at a distance of 6 meters, find the equation that expresses the relationship between the illuminance and the distance.

Solution: The general form of the equation is

$$I = \frac{k}{d^2} \qquad (\text{or} \qquad Id^2 = k)$$

To find k we insert the given values for I and d.

$$75 = \frac{k}{6^2}$$

$$75 = \frac{k}{36}$$

$$(75)(36) = k$$

$$2700 = k$$

Thus the formula is $I = \dfrac{2700}{d^2}$. ■

EXAMPLE 6 x varies inversely as y. If $x = 8$ when $y = 15$, find x when $y = 18$.

Solution: First write the equation and solve for k.

$$x = \frac{k}{y}$$

$$8 = \frac{k}{15}$$

$$120 = k$$

Now find x when $y = 18$.

$$x = \frac{120}{y} = \frac{120}{18} = 6.7 \quad \text{(to the nearest tenth)}$$ ■

Joint Variation

3 One quantity may vary jointly as a product of two or more other quantities. This type of variation is called **joint variation.**

Joint Variation

The general form of a joint variation where x varies jointly as y and z is

$$x = kyz$$

EXAMPLE 7 The area, A, of a triangle varies jointly as its base, b, and height, h. If the area of a triangle is 48 square inches when its base in 12 inches and its height is 8 inches, find the area of a triangle with a base of 15 inches and a height of 20 inches.

Solution: First write the joint variation; then solve for k.

$$A = kbh$$
$$48 = k(12)(8)$$
$$48 = k(96)$$
$$\frac{48}{96} = k$$
$$k = \frac{1}{2}$$

Now solve for the area of the given triangle.

$$A = kbh$$
$$= \frac{1}{2}(15)(20)$$
$$= 150 \text{ square inches}$$ ■

Summary of Variations

Direct	*Inverse*	*Joint*
$x = ky$	$x = \frac{k}{y}$	$x = kyz$

Combined Variation

4 Often in real-life situations one variable varies as a combination of variables. The following examples illustrate the use of **combined variations.**

EXAMPLE 8 When studying the ability of a wire to stretch or its elasticity, E, we learn that the elasticity of a wire is directly proportional to its length, L, and inversely proportional to its cross-sectional area, A. Express E in terms of L and A.

Solution: $E = \frac{kL}{A}$ ■

EXAMPLE 9 The electrostatic force, F, of attraction or repulsion between two electrical charges is jointly proportional to the two charges, q_1 and q_2, and inversely proportional to the square of the distance, d, between the two charges. Express F in terms of q_1, q_2, and d.

Solution: $F = \frac{kq_1q_2}{d^2}$ ■

EXAMPLE 10 A varies jointly as B and C and inversely as the square of D. If $A = 1$ when $B = 9$, $C = 4$ and $D = 6$, find A when $B = 8$, $C = 12$, and $D = 5$.

Solution: $A = \dfrac{kBC}{D^2}$

We must first find the constant of proportionality, k, by inserting the given values for A, B, C, and D and solving for k.

$$1 = \frac{k(9)(4)}{6^2}$$

$$1 = \frac{36k}{36}$$

$$1 = k$$

Thus the constant of proportionality equals 1. Now we find A for the given values of B, C, and D.

$$A = \frac{(1)(8)(12)}{(5)^2} = \frac{96}{25} = 3.84 \quad ■$$

Exercise Set 7.7

Use your intuition to determine if the variation between the variables given is direct or inverse.

1. The speed and distance for a constant time period.
2. The distance between two cities on a map and the actual distance between the two cities.
3. The diameter of a hose and volume of water coming from the hose.
4. A given weight and the force needed to lift that weight.
5. The cubic-inch displacement in liters and the horsepower of the engine.
6. The light illuminating an object and the distance the light is from the object.
7. The volume of a balloon and its radius.
8. The length of a board and the force applied to the center needed to break the board.
9. The shutter opening of a camera and the amount of sunlight to reach the film.
10. A person's weight (due to the earth's gravity) and his distance from the earth.
11. The number of pages a person can read in a given period of time and his reading speed.
12. The time it takes an ice cube to melt in water and the temperature of the water.
13. The time needed to get proper exposure on a film and the aperture opening of the camera lens.
14. Time to reach a certain point for a plane flying with the wind and the speed of the wind.
15. The number of calories eaten and the amount of exercise required to burn off those calories.

(a) For Exercises 16 through 33 write the variation and (b) find the quantity indicated.

16. x varies directly as y. Find x when $y = 12$ and $k = 6$.
17. C varies directly as the square of Z. Find C when $Z = 9$ and $k = \frac{3}{4}$.
18. y varies directly as R. Find y when $R = 180$ and $k = 1.7$.
19. x varies inversely as y. Find x when $y = 25$ and $k = 5$.
20. R varies inversely as W. Find R when $W = 160$ and $k = 8$.
21. L varies inversely as the square of P. Find L when $P = 4$ and $k = 100$.
22. A varies directly as B and inversely as C. Find A when $B = 12$, $C = 4$, and $k = 3$.
23. A varies jointly as R_1 and R_2 and inversely as the square of L. Find A when $R_1 = 120$, $R_2 = 8$, $L = 5$, and $k = \frac{3}{2}$.
24. T varies directly as the square of D and inversely as F. Find T when $D = 8$, $F = 15$, and $k = 12$.

25. x varies directly as y. If x is 9 when y is 18, find x when y is 36.
26. Z varies directly as W. If Z is 7 when W is 28, find Z when W is 140.
27. y varies directly as the square of R. If y is 5 when $R = 5$, find y when R is 10.
28. S varies inversely as G. If S is 12 when G is 0.4, find S when G is 5.
29. C varies inversely as J. If C is 7 when J is 0.7, find C when J is 12.
30. x varies inversely as the square of P. If $x = 10$ when P is 6, find x when $P = 20$.
31. F varies jointly as M_1 and M_2 and inversely as d. If F is 20 when $M_1 = 5$, $M_2 = 10$, and $d = 0.2$, find F when $M_1 = 10$, $M_2 = 20$, and $d = 0.4$.
32. F varies jointly as q_1 and q_2 and inversely as the square of d. If f is 8 when $q_1 = 2$, $q_2 = 8$, and $d = 4$, find f when $q_1 = 28$, $q_2 = 12$, and $d = 2$.
33. S varies jointly as I and the square of T. If S is 8 when $I = 20$ and $T = 4$, find S when $I = 2$ and $T = 2$.
34. The volume of a gas, V, varies inversely as its pressure, P. If the volume, V, is 800 cc when the pressure is 200 millimeters (mm) of mercury, find the volume when the pressure is 25 mm of mercury.
35. The amount a spring will stretch, S, varies directly with the force (or weight), F, attached to the spring. If a spring stretches 1.4 inches when 20 pounds is attached, how far will it stretch when 10 pounds is attached?
36. The pressure, P, on an object submerged in water varies directly with its depth, D. If the pressure at a depth of 50 feet is 21.6 pounds per square inch, find the pressure at a depth of 180 feet.
37. The intensity, I, of light received at a source varies inversely as the square of the distance, d, from the source. If the light intensity is 20 footcandles at 15 feet, find the light intensity at 12 feet.
38. On earth the mass of an object varies directly with its weight. If an object with a weight of 256 pounds has a mass of 8 slugs, find the mass of an object weighing 120 pounds.
39. The weight, W, of an object in the earth's atmosphere varies inversely with the square of the distance, d, between the object and the center of the earth. A 140-pound person standing on earth is approximately 4000 miles from the earth's center. Find the weight (or gravitational force of attraction) of this person at a distance 100 miles from the earth's surface.
40. The wattage rating of an appliance, W, varies jointly as the square of the current, I, and the resistance, R. If the wattage is 1 watt when the current is 0.1 ampere and the resistance is 100 ohms, find the wattage when the current is 0.4 ampere and the resistance is 250 ohms.
41. The electrical resistance of a wire, R, varies directly as its length, L, and inversely as its cross-sectional area, A. If the resistance of a wire is 0.2 ohm when the length is 200 feet and its cross-sectional area is 0.05 square inch, find the resistance of a wire whose length is 5000 feet with a cross-sectional area of 0.01 square inch.

Summary

Glossary

Algebraic fraction (or rational expression): An expression of the form p/q, where p and q are polynomials, $q \neq 0$.
Combined variation: A variation problem that involves two or more different types of variations.
Complex fraction: A fractional expression that has a fraction in its numerator or its denominator or both its numerator and denominator.
Constant of proportionality: The constant in a variation problem.
Domain of a rational expression: The set of values that can replace the variable in the expression.
Rational function: Functions of the form $f(x) = p/q$, where p and q are polynomials, $q \neq 0$.
Reduced to lowest terms: An algebraic fraction is reduced to its lowest terms when the numerator and denominator have no common factor other than 1.
Variation: An equation that relates one variable to one or more other variables using the operations of multiplication or division (or both operations).

Important Facts

Types of Variation

Direct	*Inverse*	*Joint*
$x = ky$	$x = \frac{k}{y}$	$x = kyz$

Review Exercises

[7.1] Determine the domain of each of the following.

1. $\frac{3}{x-4}$ **2.** $\frac{x}{x+1}$ **3.** $\frac{-2x}{x^2+5}$

4. $\frac{0}{(x+3)^2}$ **5.** $\frac{x+6}{x^2}$ **6.** $\frac{x^2-2}{x^2-3x-10}$

Write each expression in reduced form.

7. $\frac{x^2+xy}{x+y}$ **8.** $\frac{x^2-9}{x+3}$ **9.** $\frac{4-5x}{5x-4}$

10. $\frac{x^2+2x-3}{x^2+x-6}$ **11.** $\frac{2x^2-6x+5x-15}{2x^2+7x+5}$ **12.** $\frac{a^3-8}{a^2-4}$

[7.2] Multiply as indicated.

13. $\frac{15x^2y^3}{3z} \cdot \frac{6z^3}{5xy^3}$

14. $\frac{1}{x-2} \cdot \frac{2-x}{2}$

15. $\frac{4x+4y}{x^2y} \cdot \frac{y^3}{8x}$

16. $\frac{a-2}{a+3} \cdot \frac{a^2+4a+3}{a^2-a-2}$

17. $\frac{x^2-y^2}{x-y} \cdot \frac{x+y}{xy+x^2}$

18. $\frac{4x^2+8x-5}{2x+5} \cdot \frac{x+1}{4x^2-4x+1}$

19. $\frac{1}{a-3} \cdot \frac{a^2-2a-3}{a^2+3a+2}$

20. $\frac{2x^2+10x+12}{(x+2)^2} \cdot \frac{x+2}{x^3+5x^2+6x}$

Divide as indicated.

21. $\frac{8xy^2}{z} \div \frac{x^4y^2}{4z^2}$

22. $\frac{3x+3y}{x^2} \div \frac{x^2-y^2}{x^2}$

23. $\frac{1}{a^2+8a+15} \div \frac{3}{a+5}$

24. $(x+3) \div \frac{x^2-4x-21}{x-7}$

25. $\frac{x^2-3xy-10y^2}{6x} \div \frac{x+2y}{12x^2}$

26. $\frac{4x^2-16y^2}{9} \div \frac{(x+2y)^2}{12}$

27. $\frac{y^4-x^6}{x^3-y^2} \div (y^2-x^3)$

28. $\frac{x^3+27}{4x^2-4} \div \frac{x^2-3x+9}{(x-1)^2}$

[7.3] Add or subtract as indicated.

29. $\frac{4x}{x+2} + \frac{8}{x+2}$

30. $\frac{7x-3}{x^2+7x-30} - \frac{3x+9}{x^2+7x-30}$

31. $\dfrac{4x^2 - 11x + 4}{x - 3} - \dfrac{x^2 - 4x + 10}{x - 3}$

32. $\dfrac{6x^2 - 4x}{2x - 3} - \dfrac{-3x + 12}{2x - 3} - \dfrac{2x + 4}{2x - 3}$

33. $\dfrac{6x^2 - 4x}{2x - 3} - \dfrac{-x + 4}{2x - 3} - \dfrac{6x^2 + x - 2}{2x - 3}$

Find the least common denominator.

34. $\dfrac{6x}{x + 1} - \dfrac{3}{x}$

35. $\dfrac{9x - 3}{x + y} - \dfrac{4x + 7}{x^2 - y^2}$

36. $\dfrac{19x - 5}{x^2 + 2x - 35} + \dfrac{3x - 2}{x^2 + 9x + 14}$

37. $\dfrac{3}{(x + 2)^2} - \dfrac{6(x + 3)}{x^2 - 4} - \dfrac{4x}{x + 1}$

Add or subtract as indicated.

38. $\dfrac{4}{2x} + \dfrac{x}{x^2}$

39. $\dfrac{1}{4x} + \dfrac{6x}{xy}$

40. $\dfrac{5x}{3xy} - \dfrac{4}{x^2}$

41. $6 + \dfrac{x}{x + 2}$

42. $5 - \dfrac{3}{x + 3}$

43. $\dfrac{a + c}{c} - \dfrac{a - c}{a}$

44. $\dfrac{3}{x + 3} + \dfrac{4}{x}$

45. $\dfrac{2}{3x} - \dfrac{3x}{3x - 6}$

46. $\dfrac{x - 4}{x - 5} - \dfrac{3}{x + 5}$

47. $\dfrac{4}{x + 5} + \dfrac{6}{(x + 5)^2}$

48. $\dfrac{x + 3}{x^2 - 9} + \dfrac{2}{x + 3}$

49. $\dfrac{4}{(x + 2)(x - 3)} - \dfrac{4}{(x - 2)(x + 2)}$

50. $\dfrac{x + 2}{x^2 - x - 6} + \dfrac{x - 3}{x^2 - 8x + 15}$

51. $\dfrac{x + 5}{x^2 - 15x + 50} - \dfrac{x - 2}{x^2 - 25}$

52. $\dfrac{1}{x + 3} - \dfrac{2}{x - 3} + \dfrac{6}{x^2 - 9}$

53. $\dfrac{x - 4}{x - 5} - \dfrac{3}{x + 5} - \dfrac{10}{x^2 - 25}$

[7.4] *Simplify the complex fraction.*

54. $\dfrac{1 + \dfrac{5}{12}}{\dfrac{3}{8}}$

55. $\dfrac{4 - \dfrac{9}{16}}{1 + \dfrac{5}{8}}$

56. $\dfrac{\dfrac{15xy}{6z}}{\dfrac{3x}{z^2}}$

57. $\dfrac{36x^4y^2}{\dfrac{9xy^5}{4z^2}}$

58. $\dfrac{x + \dfrac{1}{y}}{y^2}$

59. $\dfrac{x - \dfrac{x}{y}}{\dfrac{1 + x}{y}}$

60. $\dfrac{\dfrac{4}{x} + \dfrac{2}{x^2}}{6 - \dfrac{1}{x}}$

61. $\dfrac{\dfrac{x}{x + y}}{\dfrac{x^2}{2x + 2y}}$

62. $\dfrac{a^{-1}}{a^{-2}}$

63. $\dfrac{a^{-1} + 2}{a^{-1} + \dfrac{1}{a}}$

64. $\dfrac{x^{-2} + \dfrac{1}{x}}{\dfrac{1}{x^2} - \dfrac{1}{x}}$

65. $\dfrac{\dfrac{3x}{y} - x}{\dfrac{y}{x} - 1}$

[7.5] Solve the equation.

66. $\frac{3}{x} = \frac{8}{24}$

67. $\frac{4}{a} = \frac{16}{4}$

68. $\frac{x+3}{5} = \frac{9}{5}$

69. $\frac{x}{6} = \frac{x-4}{2}$

70. $\frac{3x+4}{5} = \frac{2x-8}{3}$

71. $\frac{x}{5} + \frac{x}{2} = 14$

72. $\frac{4}{x} - \frac{1}{6} = \frac{1}{x}$

73. $\frac{1}{x-2} + \frac{1}{x+2} = \frac{1}{x^2-4}$

74. $\frac{x-3}{x-2} + \frac{x+1}{x+3} = \frac{2x^2+x+1}{x^2+x-6}$

75. $\frac{x}{x^2-9} + \frac{2}{x+3} = \frac{4}{x-3}$

Solve each problem.

76. Three resistors of 200, 400, and 1200 ohms, respectively, are wired in parallel. Find the total resistance of the circuit.

77. Two resistors are to be wired in parallel. One is to contain twice the resistance of the other. What should be the resistance of each resistor if the circuit's total resistance is to be 600 ohms?

78. What is the focal length of a mirror if the object distance is 12 centimeters and the image distance is 4 centimeters?

79. A mirror has a focal length of 10 centimeters. Find the object's distance from the lens if the image distance is twice the object's distance.

[7.6] Solve each problem.

80. It takes Dan 3 hours to mow Mr. Lee's lawn. It takes Kim 4 hours to mow the same lawn. How long will it take them working together to mow Mr. Lee's lawn?

81. A hose of $\frac{3}{4}$-inch diameter can fill a swimming pool in 7 hours. A hose of $\frac{5}{16}$-inch diameter can fill the pool in 12 hours. How long will it take to fill the pool when both hoses are on?

82. What number multiplied by the numerator and added to the denominator of the fraction $\frac{5}{8}$ makes the resulting value equal to 1?

83. When the reciprocal of twice a number is subtracted from 1, the result is the reciprocal of 3 times the number. Find the number.

84. Kit Waickman's motorboat can travel 15 miles per hour in still water. Traveling with the current of a river the boat can travel 20 miles in the same time it takes to go 10 miles against the current. Find the rate of the current.

85. A small plane and a car start from the same location, at the same time, heading toward the same town 450 miles away. The speed of the plane is 3 times the speed of the car. The plane arrives at the town 6 hours ahead of the car. Find the speed of the car and the plane.

[7.7] Find the quantity indicated.

86. A is directly proportional to B. If A is 120 when $B = 80$, find A when $B = 50$.

87. A is directly proportional to the square of C. If A is 5 when C is 5, find A when $C = 10$.

88. x is inversely proportional to y. If x is 20 when $y = 5$, find x when $y = 100$.

89. W is directly proportional to L and inversely proportional to A. If $W = 80$ when $L = 100$ and $A = 20$, find W when $L = 50$ and $A = 40$.

90. z is jointly proportional to x and y and inversely proportional to the square of r. If z is 12 when x is 20, $y = 8$, and $r = 8$, find z when $x = 10$, $y = 80$, and $r = 3$.

91. The scale of a map is 1 inch to 60 miles. How large a distance on the map represents 300 miles?

92. An electric company charges $1.62 per kilowatt-hour. What is the electric bill if 74 kilowatt-hours are used in a month?

93. The distance, d, an object drops in free fall is directly proportional to the square of the time, t. If an object falls 16 feet in 1 second, how far will an object fall in 5 seconds?

94. The area, A, of a circle varies directly with the square of its radius, r. If the area is 78.5 when the radius is 5, find the area when the radius is 8.

95. The time t, it takes for an ice cube to melt is inversely proportional to the temperature of the water it is in. If it takes an ice cube 1.7 minutes to melt in 70°F water temperature, how long will it take the same-size ice cube to melt in 50°F water?

Practice Test

1. Find the domain of $\frac{x-3}{x^2-3x-28}$.

2. Reduce to lowest terms: $\frac{x^2-5x-36}{9-x}$.

Perform the operations indicated.

3. $\frac{6x^2y^4}{4z^2} \cdot \frac{8xz^3}{9y^4}$

4. $\frac{a^2-9a+14}{a-2} \cdot \frac{a^2-4a-21}{(a-7)^2}$

5. $\frac{x^2-9y^2}{3x+6y} \div \frac{x+3y}{x+2y}$

6. $\frac{x^3+y^3}{x+y} \div \frac{x^2-xy+y^2}{x^2+y^2}$

7. $\frac{5}{x} + \frac{3}{2x^2}$

8. $\frac{x-5}{x^2-16} - \frac{x-2}{x^2+2x-8}$

9. $\frac{x+1}{4x^2-4x+1} + \frac{3}{2x^2+5x-3}$

10. *Simplify.*

$$\frac{\frac{1}{x}+\frac{1}{y}}{\frac{1}{x}-\frac{1}{y}}$$

11. *Simplify.*

$$\frac{x+\frac{x}{y}}{x^{-1}+y^{-1}}$$

12. *Solve the equation.*

$$\frac{x}{3}-\frac{x}{4}=5$$

13. *Solve the equation.*

$$\frac{x}{x-8}+\frac{6}{x-2}=\frac{x^2}{x^2-10x+16}$$

14. P varies directly as Q and inversely as R. If $P = 8$ when $Q = 4$ and $R = 10$, find P when $Q = 10$ and $R = 20$.
15. W varies jointly as P and Q and inversely as the square of T. If $W = 6$ when $P = 20$, $Q = 8$, and $T = 4$, find W if $P = 30$, $Q = 4$, and $T = 8$.
16. Kris, on his tractor, can level a 1-acre field in 8 hours. Heather, on her tractor, can level a 1-acre field in 5 hours. How long will it take them to level a 1-acre field if they work together?

See Practice Test 16.

8 Roots, Radicals, and Complex Numbers

See Section 8.6, Exercise 32

8.1 Radicals and Rational Exponents

1. *Find principal square roots.*
2. *Find cube and higher roots.*
3. *Know when the root of a number is positive, negative, or imaginary.*
4. *Evaluate radical expressions using absolute value.*
5. *Write radical expressions in exponential form.*
6. *Know that for $a > 0$, $\sqrt[n]{a^n} = a$.*

In this chapter we expand on the concept of square root introduced in Chapter 1.

In the expression $\sqrt{x}$, the $\sqrt{\ }$ is called the **radical sign.** The number or expression within the radical sign is called the **radicand.**

radical sign

$$\sqrt{x}$$

radicand

The entire expression, including the radical sign and radicand, is called the **radical expression.** Another part of the radical expression is its index. The **index** tells the "root" of the expression. Square roots have an index of 2. The index of square roots is generally not written.

$$\sqrt{x} \quad \text{means} \quad \sqrt[2]{x}$$

1 Every positive number has two square roots, a principal or positive square root and a negative square root.

> The **principal or positive square root** of a positive real number x, written $\sqrt{x}$, is that *positive* number whose square equals x.

Examples

$$\sqrt{25} = 5 \qquad \text{since } 5^2 = 5 \cdot 5 = 25$$

$$\sqrt{36} = 6 \qquad \text{since } 6^2 = 6 \cdot 6 = 36$$

$$\sqrt{\frac{4}{9}} = \frac{2}{3} \qquad \text{since } \left(\frac{2}{3}\right)^2 = \left(\frac{2}{3}\right)\left(\frac{2}{3}\right) = \frac{4}{9}$$

The negative square root of a positive real number x, written $-\sqrt{x}$, is the additive inverse or opposite of the principal square root. For example, $-\sqrt{25} = -5$ and $-\sqrt{36} = -6$. *Whenever we use the term "square root" in this book we will be referring to the principal square root.*

2 Other types of radical expressions have different indexes. For example, $\sqrt[3]{x}$ is the third or cube root of x. The index of cube roots is 3. In the expression $\sqrt[5]{xy}$, read the fifth root of xy, the index is 5 and the radicand is xy.

Radical expressions that have indices of 2, 4, 6, . . . or any even number are said to be **even roots.** Examples of even roots are $\sqrt{9}$, $\sqrt[4]{x}$, $\sqrt[12]{9x^5}$.

Radical expressions that have indices of 3, 5, 7, . . . or any odd number are said to be **odd roots.** Examples of odd roots are $\sqrt[3]{27}$, $\sqrt[5]{x}$, $\sqrt[19]{6x^2y^5}$.

Consider radicals of the form $\sqrt[n]{x}$, **where n is an even index and x is a positive real number.** The nth root of x is that *positive* number c such that $c^n = x$.

Examples

$$\sqrt{9} = 3 \qquad \text{since } 3^2 = 3 \cdot 3 = 9$$
$$\sqrt[4]{16} = 2 \qquad \text{since } 2^4 = 2 \cdot 2 \cdot 2 \cdot 2 = 16$$
$$\sqrt[4]{81} = 3 \qquad \text{since } 3^4 = 3 \cdot 3 \cdot 3 \cdot 3 = 81$$

Note that

$$-\sqrt{9} = -3$$
$$-\sqrt[4]{16} = -2$$
$$-\sqrt[4]{81} = -4$$

When considering radical expressions with even indices, the radicand must be a positive value if the number is to be real. For example, what real number is $\sqrt{-9}$ equal to? What real number when squared has a value of -9? Since the square of any real number cannot be negative, there is no real number that equals $\sqrt{-9}$. Numbers such as $\sqrt{-9}$ are called **imaginary numbers** and will be discussed shortly.

Now consider radicals of the form $\sqrt[m]{x}$, **where m is an odd index and x is any real number.** The mth root of x is that real number c such that $c^m = x$.

Examples

$$\sqrt[3]{8} = 2 \qquad \text{since } 2^3 = 2 \cdot 2 \cdot 2 = 8$$
$$\sqrt[3]{-8} = -2 \qquad \text{since } (-2)^3 = (-2)(-2)(-2) = -8$$
$$\sqrt[5]{243} = 3 \qquad \text{since } 3^5 = 3 \cdot 3 \cdot 3 \cdot 3 \cdot 3 = 243$$
$$\sqrt[5]{-243} = -3 \qquad \text{since } (-3)^5 = (-3)(-3)(-3)(-3)(-3) = -243$$

Note that

$$-\sqrt[3]{8} = -2 \qquad -\sqrt[5]{243} = -3$$
$$-\sqrt[3]{-8} = -(-2) = 2 \qquad -\sqrt[5]{-243} = -(-3) = 3$$

An odd root of a positive number is a positive number, and an odd root of a negative number is a negative number.

3 This information is summarized in Table 8.1.

Table 1

	n is even	n is odd
a is a positive real number	$\sqrt[n]{a}$ is a positive real number	$\sqrt[n]{a}$ is a positive real number
a is a negative real number	$\sqrt[n]{a}$ is not a real number (it is imaginary)	$\sqrt[n]{a}$ is a negative real number
a is 0	$\sqrt[n]{0} = 0$	$\sqrt[n]{0} = 0$

4 We can evaluate $\sqrt{a^2}$ for any real number a using absolute value.

$$\sqrt{a^2} = |a| \qquad \text{for any real number } a$$

Examples

$$\sqrt{4^2} = |4| = 4 \qquad \sqrt{7^2} = |7| = 7$$
$$\sqrt{(-4)^2} = |-4| = 4 \qquad \sqrt{(-7)^2} = |-7| = 7$$

Note that for $a > 0$, $\sqrt{a^2} = a$, and for $a < 0$, $\sqrt{a^2} = -a$. For example, $\sqrt{4^2} = 4$ and $\sqrt{(-4)^2} = -(-4) = 4$.

EXAMPLE 1 Use absolute value to evaluate each of the following.

(a) $\sqrt{5^2}$ (b) $\sqrt{(-5)^2}$ (c) $\sqrt{(-71)^2}$

Solution: (a) $\sqrt{5^2} = |5| = 5$
(b) $\sqrt{(-5)^2} = |-5| = 5$
(c) $\sqrt{(-71)^2} = |-71| = 71$ ■

EXAMPLE 2 Write each of the following as an absolute value.

(a) $\sqrt{(x+2)^2}$ (b) $\sqrt{(y-7)^2}$ (c) $\sqrt{(x^2-5x+6)^2}$

Solution: (a) $\sqrt{(x+2)^2} = |x+2|$
(b) $\sqrt{(y-7)^2} = |y-7|$
(c) $\sqrt{(x^2-5x+6)^2} = |x^2-5x+6|$ ■

If we know that a represents a nonnegative number, then we can write $\sqrt{a^2} = a$. For example, $\sqrt{4^2} = 4$ and $\sqrt{5^2} = 5$.

For the remainder of this chapter we make the assumption that all variables represent positive real numbers.

We make the foregoing assumption so that we can write many answers without absolute value signs. With this assumption, when we evaluate a radical like $\sqrt{y^2}$, we can write the answer as y rather than $|y|$.

Exponential Form of Radical Expressions

5 A radical expression of the form $\sqrt[n]{a}$ can be written as an exponential expression using the following rule.

For any nonnegative number a, and n a positive integer

$$\sqrt[n]{a} = a^{1/n}$$

Examples

$$\sqrt{6} = \sqrt[2]{6} = 6^{1/2}$$
$$\sqrt{x} = \sqrt[2]{x} = x^{1/2}$$
$$\sqrt[3]{9} = 9^{1/3}$$
$$\sqrt[3]{x} = x^{1/3}$$
$$\sqrt[4]{y} = y^{1/4}$$
$$\sqrt[5]{3} = 3^{1/5}$$

We can expand the rule so that radicals of the form $\sqrt[n]{a^m}$ can be expressed as exponential expressions.

For any nonnegative number a, and m and n integers, $n \neq 0$,

$$\sqrt[n]{a^m} = (\sqrt[n]{a})^m = a^{m/n}$$

(power: m; index: n)

Examples

$$\sqrt{3^5} = 3^{5/2} \qquad \sqrt[3]{y^5} = y^{5/3} \qquad \sqrt[4]{2^{12}} = 2^{12/4} = 2^3 = 8$$
$$\sqrt{x^6} = x^{6/2} = x^3 \qquad \sqrt[3]{z^3} = z^{3/3} = z^1 = z \qquad \sqrt[6]{y^{18}} = y^{18/6} = y^3$$
$$(\sqrt{x})^8 = x^{8/2} = x^4 \qquad (\sqrt[3]{z})^{12} = z^{12/3} = z^4 \qquad (\sqrt[3]{3})^9 = 3^{9/3} = 3^3 = 27$$

EXAMPLE 3 Write each of the following in exponential form and then simplify.

(a) $\sqrt{9^6}$ (b) $\sqrt[3]{2^{12}}$ (c) $\sqrt{x^{10}}$ (d) $\sqrt[5]{y^{15}}$

Solution:

(a) $\sqrt{9^6} = 9^{6/2} = 9^3 = 729$

(b) $\sqrt[3]{2^{12}} = 2^{12/3} = 2^4 = 16$

(c) $\sqrt{x^{10}} = x^{10/2} = x^5$

(d) $\sqrt[5]{y^{15}} = y^{15/5} = y^3$ ■

EXAMPLE 4 Write each of the following in exponential form and then simplify.

(a) $(\sqrt[3]{9})^6$ (b) $(\sqrt[4]{x})^{20}$ (c) $\sqrt[5]{x^5}$ (d) $(\sqrt[4]{x})^4$

Solution: (a) $(\sqrt[3]{9})^6 = 9^{6/3} = 9^2 = 81$

(b) $(\sqrt[4]{x})^{20} = x^{20/4} = x^5$

(c) $\sqrt[5]{x^5} = x^{5/5} = x^1 = x$

(d) $(\sqrt[4]{x})^4 = x^{4/4} = x^1 = x$ ■

6 In Example 4(c) we showed that $\sqrt[5]{x^5} = x$. Note that the index, 5, is the same as the power, 5. In the expression, $\sqrt[n]{a^n}$, the power, n in the radical matches the index, n.

$$\text{index} \longrightarrow \sqrt[n]{a^n} \longleftarrow \text{power}$$

When this happens, and $a \geq 0$, the radical simplifies to a, the base of the exponential expression in the radicand.

$$\sqrt[n]{a^n} = a^{\frac{n}{n}} = a$$

Since we are assuming that all variables represent positive real numbers, we have the following examples.

Examples

$$\sqrt{5^2} = 5 \qquad \sqrt[5]{x^5} = x$$
$$\sqrt{x^2} = x \qquad \sqrt[4]{y^4} = y$$

Note that when evaluating $\sqrt{(-2)^2}$, the property above cannot be used since $x < 0$.

$$\sqrt{(-2)^2} \neq -2^{2/2}$$
$$\neq -2^1$$
$$\neq -2$$

However, $\sqrt{(-2)^2} = |-2| = 2$.

Exercise Set 8.1

Evaluate.

1. $\sqrt{25}$
2. $\sqrt[3]{27}$
3. $\sqrt[3]{-27}$
4. $\sqrt[5]{32}$
5. $\sqrt[3]{125}$
6. $\sqrt[4]{81}$
7. $\sqrt{121}$
8. $\sqrt[6]{64}$
9. $\sqrt[3]{-8}$
10. $\sqrt[3]{216}$
11. $\sqrt{144}$
12. $\sqrt[4]{256}$
13. $\sqrt[5]{1}$
14. $\sqrt[3]{-125}$
15. $\sqrt[3]{343}$
16. $\sqrt[5]{-32}$

Use absolute value to evaluate the following.

17. $\sqrt{6^2}$
18. $\sqrt{(-6)^2}$
19. $\sqrt{(-1)^2}$
20. $\sqrt{(-17)^2}$
21. $\sqrt{(43)^2}$
22. $\sqrt{(-96)^2}$
23. $\sqrt{(147)^2}$
24. $\sqrt{(-147)^2}$
25. $\sqrt{(-83)^2}$
26. $\sqrt{(-89)^2}$
27. $\sqrt{(179)^2}$
28. $\sqrt{(213)^2}$

Write as an absolute value.

29. $\sqrt{(y-8)^2}$
30. $\sqrt{(x-7)^2}$
31. $\sqrt{(x-3)^2}$
32. $\sqrt{(3x^2-y)^2}$
33. $\sqrt{(3x+5)^2}$
34. $\sqrt{(x^2-3x+4)^2}$
35. $\sqrt{(6-3x)^2}$
36. $\sqrt{(4-5x^2)^2}$
37. $\sqrt{(y^2-4y+3)^2}$
38. $\sqrt{(x^2-3x)^2}$

Write in exponential form and then simplify. Assume all variables represent positive real numbers.

39. $\sqrt{7^2}$
40. $\sqrt{5^4}$
41. $\sqrt{3^{10}}$
42. $\sqrt[3]{5^9}$
43. $\sqrt[3]{4^3}$
44. $\sqrt[3]{2^{12}}$
45. $\sqrt[5]{8^{10}}$
46. $\sqrt[4]{7^4}$
47. $\sqrt[4]{6^8}$
48. $\sqrt[5]{4^{15}}$
49. $\sqrt[6]{2^{18}}$
50. $\sqrt{y^6}$
51. $\sqrt{z^{12}}$
52. $\sqrt{x^8}$
53. $\sqrt{y^{18}}$
54. $\sqrt{m^4}$
55. $\sqrt[3]{x^3}$
56. $\sqrt[3]{y^{12}}$
57. $\sqrt[3]{z^{21}}$
58. $\sqrt[4]{x^{20}}$
59. $\sqrt[5]{x^{30}}$
60. $\sqrt[6]{y^{18}}$
61. $\sqrt[3]{x^{60}}$
62. $\sqrt{x^2}$
63. $\sqrt[3]{y^3}$
64. $\sqrt[5]{x^5}$
65. $\sqrt[8]{x^8}$
66. $\sqrt[5]{z^5}$
67. $\sqrt[5]{x^{20}}$

68. In this section we stated that for $a \geq 0$,

$$\sqrt[n]{a^m} = a^{\frac{m}{n}}$$

Give an example to show that this does not hold true for $a < 0$.

69. Consider the expression $\sqrt[n]{x}$. Under what circumstances will this expression not be a real number?

70. Consider the expression $\sqrt[n]{x^n}$. Explain why this expression will be a real number for any real number x.

71. Consider the expression $\sqrt[n]{x^m}$. Under what circumstances will this expression not be a real number?

8.2 Multiplying and Simplifying Radicals

1. *Know the product rule for radicals.*
2. *Know the meaning of perfect factors (or perfect powers).*
3. *Simplify radicals.*
4. *Simplify a product of two radicals.*

1 We will first simplify radicals using the product rule. In the next section, we will simplify radicals using the quotient rule.

Product Rule for Radicals

For natural number n, and nonnegative real numbers a and b,

$$\sqrt[n]{a} \cdot \sqrt[n]{b} = \sqrt[n]{ab}$$

Examples of the Product Rule

$$\sqrt{60} = \begin{cases} \sqrt{1} \cdot \sqrt{60} \\ \sqrt{2} \cdot \sqrt{30} \\ \sqrt{3} \cdot \sqrt{20} \\ \sqrt{4} \cdot \sqrt{15} \\ \sqrt{5} \cdot \sqrt{12} \\ \sqrt{6} \cdot \sqrt{10} \end{cases} \qquad \sqrt[3]{60} = \begin{cases} \sqrt[3]{1} \cdot \sqrt[3]{60} \\ \sqrt[3]{2} \cdot \sqrt[3]{30} \\ \sqrt[3]{3} \cdot \sqrt[3]{20} \\ \sqrt[3]{4} \cdot \sqrt[3]{15} \\ \sqrt[3]{5} \cdot \sqrt[3]{12} \\ \sqrt[3]{6} \cdot \sqrt[3]{10} \end{cases}$$

$\sqrt{60}$ can be factored into any of these forms

$\sqrt[3]{60}$ can be factored into any of these forms

$$\sqrt{x^7} = \begin{cases} \sqrt{x} \cdot \sqrt{x^6} \\ \sqrt{x^2} \cdot \sqrt{x^5} \\ \sqrt{x^3} \cdot \sqrt{x^4} \end{cases} \qquad \sqrt[3]{x^7} = \begin{cases} \sqrt[3]{x} \cdot \sqrt[3]{x^6} \\ \sqrt[3]{x^2} \cdot \sqrt[3]{x^5} \\ \sqrt[3]{x^3} \cdot \sqrt[3]{x^4} \end{cases}$$

$\sqrt{x^7}$ can be factored into any of these forms; $\sqrt[3]{x^7}$ can be factored into any of these forms

2 To help clarify our explanations we will introduce the phrase perfect factor (or perfect power). When a radicand is an expression raised to a power, and that power is divisible by the index of the radical, we say that the radicand is a **perfect factor** (or **power**) of the radical.

Examples

$\sqrt{a^2} = a^{2 \div 2} = a^1$, thus a^2 is a perfect square factor (or power)

$\sqrt[3]{y^9} = y^{9 \div 3} = y^3$, thus y^9 is a perfect cube factor

$\sqrt[4]{x^8} = x^{8 \div 4} = x^2$, thus x^8 is a perfect fourth factor

Numbers can also be perfect factors. A number is a **perfect square** if it is the square of a natural number. A number is a **perfect cube** if it is a cube of a natural number.

1, 2, 3, 4, 5, 6, 7, 8, 9, ...	natural numbers
$1^2, 2^2, 3^2, 4^2, 5^2, 6^2, 7^2, 8^2, 9^2, \ldots$	perfect squares
1, 4, 9, 16, 25, 36, 49, 64, 81, ...	perfect square numbers
$1^3, 2^3, 3^3, 4^3, 5^3, 6^3, 7^3, 8^3, 9^3, \ldots$	perfect cubes
1, 8, 27, 64, 125, 216, 343, 512, 729, ...	perfect cube numbers

Note that the square root of any perfect square or perfect square number will be a whole number. For example,

$$\sqrt{36} = \sqrt{6^2} = 6^{2/2} = 6$$

Similarly, the cube root of any perfect cube or perfect cube number will be a whole number. For example,

$$\sqrt[3]{125} = \sqrt[3]{5^3} = 5^{3/3} = 5$$

In general for $\sqrt[n]{a}$, the expressions a^n, a^{2n}, a^{3n}, a^{4n}, and so on are perfect nth factors of the radical.

3 Now that we have introduced perfect factors, we will discuss how to simplify radicals.

To Simplify Radicals With Whole Number Radicands

1. Write the number as the product of its largest perfect factor and another factor.
2. Use the product rule to write the expression as a product of roots.
3. Find the roots of any perfect factors.

EXAMPLE 1 Simplify $\sqrt{32}$.

Solution: Since we are evaluating a square root, we look for the largest perfect square factor that divides 32. The largest perfect square factor that divides 32 is 16.

$$\begin{aligned}\sqrt{32} &= \sqrt{16 \cdot 2}\\ &= \sqrt{16}\sqrt{2}\\ &= 4\sqrt{2} \quad ■\end{aligned}$$

In Example 1 if you first believed that 4 was the largest perfect square factor of 32, you could proceed as follows:

$$\begin{aligned}\sqrt{32} &= \sqrt{4 \cdot 8}\\ &= \sqrt{4}\sqrt{8}\\ &= 2\sqrt{8} \qquad \text{4 is a perfect square factor of 8}\\ &= 2\sqrt{4}\sqrt{2}\\ &= 2 \cdot 2\sqrt{2}\\ &= 4\sqrt{2}\end{aligned}$$

Note that the final result is the same but you must work harder to find the answer.

EXAMPLE 2 Simplify $\sqrt{60}$.

Solution: $\sqrt{60} = \sqrt{4}\sqrt{15} = 2\sqrt{15}$ ■

In Example 12, $\sqrt{15}$ can be factored into $\sqrt{5}\sqrt{3}$; however, since neither of these is a perfect square factor, $\sqrt{15}$ cannot be simplified. The answer is $2\sqrt{15}$.

EXAMPLE 3 Simplify $\sqrt[3]{54}$.

The largest perfect cube factor of 54 is 27.

Solution: $\sqrt[3]{54} = \sqrt[3]{27}\sqrt[3]{2} = 3\sqrt[3]{2}$ ■

EXAMPLE 4 Simplify $\sqrt[3]{375}$.

Solution:
$$\begin{aligned}\sqrt[3]{375} &= \sqrt[3]{125}\sqrt[3]{3}\\ &= 5\sqrt[3]{3} \quad ■\end{aligned}$$

A radical containing a variable which is a perfect factor can be simplified by writing the expression in exponential form.

EXAMPLE 5 Simplify $\sqrt{x^4}$.

Solution: $\sqrt{x^4} = x^{4/2} = x^2$ ■

EXAMPLE 6 Simplify $\sqrt[3]{x^{12}}$.

Solution: $\sqrt[3]{x^{12}} = x^{12/3} = x^4$ ■

EXAMPLE 7 Simplify $\sqrt[4]{16x^{20}}$.

Solution: We can use the product rule to write

$$\begin{aligned}\sqrt[4]{16x^{20}} &= \sqrt[4]{16}\sqrt[4]{x^{20}}\\ &= 2 \cdot x^{20/4}\\ &= 2x^5\end{aligned}$$ ■

If the radicand is not a perfect factor, the radical can be simplified by using the following procedure.

To Simplify Radicals Containing Variables in the Radicand

1. Write each number and variable in the radicand as a product of its largest perfect factor and another factor.
2. Use the product rule to write the expression as a product of roots. All the perfect factors may be placed under the same radical.
3. Simplify the root containing the perfect factors.

EXAMPLE 8 Simplify $\sqrt{x^3}$.

Solution:

$$\begin{aligned}\sqrt{x^3} &= \sqrt{x^2}\sqrt{x}\\ &= x\sqrt{x}\end{aligned}$$ ■

EXAMPLE 9 Simplify $\sqrt[3]{y^{14}}$.

Solution:

$$\begin{aligned}\sqrt[3]{y^{14}} &= \sqrt[3]{y^{12}}\sqrt[3]{y^2}\\ &= y^4\sqrt[3]{y^2}\end{aligned}$$ ■

EXAMPLE 10 Simplify $\sqrt[4]{x^6y^{23}}$.

Solution: Find the highest perfect factors of x^6 and y^{23}. For an index of 4 the highest perfect factor of x^6 is x^4. The highest perfect factor of y^{23} is y^{20}.

$$\begin{aligned}\sqrt[4]{x^6y^{23}} &= \sqrt[4]{x^4 \cdot x^2 \cdot y^{20} \cdot y^3}\\ &= \sqrt[4]{x^4y^{20} \cdot x^2y^3}\\ &= \sqrt[4]{x^4y^{20}}\sqrt[4]{x^2y^3}\\ &= xy^5\sqrt[4]{x^2y^3}\end{aligned}$$ ■

EXAMPLE 11 Simplify $\sqrt[3]{54x^{17}y^{25}}$.

Solution: The highest perfect cube factor of 54 is 27. The highest perfect cube factor of x^{17} is x^{15}. The highest perfect cube factor of y^{25} is y^{24}

$$\begin{aligned}\sqrt[3]{54x^{17}y^{25}} &= \sqrt[3]{27 \cdot 2 \cdot x^{15} \cdot x^2 \cdot y^{24} \cdot y}\\ &= \sqrt[3]{27x^{15}y^{24} \cdot 2x^2y}\\ &= \sqrt[3]{27x^{15}y^{24}} \cdot \sqrt[3]{2x^2y}\\ &= 3x^5y^8\sqrt[3]{2x^2y}\end{aligned}$$ ■

HELPFUL HINT

In Example 10 we showed that

$$\sqrt[4]{x^6y^{23}} = xy^5\sqrt[4]{x^2y^3}$$

Note that the answer can be obtained by dividing the exponents on the variables in the radicand, 6 and 23, by the index, 4, and observing the quotients and remainders.

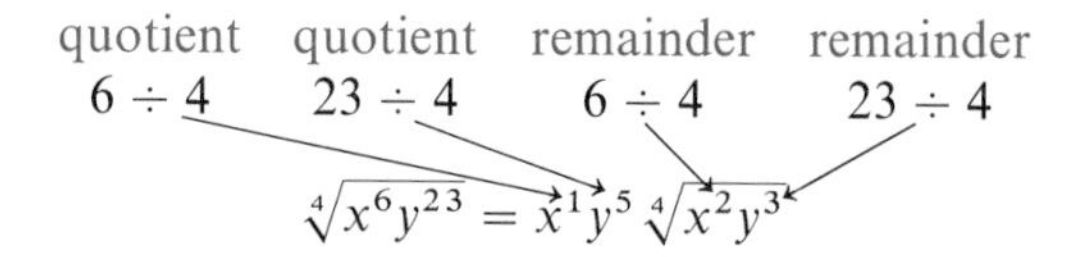

Can you explain why this procedure works? You may wish to use this procedure to work or check certain problems.

4 When we are given the product of two or more radicals, we can use the product rule to multiply the radicals together, and then simplify. Examples 12 and 13 illustrate this procedure.

EXAMPLE 12 Multiply and simplify.

(a) $\sqrt{2}\sqrt{8}$ (b) $\sqrt[3]{2x}\sqrt[3]{4x^2}$

Solution: (a) $\sqrt{2}\sqrt{8} = \sqrt{2\cdot 8}$
$= \sqrt{16} = 4$

(b) $\sqrt[3]{2x}\sqrt[3]{4x^2} = \sqrt[3]{2x\cdot 4x^2}$
$= \sqrt[3]{8x^3} = 2x$ ■

EXAMPLE 13 Multiply and simplify.

(a) $\sqrt[4]{8x^3y}\sqrt[4]{8x^6y^2}$ (b) $\sqrt[3]{5xy^4}\sqrt[3]{50x^2y^{18}}$

Solution: (a) $\sqrt[4]{8x^3y}\sqrt[4]{8x^6y^2} = \sqrt[4]{8x^3y\cdot 8x^6y^2}$
$= \sqrt[4]{64x^9y^3}$
$= \sqrt[4]{16x^8}\sqrt[4]{4xy^3}$
$= 2x^2\sqrt[4]{4xy^3}$

(b) $\sqrt[3]{5xy^4}\sqrt[3]{50x^2y^{18}} = \sqrt[3]{5xy^4\cdot 50x^2y^{18}}$
$= \sqrt[3]{250x^3y^{22}}$
$= \sqrt[3]{125x^3y^{21}}\sqrt[3]{2y}$
$= 5xy^7\sqrt[3]{2y}$ ■

Note that when simplifying a radical, the radicand of your simplified answer should not have any variable with an exponent greater than or equal to the index.

Exercise Set 8.2

Simplify each expression. Assume that all variables represent positive real numbers.

1. $\sqrt{50}$ **2.** $\sqrt{24}$ **3.** $\sqrt{40}$ **4.** $\sqrt{200}$
5. $\sqrt{32}$ **6.** $\sqrt{72}$ **7.** $\sqrt[3]{16}$ **8.** $\sqrt[3]{24}$
9. $\sqrt[3]{32}$ **10.** $\sqrt[3]{54}$ **11.** $\sqrt[4]{162}$ **12.** $\sqrt[4]{80}$
13. $\sqrt{x^3}$ **14.** $\sqrt{y^5}$ **15.** $\sqrt{x^{11}}$ **16.** $\sqrt{a^{30}}$
17. $\sqrt{b^{27}}$ **18.** $\sqrt[3]{x^5}$ **19.** $\sqrt[3]{y^7}$ **20.** $\sqrt[3]{y^{13}}$
21. $\sqrt[4]{y^9}$ **22.** $\sqrt[4]{b^{23}}$ **23.** $\sqrt[5]{x^{19}}$ **24.** $\sqrt[5]{y^{23}}$

25. $\sqrt{24x^3}$ **26.** $\sqrt{36x^5}$ **27.** $\sqrt{20x^7}$ **28.** $\sqrt[3]{24y^7}$
29. $\sqrt[3]{80x^{11}}$ **30.** $\sqrt[4]{16x^{10}}$ **31.** $\sqrt{x^3y^7}$ **32.** $\sqrt{x^5y^{12}}$
33. $\sqrt{50xy^4}$ **34.** $\sqrt[3]{x^9y^{11}z}$ **35.** $\sqrt[3]{81x^6y^8}$ **36.** $\sqrt[3]{16x^3y^6}$
37. $\sqrt[3]{54x^{12}y^{13}}$ **38.** $\sqrt[4]{x^9y^{12}z^{15}}$ **39.** $\sqrt[4]{16ab^{17}c^9}$ **40.** $\sqrt[5]{64x^{12}y^7}$
41. $\sqrt[5]{32a^2b^5}$ **42.** $\sqrt[3]{18w^{12}v^9r^{31}}$ **43.** $\sqrt[3]{32c^4w^9z}$ **44.** $\sqrt[4]{32x^8y^9z^{19}}$
45. $\sqrt[3]{81x^7y^{21}z^{50}}$ **46.** $\sqrt[3]{18x^4y^7z^{15}}$

Simplify each expression. Assume that all variables represent positive real numbers.

47. $\sqrt{5}\sqrt{5}$ **48.** $\sqrt{75}\sqrt{6}$ **49.** $\sqrt{60}\sqrt{5}$
50. $\sqrt[3]{2}\sqrt[3]{4}$ **51.** $\sqrt[3]{2}\sqrt[3]{28}$ **52.** $\sqrt[3]{3}\sqrt[3]{54}$
53. $\sqrt[4]{8}\sqrt[4]{10}$ **54.** $\sqrt{5x^2}\sqrt{8x^3}$ **55.** $\sqrt{15x^2}\sqrt{6x^5}$
56. $\sqrt{15xy^4}\sqrt{6xy^3}$ **57.** $\sqrt{14xy^2}\sqrt{3xy^3}$ **58.** $(\sqrt{6xy^2})^2$
59. $(\sqrt{4x^3y^2})^2$ **60.** $\sqrt{9x^3y^7}\sqrt{3xy^4}$ **61.** $\sqrt{20xy^4}\sqrt{6x^5y^7}$
62. $\sqrt[3]{4xy^2}\sqrt[3]{4xy^4}$ **63.** $\sqrt[3]{5xy^2}\sqrt[3]{25x^4y^{12}}$ **64.** $\sqrt[3]{9x^7y^{12}}\sqrt[3]{6x^4y}$
65. $(\sqrt[3]{2x^3y^4})^2$ **66.** $(\sqrt[3]{4x^5y^2})^2$ **67.** $(\sqrt[3]{5x^2y^6})^2$
68. $\sqrt[4]{12xy^4}\sqrt[4]{2x^3y^9z^7}$ **69.** $\sqrt[4]{3x^9y^{12}}\sqrt[4]{54x^4y^7}$ **70.** $\sqrt[5]{x^{24}y^{30}z^9}\sqrt[5]{x^{13}y^8z^7}$
71. $\sqrt[3]{2x^9y^6z}\sqrt[3]{12xy^4z^3}$ **72.** $\sqrt[4]{8x^4yz^3}\sqrt[4]{2x^2y^3z^7}$

8.3 Dividing and Simplifying Radicals

1. *Know the quotient rule for radicals.*
2. *Know when a radical is simplified.*
3. *Rationalize a denominator.*
4. *Rationalize a denominator using the conjugate of the denominator.*

1 In mathematics we sometimes need to divide one radical expression by another. To divide radicals, or to simplify radicals containing fractions, we use the quotient rule for radicals.

> **Quotient Rule for Radicals**
>
> For natural number n, and nonnegative real numbers a and b,
>
> $$\frac{\sqrt[n]{a}}{\sqrt[n]{b}} = \sqrt[n]{\frac{a}{b}} \qquad (b \neq 0)$$

Examples 1 through 3 illustrate how the quotient rule is used to simplify radical expressions.

EXAMPLE 1 Simplify each of the following.

(a) $\dfrac{\sqrt{75}}{\sqrt{3}}$ (b) $\dfrac{\sqrt[3]{24}}{\sqrt[3]{3}}$

Solution: (a) $\dfrac{\sqrt{75}}{\sqrt{3}} = \sqrt{\dfrac{75}{3}} = \sqrt{25} = 5$ (b) $\dfrac{\sqrt[3]{24}}{\sqrt[3]{3}} = \sqrt[3]{\dfrac{24}{3}} = \sqrt[3]{8} = 2$ ■

EXAMPLE 2 Simplify each of the following.

(a) $\sqrt{\frac{9}{4}}$ (b) $\sqrt[3]{\frac{8}{27}}$

Solution: (a) $\sqrt{\frac{9}{4}} = \frac{\sqrt{9}}{\sqrt{4}} = \frac{3}{2}$ (b) $\sqrt[3]{\frac{8}{27}} = \frac{\sqrt[3]{8}}{\sqrt[3]{27}} = \frac{2}{3}$ ■

EXAMPLE 3 Simplify each of the following.

(a) $\sqrt{\frac{16x^2}{8}}$ (b) $\sqrt{\frac{15xy^5}{3x^5y}}$

Solution: (a) $\sqrt{\frac{16x^2}{8}} = \sqrt{2x^2} = \sqrt{2}\sqrt{x^2} = \sqrt{2}x$ or $x\sqrt{2}$

(b) $\sqrt{\frac{15xy^5}{3x^5y}} = \sqrt{\frac{5y^4}{x^4}} = \frac{\sqrt{5y^4}}{\sqrt{x^4}} = \frac{\sqrt{y^4}\sqrt{5}}{x^2} = \frac{y^2\sqrt{5}}{x^2}$ ■

2 After you have simplified a radical expression you should check it to make sure that it is simplified as far as possible. A radical is simplified as far as possible when the following three conditions are met.

A Radical Is Simplified When the Following Are All True

1. There are no perfect factors in any radicand.
2. No radicand contains fractions.
3. There are no radicals in any denominator.

3 When the denominator of a fraction contains a radical, we generally simplify the expression by **rationalizing the denominator.** To rationalize a denominator is to remove all radicals from the denominator. Denominators are rationalized because, without a calculator, it is often easier to evaluate a fraction with a whole-number denominator than one where the denominator contains a radical.

To rationalize a denominator, multiply both the numerator and the denominator of the fraction by the denominator, or by a radical that will result in the radicand in the denominator becoming a perfect factor. The following examples illustrate the procedure to be used.

EXAMPLE 4 Simplify $\frac{1}{\sqrt{5}}$.

Solution: To simplify this expression, we must rationalize the denominator.

$$\frac{1}{\sqrt{5}} = \frac{1}{\sqrt{5}} \cdot \frac{\sqrt{5}}{\sqrt{5}} = \frac{\sqrt{5}}{\sqrt{25}} = \frac{\sqrt{5}}{5}$$ ■

In the example above, multiplying both the numerator and denominator by $\sqrt{5}$ is equivalent to multiplying the fraction by 1, which does not change the value of the original fraction.

EXAMPLE 5 Simplify each of the following.

(a) $\sqrt{\dfrac{2}{3}}$ (b) $\dfrac{x}{3\sqrt{2}}$

Solution: (a) $\sqrt{\dfrac{2}{3}} = \dfrac{\sqrt{2}}{\sqrt{3}} = \dfrac{\sqrt{2}}{\sqrt{3}} \cdot \dfrac{\sqrt{3}}{\sqrt{3}} = \dfrac{\sqrt{6}}{3}$

(b) $\dfrac{x}{3\sqrt{2}} = \dfrac{x}{3\sqrt{2}} \cdot \dfrac{\sqrt{2}}{\sqrt{2}} = \dfrac{x\sqrt{2}}{3 \cdot 2} = \dfrac{x\sqrt{2}}{6}$ ■

EXAMPLE 6 Simplify $\sqrt{\dfrac{x}{y}}$.

Solution:

$$\sqrt{\frac{x}{y}} = \frac{\sqrt{x}}{\sqrt{y}}$$

$$= \frac{\sqrt{x}}{\sqrt{y}} \cdot \frac{\sqrt{y}}{\sqrt{y}}$$

$$= \frac{\sqrt{xy}}{\sqrt{y^2}}$$

$$= \frac{\sqrt{xy}}{y}$$ ■

EXAMPLE 7 Simplify $\sqrt[3]{\dfrac{3}{5}}$.

Solution:

$$\sqrt[3]{\frac{3}{5}} = \frac{\sqrt[3]{3}}{\sqrt[3]{5}}$$

Since the denominator is a cube root, we must multiply the numerator and denominator by the cube root of an expression that will result in the product of the radicands in the denominator being a perfect cube. Multiply both the numerator and denominator by $\sqrt[3]{5^2}$.

$$\frac{\sqrt[3]{3}}{\sqrt[3]{5}} = \frac{\sqrt[3]{3}}{\sqrt[3]{5}} \cdot \frac{\sqrt[3]{5^2}}{\sqrt[3]{5^2}}$$

$$= \frac{\sqrt[3]{3} \cdot \sqrt[3]{5^2}}{\sqrt[3]{5^3}}$$

$$= \frac{\sqrt[3]{3}\sqrt[3]{25}}{5}$$

$$= \frac{\sqrt[3]{75}}{5}$$ (note that 75 has no perfect cube factors) ■

EXAMPLE 8 Simplify $\sqrt[3]{\dfrac{x}{2y^2}}$.

Solution: $\sqrt[3]{\dfrac{x}{2y^2}} = \dfrac{\sqrt[3]{x}}{\sqrt[3]{2y^2}}$

We must multiply both the numerator and denominator by the cube root of an expression that will result in the product of the radicands in the denominator being a perfect cube.

The product of the exponents in the denominator must be divisible by 3. Since the denominator is $\sqrt[3]{2y^2}$, we must multiply the expression by $\sqrt[3]{2^2y}$. Note that $2 \cdot 2^2 = 2^3$ and $y^2 \cdot y = y^3$. Multiply both the numerator and denominator by $\sqrt[3]{2^2y}$.

$$\begin{aligned}\frac{\sqrt[3]{x}}{\sqrt[3]{2y^2}} &= \frac{\sqrt[3]{x}}{\sqrt[3]{2y^2}} \cdot \frac{\sqrt[3]{2^2y}}{\sqrt[3]{2^2y}} \\ &= \frac{\sqrt[3]{x}\,\sqrt[3]{4y}}{\sqrt[3]{2^3y^3}} \\ &= \frac{\sqrt[3]{4xy}}{2y}\end{aligned}$$

■

EXAMPLE 9 Simplify $\sqrt{\dfrac{12x^3y^5}{5z}}$.

Solution: $\sqrt{\dfrac{12x^3y^5}{5z}} = \dfrac{\sqrt{12x^3y^5}}{\sqrt{5z}}$

Now simplify the numerator.

$$\begin{aligned}\frac{\sqrt{12x^3y^5}}{\sqrt{5z}} &= \frac{\sqrt{4x^2y^4}\,\sqrt{3xy}}{\sqrt{5z}} \\ &= \frac{2xy^2\sqrt{3xy}}{\sqrt{5z}}\end{aligned}$$

Now rationalize the denominator.

$$\begin{aligned}\frac{2xy^2\sqrt{3xy}}{\sqrt{5z}} &= \frac{2x^2y\sqrt{3xy}}{\sqrt{5z}} \cdot \frac{\sqrt{5z}}{\sqrt{5z}} \\ &= \frac{2x^2y\sqrt{15xyz}}{\sqrt{5^2z^2}} \\ &= \frac{2x^2y\sqrt{15xyz}}{5z}\end{aligned}$$

■

[4] When the denominator of a rational expression is a binomial which contains a radical, we again rationalize the denominator. We do this by multiplying both the numerator and the denominator of the fraction by the **conjugate** of the denominator. The conjugate of a binomial is a binomial having the same two terms with the sign of the second term changed.

Binomial	*Its conjugate*
$3 + \sqrt{2}$	$3 - \sqrt{2}$
$2\sqrt{3} - \sqrt{5}$	$2\sqrt{3} + \sqrt{5}$
$\sqrt{x} + \sqrt{y}$	$\sqrt{x} - \sqrt{y}$
$a + \sqrt{b}$	$a - \sqrt{b}$

When a binomial is multiplied by its conjugate, the outer and inner terms will sum to zero.

EXAMPLE 10 Multiply $(2 + \sqrt{3})(2 - \sqrt{3})$.

Solution: Multiply using the FOIL method.

$$\begin{aligned} &\quad\text{F}\qquad\quad\text{O}\qquad\quad\text{I}\qquad\quad\text{L} \\ &2(2) + 2(-\sqrt{3}) + 2(\sqrt{3}) + \sqrt{3}(-\sqrt{3}) = 4 - 2\sqrt{3} + 2\sqrt{3} - \sqrt{9} \\ &\qquad = 4 - \sqrt{9} \\ &\qquad = 4 - 3 \\ &\qquad = 1 \quad \blacksquare \end{aligned}$$

Note that in Example 10 we would get the same results using the difference of two squares formula, $(a + b)(a - b) = a^2 - b^2$.

$$\begin{aligned} (2 + \sqrt{3})(2 - \sqrt{3}) &= 2^2 - (\sqrt{3})^2 \\ &= 4 - 3 \\ &= 1 \end{aligned}$$

EXAMPLE 11 Multiply $(\sqrt{3} - \sqrt{5})(\sqrt{3} + \sqrt{5})$.

Solution:

$$\begin{aligned} (\sqrt{3} - \sqrt{5})(\sqrt{3} + \sqrt{5}) &= (\sqrt{3})^2 - (\sqrt{5})^2 \\ &= 3 - 5 \\ &= -2 \quad \blacksquare \end{aligned}$$

EXAMPLE 12 Simplify $\dfrac{5}{2 + \sqrt{3}}$.

Solution: To simplify this expression we must rationalize the denominator. We do this by multiplying both numerator and denominator by $2 - \sqrt{3}$, which is the conjugate of $2 + \sqrt{3}$.

$$\begin{aligned} \frac{5}{2 + \sqrt{3}} \cdot \frac{2 - \sqrt{3}}{2 - \sqrt{3}} &= \frac{5(2 - \sqrt{3})}{(2 + \sqrt{3})(2 - \sqrt{3})} \\ &= \frac{5(2 - \sqrt{3})}{4 - 3} \\ &= 5(2 - \sqrt{3}) \quad \text{or} \quad 10 - 5\sqrt{3} \quad \blacksquare \end{aligned}$$

EXAMPLE 13 Simplify $\dfrac{6}{\sqrt{5}-\sqrt{2}}$.

Solution:

$$\frac{6}{\sqrt{5}-\sqrt{2}} \cdot \frac{\sqrt{5}+\sqrt{2}}{\sqrt{5}+\sqrt{2}} = \frac{6(\sqrt{5}+\sqrt{2})}{5-2}$$

$$= \frac{\overset{2}{\cancel{6}}(\sqrt{5}+\sqrt{2})}{\underset{1}{\cancel{3}}}$$

$$= 2(\sqrt{5}+\sqrt{2}) \quad \text{or} \quad 2\sqrt{5}+2\sqrt{2}$$ ■

EXAMPLE 14 Simplify $\dfrac{x}{x+\sqrt{y}}$.

Solution: Multiply both numerator and denominator of the fraction by the conjugate of the denominator, $x-\sqrt{y}$.

$$\frac{x}{x+\sqrt{y}} \cdot \frac{x-\sqrt{y}}{x-\sqrt{y}} = \frac{x(x-\sqrt{y})}{x^2-y}$$

$$= \frac{x^2-x\sqrt{y}}{x^2-y}$$

Remember that you cannot divide out the x^2 terms because they are not factors. ■

COMMON STUDENT ERROR

The following simplifications are correct because the numbers and variables divided out are not within square roots.

Correct *Correct*

$$\frac{\overset{2}{\cancel{6}}\sqrt{2}}{\underset{1}{\cancel{3}}} = 2\sqrt{2} \qquad \frac{\cancel{x}\sqrt{2}}{\cancel{x}} = \sqrt{2}$$

An expression within a square root cannot be divided by an expression not within the square root.

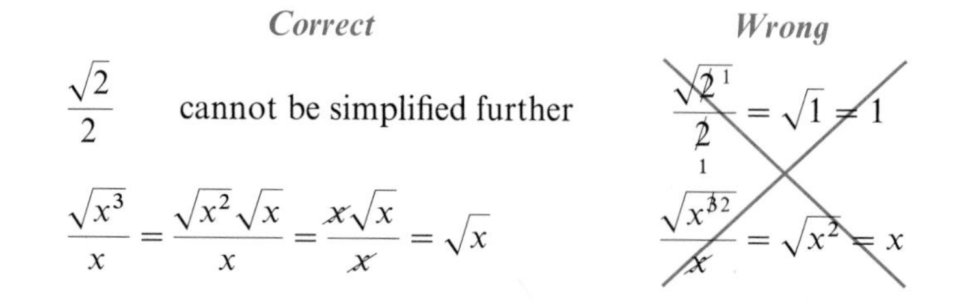

Correct *Wrong*

$\dfrac{\sqrt{2}}{2}$ cannot be simplified further

$$\frac{\sqrt{x^3}}{x} = \frac{\sqrt{x^2}\sqrt{x}}{x} = \frac{\cancel{x}\sqrt{x}}{\cancel{x}} = \sqrt{x}$$

Wrong: $\dfrac{\sqrt{\cancel{2}^{\,1}}}{\underset{1}{\cancel{2}}} = \sqrt{1} = 1$; $\dfrac{\sqrt{x^{\cancel{3}\,2}}}{\cancel{x}} = \sqrt{x^2} = x$

Exercise Set 8.3

In this exercise set assume that all variables represent positive real numbers.

Simplify each expression.

1. $\sqrt{\frac{27}{3}}$ **2.** $\sqrt{\frac{4}{25}}$ **3.** $\frac{\sqrt{3}}{\sqrt{27}}$ **4.** $\sqrt{\frac{16}{25}}$

5. $\sqrt[3]{\frac{2}{16}}$ **6.** $\frac{\sqrt[3]{108}}{\sqrt[3]{3}}$ **7.** $\frac{\sqrt{24}}{\sqrt{3}}$ **8.** $\sqrt{\frac{x}{9}}$

9. $\sqrt[3]{\frac{x^3}{27}}$ **10.** $\sqrt{\frac{x^4}{25}}$ **11.** $\frac{\sqrt[3]{2x^6}}{\sqrt[3]{16x^3}}$ **12.** $\sqrt[4]{\frac{x^4}{16}}$

13. $\sqrt{\frac{16x^4}{4}}$ **14.** $\sqrt{\frac{27x^6}{3x^2}}$ **15.** $\sqrt{\frac{2x}{8x^5}}$ **16.** $\sqrt{\frac{12xy^4}{2x^3y^4}}$

17. $\sqrt{\frac{25x^2y^5}{5x^4y}}$ **18.** $\sqrt{\frac{72x^2y^5}{8x^2y^7}}$ **19.** $\sqrt{\frac{x^4y^5}{4x^2y}}$ **20.** $\sqrt{\frac{2xy^4}{18xy^2}}$

Rationalize the denominator.

21. $\frac{1}{\sqrt{3}}$ **22.** $\frac{3}{\sqrt{3}}$ **23.** $\frac{1}{\sqrt{2}}$ **24.** $\frac{2}{\sqrt{2}}$

25. $\frac{x}{\sqrt{5}}$ **26.** $\frac{2x}{\sqrt{6}}$ **27.** $\frac{x}{\sqrt{y}}$ **28.** $\sqrt{\frac{2}{5}}$

29. $\sqrt{\frac{1}{3}}$ **30.** $\sqrt{\frac{x}{2}}$ **31.** $\sqrt{\frac{4}{5}}$ **32.** $\sqrt{\frac{x}{y}}$

33. $\sqrt{\frac{5}{8}}$ **34.** $\sqrt{\frac{x}{2y^2}}$ **35.** $\frac{2\sqrt{3}}{\sqrt{5}}$ **36.** $\frac{7\sqrt{x}}{\sqrt{98}}$

37. $\frac{2x}{\sqrt{18}}$ **38.** $\frac{2\sqrt{3}}{\sqrt{32}}$ **39.** $\sqrt{\frac{3x}{4y}}$ **40.** $\sqrt{\frac{2p}{q}}$

Rationalize the denominator.

41. $\frac{1}{\sqrt[3]{2}}$ **42.** $\frac{2}{\sqrt[3]{4}}$ **43.** $\frac{1}{\sqrt[3]{3}}$ **44.** $\sqrt[3]{\frac{1}{3}}$

45. $\frac{5}{\sqrt[3]{x}}$ **46.** $\sqrt[3]{\frac{5x}{y}}$ **47.** $\sqrt[3]{\frac{1}{4x}}$ **48.** $\frac{4x}{\sqrt[3]{5y}}$

49. $\sqrt[3]{\frac{5x}{4y^2}}$ **50.** $\frac{3}{\sqrt[4]{x}}$ **51.** $\frac{5x}{\sqrt[4]{2}}$ **52.** $\sqrt[4]{\frac{3}{2x}}$

53. $\sqrt[4]{\frac{5}{3x^3}}$ **54.** $\sqrt[4]{\frac{2x}{4y^2}}$

Simplify.

55. $\sqrt{\frac{8x^5y}{2z}}$ **56.** $\sqrt{\frac{18x^4y^6}{3z}}$ **57.** $\sqrt{\frac{5xy^4}{2z}}$ **58.** $\sqrt{\frac{24x^3y^6}{5z}}$

59. $\sqrt{\frac{20y^4z^3}{3x}}$ **60.** $\sqrt{\frac{5xy^6}{6z}}$ **61.** $\sqrt{\frac{15x^5z^7}{2y}}$ **62.** $\sqrt{\frac{25x^2y^5}{3z}}$

63. $\sqrt{\frac{18x^4y^3}{2z}}$ **64.** $\sqrt{\frac{45y^{12}z^{10}}{2x}}$ **65.** $\sqrt[3]{\frac{15x^6y^7}{z^2}}$ **66.** $\sqrt[3]{\frac{32y^{12}z^{10}}{2x}}$

Multiply as indicated.

67. $(3-\sqrt{3})(3+\sqrt{3})$
68. $(4+\sqrt{2})(4-\sqrt{2})$
69. $(\sqrt{6}+3)(\sqrt{6}-3)$
70. $(6-\sqrt{5})(6+\sqrt{5})$
71. $(\sqrt{8}-3)(\sqrt{8}+3)$
72. $(\sqrt{x}+3)(\sqrt{x}-3)$
73. $(\sqrt{x}+5)(\sqrt{x}-5)$
74. $(\sqrt{6}+x)(\sqrt{6}-x)$
75. $(\sqrt{y}-3)(\sqrt{y}+3)$
76. $(\sqrt{x}+y)(\sqrt{x}-y)$
77. $(\sqrt{x}+\sqrt{y})(\sqrt{x}-\sqrt{y})$
78. $(\sqrt{y}-x)(\sqrt{y}+x)$
79. $(x+\sqrt{y})(x-\sqrt{y})$
80. $(\sqrt{7}+\sqrt{3})(\sqrt{7}-\sqrt{3})$

Simplify.

81. $\dfrac{3}{1+\sqrt{2}}$
82. $\dfrac{1}{2+\sqrt{3}}$
83. $\dfrac{3}{\sqrt{6}-5}$
84. $\dfrac{3}{\sqrt{2}+5}$
85. $\dfrac{4}{\sqrt{2}-7}$
86. $\dfrac{2}{\sqrt{2}+\sqrt{3}}$
87. $\dfrac{\sqrt{5}}{\sqrt{5}-\sqrt{6}}$
88. $\dfrac{8}{\sqrt{5}-\sqrt{8}}$
89. $\dfrac{1}{\sqrt{17}-\sqrt{8}}$
90. $\dfrac{2}{6+\sqrt{x}}$
91. $\dfrac{5}{\sqrt{x}-3}$
92. $\dfrac{6}{4-\sqrt{y}}$
93. $\dfrac{5}{3+\sqrt{x}}$
94. $\dfrac{4}{\sqrt{x}-y}$
95. $\dfrac{\sqrt{8x}}{x+\sqrt{y}}$
96. $\dfrac{\sqrt{x}}{\sqrt{x}+\sqrt{y}}$

97. Consider the expression $\dfrac{1}{\sqrt{18}}$. Rationalize the denominator by:
(a) First simplifying $\sqrt{18}$ and then rationalizing
(b) Multiplying both numerator and denominator by $\dfrac{\sqrt{2}}{\sqrt{2}}$
(c) Multiplying both numerator and denominator by $\dfrac{\sqrt{18}}{\sqrt{18}}$.

98. Use a calculator or Appendix C to evaluate
(a) $\dfrac{\sqrt{2}}{2}$ **(b)** $\dfrac{1}{\sqrt{2}}$

99. Which is greater $\dfrac{2}{\sqrt{2}}$ or $\dfrac{3}{\sqrt{3}}$?

100. Which is greater $\dfrac{\sqrt{3}}{2}$ or $\dfrac{2}{\sqrt{3}}$?

8.4 Addition and Subtraction of Radicals

1 *Add and subtract radicals.*

1 **Like radicals** are radicals having the same radicands and index. **Unlike radicals** are radicals differing in either the radicand or the index.

Examples of like radicals	*Examples of unlike radicals*	
$\sqrt{5},\ 3\sqrt{5}$	$\sqrt{5},\ \sqrt[3]{5}$	indices differ
$5\sqrt{7},\ -2\sqrt{7}$	$\sqrt{5},\ \sqrt{7}$	radicands differ
$\sqrt{x},\ 5\sqrt{x}$	$\sqrt{x},\ \sqrt{2x}$	radicands differ
$\sqrt[3]{2x},\ -4\sqrt[3]{2x}$	$\sqrt{x},\ \sqrt[3]{x}$	indices differ
$\sqrt[4]{xy},\ -\sqrt[4]{xy}$	$\sqrt[3]{xy},\ \sqrt[3]{x^2y}$	radicands differ

Like radicals are added and subtracted in much the same way that like terms are added or subtracted. To add or subtract like radicals, add or subtract their numerical coefficients and multiply this sum or difference by the like radical.

Examples of adding like radicals

$$3\sqrt{5} + 2\sqrt{5} = (3 + 2)\sqrt{5} = 5\sqrt{5}$$
$$5\sqrt{x} - 7\sqrt{x} = (5 - 7)\sqrt{x} = -2\sqrt{x}$$
$$\sqrt[3]{4x} + 5\sqrt[3]{4x} = (1 + 5)\sqrt[3]{4x} = 6\sqrt[3]{4x}$$
$$4\sqrt{5x} - y\sqrt{5x} = (4 - y)\sqrt{5x}$$

EXAMPLE 1 Simplify each of the following.

(a) $6 + 4\sqrt{2} - \sqrt{2} + 3$ (b) $2\sqrt[3]{x} + 5x + 4\sqrt[3]{x} - 3$

Solution: (a) $6 + 4\sqrt{2} - \sqrt{2} + 3 = 3\sqrt{2} + 9$ (or $9 + 3\sqrt{2}$)

(b) $2\sqrt[3]{x} + 5x + 4\sqrt[3]{x} - 3 = 6\sqrt[3]{x} + 5x - 3$ ■

It is sometimes possible to change unlike radicals into like radicals by simplifying one or more of the radicals.

EXAMPLE 2 Simplify $\sqrt{3} + \sqrt{27}$.

Solution: Since $\sqrt{3}$ and $\sqrt{27}$ are unlike radicals, they cannot be added in their present form. We can simplify $\sqrt{27}$ to obtain like radicals.

$$\begin{aligned}\sqrt{3} + \sqrt{27} &= \sqrt{3} + \sqrt{9}\sqrt{3}\\ &= \sqrt{3} + 3\sqrt{3}\\ &= 4\sqrt{3}\end{aligned}$$ ■

To Add or Subtract Radicals

1. Simplify each radical expression.
2. Combine like radicals (if there are any).

EXAMPLE 3 Simplify $4\sqrt{24} + \sqrt{54}$.

Solution:

$$\begin{aligned}4\sqrt{24} + \sqrt{54} &= 4\sqrt{4}\cdot\sqrt{6} + \sqrt{9}\cdot\sqrt{6}\\ &= 4\cdot 2\sqrt{6} + 3\sqrt{6}\\ &= 8\sqrt{6} + 3\sqrt{6}\\ &= 11\sqrt{6}\end{aligned}$$ ■

EXAMPLE 4 Simplify $2\sqrt{45} - \sqrt{80} + \sqrt{20}$.

Solution:

$$\begin{aligned}2\sqrt{45} - \sqrt{80} + \sqrt{20} &= 2\sqrt{9}\cdot\sqrt{5} - \sqrt{16}\cdot\sqrt{5} + \sqrt{4}\cdot\sqrt{5}\\ &= 2\cdot 3\sqrt{5} - 4\sqrt{5} + 2\sqrt{5}\\ &= 6\sqrt{5} - 4\sqrt{5} + 2\sqrt{5}\\ &= 4\sqrt{5}\end{aligned}$$ ■

EXAMPLE 5 Simplify $\sqrt[3]{27} + \sqrt[3]{81} - 4\sqrt[3]{3}$.

Solution:
$$\begin{aligned}\sqrt[3]{27} + \sqrt[3]{81} - 4\sqrt[3]{3} &= 3 + \sqrt[3]{27}\sqrt[3]{3} - 4\sqrt[3]{3}\\ &= 3 + 3\sqrt[3]{3} - 4\sqrt[3]{3}\\ &= 3 - \sqrt[3]{3} \quad ■\end{aligned}$$

EXAMPLE 6 Simplify $\sqrt{x^2} - \sqrt{x^2y} + x\sqrt{y}$.

Solution:
$$\begin{aligned}\sqrt{x^2} - \sqrt{x^2y} + x\sqrt{y} &= x - \sqrt{x^2}\sqrt{y} + x\sqrt{y}\\ &= x - x\sqrt{y} + x\sqrt{y}\\ &= x \quad ■\end{aligned}$$

EXAMPLE 7 Simplify $\sqrt[3]{x^{10}y^2} - \sqrt[3]{x^4y^8}$.

Solution:
$$\begin{aligned}\sqrt[3]{x^{10}y^2} - \sqrt[3]{x^4y^8} &= \sqrt[3]{x^9} \cdot \sqrt[3]{xy^2} - \sqrt[3]{x^3y^6} \cdot \sqrt[3]{xy^2}\\ &= x^3\sqrt[3]{xy^2} - xy^2\sqrt[3]{xy^2}\end{aligned}$$

Now factor out the common factor $\sqrt[3]{xy^2}$.

$$= (x^3 - xy^2)\sqrt[3]{xy^2} \quad ■$$

EXAMPLE 8 Simplify $4\sqrt{2} - \dfrac{1}{\sqrt{8}} + \sqrt{32}$.

Solution:
$$\begin{aligned}4\sqrt{2} - \frac{1}{\sqrt{8}} + \sqrt{32} &= 4\sqrt{2} - \frac{1}{\sqrt{8}} \cdot \frac{\sqrt{2}}{\sqrt{2}} + \sqrt{16}\sqrt{2}\\ &= 4\sqrt{2} - \frac{\sqrt{2}}{\sqrt{16}} + 4\sqrt{2}\\ &= 4\sqrt{2} - \frac{\sqrt{2}}{4} + 4\sqrt{2}\\ &= \left(4 - \frac{1}{4} + 4\right)\sqrt{2}\\ &= \frac{31\sqrt{2}}{4} \quad ■\end{aligned}$$

EXAMPLE 9 Simplify $(3\sqrt{6} - 4)(2 + 5\sqrt{6})$.

Solution: Use the FOIL method to multiply; then combine like terms.

$$\begin{aligned}(3\sqrt{6})(2) + (3\sqrt{6})(5\sqrt{6}) + (-4)(2) + (-4)(5\sqrt{6}) &= 6\sqrt{6} + 15\sqrt{36} - 8 - 20\sqrt{6}\\ &= 6\sqrt{6} + 15(6) - 8 - 20\sqrt{6}\\ &= 6\sqrt{6} + 90 - 8 - 20\sqrt{6}\\ &= 82 - 14\sqrt{6} \quad ■\end{aligned}$$

EXAMPLE 10 Simplify $(3\sqrt{6} - \sqrt{3})^2$.

Solution: $(3\sqrt{6} - \sqrt{3})^2 = (3\sqrt{6} - \sqrt{3})(3\sqrt{6} - \sqrt{3})$

Now multiply using the FOIL method.

$$\begin{aligned}
&(3\sqrt{6})(3\sqrt{6}) + (3\sqrt{6})(-\sqrt{3}) + (-\sqrt{3})(3\sqrt{6}) + (-\sqrt{3})(-\sqrt{3})\\
&= 9(6) - 3\sqrt{18} - 3\sqrt{18} + 3\\
&= 54 - 3\sqrt{18} - 3\sqrt{18} + 3\\
&= 57 - 6\sqrt{18}\\
&= 57 - 6\sqrt{9}\sqrt{2}\\
&= 57 - 18\sqrt{2} \quad \blacksquare
\end{aligned}$$

COMMON STUDENT ERROR

The product rule of radicals presented in Section 8.2 is

$$\sqrt[n]{a} \cdot \sqrt[n]{b} = \sqrt[n]{ab} \qquad \textit{Correct}$$

The quotient rule of radicals presented in Section 8.3 is

$$\frac{\sqrt[n]{a}}{\sqrt[n]{b}} = \sqrt[n]{\frac{a}{b}} \qquad \textit{Correct}$$

Students often incorrectly assume similar properties exist for addition and subtraction. They do not.

Wrong

$$\sqrt[n]{a} + \sqrt[n]{b} = \sqrt[n]{a + b}$$
$$\sqrt[n]{a} - \sqrt[n]{b} = \sqrt[n]{a - b}$$

To illustrate that $\sqrt[n]{a} + \sqrt[n]{b} \neq \sqrt[n]{a + b}$, let n be a square root (index 2), $a = 9$, $b = 16$.

$$\begin{aligned}
\sqrt[n]{a} + \sqrt[n]{b} &\neq \sqrt[n]{a + b}\\
\sqrt{9} + \sqrt{16} &\neq \sqrt{9 + 16}\\
3 + 4 &\neq \sqrt{25}\\
7 &\neq 5
\end{aligned}$$

Exercise Set 8.4

In this exercise set assume all variables represent positive real numbers.

Simplify each expression.

1. $4\sqrt{3} - 2\sqrt{3}$

2. $\sqrt{5} + 2\sqrt{5}$

3. $6\sqrt[3]{7} - 8\sqrt[3]{7}$

4. $4\sqrt{10} + 6\sqrt{10} - \sqrt{10} + 2$

5. $2\sqrt{3} - 2\sqrt{3} - 4\sqrt{3} + 5$

6. $12\sqrt[3]{15} + 5\sqrt[3]{15} - 8\sqrt[3]{15}$

7. $4\sqrt{x} + \sqrt{x}$

8. $-2\sqrt{x} - 3\sqrt{x}$

9. $-\sqrt[4]{x} + 6\sqrt[4]{x} - 2\sqrt[4]{x}$

10. $3\sqrt{y} - 6\sqrt{y}$

11. $3\sqrt{y} - \sqrt{y} + 3$

12. $3\sqrt{5} - \sqrt[3]{x} + 4\sqrt{5} + 3\sqrt[3]{x}$

13. $\sqrt{x} + \sqrt{y} + x + 3\sqrt{y}$

14. $2 + 3\sqrt{y} - 6\sqrt{y} + 5$

15. $5 + 4\sqrt[3]{x} - 8\sqrt[3]{x}$

16. $5\sqrt{x} + 4 + 3\sqrt{x} + 2x - \sqrt{x}$

Simplify each expression.

17. $\sqrt{8}-\sqrt{12}$
18. $\sqrt{75}+\sqrt{108}$
19. $\sqrt{125}+\sqrt{20}$
20. $-6\sqrt{75}+4\sqrt{125}$
21. $3\sqrt{250}+5\sqrt{160}$
22. $-4\sqrt{90}+3\sqrt{40}+2\sqrt{10}$
23. $8\sqrt{45}+7\sqrt{20}+2\sqrt{5}$
24. $4\sqrt{32}-\sqrt{18}+2\sqrt{128}$
25. $\sqrt{48}+2\sqrt{75}-3\sqrt{27}$
26. $5\sqrt{8}+2\sqrt{50}-3\sqrt{72}$
27. $3\sqrt{7}+2\sqrt{63}-2\sqrt{28}$
28. $2\sqrt{5x}-3\sqrt{20x}-4\sqrt{45x}$
29. $3\sqrt{27x^2}-2\sqrt{108x^2}-\sqrt{48x^2}$
30. $3\sqrt{50x^2}-3\sqrt{72x^2}-8x\sqrt{18}$
31. $2\sqrt[3]{81}+4\sqrt[3]{24}$
32. $\sqrt[3]{54}-\sqrt[3]{16}$
33. $4\sqrt[3]{5}-5\sqrt[3]{40}$
34. $\sqrt[3]{108}+2\sqrt[3]{32}$
35. $2\sqrt[3]{16}+\sqrt[3]{54}$
36. $\sqrt[3]{27}-5\sqrt[3]{8}$
37. $4\sqrt{3x^3}-\sqrt{12x}$
38. $3\sqrt{45x^3}+\sqrt{5x}$
39. $2\sqrt[3]{x^4y^2}+3x\sqrt[3]{xy^2}$
40. $2a\sqrt{20a^3b^2}+2b\sqrt{45a^5}$
41. $x\sqrt[3]{x^2y}-\sqrt[3]{8x^5y}$
42. $2b\sqrt[4]{a^4b}+ab\sqrt[4]{16b}$
43. $2\sqrt[3]{24a^3y^4}+4a\sqrt[3]{81y^4}$
44. $3y\sqrt[4]{48x^5}-x\sqrt[4]{3x^5y^4}$
45. $\sqrt{4x^7y^5}+3x^2\sqrt{x^3y^5}-2xy\sqrt{x^5y^3}$
46. $x\sqrt[3]{27x^5y^2}-x^2\sqrt[3]{x^2y^2}+2\sqrt[3]{x^8y^2}$
47. $2\sqrt[3]{x^7y^7}-3x\sqrt[3]{x^4y^7}$
48. $\sqrt[3]{16x^9y^{10}}-2x^2y\sqrt[3]{2x^3y^7}$
49. $x\sqrt[3]{x^7y^5}-xy^2\sqrt[3]{xy^2}$
50. $2x\sqrt[3]{xy}+5y\sqrt[3]{x^4y^4}$

Simplify.

51. $\dfrac{1}{\sqrt{2}}+\dfrac{\sqrt{2}}{2}$
52. $\dfrac{1}{\sqrt{3}}+\dfrac{\sqrt{3}}{3}$
53. $\dfrac{\sqrt{6}}{2}+\dfrac{1}{\sqrt{6}}$
54. $\sqrt{3}-\dfrac{1}{\sqrt{3}}$
55. $\sqrt{6}-\sqrt{\dfrac{2}{3}}$
56. $\sqrt{\dfrac{1}{6}}+\sqrt{24}$
57. $3\sqrt{2}-\dfrac{2}{\sqrt{8}}+\sqrt{50}$
58. $4\sqrt{3}-\dfrac{3}{\sqrt{3}}+2\sqrt{18}$
59. $\sqrt{\dfrac{1}{2}}+3\sqrt{2}+\sqrt{18}$
60. $\dfrac{3}{\sqrt{18}}-2\sqrt{18}+\sqrt{\dfrac{5}{8}}$
61. $4\sqrt{x}+\dfrac{1}{\sqrt{x}}+\sqrt{\dfrac{1}{x}}$
62. $\dfrac{3}{\sqrt{y}}-\sqrt{\dfrac{9}{y}}+\sqrt{y}$

Multiply and then simplify.

63. $(\sqrt{3}+4)(\sqrt{3}+5)$
64. $(\sqrt{5}+2)(7+\sqrt{5})$
65. $(\sqrt{3}+1)(\sqrt{3}-6)$
66. $(1+\sqrt{5})(6+\sqrt{5})$
67. $(3-\sqrt{2})(4-\sqrt{8})$
68. $(4-\sqrt{2})(5+\sqrt{2})$
69. $(3\sqrt{2}-4)(\sqrt{2}+5)$
70. $(5\sqrt{6}+3)(4\sqrt{6}-2)$
71. $(\sqrt{5}+\sqrt{3})(\sqrt{5}+\sqrt{3})$
72. $(4\sqrt{3}+\sqrt{2})(\sqrt{3}-\sqrt{2})$
73. $(\sqrt{2}-\sqrt{3})(\sqrt{3}+\sqrt{8})$
74. $(\sqrt{5}+\sqrt{2})(\sqrt{2}+\sqrt{20})$
75. $(\sqrt{3}+4)^2$
76. $(2-\sqrt{3})^2$
77. $(2\sqrt{5}-3)^2$
78. $(2\sqrt{7}+\sqrt{3})^2$

79. Use a calculator or Appendix C to estimate $\sqrt{3}+3\sqrt{2}$.

80. Use a calculator or Appendix C to estimate $2\sqrt{3}+\sqrt{5}$.

81. Which is greater, $\dfrac{1}{\sqrt{3}+2}$ or $2+\sqrt{3}$? (Do not use a calculator or tables.)

82. Which is greater, $\dfrac{1}{\sqrt{3}}+\sqrt{75}$ or $\dfrac{2}{\sqrt{12}}+\sqrt{48}+2\sqrt{3}$? (Do not use a calculator or tables.)

JUST FOR FUN

1. Simplify $\dfrac{1}{\sqrt[5]{3x^7y^9}}+\dfrac{2\sqrt[5]{81x^3y}}{3x^2y^2}$

2. Simplify $\dfrac{1}{\sqrt[4]{3x^5y^6z^{13}}}$

8.5 Solving Radical Equations

1 *Solve radical equations.*

1 A **radical equation** is an equation that contains a variable in a radicand. Some examples of radical equations are:

$$\sqrt{x} = 4 \qquad \sqrt[3]{y + 4} = 9 \qquad \sqrt{x - 2} = 4 + \sqrt{x + 8}$$

To Solve Radical Equations

1. Rewrite the equation so that one radical containing a variable is isolated by itself on one side of the equation.
2. Raise each side of the equation to a power equal to the index of the radical.
3. Collect and combine like terms.
4. If the remaining equation still contains a term with a variable in a radicand, repeat steps 1 through 3.
5. Solve the resulting equation for the unknown variable.
6. Check all solutions in the original equations for extraneous roots.

Recall from Section 7.5 that an extraneous root is a number obtained when solving an equation that is not a solution to the original equation.

The following examples illustrate the procedure for solving radical equations.

EXAMPLE 1 Solve the equation $\sqrt{x} = 6$.

Solution: The square root containing the variable is already by itself on one side of the equation. Square both sides of the equation.

$$\sqrt{x} = 6$$
$$(\sqrt{x})^2 = (6)^2$$
$$x = 36$$

Check:

$$\sqrt{x} = 6$$
$$\sqrt{36} = 6$$
$$6 = 6 \qquad \text{true}$$

■

EXAMPLE 2 Solve the equation $\sqrt{x + 4} - 6 = 0$.

Solution:

$$\sqrt{x + 4} - 6 = 0$$
$$\sqrt{x + 4} = 6 \qquad \text{isolate the radical containing the variable}$$
$$(\sqrt{x + 4})^2 = 6^2 \qquad \text{square both sides of the equation}$$
$$x + 4 = 36 \qquad \text{now solve for the variable}$$
$$x = 32$$

A check will show that 32 is the solution. ■

EXAMPLE 3 Solve the equation $\sqrt[3]{x} + 4 = 6$.

Solution: Since the 4 is outside the radical, we first subtract 4 from both sides of the equation to isolate the radical.

$$\sqrt[3]{x} + 4 = 6$$
$$\sqrt[3]{x} = 2$$

Now cube both sides of the equation.

$$(\sqrt[3]{x})^3 = 2^3$$
$$x = 8$$

A check will show that 8 is the solution. ■

EXAMPLE 4 Solve the equation $\sqrt{2x - 3} = x - 3$.

Solution: $\sqrt{2x - 3} = x - 3$

Square both sides of the equation.

$$(\sqrt{2x - 3})^2 = (x - 3)^2$$
$$2x - 3 = (x - 3)(x - 3)$$
$$2x - 3 = x^2 - 6x + 9$$
$$0 = x^2 - 8x + 12$$

Now factor.

$$x^2 - 8x + 12 = 0$$
$$(x - 6)(x - 2) = 0$$
$$x - 6 = 0 \quad \text{or} \quad x - 2 = 0$$
$$x = 6 \qquad\qquad x = 2$$

Check:

$x = 6$		$x = 2$	
$\sqrt{2x - 3} = x - 3$		$\sqrt{2x - 3} = x - 3$	
$\sqrt{2(6) - 3} = 6 - 3$		$\sqrt{2(2) - 3} = 2 - 3$	
$\sqrt{9} = 3$		$\sqrt{1} = -1$	
$3 = 3$	true	$1 = -1$	false

The 6 is a solution, but 2 is not a solution to the equation. The 2 is an extraneous root (or extraneous solution). Note that 2 satisfies the equation $(\sqrt{2x - 3})^2 = (x - 3)^2$ but not the original equation $\sqrt{2x - 3} = x - 3$. ■

HELPFUL HINT

Don't forget to check your solutions in the original equation. Remember that when you raise both sides of an equation to a power, you may introduce extraneous solutions.

Consider the equation $x = 2$. Note what happens when you square both sides of the equation.

$$x = 2$$
$$x^2 = 2^2$$
$$x^2 = 4$$

Note that the equation $x^2 = 4$ has two solutions, namely, $+2$ and -2. Since the original equation $x = 2$ has only one solution, 2, we introduced the extraneous root -2.

EXAMPLE 5 Solve the equation $2x - 5\sqrt{x} - 3 = 0$.

Solution: First write the equation with the square root containing the variable by itself on one side of the equation.

$$2x - 5\sqrt{x} - 3 = 0$$
$$-5\sqrt{x} = -2x + 3$$
$$\text{or} \quad 5\sqrt{x} = 2x - 3$$

Now square both sides of the equation.

$$(5\sqrt{x})^2 = (2x - 3)^2$$
$$25x = (2x - 3)(2x - 3)$$
$$25x = 4x^2 - 12x + 9$$
$$0 = 4x^2 - 37x + 9$$
$$0 = (4x - 1)(x - 9)$$

$$4x - 1 = 0 \quad \text{or} \quad x - 9 = 0$$
$$4x = 1 \qquad\qquad x = 9$$
$$x = \frac{1}{4}$$

Check:

$$x = \frac{1}{4}$$
$$2x - 5\sqrt{x} - 3 = 0$$
$$2\left(\frac{1}{4}\right) - 5\sqrt{\frac{1}{4}} - 3 = 0$$
$$\frac{1}{2} - 5\left(\frac{1}{2}\right) - 3 = 0$$
$$-5 = 0 \quad \text{false}$$

$$x = 9$$
$$2x - 5\sqrt{x} - 3 = 0$$
$$2(9) - 5\sqrt{9} - 3 = 0$$
$$18 - 5(3) - 3 = 0$$
$$18 - 15 - 3 = 0$$
$$0 = 0 \quad \text{true}$$

The solution is 9. Note that $\frac{1}{4}$ is an extraneous root and is not a solution. ■

EXAMPLE 6 Solve the equation $3\sqrt{x-1} = 2\sqrt{2x+2}$.

Solution: Since the two radicals appear on different sides of the equation, we square both sides of the equation.

$$(3\sqrt{x-1})^2 = (2\sqrt{2x+2})^2$$
$$9(x-1) = 4(2x+2)$$
$$9x - 9 = 8x + 8$$
$$x - 9 = 8$$
$$x = 17$$

A check will show that 17 is the solution. ■

EXAMPLE 7 Solve the equation $3\sqrt[3]{x-2} = \sqrt[3]{17x-14}$.

Solution:

$$3\sqrt[3]{x-2} = \sqrt[3]{17x-14}$$
$$(3\sqrt[3]{x-2})^3 = (\sqrt[3]{17x-14})^3 \qquad \text{cube both sides of the equation}$$
$$27(x-2) = 17x - 14$$
$$27x - 54 = 17x - 14$$
$$10x - 54 = -14$$
$$10x = 40$$
$$x = 4$$

A check will show that the solution is 4. ■

EXAMPLE 8 Solve the equation $\sqrt{x+6} = \sqrt{x} + 2$.

Solution: Since one radical with a variable is isolated on one side (the left) of the equation, we begin by squaring both sides of the equation.

$$\sqrt{x+6} = \sqrt{x} + 2$$
$$(\sqrt{x+6})^2 = (\sqrt{x}+2)^2$$
$$x + 6 = (\sqrt{x}+2)(\sqrt{x}+2)$$
$$x + 6 = x + 4\sqrt{x} + 4$$

Since this equation also has a term with a variable in the radical, we must isolate the radical term.

$$x + 6 = x + 4\sqrt{x} + 4$$
$$6 = 4\sqrt{x} + 4 \qquad \text{subtract } x \text{ from both sides of the equation}$$
$$2 = 4\sqrt{x} \qquad \text{subtract 4 from both sides of the equation}$$
$$2^2 = (4\sqrt{x})^2 \qquad \text{square both sides of the equation}$$
$$4 = 16x$$
$$\frac{4}{16} = x$$
$$\frac{1}{4} = x$$

A check will show that $\frac{1}{4}$ is the solution. ■

Exercise Set 8.5

Solve each equation and then check your solution(s). If the equation has no real solution so state.

1. $\sqrt{x} = 5$
2. $\sqrt{x} = 9$
3. $\sqrt[3]{x} = 2$
4. $\sqrt[3]{x} = 4$
5. $\sqrt[4]{x} = 3$
6. $\sqrt{x-3} + 5 = 6$
7. $-\sqrt{2x+4} = -6$
8. $\sqrt{x} + 3 = 5$
9. $\sqrt[3]{2x+11} = 3$
10. $\sqrt[3]{6x-3} = 3$
11. $\sqrt[3]{3x} + 4 = 7$
12. $2\sqrt{4x-3} = 10$
13. $\sqrt{2x-3} = 2\sqrt{3x-2}$
14. $\sqrt{8x-4} = \sqrt{7x+2}$
15. $\sqrt{5x+10} = -\sqrt{3x+8}$
16. $\sqrt[4]{x+8} = \sqrt[4]{2x}$
17. $\sqrt[3]{6x+1} = \sqrt[3]{2x+5}$
18. $\sqrt[4]{3x+1} = 2$
19. $\sqrt{x^2+9x+3} = -x$
20. $\sqrt{x^2+3x+9} = x$
21. $\sqrt{m^2+4m-20} = m$
22. $\sqrt{5a+1} - 11 = 0$
23. $\sqrt{2y+3} + y = 0$
24. $\sqrt{x^2+3} = x + 1$
25. $-\sqrt{x} = 2x - 1$
26. $\sqrt{3x+4} = x - 2$
27. $\sqrt{x^2+8} = x + 2$
28. $\sqrt[3]{3x-1} + 4 = 0$
29. $\sqrt[3]{x-12} = \sqrt[3]{5x+16}$
30. $\sqrt{6x-1} = 3x$
31. $\sqrt{x+7} = 2x - 1$
32. $\sqrt[3]{4x-3} - 3 = 0$

Solve each equation. You will have to square both sides of the equation twice to eliminate all radicals (see Example 8).

33. $\sqrt{2a-3} = \sqrt{2a} - 1$
34. $\sqrt{x} + 2 = \sqrt{x+16}$
35. $\sqrt{x+1} = 2 - \sqrt{x}$
36. $\sqrt{x+3} = \sqrt{x} - 3$
37. $\sqrt{x+7} = 5 - \sqrt{x-8}$
38. $\sqrt{y+2} = 1 + \sqrt{y-3}$
39. $\sqrt{b-3} = 4 - \sqrt{b+5}$
40. $\sqrt{4x-3} = 2 + \sqrt{2x-5}$

41. Consider the equation $\sqrt{x+3} = -\sqrt{2x-1}$. Explain why this equation can have no real solution.

42. Consider the equation $-\sqrt{x^2} = \sqrt{(-x)^2}$. By studying the equation, can you determine its solution? Explain.

43. Consider the equation $\sqrt[3]{x^2} = -\sqrt[3]{x^2}$. By studying the equation, can you determine its solution? Explain.

JUST FOR FUN

Solve.

1. $\sqrt{4x+1} - \sqrt{3x-2} = \sqrt{x-5}$

2. $\dfrac{x + \sqrt{x+3}}{x - \sqrt{x+3}} = 3$

3. $\sqrt{\sqrt{x+25} - \sqrt{x}} = 5$

4. $\sqrt{\sqrt{x+9} + \sqrt{x}} = 3$

8.6 Applications of Radicals (Optional)

1. *Know the Pythagorean theorem.*
2. *Derive the distance formula.*
3. *Learn some applications of radical equations.*

In this section we will study some of the many uses of radicals. First we will introduce the Pythagorean Theorem; then we will derive the distance formula that was used in Chapter 3, and then we will study some other applications of radicals.

1 A **right triangle** is a triangle that contains a right angle. A right angle is an angle that measures 90°. A right triangle is illustrated in Fig. 8.1.

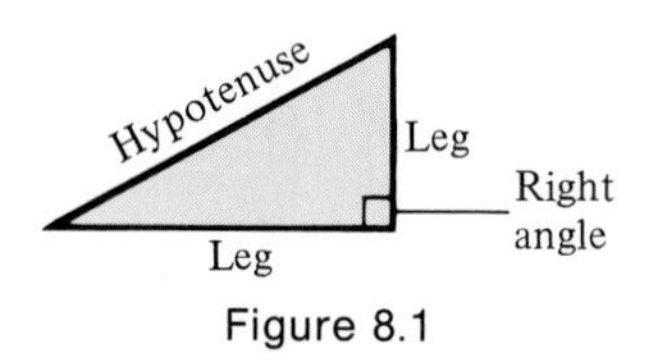

Figure 8.1

The two smaller sides of a right triangle are called the **legs** and the side opposite the right angle is called the **hypotenuse.** The Pythagorean theorem expresses the relationship between the legs of the triangle and its hypotenuse.

Pythagorean Theorem

The square of the hypotenuse of a right triangle is equal to the sum of the squares of the two legs. If a and b represent the legs and c represents the hypotenuse, then

$$a^2 + b^2 = c^2$$

c b a

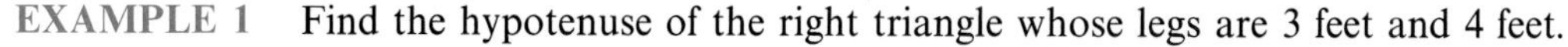

EXAMPLE 1 Find the hypotenuse of the right triangle whose legs are 3 feet and 4 feet.

Solution: It is often helpful when using the Pythagorean theorem to draw a picture of the problem before using the formula (Fig. 8.2). When drawing the picture it makes no difference which leg is called a and which leg is called b.

$$a^2 + b^2 = c^2$$
$$3^2 + 4^2 = c^2$$
$$9 + 16 = c^2$$
$$25 = c^2$$
$$\sqrt{25} = \sqrt{c^2} \quad \text{take the square root of both sides of the equation}$$
$$5 = c$$

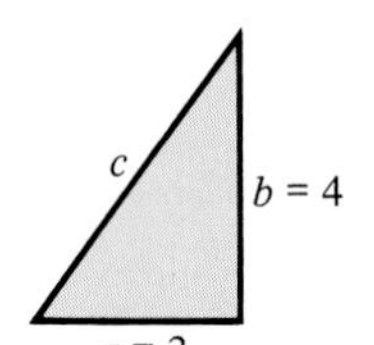

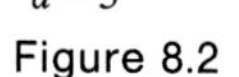
Figure 8.2

The hypotenuse is 5 feet. ■

Note that in Example 1 we obtained $\sqrt{c^2} = \sqrt{25}$. By examining this equation, we see that c can be both 5 and -5, as illustrated below.

$$\sqrt{(5)^2} = \sqrt{25} \qquad \sqrt{(-5)^2} = \sqrt{25}$$
$$\sqrt{25} = \sqrt{25} \leftarrow \text{true} \qquad \sqrt{25} = \sqrt{25} \leftarrow \text{true}$$

However, when considering lengths in physical problems, it makes no sense to talk about negative lengths. Whenever we are using the Pythagorean theorem, we will therefore use only the principal or positive root.

EXAMPLE 2 The hypotenuse of a right triangle is 8 inches. Find the second leg if one leg is 5 inches (Fig. 8.3).

Solution:

$$
\begin{aligned}
a^2 + b^2 &= c^2 \\
5^2 + b^2 &= 8^2 \\
25 + b^2 &= 64 \\
b^2 &= 39 \\
\sqrt{b^2} &= \sqrt{39} \\
b &= \sqrt{39} \quad \text{or} \quad b \approx 6.24 \text{ inches}
\end{aligned}
$$

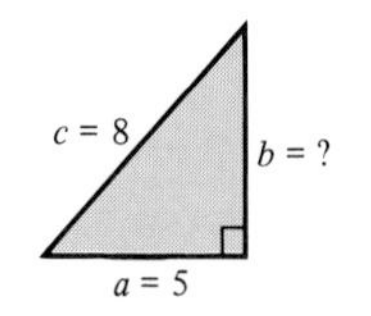

Figure 8.3

Therefore the second leg is $\sqrt{39}$ or approximately 6.24 inches. ■

Distance Formula

2 In Section 3.1 we introduced the distance formula.

> **Distance Formula**
>
> $$d = \sqrt{(x_2 - x_1)^2 + (y_2 - y_1)^2}$$

Recall that this formula can be used to find the distance between two points (x_1, y_1) and (x_2, y_2) (see Fig. 8.4).

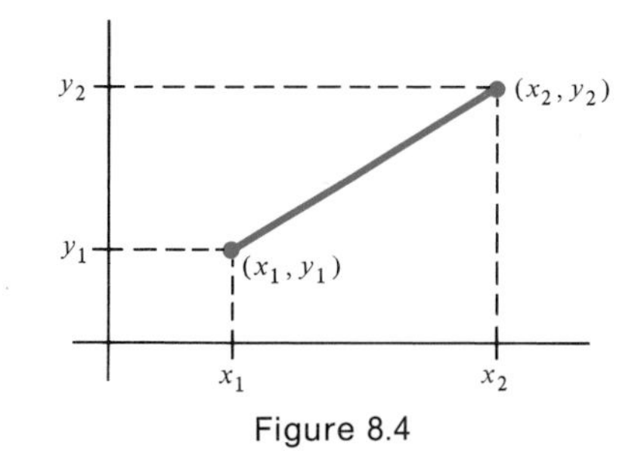

Figure 8.4

At this time we will derive the distance formula using the Pythagorean theorem. If we draw a horizontal line from point A, (x_1, y_1) and a vertical line from point B, (x_2, y_2), the two lines meet at point C, (x_2, y_1) (see Fig. 8.5).

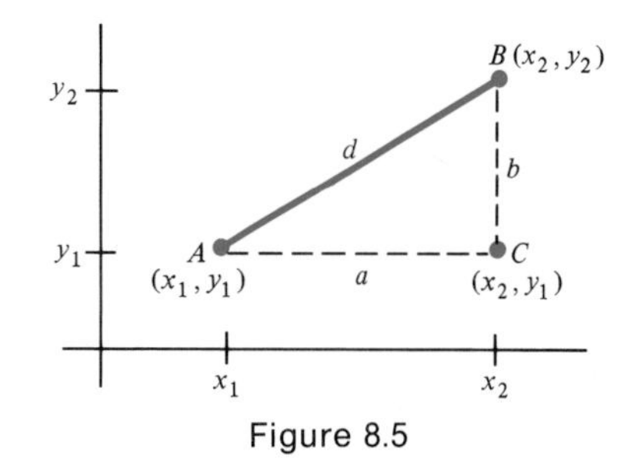

Figure 8.5

Let us designate a to be the length of the line segment from A to C, b to be the length of the line segment from B to C, and d to be the length of the line segment from A to B.

Since line segment AC is horizontal, its length is $x_2 - x_1$. Since line segment CB is vertical, its distance is $y_2 - y_1$. Line segments AC and BC are perpendicular to each other. Using the Pythagorean theorem the distance between the two given points, d, becomes

$$d^2 = a^2 + b^2$$
$$d^2 = (x_2 - x_1)^2 + (y_2 - y_1)^2$$
$$d = \sqrt{(x_2 - x_1)^2 + (y_2 - y_1)^2}$$

Note that in the distance formula above $(x_2 - x_1)^2$ and $(y_2 - y_1)^2$ will both always be nonnegative numbers. Why? Also note that when taking the square root of both sides of the equation, we use only the principal root.

3 Now we will study some of the many additional uses of radicals.

EXAMPLE 3 During the sixteenth and seventeenth centuries, Galileo Galilei did numerous experiments with objects falling freely under the influence of gravity. He showed, for example, that an object dropped from, say, 10 feet hit the ground with a higher velocity than an object dropped from 5 feet. A formula that can be used to determine the velocity of an object after it has fallen a certain distance (neglecting wind resistance) is

$$v = \sqrt{2gh}$$

where g is the acceleration due to gravity and h is the distance the object has fallen. On Earth the acceleration of gravity is approximately 32 feet per second squared.

(a) Find the velocity of an object after it has fallen 10 feet.
(b) Find the velocity of an object after it has fallen 100 feet.

Solution: (a) $v = \sqrt{2gh} = \sqrt{2(32)h} = \sqrt{64h}$

At $h = 10$ feet, $v = \sqrt{64(10)} = \sqrt{640} \approx 25.3$ feet per second

After an object has fallen 10 feet, its velocity is 25.3 feet per second.

(b) After falling 100 feet, $v = \sqrt{64(100)} = \sqrt{6400} = 80$ feet per second ■

EXAMPLE 4 The area of a triangle is $A = \frac{1}{2}bh$. If the height is not known, but instead, we know the length of each of the three sides, we can use *Hero's* formula to find the area.

$$A = \sqrt{S(S - a)(S - b)(S - c)}$$

where a, b, and c are the length of the three sides and

$$S = \frac{a + b + c}{2}$$

Use Hero's formula to find the area of a triangle with sides of 3 inches, 4 inches, and 5 inches.

Solution: The triangle is illustrated in Fig. 8.6. Let $a = 3$, $b = 4$, and $c = 5$. First find the value of S.

$$S = \frac{3+4+5}{2} = \frac{12}{2} = 6$$

Now find the area.

$$\begin{aligned} A &= \sqrt{S(S-a)(S-b)(S-c)} \\ &= \sqrt{6(6-3)(6-4)(6-5)} \\ &= \sqrt{6(3)(2)(1)} \\ &= \sqrt{36} = 6 \end{aligned}$$

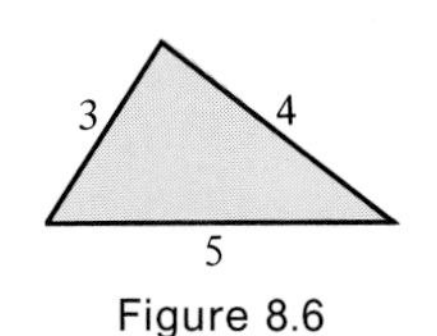

Figure 8.6

The triangle has an area of 6 square inches. ■

Exercise Set 8.6

Use the Pythagorean theorem to find the quantity indicated. You may leave your answer in square root form if a calculator with a square root key is not available and the number is not a perfect square.

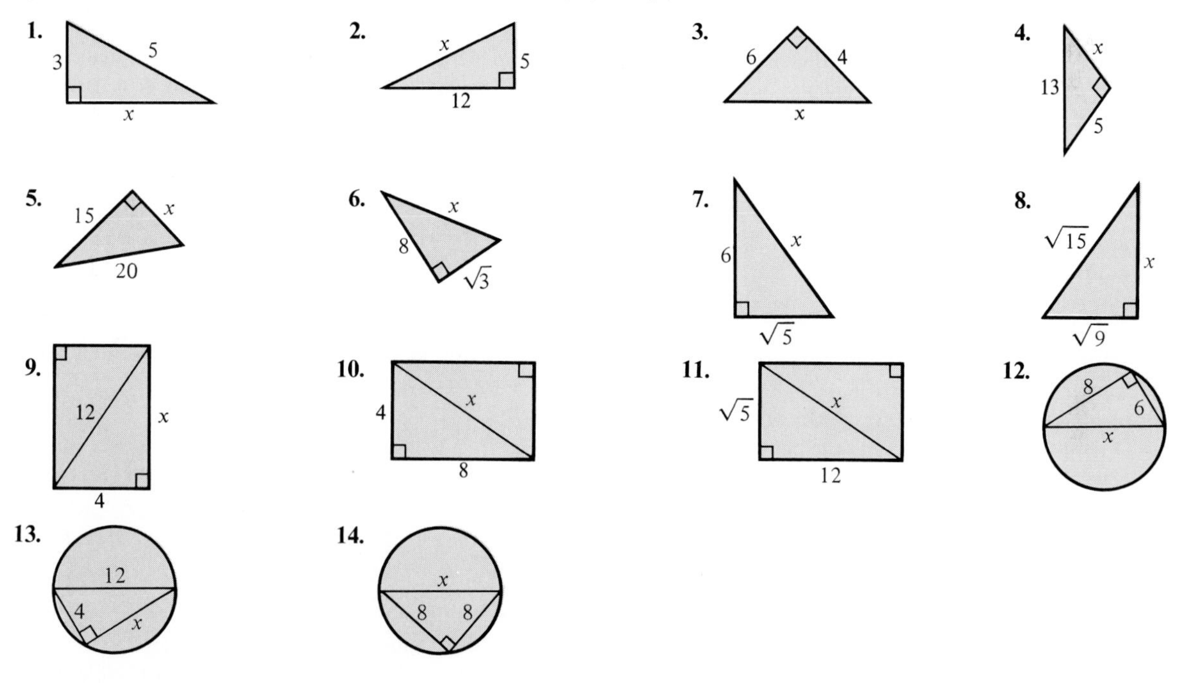

In each exercise you may leave your answer in terms of a square root if a calculator with a square root key is not available for use.

15. Find the length of the diagonal of a rectangle with a length of 12 inches and width of 5 inches.

16. A football field is 120 yards long from end zone to end zone. Find the length of the diagonal from one end zone to the other if the width of the field is 53.3 yards.

17. How long a wire does the phone company worker need to reach from the top of a 4-meter telephone pole to a point 1.5 meters from the base of the pole?

18. Ms. Song Tran places an 8-meter extension ladder against her house. The base of the ladder is 2 meters from the house. How high is the top of the ladder?

19. The length of a rectangle is 3 inches more than its width.

See Exercise 20.

If the length of the diagonal is 15 inches, find the dimensions of the rectangle.

20. A regulation baseball diamond is a square with 90 feet between bases. How far is second base from home plate?

21. Find the side of a square that has an area of 144 square inches.

22. One formula for the area of a circle is $A = \pi r^2$, where π is approximately 3.14 and r is the radius of the circle. Find the radius of a circle of area 20 square inches.

23. Find the velocity of an object after it has fallen 80 feet. Use $v = \sqrt{2gh}$. Refer to Example 3.

24. Find the velocity of an object after it has fallen 50 feet.

25. Find the area of a triangle if its three sides are 6 inches, 8 inches, and 10 inches. Use $A = \sqrt{S(S-a)(S-b)(S-c)}$. Refer to Example 4.

26. Find the area of a triangle if its three sides are 4 inches, 10 inches, and 12 inches.

27. The formula for the period of a pendulum, T, in seconds, (the time required for the pendulum to make one complete swing both back and forth) is $T = 2\pi\sqrt{L/32}$, where L is the length of the pendulum in feet. Find the period of the pendulum if its length is 8 feet. Use 3.14 as an approximation for π.

28. Find the period of a 40-foot pendulum.

29. For any planet, its "year" is the time it takes for the planet to revolve once around the sun. The number of Earth days in a given planet's year, N, is approximated by the formula

$$N = 0.2(\sqrt{R})^3$$

where R is the mean distance of the planet to the sun in millions of kilometers. Find the number of Earth days in the year of the planet Earth whose mean distance to the sun is 149.4 million kilometers.

30. Find the number of earth days in the year of the planet Mercury whose mean distance to the sun is 58 million kilometers.

31. When two forces, F_1 and F_2, pull at right angles to each other as illustrated below, the resultant, or the effective force, R, can be found by the formula $R = \sqrt{F_1^2 + F_2^2}$.

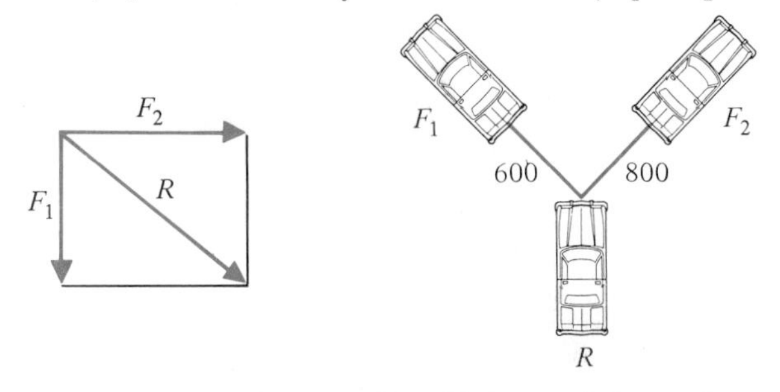

Two cars are trying to pull a third out of the mud, as illustrated. If car A is exerting a force of 600 pounds and car B is exerting a force of 800 pounds, find the resulting force on the car stuck in the mud.

32. The escape velocity, in meters per second, or the velocity needed for a spacecraft to escape a planet's gravitational field, is found by the formula $v_e = \sqrt{2gR}$, where g is the force of gravity of the planet and R is the radius of the planet in meters. Find the escape velocity for Earth where $g = 9.75$ meters per second squared and $R = 6{,}370{,}000$ meters.

33. A formula used in the study of shallow-water wave motion is $c = \sqrt{gH}$, in which c is wave velocity, H is water depth, and g is the acceleration due to gravity. Find the wave velocity if the water's depth is 10 feet. (Use $g = 32$ ft/sec^2.)

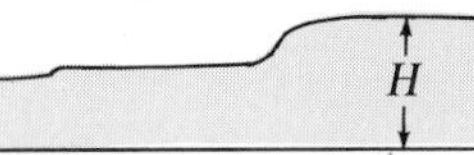

34. The length of the diagonal of a rectangular solid is given by $d = \sqrt{a^2 + b^2 + c^2}$

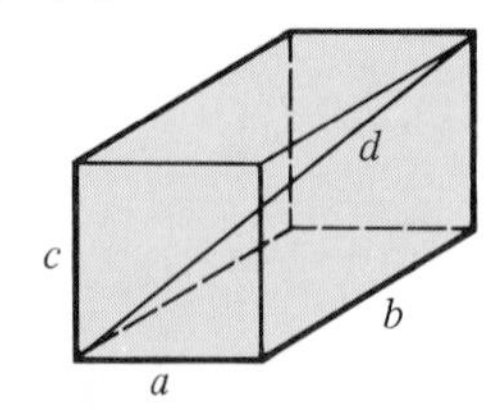

Find the length of the diagonal of a suitcase of length 37 inches, width 15 inches, and depth 9 inches.

35. A formula we will be discussing in a short while is the quadratic formula

$$x = \frac{-b \pm \sqrt{b^2 - 4ac}}{2a}$$

(a) Find x when $a = 1, b = 0, c = -4$.
(b) Find x when $a = 1, b = 1, c = -12$.
(c) Find x when $a = 2, b = 5, c = -12$.
(d) Find x when $a = -1, b = 4, c = 5$.

JUST FOR FUN

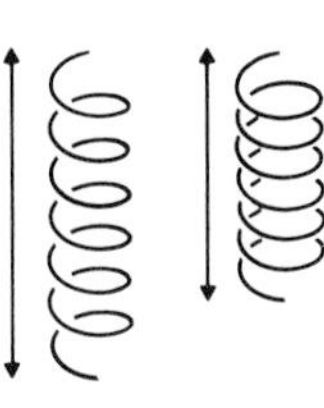

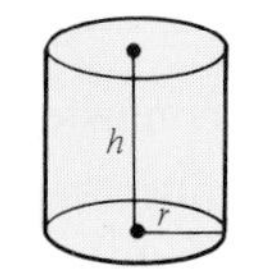

1. The force of gravity on the moon is $\frac{1}{6}$ of that on Earth. If an object falls off a rocket 100 feet above the surface of the moon, with what velocity will it strike the moon? Use $v = \sqrt{2gh}$. (See Example 3.)
2. A formula used to determine the frequency of a vibrating spring is $f = \frac{1}{2\pi}\sqrt{K/m}$, where f is the frequency of oscillation in cycles per second (also called Hertz), K is the spring stiffness constant, and m is the mass of the spring. Find the resulting frequency of a spring with a stiffness constant of 10^5 dynes/cm and a mass of 1000 grams.
3. The surface area, S, and volume, V, of a right circular cylinder of radius r and height h are given by the formulas

$$S = 2\pi r^2 + 2\pi rh \qquad V = \pi r^2 h$$

   (a) Find the surface area of the cylinder if it has a height of 10 inches and a volume of 160 cubic inches.
   (b) Find the radius if the height is 10 inches and the surface area is 160 square inches. (You may leave your answer in square root form.)

8.7 Complex Numbers

1. *Recognize an imaginary number.*
2. *Recognize a complex number.*
3. *Add and subtract complex numbers.*
4. *Multiply complex numbers.*
5. *Find the conjugate of a complex number.*
6. *Divide complex numbers.*

1 In Section 8.1 we stated the square root of negative numbers, such as $\sqrt{-4}$, are called **imaginary numbers.** There is no real number that when multiplied by itself gives -4.

$$\sqrt{-4} \neq 2 \qquad \text{since} \qquad 2 \cdot 2 = 4$$
$$\sqrt{-4} \neq -2 \qquad \text{since} \qquad (-2)(-2) = 4$$

Numbers such as -4 are called imaginary because when they were first introduced, many mathematicians refused to believe that they existed. Although they do not belong to the set of real numbers, the imaginary numbers do exist and are very useful in mathematics.

Every imaginary number has a factor of $\sqrt{-1}$. For example,

$$\sqrt{-4} = \sqrt{4}\sqrt{-1}$$
$$\sqrt{-9} = \sqrt{9}\sqrt{-1}$$
$$\sqrt{-7} = \sqrt{7}\sqrt{-1}$$

The $\sqrt{-1}$, called the **imaginary unit,** is often denoted by the letter i.

$$i = \sqrt{-1}$$

We can therefore write

$$\begin{aligned}
\sqrt{-4} &= \sqrt{4}\sqrt{-1} = 2\sqrt{-1} = 2i \\
\sqrt{-9} &= \sqrt{9}\sqrt{-1} = 3\sqrt{-1} = 3i \\
\sqrt{-7} &= \sqrt{7}\sqrt{-1} = \sqrt{7}i \quad \text{or} \quad i\sqrt{7}
\end{aligned}$$

In this book we will generally write $i\sqrt{7}$ rather than $\sqrt{7}i$ to avoid confusion with $\sqrt{7i}$.

Any number of the form bi, where b is any nonzero real number and $i = \sqrt{-1}$, is an **imaginary number.** For example, $3i$ and $i\sqrt{7}$ are imaginary numbers.

2 Now we are prepared to discuss complex numbers.

Complex Number

Every number of the form

$$a + bi$$

is a complex number.

Every real number and every imaginary number is also a complex number. A complex number has two parts: a real part and an imaginary part.

$$\underset{\text{real part}}{a} + \underset{\text{imaginary part}}{bi}$$

If $b = 0$, the complex number is a real number. If $a = 0$, the complex number is a *pure imaginary number.*

Examples of Complex Numbers

$3 + 4i$	$a = 3, b = 4$	
$5 - i\sqrt{3}$	$a = 5, b = -\sqrt{3}$	
5	$a = 5, b = 0$	(real number, $b = 0$)
$2i$	$a = 0, b = 2$	(imaginary number, $a = 0$)
$-i\sqrt{7}$	$a = 0, b = -\sqrt{7}$	(imaginary number, $a = 0$)

We stated that all real numbers and all imaginary numbers are also complex numbers. The relationship between the various sets of numbers is illustrated in Fig. 8.7.

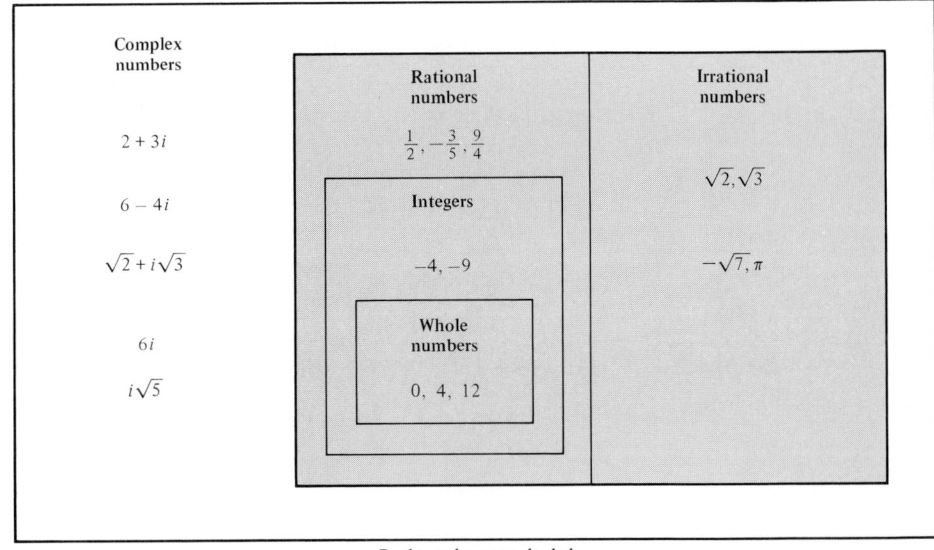

Real numbers are shaded.

Figure 8.7

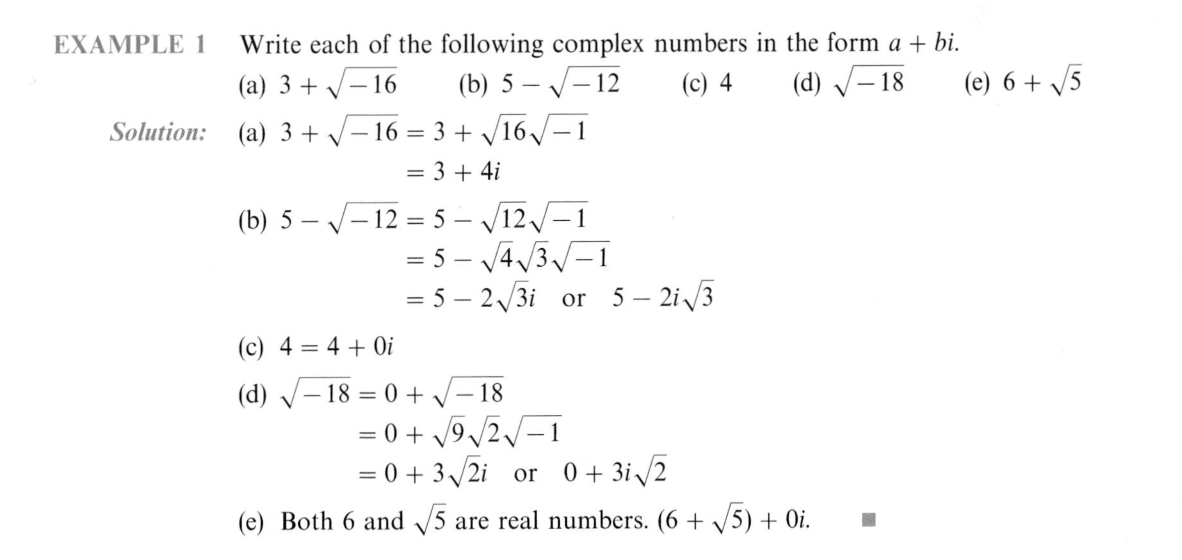

EXAMPLE 1 Write each of the following complex numbers in the form $a + bi$.

(a) $3 + \sqrt{-16}$ (b) $5 - \sqrt{-12}$ (c) 4 (d) $\sqrt{-18}$ (e) $6 + \sqrt{5}$

Solution: (a)
$$3 + \sqrt{-16} = 3 + \sqrt{16}\sqrt{-1} = 3 + 4i$$

(b)
$$5 - \sqrt{-12} = 5 - \sqrt{12}\sqrt{-1} = 5 - \sqrt{4}\sqrt{3}\sqrt{-1} = 5 - 2\sqrt{3}i \quad \text{or} \quad 5 - 2i\sqrt{3}$$

(c) $4 = 4 + 0i$

(d)
$$\sqrt{-18} = 0 + \sqrt{-18} = 0 + \sqrt{9}\sqrt{2}\sqrt{-1} = 0 + 3\sqrt{2}i \quad \text{or} \quad 0 + 3i\sqrt{2}$$

(e) Both 6 and $\sqrt{5}$ are real numbers. $(6 + \sqrt{5}) + 0i$. ■

Complex numbers can be added, subtracted, multiplied, and divided. To perform these operations we make use of the definitions that $i = \sqrt{-1}$ and

$$i^2 = -1$$

3 We will first explain how to add or subtract complex numbers. The procedures used to multiply and divide complex numbers will be explained shortly.

To Add (or Subtract) Complex Numbers

1. Change all imaginary numbers to bi form.
2. Add (or subtract) the real parts of the complex numbers.
3. Add (or subtract) the imaginary parts of the complex numbers.
4. Write the answer in the form $a + bi$.

EXAMPLE 2 Add $(4 + 13i) + (-6 - 8i)$.

Solution:

$$\begin{aligned}(4 + 13i) + (-6 - 8i) &= 4 + 13i - 6 - 8i \\ &= 4 - 6 + 13i - 8i \\ &= -2 + 5i\end{aligned}$$

EXAMPLE 3 Subtract $(-8 - 7i) - (5 - 4i)$.

Solution:

$$\begin{aligned}(-8 - 7i) - (5 - 4i) &= -8 - 7i - 5 + 4i \\ &= -13 - 3i\end{aligned}$$

EXAMPLE 4 Add $(6 - \sqrt{-8}) + (4 + \sqrt{-18})$.

Solution:

$$\begin{aligned}(6 - \sqrt{-8}) + (4 + \sqrt{-18}) &= (6 - \sqrt{8}\sqrt{-1}) + (4 + \sqrt{18}\sqrt{-1}) \\ &= (6 - \sqrt{4}\sqrt{2}\sqrt{-1}) + (4 + \sqrt{9}\sqrt{2}\sqrt{-1}) \\ &= (6 - 2i\sqrt{2}) + (4 + 3i\sqrt{2}) \\ &= 6 + 4 - 2i\sqrt{2} + 3i\sqrt{2} \\ &= 10 + i\sqrt{2}\end{aligned}$$

4

To Multiply Complex Numbers

1. Change all imaginary numbers to bi form.
2. Multiply the complex numbers as you would multiply polynomials.
3. Substitute -1 for each i^2.
4. Combine the real parts and the imaginary parts and then write the answer in $a + bi$ form.

EXAMPLE 5 Multiply $3i(5 - 2i)$.

Solution:

$$\begin{aligned}3i(5 - 2i) &= 3i(5) + (3i)(-2i) \\ &= 15i - 6i^2 \\ &= 15i - 6(-1) \\ &= 15i + 6 \quad \text{or} \quad 6 + 15i\end{aligned}$$

EXAMPLE 6 Multiply $\sqrt{-4}(\sqrt{-2}+7)$.

Solution: First write each imaginary number in bi form.

$$\begin{aligned}\sqrt{-4}(\sqrt{-2}+7) &= 2i(i\sqrt{2}+7)\\ &= (2i)(i\sqrt{2})+(2i)(7)\\ &= 2i^2\sqrt{2}+14i\\ &= 2(-1)\sqrt{2}+14i\\ &= -2\sqrt{2}+14i\end{aligned}$$ ■

COMMON STUDENT ERROR

What is $\sqrt{-4}\cdot\sqrt{-2}$?

Correct

$$\begin{aligned}\sqrt{-4}\cdot\sqrt{-2} &= 2i\cdot i\sqrt{2}\\ &= 2i^2\sqrt{2}\\ &= 2(-1)\sqrt{2}\\ &= -2\sqrt{2}\end{aligned}$$

Wrong

$$\begin{aligned}\sqrt{-4}\cdot\sqrt{-2} &= \sqrt{8}\\ &= \sqrt{4}\cdot\sqrt{2}\\ &= 2\sqrt{2}\end{aligned}$$

Recall that $\sqrt{a}\cdot\sqrt{b}=\sqrt{ab}$ for nonnegative integers a and b.

EXAMPLE 7 Multiply $(3+5i)(2-3i)$.

Solution: We may begin by multiplying using the FOIL method.

$$\begin{aligned}(3)(2)+(3)(-3i)+(5i)(2)+(5i)(-3i) &= 6-9i+10i-15i^2\\ &= 6-9i+10i-15(-1)\\ &= 6-9i+10i+15\\ &= 21+i\end{aligned}$$ ■

EXAMPLE 8 Multiply $(3-\sqrt{-8})(\sqrt{-2}+5)$.

Solution:

$$\begin{aligned}(3-\sqrt{-8})(\sqrt{-2}+5) &= (3-\sqrt{8}\sqrt{-1})(\sqrt{2}\sqrt{-1}+5)\\ &= (3-2i\sqrt{2})(i\sqrt{2}+5)\end{aligned}$$

Now use the FOIL method to multiply.

$$\begin{aligned}&= 3(i\sqrt{2})+(3)(5)+(-2i\sqrt{2})(i\sqrt{2})+(-2i\sqrt{2})(5)\\ &= 3i\sqrt{2}+15-2i^2(2)-10i\sqrt{2}\\ &= 3i\sqrt{2}+15-2(-1)(2)-10i\sqrt{2}\\ &= 3i\sqrt{2}+15+4-10i\sqrt{2}\\ &= 19-7i\sqrt{2}\end{aligned}$$ ■

5 The **conjugate** of a complex number $a + bi$ is $a - bi$. For example,

Complex Number	*Conjugate of Complex Number*
$3 + 4i$	$3 - 4i$
$1 - i\sqrt{3}$	$1 + i\sqrt{3}$
$2i$ (or $0 + 2i$)	$-2i$ (or $0 - 2i$)

When a complex number is multiplied by its conjugate using the FOIL method, the inner and outer parts will sum to zero. For example,

$$\begin{aligned}(5 + 2i)(5 - 2i) &= 25 - 10i + 10i - 4i^2 \\ &= 25 - 4i^2 \\ &= 25 - 4(-1) \\ &= 25 + 4 \\ &= 29\end{aligned}$$

6

To Divide Complex Numbers

1. Change all imaginary numbers to bi form.
2. Write the division problem as a fraction.
3. Rationalize the denominator of the fraction by multiplying both the numerator and denominator of the fraction by the conjugate of the denominator.
4. Write the answer in $a + bi$ form.

EXAMPLE 9 Simplify $\dfrac{4 + i}{i}$.

Solution: Multiply both numerator and denominator by $-i$, the conjugate of i.

$$\begin{aligned}\frac{4 + i}{i} \cdot \frac{-i}{-i} &= \frac{(4 + i)(-i)}{-i^2} \\ &= \frac{-4i - i^2}{-i^2} \qquad \text{(now substitute } i^2 = -1\text{)} \\ &= \frac{-4i - (-1)}{-(-1)} \\ &= \frac{-4i + 1}{1} = 1 - 4i\end{aligned}$$

EXAMPLE 10 Divide $\dfrac{6-5i}{2-i}$.

Solution: Multiply both numerator and denominator by $2+i$, the conjugate of $2-i$.

$$\frac{6-5i}{2-i}\cdot\frac{2+i}{2+i}=\frac{12+6i-10i-5i^2}{4-i^2}$$
$$=\frac{12-4i-5(-1)}{4-(-1)}$$
$$=\frac{17-4i}{5} \quad ■$$

Knowing that $i=\sqrt{-1}$ and $i^2=-1$ we can find other powers of i. For example,

$$i^3=i^2\cdot i=-1\cdot i=-i \qquad i^6=i^4\cdot i^2=1(-1)=-1$$
$$i^4=i^2\cdot i^2=(-1)(-1)=1 \qquad i^7=i^4\cdot i^3=1(-i)=-i$$
$$i^5=i^4\cdot i^1=1\cdot i=i \qquad i^8=i^4\cdot i^4=(1)(1)=1$$

Note that powers of i rotate through the four numbers i, -1, $-i$, 1.

Exercise Set 8.7

Write each expression as a complex number in the form $a+bi$.

1. 3
2. $\sqrt{9}$
3. $3+\sqrt{-4}$
4. $-\sqrt{5}$
5. $6+\sqrt{3}$
6. $\sqrt{-8}$
7. $\sqrt{-25}$
8. $2+\sqrt{-5}$
9. $4+\sqrt{-12}$
10. $\sqrt{-4}+\sqrt{-16}$
11. $\sqrt{-25}-2i$
12. $3+\sqrt{-72}$
13. $9-\sqrt{-9}$
14. $\sqrt{75}-\sqrt{-128}$
15. $2i-\sqrt{-80}$

Add or subtract as indicated.

16. $(3+6i)+2i$
17. $(12-6i)+(3+2i)$
18. $(6-3i)-2(2-4i)$
19. $(12+8i)-(4-5i)$
20. $(6+\sqrt{-4})+(2+7i)$
21. $(13-\sqrt{-4})-(-5+\sqrt{-9})$
22. $(7+\sqrt{5})+(2\sqrt{5}+\sqrt{-5})$
23. $(\sqrt{3}+\sqrt{2})+(3\sqrt{2}-\sqrt{-8})$
24. $(3-\sqrt{-72})+(4-\sqrt{-32})$
25. $(19+\sqrt{-147})+(\sqrt{-75})$
26. $(13+\sqrt{-108})-(\sqrt{49}-\sqrt{-48})$
27. $(\sqrt{12}+\sqrt{-49})-(\sqrt{49}-\sqrt{-12})$

Multiply as indicated.

28. $3(2+4i)$
29. $2(-3-2i)$
30. $-3(\sqrt{5}+2i)$
31. $i(6+i)$
32. $2i(2-5i)$
33. $-3i(6-4i)$
34. $\sqrt{-5}(2+3i)$
35. $\sqrt{-4}(\sqrt{3}+2i)$
36. $\sqrt{-8}(\sqrt{2}-\sqrt{-2})$
37. $\sqrt{-6}(\sqrt{3}+\sqrt{-6})$
38. $-\sqrt{-2}(3-\sqrt{-8})$
39. $(3+2i)(1+i)$
40. $(3-4i)(6+5i)$
41. $(4-6i)(3-i)$
42. $(3i+4)(2i-3)$
43. $(4+\sqrt{-3})(2+\sqrt{3})$
44. $(2-3i)(4+\sqrt{-4})$
45. $(5-\sqrt{-8})(2+\sqrt{-2})$
46. $(7-\sqrt{-5})(\sqrt{3}+2i)$

Divide as indicated.

47. $\dfrac{-5}{-3i}$
48. $\dfrac{2}{5i}$
49. $\dfrac{2+3i}{2i}$
50. $\dfrac{1+i}{-3i}$

51. $\frac{2+5i}{5i}$ **52.** $\frac{6}{2+i}$ **53.** $\frac{7}{7-2i}$ **54.** $\frac{4+2i}{1+3i}$

55. $\frac{6-3i}{4+2i}$ **56.** $\frac{4-3i}{4+3i}$ **57.** $\frac{4}{6-\sqrt{-4}}$ **58.** $\frac{5}{3+\sqrt{-5}}$

59. $\frac{\sqrt{6}}{\sqrt{3}-\sqrt{-9}}$ **60.** $\frac{\sqrt{2}}{5+\sqrt{-12}}$ **61.** $\frac{\sqrt{10}+\sqrt{-3}}{5-\sqrt{-20}}$ **62.** $\frac{12-\sqrt{-12}}{\sqrt{3}+\sqrt{-5}}$

Answer true or false.

63. Every real and every imaginary number is a complex number.
64. Every complex number is a real number.
65. The product of two pure imaginary numbers is always a real number.
66. The sum of two pure imaginary numbers is always an imaginary number.
67. The product of two complex numbers is always a real number.
68. The sum of two complex numbers is always a complex number.

8.8 The Square Root Function (Optional)

1 *Recognize square root functions.*
2 *Find the domain of square root functions.*
3 *Graph square root functions.*

1 We have discussed functions in a number of previous chapters. Here we introduce a new type of function, the square root function.

Square Root Function

$$f(x) = \sqrt{x}, \qquad x \geq 0$$

for any algebraic expression x.

2 The radicand of the square root function must be greater than or equal to 0. If the radicand is less than 0, the number becomes an imaginary number. Recall that the **domain** is the set of values that can be used for the independent variable, x. Thus the domain of $f(x) = \sqrt{x}$ (or $y = \sqrt{x}$) is all real numbers greater than or equal to 0.

EXAMPLE 1 Find the domain of $f(x) = \sqrt{x-2}$.

Solution: We must find the values of x that will result in the radicand being nonnegative. When $x \geq 2$ the radicand will be greater than or equal to 0. The domain is the set of all real numbers greater than or equal to 2.

$$D = \{x \mid x \geq 2\}$$ ■

EXAMPLE 2 Find the domain of $f(x) = \sqrt{6 - \frac{3}{5}x}$.

Solution: We must find the set of values of x that makes the radicand greater than or equal to 0. To find these values, set the radicand greater than or equal to 0 and solve for x.

$$6 - \frac{3}{5}x \geq 0$$

$$6 - 6 - \frac{3}{5}x \geq 0 - 6 \qquad \text{subtract 6 from both sides of inequality}$$

$$-\frac{3}{5}x \geq -6$$

$$\left(-\frac{5}{3}\right)\left(-\frac{3}{5}x\right) \leq -6\left(-\frac{5}{3}\right) \qquad \text{multiply both sides of inequality by } -\tfrac{5}{3} \text{ and change sense of inequality}$$

$$x \leq 10$$

Thus the domain is the set of real numbers less than or equal to 10.

$$D = \{x \mid x \leq 10\}. \quad \blacksquare$$

3 Now let us graph some square root functions.

EXAMPLE 3 Graph the function $y = \sqrt{x}$.

Solution: The domain of this function is all real numbers greater than or equal to 0. Thus when we graph we can only use values of x such that $x \geq 0$. We will substitute values for x that are perfect square numbers. The graph is illustrated in Fig. 8.8.

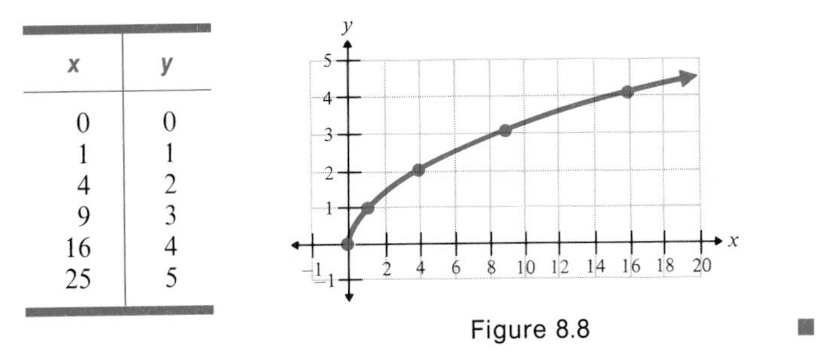

x	y
0	0
1	1
4	2
9	3
16	4
25	5

Figure 8.8 ■

Note that the graph in Fig. 8.8 is a function since it passes the vertical line test. Also note that the range of $f(x)$, or the values of y, is the set of values greater than or equal to 0, $R: \{y \mid y \geq 0\}$.

EXAMPLE 4 Graph the function $y = -\sqrt{x+3}$.

Solution: The domain is the set of values greater than or equal to -3. Therefore, when we substitute values for x we will use values that are greater than or equal to -3. We will also select values that make the radicand a perfect square. The graph is shown in Fig. 8.9.

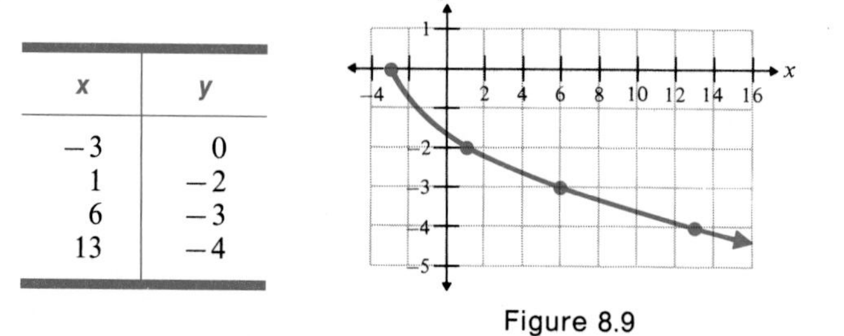

x	y
−3	0
1	−2
6	−3
13	−4

Figure 8.9 ■

Note the range of the graph in Fig. 8.9 is $\{y \mid y \leq 0\}$.

Exercise Set 8.8

Find the domain of the function.

1. $f(x) = \sqrt{x+2}$
2. $f(x) = \sqrt{x-2}$
3. $f(x) = \sqrt{4x}$
4. $f(x) = \sqrt{6-x}$
5. $f(x) = \sqrt{4-x}$
6. $f(x) = \sqrt{3x-5}$
7. $f(x) = -\sqrt{x+4}$
8. $f(x) = \sqrt{9-2x}$
9. $f(x) = -\sqrt{3x+5}$
10. $f(x) = \sqrt{5x+7}$
11. $f(x) = \sqrt{7x - \frac{1}{2}}$
12. $f(x) = \sqrt{\frac{2}{3}x - 4}$
13. $f(x) = \sqrt{\frac{5}{3} - 4x}$
14. $f(x) = \sqrt{6 - \frac{3}{2}x}$

Find the domain of the following functions. Graph the function and state its range.

15. $f(x) = \sqrt{x+4}$
16. $f(x) = \sqrt{x-4}$
17. $f(x) = \sqrt{x-2}$
18. $f(x) = \sqrt{x+2}$
19. $f(x) = -\sqrt{x+3}$
20. $f(x) = -\sqrt{x-3}$
21. $f(x) = \sqrt{3-x}$
22. $f(x) = \sqrt{5-x}$

23. $f(x) = -\sqrt{2 - x}$

24. $f(x) = \sqrt{2x + 5}$

25. $f(x) = \sqrt{2x}$

26. $f(x) = -\sqrt{2x}$

27. $f(x) = \sqrt{2x + 1}$

28. $f(x) = \sqrt{2x - 4}$

29. $f(x) = \sqrt{2x + 4}$

30. $f(x) = \sqrt{3x}$

31. Consider the square root function $y = \sqrt{x^2}$.
  (a) What is its domain?
  (b) Graph the function.
  (c) What is its range?

32. Consider the square root function $y = \sqrt{x^2 + 1}$
  (a) What is its domain?
  (b) Graph the function.
  (c) What is its range?

33. Consider the square root function $y = \sqrt{x^2 - 4}$. Is 0 in the the domain of this function? Explain.

34. Consider the square root function $y = \sqrt{4 - x^2}$. Is 0 in the domain of this function? Explain.

35. Consider the square root function $y = \sqrt{(x + 2)^2}$. What is the domain of this function?

36. Explain why the range of a function of the form $y = \sqrt{x}$ must always be $\{y | y \geq 0\}$.

37. Consider the equation $y = \sqrt[3]{x}$. Is this function a square root function? Explain. Graph the equation and determine if $y = \sqrt[3]{x}$ is a function. State the domain and range of the function or relation.

JUST FOR FUN

1. Consider the square root function $f(x) = \sqrt{4 - x^2}$.
  (a) Find the domain.
  (b) Graph the function.
  (c) Find the range.

2. Consider the square root function $f(x) = \sqrt{x^2 - 4}$.
  (a) Find the domain.
  (b) Graph the function.
  (c) Find the range.

Summary

Glossary

Complex number: A number of the form $a + bi$. Every real number and every imaginary number is a complex number.
Conjugate: The conjugate of $a + b$ is $a - b$.
Domain: The set of values that can be used for the independent variable.
Hypotenuse: The side in a right triangle opposite the right angle.
Imaginary number: A number of the form bi, $b \neq 0$.
Imaginary unit: $\sqrt{-1}$. *Note:* $\sqrt{-1} = i$.
Index: The root of an expression.
Legs of a right triangle: The two smaller sides in a right triangle.
Like radicals: Radicals that have the same index and radicand.
Perfect factor (or **perfect power**): When a radicand is an expression raised to a power, and that power is evenly divisible by the index of the radical, the radicand, including the exponent, is a perfect factor, or perfect power, of the radical.
Positive (or **principal**) **square root:** The principal square root of a positive real number x, written $\sqrt{x}$, is the positive number whose square equals x.
Radical equation: An equation that contains a variable in a radicand.
Radical expression: The radical sign and the radicand together is a radical expression.
Radical sign: $\sqrt{\ }$.
Rationalizing the denominator: To rationalize a denominator means to remove all radicals from the denominator.
Right triangle: A triangle that contains a 90° angle.
Square root function: A function of the form $f(x) = \sqrt{x}$, $x \geq 0$.

Important Facts

$$\sqrt[n]{x} = x^{\frac{1}{n}}, \qquad x \geq 0$$

$$\sqrt[n]{x^m} = (\sqrt[n]{x})^m = x^{\frac{m}{n}}, \qquad x \geq 0$$

$$\sqrt[n]{a}\sqrt[n]{b} = \sqrt[n]{ab} \qquad a \geq 0, b \geq 0$$

$$\frac{\sqrt[n]{a}}{\sqrt[n]{b}} = \sqrt[n]{\frac{a}{b}} \qquad a \geq 0, b > 0$$

Pythagorean Theorem: $a^2 + b^2 = c^2$, where a and b are legs of a right triangle and c is the hypotenuse.

Review Exercises

[8.1] Evaluate each expression.

1. $\sqrt{9}$ **2.** $\sqrt{25}$ **3.** $\sqrt[3]{-8}$ **4.** $\sqrt[4]{256}$
5. $\sqrt[3]{27}$ **6.** $\sqrt[3]{-27}$ **7.** $\sqrt{6^2}$ **8.** $\sqrt{(19)^2}$

Use absolute value to evaluate the following.

9. $\sqrt{(-7)^2}$ **10.** $\sqrt{(-93)^2}$

Write as an absolute value.

11. $\sqrt{x^2}$ **12.** $\sqrt{(x-2)^2}$ **13.** $\sqrt{(a-3)^2}$
14. $\sqrt{(x-y)^2}$ **15.** $\sqrt{(2x-3)^2}$

Write each expression in exponential form and then simplify. Assume that all variables represent positive real numbers.

16. $\sqrt{3^4}$ **17.** $\sqrt{5^6}$ **18.** $\sqrt[3]{3^6}$ **19.** $\sqrt[5]{3^{15}}$

20. $\sqrt{x^6}$ 21. $\sqrt{x^{10}}$ 22. $\sqrt[3]{y^{12}}$ 23. $\sqrt[5]{9^{10}}$

24. $\sqrt[4]{x^{20}}$ 25. $\sqrt{x^2}$ 26. $\sqrt[3]{x^3}$

[8.2] Simplify each expression. Assume that all variables represent positive real numbers.

27. $\sqrt{24}$ 28. $\sqrt{80}$ 29. $\sqrt[3]{16}$ 30. $\sqrt[3]{54}$

31. $\sqrt{8xy^4}$ 32. $\sqrt{50x^3y^7}$ 33. $\sqrt[3]{9x^6y^5}$ 34. $\sqrt[4]{16x^9y^{12}}$

35. $\sqrt[3]{125x^7y^{10}}$ 36. $\sqrt[4]{x^{12}y^{13}z^{14}}$ 37. $\sqrt[5]{32x^{12}y^7z^{17}}$ 38. $\sqrt{20}\sqrt{5}$

39. $\sqrt[3]{2}\sqrt[3]{4}$ 40. $\sqrt{5x}\sqrt{8x^5}$ 41. $\sqrt{15x^2y^3}\sqrt{3x^5y^5}$ 42. $\sqrt[3]{2x^4y^5}\sqrt[3]{16x^4y^4}$

43. $(\sqrt[3]{5x^2y^3})^2$ 44. $\sqrt[4]{8x^4y^7}\sqrt[4]{2x^5y^9}$

[8.3] Simplify each expression. Assume that all variables represent positive real numbers.

45. $\sqrt{\frac{1}{4}}$ 46. $\sqrt{\frac{36}{25}}$ 47. $\sqrt[3]{\frac{3}{81}}$ 48. $\sqrt[3]{\frac{x^3}{8}}$

49. $\frac{\sqrt[3]{2x^9}}{\sqrt[3]{16x^6}}$ 50. $\sqrt{\frac{32x^2y^5}{2x^4y}}$ 51. $\sqrt[3]{\frac{54x^3y^6}{2y^3}}$ 52. $\sqrt{\frac{75x^2y^5}{3x^4y^7}}$

Rationalize the denominator. Assume that all variables represent positive real numbers.

53. $\frac{1}{\sqrt{2}}$ 54. $\frac{1}{\sqrt{3}}$ 55. $\frac{2}{\sqrt{3}}$ 56. $\frac{x}{\sqrt{7}}$

57. $\sqrt{\frac{2}{5}}$ 58. $\sqrt{\frac{2}{y}}$ 59. $\sqrt{\frac{y}{x}}$ 60. $\sqrt{\frac{3x}{5y}}$

61. $\frac{2}{\sqrt[3]{x}}$ 62. $\sqrt[3]{\frac{3x}{5y}}$

Simplify each expression. Assume that all variables represent positive real numbers.

63. $\sqrt{\frac{3x^2}{y}}$ 64. $\sqrt{\frac{18x^4y^5}{3z}}$ 65. $\sqrt{\frac{25x^2y^5}{3z}}$ 66. $\sqrt{\frac{5x^3y^5}{2}}$

67. $\sqrt{\frac{27x^4z^5}{5y}}$ 68. $\sqrt{\frac{20y^6z^9}{6x^3}}$

Multiply as indicated, then simplify. Assume that all variables represent positive real numbers.

69. $(3-\sqrt{2})(3+\sqrt{2})$ 70. $(\sqrt{3}-\sqrt{5})(\sqrt{3}+\sqrt{5})$ 71. $(\sqrt{x}+y)(\sqrt{x}-y)$

72. $(x-\sqrt{y})(x+\sqrt{y})$ 73. $(\sqrt{3}+5)^2$ 74. $(\sqrt{5}-\sqrt{20})^2$

Simplify each expression. Assume that all variables represent positive real numbers.

75. $\frac{5}{2+\sqrt{5}}$ 76. $\frac{3}{4-\sqrt{2}}$ 77. $\frac{x}{3+\sqrt{x}}$ 78. $\frac{\sqrt{x}}{\sqrt{x}+\sqrt{y}}$

[8.4] Simplify each expression.

79. $2\sqrt{3}-\sqrt{3}+5\sqrt{3}$ 80. $\sqrt[3]{5}+2\sqrt[3]{5}-7\sqrt[3]{5}$ 81. $\sqrt[3]{x}+3\sqrt[3]{x}-2\sqrt[3]{x}$

82. $\sqrt{18}+\sqrt{32}$ 83. $\sqrt[3]{16}-\sqrt[3]{54}$ 84. $\sqrt{3}+\sqrt{27}-\sqrt{192}$

85. $\sqrt[3]{16}-5\sqrt[3]{54}+2\sqrt[3]{64}$ 86. $4\sqrt{2}-\frac{3}{\sqrt{32}}+\sqrt{50}$ 87. $\sqrt{x^3y^2}+\sqrt{x^7y^2}$

88. $3\sqrt{x^5y^6}-\sqrt{16x^7y^8}$ 89. $2\sqrt[3]{x^7y^8}-\sqrt[3]{x^4y^2}+3\sqrt[3]{x^{10}y^2}$ 90. $4\sqrt[4]{x^9y^2}-2\sqrt[4]{xy^6}+\sqrt[4]{xy^2}$

[8.5] Solve each equation; then check your solutions.

91. $\sqrt{x} = 6$

92. $\sqrt[3]{x} = 4$

93. $\sqrt{3x + 4} = 5$

94. $2 + \sqrt[3]{x} = 4$

95. $\sqrt{x^2 + 2x - 4} = x$

96. $\sqrt[3]{x - 9} = \sqrt[3]{5x + 3}$

97. $\sqrt{x^2 + 5} = x + 1$

98. $\sqrt{x + 3} = \sqrt{3x + 9}$

[8.6] Use the Pythagorean theorem to find the quantity indicated.

99.

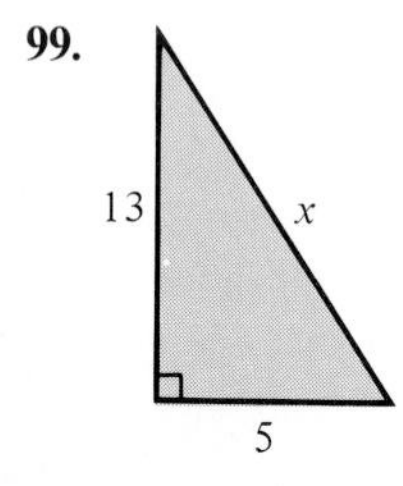

100.

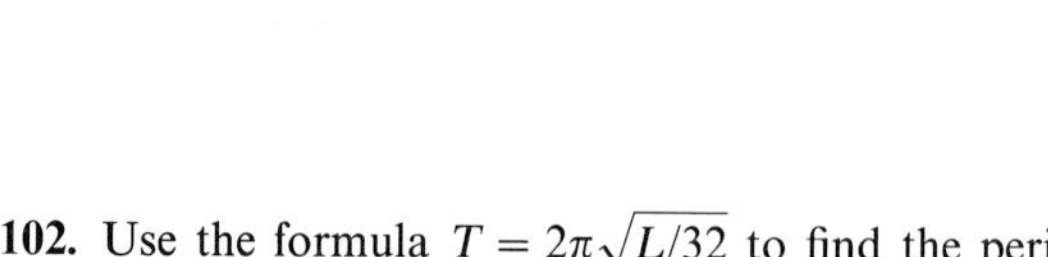

Solve each problem.

101. Use the formula $v = \sqrt{2gh}$ to find the velocity of an object after it has fallen 20 feet ($g = 32$). You may leave your answer in simplified radical form.

102. Use the formula $T = 2\pi\sqrt{L/32}$ to find the period of a pendulum (T) if its length L is 64 feet. You may leave your answer in simplified radical form.

Write each expression as a complex number in the form $a + bi$.

103. 5

104. -6

105. $2 - \sqrt{-4}$

106. $3 + \sqrt{-16}$

Add or subtract as indicated.

107. $(3 + 2i) + (4 - i)$

108. $(4 - 6i) - (3 - 4i)$

109. $(5 + \sqrt{-9}) - (3 - \sqrt{-4})$

110. $(\sqrt{3} + \sqrt{-5}) + (2\sqrt{3} - \sqrt{-7})$

Multiply as indicated.

111. $4(3 + 2i)$

112. $-2(\sqrt{3} - i)$

113. $\sqrt{8}(\sqrt{-2} + 3)$

114. $\sqrt{-6}(\sqrt{6} + \sqrt{-6})$

115. $(4 + 3i)(2 - 3i)$

116. $(6 + \sqrt{-3})(4 - \sqrt{-15})$

Divide as indicated.

117. $\dfrac{2}{3i}$

118. $\dfrac{-3}{-5i}$

119. $\dfrac{2 + \sqrt{3}}{2i}$

120. $\dfrac{5}{3 + 2i}$

121. $\dfrac{\sqrt{3}}{5 - \sqrt{-6}}$

122. $\dfrac{\sqrt{5} + 3i}{\sqrt{2} - \sqrt{-8}}$

[8.8] State the domain of each of the following functions.

123. $f(x) = \sqrt{x - 6}$

124. $f(x) = \sqrt{3x + 5}$

125. $f(x) = \sqrt{10 - 4x}$

126. $f(x) = \sqrt{\dfrac{2}{3}x - 4}$

State the domain of each function. Graph the function and state the range.

127. $f(x) = \sqrt{x}$

128. $f(x) = \sqrt{x + 2}$

129. $f(x) = \sqrt{5 - x}$

130. $f(x) = -\sqrt{x - 3}$

Practice Test

Use absolute value to evaluate.

1. $\sqrt{(-26)^2}$

Write as an absolute value.

2. $\sqrt{(3x-4)^2}$

Simplify. Assume that all variables represent positive real numbers.

3. $\sqrt{50x^5y^8}$

4. $\sqrt[3]{4x^5y^2}\,\sqrt[3]{10x^6y^8}$

5. $\sqrt{\dfrac{1}{3}}$

6. $\sqrt{\dfrac{2x^4y^5}{8z}}$

7. $\sqrt[3]{\dfrac{1}{x}}$

8. $\dfrac{\sqrt{2}}{2+\sqrt{8}}$

9. $\sqrt{27}+2\sqrt{3}-5\sqrt{75}$

10. $\sqrt[3]{8x^3y^5}+2\sqrt[3]{x^6y^8}$

11. $(\sqrt{5}-3)(2-\sqrt{8})$

Solve the equations.

12. $\sqrt{4x-3}=7$

13. $\sqrt{x^2-x-12}=x+3$

14. $\sqrt{x-15}=\sqrt{x}-3$

15. Use the Pythagorean theorem to find the quantity indicated.

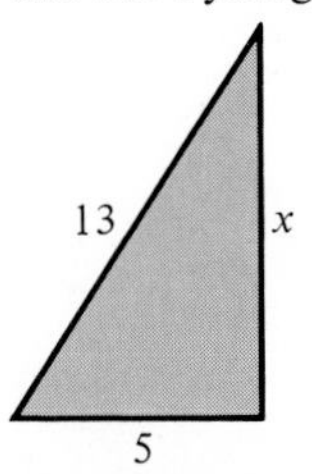

16. Multiply $(6-\sqrt{-4})(3+\sqrt{-2})$.

17. Divide $\dfrac{\sqrt{5}}{2-\sqrt{-8}}$.

*18. **(a)** State the domain of $f(x)=\sqrt{x+2}$.
(b) Graph the function and state the range.

* From an optional section.

9 Quadratic Equations and Inequalities

See Section 9.2, Exercise 57.

9.1 Solving Quadratic Equations by Completing the Square

1. *Know the square root property.*
2. *Write perfect square trinomials.*
3. *Solve quadratic equations by completing the square.*

In Section 6.5 we solved quadratic, or second-degree, equations by factoring. In this chapter we will give two techniques, completing the square and the quadratic formula, for solving quadratic equations that cannot be solved by factoring. In this section we introduce the completing the square procedure. In the next section we introduce the quadratic formula.

1 Before we can introduce completing the square you must understand the square root property. In Section 8.1 we stated that every positive number has two square roots. Thus far we have been using only the positive square root. In this section we use both the positive and negative square roots of a number.

The positive square root of 25 is 5.

$$\sqrt{25} = 5$$

The negative square root of 25 is -5.

$$-\sqrt{25} = -5$$

Notice that $5 \cdot 5 = 25$ and $(-5)(-5) = 25$. The two square roots of 25 are $+5$ and -5. A convenient way to indicate the two square roots of a number is to use the plus or minus symbol, $\pm$. For example, the square roots of 25 can be indicated ± 5, read "plus or minus 5". The equation $x^2 = 25$ has two solutions, the two roots of 25, ± 5. If you check each root you will see that each value satisfies the equation. The square root property can be used to find the solutions to equations of the form $x^2 = a$.

> **Square Root Property**
>
> If $x^2 = a$ where a is a real number, then $x = \pm\sqrt{a}$

EXAMPLE 1 Solve the equation $x^2 - 9 = 0$.

Solution: Add 9 to both sides of the equation to get the variable by itself on one side of the equation.

$$x^2 = 9$$

By the square root property.

$$x = \pm\sqrt{9}$$
$$x = \pm 3$$

Check in the original equation.

$x = 3$	$x = -3$
$x^2 - 9 = 0$	$x^2 - 9 = 0$
$3^2 - 9 = 0$	$(-3)^2 - 9 = 0$
$0 = 0$, true	$0 = 0$, true

The solutions are 3 and -3. ■

EXAMPLE 2 Solve the equation $x^2 + 5 = 65$.

Solution: Begin by subtracting 5 from both sides of the equation.

$$x^2 = 60$$
$$x = \pm\sqrt{60}$$
$$x = \pm\sqrt{4}\sqrt{15}$$
$$x = \pm 2\sqrt{15}$$

The solutions are $2\sqrt{15}$ and $-2\sqrt{15}$. ■

Not all quadratic equations have real solutions as is illustrated in Example 3.

EXAMPLE 3 Solve the equation $x^2 + 7 = 0$.

Solution:

$$x^2 + 7 = 0$$
$$x^2 = -7$$
$$x = \pm\sqrt{-7}$$
$$x = \pm i\sqrt{7}$$ ■

EXAMPLE 4 Solve the equation $(x - 4)^2 = 32$.

Solution: Begin by taking the square root of both sides of the equation.

$$(x - 4)^2 = 32$$
$$x - 4 = \pm\sqrt{32}$$
$$x = 4 \pm \sqrt{32}$$
$$x = 4 \pm \sqrt{16}\sqrt{2}$$
$$x = 4 \pm 4\sqrt{2}$$

The solutions are $4 + 4\sqrt{2}$ and $4 - 4\sqrt{2}$. ■

2 A perfect square trinomial is a trinomial that can be expressed as the square of a binomial. Some examples follow.

Perfect square trinomials	*Factors*	*Square of a binomial*
$x^2 + 8x + 16$	$= (x + 4)(x + 4)$	$= (x + 4)^2$
$x^2 - 8x + 16$	$= (x - 4)(x - 4)$	$= (x - 4)^2$
$x^2 + 10x + 25$	$= (x + 5)(x + 5)$	$= (x + 5)^2$
$x^2 - 10x + 25$	$= (x - 5)(x - 5)$	$= (x - 5)^2$

Notice each of the squared terms in the trinomials given above have a numerical coefficient of 1. When the coefficient of the squared term is 1 there is an important relationship between the coefficient of the x term and the constant. In every perfect square trinomial the constant term is the square of one-half the coefficient of the x term.

Let us examine some perfect square trinomials whose coefficients of the squared term are 1.

$$x^2 + 8x + 16 = (x + 4)^2$$

$$[\tfrac{1}{2}(8)]^2 = (4)^2$$

$$x^2 - 10x + 25 = (x - 5)^2$$

$$[\tfrac{1}{2}(-10)]^2 = (-5)^2$$

Note that when such a perfect square trinomial is written as the square of a binomial, the constant in the binomial is one-half the value of the coefficient of the x term in the perfect square trinomial. For example,

$$x^2 + 8x + 16 = (x + 4)^2$$

$$\tfrac{1}{2}(8)$$

$$x^2 - 10x + 25 = (x - 5)^2$$

$$\tfrac{1}{2}(-10)$$

3 To solve a quadratic equation by completing the square, we add (or subtract) a constant to (or from) both sides of the equation so that the remaining trinomial is a perfect square trinomial. The procedure is summarized below.

To Solve a Quadratic Equation by Completing the Square

1. Use the multiplication (or division) property if necessary to make the numerical coefficient of the squared term equal to 1.
2. Rewrite the equation with the constant by itself on the right side of the equation.
3. Take one-half the numerical coefficient of the first-degree term, square it, and add this quantity to both sides of the equation.
4. Replace the trinomial with its equivalent squared binomial.
5. Take the square root of both sides of the equation.
6. Solve for the variable.
7. Check your answers in the original equation.

EXAMPLE 5 Solve the equation $x^2 + 6x + 5 = 0$ by completing the square.

Solution: Since the numerical coefficient of the squared term is 1, step 1 is not necessary.

Step 2: Move the constant, 5, to the right side of the equation by subtracting 5 from both sides of the equation.

$$x^2 + 6x + 5 = 0$$
$$x^2 + 6x = -5$$

Step 3: Determine the square of one-half the numerical coefficient of the first-degree term.

$$\frac{1}{2}(6) = 3, \qquad 3^2 = 9$$

Add the number obtained to both sides of the equation.

$$x^2 + 6x + 9 = -5 + 9$$
$$x^2 + 6x + 9 = 4$$

Step 4: By following this procedure we produce a perfect square trinomial on the left side of the equation. The expression $x^2 + 6x + 9$ is a perfect square trinomial that can be expressed $(x + 3)^2$.

$$x^2 + 6x + 9 = 4$$

$\frac{1}{2}$ the numerical coefficient of the first-degree terms is $\frac{1}{2}(6) = +3$

$$(x + 3)^2 = 4$$

Step 5: Take the square root of both sides of the equation.

$$x + 3 = \pm 2$$

Step 6: Finally, solve for x by subtracting 3 from both sides of the equation.

$$x + 3 - 3 = -3 \pm 2$$
$$x = -3 \pm 2$$

$$x = -3 + 2 \quad \text{or} \quad x = -3 - 2$$
$$x = -1 \qquad\qquad x = -5$$

Thus the solutions are -1 and -5.

Step 7: Check both solutions in the original equation.

$$x = -1 \qquad\qquad x = -5$$
$$x^2 + 6x + 5 = 0 \qquad\qquad x^2 + 6x + 5 = 0$$
$$(-1)^2 + 6(-1) + 5 = 0 \qquad\qquad (-5)^2 + 6(-5) + 5 = 0$$
$$1 - 6 + 5 = 0 \qquad\qquad 25 - 30 + 5 = 0$$
$$0 = 0, \quad \text{true} \qquad\qquad 0 = 0, \quad \text{true}$$

■

EXAMPLE 6 Solve the equation $x^2 = 3x + 18$ by completing the square.

Solution: Move all terms except the constant to the left side of the equation.

$$x^2 = 3x + 18$$
$$x^2 - 3x = 18$$

Take half the numerical coefficient of the x term, square it, and add this product to both sides of the equation.

$$\frac{1}{2}(-3) = -\frac{3}{2}, \qquad \left(-\frac{3}{2}\right)^2 = \frac{9}{4}$$
$$x^2 - 3x + \frac{9}{4} = 18 + \frac{9}{4}$$
$$\left(x - \frac{3}{2}\right)^2 = 18 + \frac{9}{4}$$
$$\left(x - \frac{3}{2}\right)^2 = \frac{72}{4} + \frac{9}{4}$$
$$\left(x - \frac{3}{2}\right)^2 = \frac{81}{4}$$
$$x - \frac{3}{2} = \pm\sqrt{\frac{81}{4}}$$
$$x - \frac{3}{2} = \pm\frac{9}{2}$$
$$x = \frac{3}{2} \pm \frac{9}{2}$$
$$x = \frac{3}{2} + \frac{9}{2} \quad \text{or} \quad x = \frac{3}{2} - \frac{9}{2}$$
$$x = \frac{12}{2} = 6 \qquad\qquad x = -\frac{6}{2} = -3$$

In the following examples we will not illustrate some of the intermediate steps.

EXAMPLE 7 Solve the equation $x^2 - 6x + 17 = 0$.

Solution:

$$x^2 - 6x + 17 = 0$$
$$x^2 - 6x = -17$$
$$x^2 - 6x + 9 = -17 + 9$$
$$(x - 3)^2 = -8$$
$$x - 3 = \pm\sqrt{-8}$$
$$x - 3 = \pm 2i\sqrt{2}$$
$$x = 3 \pm 2i\sqrt{2}$$

The solutions are $3 + 2i\sqrt{2}$ and $3 - 2i\sqrt{2}$. Note that the solutions to the equation $x^2 - 6x + 17 = 0$ are not real. The solutions are complex numbers.

EXAMPLE 8 Solve the equation $3m^2 - 6m - 24 = 0$ by completing the square.

Solution: To solve an equation by completing the square, the numerical coefficient of the squared term must be 1. Since the numerical coefficient of the squared term is 3, we multiply both sides of the equation by $\frac{1}{3}$ to make the numerical coefficient of the squared term equal to 1.

$$3m^2 - 6m - 24 = 0$$
$$\frac{1}{3}(3m^2 - 6m - 24) = \frac{1}{3}(0)$$
$$m^2 - 2m - 8 = 0$$

Now proceed as before.

$$m^2 - 2m = 8$$
$$m^2 - 2m + 1 = 8 + 1$$
$$(m - 1)^2 = 9$$
$$m - 1 = \pm 3$$
$$m = 1 \pm 3$$

$$m = 1 + 3 \quad \text{or} \quad m = 1 - 3$$
$$m = 4 \qquad\qquad m = -2$$

Exercise Set 9.1

Solve each equation by completing the square.

1. $x^2 + 2x - 3 = 0$
2. $x^2 - 6x + 8 = 0$
3. $x^2 - 4x - 5 = 0$
4. $x^2 + 8x + 12 = 0$
5. $x^2 + 3x + 2 = 0$
6. $x^2 + 4x - 32 = 0$
7. $x^2 - 8x + 15 = 0$
8. $x^2 - 9x + 14 = 0$
9. $x^2 + 2x + 15 = 0$
10. $x^2 + 5x + 4 = 0$
11. $x^2 = -5x - 6$
12. $x^2 - 2x + 4 = 0$
13. $x^2 + 9x + 18 = 0$
14. $x^2 - 9x + 18 = 0$
15. $x^2 = 15x - 56,$
16. $x^2 = 3x + 28$
17. $-4x = -x^2 + 12$
18. $x^2 + 3x + 6 = 0$
19. $x^2 + 2x - 6 = 0$
20. $x^2 - 4x + 2 = 0$
21. $6x + 6 = -x^2$
22. $x^2 - x - 3 = 0$
23. $-x^2 + 5x = -8$
24. $x^2 + 3x - 6 = 0$
25. $2x^2 + 4x = 0$
26. $2x^2 - 6x = 0$
27. $12x^2 - 4x = 0$
28. $6x^2 = 9x$
29. $2x^2 + 4x - 6 = 0$
30. $2x^2 + 2x - 24 = 0$
31. $2x^2 + 18x + 4 = 0$
32. $2x^2 = 8x + 90$
33. $3x^2 + 33x + 72 = 0$
34. $4x^2 = -28x + 32$

35. $2x^2 + 4x + 3 = 0$

36. $3x^2 - 8x + 4 = 0$

37. $-3x^2 + 6x = 6$

38. $2x^2 - x = -5$

39. $x^2 + \frac{3}{5}x - \frac{1}{2} = 0$

40. $x^2 - \frac{2}{3}x - \frac{1}{5} = 0$

41. $3x^2 + \frac{1}{2}x = -4$

42. $2x^2 - \frac{1}{3}x = -2$

Solve each problem.

43. The product of two consecutive positive odd integers is 63. Find the two odd integers.
44. The larger of two integers is 2 more than twice the smaller. Find the two numbers if their product is 12.
45. The length of a rectangle is 2 feet more than twice its width. Find the dimensions of the rectangle if the area is 60 square feet.
46. Find the length of the side of a square whose diagonal is 10 feet longer than the length of its side.
47. Find the length of the side of a square whose diagonal is 12 feet longer than the length of a side.

JUST FOR FUN

Write each equation in the form $a(x - h)^2 + b(y - k)^2 = c$ by completing the square twice: once for the x values and once for the y values.

1. $x^2 + 4x + y^2 - 6y = 3$
2. $4x^2 + 9y^2 - 48x + 72y + 144 = 0$
3. $x^2 - 4y^2 + 4x - 16y - 28 = 0$

9.2 Solving Quadratic Equations by the Quadratic Formula

1. *Derive the quadratic formula.*
2. *Use the quadratic formula to solve equations.*
3. *Use the discriminant to determine the number of solutions to a quadratic equation.*

1 The quadratic formula can be used to solve any quadratic equation. In fact, it is the most useful and most versatile method of solving quadratic equations.

The standard form of a quadratic equation is $ax^2 + bx + c = 0$, where a is the numerical coefficient of the squared term, b is the numerical coefficient of the first-degree term, and c is the constant.

Quadratic equation in standard form	*Values of a, b, and c*
$x^2 - 3x + 4 = 0$	$a = 1, \quad b = -3, \quad c = 4$
$3x^2 - 4 = 0$	$a = 3, \quad b = 0, \quad c = -4$
$-5x^2 + 3x = 0$	$a = -5, \quad b = 3, \quad c = 0$

We can derive the quadratic formula by starting with a quadratic equation in standard form and completing the square as discussed in the preceding section.

$ax^2 + bx + c = 0$	
$ax^2 + bx = -c$	subtract c from both sides of equation
$x^2 + \frac{b}{a}x = \frac{-c}{a}$	divide both sides of equation by a
$x^2 + \frac{b}{a}x + \frac{b^2}{4a^2} = \frac{-c}{a} + \frac{b^2}{4a^2}$	take 1/2 of b/a to get $b/2a$, square this expression to get $b^2/4a^2$, and add this expression to both sides of the equation
$\left(x + \frac{b}{2a}\right)^2 = \frac{-c}{a} + \frac{b^2}{4a^2}$	express the left side of the equation as the square of a binomial
$\left(x + \frac{b}{2a}\right)^2 = \frac{4a}{4a} \cdot \frac{-c}{a} + \frac{b^2}{4a^2}$	obtain a common denominator so that the fractions can be added
$\left(x + \frac{b}{2a}\right)^2 = \frac{-4ac + b^2}{4a^2}$	
$\left(x + \frac{b}{2a}\right)^2 = \frac{b^2 - 4ac}{4a^2}$	
$x + \frac{b}{2a} = \pm\sqrt{\frac{b^2 - 4ac}{4a^2}}$	take the square root of both sides of the equation
$x + \frac{b}{2a} = \pm\frac{\sqrt{b^2 - 4ac}}{2a}$	
$x = \frac{-b}{2a} \pm \frac{\sqrt{b^2 - 4ac}}{2a}$	subtract $b/2a$ from both sides of the equation
$x = \frac{-b \pm \sqrt{b^2 - 4ac}}{2a}$	quadratic formula

2

To Solve a Quadratic Equation by the Quadratic Formula

1. Write the quadratic equation in standard form, $ax^2 + bx + c = 0$, and determine the numerical values for a, b, and c.
2. Substitute the values for a, b, and c in the quadratic formula and then evaluate the formula to obtain the solution.

The quadratic formula

$$x = \frac{-b \pm \sqrt{b^2 - 4ac}}{2a}$$

EXAMPLE 1 Solve the equation $x^2 + 2x - 8 = 0$ using the quadratic formula.

Solution: $a = 1$, $b = 2$, $c = -8$.

$$x = \frac{-b \pm \sqrt{b^2 - 4ac}}{2a}$$

$$x = \frac{-(2) \pm \sqrt{(2)^2 - 4(1)(-8)}}{2(1)}$$

$$= \frac{-2 \pm \sqrt{4 + 32}}{2}$$

$$= \frac{-2 \pm \sqrt{36}}{2}$$

$$= \frac{-2 \pm 6}{2}$$

$$x = \frac{-2 + 6}{2} \quad \text{or} \quad x = \frac{-2 - 6}{2}$$

$$= \frac{4}{2} = 2 \qquad\qquad = \frac{-8}{2} = -4$$

The solutions are 2 and -4. ■

COMMON STUDENT ERROR

The entire numerator of the quadratic formula must be divided by $2a$.

Correct

$$x = \frac{-b \pm \sqrt{b^2 - 4ac}}{2a}$$

Wrong (crossed out)

$$x = -b \pm \frac{\sqrt{b^2 - 4ac}}{2a}$$

$$= \frac{-b}{2a} \pm \sqrt{b^2 - 4ac}$$

EXAMPLE 2 Solve the equation $2x^2 + 4x - 5 = 0$ using the quadratic formula.

Solution: $a = 2$, $b = 4$, $c = -5$.

$$x = \frac{-b \pm \sqrt{b^2 - 4ac}}{2a}$$

$$x = \frac{-4 \pm \sqrt{(4)^2 - 4(2)(-5)}}{2(2)}$$

$$= \frac{-4 \pm \sqrt{16 + 40}}{4}$$

$$= \frac{-4 \pm \sqrt{56}}{4}$$

$$= \frac{-4 \pm 2\sqrt{14}}{4}$$

Now factor out a 2 from both terms in the numerator; then divide out the common factor.

$$x = \frac{\overset{1}{\cancel{2}}(-2 \pm \sqrt{14})}{\underset{2}{\cancel{4}}}$$

$$= \frac{-2 \pm \sqrt{14}}{2}$$

Thus the answers are $\dfrac{-2 + \sqrt{14}}{2}$ and $\dfrac{-2 - \sqrt{14}}{2}$. ■

COMMON STUDENT ERROR

Many students use the quadratic formula correctly until the last step, where they make an error. Below are illustrated both the correct and incorrect procedures for simplifying an answer.

When *both* terms in the numerator, *and* the denominator, have a common factor, that common factor may be divided out, as illustrated below.

Correct

$$\frac{2 + 4\sqrt{3}}{2} = \frac{\overset{1}{\cancel{2}}(1 + 2\sqrt{3})}{\underset{1}{\cancel{2}}} = 1 + 2\sqrt{3}$$

$$\frac{6 + 3\sqrt{3}}{6} = \frac{\overset{1}{\cancel{3}}(2 + \sqrt{3})}{\underset{2}{\cancel{6}}} = \frac{2 + \sqrt{3}}{2}$$

Below are some common errors. Study them carefully so you will not make them. Can you explain why each of the following procedures is wrong?

Wrong

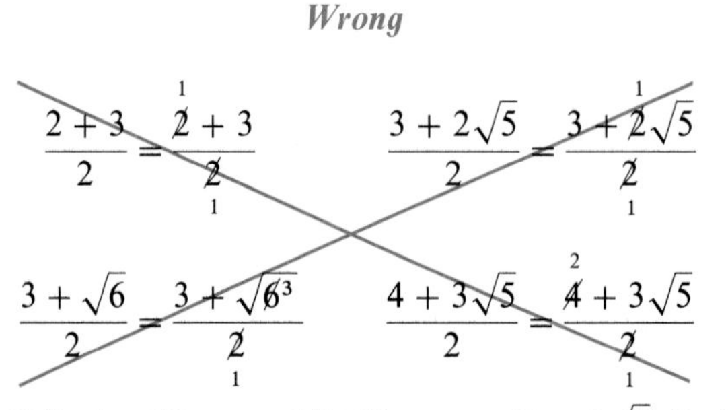

Note that $(2 + 3)/2$ simplifies to $5/2$. However $(3 + 2\sqrt{5})/2$, $(3 + \sqrt{6})/2$ and $(4 + 3\sqrt{5})/2$ cannot be simplified any further.

EXAMPLE 3 Solve the equation $-p^2 - 2p + 8 = 0$ using the quadratic formula.

Solution: Do not let the change in variable worry you. The quadratic formula is used in exactly the same way as before.

$$a = -1, \qquad b = -2, \qquad c = 8$$

$$\begin{aligned} p &= \frac{-b \pm \sqrt{b^2 - 4ac}}{2a} \\ &= \frac{-(-2) \pm \sqrt{(-2)^2 - 4(-1)(8)}}{2(-1)} \\ &= \frac{2 \pm \sqrt{4 + 32}}{-2} \\ &= \frac{2 \pm \sqrt{36}}{-2} \\ &= \frac{2 \pm 6}{-2} \end{aligned}$$

$$p = \frac{2 + 6}{-2} = \frac{8}{-2} = -4 \quad \text{or} \quad p = \frac{2 - 6}{-2} = \frac{-4}{-2} = 2$$

Thus the solutions are -4 and 2. ■

EXAMPLE 4 Solve the quadratic equation $2x^2 + 5x = -6$.

Solution: $2x^2 + 5x + 6 = 0$

$$a = 2, \qquad b = 5, \qquad c = 6$$

$$\begin{aligned} x &= \frac{-b \pm \sqrt{b^2 - 4ac}}{2a} \\ &= \frac{-5 \pm \sqrt{(5)^2 - 4(2)(6)}}{2(2)} \\ &= \frac{-5 \pm \sqrt{25 - 48}}{4} \\ &= \frac{-5 \pm \sqrt{-23}}{4} \end{aligned}$$

The solutions are $\frac{-5 + i\sqrt{23}}{4}$ and $\frac{-5 - i\sqrt{23}}{4}$. Note that neither solution is a real solution. ■

EXAMPLE 5 Solve the equation $x^2 + \frac{2}{5}x - \frac{1}{3} = 0$ using the quadratic formula.

Solution: We could solve this equation using the quadratic formula with $a = 1$, $b = \frac{2}{5}$, and $c = -\frac{1}{3}$.

When a quadratic equation contains fractions, it is generally easier to begin by multiplying both sides of the equation by the least common denominator. In this example the least common denominator is 15.

$$15\left(x^2 + \frac{2}{5}x - \frac{1}{3}\right) = 15(0)$$

$$15x^2 + 6x - 5 = 0$$

Now use the quadratic formula with $a = 15$, $b = 6$, $c = -5$.

$$x = \frac{-b \pm \sqrt{b^2 - 4ac}}{2a}$$

$$= \frac{-6 \pm \sqrt{6^2 - 4(15)(-5)}}{2(15)}$$

$$= \frac{-6 \pm \sqrt{36 + 300}}{30}$$

$$= \frac{-6 \pm \sqrt{336}}{30}$$

$$= \frac{-6 \pm \sqrt{16}\sqrt{21}}{30}$$

$$= \frac{-6 \pm 4\sqrt{21}}{30}$$

$$= \frac{\overset{1}{\cancel{2}}(-3 \pm 2\sqrt{21})}{\underset{15}{\cancel{30}}}$$

$$= \frac{-3 \pm 2\sqrt{21}}{15}$$

The solutions are $\dfrac{-3 + 2\sqrt{21}}{15}$ and $\dfrac{-3 - 2\sqrt{21}}{15}$.

The same solution could be obtained using the fractional values stated earlier. ■

When we are given a quadratic equation where all the numerical coefficients have a common factor, it is often worthwhile to factor the quadratic equation before using the quadratic formula. For example, consider the equation $15x^2 + 30x - 45 = 0$. In this equation $a = 15$, $b = 30$, $c = -45$. If we use the quadratic formula we would eventually obtain $x = 1$ and $x = -3$ as solutions. By factoring the equation before using the formula we get:

$$15x^2 + 30x - 45 = 0$$

$$15(x^2 + 2x - 3) = 0$$

If we consider $x^2 + 2x - 3 = 0$, then $a = 1$, $b = 2$, and $c = -3$. If we use these new values of a, b, and c in the quadratic formula we will obtain the identical solution, namely, $x = 1$ and $x = -3$. However the calculations with these smaller values of a, b, and c are greatly simplified. Solve both equations now using the quadratic formula to convince yourself. Why can this procedure always be used to obtain the correct answer?

3 The expression under the square root sign in the quadratic formula is called the **discriminant.**

$$\underbrace{b^2 - 4ac}_{\text{discriminant}}$$

The discriminant gives the number and type of solutions to a quadratic equation.

For a quadratic equation of the form $ax^2 + bx + c = 0$, $a \neq 0$:

If $b^2 - 4ac > 0$, then the quadratic equation has two distinct real solutions.
If $b^2 - 4ac = 0$, then the quadratic equation has a single real solution (also called a double root).
If $b^2 - 4ac < 0$, then the quadratic equation has no real solution.

EXAMPLE 6 (a) Find the discriminant of the equation $x^2 - 8x + 16 = 0$.
(b) Use the quadratic formula to find the solution.

Solution: (a) $a = 1$, $b = -8$, $c = 16$.

$$\begin{aligned} b^2 - 4ac &= (-8)^2 - 4(1)(16) \\ &= 64 - 64 \\ &= 0 \end{aligned}$$

Since the discriminant equals zero, there is a single real solution.

(b) $$\begin{aligned} x &= \frac{-b \pm \sqrt{b^2 - 4ac}}{2a} \\ &= \frac{-(-8) \pm \sqrt{0}}{2(1)} \\ &= \frac{8 \pm 0}{2} \\ &= \frac{8}{2} = 4 \end{aligned}$$

The only solution is 4. ■

EXAMPLE 7 Without actually finding the solutions, determine if the following equations have two distinct real solutions, a single real solution, or no real solution.

(a) $2x^2 - 4x + 6 = 0$
(b) $x^2 - 5x - 8 = 0$
(c) $4x^2 - 12x = -9$

Solution: We use the discriminant of the quadratic formula to answer these questions.

(a) $b^2 - 4ac = (-4)^2 - 4(2)(6) = 16 - 48 = -32$
Since the discriminant is negative, this equation has no real solution.
(b) $b^2 - 4ac = (-5)^2 - 4(1)(-8) = 25 + 32 = 57$
Since the discriminant is positive, this equation has two distinct real solutions.
(c) First rewrite $4x^2 - 12x = -9$ as $4x^2 - 12x + 9 = 0$.

$$b^2 - 4ac = (-12)^2 - 4(4)(9) = 144 - 144 = 0$$

Since the discriminant is zero, this equation has a single real solution. ■

Exercise Set 9.2

Determine whether each equation has two distinct real solutions, a single real solution, or no real solution.

1. $x^2 + 4x - 3 = 0$
2. $3x^2 + x + 3 = 0$
3. $2x^2 - 4x + 7 = 0$
4. $-2x^2 + x - 8 = 0$
5. $5x^2 + 3x - 7 = 0$
6. $2x^2 = 16x - 32$
7. $4x^2 - 24x = -36$
8. $5x^2 - 4x = 7$
9. $2x^2 - 8x + 5 = 0$
10. $3x^2 - 5x - 9 = 0$
11. $-3x^2 + 5x - 8 = 0$
12. $x^2 + 4x - 8 = 0$
13. $x^2 + 7x - 3 = 0$
14. $2x^2 - 6x + 9 = 0$
15. $4x^2 - 9 = 0$
16. $6x^2 - 5x = 0$

Use the quadratic formula to solve the equation.

17. $x^2 - 3x + 2 = 0$
18. $x^2 + 6x + 8 = 0$
19. $x^2 - 9x + 20 = 0$
20. $x^2 - 3x - 10 = 0$
21. $x^2 + 5x - 24 = 0$
22. $x^2 - 6x = -5$
23. $x^2 = 13x - 36$
24. $x^2 - 36 = 0$
25. $x^2 - 25 = 0$
26. $x^2 - 6x = 0$
27. $x^2 - 3x = 0$
28. $z^2 - 17z + 72 = 0$
29. $4p^2 - 7p + 13 = 0$
30. $2x^2 - 3x + 2 = 0$
31. $2y^2 - 7y + 4 = 0$
32. $2x^2 - 7x = -5$
33. $6x^2 = -x + 1$
34. $4r^2 + r - 3 = 0$
35. $2x^2 - 4x - 1 = 0$
36. $3w^2 - 4w + 5 = 0$
37. $2s^2 - 4s + 3 = 0$
38. $x^2 - 7x + 3 = 0$
39. $4x^2 = x + 5$
40. $x^2 - 2x - 1 = 0$
41. $2x^2 - 7x = 9$
42. $-x^2 + 2x + 15 = 0$
43. $-2x^2 + 11x - 15 = 0$
44. $6x^2 - 5x + 9 = 0$
45. $2x^2 + x + 3 = 0$
46. $9x^2 + 6x + 1 = 0$
47. $2x^2 + 6x = 0$
48. $3x^2 - 5x = 0$

49. $4x^2 - 7 = 0$

50. $-2x^2 = -10$

51. $\frac{x^2}{2} + 2x + \frac{2}{3} = 0$

52. $x^2 - \frac{x}{5} - \frac{1}{3} = 0$

53. $x^2 - \frac{11}{3}x - \frac{10}{3} = 0$

54. $x^2 - \frac{7}{6}x + \frac{2}{3} = 0$

55. Twice the square of a positive number increased by 3 times the number is 14. Find the number.

56. Three times the square of a positive number decreased by twice the number is 21. Find the number.

57. The length of a rectangular garden is 2 feet less than 3 times its width. Find the length and width if the area of the garden is 21 square feet.

58. Lora Moore wishes to form a rectangular region along a river bank by constructing fencing as illustrated in the diagram.

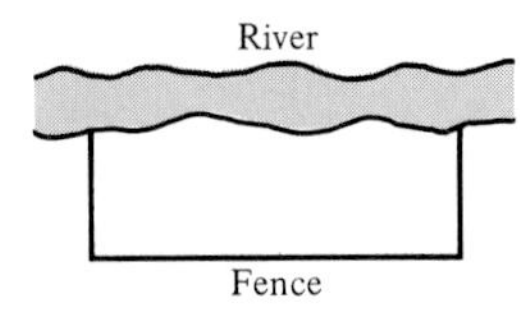

If she has only 400 feet of fencing and wishes to enclose an area of 15,000 square feet, find the dimensions of the rectangular region.

59. John Williams, a professional photographer, has a 6-inch by 8-inch photo. He wishes to reduce the photo by the same amount on each side so that the resulting photo will have half the area of the original photo. By how much will he have to reduce the length of each side?

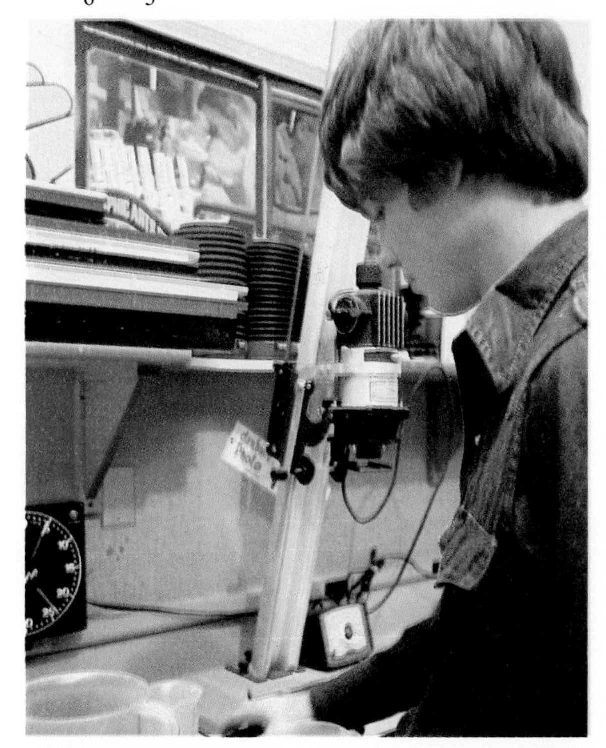

See Exercise 59.

JUST FOR FUN

Solve each of the following using the quadratic formula.

1. $x^2 - \sqrt{5}x - 10 = 0$
2. $x^2 + 5\sqrt{6}x + 36 = 0$
3. $a^4 - 5a^2 + 4 = 0$
4. By increasing the speed of his jet by 200 miles per hour, Jack can decrease the time needed to cover 4000 miles by 1 hour. What is its speed?
5. A metal cube expands when heated. If each edge is increased 0.20 millimeter after being heated and the total volume increases by 6 cubic millimeters, find the original length of a side of the cube.

9.3 Quadratic and Other Inequalities in One Variable

1 *Solve quadratic inequalities using a sign graph.*

2 *Solve other types of inequalities.*

When the equal sign in a quadratic equation of the form $ax^2 + bx + c = 0$ is replaced by an inequality sign, we get a quadratic inequality. Some examples of quadratic inequalities are

$$x^2 + x - 12 \geq 0 \qquad 2x^2 - 9x - 5 < 0$$

The solution to a quadratic inequality is the set of all values that make the inequality a true statement. For example, if we substitute 5 for x in $x^2 + x - 12 > 0$, we obtain

$$x^2 + x - 12 > 0$$
$$5^2 + 5 - 12 > 0$$
$$25 + 5 - 12 > 0$$
$$18 > 0, \qquad \text{true}$$

Since the inequality is true when x is 5, 5 satisfies the inequality. However, the solution is not only 5, for there are other values that satisfy (or are solutions to) the inequality. Does 4 satisfy the inequality? Does 2 satisfy the inequality?

1 A number of methods can be used to find the solution to quadratic inequalities. We will study only the method that uses a **sign graph.** Consider the inequality $x^2 + x - 12 > 0$. The first step in obtaining the solution is to factor the left side of the inequality

$$x^2 + x - 12 > 0$$
$$(x + 4)(x - 3) > 0$$

We must now determine the values of x that make the product of the factors $(x + 4)$ and $(x - 3)$ greater than 0, or positive. The product of two positive numbers is a positive number, and the product of two negative numbers is also a positive number. Therefore we must find the values of x which make both factors positive numbers, or both factors negative numbers. One way to do this is to use number lines. Consider the factor $x - 3$. For values of x less than 3 the value of $x - 3$ will be negative, and for values of x greater than 3 the value of $x - 3$ will be positive. This is illustrated in Fig. 9.1.

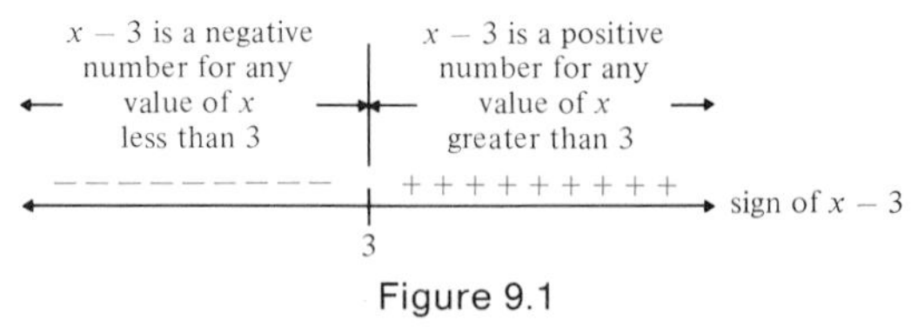

Figure 9.1

Now consider the factor $x + 4$. This factor will be negative when x is less than -4 and positive when x is greater than -4; see Fig. 9.2.

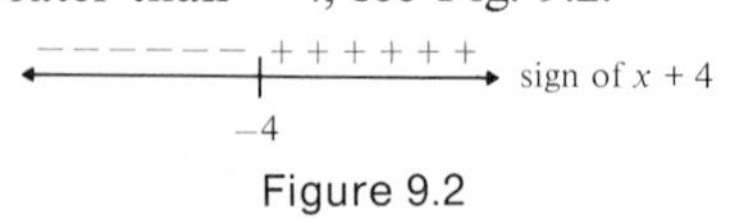

Figure 9.2

If we draw the two number lines together and draw a dashed vertical line through the values -4 and 3 we get the sign graph in Fig. 9.3. Note that when we place the two lines together the -4 is marked to the left of 3 since -4 is less than 3. Also note that the two vertical lines divide the sign graph into three vertical regions labeled a, b, and c.

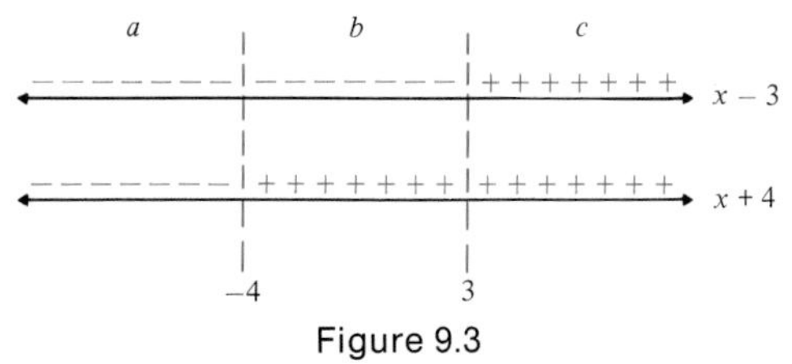

Figure 9.3

Next determine the sign of the product of the numbers in each region. If the two signs in a given vertical region are the same $(-)(-)$ or $(+)(+)$ the product will be positive. If the signs in a given vertical region are different $(+)(-)$ or $(-)(+)$ the product will be negative, see Fig. 9.4.

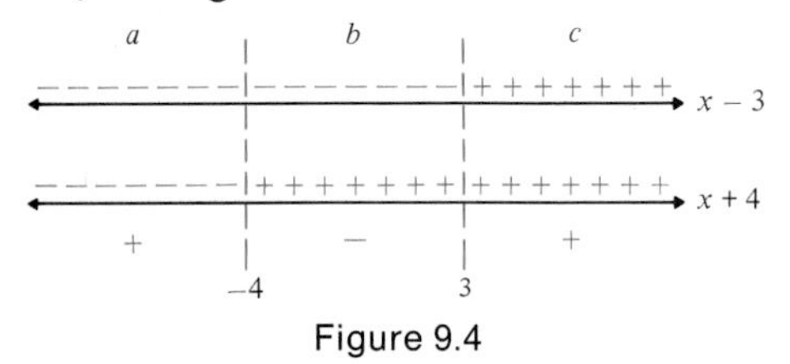

Figure 9.4

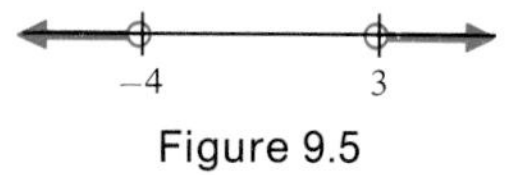

Figure 9.5

Since we are solving the inequality $x^2 + x - 12 > 0$ the solutions are the numbers in regions a and c. Any number less than -4 or greater than 3 is a solution to the inequality. The solution is illustrated on the number line in Fig. 9.5.

The open circles at -4 and 3 indicate those values are not part of the solution.

EXAMPLE 1 Graph the solution to $x^2 + 9x \leq -18$ on the number line.

Solution: First write the quadratic inequality in the form $x^2 + 9x + 18 \leq 0$; then factor.

$$x^2 + 9x + 18 \leq 0$$

$$(x + 6)(x + 3) \leq 0$$

Now draw a sign graph using one number line for each factor, see Fig. 9.6.

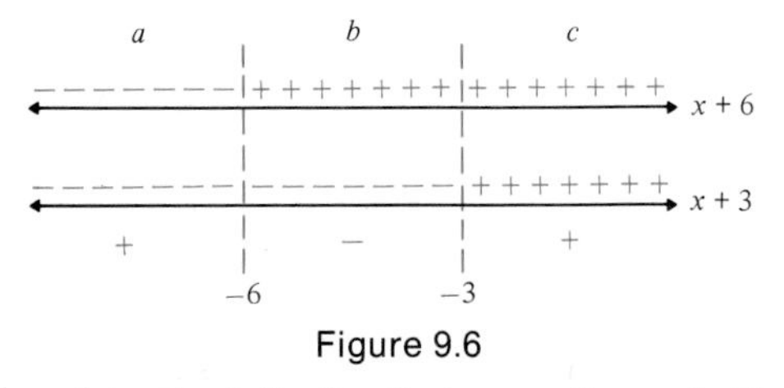

Figure 9.6

Note the -6 was placed to the left of -3 since $-6 < -3$. Since the inequality we are evaluating is $(x + 6)(x + 3) \leq 0$ the solution will be the numbers in region b, where the product of the factors is negative. The solution is illustrated in Fig. 9.7. Note the darkened dots at -6 and -3. They indicate that the -6 and -3 are a part of the solution. ■

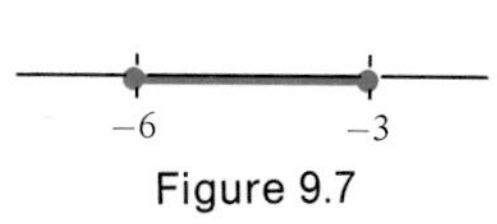

Figure 9.7

EXAMPLE 2 Graph the solution to the inequality $x^2 - 10x + 25 \geq 0$ on the real number line.

Solution:

$$x^2 - 10x + 25 \geq 0$$

$$(x - 5)(x - 5) \geq 0$$

or

$$(x - 5)^2 \geq 0$$

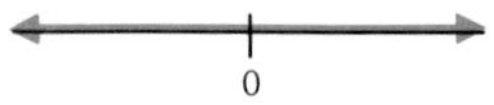

0

Figure 9.8

This is a special case in which both factors are the same. Since any real number squared will be greater than or equal to 0, the solution to this inequality is all real numbers; see Fig. 9.8. ■

EXAMPLE 3 Graph the solution to the inequality $6x^2 + 7x - 3 \geq 0$ on the number line.

Solution:

$$6x^2 + 7x - 3 \geq 0$$

$$(2x + 3)(3x - 1) \geq 0$$

To find out where each factor equals 0 we can set each factor equal to 0 and solve.

$$2x + 3 = 0 \qquad 3x - 1 = 0$$

$$2x = -3 \qquad 3x = 1$$

$$x = -3/2 \qquad x = 1/3$$

Now construct a sign graph, Fig. 9.9.

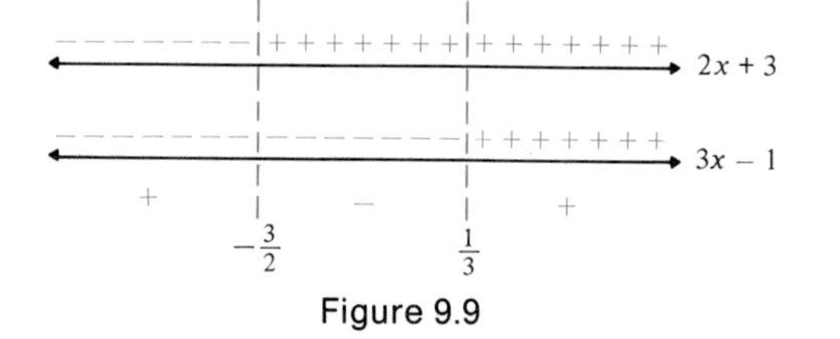

Figure 9.9

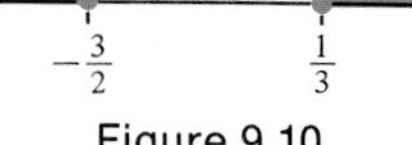
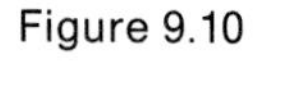

Figure 9.10

The solution is all values less than or equal to $-3/2$ or all values greater than or equal to $1/3$; see Fig. 9.10. ■

2 This procedure used to solve quadratic inequalities can be used to solve other types of inequalities as illustrated in the following examples.

EXAMPLE 4 Solve the inequality $(x - 2)(x + 3)(x + 5) < 0$ and graph the solution on the number line.

Solution: We will again use a sign graph. However since there are three factors we will need three number lines; see Fig. 9.11.

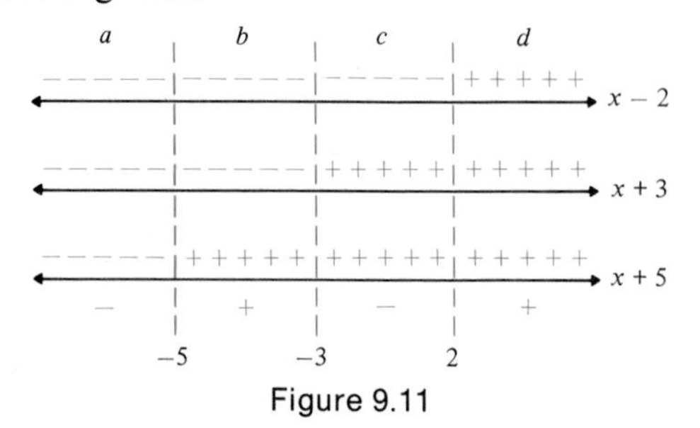

Figure 9.11

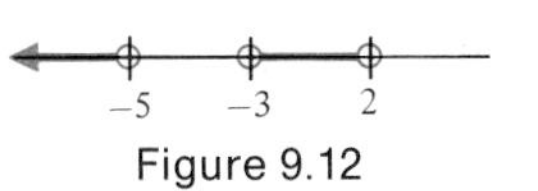

Figure 9.12

The products in regions a and c will be negative for the product of an odd number of negative factors is negative. The products in regions b and d will be positive for the product of an even number of negative factors is positive. The solution is illustrated in Fig. 9.12. ■

EXAMPLE 5 Solve the inequality $\frac{x+3}{x-4} \leq 0$ and graph the solution on the number line.

Solution: We will use basically the same procedure as in previous examples. Recall that a quotient of two numbers with like signs is positive. Construct a sign graph, Fig. 9.13.

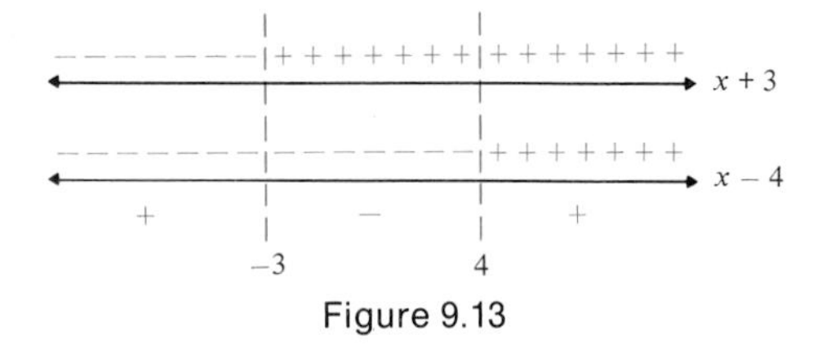

Figure 9.13

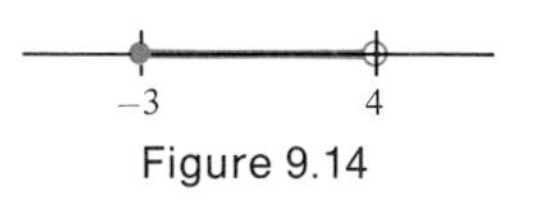

Figure 9.14

The solution is indicated on the number line in Fig. 9.14. Note that the darkened circle at the 3 indicates that the 3 is part of the solution. When $x = 3$ the quotient is equal to 0. However since we cannot divide by 0 the 4 is not part of the solution. Thus, we place an open dot at 4. ■

EXAMPLE 6 Solve the inequality $\frac{6}{x-2} > 4$.

Solution: Your first reaction might be to multiply both sides of the inequality by $x - 2$. This is wrong. Do you know why? Remember that if we multiply both sides of an inequality by a negative number, we must change the sense (or direction) of the equality symbol. Since we do not know the value of x, we do not know whether $x - 2$ represents a positive or negative number. To solve this inequality, subtract 4 from both sides of the inequality. Then rewrite 4 with a denominator of $x - 2$.

$$\frac{6}{x-2} > 4$$

$$\frac{6}{x-2} - 4 > 0$$

$$\frac{6}{x-2} - \frac{4(x-2)}{x-2} > 0$$

$$\frac{6-4(x-2)}{x-2} > 0$$

$$\frac{6-4x+8}{x-2} > 0$$

$$\frac{-4x+14}{x-2} > 0$$

We can now multiply both sides of the inequality by -1 to make the x term in the numerator positive. Remember when we multiply an inequality by a negative number we must change the sense of the inequality.

$$-1\left(\frac{-4x + 14}{x - 2}\right) < -1(0)$$

$$\frac{4x - 14}{x - 2} < 0$$

Now determine the solution using a sign graph, Fig. 9.15.

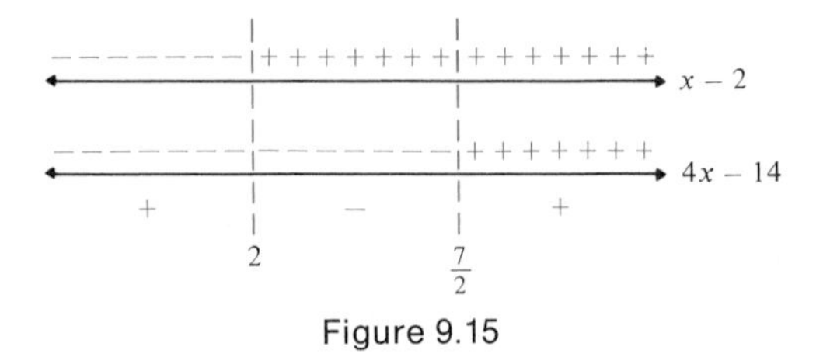

Figure 9.15

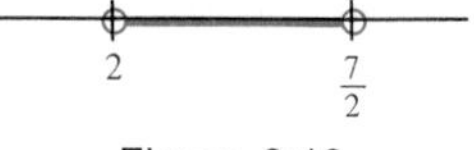

Figure 9.16

The solution is graphed on the number line in Fig. 9.16. ■

EXAMPLE 7 Graph the solution to $\dfrac{(x-3)(x+4)}{x+1} \geq 0.$

Solution: When a problem involves both multiplication and division, any combination of an odd number of negative factors will result in the answer being negative. An even number of negative factors will result in a positive answer. Set up a sign graph with one number line for each factor as illustrated in Fig. 9.17.

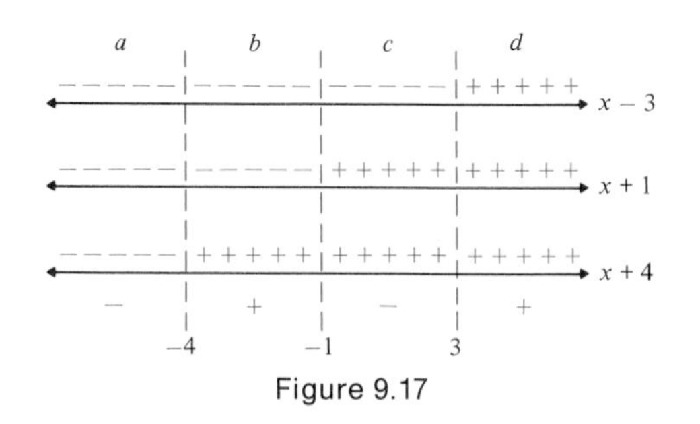

Figure 9.17

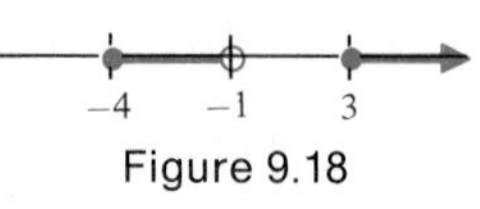

Figure 9.18

The answer is illustrated in Fig. 9.18.
Notice -1 is not part of the solution since we cannot divide by 0. ■

Exercise Set 9.3

Solve each inequality and graph the solution on the number line.

1. $x^2 - 3x - 10 \geq 0$
2. $x^2 + 8x + 7 < 0$
3. $x^2 + 4x > 0$
4. $x^2 - 5x \geq 0$
5. $x^2 - 16 < 0$
6. $x^2 - 25 \leq 0$
7. $x^2 + 9x + 20 \geq 0$
8. $x^2 - 13x + 42 > 0$
9. $x^2 - 6x \leq 27$
10. $x^2 \geq -9x - 14$
11. $x^2 \leq 2x + 35$
12. $x^2 \geq -4x$
13. $x^2 < 36$
14. $x^2 + 10x \leq -21$
15. $2x^2 + 10x + 12 > 0$
16. $3x^2 - 21x - 24 \leq 0$
17. $2x^2 - 7x - 15 \leq 0$
18. $6x^2 > 5x + 6$
19. $4x^2 - 11x \leq 20$
20. $5x^2 \leq -21x - 4$
21. $2y^2 + y < 15$
22. $3a^2 - 4a + 1 > 0$
23. $3x^2 + x - 10 \geq 0$
24. $2y^2 - 5y + 2 \geq 0$

Solve each inequality and graph the solution on the number line.

25. $(x - 1)(x + 1)(x + 4) > 0$
26. $(x - 3)(x + 2)(x + 5) \leq 0$
27. $(x - 4)(x - 1)(x + 3) \leq 0$
28. $(2x - 4)(x + 3)(x + 6) > 0$
29. $x(x - 3)(2x + 6) \geq 0$
30. $(x - 3)(x + 4)(x - 2) \leq 0$
31. $(2x + 6)(3x - 6)(x + 6) > 0$
32. $(2x - 1)(x + 5)(3x + 6) \geq 0$
33. $(x + 2)(x + 2)(x - 3) \geq 0$
34. $(x + 3)^2(x - 5) \leq 0$
35. $(x + 3)(x + 3)(x - 4) < 0$
36. $x(x - 5)(x + 5) > 0$

Solve each inequality and graph the solution on the number line.

37. $\frac{x + 3}{x - 1} > 0$
38. $\frac{x - 5}{x + 2} < 0$
39. $\frac{x - 4}{x - 1} \leq 0$
40. $\frac{x + 6}{x + 2} \geq 0$
41. $\frac{2x - 4}{x - 1} < 0$
42. $\frac{3x + 6}{x + 4} \geq 0$
43. $\frac{3x + 6}{2x - 1} \geq 0$
44. $\frac{3x + 4}{2x - 1} < 0$
45. $\frac{x + 4}{x - 4} \leq 0$
46. $\frac{x + 3}{x} \geq 0$
47. $\frac{4x - 2}{2x - 4} > 0$
48. $\frac{3x + 5}{x - 2} \leq 0$

Solve each inequality and graph the solution on the number line.

49. $\frac{(x + 2)(x - 4)}{x + 6} < 0$
50. $\frac{(x - 3)(x - 6)}{x + 4} \geq 0$
51. $\frac{(x - 6)(x - 1)}{x - 3} \geq 0$
52. $\frac{x + 6}{(x - 2)(x + 4)} > 0$
53. $\frac{x - 6}{(x + 4)(x - 1)} \leq 0$
54. $\frac{x}{(x + 3)(x - 3)} \leq 0$
55. $\frac{(x - 3)(2x + 5)}{(x - 6)} > 0$
56. $\frac{x(x - 3)}{2x + 6} < 0$
57. $\frac{(x + 2)(2x - 3)}{x} \geq 0$
58. $\frac{(x - 4)(3x - 2)}{x + 2} \leq 0$

Solve each inequality. Graph the solution on the number line.

59. $\frac{2}{x - 3} \geq -1$
60. $\frac{3}{x - 1} \leq -1$
61. $\frac{4}{x - 2} \geq 2$
62. $\frac{2}{2a - 1} > 2$

63. $\frac{2x-5}{x-4} \le 1$ **64.** $\frac{2x}{x+1} > 1$ **65.** $\frac{w}{3w-2} > -2$ **66.** $\frac{x-1}{2x+6} \le -3$

67. $\frac{x}{3x-1} < -1$ **68.** $\frac{4x-5}{x+3} < 3$ **69.** $\frac{x+8}{x+2} > 1$ **70.** $\frac{4x+2}{2x-3} \ge 2$

71. What is the solution to the inequality $(x+3)^2(x-1)^2 \ge 0$? Explain your answer.

72. What is the solution to the inequality $x^2(x-3)^2(x+4)^2 \le 0$? Explain your answer.

73. What is the solution to the inequality $x^2(x-3)^2(x+4)^2 < 0$? Explain your answer.

74. What is the solution to the inequality $\frac{x^2}{(x+1)^2} \ge 0$? Explain your answer.

75. What is the solution to the inequality $\frac{x^2}{(x-3)^2} > 0$? Explain your answer.

JUST FOR FUN

Solve the following inequalities. Graph the solution on the number line.

1. $(x+1)(x-3)(x+5)(x+9) \ge 0$ **2.** $\frac{(x-4)(x+2)}{x(x+6)} \ge 0$

9.4 Quadratic Functions

1 *Identify quadratic functions.*

2 *Derive the equation for the axis of symmetry, and find the vertex of a parabola.*

3 *Find the roots of a quadratic equation.*

4 *Sketch graphs of quadratic equations.*

5 *Write a quadratic equation given its roots.*

6 *Apply quadratic equations to some practical situations.*

1 We introduced quadratic functions in Section 5.8. In this section we study quadratic functions in more depth.

Quadratic Function

$$f(x) = ax^2 + bx + c, \qquad a \ne 0$$

Since y may replace $f(x)$, equations of the form $y = ax^2 + bx + c$, $a \ne 0$, are also quadratic functions. Recall from Section 5.8 that every quadratic equation of the form above, when graphed, will be a **parabola.** The graph of $y = ax^2 + bx + c$ will have one of the general forms indicated in Fig. 9.19. Note that both parabolas are functions since they pass the vertical line test.

When a quadratic equation is in the form given above, the sign of the numerical coefficient of the squared term, a, will determine whether the parabola will open upward (Fig. 9.19a) or downward (Fig. 9.19b). When the coefficient of the squared term is positive, the parabola will open upward, and when the coefficient of the squared term is negative, the parabola will open downward. The **vertex** is the lowest point on a parabola that opens upward and the highest point on a parabola that opens downward.

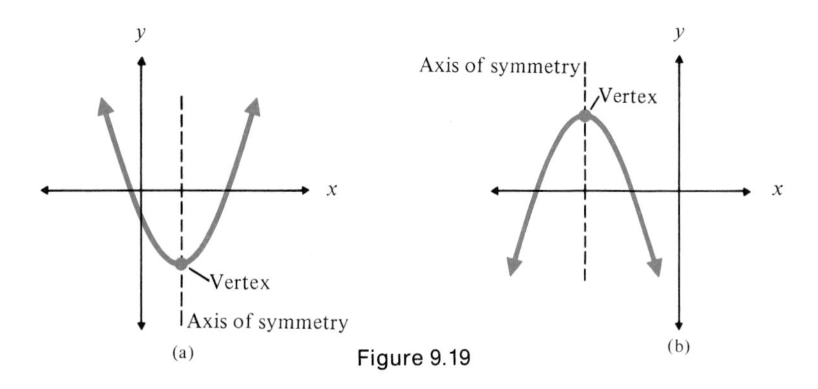

Figure 9.19

2 Graphs of quadratic equations of the form $y = ax^2 + bx + c$ will have **symmetry** about a line through the vertex. This means that if we fold the paper along this imaginary line, called the **axis of symmetry,** the two halves of the graph would coincide.

Axis of Symmetry of Parabola

For an equation of the form $y = ax^2 + bx + c$, $a \neq 0$, the axis of symmetry of the parabola is

$$x = \frac{-b}{2a}$$

At this point we have the knowledge to see where this formula comes from. We will derive the formula $x = \frac{-b}{2a}$, and the coordinates of the vertex of a parabola, by beginning with a quadratic equation in standard form and completing the square on the first two terms.

$$y = ax^2 + bx + c$$

$$y = a\left(x^2 + \frac{b}{a}x\right) + c$$

$$y = a\left[x^2 + \frac{b}{a}x + \left(\frac{b}{2a}\right)^2\right] + c - a\left(\frac{b}{2a}\right)^2$$

$$y = a\left(x + \frac{b}{2a}\right)^2 + c - \cancel{a}\left(\frac{b^2}{4a^2}\right)$$

$$y = a\left(x + \frac{b}{2a}\right)^2 + c\left(\frac{4a}{4a}\right) - \frac{b^2}{4a}$$

$$y = a\left(x + \frac{b}{2a}\right)^2 + \frac{4ac}{4a} - \frac{b^2}{4a}$$

$$y = a\left(x + \frac{b}{2a}\right)^2 + \frac{4ac - b^2}{4a}$$

Since a, b, and c are all constants, the expression $\frac{4ac - b^2}{4a}$ must be a constant. The expression $\left(x + \frac{b}{2a}\right)^2$ will always be a number greater than or equal to 0. The minimum value of $\left(x + \frac{b}{2a}\right)^2$ will be 0 when $x = \frac{-b}{2a}$. If a is positive, then $y = a\left(x + \frac{b}{2a}\right)^2 + \frac{4ac - b^2}{4a}$ will have a minimum of $\frac{4ac - b^2}{4a}$ when $x = \frac{-b}{2a}$. Thus the axis of symmetry, upon which the vertex lies, is $x = -\frac{b}{2a}$.

By observing the equation above we can also see that the coordinates of the vertex of a parabola are as given below.

Coordinates of Vertex of a Parabola

For an equation of the form $y = ax^2 + bx + c$, $a \neq 0$, the coordinates of the vertex of the parabola are

$$\left(\frac{-b}{2a}, \frac{4ac - b^2}{4a}\right)$$

Roots of an Equation

3 The value or values of x where the graph crosses the x axis (the x intercepts) are called the **roots** of the equation. Such points must have a y value of zero (see Fig. 9.20).

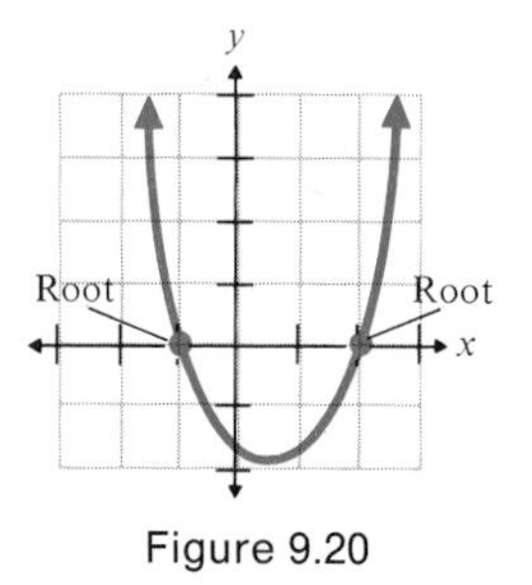

Figure 9.20

A graph of a quadratic equation of the form $y = ax^2 + bx + c$ will have two distinct real roots (Fig. 9.21a), a double root (Fig. 9.21b), or no real (two imaginary) roots (Fig. 9.21c). In Section 9.2 we mentioned that when the discriminant $b^2 - 4ac$ is greater than zero, there are two distinct real solutions; when equal to zero, there is a single real solution (also called a double root); and when less than zero, there is no real solution. This concept is illustrated in Fig. 9.21.

The roots of an equation are the points of intersection of the graph with the x axis. The value of y at the points of intersection must always be 0. Therefore, to find the roots of an equation algebraically, set $y = 0$ and solve the resulting equation for x.

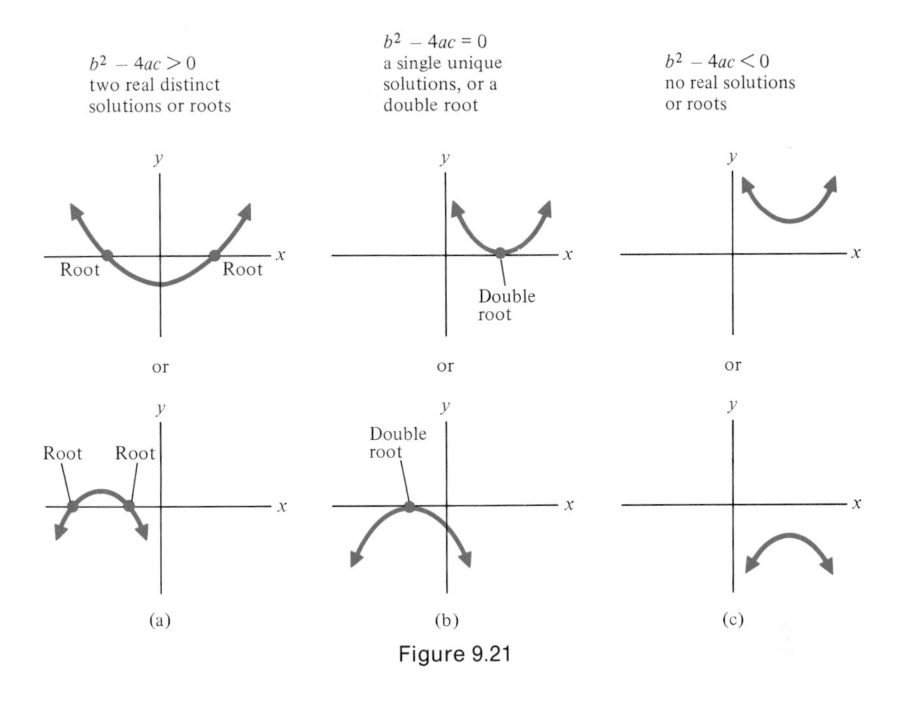

Figure 9.21

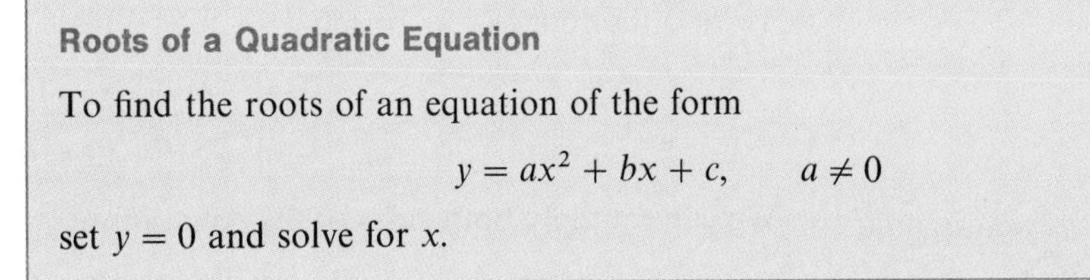

Roots of a Quadratic Equation

To find the roots of an equation of the form

$$y = ax^2 + bx + c, \qquad a \neq 0$$

set $y = 0$ and solve for x.

If we set $y = 0$ when finding the roots, we get an equation of the form

$$ax^2 + bx + c = 0$$

We learned earlier in this chapter that equations of this form may be solved (1) by factoring, if the trinomial is factorable; (2) by completing the square; or (3) by the quadratic formula.

The roots may also be found by graphing the equation and finding the point(s) of intersection of the graph with the x axis. However, this method may result in an inaccurate answer since you may only be able to estimate the points of intersection.

4 We can sketch the graph of a quadratic equation by noticing whether the graph opens upward or downward, finding the vertex, finding the roots, and finding the y intercept. Recall that to find the y intercept, set $x = 0$ and solve for y. Example 1 illustrates how a quadratic function may be sketched.

EXAMPLE 1 Consider the equation $y = -x^2 + 8x - 12$.

(a) Determine whether the parabola opens upward or downward.
(b) Find the y intercept.
(c) Find the vertex.
(d) Find the roots (if they exist).
(e) Sketch the graph.

Solution: (a) Since a is -1, which is less than 0, the parabola opens downward.

(b) To find the y intercept set $x = 0$ and solve for y.

$$y = -(0)^2 + 8(0) - 12 = -12$$

The y intercept is -12.

(c)
$$x = \frac{-b}{2a} = \frac{-8}{2(-1)} = 4$$
$$y = \frac{4ac - b^2}{4a} = \frac{4(-1)(-12) - 8^2}{4(-1)} = \frac{48 - 64}{-4} = 4$$

The vertex is at (4, 4). The y value of the vertex could also be found by substituting 4 for x in the original equation and finding the corresponding value of y, 4.

(d) To find the roots we set $y = 0$

$$0 = -x^2 + 8x - 12$$
$$\text{or} \quad -x^2 + 8x - 12 = 0$$

We can multiply both sides of the equation by -1, then factor.

$$-1(-x^2 + 8x - 12) = -1(0)$$
$$x^2 - 8x + 12 = 0$$
$$(x - 6)(x - 2) = 0$$
$$x - 6 = 0 \quad \text{or} \quad x - 2 = 0$$
$$x = 6 \quad \text{or} \quad x = 2$$

Thus the roots are 2 and 6. The roots could be found by the quadratic formula (or by completing the square). Why not, at this time, find the roots by the quadratic formula?

(e) Now we use all this information to sketch the graph (Figure 9.22). ■

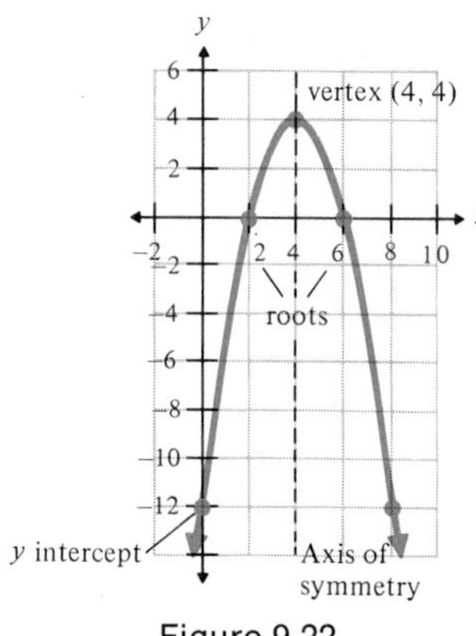

Figure 9.22

HELPFUL HINT

> In Example 1(d) we multiplied both sides of the equation $0 = -x^2 + 8x - 12$ by -1 to make the factoring easier. Since $-1(0)$ equals 0, we can do this. When given an equation like $y = -x^2 + 8x - 12$, you *cannot* just multiply the right side of the equation by -1 to get $y = x^2 - 8x + 12$. Note that $y = -x^2 + 8x - 12$ is a parabola that opens downward while $y = x^2 - 8x + 12$ is a parabola that opens upward. These are two different functions and they will have two different graphs.

EXAMPLE 2 Consider the equation $y = 2x^2 + 3x + 4$.

(a) Determine whether the parabola opens upward or downward.
(b) Find the y intercept.
(c) Find the vertex.
(d) Find the roots if they exist.
(e) Sketch the graph.

Solution: (a) Since a is 2, which is greater than 0, the parabola opens upward.

(b) $y = 2(0)^2 + 3(0) + 4 = 4$

The y intercept is 4.

(c)
$$x = -\frac{b}{2a} = \frac{-3}{2(2)} = \frac{-3}{4} \qquad y = \frac{4ac - b^2}{4a} = \frac{4(2)(4) - 3^2}{4(2)} = \frac{32 - 9}{8} = \frac{23}{8} \text{ or } 2\frac{7}{8}$$

The vertex is $(-\frac{3}{4}, \frac{23}{8})$.

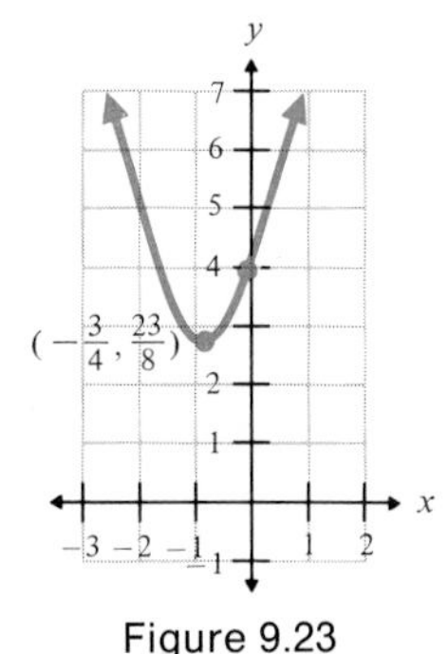

Figure 9.23

(d) $0 = 2x^2 + 3x + 4$

This trinomial cannot be factored. To determine if this equation has any real roots, we will determine the discriminant.

$$b^2 - 4ac = 3^2 - 4(2)(4) = 9 - 32 = -23$$

Since the discriminant is less than 0, this equation has no real roots. We should have expected this answer for the y value of the vertex is a positive number, and therefore above the x axis. Since the parabola opens upward it cannot intersect the x axis.

(e) The graph is sketched in Fig. 9.23. ■

Note that when sketching by this procedure, the exact curve of the graph may be slightly inaccurate, for we are not plotting point by point. However, for our needs a sketch is generally sufficient.

5 If we are given the roots of an equation, we can find the equation. This procedure is illustrated in Example 3.

EXAMPLE 3 Write an equation in two variables that has roots of -3 and 2.

Solution: If the roots are -3 and 2, the factors must be $(x + 3)$ and $(x - 2)$. Therefore, the equation is

$$y = (x + 3)(x - 2)$$

$$\text{or} \quad y = x^2 + x - 6$$ ■

In Example 3 we gave the answer as $y = (x + 3)(x - 2)$. Many other equations have roots -3 and 2, in fact, any equation of the form $y = a(x + 3)(x - 2)$ for any real number a will also have roots of -3 and 2. Can you explain why this is true?

6 Here are some applications of quadratic functions.

EXAMPLE 4 Consider the squares in Fig. 9.24. The number of complete squares below (or above) the diagonal of a square whose length and width are divided into N equal parts can be determined by the formula

$$C = \frac{1}{2}(N^2 - N)$$

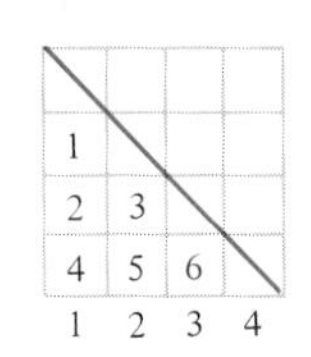

(a)

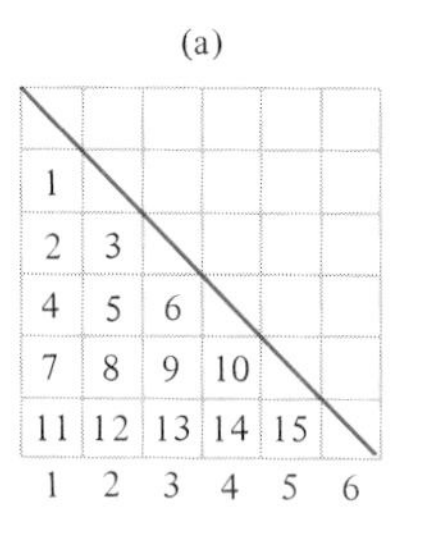

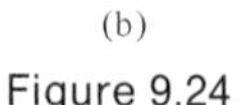

(b)

Figure 9.24

For example, in Fig. 9.24a, $N = 4$ and the number of complete squares below (or above) the diagonal is

$$C = \frac{1}{2}(N^2 - N) = \frac{1}{2}(4^2 - 4) = \frac{1}{2}(12) = 6$$

In Figure 9.24b, $N = 6$ and the number of complete squares below (or above) the diagonal is

$$C = \frac{1}{2}(N^2 - N) = \frac{1}{2}(6^2 - 6) = \frac{1}{2}(30) = 15$$

(a) Use the formula $C = \frac{1}{2}(N^2 - N)$ to determine the number of complete squares below the diagonal if the length and width of a square are divided into 12 equal parts.
(b) If there are 36 complete squares below the diagonal, determine the number of equal units in the length and width of the square.

Solution: (a) $N = 12$.

$$C = \frac{1}{2}(N^2 - N) = \frac{1}{2}(12^2 - 12) = \frac{1}{2}(132) = 66$$

(b) We are told that $C = 36$ and we must find N.

$$C = \frac{1}{2}(N^2 - N)$$

$$36 = \frac{1}{2}(N^2 - N)$$

$$72 = N^2 - N$$

or

$$N^2 - N - 72 = 0$$

$$(N - 9)(N + 8) = 0$$

$$N - 9 = 0 \quad \text{or} \quad N + 8 = 0$$

$$N = 9 \qquad\qquad N = -8$$

Since N must be a positive number, the only possible answer to our question is 9. Thus the length and width of the square are divided into nine equal parts. ■

EXAMPLE 5 When an object is projected in an upward direction from ground level, its distance (or height), d, from the ground at any time, t, is determined by the formula $d = v_0t - \frac{1}{2}gt^2$, where v_0 is the initial velocity with which the object is projected and g is the force of gravity. The gravity of Earth is 32 feet per second squared. If an object is projected upward with an initial velocity of 128 feet per second,

(a) Determine the height at $t = 1$ second.
(b) Sketch a graph of distance versus time.
(c) What is the maximum height the object will reach?
(d) At what time will the object reach its maximum height?
(e) At what time will the object strike the ground?

Solution: (a) The initial velocity, v_0, is 128 and gravity, g, is 32.

$$d = v_0 t - \frac{1}{2} g t^2$$
$$= 128t - \frac{1}{2}(32)t^2$$
$$= 128t - 16t^2$$

When $t = 1$ second, the height is

$$d = 128(1) - 16(1)^2$$
$$= 128 - 16 = 112 \text{ feet}$$

(b) Find the vertex of $d = 128t - 16t^2$ which can be written $d = -16t^2 + 128t$.

$$t = \frac{-b}{2a} = \frac{-128}{-32} = 4$$

When $t = 4$, $d = -16(4)^2 + 128(4) = 256$. Therefore, the vertex is at (4, 256).
To find the roots, set $d = 0$.

$$d = -16t^2 + 128t$$
$$0 = -16t^2 + 128t$$
$$0 = -16t(t - 8)$$
$$-16t = 0 \quad \text{or} \quad t - 8 = 0$$
$$t = 0 \qquad\qquad t = 8$$

Thus the roots are at 0 and 8. The graph is sketched in Fig. 9.25

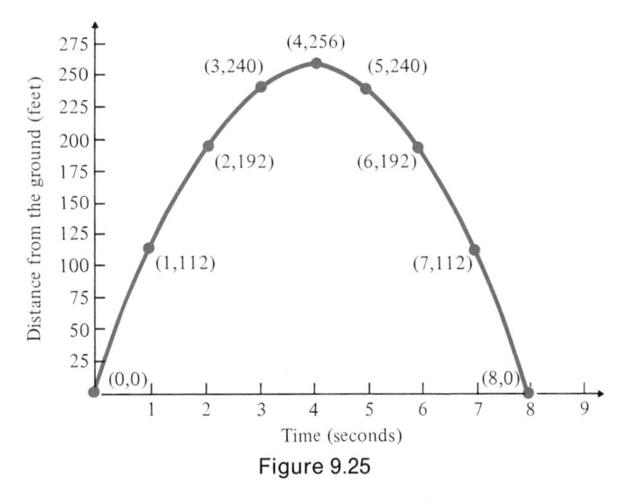

Figure 9.25

(c) The maximum height obtained by the object is 256 feet.

(d) By observation of the graph we see that the object reaches its maximum height of 256 feet at $t = 4$ seconds.

(e) The object strikes the ground at $t = 8$ seconds. Note that it takes 4 seconds for the object to reach its maximum height and 4 seconds to return to the ground. Also note that in part (b) we found the roots to be 0 and 8. Note the object is on the ground at $t = 0$ seconds and $t = 8$ seconds. ■

Maximum/Minimum Problems

In Fig. 9.19 we can see that a parabola that opens upward has a minimum value and a parabola that opens downward has a maximum value. The **maximum or minimum value** of a parabola of the form $y = ax^2 + bx + c$ will be the y coordinate of the vertex. Therefore, the maximum or minimum value of a problem that can be represented as a quadratic equation will be

$$y = \frac{4ac - b^2}{4a}$$

The maximum or minimum value will occur at

$$x = \frac{-b}{2a}$$

EXAMPLE 6 When a cannon is fired at a certain angle, the distance of the shell above the ground, h, in meters, at time, t, in seconds, is given by the formula $h = -4.9t^2 + 24t + 5$. Find the maximum height attained by the shell.

Solution: Since this formula is a quadratic function, its graph will be a parabola. Since $a = -4.9$, the parabola opens downward and the function has a maximum value. We can use $y = (4ac - b^2)/4a$ to find the maximum height.

$$\begin{aligned} y &= \frac{4ac - b^2}{4a} \\ &= \frac{4(-4.9)(5) - (24)^2}{4(-4.9)} \\ &= \frac{-98 - 576}{-19.6} \\ &= \frac{-674}{-19.6} \\ &\approx 34.4 \text{ meters} \end{aligned}$$

Thus the maximum height obtained by the shell is about 34.4 meters. The shell reaches this maximum height at

$$t = \frac{-b}{2a} = \frac{-24}{2(-4.9)} \approx 2.4 \text{ seconds}$$ ■

Exercise Set 9.4

(a) Determine whether the parabola opens upward or downward, (b) find the y intercept, (c) find the vertex, and (d) find the roots (if they exist). You may need to use Appendix C or a calculator to find an approximate value for the roots if they are irrational. (e) Sketch the graph.

1. $y = x^2 + 8x + 15$

2. $y = x^2 + 2x - 3$

3. $f(x) = x^2 - 6x + 4$

4. $y = -x^2 + 4x - 5$

5. $y = x^2 + 6x + 9$

6. $y = 2x^2 + 4x + 2$

7. $f(x) = x^2 - 4x + 4$

8. $y = -2x^2 + 4x - 8$

9. $y = 2x^2 - x - 6$

10. $y = -3x^2 + 6x - 9$

11. $y = 2x^2 - 3x - 9$

12. $f(x) = 3x^2 - 5x - 12$

13. $y = 3x^2 + 4x + 3$

14. $f(x) = -3x^2 - 2x - 6$

15. $f(x) = -2x^2 - 6x + 4$

16. $f(x) = 2x^2 + x - 6$

17. $y = x^2 + 4$

18. $y = -x^2 + 4$

19. $y = -9x^2 + 4$

20. $y = x^2 + 4x$

21. $y = -x^2 + 6x$

22. $f(x) = 3x^2 + 10x$

23. $f(x) = -5x^2 + 5$

24. $f(x) = 2x^2 - 6x + 4$

25. $y = 3x^2 + 4x - 6$

26. $f(x) = -x^2 + 3x - 5$

27. $y = -x^2 + 3x + 6$

28. $f(x) = -2x^2 - 6x + 5$

29. $f(x) = -4x^2 + 6x - 9$

30. $y = -2x^2 + 5x + 4$

Write an equation in two variables that has the roots given.

31. $4, 6$

32. $-2, 5$

33. $3, -4$

34. $0, 4$

35. $2, -3$

36. $-1, -6$

37. $2, 2$

38. $\frac{1}{2}, 3$

39. $-2, \frac{2}{3}$

40. $-\frac{3}{5}, \frac{2}{3}$

41. $-\frac{1}{2}, \frac{2}{3}$

42. $\frac{3}{5}, \frac{1}{4}$

See Exercise 48.

43. Use the formula given in Example 4, $C = \frac{1}{2}(N^2 - N)$, to determine the number of whole squares below the diagonal if the length and width of a square are divided into eight equal parts.

44. The numbers 1 through 144 form a 12 by 12 square array of numbers. The number 1 is in the upper left-hand corner and the number 144 is in the lower right-hand corner. If a diagonal is drawn from number 1 to number 144, how many numbers lie below and above the diagonal?

45. Quentin has planted a 6-foot-tall willow tree. Its height, h, in feet, t years after being planted can be estimated by the formula $h = -0.3t^2 + 6.5t + 6$, $t \leq 10$. Find the height of the tree after 8 years.

46. The number of centimeters that a specific spring will stretch, s, when a mass, m, in kilograms, is attached to it can be found by the formula $s = 3.7m - 0.5m^2$ (for $m \leq 7$ kg). How far will the spring stretch if an object with a mass of 4 kilograms is attached?

47. A person of weight w on the end of a diving board causes it to dip d inches. The relationship between the weight on the board and the dip is $d = 0.00003w^2 + 0.05w$. How much will the board dip if a 200-pound person stands on the tip of the board?

48. The pressure in pounds per square inch within an aerosol spray can diminishes with the amount of time the spray is applied. The pressure, p, inside a spray can after t minutes of application is found by the formula $p = 120 - 4.3t^2$. Find the pressure within the can after 2 minutes.

49. An object is projected upward with an initial velocity of 192 feet per second. Use the formula given in Example 5, $d = v_0t - \frac{1}{2}gt^2$, to:

(a) Find the object's distance from the ground in 3 seconds.
(b) Make a graph of distance versus time.
(c) What is the maximum height the object will reach?
(d) At what time will it reach its maximum height?
(e) At what time will the object strike the ground?

50. An object is projected upward with an initial velocity of 160 feet per second. Answer the questions asked in Exercise 49.

51. The Rochester Philharmonic is trying to determine the price to charge for concert tickets. If the price is too low, they will not make enough money to cover expenses, and if their price is too high, not enough people will wish to pay the price of a ticket. They estimate their total income, I, in hundreds of dollars, per show, can be approximated by the formula $I = -x^2 + 24x - 44$, $0 \leq x \leq 24$, where x is the cost of a ticket.

(a) Draw a graph of income versus the cost of a ticket.
(b) Determine the minimum cost of a ticket for the Philharmonic to break even.
(c) Determine the maximum cost of a ticket that the Philharmonic can charge and break even.
(d) How much should they charge if they are to receive the maximum income?
(e) Find the maximum income.

52. A company earns a weekly profit of P dollars according to the formula $P = -0.4x^2 + 80x - 200$. Find the number of items the company must sell each week to obtain the largest profit. Find the maximum profit.

53. Ramon throws a ball into the air with an initial velocity of 32 feet per second. The height of the ball at any time t is given by the formula $h = 32t - 16t^2$. At what time does the ball reach its maximum height? What is the maximum height?

54. When an object is projected upward with a velocity of 64 feet per second from the top of a 160-foot-tall building, its distance from the ground, h, at any time, t, is $h = -16t^2 + 64t + 160$. Find the maximum height obtained by the object.

JUST FOR FUN

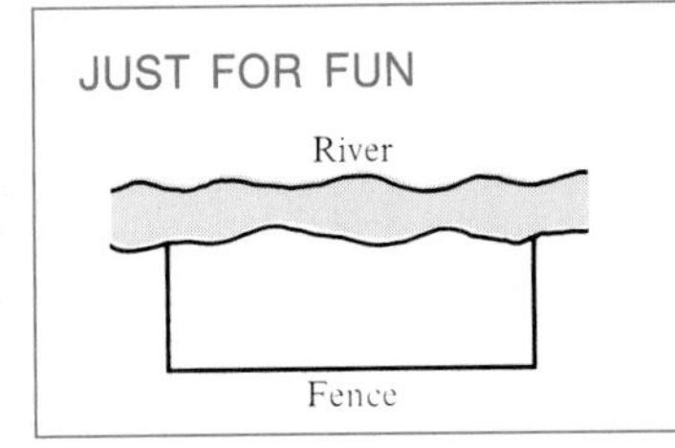

1. What are the dimensions of the rectangle with the largest area that can be enclosed with 64 feet of fence?
2. A landscaper wishes to make a rectangular region along the bank of a river. She need only fence on three sides, as illustrated. What dimensions should she use if she wishes to maximize the rectangle's area and has only 100 feet of fencing?

Summary

Glossary

Axis of symmetry: Imaginary line about which a parabola is symmetric.
Discriminant: For a quadratic equation of the form $ax^2 + bx + c = 0$ the discriminant is $b^2 - 4ac$.
Parabola: The shape of the graph of a quadratic function.
Perfect square trinomial: A trinomial that can be expressed as the square of a binomial.
Quadratic function: Functions of the form $f(x) = ax^2 + bx + c$, $a \neq 0$.
Root of a quadratic equation: The solution to a quadratic equation of the form $ax^2 + bx + c = 0$, $a \neq 0$. When a quadratic function in two variables is graphed, the roots of the equation are the point(s) of intersection of the graph with the x axis.
Solution to a quadratic equation: The value or values of the variable that make the equation a true statement.
Standard form of a quadratic equation: $ax^2 + bx + c = 0$, $a \neq 0$.
Vertex: The lowest point on a parabola that opens upward or the highest point on a parabola that opens downward.

Important Facts

Quadratic Formula $x = \dfrac{-b \pm \sqrt{b^2 - 4ac}}{2a}$

Discriminant $b^2 - 4ac$

If $b^2 - 4ac > 0$, the quadratic equation has two distinct real solutions.
If $b^2 - 4ac = 0$, the quadratic equation has one real solution.
If $b^2 - 4ac < 0$, the quadratic equation has no real solution.

Vertex at $\left(\dfrac{-b}{2a}, \dfrac{4ac - b^2}{4a}\right)$ for an equation of the form $y = ax^2 + bx + c$.

Review Exercises

[9.1] Solve each equation by completing the square.

1. $x^2 - 10x + 16 = 0$
2. $x^2 - 8x + 15 = 0$
3. $x^2 - 14x + 13 = 0$
4. $x^2 + x - 6 = 0$
5. $x^2 - 3x - 54 = 0$
6. $x^2 = -5x + 6$
7. $x^2 + 2x - 5 = 0$
8. $x^2 - 3x + 8 = 0$
9. $2x^2 - 8x = -64$
10. $2x^2 - 4x = 30$
11. $4x^2 - 2x + 12 = 0$
12. $-x^2 - 6x + 10 = 0$

[9.2] Determine whether the equation has two distinct real solutions, a single real solution, or no real solution.

13. $3x^2 - 4x - 20 = 0$
14. $-3x^2 + 4x = 9$
15. $2x^2 + 6x + 7 = 0$
16. $x^2 - x + 8 = 0$
17. $x^2 - 12x = -36$
18. $3x^2 - 4x + 5 = 0$
19. $-3x^2 - 4x + 8 = 0$
20. $x^2 - 9x + 6 = 0$

Solve by the quadratic formula.

21. $x^2 - 9x + 14 = 0$
22. $x^2 + 7x - 30 = 0$
23. $x^2 = 7x - 10$
24. $5x^2 - 7x = 6$
25. $x^2 - 18 = 7x$
26. $x^2 - x + 30 = 0$
27. $6x^2 + x - 15 = 0$
28. $2x^2 + 4x - 3 = 0$
29. $-2x^2 + 3x + 6 = 0$
30. $x^2 - 6x + 7 = 0$
31. $3x^2 - 4x + 6 = 0$
32. $3x^2 - 6x - 8 = 0$
33. $2x^2 + 3x = 0,\quad 0,$
34. $2x^2 - 5x = 0$

[9.1, 9.2] Find the solution to the quadratic equation by the method of your choice.

35. $x^2 - 11x + 24 = 0$
36. $x^2 - 16x + 63 = 0$
37. $x^2 = -3x + 40$
38. $x^2 + 6x = 27$
39. $x^2 - 4x - 60 = 0$
40. $x^2 - x + 42 = 0$
41. $x^2 + 11x + 12 = 0$
42. $x^2 = 25$
43. $x^2 + 6x = 0$
44. $2x^2 + 5x = 3$
45. $3x^2 = 9x - 10$
46. $6x^2 + 5x = 6$
47. $x^2 + 3x - 6 = 0$
48. $3x^2 - 11x + 10 = 0$
49. $-3x^2 - 5x + 8 = 0$
50. $-2x^2 + 6x = -9$
51. $2x^2 - 5x = 0$
52. $3x^2 + 5x = 0$
53. $x^2 + \frac{5x}{4} = \frac{3}{8}$
54. $x^2 = \frac{5}{6}x + \frac{25}{6}$

[9.3] Graph the solution to each inequality on the number line.

55. $x^2 + 6x + 5 \geq 0$
56. $x^2 + 2x - 15 \leq 0$
57. $x^2 \leq 11x - 30$
58. $2x^2 + 6x > 0$
59. $3x^2 + 8x > 16$
60. $4x^2 - 16 \leq 0$
61. $5x^2 - 25 > 0$
62. $9x^2 > 25$

Solve each inequality and graph the solution on the number line.

63. $\frac{x+2}{x-3} > 0$
64. $\frac{x-5}{x+2} \leq 0$
65. $\frac{2x-4}{x+1} \geq 0$
66. $\frac{3x+5}{x-6} < 0$
67. $(x+3)(x+1)(x-2) > 0$
68. $x(x-3)(x-5) \leq 0$
69. $(x+4)(x-1)(x-3) \geq 0$
70. $x(x+2)(x+5) < 0$
71. $\frac{x(x-4)}{x+2} > 0$
72. $\frac{(x-2)(x-5)}{x+3} < 0$
73. $\frac{x-3}{(x+2)(x-5)} \geq 0$
74. $\frac{x(x-5)}{x+3} \leq 0$

Solve each inequality and graph the solution on the number line.

75. $\frac{3}{x+4} \geq -1$
76. $\frac{2x}{x-2} \leq 1$
77. $\frac{2x+3}{x-5} < 2$
78. $\frac{4x}{x+5} \leq 3$

[9.4] (a) Determine whether the parabola opens upward or downward, (b) find the y intercept, (c) find the vertex, (d) find the roots if they exist, and (e) sketch the graph.

79. $y = x^2 + 6x$
80. $y = x^2 + 2x - 8$
81. $y = 2x^2 + 4x - 16$
82. $y = -x^2 - 9$
83. $y = -2x^2 - x + 15$
84. $y = x^2 + 3x + 8$

Write an equation in two variables that has the given roots.

85. $3, -2$

86. $\frac{2}{3}, -3$

87. $-3, -3$

88. 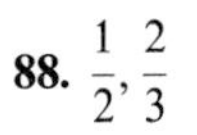 $\frac{1}{2}, \frac{2}{3}$

[9.1–9.2]

89. The product of two consecutive positive integers is 90. Find the integers.

90. The larger of two positive numbers is 4 greater than the smaller. Find the two numbers if their product is 45.

91. The length of a rectangle is 1 inch less than twice its width. Find the sides of the rectangle if its area is 66 square inches.

92. The value, V, of a wheat crop per acre, in dollars, d days after planting is given by the formula $V = 12d - 0.05d^2$, $20 < d < 80$. Find the value of an acre of wheat after it has been planted 50 days.

93. The distance an object is from the ground, in feet, y seconds after being dropped from an airplane is given by the formula $d = -16t^2 + 1800$. Find the distance the object is from the ground 3 seconds after it has been dropped.

See Exercise 93.

94. If an object is dropped from the top of a 100-foot-tall building, its height above the ground, h, at any time, t, can be found by the formula $h = -16t^2 + 100$, $t < 2.25$ seconds. Find the height of the object at 2 seconds.

[9.4]

95. When a cannon is fired at a certain angle the height of its shell above the ground, in feet, at any time, t, can be calculated by the formula $h = -5t^2 + 26t + 8$.

(a) Find the time at which the shell reaches its maximum height.

(b) Find the maximum height.

96. Ruben throws a rock upward from the top of a building 64 feet tall. The rock's distance from the ground, d, in feet, can be found by the formula $d = -16t^2 + 32t + 64$.

(a) Find the time, t, at which the rock attains its maximum height.

(b) Find the maximum height obtained by the rock.

Practice Test

Solve by completing the square.

1. $x^2 = -x + 12$

2. $4x^2 + 8x = -12$

Solve by the quadratic formula.

3. $x^2 - 5x - 6 = 0$

4. $x^2 + 5 = -8x$

Solve by the method of your choice.

5. $3x^2 - 5x = 0$

6. $-2x^2 = 9x - 5$

7. Determine whether the following equation has two distinct real solutions, a single unique solution, or no real solution: $5x^2 = 4x + 2$.

Graph the solution to the inequality on the number line.

8. $x^2 - x \geq 42$

9. $\dfrac{(x + 3)(x - 4)}{x + 1} \geq 0$

10. $\dfrac{x + 3}{x + 2} \leq -1$

In problems 11 and 12, (a) determine whether the parabola opens upward or downward, (b) find the y intercept, (c) find the vertex. (d) find the roots if they exist, and (e) sketch the graph.

11. $y = x^2 - 2x - 8$

12. $y = -2x^2 - 3x + 9$

13. Write an equation in two variables that has roots $-6, \frac{1}{2}$.

14. The length of a rectangle is 4 feet greater than twice its width. Find the length and width of the rectangle if the area of the rectangle is 48 square feet.

15. Kerry throws a ball upward from the top of a building. The distance, d, of the ball from the ground at any time t is $d = -16t^2 + 65t + 80$.

(a) Find the time the object reaches its maximum height.

(b) Find the maximum height.

10 Conic Sections

See Section 10.3.

10.1 The Circle

1. *Identify and describe the conic sections.*
2. *Graph circles with center at the origin.*
3. *Graph circles with center at (h, k).*

Conic Sections

1 In Chapter 9 we discussed parabolas. A parabola is one type of conic section. Parabolas will be discussed further in Section 10.3. Other conic sections are circles, ellipses, and hyperbolas. Each of these shapes is called a conic section because each can be made by slicing a cone and observing the shape of the resulting slice. The methods used to slice the cone to obtain each individual conic section are illustrated in Fig. 10.1.

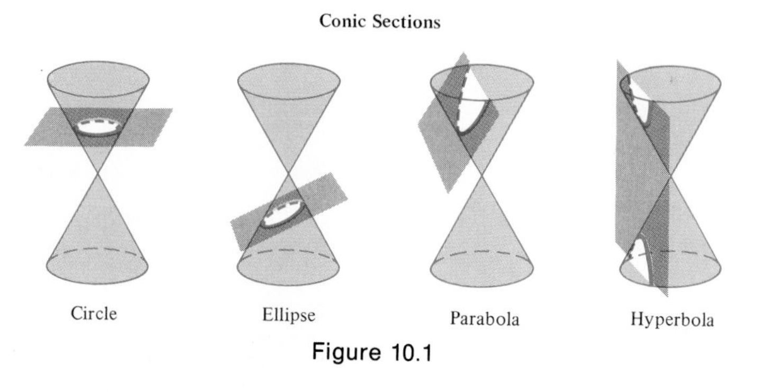

Figure 10.1

The Circle

2 A **circle** may be defined as the set of points in a plane equidistant from a fixed point called its **center** (Fig. 10.2). The **standard form** of the equation of a circle with its center at the origin is $x^2 + y^2 = r^2$, where r is the radius.

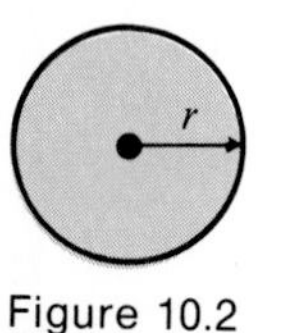

Figure 10.2

Circle with Center at the Origin and Radius r

$$x^2 + y^2 = r^2$$

For example, $x^2 + y^2 = 16$ is a circle with center at the origin and a radius of 4, and $x^2 + y^2 = 7$ is a circle with center at the origin and a radius of $\sqrt{7}$. Note that $4^2 = 16$ and $(\sqrt{7})^2 = 7$.

EXAMPLE 1 Sketch the graph of $x^2 + y^2 = 36$.

Solution: If we rewrite the equation as

$$x^2 + y^2 = 6^2$$

we see that the radius is 6. The graph is illustrated in Fig. 10.3.

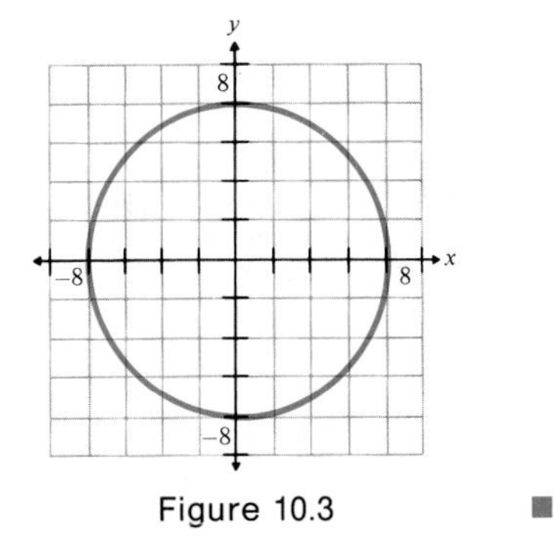

Figure 10.3 ■

3 The standard form of a circle with center at (h, k) and radius r can be derived using the distance formula discussed in Section 8.6. Let (h, k) be the center of the circle and let (x, y) be any point on the circle (see Fig. 10.4).

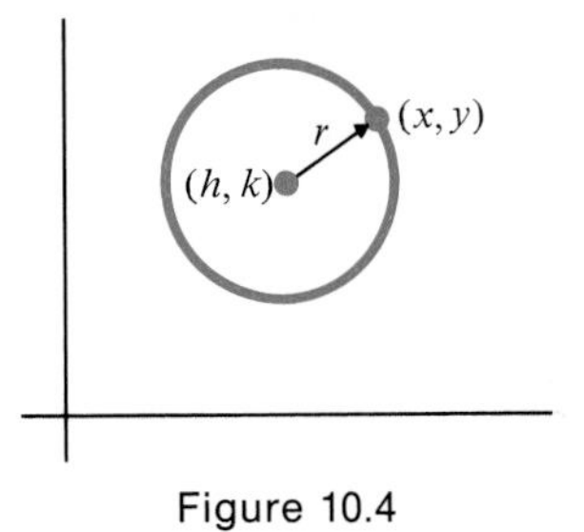

Figure 10.4

If the radius r represents the distance between points (x, y) and (h, k), then by the distance formula

$$r = \sqrt{(x - h)^2 + (y - k)^2}$$

We now square both sides of the equation to obtain the standard form of a circle with center at (h, k) and radius r.

$$r^2 = (x - h)^2 + (y - k)^2$$

Circle with Center at (*h*, *k*) and Radius *r*

$$(x - h)^2 + (y - k)^2 = r^2$$

EXAMPLE 2 (a) Determine the equation of the circle with center at $(3, -1)$ with radius 4.
(b) Sketch the circle.

Solution: (a) The center is $(3, -1)$. Thus h has a value of 3 and k is -1. The radius, r, is 4.

$$(x - h)^2 + (y - k)^2 = r^2$$
$$(x - 3)^2 + [y - (-1)]^2 = 4^2$$
$$(x - 3)^2 + (y + 1)^2 = 16$$

(b)

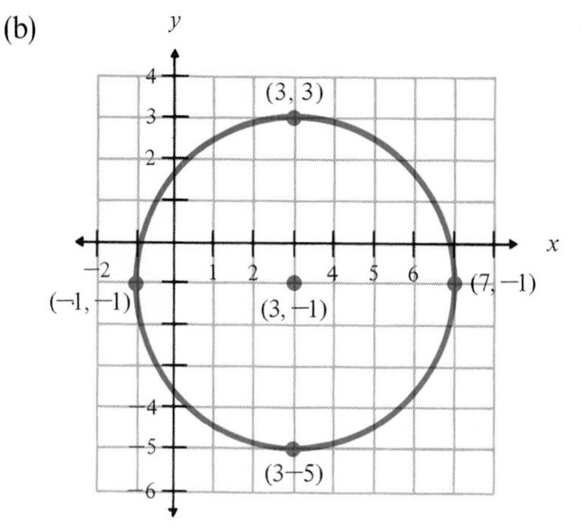

Note that each point on the circle is four units from the center. ■

EXAMPLE 3 Determine the equation of the circle shown in Fig. 10.5.

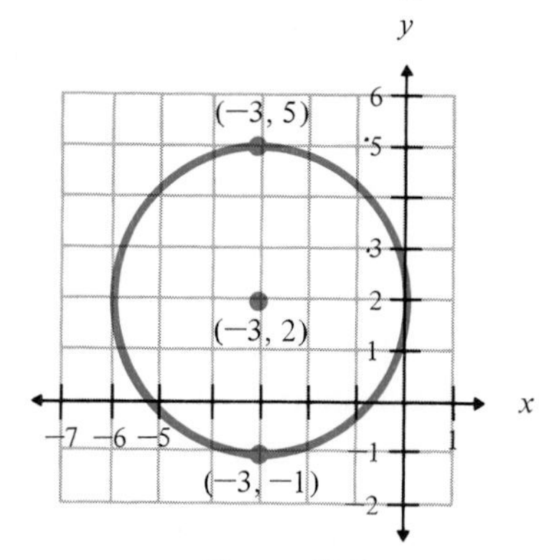

Figure 10.5

Solution: The center is $(-3, 2)$ and the radius is 3. The equation is therefore

$$[x - (-3)]^2 + (y - 2)^2 = 3^2$$
$$(x + 3)^2 + (y - 2)^2 = 9 \quad ■$$

EXAMPLE 4 (a) Show that the equation $x^2 + y^2 + 6x - 2y + 6 = 0$ is an equation of a circle by using the procedure for completing the square discussed in Section 9.1 to rewrite this equation in standard form.
(b) Determine the center and radius of the circle and then sketch the circle.

Solution: (a) First rewrite the equation, placing all the terms containing like variables together.

$$x^2 + 6x + y^2 - 2y + 6 = 0$$

Then move the constant to the right side of the equation.

$$x^2 + 6x + y^2 - 2y = -6$$

Now we complete the square twice, once for each variable. We will first work with the variable x.

$$x^2 + 6x \quad + y^2 - 2y = -6$$
$$x^2 + 6x + 9 + y^2 - 2y = -6 + 9$$

Now work with the variable y.

$$x^2 + 6x + 9 + y^2 - 2y + 1 = -6 + 9 + 1$$

or

$$\underbrace{x^2 + 6x + 9}_{(x+3)^2} + \underbrace{y^2 - 2y + 1}_{(y-1)^2} = 4$$
$$(x + 3)^2 + (y - 1)^2 = 4$$
$$(x + 3)^2 + (y - 1)^2 = 2^2$$

(b) The center of the circle is at $(-3, 1)$ and the radius is 2. The circle is sketched in Fig. 10.6.

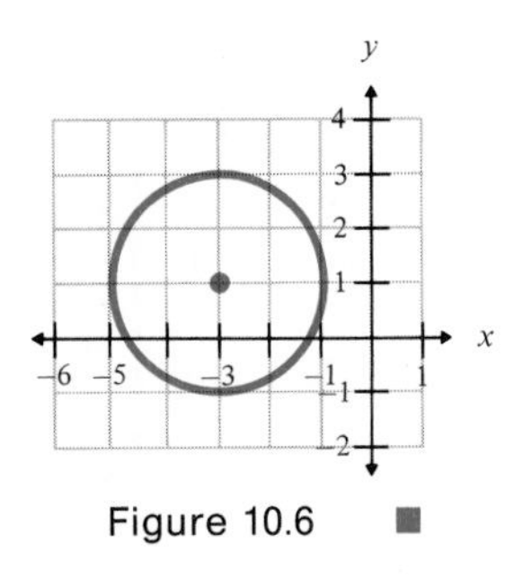

Figure 10.6 ■

Exercise Set 10.1

Write the equation of the circle with the given center and radius; then sketch the graph of the equation.

1. Center (0, 0), radius 3

2. Center (0, 0), radius 5

3. Center (3, 0), radius 1

4. Center $(0, -2)$, radius 7

5. Center $(-6, 5)$, radius 5

6. Center $(-4, -1)$, radius 4

7. Center (4, 7), radius $\sqrt{8}$

8. Center $(0, -2)$, radius $\sqrt{12}$

Sketch the graph of each equation.

9. $x^2 + y^2 = 16$

10. $x^2 + y^2 = 9$

11. $x^2 + y^2 = 3$

12. $x^2 + y^2 = 10$

13. $x^2 + (y - 3)^2 = 4$

14. $(x + 4)^2 + y^2 = 25$

15. $(x - 2)^2 + (y + 3)^2 = 16$

16. $(x + 8)^2 + (y + 2)^2 = 9$

17. $(x + 1)^2 + (y - 4)^2 = 36$

Determine the equation of the circle.

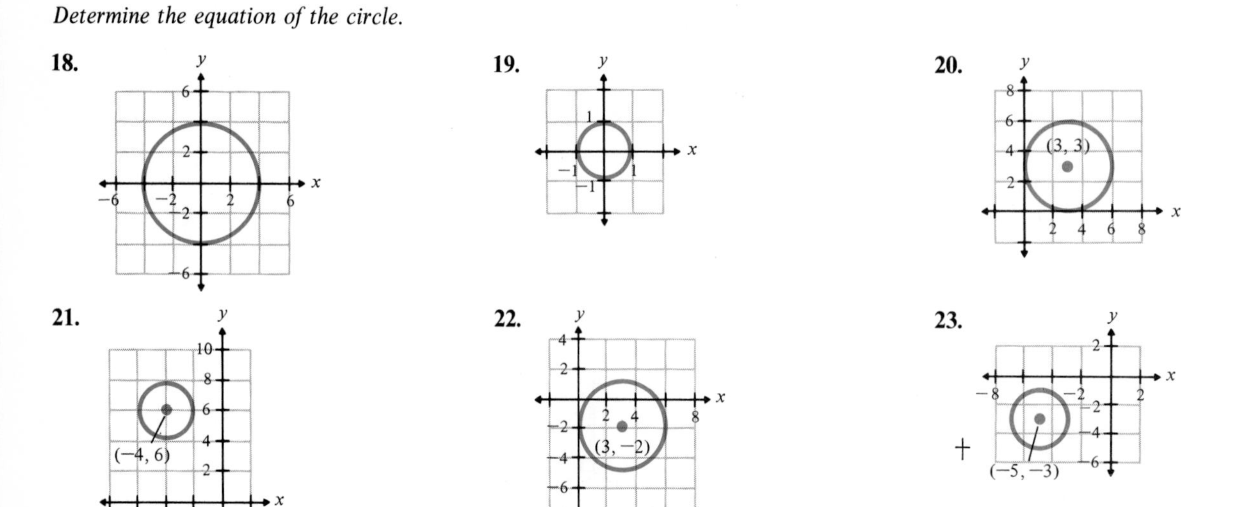

Use the procedure illustrated in Example 4 to write the equation in standard form; then sketch the graph.

24. $x^2 + 4x + y^2 - 12 = 0$

25. $x^2 + y^2 + 10y - 75 = 0$

26. $x^2 + y^2 - 4y = 0$

27. $x^2 + 8x - 9 + y^2 = 0$

28. $x^2 + y^2 + 6x - 4y + 9 = 0$

29. $x^2 + y^2 + 2x - 4y - 4 = 0$

30. $x^2 + y^2 + 4x - 6y - 3 = 0$

31. $x^2 + y^2 + 6x - 2y + 6 = 0$

32. $x^2 + y^2 + 8x - 4y + 4 = 0$

33. $x^2 + y^2 - 8x + 2y + 13 = 0$

10.2 The Ellipse

1 *Graph ellipses.*

1 An **ellipse** may be defined as a set of points in a plane, the sum of whose distances from two fixed points is a constant. The two fixed points are called the **foci** (each is a focus) of the ellipse (see Fig. 10.7).

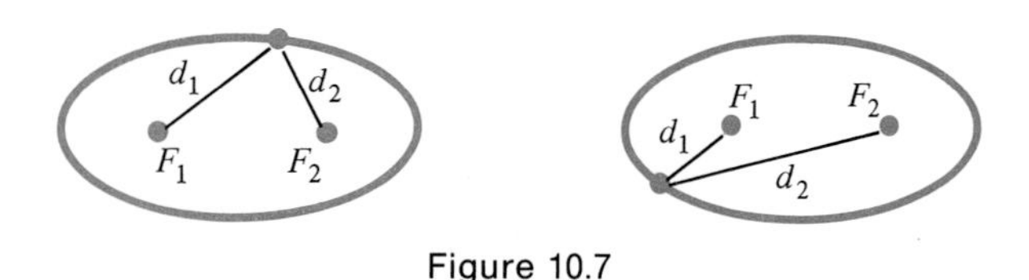

Figure 10.7

We can construct an ellipse using a loop of string and two thumbtacks. Place the two thumbtacks fairly close together (Fig. 10.8). Then place the loop of string around the two thumbtacks. With a pencil or pen pull the string taut, and while keeping the string taut, draw the ellipse by moving the pencil around the thumbtacks.

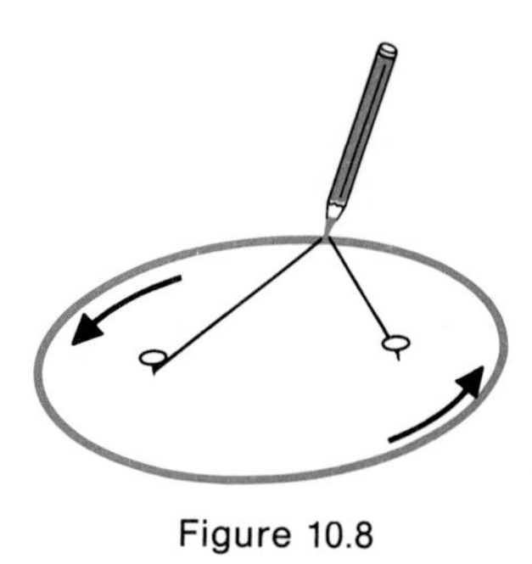

Figure 10.8

The standard form of an ellipse with its center at the origin (Fig. 10.9) is given below.

Ellipse with Center at the Origin

$$\frac{x^2}{a^2} + \frac{y^2}{b^2} = 1$$

where a and $-a$ are the x intercepts and b and $-b$ are the y intercepts.

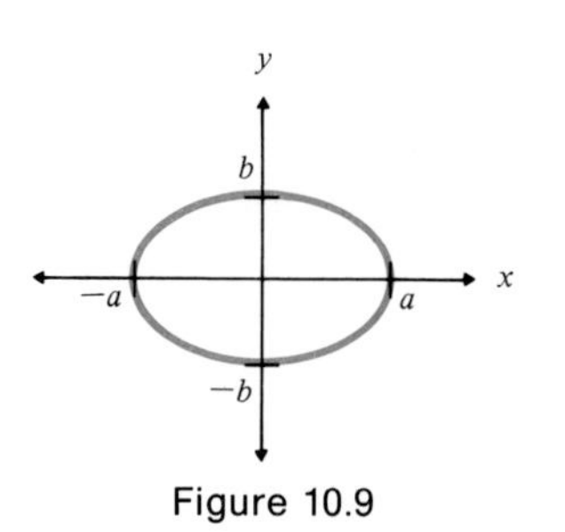

Figure 10.9

In Fig. 10.9 the x axis is the longer or **major axis** and the y axis is the shorter or **minor axis** of the ellipse.

EXAMPLE 1 Sketch the graph of $\frac{x^2}{9} + \frac{y^2}{4} = 1$.

Solution: We can rewrite the equation as

$$\frac{x^2}{3^2} + \frac{y^2}{2^2} = 1$$

Thus $a = 3$ and the x intercepts are ± 3. Since $b = 2$, the y intercepts are ± 2. The ellipse is illustrated in Fig. 10.10.

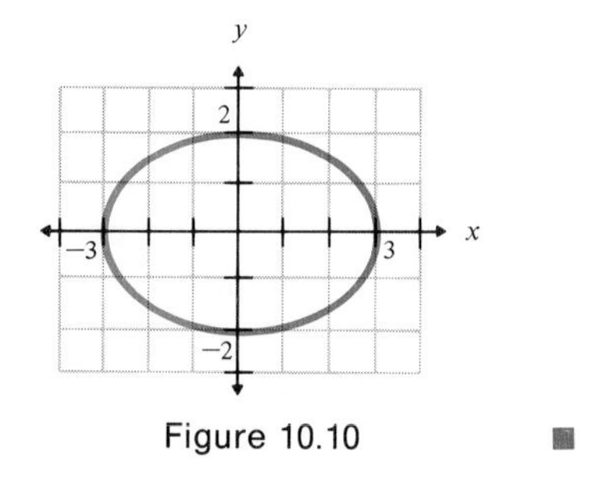

Figure 10.10 ■

An equation may be camouflaged so that it may not be obvious that its graph is an ellipse. This is illustrated in Example 2.

EXAMPLE 2 Sketch the graph of $20x^2 + 9y^2 = 180$.

Solution: If we divide both sides of the equation by 180, we obtain an equation that we can recognize as an ellipse.

$$\frac{20x^2 + 9y^2}{180} = \frac{180}{180}$$

$$\frac{20x^2}{180} + \frac{9y^2}{180} = 1$$

$$\frac{x^2}{9} + \frac{y^2}{20} = 1$$

The equation can now be recognized as an ellipse in standard form.

$$\frac{x^2}{a^2} + \frac{y^2}{b^2} = 1$$

Since $a^2 = 9$, $a = 3$. We know that $b^2 = 20$; thus $b = \sqrt{20}$ (or approximately 4.47). The x intercepts are at ± 3. The y intercepts are at $\pm \sqrt{20}$. The graph is illustrated in Fig. 10.11. Note that the y axis is the major axis of the ellipse.

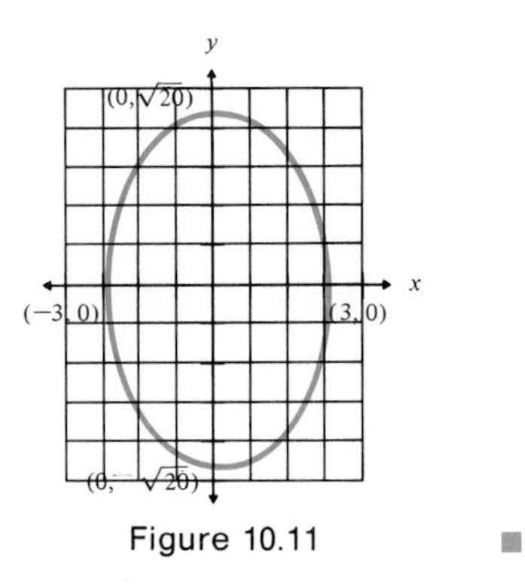

Figure 10.11

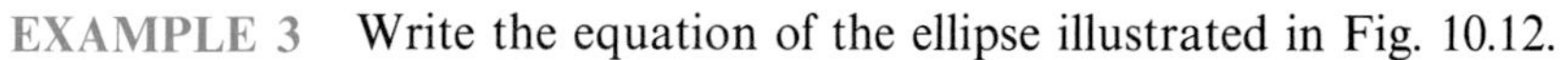

EXAMPLE 3 Write the equation of the ellipse illustrated in Fig. 10.12.

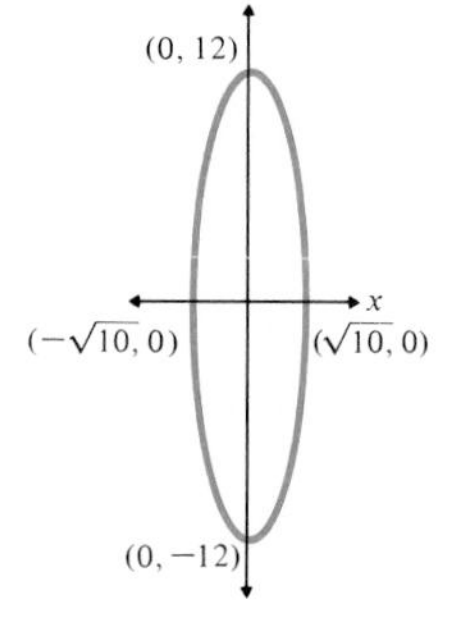

Figure 10.12

Solution: The x intercepts are $\pm\sqrt{10}$; thus $a = \sqrt{10}$ and $a^2 = 10$. The y intercepts are ± 12; thus $b = 12$ and $b^2 = 144$.

$$\frac{x^2}{a^2} + \frac{y^2}{b^2} = 1$$

$$\frac{x^2}{10} + \frac{y^2}{144} = 1$$

Exercise Set 10.2

Sketch the graph of each equation.

1. $\frac{x^2}{4} + \frac{y^2}{1} = 1$

2. $\frac{x^2}{9} + \frac{y^2}{4} = 1$

3. $\frac{x^2}{4} + \frac{y^2}{9} = 1$

4. $\frac{x^2}{25} + \frac{y^2}{9} = 1$

5. $\frac{x^2}{16} + \frac{y^2}{25} = 1$

6. $\frac{x^2}{9} + \frac{y^2}{121} = 1$

7. $9x^2 + 12y^2 = 108$

8. $9x^2 + 4y^2 = 36$

9. $25x^2 + 16y^2 = 400$

10. $x^2 + 36y^2 = 36$

11. $9x^2 + 25y^2 = 225$

12. $x^2 + 2y^2 = 8$

JUST FOR FUN

The standard form of an ellipse with center at (h, k) is

$$\frac{(x-h)^2}{a^2} + \frac{(y-k)^2}{b^2} = 1$$

where a and b are distances from the center to the vertices as shown.

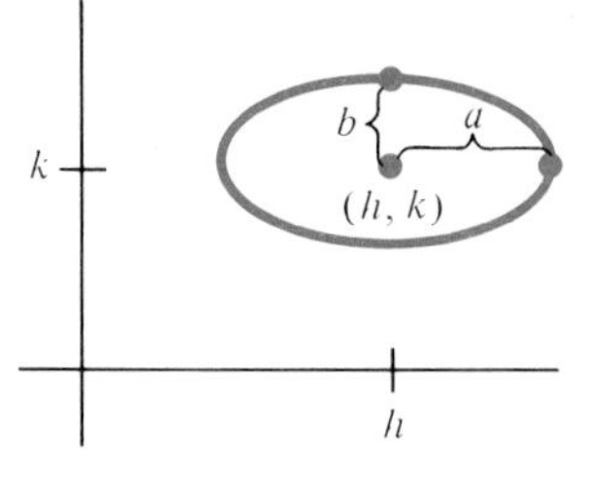

1. Sketch a graph of the equation

$$\frac{x^2}{16} + \frac{(y-2)^2}{9} = 1.$$

2. Sketch the graph of

$$\frac{(x-4)^2}{9} + \frac{(y+3)^2}{25} = 1.$$

3. Write the following equation in standard form. Determine the center of the ellipse and then sketch the ellipse.

$$x^2 + 4y^2 - 4x - 8y - 92 = 0$$

10.3 The Parabola

1 *Derive the equation* $y = a(x-h)^2 + k$.

2 *Graph parabolas of the forms* $y = a(x-h)^2 + k$ *and* $x = a(y-k)^2 + h$.

3 *Convert equations from* $y = ax^2 + bx + c$ *form to* $y = a(x-h)^2 + k$ *form.*

1 We have discussed the parabola in Chapters 5 and 9. In this section we will discuss the parabola further. In Section 9.4 we began with a quadratic equation of the form $y = ax^2 + bx + c$. By completing the square we obtained

$$y = a\left(x + \frac{b}{2a}\right)^2 + \frac{4ac - b^2}{4a}$$

We then showed that the coordinates of the vertex of the parabola are

$$\left(\frac{-b}{2a}, \frac{4ac - b^2}{4a}\right)$$

Since a, b, and c are all constants, the expressions $-b/2a$ and $(4ac - b^2)/4a$, will also be constants. If we let $h = -b/2a$ and $k = (4ac - b^2)/4a$, then by substitution we get

$$y = a\left[x - \left(\frac{-b}{2a}\right)\right]^2 + \frac{4ac - b^2}{4a}$$

$$y = a(x - h)^2 + k$$

We can therefore reason that an equation of the form $y = a(x - h)^2 + k$ will be a parabola with its vertex at the point (h, k). If a in the equation $y = a(x - h)^2 + k$ is a positive number, the parabola will open upward, and if a is a negative number the parabola will open downward.

Parabolas can also open to the right or left. The graph of an equation of the form $x = a(y - k)^2 + h$ will be a parabola whose vertex is at the point (h, k). If a is a positive number the parabola will open to the right, and if a is a negative number the parabola will open to the left. The four different forms of a parabola are shown in Fig. 10.13 below.

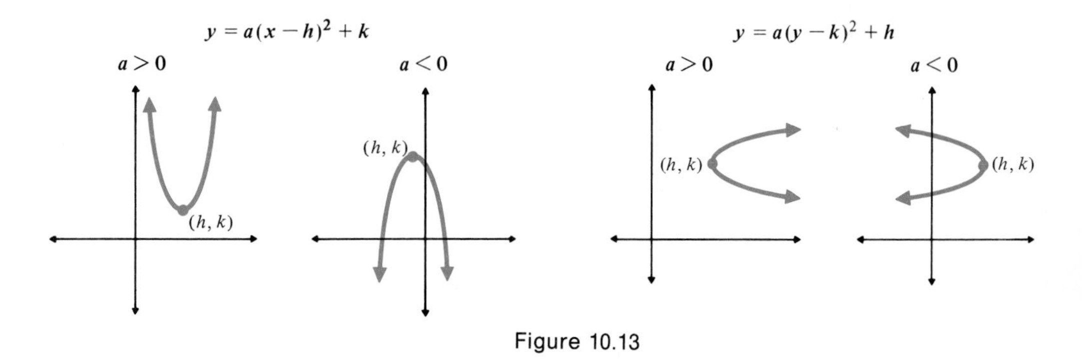

Figure 10.13

Parabola with Vertex at (*h*, *k*)

(a) $y = a(x - h)^2 + k,\ a > 0$ (opens upward)
(b) $y = a(x - h)^2 + k,\ a < 0$ (opens downward)
(c) $x = a(y - k)^2 + h,\ a > 0$ (opens to the right)
(d) $x = a(y - k)^2 + h,\ a < 0$ (opens to the left)

Note that equations of the form $y = a(x - h)^2 + k$ are functions since their graphs pass the vertical line test. However equations of the form $x = a(y - k)^2 + h$ are not functions since their graphs do not pass the vertical line test.

2 Now we will sketch some parabolas.

EXAMPLE 1 Sketch the graph of $y = (x - 2)^2 + 3$.

Solution: The graph opens upward since the equation is of the form $\mathbf{y = a(x - h)^2 + k}$ and $a = 1$, which is greater than 0. The vertex is at (2, 3), see Fig. 10.14. Note that when $x = 0$, the y intercept is $(0 - 2)^2 + 3 = 4 + 3$ or 7.

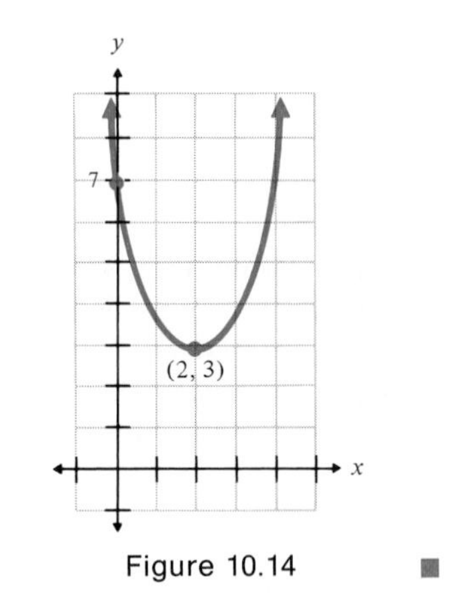

Figure 10.14 ■

EXAMPLE 2 Sketch the graph of $x = -2(y + 4)^2 - 1$.

Solution: The graph opens to the left since the equation is of the form $\mathbf{x = a(y - k)^2 + h}$ and $a = -2$, which is less than 0 (Fig. 10.15). The equation can be expressed as $x = -2[y - (-4)]^2 - 1$. Thus $h = -1$ and $k = -4$. The vertex of the graph is $(-1, -4)$. When $y = 0$ we see that the x intercept is at $-2(0 + 4)^2 - 1 = -2(16) - 1$ or -33.

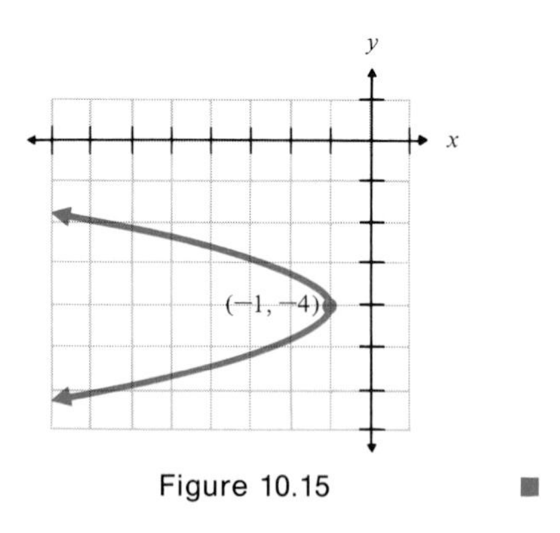

Figure 10.15 ■

3 In Examples 3 and 4 we will convert an equation from $y = ax^2 + bx + c$ form to $y = a(x - h)^2 + k$ form before graphing.

EXAMPLE 3 (a) Write the equation $y = x^2 - 6x + 8$ in $y = a(x - h)^2 + k$ form.
(b) Sketch the graph of $y = x^2 - 6x + 8$.

Solution: (a) We convert $y = x^2 - 6x + 8$ to $y = a(x - h)^2 + k$ form by completing the square.

$$y = x^2 - 6x + 8$$

Take one-half the coefficient of the x term; then square it.

$$\frac{-6}{2} = -3, \qquad (-3)^2 = 9$$

Now add $+9$ and -9 to the equation to obtain

$$y = x^2 - 6x + 9 - 9 + 8$$

By doing this we have created a perfect square trinomial plus some constant.

$$y = \underbrace{x^2 - 6x + 9}_{} \underbrace{- 9 + 8}_{}$$
$$y = (x - 3)^2 \quad -1$$

(b) The vertex of the parabola is at $(3, -1)$ and the parabola opens upward since $a = 1$. The y intercept is at 8. The graph is sketched in Fig. 10.16.

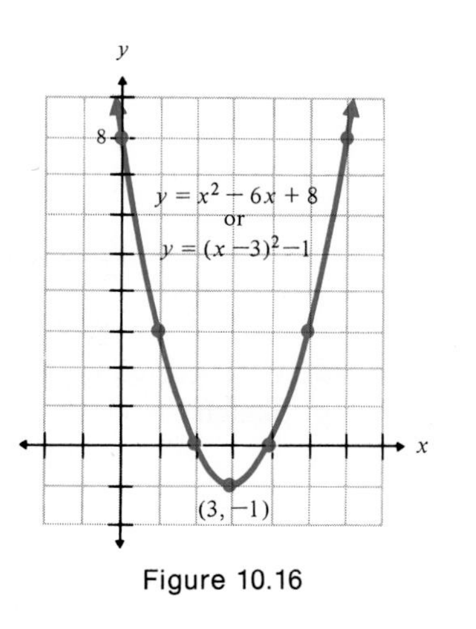

Figure 10.16 ■

EXAMPLE 4 (a) Write the equation $y = 2x^2 + 4x - 6$ in $y = a(x - h)^2 + k$ form.
(b) Sketch the graph of $y = 2x^2 + 4x - 6$.

Solution: (a) First factor 2 from the two terms containing the variable to make the coefficient of the squared term equal to 1. Do not factor 2 from the constant, -6.

$$y = 2x^2 + 4x - 6$$
$$y = 2(x^2 + 2x) - 6$$

Now complete the square by taking one-half the coefficient of the x term and squaring it.

$$\frac{2}{2} = 1, \qquad 1^2 = 1$$

Now add $+1$ inside the parentheses. Since the terms inside the parentheses are multiplied by 2, we are really adding 2(1) or 2. Therefore, we must also add a -2 to the right of the parentheses. In doing this we are not changing the equation since we are adding 2 and -2, which sums to zero.

$$y = 2(x^2 + 2x + 1) - 2 - 6$$
$$y = 2(x + 1)^2 - 8$$

(b) This parabola opens upward since $a = 2$, which is greater than 0. Its vertex is at $(-1, -8)$. The y intercept is at $2(0 + 1)^2 - 8 = 2(1) - 8$ or -6 (see Fig. 10.17).

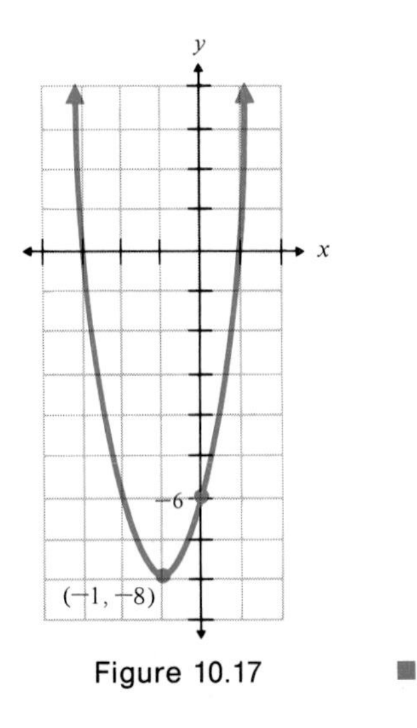

Figure 10.17 ■

EXAMPLE 5 (a) Write the equation $x = -2y^2 + 4y + 5$ in $x = a(y - k)^2 + h$ form.
(b) Sketch the graph of $x = -2y^2 + 4y + 5$.

Solution: (a)

$$x = -2y^2 + 4y + 5$$
$$x = -2(y^2 - 2y) + 5$$
$$x = -2(y^2 - 2y + 1) + 2 + 5$$
$$x = -2(y - 1)^2 + 7$$

(b) Since $a < 0$, this parabola opens to the left. Note that when $y = 0$, $x = 5$. Therefore the x intercept is 5. The vertex of the parabola is (7, 1). The graph is shown in Fig. 10.18.

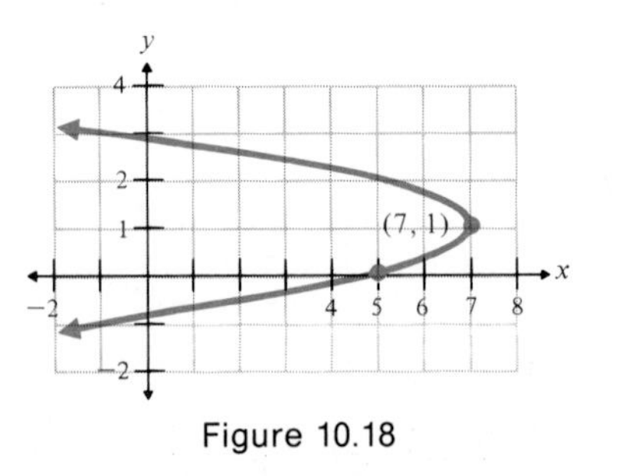

Figure 10.18 ■

Exercise Set 10.3

In Exercises 1 through 12, sketch the graph of each equation.

1. $y = (x - 2)^2 + 3$

2. $y = (x + 1)^2 - 2$

3. $y = -(x - 3)^2 - 6$

4. $x = (y + 6)^2 + 1$

5. $x = (y - 4)^2 - 3$

6. $x = -(y - 5)^2 + 4$

7. $y = 2(x + 6)^2 - 4$

8. $y = -3(x - 5)^2 + 3$

9. $x = -5(y + 3)^2 - 6$

10. $x = -(y - 7)^2 + 8$

11. $y = -2\left(x + \frac{1}{2}\right)^2 + 6$

12. $y = -\left(x - \frac{5}{2}\right)^2 + \frac{1}{2}$

Write each equation in the form $y = a(x - h)^2 + k$ or $x = a(y - k)^2 + h$, and then sketch the graph of the equation.

13. $y = x^2 + 2x$

14. $y = x^2 - 4x$

15. $x = y^2 + 6y$

16. $y = x^2 - 6x + 8$

17. $y = x^2 + 2x - 15$

18. $x = y^2 + 8y + 7$

19. $x = -y^2 + 6y - 9$

20. $y = -x^2 + 4x - 4$

21. $y = x^2 + 7x + 10$

22. $x = -y^2 + 3y - 4$

23. $x = -y^2 - 4y$

24. $y = 2x^2 - 4x - 4$

25. $x = 3y^2 - 12y - 36$

10.4 The Hyperbola

1. *Identify hyperbolas.*
2. *Graph hyperbolas in standard form using asymptotes.*
3. *Graph hyperbolas in nonstandard form.*
4. *Review conic sections.*
5. *Identify nonstandard forms of conic sections.*

1 A **hyperbola** is the set of all points in a plane the difference of whose distances from two fixed points (called foci) is a constant. A hyperbola looks like a pair of parabolas (Fig. 10.19). However, the shapes are actually quite different. A hyperbola has two **vertices.** The point halfway between the two vertices is the **center** of the hyperbola.

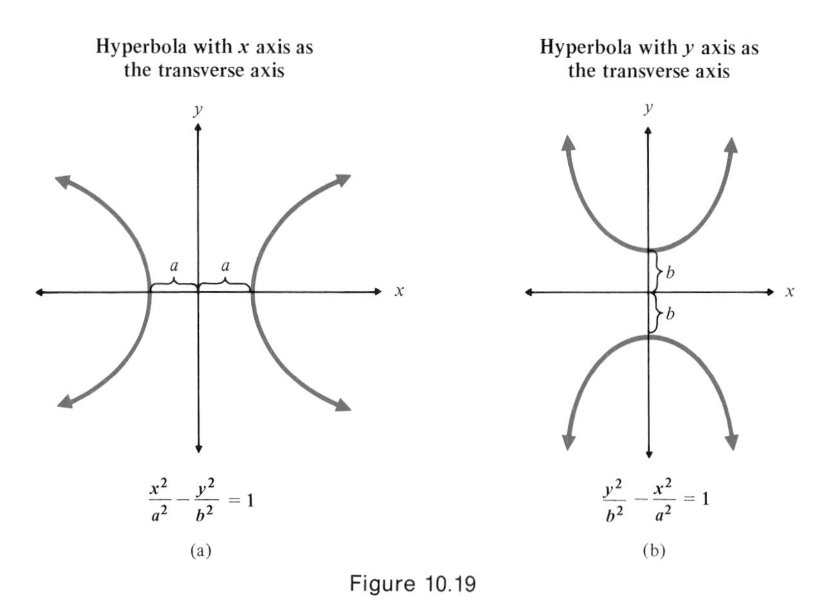

Figure 10.19

Also given in Fig. 10.19 is the standard form of the equation for each hyperbola. The axis through the center that intersects the vertices is called the **transverse axis.** The axis perpendicular to the transverse axis is the **conjugate axis.** In Fig. 10.19a the x axis is the transverse axis and the y axis is the conjugate axis. Note both vertices are a units from the origin. In Fig. 10.19b the y axis is the transverse axis and the x axis is the conjugate axis. Note both vertices are b units from the origin. Note that in the standard form of the equation the denominator of the x^2 term is always a^2 and the denominator of the y^2 term is always b^2.

Note that a hyperbola whose axes are parallel to the coordinate axes has either x intercepts (Fig. 10.19a) or y intercepts (Fig. 10.19b) but not both. When written in standard form, the intercepts will be on the axis indicated by the variable with the positive coefficient. For example, $\frac{x^2}{25} - \frac{y^2}{9} = 1$ will intersect the x axis, and $\frac{y^2}{9} - \frac{x^2}{25} = 1$ will intersect the y axis. In either case the intercepts will be the positive and the negative square root of the denominator of the positive term (which will be the first term when written in standard form). Thus $\frac{x^2}{25} - \frac{y^2}{9} = 1$ has x intercepts of ± 5 and $\frac{y^2}{9} - \frac{x^2}{25} = 1$ has y intercepts of ± 3. Note that the intercepts are the vertices.

2 **Asymptotes** can be used as an aid in graphing hyperbolas. The asymptotes are two straight lines that go through the center of the hyperbola. As the values of x and y get larger, the graph of the hyperbola will approach the asymptotes. The asymptotes are not a part of the hyperbola. They are only an aid in graphing the hyperbola. The equations of the asymptotes are determined using a and b. The equations of the asymptotes of a hyperbola whose center is the origin are:

$$y = \frac{b}{a}x \quad \text{and} \quad y = -\frac{b}{a}x$$

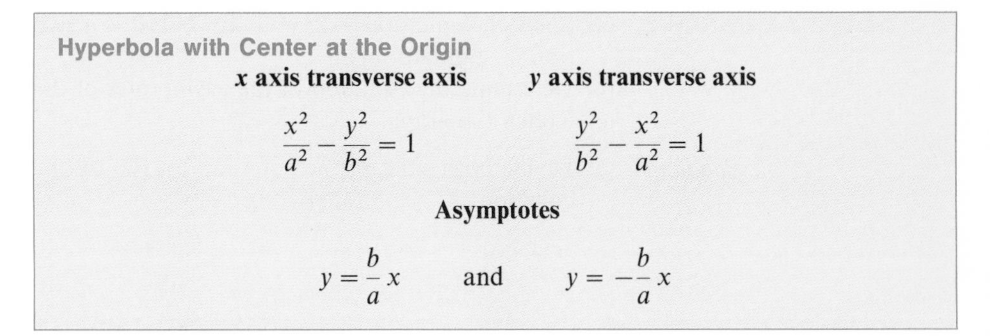

Hyperbola with Center at the Origin

x axis transverse axis	**y axis transverse axis**
$\dfrac{x^2}{a^2} - \dfrac{y^2}{b^2} = 1$	$\dfrac{y^2}{b^2} - \dfrac{x^2}{a^2} = 1$

Asymptotes

$$y = \frac{b}{a}x \qquad \text{and} \qquad y = -\frac{b}{a}x$$

EXAMPLE 1 (a) Determine the equations of the asymptotes of the hyperbola with equation

$$\frac{x^2}{9} - \frac{y^2}{16} = 1$$

(b) Sketch the hyperbola using the asymptotes as an aid.

Solution: (a) The value of a^2 is 9; the positive square root of 9 is 3. The value of b^2 is 16; the positive square root of 16 is 4. The asymptotes are:

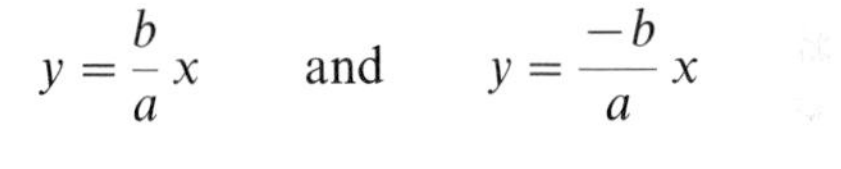

$$y = \frac{b}{a}x \qquad \text{and} \qquad y = \frac{-b}{a}x$$

or

$$y = \frac{4}{3}x \qquad \text{and} \qquad y = \frac{-4}{3}x$$

(b) To graph the hyperbola, we first graph the asymptotes as illustrated in Fig. 10.20. Since the x term is positive, the graph intersects the x axis. Since the denominator of the positive term is 9, the vertices will be at 3 and -3. Now draw the hyperbola by letting the hyperbola approach its asymptotes (Fig. 10.21). Note that the asymptotes are drawn using dashed lines since they are not part of the hyperbola. They are used merely to help sketch the graph.

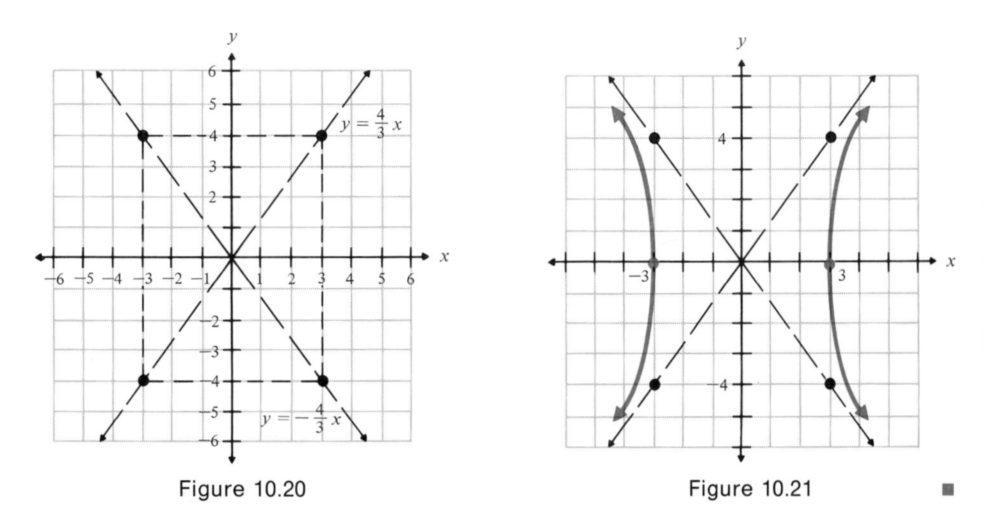

Figure 10.20 Figure 10.21 ■

EXAMPLE 2 (a) Show the equation $-25x^2 + 4y^2 = 100$ is a hyperbola by expressing the equation in standard form.
(b) Determine the equation of the asymptotes of the graph.
(c) Sketch the graph.

Solution: (a) Divide both sides of the equation by 100 to obtain a 1 on the right side of the equation.

$$\frac{-25x^2 + 4y^2}{100} = \frac{100}{100}$$

$$\frac{-25x^2}{100} + \frac{4y^2}{100} = 1$$

$$\frac{-x^2}{4} + \frac{y^2}{25} = 1$$

Rewriting the equation in standard form (positive term first), we get

$$\frac{y^2}{25} - \frac{x^2}{4} = 1$$

(b) The equations of the asymptotes are:

$$y = \frac{5}{2}x \qquad \text{and} \qquad y = \frac{-5}{2}x$$

(c) The graph intersects the y axis at 5 and -5. Figure 10.22a illustrates the asymptotes and Fig. 10.22b illustrates the graph of the hyperbola.

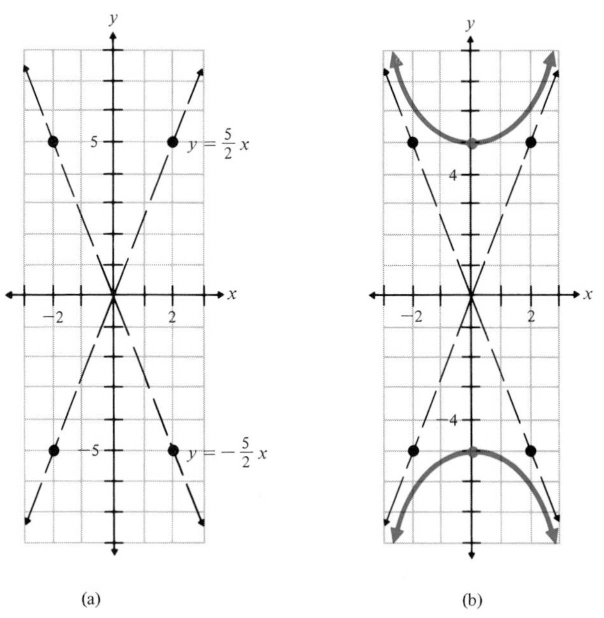

Figure 10.22 ■

Nonstandard Form of the Hyperbola

3 Another form of the hyperbola is $xy = c$, where c is a nonzero constant. The following equations are examples of hyperbolas in nonstandard form:

$$xy = 4, \qquad xy = -6, \qquad x = \frac{1}{y}, \qquad y = -\frac{3}{x}$$

Note that $x = \frac{1}{y}$ is equivalent to $xy = 1$ and $y = \frac{-3}{x}$ is equivalent to $xy = -3$.

Equations of the form $xy = c$ will be hyperbolas with the x and y axes as asymptotes.

EXAMPLE 3 (a) Sketch the graph of $xy = 6$.
(b) Sketch the graph of $xy = -6$.

Solution: (a) $xy = 6$ or $y = \frac{6}{x}$.

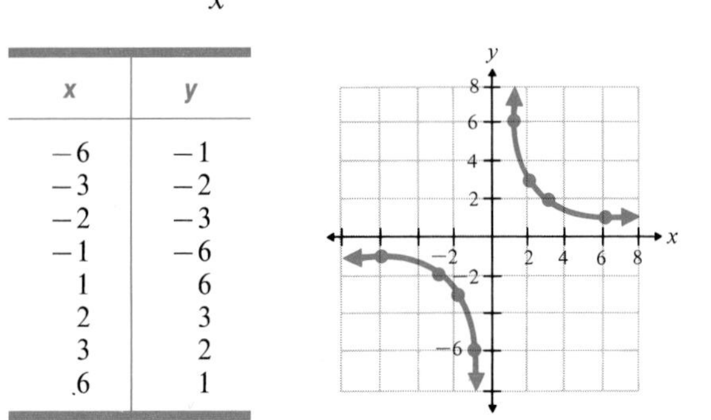

x	y
−6	−1
−3	−2
−2	−3
−1	−6
1	6
2	3
3	2
6	1

(b) $xy = -6$ or $y = \frac{-6}{x}$.

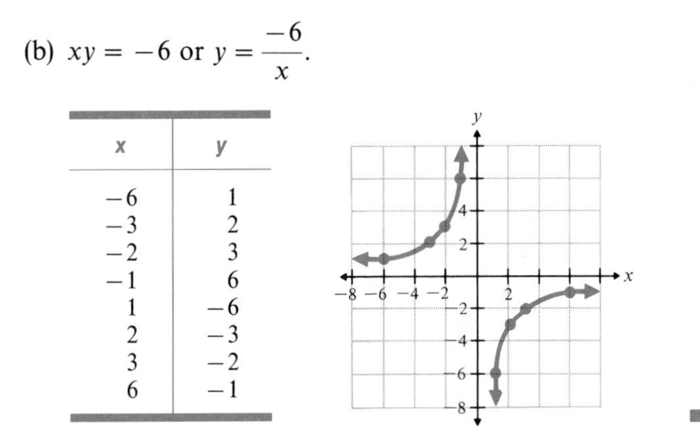

x	y
−6	1
−3	2
−2	3
−1	6
1	−6
2	−3
3	−2
6	−1

■

Summary of Conic Sections

4 The following chart summarizes conic sections.

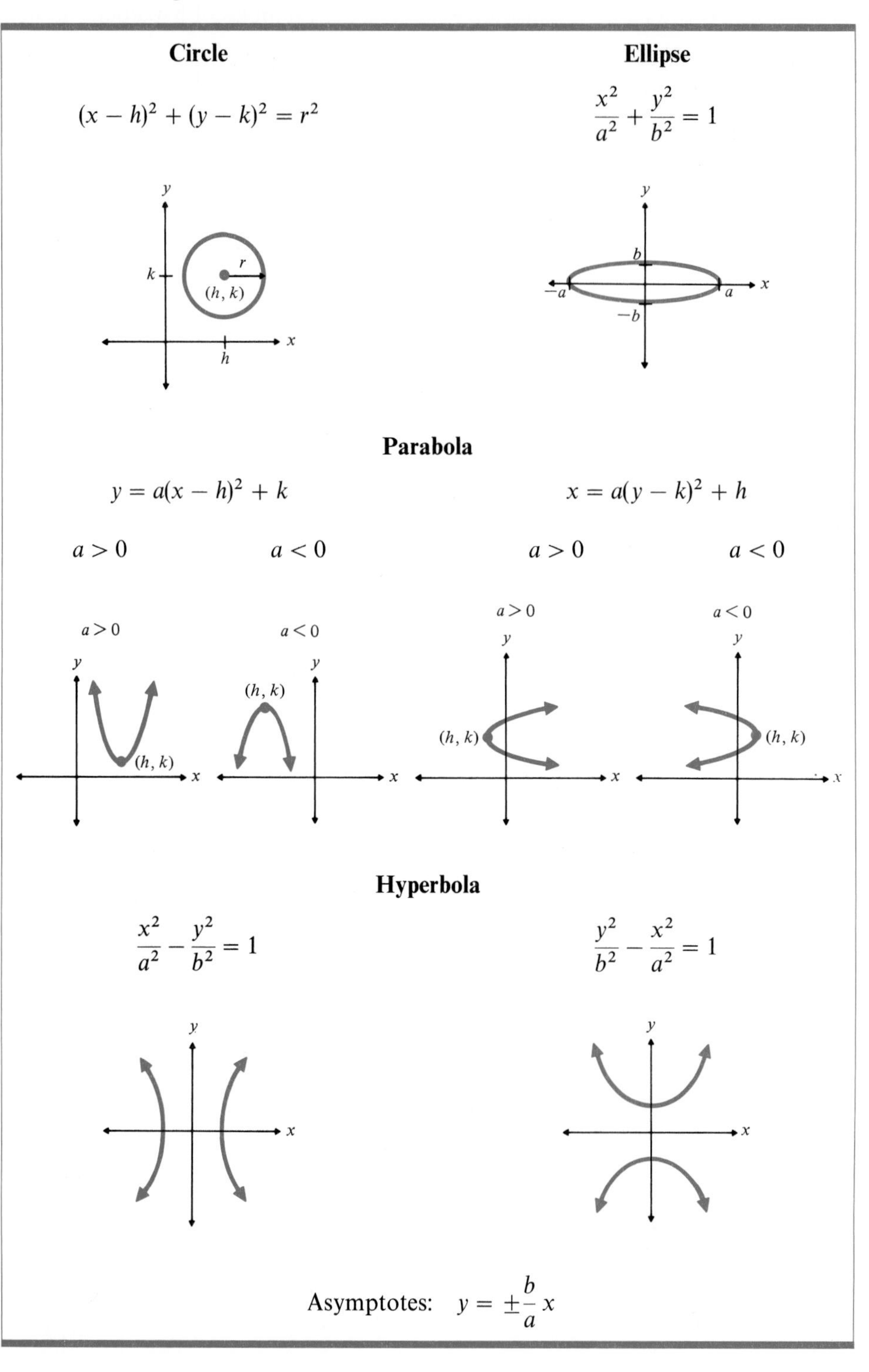

Nonstandard Forms of Conic Sections

5 Often, a conic section will be given in a nonstandard form. Conic sections given in nonstandard form can be recognized with a little practice. The circle, ellipse, and hyperbola can be discussed using the equation

$$ax^2 + by^2 = c^2$$

If a and b are both positive, and $a = b$, the equation is a circle. For example, $4x^2 + 4y^2 = 16$ is a circle.

$$4x^2 + 4y^2 = 16$$

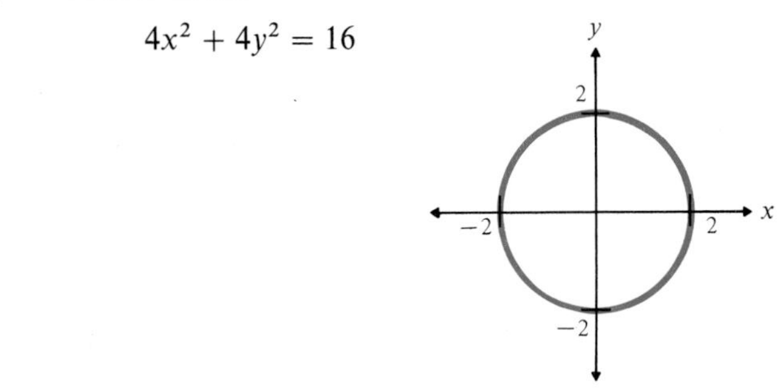

If a and b are both positive and $a \neq b$, the equation is an ellipse. If $a < b$, the major axis is the x axis. If $a > b$, the major axis is the y axis. For example,

$$4x^2 + 9y^2 = 36 \qquad\qquad 9x^2 + 4y^2 = 36$$

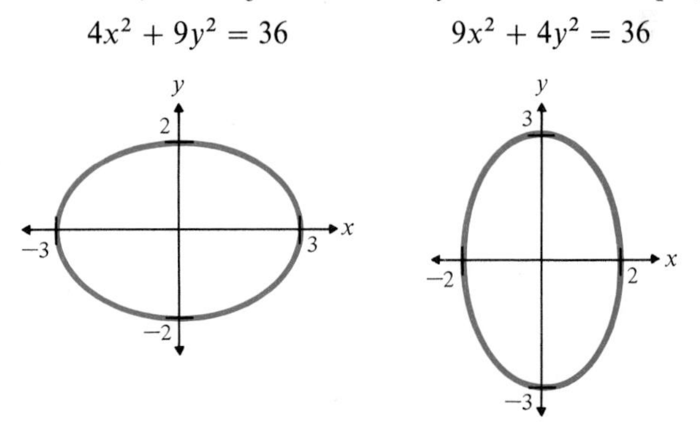

If a and b have opposite signs, the equation is a hyperbola. If $a > 0$, the hyperbola will intersect the x axis. If $b > 0$, the hyperbola will intersect the y axis. For example,

$$4x^2 - 9y^2 = 36 \qquad\qquad -4x^2 + 9y^2 = 36$$

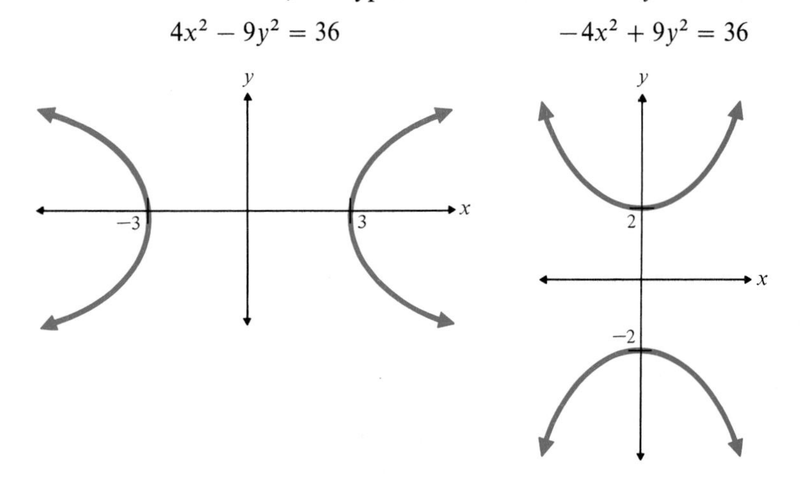

Nonstandard forms of parabolas are $y = ax^2 + bx + c$ (parabola opens up when $a > 0$, down when $a < 0$), and $x = ay^2 + by + c$ (parabola opens right when $a > 0$, left when $a < 0$).

Exercise Set 10.4

Determine the equations of the asymptotes, and then sketch the graph of the equation.

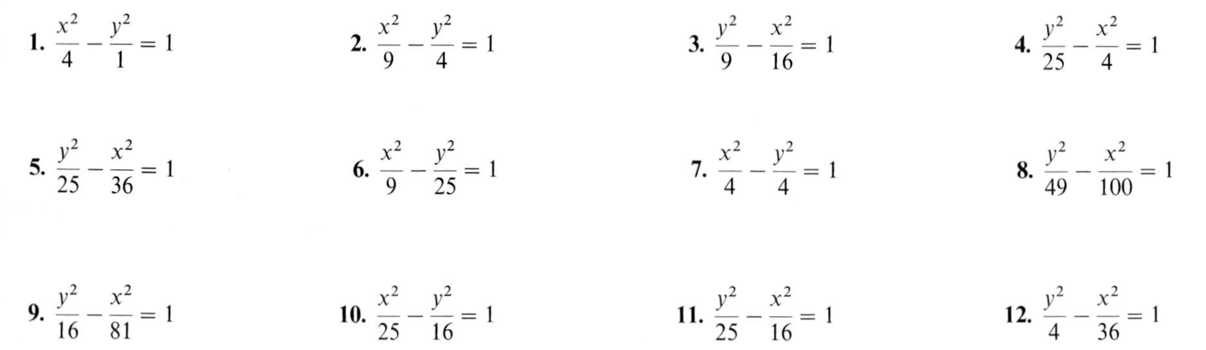

1. $\frac{x^2}{4} - \frac{y^2}{1} = 1$ **2.** $\frac{x^2}{9} - \frac{y^2}{4} = 1$ **3.** $\frac{y^2}{9} - \frac{x^2}{16} = 1$ **4.** $\frac{y^2}{25} - \frac{x^2}{4} = 1$

5. $\frac{y^2}{25} - \frac{x^2}{36} = 1$ **6.** $\frac{x^2}{9} - \frac{y^2}{25} = 1$ **7.** $\frac{x^2}{4} - \frac{y^2}{4} = 1$ **8.** $\frac{y^2}{49} - \frac{x^2}{100} = 1$

9. $\frac{y^2}{16} - \frac{x^2}{81} = 1$ **10.** $\frac{x^2}{25} - \frac{y^2}{16} = 1$ **11.** $\frac{y^2}{25} - \frac{x^2}{16} = 1$ **12.** $\frac{y^2}{4} - \frac{x^2}{36} = 1$

Write each equation in standard form, determine the equation of the asymptotes, and then sketch the graph.

13. $16x^2 - 4y^2 = 64$ **14.** $25x^2 - 16y^2 = 400$ **15.** $9y^2 - x^2 = 9$ **16.** $4y^2 - 25x^2 = 100$

17. $4y^2 - 36x^2 = 144$ **18.** $x^2 - 25y^2 = 25$ **19.** $25x^2 - 4y^2 = 100$ **20.** $25x^2 - 9y^2 = 225$

21. $81x^2 - 9y^2 = 729$ **22.** $64y^2 - 25x^2 = 1600$

Graph each hyperbola.

23. $xy = 10$ **24.** $xy = 1$ **25.** $xy = -8$

26. $xy = -4$ **27.** $y = \frac{12}{x}$ **28.** $y = -\frac{3}{x}$

Indicate whether the equation when graphed will be a parabola, circle, ellipse, or hyperbola.

29. $y = 6x^2 + 4x + 3$
30. $4x^2 + 4y^2 = 16$
31. $5x^2 - 5y^2 = 25$
32. $9x^2 - 16y^2 = 36$
33. $x = y^2 + 6y - 7$
34. $x^2 - 4y^2 = 36$
35. $-2x^2 + 4y^2 = 16$
36. $3x^2 + 3y^2 = 12$
37. $5x^2 + 10y^2 = 12$
38. $9x^2 + 16y^2 = 144$
39. $x = 3y^2 - y + 7$
40. $6x^2 - 9y^2 = 36$
41. $6x^2 + 6y^2 = 36$
42. $-y^2 + 4x^2 = 16$
43. $-3x^2 - 3y^2 = -27$
44. $-6x^2 + 2y^2 = -6$
45. $-6y^2 + x^2 = -9$

JUST FOR FUN

The standard forms of a hyperbola with center at (h, k) are given below.

$$\frac{(x-h)^2}{a^2} - \frac{(y-k)^2}{b^2} = 1 \qquad \frac{(y-k)^2}{b^2} - \frac{(x-h)^2}{a^2} = 1$$

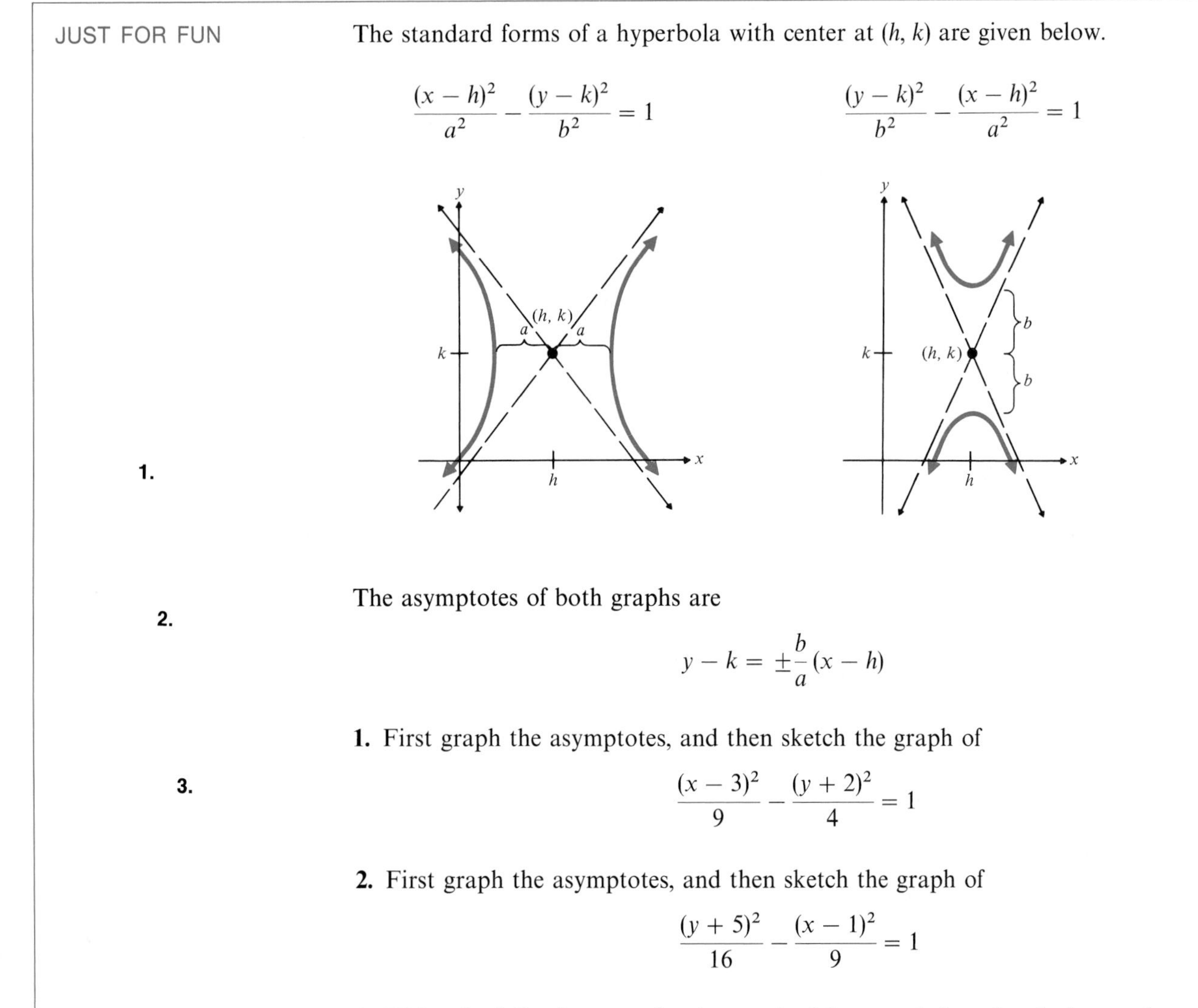

The asymptotes of both graphs are

$$y - k = \pm\frac{b}{a}(x - h)$$

1. First graph the asymptotes, and then sketch the graph of

$$\frac{(x-3)^2}{9} - \frac{(y+2)^2}{4} = 1$$

2. First graph the asymptotes, and then sketch the graph of

$$\frac{(y+5)^2}{16} - \frac{(x-1)^2}{9} = 1$$

3. Write the following equation in standard form, and then sketch the graph.

$$y^2 - 4x^2 + 2y + 8x - 7 = 0$$

10.5 Nonlinear Systems of Equations

1 *Identify a nonlinear system of equations.*

2 *Solve a nonlinear system using substitution.*

3 *Solve a nonlinear system using addition.*

1 In Chapter 4 we discussed systems of linear equations. Here, we discuss nonlinear systems of equations. A **nonlinear system of equations** is a system of equations containing at least one that is not a linear equation (that is, one whose graph is not a straight line).

The solution to a system of equations is the point or points that satisfy all equations in the system. Consider the system of equations

$$\begin{aligned} x^2 + y^2 &= 25 \\ 3x + 4y &= 0 \end{aligned}$$

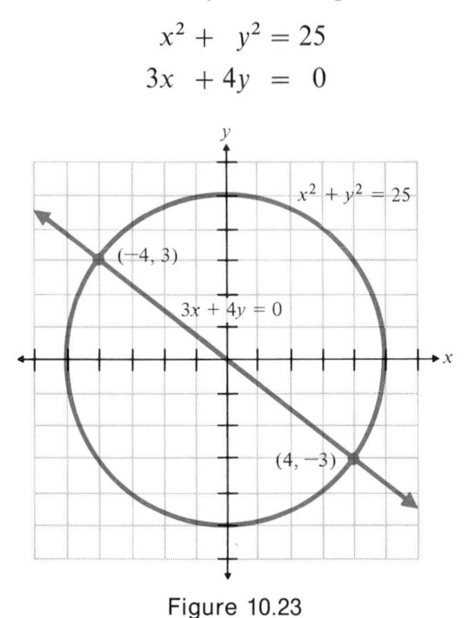

Figure 10.23

Both equations are graphed on the same set of axes in Fig. 10.23. Note that the graphs appear to intersect at the points $(-4, 3)$ and $(4, -3)$. The check shows these points satisfy all equations in the system and are therefore solutions to the system.

Check:

$(-4, 3)$

$$\begin{aligned} x^2 + y^2 &= 25 \\ (-4)^2 + 3^2 &= 25 \\ 16 + 9 &= 25 \\ 25 &= 25, \quad \text{true} \end{aligned} \qquad \begin{aligned} 3x + 4y &= 0 \\ 3(-4) + 4(3) &= 0 \\ -12 + 12 &= 0 \\ 0 &= 0, \quad \text{true} \end{aligned}$$

Check:

$(4, -3)$

$$\begin{aligned} 4^2 + (-3)^2 &= 25 \\ 16 + 9 &= 25 \\ 25 &= 25, \quad \text{true} \end{aligned} \qquad \begin{aligned} 3(4) + 4(-3) &= 0 \\ 12 - 12 &= 0 \\ 0 &= 0, \quad \text{true} \end{aligned}$$

2 The graphical procedure for solving a system of equations may result in an inaccurate result since you will have to estimate the point or points of intersection. An exact answer may be obtained algebraically.

To solve a system of equations algebraically, we often solve one or more of the equations for one of the variables and then use the substitution principle. This procedure is illustrated in Examples 1 through 3.

EXAMPLE 1 Solve the system of equations algebraically using substitution.

$$x^2 + y^2 = 25$$
$$3x + 4y = 0$$

Solution: Solve the linear equation, $3x + 4y = 0$, for either x or y. We will select to solve for y.

$$3x + 4y = 0$$
$$4y = -3x$$
$$y = -\frac{3x}{4}$$

Now substitute $-3x/4$ for y in the equation $x^2 + y^2 = 25$ and solve for the remaining variable, x.

$$x^2 + y^2 = 25$$
$$x^2 + \left(\frac{-3x}{4}\right)^2 = 25$$
$$x^2 + \frac{9x^2}{16} = 25$$
$$16\left(x^2 + \frac{9x^2}{16}\right) = 16(25)$$
$$16x^2 + 9x^2 = 400$$
$$25x^2 = 400$$
$$x^2 = \frac{400}{25} = 16$$
$$x = \pm\sqrt{16} = \pm 4$$

Next, find the corresponding value of y for each value of x by substituting each value of x (one at a time) in the equation solved for y.

$$x = 4 \qquad\qquad x = -4$$
$$y = -\frac{3x}{4} \qquad\qquad y = \frac{-3x}{4}$$
$$y = \frac{-3(4)}{4} \qquad\qquad y = \frac{-3(-4)}{4}$$
$$y = -3 \qquad\qquad y = 3$$

Solutions: $(4, -3)$ and $(-4, 3)$

This answer checks with the answer obtained graphically. ■

Note that our objective in the substitution method is to make a substitution that will result in obtaining a single equation containing only one variable.

EXAMPLE 2 Solve the following system of equations using substitution.

$$y = x^2 - 3$$
$$x^2 + y^2 = 9$$

Solution: Since both equations contain an x^2 term, we will solve one of the equations for x^2.

$$y = x^2 - 3$$
$$y + 3 = x^2$$

Now substitute $y + 3$ for x^2 in the equation $x^2 + y^2 = 9$.

$$x^2 + y^2 = 9$$
$$(y + 3) + y^2 = 9$$
$$y^2 + y + 3 = 9$$
$$y^2 + y - 6 = 0$$
$$(y + 3)(y - 2) = 0$$

$$y + 3 = 0 \quad \text{or} \quad y - 2 = 0$$
$$y = -3 \qquad\qquad y = 2$$

Now find the corresponding values of x.

$$y = -3 \qquad\qquad y = 2$$
$$y = x^2 - 3 \qquad\qquad y = x^2 - 3$$
$$-3 = x^2 - 3 \qquad\qquad 2 = x^2 - 3$$
$$0 = x^2 \qquad\qquad 5 = x^2$$
$$0 = x \qquad\qquad \pm\sqrt{5} = x$$

Solutions: $(0, -3)$ $\quad (\sqrt{5}, 2), (-\sqrt{5}, 2)$

This system has three solutions: $(0, -3)$, $(\sqrt{5}, 2)$, and $(-\sqrt{5}, 2)$. ■

Note that in Example 2 the equation $y = x^2 - 3$ is a parabola and $x^2 + y^2 = 9$ is a circle. The graphs of both equations are illustrated in Fig. 10.24.

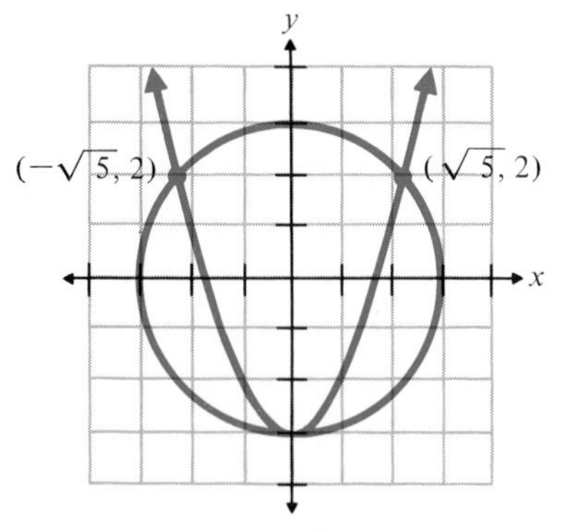

Figure 10.24

HELPFUL HINT

> Students will sometimes solve for one variable and assume that they have found the answer. Remember that the solution, if one exists, consists of one or more *ordered pairs*.

3 We can often solve systems of equations more easily using the addition method that was discussed in Section 4.1. As with the substitution method, our objective will be to obtain a single equation containing only one variable.

EXAMPLE 3 Solve the system of equations by the addition method.

$$\begin{aligned} x^2 + y^2 &= 9 \\ 2x^2 - y^2 &= -6 \end{aligned}$$

Solution: If we add the two equations, we will obtain one equation containing only one variable.

$$\begin{aligned} x^2 + y^2 &= 9 \\ 2x^2 - y^2 &= -6 \\ \hline 3x^2 \quad &= 3 \\ x^2 &= 1 \\ x &= \pm 1 \end{aligned}$$

Now solve for the variable y by substituting $x = \pm 1$ in either of the original equations.

$$\begin{array}{ll} x = 1 & x = -1 \\ x^2 + y^2 = 9 & x^2 + y^2 = 9 \\ 1^2 + y^2 = 9 & (-1)^2 + y^2 = 9 \\ 1 + y^2 = 9 & 1 + y^2 = 9 \\ y^2 = 8 & y^2 = 8 \\ y = \pm\sqrt{8} = \pm 2\sqrt{2} & y = \pm\sqrt{8} = \pm 2\sqrt{2} \end{array}$$

Solutions: $(1, 2\sqrt{2}), (1, -2\sqrt{2})$ $\qquad (-1, 2\sqrt{2}), (-1, -2\sqrt{2})$

There are four solutions to this system of equations: $(1, 2\sqrt{2})$, $(1, -2\sqrt{2})$, $(-1, 2\sqrt{2})$, and $(-1, -2\sqrt{2})$. The graph of the system is illustrated in Fig. 10.25.

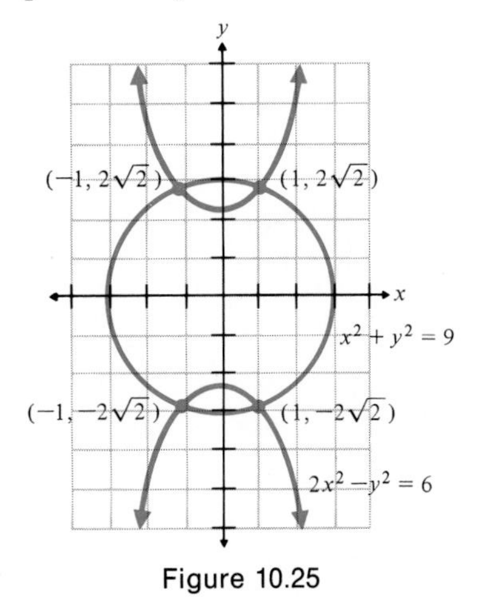

Figure 10.25 ■

It is possible that a system of equations has no real solution (therefore, the graphs do not intersect). Example 4 illustrates such a case.

EXAMPLE 4 Solve the system of equations using the addition method.

$$x^2 + 4y^2 = 16$$
$$x^2 + y^2 = 1$$

Solution: Recall from Section 4.1 that in this text we place an equation within brackets, [ ], to indicate that the entire equation is to be multiplied by the value to the left of the brackets.

$$\begin{aligned} x^2 + 4y^2 &= 16 \\ -1[x^2 + y^2 &= 1] \end{aligned} \quad \text{gives} \quad \begin{aligned} x^2 + 4y^2 &= 16 \\ -x^2 - y^2 &= -1 \\ \hline 3y^2 &= 15 \\ y^2 &= 5 \\ y &= \pm\sqrt{5} \end{aligned}$$

Now solve for x.

$$\begin{aligned} y &= \sqrt{5} \\ x^2 + y^2 &= 1 \\ x^2 + (\sqrt{5})^2 &= 1 \\ x^2 + 5 &= 1 \\ x^2 &= -4 \\ x &= \pm\sqrt{-4} \\ x &= \pm 2i \end{aligned} \qquad \begin{aligned} y &= -\sqrt{5} \\ x^2 + y^2 &= 1 \\ x^2 + (-\sqrt{5})^2 &= 1 \\ x^2 + 5 &= 1 \\ x^2 &= -4 \\ x &= \pm\sqrt{-4} \\ x &= \pm 2i \end{aligned}$$

Since x is an imaginary number for both values of y, this system of equations has no real solution. The graphs of the two equations are illustrated in Fig. 10.26.

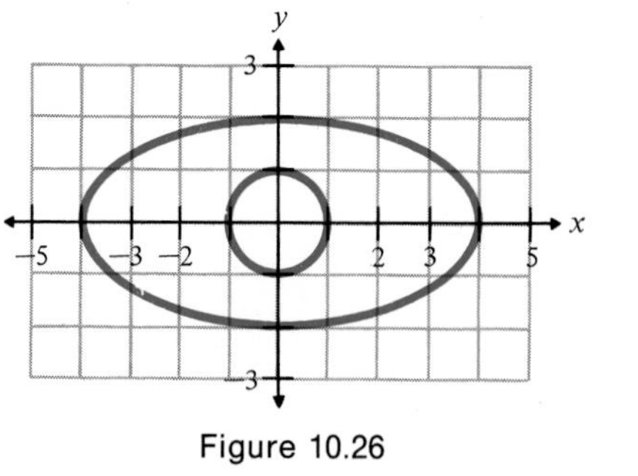

Figure 10.26 ■

EXAMPLE 5 Solve the system of equations by the addition method.

$$6x^2 + y^2 = 10$$
$$2x^2 + 4y^2 = 40$$

Solution: We can select to eliminate either the variable x or the variable y. In this example we will eliminate the variable x.

$$\begin{aligned} 6x^2 + y^2 &= 10 \\ -3[2x^2 + 4y^2 &= 40] \end{aligned} \quad \text{gives} \quad \begin{aligned} 6x^2 + y^2 &= 10 \\ -6x^2 - 12y^2 &= -120 \\ \hline -11y^2 &= -110 \\ y^2 &= \frac{-110}{-11} = 10 \\ y &= \pm\sqrt{10} \end{aligned}$$

Now find the value of x by substituting $y = \pm\sqrt{10}$ in either of the original equations.

$y = \sqrt{10}$	$y = -\sqrt{10}$
$6x^2 + y^2 = 10$	$6x^2 + y = 10$
$6x^2 + (\sqrt{10})^2 = 10$	$6x^2 + (-\sqrt{10})^2 = 10$
$6x^2 + 10 = 10$	$6x^2 + 10 = 10$
$6x^2 = 0$	$6x^2 = 0$
$x^2 = 0$	$x^2 = 0$
$x = 0$	$x = 0$

Solutions: $(0, \sqrt{10})$ $(0, -\sqrt{10})$ ■

Exercise Set 10.5

Solve the system of equations using the substitution method.

1. $x^2 + y^2 = 4$, $x - 2y = 4$
2. $x^2 + y^2 = 9$, $x + 2y - 3 = 0$
3. $y = x^2 - 5$, $3x + 2y = 10$
4. $x + y = 4$, $x^2 - y^2 = 4$
5. $2x^2 - y^2 = -8$, $x - y = 6$
6. $y^2 = -x + 4$, $x^2 + y^2 = 6$
7. $x^2 - 4y^2 = 16$, $x^2 + y^2 = 1$
8. $2x^2 + y^2 = 16$, $x^2 - y^2 = -4$
9. $xy = 4$, $6x - y = 5$
10. $xy = 4$, $y = 5 - x$
11. $y = x^2 - 3$, $x^2 + y^2 = 9$
12. $x^2 + y^2 = 25$, $x - 3y = -5$

Solve the system of equations using the addition method.

13. $x^2 - y^2 = 4$, $x^2 + y^2 = 4$
14. $x^2 + y^2 = 25$, $x^2 - 2y^2 = 7$
15. $x^2 + y^2 = 13$, $2x^2 + 3y^2 = 30$
16. $3x^2 - y^2 = 4$, $x^2 + 4y^2 = 10$
17. $4x^2 + 9y^2 = 36$, $2x^2 - 9y^2 = 18$
18. $5x^2 - 2y^2 = -13$, $3x^2 + 4y^2 = 39$
19. $2x^2 + 3y^2 = 21$, $x^2 + 2y^2 = 12$
20. $2x^2 + y^2 = 11$, $x^2 + 3y^2 = 28$
21. $-x^2 - 2y^2 = 6$, $5x^2 + 15y^2 = 20$
22. $x^2 - 2y^2 = 7$, $x^2 + y^2 = 34$
23. $x^2 + y^2 = 9$, $16x^2 - 4y^2 = 64$
24. $3x^2 + 4y^2 = 35$, $2x^2 + 5y^2 = 42$

10.6 Second-Degree Inequalities in Two Variables

1 *Graph second-degree inequalities.*

2 *Solve a system of inequalities graphically.*

1 In Sections 10.1 through 10.5 we graphed second-degree equations such as $x^2 + y^2 = 25$. Now we will graph second-degree inequalities, such as $x^2 + y^2 \leq 25$. To graph second-degree inequalities we use a procedure similar to the one we used to graph linear inequalities. That is, graph the equation using a solid line if the inequality is $\leq$ or $\geq$, and a dashed line if the inequality is $<$ or $>$, and then shade in the region that satisfies the inequality. Examples 1 and 2 illustrate this procedure.

EXAMPLE 1 Graph the inequality.

$$\frac{x^2}{4} + \frac{y^2}{9} \geq 1.$$

Solution: First graph the equation $\frac{x^2}{4} + \frac{y^2}{9} = 1$. Use a solid line when drawing the ellipse since the inequality contains $\geq$ (Fig. 10.27a). Next select a test point not on the graph and test it in the original inequality. We will select the test point (0, 0).

$$\frac{x^2}{4} + \frac{y^2}{9} \geq 1$$

$$\frac{0^2}{4} + \frac{0^2}{9} \geq 1$$

$$0 + 0 \geq 1$$

$$0 \geq 1, \qquad \text{false}$$

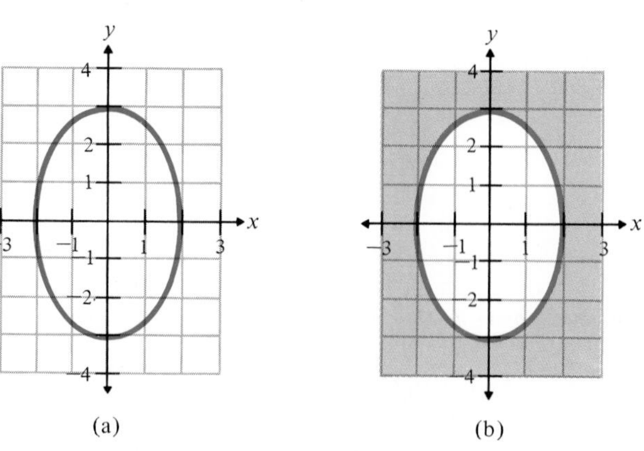

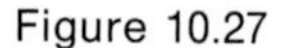
Figure 10.27

The point (0, 0), which is within the ellipse, is not a solution. Thus all the points outside the ellipse will satisfy the inequality. Shade in this outer area (Fig. 10.27b). The shaded area, and the ellipse itself, is the solution to the inequality. ■

EXAMPLE 2 Graph the inequality.

$$\frac{y^2}{25} - \frac{x^2}{9} < 1$$

Solution: Graph the equation $\frac{y^2}{25} - \frac{x^2}{9} = 1$. Use a dashed line when drawing the hyperbola since the inequality contains $<$ (Fig. 10.28a). Next select the test point. We will use (0, 0).

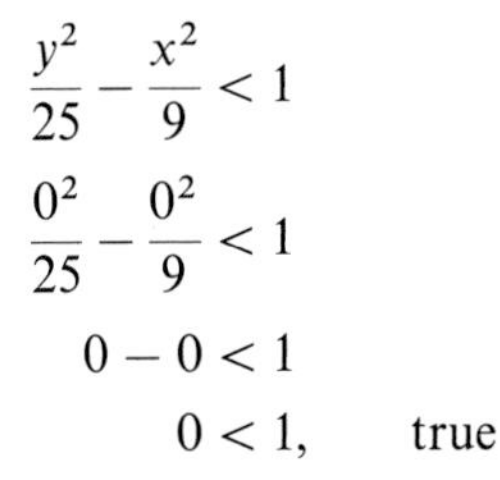

$$\frac{y^2}{25} - \frac{x^2}{9} < 1$$

$$\frac{0^2}{25} - \frac{0^2}{9} < 1$$

$$0 - 0 < 1$$

$$0 < 1, \qquad \text{true}$$

Since (0, 0) is a solution, we shade in the region containing the point (0, 0). The solution is indicated in Fig. 10.28b.

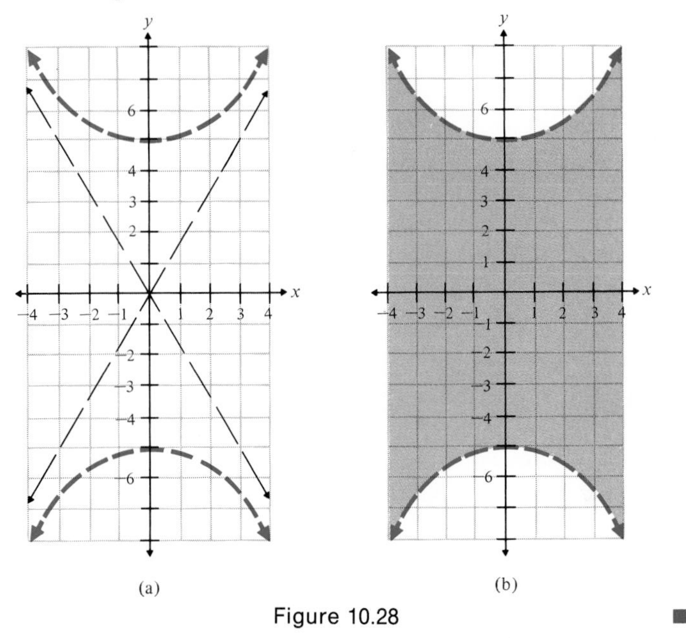

Figure 10.28 ■

2 Now we will find the graphical solution to a system of inequalities in which at least one inequality is not linear. The solution to a system of inequalities is the set of ordered pairs that satisfy all inequalities in the system.

> **To Solve a System of Inequalities Graphically**
>
> **1.** Graph one inequality.
>
> **2.** On the same set of axes, draw the second inequality. Use a different type of shading than was used in the first inequality.
>
> **3.** The solution is the area containing the shaded area from both inequalities.

EXAMPLE 3 Solve the system of inequalities graphically.

$$x^2 + y^2 < 25$$
$$2x + y \geq 4$$

Solution: First graph the inequality $x^2 + y^2 < 25$. The inner region of the circle satisfies the inequality (Fig. 10.29). On the same set of axes, graph the second inequality $2x + y \geq 4$. The solution to the system is the area containing both types of shading and that part of the straight line within the boundaries of the circle (see Fig. 10.30.).

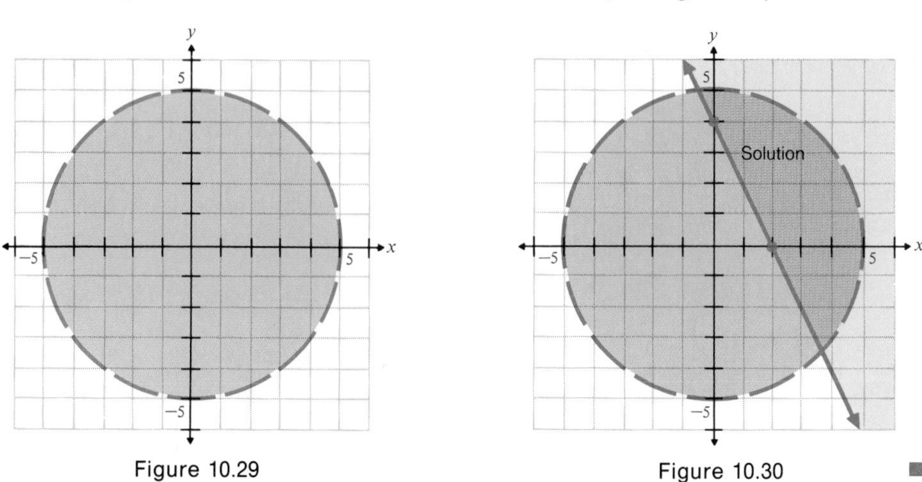

Figure 10.29

Figure 10.30 ■

EXAMPLE 4 Solve the system of inequalities graphically.

$$\frac{x^2}{4} - \frac{y^2}{9} \leq 1$$
$$y > (x + 2)^2 - 4$$

Solution: Graph each inequality on the same set of axes (Fig. 10.31). The solution is the area containing both types of shading and the solid line within the parabola.

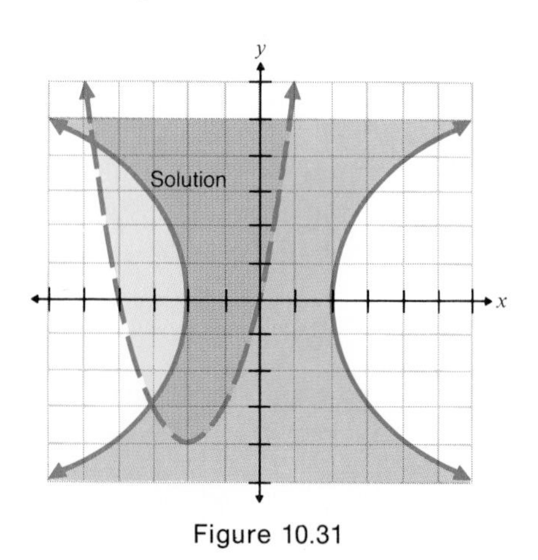

Figure 10.31 ■

Exercise Set 10.6

Graph each system of inequalities.

1. $x^2 + y^2 \geq 16$
$y + x < 5$

2. $\frac{x^2}{25} - \frac{y^2}{9} < 1$
$2x + 3y > 6$

3. $4x^2 + y^2 > 16$
$y \geq 2x + 2$

4. $x^2 - y^2 > 1$
$\frac{x^2}{9} + \frac{y^2}{4} \leq 1$

5. $x^2 + y^2 \leq 36$
$y < (x + 1)^2 - 5$

6. $y \geq x^2 - 2x + 1$
$x^2 + y^2 > 4$

7. $\frac{y^2}{16} - \frac{x^2}{4} \geq 1$
$\frac{x^2}{4} - \frac{y^2}{1} < 1$

8. $4x^2 + 9y^2 \leq 36$
$x > (y - 2)^2 + 1$

9. $xy \leq 6$
$2x - y \leq 8$

10. $25x^2 - 4y^2 > 100$
$5x + 3y \leq 15$

11. $(x - 3)^2 + (y + 2)^2 \geq 16$
$y \leq 4x - 2$

12. $x^2 + (y - 3)^2 \leq 25$
$y > (x - 2)^2 + 3$

JUST FOR FUN

Graph the following systems of inequalities.

1. $y > 4x - 6$
$x^2 + y^2 \geq 36$
$2x + y \leq 8$

2. $x^2 + y^2 > 25$
$\frac{x^2}{25} + \frac{y^2}{9} > 1$
$2x - 3y < 12$

Summary

Glossary

Asymptotes of a hyperbola: Two lines through the center of the hyperbola that are used as an aid in graphing the hyperbola.

Circle: The set of points in a plane that are equidistant from a fixed point called the center.

Conic sections: Circles, ellipses, hyperbolas, and parabolas.

Conjugate axis of a hyperbola: The axis perpendicular to the transverse axis.

Ellipse: The set of points in a plane the sum of whose distance from two fixed points, called **foci**, is a constant.

Hyperbola: The set of points in a plane the difference of whose distance from two fixed points, called **foci**, is a constant.

Major axis of ellipse: The longer axis through the center of the ellipse.

Minor axis of ellipse: The shorter axis through the center of the ellipse.

Nonlinear system of equations: A system of equations containing at least one equation that is not a linear equation.

Parabola: The shape of the graph of a quadratic equation of the form $y = a(x - h)^2 + k$ or $x = a(y - k)^2 + h$.

Transverse axis of a hyperbola: The axis through the center of the hyperbola that intersects the vertices.

Important Facts

Circle

$(x - h)^2 + (y - k)^2 = r^2$
center at (h, k)
radius r

Ellipse

$$\frac{x^2}{a^2} + \frac{y^2}{b^2} = 1$$
center at $(0, 0)$

Parabola

$y = a(x - h)^2 + k$
vertex at (h, k)
opens up when $a > 0$
opens down when $a < 0$
$x = a(y - k)^2 + h$
vertex at (h, k)
opens right when $a > 0$
opens left when $a < 0$

Hyperbola

$$\frac{x^2}{a^2} - \frac{y^2}{b^2} = 1$$
x axis transverse axis
$$\frac{y^2}{b^2} - \frac{x^2}{a^2} = 1$$
y axis transverse axis
asymptotes:
$$y = \pm\frac{b}{a}x$$

Review Exercises

[10.1] Write the equation of the circle in standard form; then sketch the graph of the circle.

1. Center $(0, 0)$, radius 5

2. Center $(-3, 4)$, radius 3

3. Center $(4, 2)$, radius $\sqrt{8}$

Sketch the graph of each equation.

4. $x^2 + y^2 = 25$

5. $(x - 2)^2 + (y + 3)^2 = 25$

6. $(x + 1)^2 + (y + 3)^2 = 6$

Determine the equation of the circle.

7.

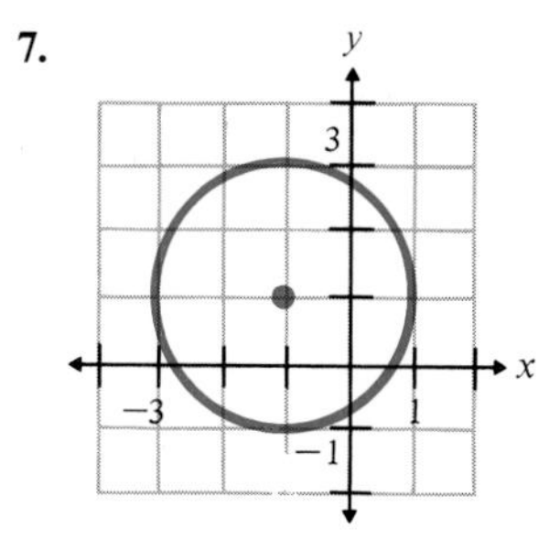

8.

Write the equation in standard form; then sketch the graph.

9. $x^2 + y^2 - 4y = 0$

10. $x^2 + y^2 - 2x + 6y + 1 = 0$

11. $x^2 - 8x + y^2 - 10y + 40 = 0$

12. $x^2 + y^2 - 4x + 10y + 17 = 0$

[10.2] Sketch the graph of each equation.

13. $\frac{x^2}{9} + \frac{y^2}{4} = 1$

14. $\frac{x^2}{16} + \frac{y^2}{1} = 1$

15. $\frac{x^2}{9} + \frac{y^2}{64} = 1$

16. $4x^2 + 9y^2 = 36$

17. $9x^2 + 16y^2 = 144$

18. $16x^2 + y^2 = 16$

[10.3] Graph each equation.

19. $y = (x - 3)^2 + 4$

20. $y = (x + 4)^2 - 5$

21. $x = (y - 1)^2 + 4$

22. $x = -2(y + 4)^2 - 3$

23. $y = -3(x + 5)^2$

Write each equation in the form $y = a(x - h)^2 + k$ or $x = a(y - k)^2 + h$; then graph the equation.

24. $y = x^2 - 6x$

25. $y = x^2 - 2x - 3$

26. $x = -y^2 - 2y + 8$

27. $x = y^2 + 5y + 4$

28. $y = 2x^2 - 8x - 24$

[10.4] Determine the equations of the asymptotes; then sketch the graph.

29. $\frac{x^2}{4} - \frac{y^2}{9} = 1$

30. $\frac{y^2}{16} - \frac{x^2}{4} = 1$

31. $\frac{y^2}{9} - \frac{x^2}{25} = 1$

32. $\frac{x^2}{4} - \frac{y^2}{36} = 1$

Write each equation in standard form, determine the equations of the asymptotes; then sketch the graph.

33. $9y^2 - 4x^2 = 36$

34. $x^2 - 16y^2 = 16$

35. $25x^2 - 16y^2 = 400$

36. $49y^2 - 9x^2 = 441$

Graph each equation.

37. $xy = 6$

38. $xy = -8$

39. $y = \frac{3}{x}$

40. $y = -\frac{10}{x}$

Identify the graph as a circle, ellipse, parabola, or hyperbola.

41. $\frac{x^2}{4} - \frac{y^2}{16} = 1$

42. $16x^2 + 9y^2 = 144$

43. $(x-4)^2 + (y+2)^2 = 16$

44. $x^2 - 25y^2 = 25$

45. $\frac{x^2}{64} + \frac{y^2}{9} = 1$

46. $y = (x - 2)^2 - 5$

47. $4x^2 + 9y^2 = 36$

48. $y = x^2 + 2x - 3$

49. $x = -2(y+3)^2 - 1$

50. $25y^2 - 9x^2 = 225$

[10.5] Solve the systems of equations using the substitution method.

51. $x^2 + y^2 = 9$
$y = 3x + 9$

52. $xy = 5$
$3x + y = 16$

53. $x^2 + y^2 = 4$
$x^2 - y^2 = 4$

54. $x^2 + 4y^2 = 4$
$x^2 - 6y^2 = 12$

Solve the systems of equations using the addition method.

55. $x^2 + y^2 = 16$
$x^2 - y^2 = 16$

56. $x^2 + y^2 = 25$
$x^2 - 2y^2 = -2$

57. $-4x^2 + y^2 = -12$
$8x^2 + 2y^2 = -8$

58. $-2x^2 - 3y^2 = -6$
$5x^2 + 4y^2 = 15$

[10.6] Graph each system of nonlinear inequalities.

59. $2x + y \geq 6$
$x^2 + y^2 < 9$

60. $xy > 5$
$y < 3x + 4$

61. $4x^2 + 9y^2 \leq 36$
$x^2 + y^2 > 25$

62. $\dfrac{x^2}{4} - \dfrac{y^2}{9} > 1$
$y \geq (x - 3)^2 - 4$

Practice Test

1. Write the equation of the circle with center at $(-3, -1)$ and radius 9. Then sketch the circle.

2. Write the equation in standard form: then sketch the graph

$$x^2 + y^2 - 2x - 6y + 1 = 0$$

3. Sketch the graph of $9x^2 + 16y^2 = 144$.

4. Sketch the graph of $y = -2(x - 3)^2 - 9$.

5. Sketch the graph of $\dfrac{y^2}{25} - \dfrac{x^2}{1} = 1$.

6. Sketch the graph of $y = \dfrac{8}{x}$.

7. Solve the system of equations by the method your choice.

$$x^2 + y^2 = 16$$
$$2x^2 - y^2 = 2$$

8. Sketch the system of inequalities

$$\frac{x^2}{9} - \frac{y^2}{25} < 1$$
$$x^2 + y^2 \leq 4$$

11 Exponential and Logarithmic Functions

See Section 11.2, Example 11.

11.1 Inverse Functions

1. *Identify one-to-one functions.*
2. *Find the inverse function of a set of ordered pairs.*
3. *Find inverse functions.*

---

1 Before we can discuss exponential and logarithmic functions we must have an understanding of *inverse functions*. We will discuss inverse functions shortly, but first we need to explain what is meant by *one-to-one functions*.

Consider the function $f(x) = x^2$, see Fig. 11.1.

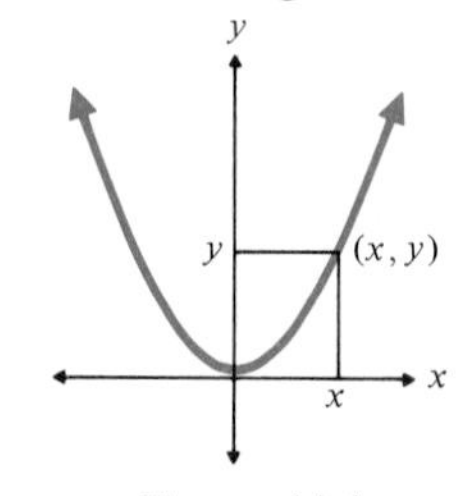

Figure 11.1

Note that it is a function since it passes the vertical line test. For each value of x there is a unique value of y. Does each value of y also have a unique value of x? The answer is no, see Fig. 11.2.

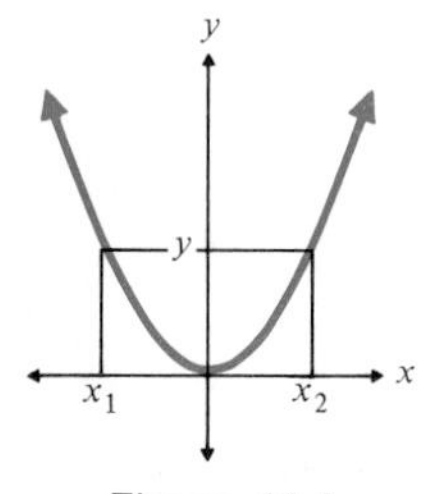

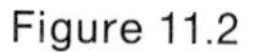

Figure 11.2

Note that for the indicated value of y there are two values of x, x_1 and x_2. If we limit the domain of $f(x) = x^2$ to values of x greater than or equal to 0, then each x value has a unique y value and each y value has a unique x value, see Fig. 11.3.

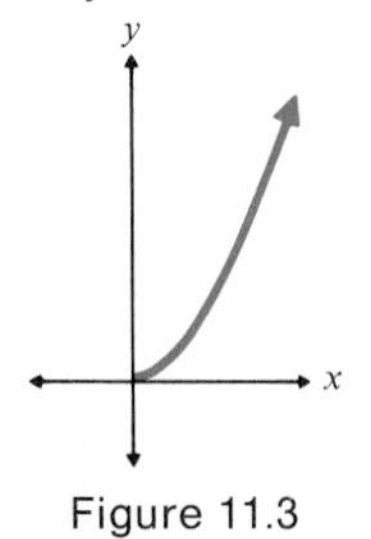

Figure 11.3

The function $f(x) = x^2$, $x \geq 0$, is an example of a one-to-one function. A **one-to-one function** is a function where each y value has a unique x value. For a function to be a one-to-one function, it must pass not only a **vertical line test** (the test to ensure that it is a function) but also a **horizontal line test** (to test the one-to-one criteria).

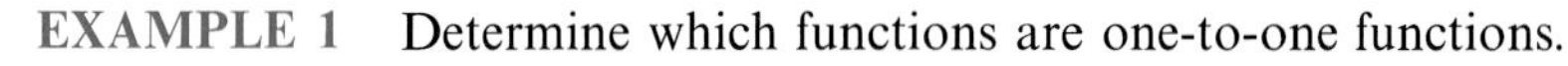

EXAMPLE 1 Determine which functions are one-to-one functions.

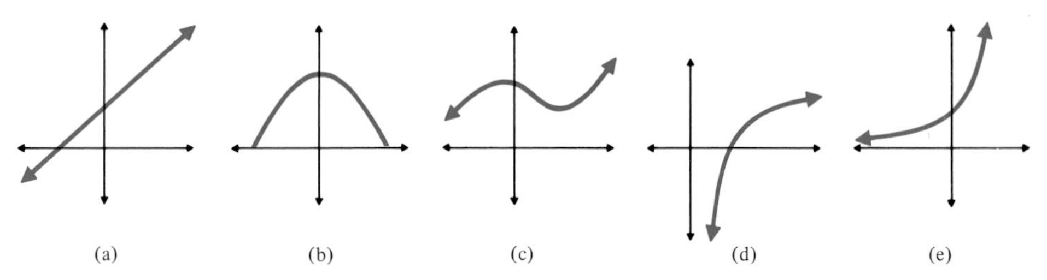

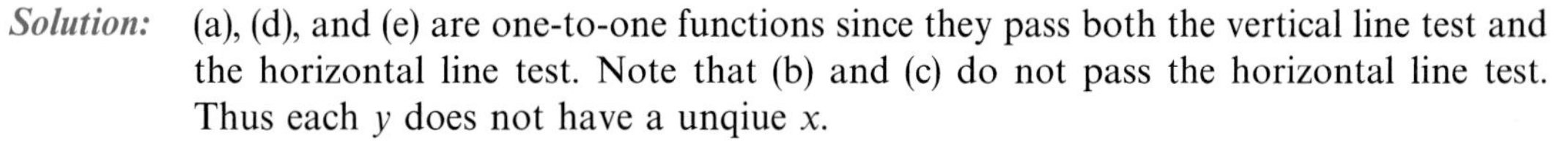

Solution: (a), (d), and (e) are one-to-one functions since they pass both the vertical line test and the horizontal line test. Note that (b) and (c) do not pass the horizontal line test. Thus each y does not have a unqiue x.

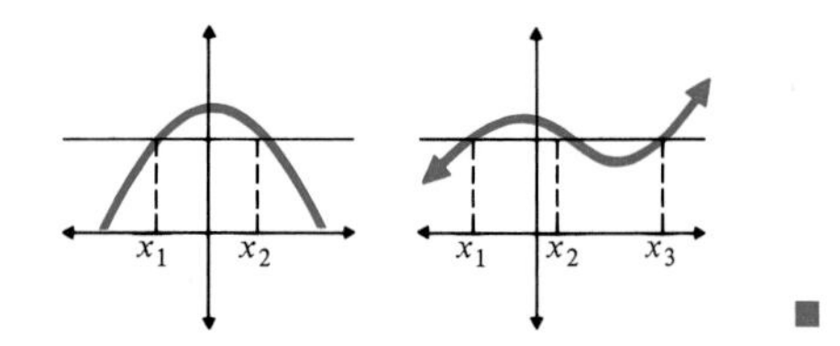

■

2 Now that we have discussed one-to-one functions we can introduce inverse functions. If a function is one-to-one, its **inverse function** may be obtained by interchanging the first and second coordinates in each ordered pair of the function. Thus for each ordered pair (x, y) in the function, the ordered pair (y, x) will be in the inverse function. For example,

Function:	$\{(1, 4), (2, 0), (3, 7), (-2, 1), (-1, -5)\}$
Inverse function:	$\{(4, 1), (0, 2), (7, 3), (1, -2), (-5, -1)\}$

Note that the domain of the function becomes the range of the inverse function and the range of the function is the domain of the inverse function.

If we graph the points in the function and the points in the inverse function (Fig. 11.4), we see that the points are symmetric about the line $y = x$.

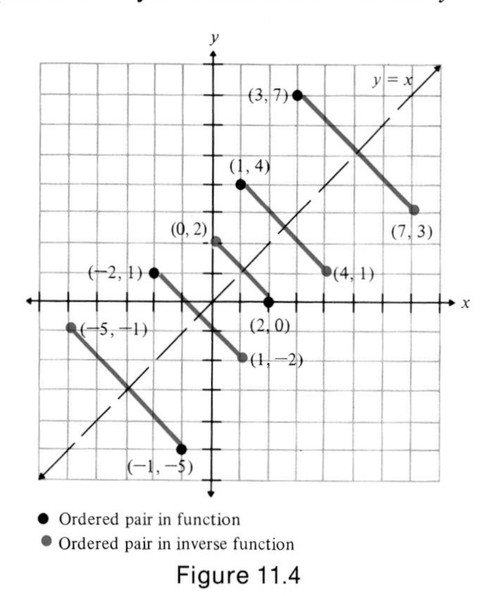

Figure 11.4

If $f(x)$ is used to represent a function, the notation $f^{-1}(x)$ represents the inverse function of $f(x)$. Note the -1 in the notation is *not* an exponent.

Inverse Function

If $f(x)$ is a one-to-one function with ordered pairs of the form (x, y), then its inverse function, $f^{-1}(x)$, will be a one-to-one function with ordered pairs of the form (y, x).

3 When a one-to-one function is given in the form of an equation, its inverse function can be found by following this procedure:

To Find the Inverse Function of a One-to-One Function of the Form $y = f(x)$

1. Interchange the two variables x and y.
2. Solve the equation for y. The resulting equation will be the inverse function.

The following example will illustrate the procedure.

EXAMPLE 2 (a) Find the inverse function of $y = f(x) = 4x + 2$.
(b) On the same set of axes, graph $f(x)$ and $f^{-1}(x)$.

Solution: (a) $y = 4x + 2$.

First interchange x and y.

$$x = 4y + 2$$

Now solve for y.

$$x - 2 = 4y$$

$$\frac{x-2}{4} = y$$

$$y = f^{-1}(x) = \frac{x-2}{4}$$

(b)

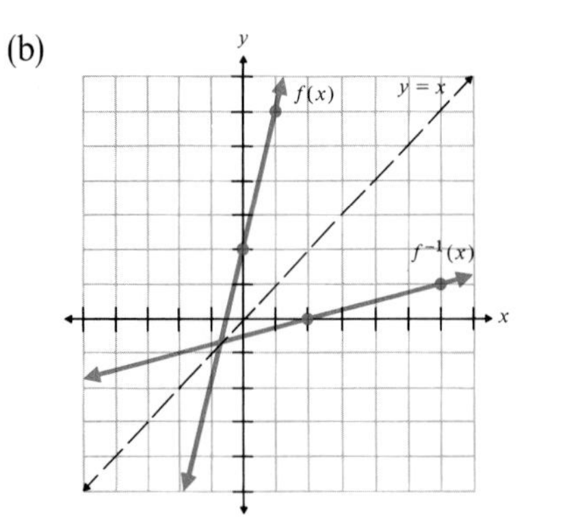

Note the symmetry of $f(x)$ and $f^{-1}(x)$ about the line $y = x$. ■

EXAMPLE 3 (a) Find the inverse function of $y = f(x) = \dfrac{-3x - 2}{4}$.

(b) On the same set of axes, graph $f(x)$ and $f^{-1}(x)$.

Solution: (a) $y = \dfrac{-3x - 2}{4}$.

Interchange x and y; then solve for y.

$$x = \frac{-3y - 2}{4}$$

$$4x = -3y - 2$$

$$4x + 2 = -3y$$

$$\frac{4x + 2}{-3} = y$$

$$\text{or} \quad y = \frac{-4x - 2}{3}$$

Thus $f^{-1}(x) = \dfrac{-4x - 2}{3}$.

(b)

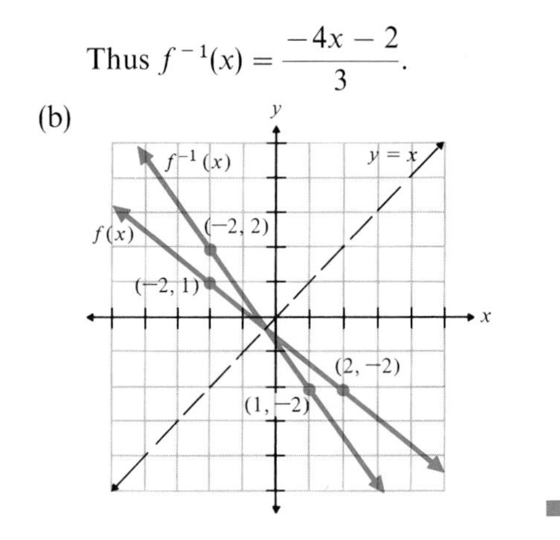

■

Exercise Set 11.1

In Exercises 1 through 10, determine whether each function is a one-to-one function.

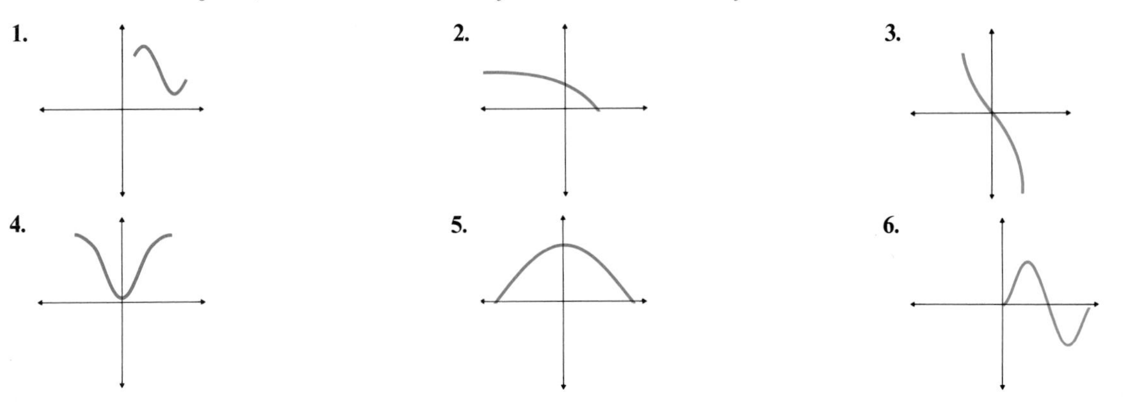

7. $\{(-2, 4), (3, -7), (5, 3), (-6, 0)\}$

8. $\{(-4, 2), (2, 3), (4, 1), (0, 4)\}$

9. $\{(-4, 2), (5, 3), (0, 2), (3, 7)\}$

10. $\{(-4, 5), (1, 4), (-3, 5), (4, 2)\}$

Give the domain and range of $f(x)$ and $f^{-1}(x)$.

11. Function: $\{(4, 0), (9, 3), (2, 7), (-1, 6), (-2, 4)\}$

12. Function: $\left\{(-2, -3), (-4, 0), (5, 3), (6, 2), \left(2, \frac{1}{2}\right)\right\}$

13. Function: $\{(1.7, 3), (-2.9, 4), (5.7, -3.4), (0, 9.76)\}$

14. Function: $\left\{(2.3, -2), \left(0, \frac{3}{8}\right), \left(-\frac{5}{3}, -\frac{1}{2}\right), (\sqrt{3}, 4)\right\}$

Find $f^{-1}(x)$ and graph $f(x)$ and $f^{-1}(x)$ on the same set of axes.

15. $y = f(x) = 2x + 8$

16. $y = f(x) = -3x + 6$

17. $y = f(x) = -3x - 10$

18. $y = f(x) = \frac{1}{2}x + 3$

19. $y = f(x) = 2x - \frac{3}{5}$

20. $y = f(x) = \frac{x + 3}{6}$

21. $y = f(x) = 6 - 3x$

22. $y = f(x) = \frac{2 - 5x}{3}$

23. $y = f(x) = \frac{6 + 4x}{3}$

24. $y = f(x) = \frac{-3}{5}x + \frac{1}{2}$

25. $y = f(x) = -\frac{2}{3} + \frac{5}{8}x$

26. $y = f(x) = -\frac{9}{4}x + 2$

JUST FOR FUN

Find $f^{-1}(x)$ and graph $f(x)$ and $f^{-1}(x)$ on the same set of axes.

Use the fact that the domain of $f(x)$ is the range of $f^{-1}(x)$ and the range of $f(x)$ is the domain of $f^{-1}(x)$ to graph $f^{-1}(x)$.

1. $y = \sqrt{x^2 - 9},\ x \geq 3$

2. $y = \sqrt{x^2 - 9},\ x \leq -3$

11.2 Exponential and Logarithmic Functions

1. *Graph exponential functions.*
2. *Evaluate exponential equations.*
3. *Change an exponential equation to logarithmic form.*
4. *Graph logarithmic functions.*
5. *Graph inverse functions of the form $y = a^x$ and $y = \log_a x$ on the same set of axes.*
6. *Solve logarithmic equations.*

1 An **exponential function** is a function of the form $y = f(x) = a^x$, where a is a positive real number not equal to 1. Examples of exponential functions are:

$$y = 2^x, \qquad y = 5^x, \qquad y = \left(\frac{1}{2}\right)^x$$

Exponential Function

$$f(x) = a^x$$

for any real number $a > 0$ and $a \neq 1$ is an exponential function.

Exponential functions can be graphed by selecting values for x, finding the corresponding values of y [or $f(x)$], and plotting the points.

EXAMPLE 1 Graph the exponential function $y = 2^x$. State the domain and range of the function.

Solution: First construct a table of values.

x	-3	-2	-1	0	1	2	3	4
y	$\frac{1}{8}$	$\frac{1}{4}$	$\frac{1}{2}$	1	2	4	8	16

Now plot these points and connect them with a smooth curve (Fig. 11.5).

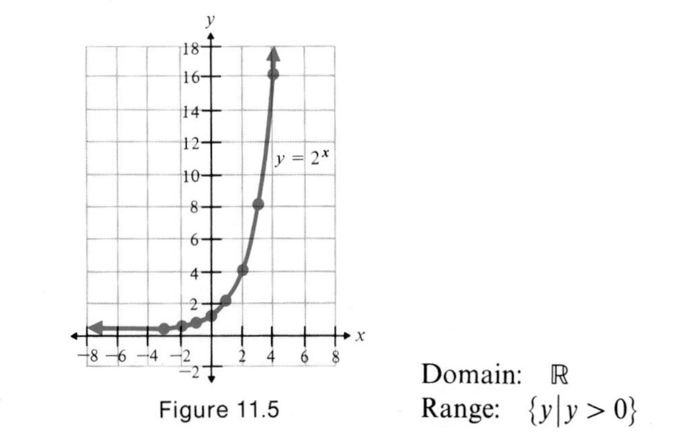

Figure 11.5

Domain: $\mathbb{R}$
Range: $\{y | y > 0\}$

The domain of this function is all real numbers, $\mathbb{R}$. The range is the set of values greater than 0. Why can y never equal 0? ■

EXAMPLE 2 Graph $y = \left(\frac{1}{2}\right)^x$. State the domain and range of the function.

Solution: We will construct a table of values and plot the curve (Fig. 11.6).

x	-3	-2	-1	0	1	2	3	4
y	8	4	2	1	$\frac{1}{2}$	$\frac{1}{4}$	$\frac{1}{8}$	$\frac{1}{16}$

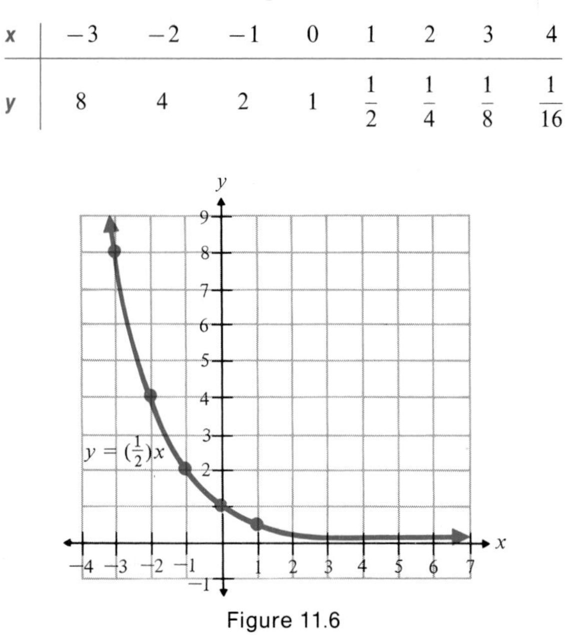

Figure 11.6

The domain is the set of real numbers. The range is $\{y|y > 0\}$. ■

Note that the graphs in Figures 11.5 and 11.6 are both one-to-one functions. Both graphs appear to be horizontal along the x axis, but they really are not. In Fig. 11.5 when $x = -2$, $y = \frac{1}{4}$, and when $x = -3$, $y = \frac{1}{8}$, and so on. As the values of x decrease, the values of y get closer to 0. When $x = -10$, $y = \frac{1}{1024}$ or about 0.001.

Exponential functions of the form $y = a^x$ will have a shape similar to that in Fig. 11.5 when $a > 1$ and a shape similar to Fig. 11.6 when $0 < a < 1$. Note that $y = 1^x$ is not a one-to-one function, so we exclude it from our discussion of exponential functions.

2 An **exponential equation** is one that has a variable as an exponent. Exponential equations are often used in applications describing growth and decay of some quantities. Example 3 illustrates the use of an exponential equation in the study of heredity.

EXAMPLE 3 The number of gametes, g, in a certain species of plant is determined by the equation $g = 2^n$, where n is the number of cells in the species. Determine the number of gametes in a species that has 12 cells.

Solution: By evaluating 2^{12} on a calculator we can determine that the species with 12 cells has 4096 gametes. ■

EXAMPLE 4 The formula $A = p(1 + r)^n$ is called the **compound interest formula.** When interest is compounded periodically (yearly, monthly, daily, etc.), this formula can be used to find the amount, A, in the account after n periods. In the formula, p is the principal, or the original amount invested, r is the interest rate per compounding period, and n is the number of compounding periods. For example, suppose that \$10,000 is invested at 8% interest compounded quarterly for 6 years. Then $P = 10{,}000$; the interest rate per period, r, is $\frac{8\%}{4}$ or 2%; and the number of periods, n, is $6 \cdot 4$ or 24. Find the amount in the account after 6 years for these given values.

Solution:

$$\begin{aligned} A &= p(1 + r)^n \\ &= 10{,}000(1 + .02)^{24} \\ &= 10{,}000(1.02)^{24} \\ &= 10{,}000(1.61) \quad \text{from a calculator (to the nearest hundredth).} \\ &= 16{,}100 \end{aligned}$$

Therefore, the original \$10,000 investment has grown to \$16,100. ■

EXAMPLE 5 Carbon 14 dating is used by scientists to find the age of fossils and other items. The formula used in carbon dating is:

$$A = A_0 \cdot 2^{\frac{-t}{5600}}$$

where A_0 represents the original amount of carbon 14 present and A represents the amount of carbon 14 present after t years. If 500 grams of carbon 14 are present originally, how many grams will remain after 2000 years?

Solution:

$$\begin{aligned} A &= A_0 \cdot 2^{\frac{-t}{5600}} \\ &= 500(2)^{\frac{-2000}{5600}} \\ &= 500(2)^{-0.36} \\ &= 500(0.78) \\ &= 390 \text{ grams} \end{aligned}$$ ■

3 Consider the exponential function $y = 2^x$. To find the inverse function we interchange the x and y and solve the equation for y. Interchanging x and y gives the equation $x = 2^y$, which is the inverse of $y = 2^x$. But at this time we have no way of solving the equation $x = 2^y$ for y. To solve this equation for y we introduce a new definition.

Definition of Logarithm

For all positive numbers a, where $a \neq 1$,

$$\mathbf{y = \log_a x \quad \text{means} \quad x = a^y}$$

In the expression $y = \log_a x$, the word "log" is an abbreviation for the word "logarithm." $y = \log_a x$ is read "y is the logarithm of x to the base a." The letter y represents the logarithm, the letter a represents the base, and the letter x represents the number.

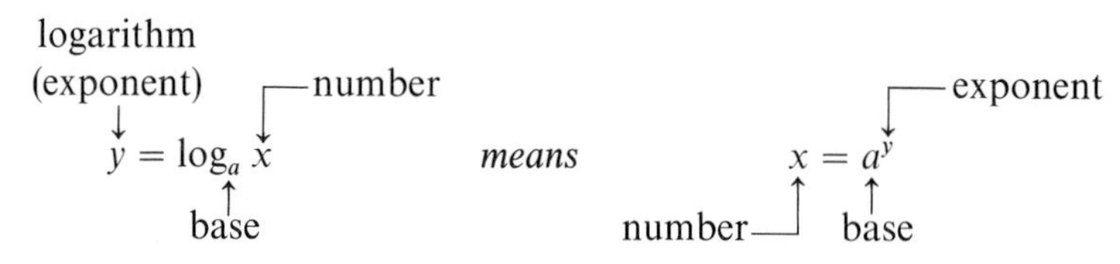

In words, the logarithm of the number x to the base a is the exponent to which the base a must be raised to obtain the number x. For example,

$$2 = \log_{10} 100 \qquad \text{means} \qquad 100 = 10^2$$

The logarithm is 2, the base is 10, and the number is 100. The logarithm, 2, is the exponent to which the base, 10, must be raised to give the number, 100. Note $10^2 = 100$.

Following are some examples of how an exponential expression can be converted to a logarithmic expression.

Exponential Form	*Logarithmic Form*		
$10^0 = 1$	$0 = \log_{10} 1$	or	$\log_{10} 1 = 0$
$4^2 = 16$	$2 = \log_4 16$	or	$\log_4 16 = 2$
$\left(\frac{1}{2}\right)^5 = \frac{1}{32}$	$5 = \log_{1/2}\left(\frac{1}{32}\right)$	or	$\log_{1/2}\left(\frac{1}{32}\right) = 5$
$5^{-2} = \frac{1}{25}$	$-2 = \log_5\left(\frac{1}{25}\right)$	or	$\log_5\left(\frac{1}{25}\right) = -2$

Now let us do a few examples involving conversion from exponential to logarithmic form, and vice versa.

EXAMPLE 6 Write each of the following in logarithmic form.

(a) $3^4 = 81$ (b) $\left(\frac{1}{5}\right)^3 = \frac{1}{125}$ (c) $2^{-4} = \frac{1}{16}$

Solution: (a) $\log_3 81 = 4$ (b) $\log_{1/5}\left(\frac{1}{125}\right) = 3$ (c) $\log_2\left(\frac{1}{16}\right) = -4$ ■

EXAMPLE 7 Write each of the following logarithms in exponential form.

(a) $\log_6 36 = 2$ (b) $\log_3 9 = 2$ (c) $\log_{1/3}\left(\frac{1}{81}\right) = 4$

Solution: (a) $6^2 = 36$ (b) $3^2 = 9$ (c) $\left(\frac{1}{3}\right)^4 = \frac{1}{81}$ ■

EXAMPLE 8 Write each logarithm in exponential form; then find the missing value.

(a) $y = \log_5 25$ (b) $2 = \log_a 16$ (c) $3 = \log_{1/2} x$

Solution: (a) $5^y = 25$

Since $5^2 = 25$, $y = 2$.

(b) $a^2 = 16$,

Since $4^2 = 16$, $a = 4$.

(c) $\left(\frac{1}{2}\right)^3 = x$

Since $\left(\frac{1}{2}\right)^3 = \frac{1}{8}$, $x = \frac{1}{8}$. ■

4 Now that we know how to convert from exponential form to logarithmic form and vice versa, we will graph logarithmic functions. To graph a logarithmic function, change it to exponential form. This procedure is illustrated in Examples 9 and 10.

EXAMPLE 9 Graph $y = \log_2 x$. State the domain and range of the function.

Solution: $y = \log_2 x$ means $x = 2^y$.

Using $x = 2^y$ construct a table of values. The table will be easier to develop by selecting values for y and finding the corresponding values for x.

x	$\frac{1}{8}$	$\frac{1}{4}$	$\frac{1}{2}$	1	2	3	8	16
y	−3	−2	−1	0	1	2	3	4

Now draw the graph (Fig. 11.7). The domain, the x values, is $\{x|x > 0\}$. The range, the y values, is all real numbers, $\mathbb{R}$.

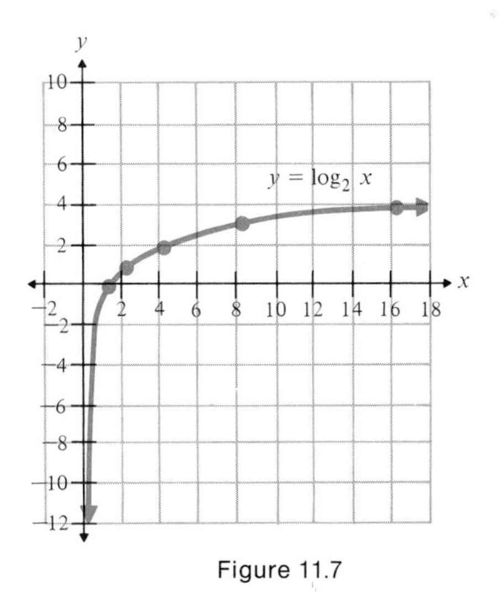

Figure 11.7 ■

EXAMPLE 10 Graph $y = \log_{1/2} x$. State the domain and the range of the function.

Solution: $y = \log_{1/2} x$ means $x = (\frac{1}{2})^y$. Construct a table of values, by selecting values for y and finding the corresponding values of x.

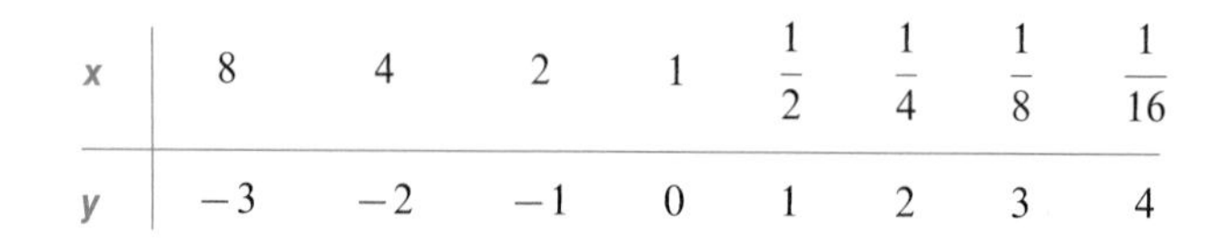

x	8	4	2	1	$\frac{1}{2}$	$\frac{1}{4}$	$\frac{1}{8}$	$\frac{1}{16}$
y	-3	-2	-1	0	1	2	3	4

The graph is illustrated in Fig. 11.8. The domain is $\{x | x > 0\}$. The range is all real numbers, $\mathbb{R}$.

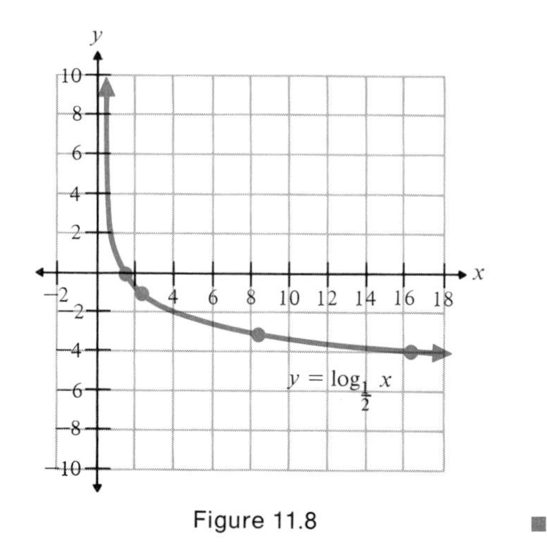

Figure 11.8

■

If we study the domains in Examples 9 and 10 we see that both $y = \log_2 x$ and $y = \log_{1/2} x$ have domains $\{x | x > 0\}$. In fact, **for any expression of the form $y = \log_a x$, the domain will be $\{x | x > 0\}$.** Also note that the graphs in Examples 9 and 10 are both one-to-one functions.

5 Earlier we stated that functions of the form $y = 2^x$ and $x = 2^y$ are inverse functions. Since $x = 2^y$ means $y = \log_2 x$, the functions $y = 2^x$ and $y = \log_2 x$ are inverse functions. In fact, for any $a > 0$, $a \neq 1$, the functions $y = a^x$ and $y = \log_a x$ will be inverse functions.

The graphs of $y = 2^x$ and $y = \log_2 x$ are illustrated in Fig. 11.9. Note that they are symmetric with respect to the line $y = x$. Since these graphs are inverses of each other, the symmetry is expected.

The graphs of $y = (\frac{1}{2})^x$ and $y = \log_{1/2} x$ are illustrated in Fig. 11.10. They are also inverses of each other and are symmetric with respect to the line $y = x$.

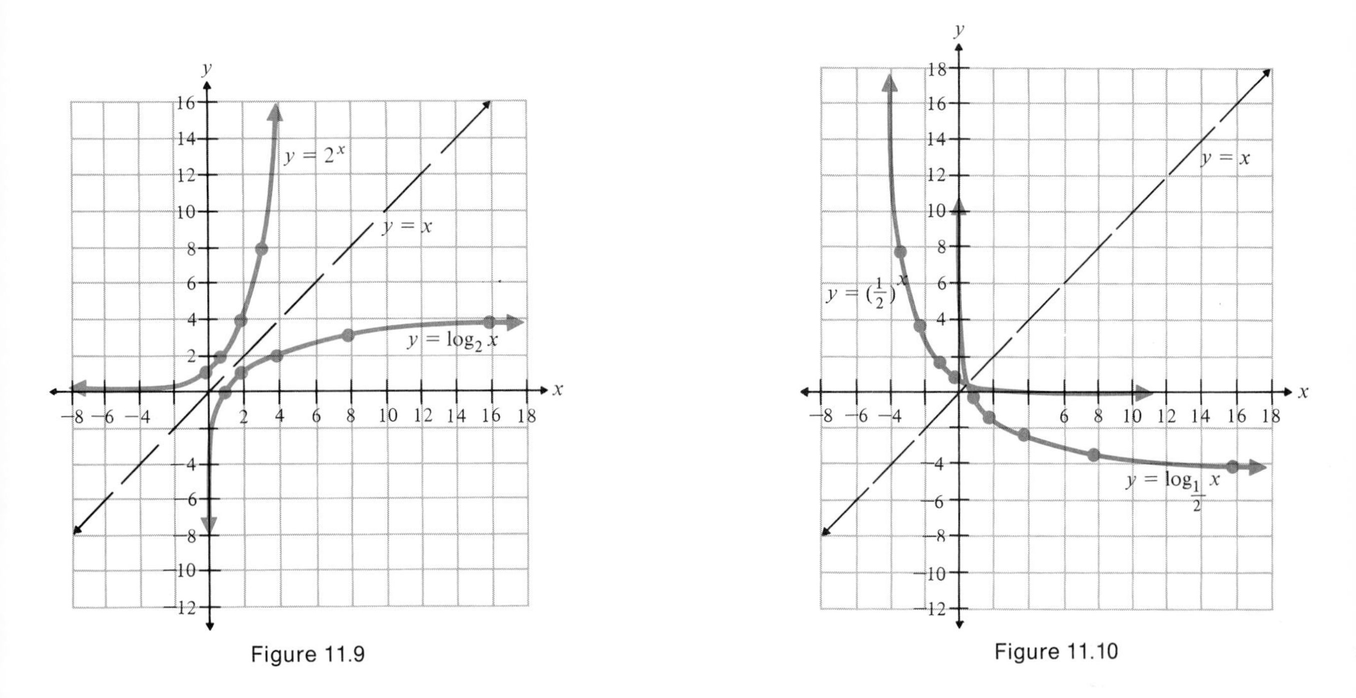

Figure 11.9

Figure 11.10

6 When finding the logarithm of an expression, the expression is called the **argument** of the logarithm. For example, in log 3 the 3 is the argument, and in $\log(2x + 4)$, the $2x + 4$ is the argument. When the argument contains a variable, we assume that the argument represents a number that is greater than 0. *Remember only the logarithms of positive numbers exist.*

A **logarithmic equation** is one in which a variable appears in the argument of some logarithm. Exponential and logarithmic equations are discussed in more detail in Section 11.4. One application of logarithms is to measure the intensity of earthquakes. Example 11 illustrates this use of logarithms.

EXAMPLE 11 The magnitude of an earthquake on the Richter Scale, R, is given by the formula

$$R = \log_{10} I$$

The I in the formula represents the number of times greater (or more intense) the earthquake is than the smallest measurable activity that can be measured on the seismograph.

(a) If an earthquake measures 4 on the Richter Scale, how many times more intense is it than the smallest measurable activity?

(b) How many times more intense is an earthquake that measures 5 than an earthquake that measures 4?

Solution: (a)

$$R = \log_{10} I$$
$$4 = \log_{10} I$$

or

$$10^4 = I$$
$$10{,}000 = I$$

Therefore, an earthquake that measures 4 is 10,000 times greater than the smallest measurable activity.

(b)

$$5 = \log_{10} I$$
$$10^5 = I$$
$$100{,}000 = I$$

Since (10,000)(10) = 100,000, an earthquake that measures 5 is 10 times more intense than an earthquake that measures 4. ■

Exercise Set 11.2

Graph the exponential function.

1. $y = 2^x$

2. $y = 3^x$

3. $y = \left(\frac{1}{2}\right)^x$

4. $y = \left(\frac{1}{3}\right)^x$

5. $y = 4^x$

6. $y = 5^x$

Graph the logarithmic function.

7. $y = \log_2 x$

8. $y = \log_3 x$

9. $y = \log_{1/2} x$

10. $y = \log_{1/3} x$

11. $y = \log_5 x$

12. $y = \log_{1/4} x$

Graph each pair of functions on the same set of axes.

13. $y = 2^x$, $y = \log_{1/2} x$

14. $y = \left(\frac{1}{2}\right)^x$, $y = \log_2 x$

15. $y = 2^x$, $y = \log_2 x$

16. $y = \left(\frac{1}{2}\right)^x$, $y = \log_{1/2} x$

Write each expression in logarithmic form.

17. $2^3 = 8$

18. $5^2 = 25$

19. $3^5 = 243$

20. $9^{1/2} = 3$

21. $8^{1/3} = 2$

22. $\left(\frac{1}{2}\right)^5 = \frac{1}{32}$

23. $\left(\frac{1}{4}\right)^2 = \frac{1}{16}$

24. $2^{-3} = \frac{1}{8}$

25. $5^{-2} = \frac{1}{25}$

26. $4^3 = 64$

27. $4^{-3} = \frac{1}{64}$

28. $(64)^{1/3} = 4$

29. $16^{-1/2} = \frac{1}{4}$

30. $36^{1/2} = 6$

31. $8^{-1/3} = \frac{1}{2}$

32. $81^{-1/4} = \frac{1}{3}$

Write each expression in exponential form.

33. $\log_2 8 = 3$

34. $\log_3 9 = 2$

35. $\log_4 64 = 3$

36. $\log_{1/3}\left(\frac{1}{9}\right) = 2$

37. $\log_{1/2}\left(\frac{1}{16}\right) = 4$

38. $\log_4\left(\frac{1}{16}\right) = -2$

39. $\log_5\left(\frac{1}{125}\right) = -3$

40. $\log_9 3 = \frac{1}{2}$

41. $\log_{125} 5 = \frac{1}{3}$

42. $\log_8\left(\frac{1}{64}\right) = -2$

43. $\log_{27}\left(\frac{1}{3}\right) = -\frac{1}{3}$

44. $\log_{10} 100 = 2$

45. $\log_{10} 1000 = 3$

46. $\log_6 216 = 3$

Write each logarithm in exponential form; then find the missing value.

47. $y = \log_6 36$

48. $y = \log_3 27$

49. $3 = \log_2 x$

50. $5 = \log_3 x$

51. $y = \log_{64} 8$

52. $\frac{1}{2} = \log_{100} x$

53. $3 = \log_a 64$

54. $2 = \log_a 16$

55. $4 = \log_{1/2} x$

56. $3 = \log_{1/4} x$

57. $5 = \log_a 32$

58. $y = \log_2\left(\frac{1}{4}\right)$

59. $\frac{1}{3} = \log_8 x$

60. $-\frac{1}{4} = \log_{16} x$

61. Use the formula $g = 2^n$ to determine the number of gametes the specified plant has if it has 8 cells. See Example 3.

62. If José invests \$5000 at 10% interest compounded quarterly, find the amount after 4 years. Use $A = p(1 + r)^n$. See Example 4.

63. If Marsha invests \$8000 at 12% interest compounded quarterly, find the amount after 5 years.

64. If 12 grams of carbon 14 are originally present in a certain animal bone, how much will remain at the end of 1000 years? Use $A = A_0 \cdot 2^{-t/5600}$. See Example 5.

65. If 60 grams of carbon 14 are originally present in the fossil Jonas found at the archaeological site, how much will remain after 10,000 years?

66. The expected future population of Ackworth, which presently has 2000 residents, can be approximated by the formula $y = 2000\,(1.2)^{.1x}$, where x is the number of years in the future. Find the expected population of the town in: **(a)** 10 years **(b)** 100 years

67. The amount of a radioactive substance present, in grams, at time t is given by the formula $y = 80(2)^{-0.4t}$, where t is measured in years. Find the number of grams present at: **(a)** 10 years **(b)** 100 years

68. The number of a certain type of bacteria present in a culture is determined by the equation $y = 5000(3)^x$, where x is the number of days the culture has been growing. Find the number of bacteria in: **(a)** 5 days **(b)** 7 days

69. If an earthquake has a magnitude of 7 on the Richter Scale, how many times greater is the earthquake than the smallest measurable activity? Use $R = \log_{10} I$. See Example 11.

70. How many times greater is an earthquake that measures 3 on the Richter Scale than an earthquake that measures 1?

71. For the logarithmic function $y = \log_a x$,
(a) What are the restrictions on a?
(b) What is the domain of the function?
(c) What is the range of the function?

72. For the logarithmic function $y = \log_a(x - 3)$, what must be true about x? Explain your answer.

11.3 Properties of Logarithms

1. *Use the product rule for logarithms.*
2. *Use the quotient rule for logarithms.*
3. *Use the power rule for logarithms.*

1 To be able to do calculations using logarithms, you must have an understanding of their properties. We use these properties in Section 11.4 when we solve exponential and logarithmic equations. The first property we discuss is the product rule for logarithms.

Product Rule For Logarithms

For positive real numbers x, y, and a, $a \neq 1$,

$$\log_a xy = \log_a x + \log_a y \qquad \text{Property 1}$$

To prove this property, let $\log_a x = m$ and $\log_a y = n$. Now write each logarithm in exponential form.

$$\log_a x = m \quad \text{means} \quad a^m = x$$
$$\log_a y = n \quad \text{means} \quad a^n = y$$

By substitution and using the rules of exponents, we see that

$$xy = a^m \cdot a^n = a^{m+n}$$

We can now convert $xy = a^{m+n}$ back to logarithmic form.

$$xy = a^{m+n} \quad \text{means} \quad \log_a xy = m + n$$

Finally substituting $\log_a x$ for m, and $\log_a y$ for n, we obtain

$$\log_a xy = \log_a x + \log_a y$$

which is property 1.

Examples of Property 1

$$\log_3 5 \cdot 7 = \log_3 5 + \log_3 7$$
$$\log_4 3x = \log_4 3 + \log_4 x$$
$$\log_8 x^2 = \log_8 (x \cdot x) = \log_8 x + \log_8 x$$

2 Now we give the Quotient Rule for logarithms, which we refer to as property 2.

Quotient Rule for Logarithms

For positive real numbers x, y, and a, $a \neq 1$,

$$\log_a \frac{x}{y} = \log_a x - \log_a y \qquad \text{Property 2}$$

Property 2 and other properties we will discuss can be proved in a manner similar to property 1.

Examples of Property 2

$$\log_3 \frac{12}{4} = \log_3 12 - \log_3 4$$

$$\log_6 \frac{x}{3} = \log_6 x - \log_6 3$$

$$\log_5 \frac{x}{x+2} = \log_5 x - \log_5 (x+2)$$

3 The last property we discuss in this section is the power rule for logarithms.

Power Rule for Logarithms

If x and a are positive real numbers, $a \neq 1$, and n is any real number, then

$$\log_a x^n = n \log_a x \qquad \text{Property 3}$$

Examples of Property 3

$$\log_2 4^3 = 3 \log_2 4$$

$$\log_{10} x^2 = 2 \log_{10} x$$

$$\log_5 \sqrt{12} = \log_5 (12)^{1/2} = \frac{1}{2} \log_5 12$$

$$\log_8 \sqrt[5]{x+3} = \log_8 (x+3)^{1/5} = \frac{1}{5} \log_8 (x+3)$$

EXAMPLE 1 Use properties 1 through 3 to expand each of the following.

(a) $\log_8 \frac{27}{43}$

(b) $\log_4 (64)(180)$

(c) $\log_{10} (32)^{1/5}$

Solution: (a) $\log_8 \frac{27}{43} = \log_8 27 - \log_8 43$

(b) $\log_4 (64)(180) = \log_4 64 + \log_4 180$

(c) $\log_{10} (32)^{1/5} = \frac{1}{5} \log_{10} 32$ ■

Quite often we will have to use two or more of these properties in the same problem.

EXAMPLE 2 Expand each of the following.

(a) $\log_{10} 4(x+2)^3$ (b) $\log_5 \dfrac{(4-x)^2}{3}$

(c) $\log_5 \left(\dfrac{4-x}{3}\right)^2$ (d) $\log_5 \dfrac{[x(x+4)]^3}{2}$

Solution: (a)
$$\log_{10} 4(x+2)^3 = \log_{10} 4 + \log_{10} (x+2)^3$$
$$= \log_{10} 4 + 3 \log_{10} (x+2)$$

(b)
$$\log_5 \frac{(4-x)^2}{3} = \log_5 (4-x)^2 - \log_5 3$$
$$= 2 \log_5 (4-x) - \log_5 3$$

(c)
$$\log_5 \left(\frac{4-x}{3}\right)^2 = 2 \log_5 \left(\frac{4-x}{3}\right)$$
$$= 2[\log_5 (4-x) - \log_5 3]$$
$$= 2 \log_5 (4-x) - 2 \log_5 3$$

(d)
$$\log_5 \frac{[x(x+4)]^3}{2} = \log_5 [x(x+4)]^3 - \log_5 2$$
$$= 3 \log_5 x(x+4) - \log_5 2$$
$$= 3[\log_5 x + \log_5 (x+4)] - \log_5 2$$
$$= 3 \log_5 x + 3 \log_5 (x+4) - \log_5 2$$ ■

Note that property 1 can be expanded to evaluate the product of 3 or more quantities. Thus, for example, $\log_5 xyz = \log_5 x + \log_5 y + \log_5 z$.

EXAMPLE 3 Write each of the following as the logarithm of a single expression.

(a) $3 \log_8 (x+2) - \log_8 x$

(b) $[\log_7 (x+1) + 2 \log_7 (x+4)] - 3 \log_7 (x-5)$

Solution: (a)
$$3 \log_8 (x+2) - \log_8 x = \log_8 (x+2)^3 - \log_8 x$$
$$= \log_8 \frac{(x+2)^3}{x}$$

(b)
$$[\log_7 (x+1) + 2 \log_7 (x+4)] - 3 \log_7 (x-5)$$
$$= [\log_7 (x+1) + \log_7 (x+4)^2] - \log_7 (x-5)^3$$
$$= \log_7 (x+1)(x+4)^2 - \log_7 (x-5)^3$$
$$= \log_7 \frac{(x+1)(x+4)^2}{(x-5)^3}$$ ■

COMMON STUDENT ERROR

The correct rules are:

$$\log(A \cdot B) = \log A + \log B$$

$$\log \frac{A}{B} = \log A - \log B$$

A common mistake made by students is to use the following *incorrect procedures.*

Wrong	*Wrong*
$\log(A + B) = \log A + \log B$	$\frac{\log A}{\log B} = \log A - \log B$
$\log(A - B) = \log A - \log B$	$\frac{\log A}{\log B} = \log \frac{A}{B}$
$\log(A \cdot B) = (\log A)(\log B)$	

For example,

$$\log(x + 2) \neq \log x + \log 2 \qquad \frac{\log x}{\log 2} \neq \log x - \log 2$$

$$\log(3x) \neq (\log 3)(\log x) \qquad \frac{\log 10}{\log x} \neq \log \frac{10}{x}$$

Exercise Set 11.3

Use properties 1 through 3 to expand each of the following.

1. $\log_3 7 \cdot 12$
2. $\log_5 8 \cdot 29$
3. $\log_8 7(x + 3)$
4. $\log_{10} x(x - 2)$
5. $\log_4 \frac{15}{7}$
6. $\log_9 \frac{x}{12}$
7. $\log_{10} \left(\frac{\sqrt{x}}{x - 3}\right)$
8. $\log_5 3^{12}$
9. $\log_8 x^4$
10. $\log_5 (x + 4)^3$
11. $\log_{10} 3(8^2)$
12. $\log_8 x^2(x - 2)$
13. $\log_4 \left(\frac{x^5}{x + 4}\right)$
14. $\log_{10} (x - 3)^2 x^3$
15. $\log_{10} \left[\frac{x^4}{(x + 2)^3}\right]$
16. $\log_7 x^2(x - 2)$
17. $\log_8 \left[\frac{x(x - 6)}{x^3}\right]$
18. $\log_{10} \left(\frac{x}{6}\right)^2$
19. $\log_{10} \left(\frac{2x}{3}\right)$
20. $\log_5 \left[\frac{(3x)^2}{x + 5}\right]$

Write each of the following as a logarithm of a single expression.

21. $2 \log_{10} x - \log_{10} (x - 2)$
22. $3 \log_8 x + 2 \log_8 (x + 1)$
23. $2(\log_5 x - \log_5 4)$
24. $2[\log_6 (x - 1) - \log_6 3]$
25. $[\log_{10} x + \log_{10} (x - 4)] - \log (x + 1)$
26. $[2 \log_5 x + \log_5 (x - 4)] + \log_5 (x - 2)$
27. $\frac{1}{2} [\log_7 (x - 2) - \log_7 x]$
28. $[5 \log_7 (x + 3) + 2 \log_7 (x - 1)] - \frac{1}{2} \log_7 x$
29. $(2 \log_9 5 - 4 \log_9 6) + \log_9 3$
30. $5 \log_6 (x + 3) - [2 \log_6 (x - 4) + 3 \log_6 x]$
31. $4 \log_6 3 - [2 \log_6 (x + 3) + 4 \log_6 x]$
32. $2 \log_7 (x - 6) + 3 \log_7 (x + 3) - [5 \log_7 2 + 3 \log_7 (x - 2)]$

11.4 Common Logarithms

1. *Identify common logarithms.*
2. *Find common logarithms of powers of 10.*
3. *Find common logarithms.*
4. *Find antilogarithms.*
5. *Perform calculations using logarithms.*

1 The properties discussed in Section 11.3 can be used with any valid base (greater than 0 and not equal to 1). However, since we are used to working in base 10, we will often use the base 10 when computing with logarithms. Base 10 logarithms are called **common logarithms.** When we are working with common logarithms it is not necessary to list the base. Thus $\log x$ is the same as $\log_{10} x$.

The properties of logarithms written as common logarithms follow, for positive real numbers x and y, and any real number, n.

1. $\log xy = \log x + \log y$

2. $\log \frac{x}{y} = \log x - \log y$

3. $\log x^n = n \log x$

2 In Chapter 5 we learned that

$$10^{-2} = \frac{1}{10^2} = \frac{1}{100} = 0.01$$
$$10^{-1} = \frac{1}{10^1} = \frac{1}{10} = 0.1$$
$$10^0 = 1$$
$$10^1 = 10$$
$$10^2 = 100$$

We could enlarge this chart by using exponents less than -2 or greater than 2.

Note that the number 1 can be expressed as 10^0 and the number 10 can be expressed as 10^1. Since, for example, the number 5 is between 1 and 10, it must be a number between 10^0 and 10^1.

$$1 < 5 < 10$$
$$10^0 < 5 < 10^1$$

The number 5 can be expressed as the base 10 raised to an exponent between 0 and 1. The number 5 is approximately equal to $10^{0.69897}$. The common logarithm of 5 is approximately 0.69897.

$$\log 5 \approx 0.69897$$

The common logarithm of a number is the exponent to which the base 10 is raised to obtain the number.

Now consider the number 50.

$$10 < 50 < 100$$
$$10^1 < 50 < 10^2$$

The number 50 can be expressed as the base 10 raised to an exponent between 1 and 2. The number $50 \approx 10^{1.69897}$; thus $\log 50 \approx 1.69897$.

We will shortly be able to determine that $\log 500 \approx 2.69897$, $\log 5000 \approx 3.69897$, and so on.

To compute successfully with logarithms, you must understand how to use **scientific notation.** If you have forgotten how to change to and from scientific notation form, review Section 5.1 now.

Property 4 is useful in finding common logarithms.

For any real number n

$$\log 10^n = n \qquad \text{Property 4}$$

To help explain property 4, consider log 10. To what is log 10 equal? Let $\log 10 = b$; then

$$\log 10 = b$$
$$10^b = 10$$
$$10^b = (10)^1$$
$$b = 1$$

Therefore, $\log 10 = 1$.

Now consider $\log 10^2$, $\log 10^5$, and $\log 10^{-3}$.

$$\log 10^2 = 2 \log 10 \qquad \log 10^5 = 5 \log 10 \qquad \log 10^{-3} = -3 \log 10$$
$$= 2(1) \qquad = 5(1) \qquad = -3(1)$$
$$= 2 \qquad = 5 \qquad = -3$$

Other Examples

$$\log 10^0 = 0 \qquad \log 10^{-9} = -9$$
$$\log 10^{12} = 12 \qquad \log 10^{0.5} = 0.5$$

3 To find logarithms of numbers we can use a calculator. The procedure to follow to evaluate logarithms on a calculator is given at the end of this section. If a calculator is not available, we can find logarithms of numbers with the use of Appendix D. Appendix D lists the logarithms of numbers from 1.00 to 9.99 inclusive. Almost all logarithms are irrational numbers and can only be approximated when written in decimal form. The logarithms given in Appendix D have been rounded to four decimal places. Although the logarithms are only close approximations, it is customary to write logarithms using an equal sign. If greater accuracy is desired, more exact tables are available.

To explain how to find logarithms using Appendix D, consider log 6.37. Since we are finding the logarithm of a number between 1.00 and 9.99, we can find the log directly from Appendix D. A portion of the appendix is illustrated in Table 1.

Table 1

	0	1	2	3	4	5	6	7	8	9
6.0	.7782	.7789	.7797	.7803	.7810	.7818	.7825	.7832	.7839	.7846
6.1	.7853	.7860	.7868	.7875	.7882	.7889	.7896	.7903	.7910	.7917
6.2	.7924	.7931	.7938	.7945	.7952	.7959	.7966	.7973	.7980	.7987
6.3	.7993	.8000	.8007	.8014	.8021	.8028	.8035	.8041	.8048	.8055
6.4	.8062	.8069	.8075	.8082	.8089	.8096	.8102	.8109	.8116	.8122

To find log 6.37, find the first two digits 6.3, in the left-hand column. Next find the digit in the hundredths position, 7, in the top row. If a horizontal line is drawn from the 6.3 and a vertical line is drawn from the 7, the two lines meet at .8041. This number is circled in the table. Thus log 6.37 = 0.8041. Note that $10^{.8041} \approx 6.37$.

EXAMPLE 1 Find the following logarithms using Appendix D.

(a) log 7.13 (b) log 2.09

Solution: (a) log 7.13 = 0.8531
(b) log 2.09 = 0.3201 ■

To find logarithms of numbers not between 1.00 and 9.99, we first write the number using scientific notation. Consider log 637.

$$\log 637 = \log (6.37 \cdot 10^2)$$

By property 1 we can write

$$\log 637 = \log 6.37 + \log 10^2$$

By property 4 we know that $\log 10^2 = 2$. Thus

$$\begin{aligned} \log 637 &= \log 6.37 + 2 \\ &= 0.8041 + 2 \qquad \text{(from table)} \\ &= 2.8041 \end{aligned}$$

Thus log 637 = 2.8041. Note that $10^{2.8041} \approx 637$.

The logarithm 2.8014 consists of two parts: the integer part, called the **characteristic,** and the decimal part, called the **mantissa.**

$$\log 637 = \underbrace{2}_{\text{characteristic}}.\underbrace{8014}_{\text{mantissa}}$$

When finding the logarithm of a number, the characteristic of a logarithm is the exponent on the base 10 when the number is written in scientific notation. The characteristic will be the number of places the decimal is moved in order to write the number in scientific notation. (Remember that when the decimal is moved to the left, the exponent is positive; to the right, negative.)

Let us now find some other logarithms.

EXAMPLE 2 Find log 40,300.

Solution: First write the number using scientific notation.

$$40{,}300 = 4.03 \times 10^4$$
$$\log 40{,}300 = \log 4.03 + \log 10^4$$
$$= \underbrace{0.6053}_{\text{From Appendix D}} + \underbrace{4}_{\text{by property 4}}$$

From Appendix D, — by property 4

$$\log 40{,}300 = 4.6053$$ ■

EXAMPLE 3 Find log 8,390,000.

Solution:

$$8{,}390{,}000 = 8.39 \times 10^6$$
$$\log 8{,}390{,}000 = \log 8.39 + \log 10^6$$
$$= 0.9238 + 6$$
$$= 6.9238$$ ■

EXAMPLE 4 Find log 0.0538.

Solution:

$$0.0538 = 5.38 \times 10^{-2}$$
$$\log 0.0538 = \log 5.38 + \log 10^{-2}$$
$$= 0.7308 - 2$$

When the characteristic is a negative number, we generally do not subtract the characteristic from the mantissa because it would change the logarithm to a negative number. For example, if we subtracted 2 from 0.7308, we would get -1.2692. Often when the characteristic is negative, we write the logarithm in an equivalent form. For example, we can write

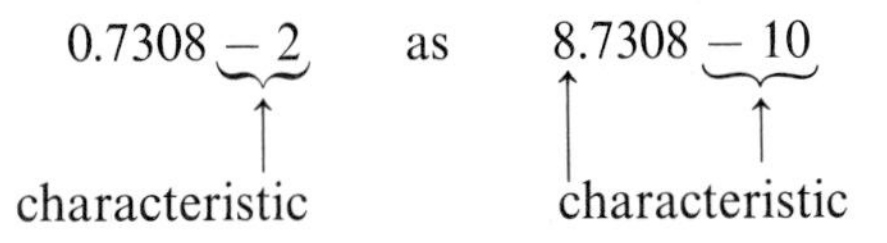

Note that the characteristic -2 is equal to the characteristic $8 - 10$. Thus $\log 0.0538 = 8.7308 - 10$. ■

EXAMPLE 5 Find log 0.000392.

Solution:

$$0.000392 = 3.92 \times 10^{-4}$$
$$\log 0.000392 = \log 3.92 + \log 10^{-4}$$
$$= 0.5933 - 4$$
$$= 6.5933 - 10$$

Note that $-4 = 6 - 10$. ■

HELPFUL HINT

The characteristic of a logarithm can be found by counting the number of places the decimal point must be moved to obtain a number greater than or equal to 1 and less than 10. If the decimal point is moved to the left, the characteristic is positive; to the right, negative.

Examples

log 62800—the characteristic is the 4 since the decimal point is moved 4 places to the left, 62800

log 0.00000418—the characteristic is -6 since the decimal point is moved 6 places to the right, 0.00000418

4 We also need to know how to find the **antilogarithm** (abbreviated antilog) of a number. If the logarithm of N is L, then N is the antilogarithm of L.

$$\text{If} \quad \log N = L \quad \text{then} \quad N = \text{antilog } L$$

In Example 2 we found log 40,300 = 4.6053, thus 40,300 = antilog 4.6053. In Example 4 we found log 0.0538 = 8.7308 − 10, thus 0.0538 = antilog 8.7308 − 10. To find the antilogarithm we reverse the procedure used in finding the logarithm. When finding an antilog we are converting a logarithm, or exponent, back into a number.

To explain how to find antilogs, consider the following problem:

$$\log N = 3.7619$$
$$N = \text{antilog } 3.7619$$

We must find the number whose logarithm is 3.7619. The logarithm consists of the characteristic, 3, and the mantissa, 0.7619. To find N, the antilog of 3.7619, first look up the mantissa 0.7619 in the body of Appendix D. The portion of the table containing 0.7619 is reproduced below.

	0	1	2	3	4	5	6	7	8	9
5.5								.7459	.7466	.7474
5.6								.7536	.7543	.7551
5.7								.7612	.7619	.7627
5.8								.7686	.7694	.7701
5.9								.7760	.7767	.7774

We see that the mantissa 0.7619 yields the number 5.78. Since the characteristic is 3, which represents 10^3, write

$$5.78 \times 10^3 = 5780$$

Thus antilog 3.7619 = 5780. The process above can be summarized as follows

$$\log N = 3.7619$$
$$N = \text{antilog } 3.7619$$
$$N = 5780$$

To find antilogarithm, look up the mantissa in the body of Appendix D. This will give a number between 1.00 and 9.99. Then move the decimal point to the left or right as indicated by the characteristic. When the characteristic is positive, the decimal point moves to the right; when the characteristic is negative, the decimal point moves to the left.

Examples

antilog 3.7619 = 5780.
move decimal three places to right
Thus antilog 3.7619 = 5780.

Thus antilog 8.7619 − 10 = 0.0578.
characteristic is −2; move decimal two places to left
Thus antilog 8.7619 − 10 = 0.0578.

EXAMPLE 6 Find antilog 4.9350.

Solution: Look up the mantissa 0.9350 in Appendix D. The number obtained from the borders is 8.61. Since the characteristic is 4, we move the decimal point four places to the right to obtain 86,100.

$$\text{antilog } 4.9350 = 86{,}100 \quad ■$$

EXAMPLE 7 $\log N = 6.5877 - 10$, find N.

Solution:

$$\log N = 6.5877 - 10$$
$$N = \text{antilog } 6.5877 - 10$$
$$N = 0.000387 \quad ■$$

☐ *Calculator Corner*

Only a few years ago problems involving logarithms were often done on a slide rule. Nowadays, problems involving logarithms are greatly simplified with the use of calculators. Calculators that can work with logarithms contain a [log] key.

To find the logarithm of a given number, enter the number, then press the [log] key. For example, to find the log 400, you press

400 [log]

After you press [log], the answer 2.60206 will be displayed. Thus log 400 = 2.60206.

To find log 0.0538, press

0.0538 [log]

After you press [log], the answer −1.2692177 will appear. Thus log 0.0538 = −1.2692177.

On the calculator, log 0.0538 has a value of approximately −1.2692. When we find log 0.0538 using the log table, we find that log 0.0538 has a value of approximately 0.7308 − 2 or 8.7308 − 10. The number −1.2692 is equivalent to both 0.7308 − 2 and 8.7308 − 10. For example, 0.7308 − 2.0000 = −1.2692.

Antilogs can also be found on the calculator. To find the antilog of a number, on certain calculators, use the combination of inverse and log keys. For example, to find antilog 3.7619, press

3.7619 [inv] [log]

The answer, approximately 5779.6295, will then be displayed. Thus antilog 3.7619 = 5779.6295.

To find antilog −1.2692, press

1.2692 [+/−] [inv] [log]

The answer, approximately 0.0538, will be displayed. Thus antilog −1.2692 = 0.0538.

5 Before the development of the calculator logarithms were often used to perform certain calculations. The following examples illustrate how certain arithmetic problems can be done using logarithms.

EXAMPLE 8 Divide $\dfrac{704}{0.0838}$ using logarithms.

Solution: Call the fraction N. Then take the logarithm of both sides of the equation.

$$N = \frac{704}{0.0838}$$

$$\log N = \log \frac{704}{0.0838}$$

$$\begin{aligned}\log N &= \log 704 - \log 0.0838 \qquad \text{(property 2)}\\ &= 2.8476 - (8.9232 - 10)\\ &= 2.8476 - 8.9232 + 10\\ &= 12.8476 - 8.9232 = 3.9244\end{aligned}$$

$$N = \text{antilog } 3.9244 \approx 8400$$

Thus $\dfrac{704}{0.0838} \approx 8400$. ∎

EXAMPLE 9 Evaluate $(4.08)^5$ using logarithms.

Solution:

$$N = (4.08)^5$$
$$\log N = \log (4.08)^5$$
$$\log N = 5 \log 4.08 \qquad \text{(property 3)}$$
$$\log N = 5(0.6107)$$
$$\log N = 3.0535$$
$$N = \text{antilog } 3.0535$$
$$N \approx 1130$$

Thus $(4.08)^5 \approx 1130$. ■

EXAMPLE 10 Evaluate $\sqrt[3]{0.574}$ using logarithms.

Solution:

$$N = \sqrt[3]{0.574}$$
$$\log N = \log \sqrt[3]{0.574}$$
$$= \log (0.574)^{1/3}$$
$$= \frac{1}{3} \log 0.574$$
$$= \frac{1}{3}(9.7589 - 10)$$

Since we wish the number being subtracted, the 10, to be divisible by 3, we write $9.7589 - 10$ as $29.7589 - 30$. Note that both numbers have a characteristic of -1.

$$\log N = \frac{1}{3}(29.7589 - 30)$$
$$\log N = 9.9196 - 10$$
$$N = \text{antilog } 9.9196 - 10$$
$$N \approx 0.831$$

Thus $\sqrt[3]{0.574} \approx 0.831$. ■

Exercise Set 11.4

Find the common logarithm of the number.

1. 870
2. 36
3. 8
4. 19,200
5. 1000
6. 0.00152
7. 0.0000857
8. 27,700
9. 100
10. 0.000835
11. 1.74
12. 3.75
13. 0.375
14. 0.0000375
15. 0.00872
16. 960
17. 102
18. 8.92
19. 0.00128
20. 73,700,000

Find the antilog of the logarithm.

21. 0.5416
22. 2.6464
23. 2.3201
24. 5.8149
25. 8.9415 – 10
26. 7.6618 – 10
27. 0.0000
28. 5.5922
29. 2.5011
30. 0.5694 – 5

Find the number N.

31. $\log N = 2.0000$
32. $\log N = 1.6730$
33. $\log N = 6.8960 - 10$
34. $\log N = 1.9330$
35. $\log N = 3.8202$
36. $\log N = 2.7404$
37. $\log N = 0.9400 - 2$
38. $\log N = 8.8531 - 10$
39. $\log N = 1.1903$
40. $\log N = 9.6314 - 10$
41. $\log N = 4.8537$
42. $\log N = 1.5159$
43. $\log N = 8.3909 - 10$
44. $\log N = 7.8082 - 10$

Evaluate each of the following using logarithms. (Note answers were found using Appendix D. If you obtain log values using a calculator your answers may differ.)

45. $(37)(109)$
46. $(4.62)(943)$
47. $(18.3)(0.962)$
48. $(0.162)(18{,}000)$
49. $(0.00314)(0.164)$
50. $\dfrac{986}{3.17}$
51. $\dfrac{14{,}200}{82.7}$
52. $\dfrac{0.0483}{0.000817}$
53. $\dfrac{786}{19{,}600}$
54. $(8)^4$
55. $(11.6)^5$
56. $\sqrt{97.6}$
57. $\sqrt[3]{85{,}900}$
58. $(0.0276)^3$
59. $(4)(8)^3$
60. $\dfrac{714}{\sqrt{37.2}}$
61. $\dfrac{(708)(0.00762)}{0.0752}$
62. $\dfrac{(13.2)(0.0074)}{83{,}400}$

JUST FOR FUN

Evaluate using logarithms

1. $\dfrac{\sqrt[3]{87.3}}{(9.02)(12.8)}$

2. $\dfrac{(0.763)^2\sqrt[3]{0.00183}}{76}$

11.5 Exponential and Logarithmic Equations

1 *Solve exponential and logarithmic equations.*

2 *Solve some practical problems using exponential and logarithmic equations.*

1 In Section 11.2 we introduced exponential and logarithmic equations. In this section we give more examples of their use and discuss further procedures for solving such equations.

To solve exponential and logarithmic equations, we often use Properties 5a through 5d.

To Solve Exponential and Logarithmic Equations

We may use these properties: Properties 5a–5d

a. If $x = y$, then $a^x = a^y$.
b. If $a^x = a^y$, then $x = y$.
c. If $x = y$, then $\log x = \log y \quad (x > 0, y > 0)$.
d. If $\log x = \log y$, then $x = y \quad (x > 0, y > 0)$.

EXAMPLE 1 Solve the equation $8^x = \frac{1}{2}$.

Solution:

$$8^x = \frac{1}{2}$$
$$(2^3)^x = \frac{1}{2}$$
$$(2^3)^x = 2^{-1}$$
$$2^{3x} = 2^{-1}$$

Using property 5b we can write

$$3x = -1$$
$$x = -\frac{1}{3} \quad ■$$

When both sides of the exponential equation cannot be written as a power of the same base, we often begin by taking the logarithm of both sides of the equation as illustrated in Example 2.

EXAMPLE 2 Solve the equation $5^n = 20$.

Solution: Take the logarithm of both sides of the equation and solve for the variable n.

$$\log 5^n = \log 20$$
$$n \log 5 = \log 20$$
$$n = \frac{\log 20}{\log 5}$$
$$= \frac{1.3010}{0.6990}$$
$$n \approx 1.861 \quad ■$$

Some logarithmic equations can be solved by expressing the equation in exponential form. **It is necessary to check logarithmic equations for extraneous roots.** When checking an answer if you obtain the logarithm of a nonpositive number, that answer is not a solution to the equation.

EXAMPLE 3 Solve the equation $\log_2 (x + 1)^3 = 4$.

Solution: Writing the equation in exponential form gives

$$(x + 1)^3 = 2^4$$
$$(x + 1)^3 = 16$$
$$x + 1 = \sqrt[3]{16}$$
$$x = -1 + \sqrt[3]{16}$$

Check:

$$\begin{aligned}
\log_2 (x+1)^3 &= 4\\
\log_2 [(-1+\sqrt[3]{16})+1]^3 &= 4\\
\log_2 (\sqrt[3]{16})^3 &= 4\\
\log_2 16 &= 4\\
2^4 &= 16\\
16 &= 16, \qquad \text{true}
\end{aligned}$$

■

Other logarithmic equations can be solved using the properties of logarithms given in earlier sections.

EXAMPLE 4 Solve the equation $\log(3x+2) + \log 9 = \log(x+5)$.

Solution:

$$\begin{aligned}
\log(3x+2) + \log 9 &= \log(x+5)\\
\log(3x+2)(9) &= \log(x+5)\\
(3x+2)(9) &= (x+5) \qquad \text{Property 5d}\\
27x + 18 &= x + 5\\
26x + 18 &= 5\\
26x &= -13\\
x &= -\frac{1}{2}
\end{aligned}$$

Check:

$$\begin{aligned}
\log(3x+2) + \log 9 &= \log(x+5)\\
\log\left[3\left(-\frac{1}{2}\right)+2\right] + \log 9 &= \log\left(-\frac{1}{2}+5\right)\\
\log\left(-\frac{3}{2}+2\right) + \log 9 &= \log\frac{9}{2}\\
\log\frac{1}{2} + \log 9 &= \log\frac{9}{2}\\
\log\left(\frac{1}{2}\right)9 &= \log\frac{9}{2}\\
\log\frac{9}{2} &= \log\frac{9}{2}, \qquad \text{true}
\end{aligned}$$

Thus the solution is $-\frac{1}{2}$. ■

EXAMPLE 5 Solve the equation $\log x + \log (x + 1) = \log 12$.

Solution:

$$\log x + \log (x + 1) = \log 12$$
$$\log x(x + 1) = \log 12$$
$$x(x + 1) = 12$$
$$x^2 + x = 12$$
$$x^2 + x - 12 = 0$$
$$(x + 4)(x - 3) = 0$$
$$x + 4 = 0 \quad \text{or} \quad x - 3 = 0$$
$$x = -4 \quad \text{or} \quad x = 3$$

Check:

$x = 3$

$$\log x + \log (x + 1) = \log 12$$
$$\log 3 + \log 4 = \log 12$$
$$\log (3)(4) = \log 12$$
$$\log 12 = \log 12, \text{ true}$$

$x = -4$

$$\log x + \log (x + 1) = \log 12$$
$$\log (-4) + \log (-3) = \log 12$$

↑ ↑ logarithms of negative numbers are not real numbers

Thus -4 is an extraneous solution. The only solution is 3. ■

EXAMPLE 6 Solve the equation $\log (3x - 5) - \log 5x = 1.23$.

Solution:

$$\log \left[\frac{3x - 5}{5x}\right] = 1.23 \qquad \text{by property 2}$$
$$\frac{3x - 5}{5x} = \text{antilog } 1.23$$
$$\frac{3x - 5}{5x} = 17.0$$
$$3x - 5 = 5x(17.0)$$
$$3x - 5 = 85x$$
$$-5 = 82x$$
$$x = -\frac{5}{82} \approx -0.061$$

Check:

$$\log (3x - 5) - \log 5x = 1.23$$
$$\log [3(-0.061) - 5] - \log [(5)(-0.061)] = 1.23$$
$$\log (-5.183) - \log (-0.305) = 1.23$$

Since we are evaluating the logs of negative numbers, -0.061 is an extraneous solution. Thus the answer to this equation is no solution, or the empty set, { }. ■

2 In Section 11.2 we introduced the compound interest formula, $A = P(1 + r)^n$. We found A at that time by using a calculator. The value of A can also be found by using logarithms as illustrated in Example 7.

EXAMPLE 7 The amount of money, A, accumulated in a savings account for a given principal, P, interest rate r, and number of compounding periods, n, can be found by the formula

$$A = P(1 + r)^n$$

For example, if \$1000 is invested in a savings account giving 8% interest compounded annually for 5 years, the amount accumulated at the end of 5 years is

$$A = 1000(1 + 0.08)^5$$

Use logarithms to find the amount accumulated.

Solution:

$$\begin{aligned}
\log A &= \log 1000(1 + 0.08)^5 \\
&= \log 1000 + \log (1 + 0.08)^5 \\
&= \log 1000 + \log (1.08)^5 \\
&= \log 1000 + 5 \log 1.08 \\
&= 3.00 + 5(0.0334) \\
&= 3.00 + 0.167 \\
&= 3.167 \\
A &= \text{antilog } 3.167 \\
A &\approx 1470
\end{aligned}$$

In 5 years the \$1000 would grow to about \$1470. This includes the \$1000 principal and \$470 interest. ■

EXAMPLE 8 If there are initially 1000 bacteria in a culture, and the number of bacteria double each hour, the number of bacteria after t hours can be found by the formula

$$N = 1000(2^t)$$

How long will it take for the culture to grow to 30,000 bacteria?

Solution:

$$\begin{aligned}
N &= 1000(2^t) \\
30{,}000 &= 1000(2^t)
\end{aligned}$$

Now take the log of both sides of the equation.

$$\begin{aligned}
\log 30{,}000 &= \log 1000(2^t) \\
\log 30{,}000 &= \log 1000 + \log 2^t \\
\log 30{,}000 &= \log 1000 + t(\log 2) \\
4.4771 &= 3.000 + t(0.3010) \\
4.4771 &= 3.000 + 0.3010t \\
1.4771 &= 0.3010t \\
\frac{1.4771}{0.3010} &= t \\
4.91 &\approx t
\end{aligned}$$

In approximately 4.91 hours there will be 30,000 bacteria in the culture. ■

Exercise Set 11.5

Solve the exponential equation. Use a calculator if available and round answer to nearest hundredth. Note answers may be slightly different if tables are used, due to roundoff error.

1. $3^x = 243$ **2.** $16^x = \frac{1}{4}$ **3.** $5^x = 125$ **4.** $8^x = 60$

5. $1.05^x = 15$ **6.** $4^{x-1} = 20$ **7.** $1.63^{x+1} = 25$ **8.** $2^{2x-1} = 50.6$

Find the solutions to the logarithmic equation. Round the answer to the nearest hundredth. If the equation has no real solution so state.

9. $\log_4 (x + 1)^3 = 3$ **10.** $\log_3 (x - 2)^2 = 2$ **11.** $\log_2 (x + 4)^2 = 4$
12. $\log_2 x + \log_2 3 = 1$ **13.** $\log (2x - 3)^3 = 3$ **14.** $\log_3 3 + \log_3 x = 3$
15. $\log (x + 2) = \log (3x - 1)$ **16.** $\log x = \log(1 - x)$ **17.** $\log (3x - 1) + \log 4 = \log (9x + 2)$
18. $\log (x + 3) + \log x = \log 4$ **19.** $\log x + \log (3x - 5) = \log 2$ **20.** $\log (x + 4) - \log x = \log (x + 1)$
21. $\log x + \log 4 = .56$ **22.** $\log (x + 4) - \log x = 1.22$ **23.** $2 \log x - \log 4 = 2$
24. $\log 6000 - \log (x + 2) = 3.15$

Find the solution. Use a calculator if available and round the answer to the nearest hundredth.

25. In astronomy, a formula used to find the diameter, in kilometers, of minor planets (also called asteroids) is $\log d = 3.7 - 0.2g$ where g is a quantity called the absolute magnitude of the minor planet. Find the diameter of a minor planet that has an absolute magnitude of **(a)** 11; **(b)** 20.

26. Find the amount accumulated if Anne puts \$1200 in a savings account offering 10% interest compounded annually for 5 years. Use $A = P(1 + r)^n$. See Example 7.

27. Find the amount of money accumulated if Doang Nhu invests \$15,000 in a savings account giving 8% interest compounded annually for 20 years.

28. If the initial amount of bacteria in the culture in Example 8 is 4500, how long will it take for the number of bacteria in the culture to reach 50,000? Use $N = 1000(2^t)$.

29. If after 4 hours the culture in Example 8 contains 2224 bacteria, how many bacteria were present initially?

30. The amount, A, of 200 grams of a certain radioactive material remaining after t years can be found by the equation $A = 200(0.800)^t$. After how long will there be 40 grams of the material remaining?

31. A machine or automobile purchased for business can be depreciated to reduce income tax. The value of the machine at the end of its useful life is called its scrap value. When the machine depreciates by a constant percentage annually, its scrap value, S, is

$$S = c(1 - r)^n$$

where c is the original cost, r is the annual rate of depreciation as a decimal, and n is the useful life in years.

Find the scrap value of a machine that costs \$50,000, has a useful life of 12 years, and has an annual depreciation rate of 15%.

32. If the machine in Exercise 31 costs \$100,000, has a useful life of 15 years, and has an annual depreciation rate of 8%, find its scrap value.

33. The power gain, P, of an amplifier is defined to be

$$P = \log 10 \frac{P_{out}}{P_{in}}$$

where P_{out} is the power output in watts and P_{in} is the power input in watts. If an amplifier has an output power of 12.6 watts and an input power of 0.146 watts, find the power gain.

34. Measured by the Richter Scale, the magnitude, R, of an earthquake of intensity I is defined by

$$R = \log I$$

where I is the number of times greater (or more intense) the earthquake is than the minimum level for comparison.

(a) How many times more intense was the 1906 San Francisco earthquake which measured 8.25 on the Richter Scale than the minimum level for comparison?

(b) How many times more intense is an earthquake that measures 6.4 on the Richter Scale than one that measures 4.7?

35. The decibel scale is used to measure the magnitude of sound. The magnitude of a sound, d, in decibels, is defined to be

$$d = 10 \log I$$

where I is the number of times greater (or more intense) the given sound is than the minimum intensity of audible sound. An airplane engine (nearby) measures 120 decibels.

(a) How many times the minimum level of audible sound is the airplane engine?

(b) A busy city street has an intensity of 70 decibels. How many times greater is the intensity of the sound of the airplane engine than the sound of the city street?

JUST FOR FUN

1. The pH is a measure of the acidity or alkalinity of a solution. The pH of water, for example, is 7. In general acids have pH numbers less than 7 while alkaline solutions have pH numbers greater than 7. The pH of a solution is defined to be

$$\text{pH} = -\log\,[H_3O^+]$$

where H_3O^+ represents the hydronium ion concentration of the solution. Find the pH of a solution with a hydronium ion concentration of 2.8×10^{-3}.

Summary

Glossary

Antilogarithm: If $\log N = L$, then $N = \text{antilog } L$.
Argument of a logarithm: When finding the logarithm of an expression, the expression is called the argument. In $y = \log(x + 3)$, the $x + 3$ is the argument.
Characteristic of a logarithm: The integer part of a logarithm.
Common logarithm: Logarithms to the base 10 are common logarithms.
Exponential equation: An equation that has a variable as an exponent.
Exponential function: A function of the form $f(x) = a^x$, $a > 0$, $a \neq 1$.
Logarithm: $y = \log_a x$ means $x = a^y$, $a > 0$, $a \neq 1$.
Logarithmic equation: An equation in which a variable appears in the argument of some logarithm.
Logarithmic functions: Equations of the form $y = \log_a x$.
Mantissa of a logarithm: The decimal part of a logarithm.

Important Facts

$y = a^x$ and $y = \log_a x$ are inverse functions.
The domain of a logarithm of the form $y = \log_a x$ is $x > 0$.

Properties of Logarithms

1. $\log_a xy = \log_a x + \log_a y$
2. $\log_a \dfrac{x}{y} = \log_a x - \log_a y$
3. $\log_a x^n = n \log_a x$
4. $\log 10^n = n$

To solve exponential and logarithmic equations, we may use these properties:

1. If $x = y$, then $a^x = a^y$.
2. If $a^x = a^y$, then $x = y$.
3. If $x = y$, then $\log x = \log y \quad (x > 0,\ y > 0)$.
4. If $\log x = \log y$, then $x = y \quad (x > 0,\ y > 0)$.

Review Exercises

[11.1] Determine which functions are one-to-one.

1. 2. 3. 4.

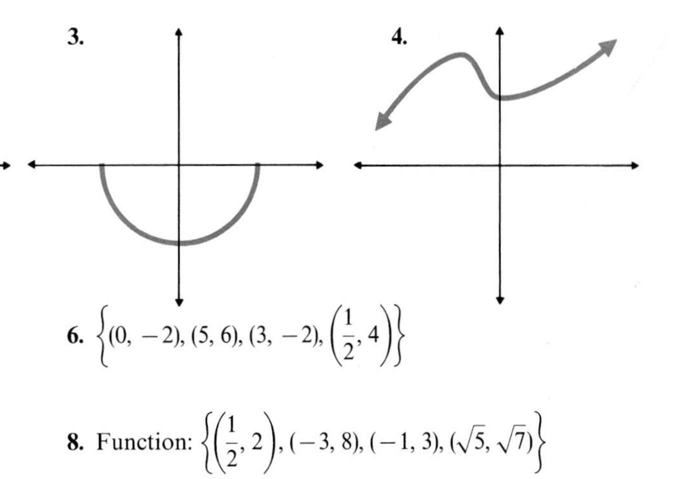

5. $\{(2, 3), (4, 0), (-5, 7), (3, 8)\}$

6. $\left\{(0, -2), (5, 6), (3, -2), \left(\frac{1}{2}, 4\right)\right\}$

Give the domain and range of $f(x)$ and $f^{-1}(x)$.

7. Function: $\{(5, 3), (6, 2), (-4, -3), (0, 7)\}$

8. Function: $\left\{\left(\frac{1}{2}, 2\right), (-3, 8), (-1, 3), (\sqrt{5}, \sqrt{7})\right\}$

Find $f^{-1}(x)$ and graph $f(x)$ and $f^{-1}(x)$ on the same set of axes.

9. $y = f(x) = 4x - 2$

10. $y = f(x) = -3x - 5$

11. $y = f(x) = \dfrac{2x + 5}{3}$

12. $y = f(x) = \dfrac{3}{5} - \dfrac{2}{3}x$

[11.2] Graph the following functions.

13. $y = 2^x$

14. $y = \left(\dfrac{1}{2}\right)^x$

15. $y = \log_2 x$

16. $y = \log_{1/2} x$

17. On the same set of axes graph

$$y = 3^x, \qquad y = \log_3 x$$

Write in logarithmic form.

18. $4^2 = 16$

19. $8^{1/3} = 2$

20. $6^{-2} = \dfrac{1}{36}$

21. $25^{1/2} = 5$

Write in exponential form.

22. $\log_5 25 = 2$

23. $\log_{1/3}\left(\dfrac{1}{9}\right) = 2$

24. $\log_3\left(\dfrac{1}{9}\right) = -2$

25. $\log_2 32 = 5$

Write in exponential form and find the missing value.

26. $3 = \log_4 x$

27. $2 = \log_4 x$

28. $3 = \log_a 8$

29. $-3 = \log_{1/4} x$

[11.3] Use the properties of logarithms to expand each expression.

30. $\log_8 \sqrt{12}$

31. $\log (x - 8)^5$

32. $\log \frac{2(x-3)}{x}$

33. $\log \frac{x^4}{39(2x+8)}$

Write as the logarithm of a single expression.

34. $\log_5 (x - 2) - 2 \log_5 x$

35. $2 \log x - 3 \log (x + 1)$

36. $3[\log x + \log 2] - \log x$

37. $[2 \log_8 (x + 3) + 4 \log_8 (x - 1)] - \frac{1}{2} \log_8 x$

[11.4] Use a calculator or Appendix D to find the common logarithm.

38. 8200

39. 0.000716

40. 0.00189

41. 17,600

Use a calculator or Appendix D to find the antilog.

42. 1.7528

43. 2.9186

44. 8.9009 – 10

45. 0.6253 – 2

Use a calculator or Appendix D to find N.

46. $\log N = 2.3304$

47. $\log N = 8.7738 - 10$

Evaluate using logarithms. The answers given were obtained using the log table. If you obtain your log values using a calculator your answer may differ.

48. (2040)(0.00738)

49. $\frac{2.93}{0.00174}$

50. $(183)^3$

51. $\sqrt{13{,}600}$

52. $\frac{(516)(384)}{0.076}$

[11.5] Solve the exponential equation.

53. $4^x = 37$

54. $(3.2)^x = 187$

55. $(10.9)^{x+1} = 492$

56. $49^x = \frac{1}{7}$

Solve the logarithmic equation.

57. $\log_5 (x + 2) = 3$

58. $\log (3x + 2) = \log 300$

59. $\log x - \log (3x - 5) = \log 2$

60. $\log_3 x + \log_3 (2x + 1) = 1$

61. Find the amount of money accumulated if Mrs. Elwood puts $12,000 in a savings account yielding 10% interest per year for 8 years. Use $A = p(1 + r)^n$.

62. Plutonium is a radioactive element that decays, or disintegrates, over time. If there are originally 1000 mg of plutonium, the amount remaining, R, after t years is

$$R = 1000(0.5)^{0.000041t}$$

Calculate the amount of plutonium present after 20,000 years.

Practice Test

1. **(a)** Find the inverse function of $f(x)$ and **(b)** sketch $f(x)$ and $f^{-1}(x)$ on the same set of axes.

$$y = f(x) = 2x + 4$$

Write in exponential form and find the missing value.

6. $4 = \log_2 x$

7. $y = \log_{27} 3$

8. Expand $\log_3 \left[\dfrac{x(x-4)}{x^2}\right]$.

9. Write as the logarithm of a single expression.

$$[3 \log_8 (x-4) + 2 \log_8 (x+1)] - \frac{1}{2} \log_8 x$$

2. Graph $y = 2^x$.
3. Graph $y = \log_2 x$.
4. Write $4^{-3} = \frac{1}{64}$ in logarithmic form.
5. Write $\log_3 243 = 5$ in exponential form.

10. Find log 4620.
11. Find log 0.000638.
12. Log $N = 7.6998 - 10$; find N.
13. Solve for x: $3^x = 123$.
14. Solve for x: $\log 4x = \log (x + 3) + \log 2$.
15. What amount of money accumulates if Say-Chun puts \$1500 in a savings account yielding 12% interest per year for 10 years? Use $A = p(1 + r)^n$.

12 Sequences, Series, and the Binomial Theorem

See Section 12.3, Exercise 36.

12.1 Sequences and Series

1. *Find the terms of a sequence given the general term.*
2. *Write a series.*
3. *Find partial sums of a series.*

1 Consider the following two lists of numbers

$$5, 10, 15, 20, 25, 30, \ldots$$
$$2, 4, 8, 16, 32, \ldots$$

The three dots at the ends of the lists indicate that the lists continue in the same manner. Can you determine the next two numbers in the first list? In the second list? The next numbers in the first list would seem to be 35 and 40. In the second list of numbers they would seem to be 64 and 128. The two lists of numbers above are examples of sequences. A **sequence (or progression)** of numbers is a list of numbers arranged in a specified order. Consider the sequence 5, 10, 15, 20, 25, 30, ... The first term is 5. We indicate this by writing $a_1 = 5$. Since the second term is 10, $a_2 = 10$ and so on.

> An **infinite sequence** is a function whose domain is the set of natural numbers.

Consider the infinite sequence 5, 10, 15, 20, 25, 30, ...

$$\begin{array}{ll} \text{Domain:} & \{1,\ 2,\ 3,\ 4,\ 5,\ 6,\ 7,\ \ldots,\ n,\ \ldots\} \\ & \ \ \downarrow\ \ \downarrow\ \ \downarrow\ \ \downarrow\ \ \downarrow\ \ \downarrow\ \ \downarrow\qquad\ \ \downarrow \\ \text{Range:} & \{5,\ 10,\ 15,\ 20,\ 25,\ 30,\ 35,\ \ldots,\ 5n,\ \ldots\} \end{array}$$

Note that the terms of the sequence 5, 10, 15, 20, . . . , are found by multiplying each natural number by 5. For any natural number n, the corresponding term in the sequence is $5 \cdot n$ or $5n$. The **general term of the sequence,** a_n, which defines the sequence, is $a_n = 5n$.

$$a_n = f(n) = 5n$$

To find the twelfth term of the sequence substitute 12 for n in the general term of the sequence, $a_{12} = 5 \cdot 12 = 60$. Thus, the twelfth term of the sequence is 60. Note that the terms in the sequence are the function values, or the range of the function. When writing the sequence we do not use set braces. The general form of a sequence is:

$$a_1, a_2, a_3, a_4, \ldots, a_n, \ldots$$

For the infinite sequence $2, 4, 8, 16, 32, \ldots, 2^n, \ldots$ we can write

$$a_n = f(n) = 2^n$$

Notice when $n = 1$, $a_1 = 2^1 = 2$, when $n = 2$, $a_2 = 2^2 = 4$, when $n = 3$, $a_3 = 2^3 = 8$, when $n = 4$, $a_4 = 2^4 = 16$, and so on. What is the seventh term of this sequence? The answer is $a_7 = 2^7 = 128$.

A sequence may also be finite.

> A **finite sequence** is a function whose domain includes only the first n natural numbers.

A finite sequence has only a finite number of terms.

Examples of finite sequences

5, 10, 15, 20	domain is $\{1, 2, 3, 4\}$
2, 4, 8, 16, 32	domain is $\{1, 2, 3, 4, 5\}$

EXAMPLE 1 Write the finite sequence defined by $a_n = 2n + 1$, for $n = 1, 2, 3, 4$.

Solution:

$$a_1 = 2(1) + 1 = 3$$
$$a_2 = 2(2) + 1 = 5$$
$$a_3 = 2(3) + 1 = 7$$
$$a_4 = 2(4) + 1 = 9$$

Thus the sequence is 3, 5, 7, 9. ■

EXAMPLE 2 Given $a_n = (2n + 3)/n^2$, find the following.

(a) The first term in the sequence.
(b) The third term in the sequence.
(c) The fifth term in the sequence.

Solution:

(a) When $n = 1$, $a_1 = \dfrac{2(1) + 3}{1^2} = \dfrac{5}{1} = 5$

(b) When $n = 3$, $a_3 = \dfrac{2(3) + 3}{3^2} = \dfrac{9}{9} = 1$

(c) When $n = 5$, $a_5 = \dfrac{2(5) + 3}{5^2} = \dfrac{13}{25}$ ■

Note that in Example 2 since there is no restriction on n, a_n is the general term of an infinite sequence.

2 A **series** is the sum of a sequence. A series may be finite or infinite depending on whether the sequence it is based on is finite or infinite. An example of a finite sequence and finite series is illustrated below.

a_1, a_2, a_3, a_4, a_5	**finite sequence**
$a_1 + a_2 + a_3 + a_4 + a_5$	**finite series**

Examples of an infinite sequence and an infinite series are

$a_1, a_2, a_3, a_4, a_5, \ldots, a_n, \ldots$	**infinite sequence**
$a_1 + a_2 + a_3 + a_4 + a_5 + \cdots + a_n + \cdots$	**infinite series**

EXAMPLE 3 Write the first eight terms of the sequence; then write the series that represents the sum of that sequence if:

(a) $a_n = \left(\frac{1}{2}\right)^n$ (b) $a_n = \left(\frac{1}{2}\right)^{n-1}$

Solution: (a) We begin with $n = 1$; thus the first eight terms of the sequence whose general term is $a_n = (\frac{1}{2})^n$ are:

$$\left(\frac{1}{2}\right)^1, \left(\frac{1}{2}\right)^2, \left(\frac{1}{2}\right)^3, \left(\frac{1}{2}\right)^4, \left(\frac{1}{2}\right)^5, \left(\frac{1}{2}\right)^6, \left(\frac{1}{2}\right)^7, \left(\frac{1}{2}\right)^8$$

or

$$\frac{1}{2}, \frac{1}{4}, \frac{1}{8}, \frac{1}{16}, \frac{1}{32}, \frac{1}{64}, \frac{1}{128}, \frac{1}{256}$$

The series that represents the sum of the sequence is

$$\frac{1}{2} + \frac{1}{4} + \frac{1}{8} + \frac{1}{16} + \frac{1}{32} + \frac{1}{64} + \frac{1}{128} + \frac{1}{256}$$

(b) We again begin with $n = 1$; thus the first eight terms of the sequence whose general term is $a_n = (\frac{1}{2})^{n-1}$ are:

$$\left(\frac{1}{2}\right)^{1-1}, \left(\frac{1}{2}\right)^{2-1}, \left(\frac{1}{2}\right)^{3-1}, \left(\frac{1}{2}\right)^{4-1}, \left(\frac{1}{2}\right)^{5-1}, \left(\frac{1}{2}\right)^{6-1}, \left(\frac{1}{2}\right)^{7-1}, \left(\frac{1}{2}\right)^{8-1}$$

or

$$\left(\frac{1}{2}\right)^0, \left(\frac{1}{2}\right)^1, \left(\frac{1}{2}\right)^2, \left(\frac{1}{2}\right)^3, \left(\frac{1}{2}\right)^4, \left(\frac{1}{2}\right)^5, \left(\frac{1}{2}\right)^6, \left(\frac{1}{2}\right)^7$$

or

$$1, \frac{1}{2}, \frac{1}{4}, \frac{1}{8}, \frac{1}{16}, \frac{1}{32}, \frac{1}{64}, \frac{1}{128}$$

The series that represents the sum of this sequence is:

$$1 + \frac{1}{2} + \frac{1}{4} + \frac{1}{8} + \frac{1}{16} + \frac{1}{32} + \frac{1}{64} + \frac{1}{128}$$ ■

3 A **partial sum of a series** is the sum of a finite number of consecutive terms of the series, beginning with the first term.

$$\begin{aligned}
s_1 &= a_1 && \text{(first partial sum)}\\
s_2 &= a_1 + a_2 && \text{(second partial sum)}\\
s_3 &= a_1 + a_2 + a_3 && \text{(third partial sum)}\\
&\ \vdots\\
s_n &= a_1 + a_2 + a_3 + \cdots + a_n && \text{(}n\text{th partial sum)}
\end{aligned}$$

EXAMPLE 4 Given the infinite sequence defined by $a_n = (1 + n^2)/n$, find the following.
(a) The first partial sum.
(b) The third partial sum.

Solution: (a) $s_1 = a_1 = \frac{1 + 1^2}{1} = \frac{1 + 1}{1} = 2$

(b) $$\begin{aligned} s_3 &= a_1 + a_2 + a_3 \\ &= \frac{1 + 1^2}{1} + \frac{1 + 2^2}{2} + \frac{1 + 3^2}{3} \\ &= 2 + \frac{5}{2} + \frac{10}{3} \\ &= \frac{12}{6} + \frac{15}{6} + \frac{20}{6} \\ &= \frac{47}{6} \quad \text{or} \quad 7\frac{5}{6} \end{aligned}$$ ■

Exercise Set 12.1

Write the first five terms of the sequence whose nth term is shown.

1. $a_n = 2n$
2. $a_n = 2n + 3$
3. $a_n = \frac{n + 5}{n}$
4. $a_n = \frac{n}{n^2}$
5. $a_n = \frac{1}{n}$
6. $a_n = n^2 - n$
7. $a_n = \frac{n + 2}{n + 1}$
8. $a_n = \frac{n + 2}{n + 3}$
9. $a_n = (-1)^n$
10. $a_n = (-1)^{2n}$
11. $a_n = (-2)^{n+1}$
12. $a_n = n(n + 2)$

Find the indicated term of the sequence whose nth term is shown.

13. $a_n = 2n + 3$, twelfth term
14. $a_n = 2^n$, seventh term
15. $a_n = 2n - 4$, fifth term
16. $a_n = (-1)^n$, eighth term
17. $a_n = (-2)^n$, fourth term
18. $a_n = n(n + 5)$, eighth term
19. $a_n = \frac{n^2}{(2n + 1)}$, ninth term
20. $a_n = \frac{n(n + 1)}{n^2}$, tenth term

Find the first and third partial sums, s_1 and s_3, for the given sequences.

21. $a_n = 2n + 5$
22. $a_n = \frac{3n}{n + 2}$
23. $a_n = 2^n + 1$
24. $a_n = \frac{n - 1}{n + 1}$
25. $a_n = (-1)^{2n}$
26. $a_n = \frac{2n^2}{n + 1}$
27. $a_n = \frac{n^2}{2}$
28. $a_n = \frac{n + 3}{2n}$

Write the next three terms of each sequence.

29. $2, 4, 8, 16, 32, \ldots$
30. $\frac{1}{2}, \frac{1}{3}, \frac{1}{4}, \frac{1}{5}, \ldots$
31. $3, 5, 7, 9, 11, 13, \ldots$
32. $5, 10, 15, 20, 25, \ldots$
33. $1, \frac{1}{2}, \frac{1}{3}, \frac{1}{4}, \frac{1}{5}, \ldots$
34. $\frac{2}{3}, \frac{3}{4}, \frac{4}{5}, \frac{5}{6}, \frac{6}{7}, \ldots$

35. $-1, 1, -1, 1, -1, \ldots$

36. $-10, -20, -30, -40, \ldots$

37. $1, \frac{1}{3}, \frac{1}{9}, \frac{1}{27}, \ldots$

38. $\frac{1}{3}, \frac{2}{3}, \frac{3}{3}, \frac{4}{3}, \ldots$

39. $1, -\frac{1}{2}, \frac{1}{4}, -\frac{1}{8}, \ldots$

40. $\frac{2}{3}, \frac{1}{3}, \frac{1}{6}, \frac{1}{12}, \ldots$

41. $7, -1, -9, -17, \ldots$

42. $37, 32, 27, 22, \ldots$

12.2 Arithmetic Sequences and Series

1 *Find the common difference in an arithmetic sequence.*

2 *Find the nth term of an arithmetic sequence.*

3 *Find the nth partial sum of an arithmetic series.*

1 A sequence in which each term after the first differs from the preceding term by a constant amount is called an **arithmetic sequence** or **arithmetic progression.** The constant amount by which each pair of successive terms differs is called the **common difference,** d. The common difference can be found by subtracting any term from the term that directly follows it.

$$1, 4, 7, 10, 13, 16, \ldots \qquad d = 4 - 1 = 3$$

$$-7, -2, 3, 8, 13, \ldots \qquad d = -2 - (-7) = -2 + 7 = 5$$

$$\frac{7}{2}, \frac{2}{2}, -\frac{3}{2}, -\frac{8}{2}, -\frac{13}{2}, -\frac{18}{2}, \ldots \qquad d = \frac{2}{2} - \frac{7}{2} = -\frac{5}{2}$$

EXAMPLE 1 Write the first five terms of the arithmetic sequence with the first term 6 and the common difference 3.

Solution: 6, 9, 12, 15, 18 ■

EXAMPLE 2 Write the first six terms of the arithmetic sequence with the first term 3 and the common difference -2.

Solution: $3, 1, -1, -3, -5, -7$ ■

2 In general, an arithmetic sequence with the first term a_1 and the common difference d will have the following terms:

$$a_1 = a_1, \quad a_2 = a_1 + d, \quad a_3 = a_1 + 2d, \quad a_4 = a_1 + 3d, \quad \text{and so on.}$$

If we continue this process we can see that the nth term, a_n, can be found by

nth Term of an Arithmetic Sequence

$$a_n = a_1 + (n - 1)d$$

EXAMPLE 3 (a) Write an expression for the general (or nth) term, a_n, of the arithmetic sequence with a first term of -3 and a difference of 4.
(b) Find the twelfth term of the sequence.

Solution: (a) The nth term of the sequence is $a_n = a_1 + (n-1)d$. Substituting $a_1 = -3$ and $d = 4$ we obtain

$$a_n = a_1 + (n-1)d = -3 + (n-1)4 = -3 + 4(n-1)$$

Thus $a_n = -3 + 4(n-1)$.

(b) $a_n = -3 + 4(n-1)$

$$a_{12} = -3 + 4(12-1) = -3 + 4(11) = -3 + 44 = 41$$

The twelfth term in the sequence will be 41. ■

EXAMPLE 4 Find the number of terms in the arithmetic sequence 5, 9, 13, 17, . . . , 41.

Solution: The first term, a_1, is 5; the nth term is 41; and the common difference, d, is 4.

$$\begin{aligned} a_n &= a_1 + (n-1)d \\ 41 &= 5 + (n-1)4 \\ 41 &= 5 + 4n - 4 \\ 41 &= 4n + 1 \\ 40 &= 4n \\ 10 &= n \end{aligned}$$

Thus the sequence has a total of 10 terms. ■

3 An **arithmetic series** is the sum of the terms of an arithmetic sequence. A finite arithmetic series can be written

$$s_n = a_1 + (a_1 + d) + (a_1 + 2d) + (a_1 + 3d) + \cdots + (a_n - 2d) + (a_n - d) + a_n$$

If we consider the last term as a_n, the term before the last term will be $a_n - d$, the second before the last term will be $a_n - 2d$, and so on.

A formula for the nth partial sum, s_n, can be obtained by adding the reverse of s_n to itself.

$$\begin{aligned} s_n &= a_1 + (a_1 + d) + (a_1 + 2d) + \cdots + (a_n - 2d) + (a_n - d) + a_n \\ s_n &= a_n + (a_n - d) + (a_n - 2d) + \cdots + (a_1 + 2d) + (a_1 + d) + a_1 \\ \hline 2s_n &= (a_1 + a_n) + (a_1 + a_n) + (a_1 + a_n) + \cdots + (a_1 + a_n) + (a_1 + a_n) + (a_1 + a_n) \end{aligned}$$

Since the right side of the equation contains n terms of $(a_1 + a_n)$, we can write

$$2s_n = n(a_1 + a_n)$$

Therefore,

nth Partial Sum of an Arithmetic Series

$$s_n = \frac{n(a_1 + a_n)}{2}$$

EXAMPLE 5 Find the sum of the first 25 natural numbers.

Solution: The arithmetic sequence is 1, 2, 3, 4, 5, 6, . . . , 25. The first term, a_1, is 1; the last term, a_n, is 25. There are 25 terms; thus $n = 25$.

$$s_n = \frac{n(a_1 + a_n)}{2} = \frac{25(1 + 25)}{2} = \frac{25(26)}{2} = 25(13) = 325$$

The sum of the first 25 natural numbers is 325. Thus $s_{25} = 325$. ■

EXAMPLE 6 The first term of an arithmetic sequence is 4, and the last term is 31. If $s_n = 175$, find the number of terms in the sequence and the common difference.

Solution: $a_1 = 4$, $a_n = 31$, and $s_n = 175$.

$$s_n = \frac{n(a_1 + a_n)}{2}$$

$$175 = \frac{n(4 + 31)}{2}$$

$$175 = \frac{35n}{2}$$

$$350 = 35n$$

$$10 = n$$

Thus there are 10 terms in the sequence. We can now find the common difference.

$$a_n = a_1 + (n - 1)d$$

$$31 = 4 + (10 - 1)d$$

$$31 = 4 + 9d$$

$$27 = 9d$$

$$3 = d$$

The common difference is 3. The sequence is 4, 7, 10, 13, 16, 19, 22, 25, 28, 31. ■

Examples 7 and 8 illustrate some applications of arithmetic sequences and series.

EXAMPLE 7 Donna Stansell is given a starting salary of \$25,000 and is promised a \$1200 raise after each of the next 8 years. Find her salary during her eighth year of work.

Solution: Her salaries after the first few years would be:

$$25{,}000,\ 26{,}200,\ 27{,}400,\ 28{,}600, \ldots$$

Since we are adding a constant amount each year this is an arithmetic sequence. The general term of an arithmetic sequence is $a_n = a_1 + (n - 1)d$. In this example $a_1 = 25{,}000$ and $d = 1200$. Thus for $n = 8$, Donna's salary would be:

$$\begin{aligned} a_8 &= 25{,}000 + (8 - 1)1200 \\ &= 25{,}000 + 7(1200) \\ &= 25{,}000 + 8400 \\ &= 33{,}400 \end{aligned}$$

If we listed all the salaries for the 8-year period they would be:

$$25{,}000,\ 26{,}200,\ 27{,}400,\ 28{,}600,\ 29{,}800,\ 31{,}000,\ 32{,}200,\ 33{,}400$$ ■

EXAMPLE 8 Each swing of a pendulum is 3 inches shorter than its preceding swing. The first swing is 8 feet.

(a) Find the length of the twelfth swing.
(b) Determine the total distance traveled by the pendulum during the first 12 swings.

Solution: (a) Since each swing is decreasing by a constant amount this problem can be represented as an arithmetic series. Since the first swing is given in feet and the decrease in swing in inches, we will change 3 inches to 0.25 feet ($3 \div 12 = 0.25$). The twelfth swing can be considered a_{12}.

$$\begin{aligned} a_n &= a_1 + (n - 1)d \\ a_{12} &= 8 + (12 - 1)(-0.25) \\ a_{12} &= 8 + 11(-0.25) \\ &= 8 - 2.75 \\ &= 5.25 \text{ feet} \end{aligned}$$

The twelfth swing will travel 5.25 feet. Note the difference, d, is negative since the distance is decreasing with each swing.

(b) The total distance traveled during the first 12 swings can be found using the formula $s_n = \dfrac{n(a_1 + a_n)}{2}$.

$$s_{12} = \frac{12(a_1 + a_{12})}{2}$$

$$s_{12} = \frac{12(8 + 5.25)}{2} = \frac{12(13.25)}{2} = 6(13.25) = 79.5 \text{ feet}$$

The pendulum travels a total of 79.5 feet during its first 12 swings. ■

Exercise Set 12.2

Write the first five terms of the arithmetic sequence with the given first term and common difference. Write the expression for the general (or nth) term, a_n, of the arithmetic sequence.

1. $a_1 = 3, d = 4$
2. $a_1 = 8, d = 2$
3. $a_1 = -5, d = 2$
4. $a_1 = -8, d = -3$
5. $a_1 = \frac{1}{2}, d = \frac{3}{2}$
6. $a_1 = -\frac{5}{3}, d = -\frac{1}{3}$
7. $a_1 = 100, d = -5$
8. $a_1 = \frac{5}{4}, d = -\frac{3}{4}$

Find the desired quantity of the arithmetic sequence.

9. $a_1 = 4, d = 3$; find a_7
10. $a_1 = 8, d = -2$; find a_6
11. $a_1 = -6, d = -1$; find a_{18}
12. $a_1 = -15, d = 3$; find a_{20}
13. $a_1 = -2, d = \frac{5}{3}$; find a_{10}
14. $a_1 = 5, a_8 = -21$; find d
15. $a_1 = 3, a_9 = 19$; find d
16. $a_1 = \frac{1}{2}, a_7 = \frac{19}{2}$; find d
17. $a_1 = 4, a_n = 28, d = 3$; find n
18. $a_1 = -2, a_n = -20, d = -3$; find n
19. $a_1 = -\frac{7}{3}, a_n = -\frac{17}{3}, d = -\frac{2}{3}$; find n
20. $a_1 = 100, a_n = 60, d = -8$; find n

Find the sum, s_n, and common difference, d.

21. $a_1 = 1, a_n = 19, n = 10$
22. $a_1 = -5, a_n = 13, n = 7$
23. $a_1 = \frac{3}{5}, a_n = 2, n = 8$
24. $a_1 = 12, a_n = -23, n = 8$
25. $a_1 = \frac{12}{5}, a_n = \frac{28}{5}, n = 5$
26. $a_1 = -3, a_n = 15.5, n = 6$
27. $a_1 = 7, a_n = 67, n = 11$

Write the first four terms of each sequence; then find a_{10} and s_{10}.

28. $a_1 = 5, d = 3$
29. $a_1 = -4, d = -2$
30. $a_1 = \frac{7}{2}, d = \frac{5}{2}$
31. $a_1 = -8, d = -5$
32. $a_1 = -15, d = 4$
33. $a_1 = 100, d = -7$
34. $a_1 = 35, d = 3$
35. $a_1 = \frac{9}{5}, d = \frac{3}{5}$

Find the number of terms in each sequence and s_n.

36. $1, 4, 7, 10, \ldots, 43$
37. $-8, -6, -4, -2, \ldots, 42$
38. $-9, -5, -1, 3, \ldots, 27$
39. $\frac{1}{2}, \frac{2}{2}, \frac{3}{2}, \frac{4}{2}, \frac{5}{2}, \ldots, \frac{17}{2}$
40. $-\frac{5}{6}, -\frac{7}{6}, -\frac{9}{6}, -\frac{11}{6}, \ldots, -\frac{21}{6}$
41. $7, 14, 21, 28, \ldots, 63$
42. $-12, -16, -20, \ldots, -52$
43. $9, 12, 15, 18, \ldots, 93$

44. Mr. Baudean is given a starting salary of $20,000 and is told he will receive a $1000 raise at the end of each year for the next 5 years.
- **(a)** Write a sequence showing his salary for the next 5 years.
- **(b)** What is the general term for this sequence?
- **(c)** If this procedure is extended, find his salary after 12 years.

45. Find the sum of the first 1000 positive integers.

46. Find the sum of the numbers between 50 and 200 inclusive.

47. Determine how many numbers between 7 and 1610 are divisible by 6.

48. An object falls 16.0 feet during the first second, 48.0 feet during the second second, 80.0 feet during the third second, etc.

(a) How far will it fall in its tenth second?
(b) How far will the object fall, in total, during the first 10 seconds?

49. Each time a ball bounces the height attained is 6 inches less than the previous height reached. If its first bounce reaches a height of 6 feet, find the height attained on the eleventh bounce.

50. If you are given $1 on January 1, $2 on January 2, $3 on the 3rd, etc., how much money will you have received in total by January 31?

51. Jack piles logs so that there are 20 logs in the bottom layer, and each layer contains one log less than the layer below it. How many logs are on the pile?

12.3 Geometric Sequences and Series

1. *Find the common ratio in a geometric sequence.*
2. *Find the nth term of a geometric sequence.*
3. *Find the nth partial sum of a geometric series.*

1 A **geometric sequence** (or **geometric progression**) is a sequence in which each term after the first is the same multiple of the preceding term. The common multiple is called the **common ratio.** The common ratio, r, in any geometric sequence can be found by taking any term, except the first, and dividing that term by the preceding term. Consider the geometric sequence

$$1, 3, 9, 27, 81, \ldots, 3^{n-1}, \ldots$$

The common ratio is 3 since $3 \div 1 = 3$ (or $9 \div 3 = 3$, etc.).

The common ratio of the geometric sequence

$$4, 8, 16, 32, 64, \ldots, 4(2^{n-1}), \ldots \text{ is } 2.$$

The common ratio of the geometric sequence

$$3, 12, 48, 192, 576, \ldots, 3(4^{n-1}), \ldots \text{ is } 4.$$

The common ratio of the geometric sequence

$$7, \frac{7}{2}, \frac{7}{4}, \frac{7}{8}, \frac{7}{16}, \ldots, 7\left(\frac{1}{2}\right)^{n-1}, \ldots \text{ is } \frac{1}{2}.$$

EXAMPLE 1 Determine the first five terms of the geometric sequence if $a_1 = 4$ and $r = \frac{1}{2}$.

Solution: $a_1 = 4, \quad a_2 = 4 \cdot \frac{1}{2} = 2, \quad a_3 = 2 \cdot \frac{1}{2} = 1, \quad a_4 = 1 \cdot \frac{1}{2} = \frac{1}{2}, \quad a_5 = \frac{1}{2} \cdot \frac{1}{2} = \frac{1}{4}$

Thus the first five terms of the geometric sequence are:

$$4, 2, 1, \frac{1}{2}, \frac{1}{4}$$ ■

2 In general a geometric sequence with first term a_1 and common ratio r has the following terms:

a_1,	a_1r,	a_1r^2,	a_1r^3,	$a_1r^4, \ldots,$	$a_1r^{n-1}, \ldots$
↑	↑	↑	↑	↑	↑
1st term	2nd term	3rd term	4th term	5th term	nth term

By observing the preceding geometric sequence we can see that the nth term of a geometric sequence is:

> **nth Term of a Geometric Sequence**
>
> $$a_n = a_1 r^{n-1}$$

EXAMPLE 2 (a) Write an expression for the general (or nth) term, a_n, of the geometric sequence with $a_1 = 3$ and $r = -2$.
(b) Find the twelfth term of this sequence.

Solution: (a) The nth term of the sequence is $a_n = a_1 n^{r-1}$. Substituting $a_1 = 3$ and $r = -2$ we obtain

$$a_n = a_1 r^{n-1} = 3(-2)^{n-1}$$

Thus $a_n = 3(-2)^{n-1}$.

(b)

$$a_n = 3(-2)^{n-1}$$
$$a_{12} = 3(-2)^{12-1} = 3(-2)^{11} = 3(-2048) = -6144$$

The twelfth term of the sequence is -6144. The first twelve terms of the sequence are 3, -6, 12, -24, 48, -96, 192, -384, 768, -1536, 3072, -6144. ■

EXAMPLE 3 Find r and a_1 for the geometric sequence with $a_2 = 24$ and $a_5 = 648$.

Solution: The sequence looks like

$$\underline{\quad},\ \underset{\substack{\uparrow \\ a_2}}{24},\ \underline{\quad},\ \underline{\quad},\ \underset{\substack{\uparrow \\ a_5}}{648}$$

If we assume that a_2 is the first term of a sequence with the same common ratio, we obtain

$$\underset{\substack{\uparrow \\ \text{1st} \\ \text{term}}}{24},\ \underline{\quad},\ \underline{\quad},\ \underset{\substack{\uparrow \\ \text{4th} \\ \text{term}}}{648}$$

Now use the formula

$$a_n = a_1 r^{n-1}$$
$$648 = 24r^{4-1}$$
$$648 = 24r^3$$
$$\frac{648}{24} = r^3$$
$$27 = r^3$$
$$3 = r$$

Thus the common ratio is 3.

The first term of the original sequence must be $24 \div 3$ or 8. Thus $a_1 = 8$. The first term could also be found using

$$\begin{aligned} a_n &= a_1 r^{n-1} \\ 648 &= a_1 3^4 \\ 648 &= a_1(81) \\ \frac{648}{81} &= a_1 \\ 8 &= a_1 \end{aligned}$$

■

3 A **geometric series** is the sum of the terms of a geometric sequence. The sum of the first n terms, s_n, of a geometric sequence can be expressed as

$$s_n = a_1 + a_1 r + a_1 r^2 + a_1 r^3 + \cdots + a_1 r^{n-1}$$

If we multiply both sides of the equation by r, we obtain

$$rs_n = a_1 r + a_1 r^2 + a_1 r^3 + \cdots + a_1 r^n$$

Now subtract the second equation from the first.

$$\begin{aligned} s_n &= a_1 + a_1 r + a_1 r^2 + a_1 r^3 + \cdots + a_1 r^{n-1} \\ rs_n &= \phantom{a_1 +{}} a_1 r + a_1 r^2 + a_1 r^3 + \cdots + a_1 r^{n-1} + a_1 r^n \\ \hline s_n - rs_n &= a_1 \phantom{+ a_1 r + a_1 r^2 + a_1 r^3 + \cdots + a_1 r^{n-1}} - a_1 r^n \end{aligned}$$

or

$$\begin{aligned} s_n - rs_n &= a_1 - a_1 r^n \\ s_n(1 - r) &= a_1 - a_1 r^n \\ s_n &= \frac{a_1 - a_1 r^n}{1 - r} \end{aligned}$$

Thus we have the following formula for the nth partial sum of a geometric series.

nth Partial Sum of a Geometric Series

$$s_n = \frac{a_1(1 - r^n)}{1 - r}, \quad r \neq 1$$

EXAMPLE 4 Find the seventh partial sum of a geometric series whose first term is 8 and whose common ratio is $\frac{1}{2}$.

Solution:

$$s_n = \frac{a_1(1 - r^n)}{1 - r}$$

$$s_7 = \frac{8\left[1 - \left(\frac{1}{2}\right)^7\right]}{1 - \frac{1}{2}} = \frac{8\left(1 - \frac{1}{128}\right)}{\frac{1}{2}} = \frac{8\left(\frac{127}{128}\right)}{\frac{1}{2}} = \frac{127}{16} \cdot \frac{2}{1} = \frac{127}{8}$$

Thus $s_7 = \frac{127}{8}$. ■

EXAMPLE 5 Given $s_n = 93$, $a_1 = 3$, $r = 2$, find n.

Solution:

$$s_n = \frac{a_1(1 - r^n)}{1 - r}$$

$$93 = \frac{3(1 - 2^n)}{1 - 2}$$

$$93 = \frac{3(1 - 2^n)}{-1}$$

$$-93 = 3(1 - 2^n)$$

$$-31 = 1 - 2^n$$

$$-32 = -2^n$$

$$32 = 2^n$$

$$2^5 = 2^n$$

Therefore, $n = 5$. ■

Examples 6 and 7 illustrate useful applications of geometric sequences and series.

EXAMPLE 6 A certain substance decomposes and loses 20% of its weight each hour. If there are originally 300 grams of the substance, how much remains after 7 hours?

Solution: This problem can be considered as a geometric sequence since the substance is decreasing by a certain rate (or percent) each hour. Often when working with a sequence it is helpful to write out the first few terms of the sequence. In this problem since we are concerned with the amount of the substance *remaining*, the rate, r, is 100% − 20% = 80% or 0.80. To obtain the terms in the sequence, the preceding term is multiplied by 0.80 giving the amount of the substance left in each succeeding hour.

amount remaining at beginning of			
1st hour	2nd hour	3rd hour	4th hour
$a_1 = 300$	$a_2 = 300(0.80)$	$a_3 = 300(0.80)^2$	$a_4 = 300(0.80)^3$

In this geometric sequence $a_1 = 300$ and $r = 0.8$. In general the amount of substance remaining *at the beginning* of the nth hour is $a_n = 300(0.80)^{n-1}$.

We are asked to find the amount remaining after 7 hours. Thus we must therefore find a_8 since the amount remaining after 7 hours is the same as the amount remaining at the beginning of the eighth hour.

$$\begin{aligned} a_8 &= 300(0.8)^{8-1} \\ &= 300(0.8)^7 \\ &= 300(0.2097) \\ &= 62.91 \text{ grams} \end{aligned}$$

Note that $a_n = 300(0.80)^n$ could also be used to find the amount *remaining after n* hours. Thus after 7 hours, $a_7 = 300(0.80)^7$. In explaining the solution to this example we used the form $a_n = a_1r^{n-1}$ rather than $a_n = a_1r^n$ since the first form was presented in the section. ■

EXAMPLE 7 Mary Foster invests \$1000 at 8% interest compounded annually in a savings account. Determine the amount in her account at the end of 6 years.

Solution: At the beginning of the second year the amount is $1000 + 0.08(1000) = 1000(1 + 0.8) = 1000(1.08)$. At the beginning of the third year this new amount increases by 8% to give $1000(1.08)^2$, and so on.

amount in account at beginning of

1st year	2nd year	3rd year	4th year
$a_1 = 1000$	$a_2 = 1000(1.08)$	$a_3 = 1000(1.08)^2$	$a_4 = 1000(1.08)^3$

Note this is a geometric sequence with $a_1 = 1000$ and $r = 1.08$. Since we are seeking the amount at the end of 6 years, which is the same as the beginning of the seventh year, we will find a_7.

$$\begin{aligned} a_n &= 1000(1.08)^{n-1} \\ a_7 &= 1000(1.08)^{7-1} \\ &= 1000(1.08)^6 \\ &= 1000(1.58687) \\ &= 1586.87 \end{aligned}$$

After 6 years the amount in the account is \$1586.87. The amount of interest gained is \$1586.87 − \$1000 = \$586.87. ■

Exercise Set 12.3

Determine the first five terms of the geometric sequence.

1. $a_1 = 5, r = 3$

2. $a_1 = 6, r = \frac{1}{2}$

3. $a_1 = 90, r = \frac{1}{3}$

4. $a_1 = -12, r = -1$

5. $a_1 = -15, r = -2$

6. $a_1 = 1, r = -\frac{1}{2}$

7. $a_1 = 3, r = \frac{3}{2}$

8. $a_1 = 60, r = \frac{1}{3}$

Find the indicated term of the geometric sequence.

9. $a_1 = 5, r = 2$; find a_6

10. $a_1 = -12, r = \frac{1}{2}$; find a_{10}

11. $a_1 = 18, r = 3$; find a_7

12. $a_1 = -20, r = -2$; find a_{10}

13. $a_1 = 2, r = \frac{1}{2}$; find a_8

14. $a_1 = 5, r = \frac{2}{3}$; find a_9

15. $a_1 = -3, r = -2$; find a_{12}

16. $a_1 = 80, r = \frac{1}{3}$; find a_{12}

Find the sum indicated.

17. $a_1 = 3, r = 4$; find s_5

18. $a_1 = 9, r = \frac{1}{2}$; find s_6

19. $a_1 = 80, r = 2$; find s_7

20. $a_1 = 1, r = -2$; find s_{12}

21. $a_1 = -30, r = -\frac{1}{2}$; find s_9

22. $a_1 = \frac{3}{5}, r = 3$; find s_7

23. $a_1 = -9, r = \frac{2}{5}$; find s_5

24. $a_1 = 35, r = \frac{1}{5}$; find s_{12}

Find the common ratio, r; then write an expression for the general (or nth) terms, a_n, for the geometric sequence.

25. $5, \frac{5}{2}, \frac{5}{4}, \frac{5}{8}, \ldots$

26. $3, 9, 27, 81, \ldots$

27. $2, -6, 18, -54, \ldots$

28. $\frac{3}{4}, \frac{6}{12}, \frac{12}{36}, \frac{24}{108}$

29. $-1, -3, -9, -18, \ldots$

30. $\frac{4}{3}, \frac{8}{3}, \frac{16}{3}, \frac{32}{3}, \ldots$

In the following exercises you may wish to use a calculator or logarithms.

31. In a geometric series $a_3 = 28$ and $a_5 = 112$; find r and a_1.

32. In a geometric series $a_2 = 27$ and $a_5 = 1$; find r and a_1.

33. In a geometric series $a_2 = 15$ and $a_5 = 405$; find r and a_1.

34. In a geometric series $a_2 = 12$ and $a_5 = -324$; find r and a_1.

35. Your salary increases at a rate of 15% per year If your initial salary is \$20,000, what will your salary be at the start of your twenty-fifth year?

36. A ball, when dropped, rebounds to four-fifths of its original height. How high will the ball rebound after the fourth bounce if it is dropped from a height of 30 feet?

37. A substance loses half its mass each day. If there are initially 300 grams of the substance, find:

(a) The number of days after which only 37.5 grams of the substance remain.

(b) The amount of the substance remaining after 8 days.

38. A certain type of bacteria doubles every hour. If there are initially 1000 bacteria, how long will it take for the number of bacteria to reach 64,000?

39. The population of the United States is 217.3 million. If the population grows at a rate of 6% per year, find:

(a) The population in 12 years.

(b) The number of years for the population to double.

40. A piece of farm equipment that costs \$75,000 new decreases in value by 15% per year. Find the value of the equipment after 4 years.

41. The amount of light filtering through a lake diminishes by $\frac{1}{2}$ for each meter of depth.

(a) Write a sequence indicating the amount of light remaining at depths of 1, 2, 3, 4, and 5 meters.

(b) What is the general term for this sequence?

(c) What is the remaining light at a depth of 7 meters.

42. You invest \$10,000 in a savings account paying 6% interest annually. Find the amount in your account at the end of 8 years.

43. A tracer dye is injected into Mark for medical reasons. After each hour, $\frac{2}{3}$ of the previous hour's dye remains. How much dye remains in Mark's system after 10 hours?

44. If you start with \$1 and double your money each day, how many days will it take to surpass \$1,000,000?

45. One method of depreciating an item on an income tax return is the declining balance method. With this method a given percentage of the original value of the item is depreciated each year. Suppose an item has a 5-year life and is depreciated using the declining balance method. Then at the end of its first year it loses $\frac{1}{5}$ of its value and $\frac{4}{5}$ of its value remains. At the end of the second year it loses $\frac{1}{5}$ of the remaining $\frac{4}{5}$ of its value, and so on. A car has a 5-year life expectancy and costs \$9800.

(a) Write a sequence showing the value of the car that remains for each of the first 3 years.

(b) What is the general term of this sequence?

(c) Find its value at the end of 5 years.

JUST FOR FUN

1. In Exercise Set 11.5, number 31, a formula for scrap value was given. The scrap value, S, is found by

$$S = c(1 - r)^n$$

where c is the original cost, r is the annual depreciation rate, and n is the number of years the object is depreciated.

(a) If you have not already done so, do Exercise 45 of this section to find the value of the car remaining at the end of 5 years.

(b) Use the formula above to find the scrap value of the car at the end of 5 years and compare this answer with the answer found in (a) above.

2. Find the sum of the sequence 1, 2, 4, 8, . . . , 1048576 and the number of terms in the sequence.

12.4 Infinite Geometric Series

1 *Identify infinite geometric series.*

2 *Find the sum of an infinite geometric series.*

1 All the geometric sequences that we have examined thus far have been finite since they have had a last term. The following sequence is an example of an infinite geometric sequence.

$$1, \frac{1}{2}, \frac{1}{4}, \frac{1}{8}, \frac{1}{16}, \ldots, \left(\frac{1}{2}\right)^{n-1}, \ldots$$

Note that the three dots at the end of the sequence indicate that the sequence continues indefinitely in the same manner. The sum of the terms in an infinite geometric sequence form an **infinite geometric series.** For example,

$$1 + \frac{1}{2} + \frac{1}{4} + \frac{1}{8} + \frac{1}{16} + \cdots + \left(\frac{1}{2}\right)^{n-1} + \cdots$$

is an infinite geometric series. Let's find some partial sums.

Sum of first two terms, s_2: $1 + \frac{1}{2} = \frac{3}{2} = 1.5$

Sum of first three terms, s_3: $1 + \frac{1}{2} + \frac{1}{4} = \frac{7}{4} = 1.75$

Sum of first four terms, s_4: $1 + \frac{1}{2} + \frac{1}{4} + \frac{1}{8} = \frac{15}{8} = 1.875$

Sum of first five terms, s_5: $1 + \frac{1}{2} + \frac{1}{4} + \frac{1}{8} + \frac{1}{16} = \frac{31}{16} = 1.9375$

Sum of first six terms, s_6: $1 + \frac{1}{2} + \frac{1}{4} + \frac{1}{8} + \frac{1}{16} + \frac{1}{32} = \frac{63}{32} = 1.96875$

Note that since each term of the geometric sequence is smaller than the preceding one, each additional term adds less and less to the sum. Also note that the sum seems to be getting closer and closer to 2.00.

2 Consider the formula for the sum of an n-term geometric series:

$$s_n = \frac{a_1(1 - r^n)}{1 - r}, \qquad r \neq 1$$

What happens to r^n if $|r| < 1$ and n gets larger and larger? Suppose that $r = \frac{1}{2}$; then

$$\left(\frac{1}{2}\right)^1 = \frac{1}{2} = 0.5$$
$$\left(\frac{1}{2}\right)^2 = \frac{1}{4} = 0.25$$
$$\left(\frac{1}{2}\right)^3 = \frac{1}{8} = 0.125$$
$$\left(\frac{1}{2}\right)^{20} \approx 0.000001$$

We can see that when $|r| < 1$ the value of r^n gets exceedingly close to 0 as n gets larger and larger. Thus when considering the sum of an infinite series, symbolized s_∞, the expression r^n approaches 0 when $|r| < 1$. Therefore, replacing r^n with a 0 in the formula $s_n = \dfrac{a_1(1 - r^n)}{1 - r}$ gives

Sum of an Infinite Series

$$s_\infty = \frac{a_1}{1 - r}, \qquad |r| < 1$$

EXAMPLE 1 Find the sum of the terms of the infinite sequence $1, \frac{1}{2}, \frac{1}{4}, \frac{1}{16}, \ldots$.

Solution: $a_1 = 1$ and $r = \frac{1}{2}$.

$$s_\infty = \frac{a_1}{1 - r} = \frac{1}{1 - \frac{1}{2}} = \frac{1}{\frac{1}{2}} = 2$$

Thus $1 + \frac{1}{2} + \frac{1}{4} + \frac{1}{8} + \frac{1}{16} + \cdots + (\frac{1}{2})^{n-1} + \cdots = 2$. ■

EXAMPLE 2 Find the sum of the infinite geometric series

$$5 - 2 + \frac{4}{5} - \frac{8}{25} + \frac{16}{125} - \frac{32}{625} + \cdots$$

Solution: The terms of the corresponding sequence are:

$$5, -2, \frac{4}{5}, -\frac{8}{25}, \ldots$$

$$r = -2 \div 5 = -\frac{2}{5} \quad \text{and} \quad a_1 = 5$$

$$s_\infty = \frac{a_1}{1 - r} = \frac{5}{1 - \left(-\frac{2}{5}\right)} = \frac{5}{1 + \frac{2}{5}}$$

$$= \frac{5}{\frac{7}{5}}$$

$$= \frac{25}{7}$$ ■

EXAMPLE 3 Write a fraction equivalent to 0.343434 . . . .

Solution: We can write this decimal as

$$0.34 + 0.0034 + 0.000034 + \cdots + (0.34)(0.01)^{n-1} + \cdots$$

This is an infinite geometric series with $r = 0.01$. Since $|r| < 1$,

$$s_\infty = \frac{a_1}{1 - r} = \frac{0.34}{1 - 0.01} = \frac{0.34}{0.99} = \frac{34}{99}$$

If you divide 34 by 99 on a calculator, you will see .34343434 displayed. ■

EXAMPLE 4 Each swing of a certain pendulum travels 90% of its previous swing. For example, if the swing to the right is 10 feet, then the swing back to the left is $0.9 \cdot 10 = 9$ feet (see Fig. 12.1). If the first swing is 10 feet long, determine the total distance traveled by the pendulum by the time it comes to rest.

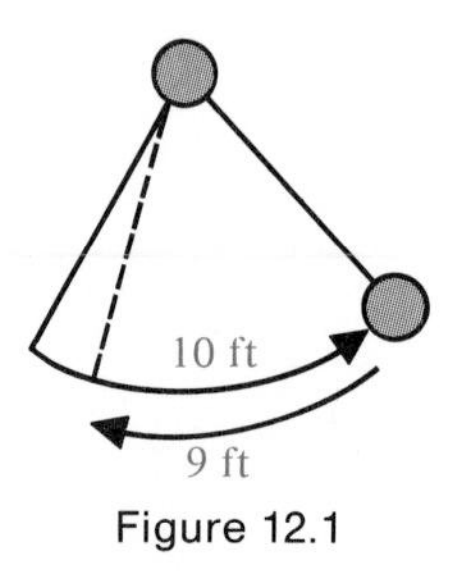

Figure 12.1

Solution: This problem may be considered an infinite geometric series with $a_1 = 10$ and $r = 0.9$. We can therefore use the formula $s_\infty = \dfrac{a_1}{1-r}$ to find the total distance traveled by the pendulum.

$$s_\infty = \frac{a_1}{1-r} = \frac{10}{1-0.9} = \frac{10}{0.1} = 100 \text{ feet} \quad \blacksquare$$

What is the sum of a geometric series where $|r| > 1$? Consider the geometric sequence where $a_1 = 1$ and $r = 2$.

$$1, 2, 4, 8, 16, 32, \ldots, 2^{n-1}, \ldots$$

The sum of its terms is

$$1 + 2 + 4 + 8 + 16 + 32 + \cdots + 2^{n-1}$$

What is the sum of this series? As n gets larger and larger, the sum gets larger and larger. We therefore say that the sum "does not exist." For $|r| > 1$, the sum of an infinite geometric series does not exist.

Exercise Set 12.4

Find the sum of the terms in each sequence.

1. $6, 3, \frac{3}{2}, \frac{3}{4}, \frac{3}{8}, \ldots$

2. $\frac{1}{3}, \frac{1}{9}, \frac{1}{27}, \frac{1}{81}, \ldots$

3. $5, 2, \frac{4}{5}, \frac{8}{25}, \ldots$

4. $-\frac{4}{3}, -\frac{4}{9}, -\frac{4}{27}, -\frac{4}{81}, \ldots$

5. $\frac{1}{3}, \frac{4}{15}, \frac{16}{75}, \frac{64}{375}, \ldots$

6. $6, -2, \frac{2}{3}, -\frac{2}{9}, \frac{2}{27}, \ldots$

7. $9, -1, \frac{1}{9}, -\frac{1}{81}, \ldots$

8. $\frac{5}{3}, 1, \frac{3}{5}, \frac{9}{25}, \ldots$

Find the sum of each infinite series.

9. $1 + \frac{1}{2} + \frac{1}{4} + \frac{1}{8} + \cdots$

10. $4 + 2 + 1 + \frac{1}{2} + \cdots$

11. $8 + \frac{16}{3} + \frac{32}{9} + \frac{64}{27} + \cdots$

12. $10 - 5 + \frac{5}{2} - \frac{5}{4} + \cdots$

13. $-60 + 20 - \frac{20}{3} + \frac{20}{9} - \cdots$

14. $4 + \frac{8}{3} + \frac{16}{9} + \frac{32}{27} + \cdots$

15. $-12 - \frac{12}{5} - \frac{12}{25} - \frac{12}{125} - \cdots$

16. $5 - 1 + \frac{1}{5} - \frac{1}{25} + \cdots$

Write a fraction equivalent to the repeating decimal.

17. 0.2727 . . .

18. 0.454545 . . .

19. 0.5555 . . .

20. 0.375375 . . .

21. 0.515151 . . .

22. 0.742742 . . .

23. Each swing of a pendulum travels 80% of its previous swing. If the first swing is 8 feet long, determine the total distance traveled by the pendulum by the time it comes to rest.

JUST FOR FUN

1. A ball is dropped from a height of 10 feet. The ball bounces to a height of 9 feet. On each successive bounce the ball reaches a height of 90% of its previous height. Find the total vertical distance traveled by the ball when it comes to rest (see figure below).

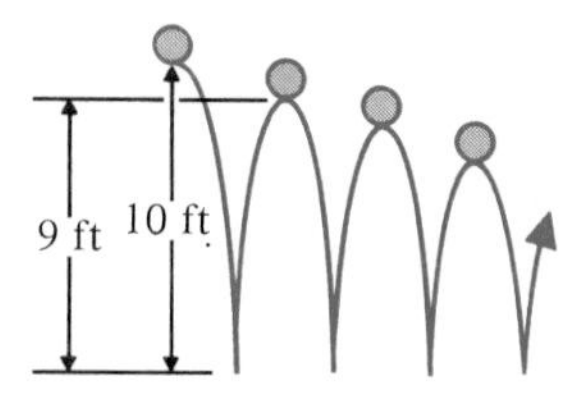

2. A particle follows the path indicated by the wave shown. Find the total vertical distance traveled by the particle.

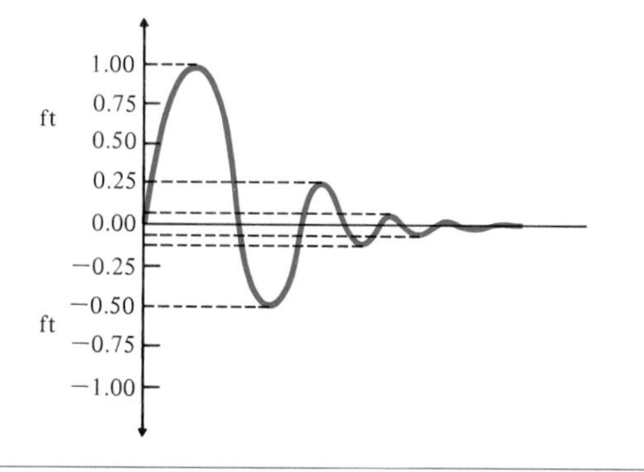

12.5 The Binomial Theorem (Optional)

1. *Evaluate factorials.*
2. *Use Pascal's triangle to find the numerical coefficients of an expanded binomial.*
3. *Use the binomial formula to expand binomials.*

1 To understand the binomial theorem, you must have an understanding of what **factorials** are. The notation $n!$ is read "n factorial" and is defined by

n Factorial

$$n! = n(n-1)(n-2)(n-3)\cdots(1)$$

for any positive integer n.

For example,

$$6! = 6 \cdot 5 \cdot 4 \cdot 3 \cdot 2 \cdot 1 = 720$$
$$4! = 4 \cdot 3 \cdot 2 \cdot 1 = 24$$
$$8! = 8 \cdot 7 \cdot 6 \cdot 5 \cdot 4 \cdot 3 \cdot 2 \cdot 1 = 40{,}320$$

Note that *0! is defined to be 1.*

Using direct multiplication, we can obtain the following expansions of the binomial $a + b$:

$$(a + b)^0 = 1$$
$$(a + b)^1 = a + b$$
$$(a + b)^2 = a^2 + 2ab + b^2$$
$$(a + b)^3 = a^3 + 3a^2b + 3ab^2 + b^2$$
$$(a + b)^4 = a^4 + 4a^3b + 6a^2b^2 + 4ab^3 + b^4$$
$$(a + b)^5 = a^5 + 5a^4b + 10a^3b^2 + 10a^2b^3 + 5ab^4 + b^5$$
$$(a + b)^6 = a^6 + 6a^5b + 15a^4b^2 + 20a^3b^3 + 15a^2b^4 + 6ab^5 + b^6$$

Note that when expanding a binomial of the form $(a + b)^n$:

1. There are $n + 1$ terms in the expansion.
2. The first term is a^n and the last term is b^n.
3. Progressing from the term on the left to the term on the right, the exponents on a decrease by 1 from term to term while the exponents on b increase by 1 from term to term.
4. The sum of the exponents in any one term is n.
5. The coefficients of the terms equidistant from the ends are the same.

2 If we examine just the variables in $(a + b)^5$, we have a^5, a^4b, a^3b^2, a^2b^3, ab^4, and b^5. The numerical coefficients of each term can be found by using **Pascal's triangle.**

Pascal's triangle

$n = 0$	1
$n = 1$	1 1
$n = 2$	1 2 1
$n = 3$	1 3 3 1
$n = 4$	1 4 6 4 1
$n = 5$	1 5 10 10 5 1
$n = 6$	1 6 15 20 15 6 1

Let us examine row 5 ($n = 4$) and row 6 ($n = 5$).

$$1 + 4 + 6 + 4 + 1$$

$$1 \quad 5 \quad 10 \quad 10 \quad 5 \quad 1$$

Note that the first and last numbers in each row are 1 and the inner numbers are obtained by adding the two numbers in the row above (to the right and left). The numerical coefficients of $(a + b)^5$ are 1, 5, 10, 10, 5, and 1. Combining the numerical coefficients and the variables, we obtain

$$(a + b)^5 = a^5 + 5a^4b + 10a^3b^2 + 10a^2b^3 + 5ab^4 + b^5$$

This method of expanding a binomial is not practical when n is large. A more practical procedure is to use the binomial formula.

3 The binomial formula is a convenient method for expanding any binomial to any positive integral power.

Binomial Formula

$$(a + b)^n = a^n + na^{n-1}b + \frac{n(n-1)}{2!}a^{n-2}b^2 + \frac{n(n-1)(n-1)}{3!}a^{n-3}b^3 + \cdots + b^n$$

For example,

$$\begin{aligned}(a + b)^5 &= a^5 + 5a^{5-1}b + \frac{5 \cdot 4}{2 \cdot 1}a^{5-2}b^2 + \frac{5 \cdot 4 \cdot 3}{3 \cdot 2 \cdot 1}a^{5-3}b^3 + \frac{5 \cdot 4 \cdot 3 \cdot 2}{4 \cdot 3 \cdot 2 \cdot 1}a^{5-4}b^4 + b^5 \\ &= a^5 + 5a^4b + 10a^3b^2 + 10a^2b^3 + 5ab^4 + b^5\end{aligned}$$

EXAMPLE 1 Use the binomial formula to expand $(2x + 3)^6$.

Solution: If we use $2x$ for a and 3 for b, we obtain

$$\begin{aligned}(2x + 3)^6 &= (2x)^6 + 6(2x)^5(3) + \frac{6 \cdot 5}{2 \cdot 1}(2x)^4(3)^2 + \frac{6 \cdot 5 \cdot 4}{3 \cdot 2 \cdot 1}(2x)^3(3)^3 \\ &\quad + \frac{6 \cdot 5 \cdot 4 \cdot 3}{4 \cdot 3 \cdot 2 \cdot 1}(2x)^2(3)^4 + \frac{6 \cdot 5 \cdot 4 \cdot 3 \cdot 2}{5 \cdot 4 \cdot 3 \cdot 2 \cdot 1}(2x)(3)^5 + (3)^6 \\ &= 64x^6 + 6(32x^5)(3) + 15(16x^4)(9) + 20(8x^3)(27) \\ &\quad + 15(4x^2)(81) + 6(2x)(243) + 729 \\ &= 64x^6 + 576x^5 + 2160x^4 + 4320x^3 + 4860x^2 + 2916x + 729\end{aligned}$$ ■

EXAMPLE 2 Use the binomial formula to expand $(5x - 2y)^4$.

Solution: Write $(5x - 2y)^4$ as $[5x + (-2y)]^4$. Use $5x$ in place of a and $-2y$ in place of b in the binomial formula.

$$\begin{aligned}(5x - 2y)^4 &= (5x)^4 + 4(5x)^3(-2y) + \frac{4 \cdot 3}{2 \cdot 1}(5x)^2(-2y)^2 \\ &\quad + \frac{4 \cdot 3 \cdot 2}{3 \cdot 2 \cdot 1}(5x)(-2y)^3 + (-2y)^4 \\ &= 625x^4 - 1000x^3y + 600x^2y^2 - 160xy^3 + 16y^4\end{aligned}$$ ■

EXAMPLE 3 Use the binomial formula to evaluate $(1.06)^6$ correct to four decimal places.

Solution: Write $(1.06)^6$ as $(1 + 0.06)^6$. Then use the binomial formula to evaluate.

$$\begin{aligned}(1 + 0.06)^6 &= 1^6 + 6(1)^5(0.06) + \frac{6 \cdot 5}{2 \cdot 1}(1)^4(0.6)^2 + \frac{6 \cdot 5 \cdot 4}{3 \cdot 2 \cdot 1}(1)^3(0.06)^3 \\ &\quad + \frac{6 \cdot 5 \cdot 4 \cdot 3}{4 \cdot 3 \cdot 2 \cdot 1}(1)^2(0.06)^4 + \cdots \\ &= 1 + 6(1)(0.06) + (15)(1)(0.0036) + (20)(1)(0.000216) \\ &\quad + (15)(1)(0.000013) + \cdots \\ &= 1 + 0.36 + 0.054 + 0.00432 + 0.0001944 \\ &= 1.4185144 \\ &= 1.4185\end{aligned}$$

Thus $(1.06)^6 = 1.4185$ to four places. We have stopped at five terms. The next term would contain at least five zeros following the decimal point. Therefore we need not go any farther. ■

Exercise Set 12.5

Use the binomial formula to expand each expression.

1. $(x + 4)^3$ **2.** $(2x + 3)^3$ **3.** $(a - b)^4$ **4.** $(2r + s^2)^4$

5. $(3a - b)^5$ **6.** $(x + 2y)^5$ **7.** $\left(2x + \frac{1}{2}\right)^4$ **8.** $\left(\frac{2}{3}x + \frac{3}{2}\right)^4$

9. $\left(\frac{x}{2} - 3\right)^4$ **10.** $(3x^2 + y)^5$

Write the first four terms of each expansion.

11. $(x + y)^{10}$ **12.** $(2x + 3)^8$ **13.** $(3x - y)^7$ **14.** $(3p + 2q)^{11}$

15. $(x^2 - 3y)^8$ **16.** $\left(2x + \frac{y}{5}\right)^9$

17. Evaluate $(1.04)^5$ correct to three decimal places.
18. Evaluate $(2.02)^6$ correct to three decimal places.
19. Evaluate $(1.01)^7$ correct to three decimal places.

JUST FOR FUN

1. Use the binomial theorem to approximate the value of $\sqrt[3]{1004}$ to three decimal places.
2. Use the binomial theorem to approximate the value of $\sqrt[4]{80}$ to three decimal places.
3. Use the binomial theorem to approximate the value of $(1.02)^{-4}$ to three decimal places.

Summary

Glossary

Arithmetic sequence (or **arithmetic progression**): A sequence in which each term after the first differs from the preceding term by a constant amount.
Arithmetic series: The sum of the terms of an arithmetic sequence.
Common difference: The amount by which each pair of terms differs in an arithmetic sequence.
Common ratio: The common multiple in a geometric series.
Finite sequence: A sequence that has a last term.
General term of a sequence: An expression that defines the nth term of the sequence.
Geometric series (or **geometric progression**): A sequence in which each term after the first is the same multiple of the preceding term.
Infinite geometric series: The sum of the terms in an infinite geometric sequence.
Infinite sequence: A sequence that has no last term.
Partial sum of a series: The sum of a finite number of consecutive terms of a series, beginning with the first term.
Sequence (or **progression**) A list of numbers arranged in a specific order.
Series: The sum of a sequence.

Important Facts

***n* Factorial:** $n! = n(n-1)(n-2)\cdots(2)(1)$

***n*th Term of an Arithmetic Sequence** $a_n = a_1 + (n-1)d$

***n*th Partial Sum of an Arithmetic Series** $s_n = \dfrac{n(a_1 + a_n)}{2}$

***n*th Term of a Geometric Sequence** $a_n = a_1 r^{n-1}$

***n*th Partial Sum of a Geometric Series** $s_n = \dfrac{a_1(1-r^n)}{1-r}, \; r \neq 1$

Sum of an Infinite Series $s_\infty = \dfrac{a_1}{1-r}, \; |r| < 1$

Binomial Formula

$$(a+b)^n = a^n + na^{n-1}b + \frac{n(n-1)}{2!}a^{n-2}b^2 + \frac{n(n-1)(n-2)}{3!}a^{n-3}b^3 + \cdots + b^n$$

Review Exercises

[12.1] Write the first five terms of the sequence.

1. $a_n = n + 2$ **2.** $a_n = \dfrac{1}{n}$ **3.** $a^n = n(n+1)$ **4.** $a_n = \dfrac{n^2}{n+4}$

Find the indicated term of the sequence.

5. $a_n = 3n + 4$, seventh term **6.** $a_n = (-1)^n + 3$, seventh term

7. $a_n = \dfrac{n+7}{n^2}$, ninth term **8.** $a_n = (n)(n-3)$, eleventh term

Find the first and third partial sums, s_1 and s_3.

9. $a_n = 3n + 2$ **10.** $a_n = 2n^2$ **11.** $a_n = \dfrac{n+3}{n+2}$ **12.** $a_n = (-1)^n(n+2)$

Write the next three terms of each sequence; then write an expression for the general term, a_n.

13. $1, 2, 4, 8, \ldots$

14. $-8, 4, -2, 1, \ldots$

15. $\frac{2}{3}, \frac{4}{3}, \frac{8}{3}, \frac{16}{3}, \ldots$

16. $9, 6, 3, 0, \ldots$

17. $-1, 1, -1, 1, \ldots$

18. $6, 12, 18, 24, \ldots$

[12.2] Write the first five terms of the arithmetic sequence with first term and common difference as given.

19. $a_1 = 5, d = 2$

20. $a_1 = \frac{1}{2}, d = -2$

21. $a_1 = -12, d = -\frac{1}{2}$

22. $a_1 = -100, d = \frac{1}{5}$

Find the desired quantity of the arithmetic sequence.

23. $a_1 = 2, d = 3$; find a_9

24. $a_1 = -12, d = -\frac{1}{2}$; find a_7

25. $a_1 = 50, a_5 = 34$; find d

26. $a_1 = -3, a_7 = 0$; find d

27. $a_1 = 12, a_n = -13, d = -5$; find n

28. $a_1 = 80, a_n = 152, d = 12$; find n

Find s_n and d for each arithmetic sequence.

29. $a_1 = 7, a_n = 21, n = 8$

30. $a_1 = -12, a_n = -48, n = 7$

31. $a_1 = \frac{3}{5}, a_n = 3, n = 7$

32. $a_1 = -\frac{10}{3}, a_n = -6, n = 9$

Write the first four terms of each arithmetic sequence; then find a_{10} and s_{10}.

33. $a_1 = 2, d = 4$

34. $a_1 = -8, d = -3$

35. $a_1 = \frac{5}{6}, d = \frac{2}{3}$

36. $a_1 = -80, d = 4$

Find the number of terms in each arithmetic sequence and s_n.

37. $3, 8, 13, \ldots, 53$

38. $-16, -11, -6, -1, \ldots, 24$

39. $\frac{6}{10}, \frac{9}{10}, \frac{12}{10}, \frac{15}{10}, \ldots, \frac{36}{10}$

40. $-5, 0, 5, 10, \ldots, 85$

[12.3] Determine the first five terms of each geometric sequence.

41. $a_1 = 5, r = 2$

42. $a_1 = -12, r = \frac{1}{2}$

43. $a_1 = 20, r = -\frac{2}{3}$

44. $a_1 = -100, r = \frac{1}{5}$

Find the indicated term of the geometric sequence.

45. $a_1 = 12, r = \frac{1}{3}$; find a_7

46. $a_1 = 25, r = 2$; find a_9

47. $a_1 = -8, r = -2$; find a_9

48. $a_1 = \frac{5}{12}, r = \frac{2}{3}$; find a_8

Find the indicated sum.

49. $a_1 = 12, r = 2$; find s_8

50. $a_1 = \frac{3}{5}, r = \frac{5}{3}$; find s_7

51. $a_1 = -84, r = -\frac{1}{4}$; find s_5

52. $a_1 = 9, r = \frac{3}{2}$; find s_9

Find the common ratio, r, then write an expression for the general term a_n, for the geometric sequences.

53. $6, 12, 24, \ldots$

54. $8, \frac{8}{3}, \frac{8}{9}, \ldots$

55. $-4, -20, -100, \ldots$

56. $\frac{9}{5}, \frac{18}{15}, \frac{36}{45}, \ldots$

[12.4] Find the sum of the terms in each infinite sequence.

57. $7, \frac{7}{2}, \frac{7}{4}, \frac{7}{8}, \ldots$

58. $-8, \frac{8}{3}, -\frac{8}{9}, \frac{8}{27}, \ldots$

59. $-5, -\frac{10}{3}, -\frac{20}{9}, -\frac{40}{27}, \ldots$

60. $\frac{7}{2}, 1, \frac{2}{7}, \frac{4}{49}, \ldots$

Find the sum of each infinite series.

61. $2 + 1 + \frac{1}{2} + \frac{1}{4} + \cdots$

62. $7 + \frac{7}{3} + \frac{7}{9} + \frac{7}{27} + \cdots$

63. $-12 - \frac{24}{3} - \frac{48}{9} - \frac{96}{27} - \cdots$

64. $5 - 1 + \frac{1}{5} - \frac{1}{25} + \cdots$

[12.4] Write a fraction equivalent to the repeating decimal.

65. 0.5252 . . .

66. 0.375375 . . .

[12.5] Use the binomial formula to expand the following.

67. $(3x + y)^4$

68. $(2x - 3y^2)^3$

Write the first four terms of the expansion.

69. $(x - 2y)^9$

70. $(2a^2 + 3b)^8$

[12.2–12.4]

71. Find the sum of the integers between 100 and 200.

72. Prof. Gayvert is offered a job with a starting salary of \$30,000 with the agreement that his salary will increase \$1000 at the end of each of the next seven years.
- **(a)** Write a sequence showing his salary for the first five years.
- **(b)** What is the general term of this sequence?
- **(c)** If this process were continued, what would his salary be after nine years?

73. You begin with \$100, double that to get \$200, double that again to get \$400, and so on. How much will you have after you perform this process 10 times?

74. If the inflation rate remains constant at 15% per year (each year the cost of living is 15% greater than the previous year), how much would an object that presently costs \$200 cost after 12 years?

75. Each successive swing of a pendulum travels 92% of the length of its previous swing. If the first swing is 8 feet in length, find the distance traveled by the pendulum by the time it comes to rest.

Practice Test

1. Write the first five terms of the sequence with $a_n = \frac{n+2}{n^2}$.

2. Find the first and third partial sums of $a_n = \frac{2n+3}{n}$.

3. Write the general term for the arithmetic sequence

$$\frac{1}{3}, \frac{2}{3}, \frac{3}{3}, \frac{4}{3}, \ldots$$

4. Write the general term for the geometric sequence 5, 10, 20, 40, . . .

Write the first four terms of the sequence.

5. $a_1 = 12, d = -3$

6. $a_1 = \frac{5}{8}, r = \frac{2}{3}$

7. Find a_8 when $a_1 = 100$ and $d = -12$.

8. Find s_8 when $a_1 = 3$, $a_8 = -11$ and $d = -2$.

9. Find the number of terms in the arithmetic sequence

$$-4, -16, -28, \ldots, -148$$

10. Find a_7 when $a_1 = 8$ and $r = \frac{2}{3}$.

11. Find s_7 when $a_1 = \frac{3}{5}$ and $r = -5$.

12. Find the common ratio and write an expression for the general term of the sequence.

$$12, 6, 3, \frac{3}{2}, \ldots$$

13. Find the sum of the infinite geometric series

$$3 + \frac{6}{3} + \frac{12}{9} + \frac{24}{27} + \cdots$$

14. Use the binomial formula to expand $(x + 2y)^4$.

Appendixes

A

A Review of Fractions

To be successful in algebra you must have a thorough understanding of fractions. This appendix gives a brief review of addition, subtraction, multiplication, and division of fractions. For a more complete explanation of fractions, see any arithmetic text.

The top number of a fraction is called the **numerator.** The bottom number is called the **denominator.**

$$\frac{3}{5} \begin{array}{l} \leftarrow \text{numerator} \\ \leftarrow \text{denominator} \end{array}$$

Multiplication of Fractions

The product of two or more fractions is obtained by multiplying their numerators together, then multiplying their denominators together as follows:

$$\frac{a}{b} \cdot \frac{c}{d} = \frac{a \cdot c}{b \cdot d}$$

EXAMPLE 1 (a) $\frac{3}{5} \cdot \frac{4}{7} = \frac{3 \cdot 4}{5 \cdot 7} = \frac{12}{35}$

(b) $\frac{5}{12} \cdot \frac{3}{10} = \frac{5 \cdot 3}{10 \cdot 12} = \frac{15}{120} = \frac{1}{8}$ ■

In Example 1(b) the fraction $\frac{15}{120}$ was reduced to $\frac{1}{8}$. To help avoid having to reduce fractions after multiplication, we often divide out common factors. When any numerator and any denominator in a **multiplication problem** have a common factor, divide both the numerator and the denominator by the common factor prior to multiplying. This process is illustrated in Example 2.

EXAMPLE 2 $\frac{5}{12} \cdot \frac{3}{10} = \frac{\overset{1}{\cancel{5}}}{12} \cdot \frac{3}{\underset{2}{\cancel{10}}}$ divide both 5 and 10 by 5

$= \frac{\overset{1}{\cancel{5}}}{\underset{4}{\cancel{12}}} \cdot \frac{\overset{1}{\cancel{3}}}{\underset{2}{\cancel{10}}}$ divide both 3 and 12 by 3

$= \frac{1 \cdot 1}{4 \cdot 2} = \frac{1}{8}$ ■

EXAMPLE 3 $\frac{25}{36} \cdot \frac{8}{15} = \frac{\overset{5}{\cancel{25}}}{36} \cdot \frac{8}{\underset{3}{\cancel{15}}}$ divide both 25 and 15 by 5

$= \frac{\overset{5}{\cancel{25}}}{\underset{9}{\cancel{36}}} \cdot \frac{\overset{2}{\cancel{8}}}{\underset{3}{\cancel{15}}}$ divide both 8 and 36 by 4

$= \frac{5 \cdot 2}{9 \cdot 3} = \frac{10}{27}$ ■

Division of Fractions

To divide fractions, invert the divisor and proceed as in multiplication.

$$\frac{a}{b} \div \frac{c}{d} = \frac{a}{b} \cdot \frac{d}{c} = \frac{a \cdot d}{b \cdot c}$$

EXAMPLE 4 $\frac{3}{8} \div \frac{5}{9} = \frac{3}{8} \cdot \frac{9}{5} = \frac{27}{40}$ ■

EXAMPLE 5 $\frac{7}{15} \div \frac{3}{5} = \frac{7}{\underset{3}{\cancel{15}}} \cdot \frac{\overset{1}{\cancel{5}}}{3} = \frac{7}{9}$ ■

EXAMPLE 6 $4 \div \frac{3}{5} = \frac{4}{1} \cdot \frac{5}{3} = \frac{20}{3}$ ■

EXAMPLE 7 $\frac{9}{16} \div 6 = \frac{9}{16} \div \frac{6}{1} = \frac{\overset{3}{\cancel{9}}}{16} \cdot \frac{1}{\underset{2}{\cancel{6}}} = \frac{3}{32}$ ■

Addition and Subtraction of Fractions

Only fractions with the same denominators can be added or subtracted. To add (or subtract) fractions with the same denominators, add (or subtract) the numerators while maintaining the common denominator.

$$\frac{a}{c}+\frac{b}{c}=\frac{a+b}{c}, \qquad \frac{a}{c}-\frac{b}{c}=\frac{a-b}{c}$$

EXAMPLE 8 (a) $\frac{3}{7}+\frac{2}{7}=\frac{3+2}{7}=\frac{5}{7}$

(b) $\frac{8}{12}-\frac{5}{12}=\frac{8-5}{12}=\frac{3}{12}=\frac{1}{4}$ ■

To add or subtract fractions with different denominators, rewrite the fractions so that they have the same, or a common, denominator; then proceed as above.

EXAMPLE 9 (a) $\frac{3}{4}+\frac{5}{6}=\left(\frac{3}{4}\cdot\frac{3}{3}\right)+\left(\frac{5}{6}\cdot\frac{2}{2}\right)=\frac{9}{12}+\frac{10}{12}=\frac{19}{12}$

(b) $\frac{5}{12}-\frac{7}{18}=\left(\frac{5}{12}\cdot\frac{3}{3}\right)-\left(\frac{7}{18}\cdot\frac{2}{2}\right)=\frac{15}{36}-\frac{14}{36}=\frac{1}{36}$ ■

EXAMPLE 10 $4+\frac{5}{7}=\frac{4}{1}+\frac{5}{7}=\frac{4}{1}\cdot\frac{7}{7}+\frac{5}{7}=\frac{28}{7}+\frac{5}{7}=\frac{33}{7}$ ■

COMMON STUDENT ERROR

Dividing out common factors can be be performed only when multiplying fractions; it cannot be performed when adding or subtracting fractions.

Correct

$$\frac{\overset{1}{\cancel{3}}}{5}\cdot\frac{4}{\underset{3}{\cancel{9}}}=\frac{1\cdot 4}{5\cdot 3}=\frac{4}{15}$$

$$\frac{\overset{1}{\cancel{4}}\cdot 7}{\underset{2}{\cancel{8}}}=\frac{1\cdot 7}{2}=\frac{7}{2}$$

Wrong

$$\frac{\overset{1}{\cancel{3}}}{5}+\frac{\overset{1}{\cancel{1}}}{\underset{1}{\cancel{3}}}$$

$$\frac{\overset{1}{\cancel{4}}+7}{\underset{2}{\cancel{8}}}$$

cannot divide out common factors when adding or subtracting

B Geometric Formulas

AREAS AND PERIMETERS

Figure	Sketch	Area	Perimeter
Square		$A = s^2$	$P = 4s$
Rectangle		$A = lw$	$P = 2l + 2w$
Parallelogram		$A = lh$	$P = 2l + 2w$
Trapezoid		$A = \frac{1}{2}h(b_1 + b_2)$	$P = s_1 + s_2 + b_1 + b_2$
Triangle		$A = \frac{1}{2}bh$	$P = s_1 + s_2 + b$

AREA AND CIRCUMFERENCE OF CIRCLE

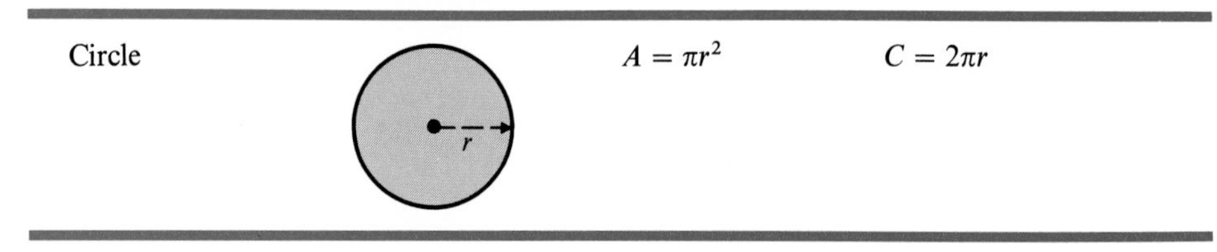

Circle		$A = \pi r^2$	$C = 2\pi r$

VOLUME AND SURFACE AREA OF THREE-DIMENSIONAL FIGURES

Figure	Sketch	Volume	Surface Area
Rectangular solid	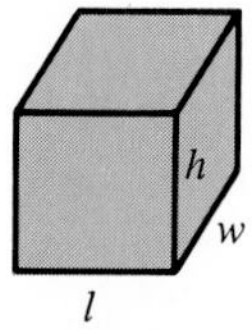 h, w, l	$V = lwh$	$s = 2lh + 2wh + 2wl$
Right circular cylinder	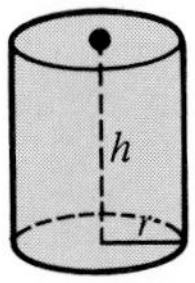 h, r	$V = \pi r^2 h$	$s = 2\pi rh + 2\pi r^2$
Sphere	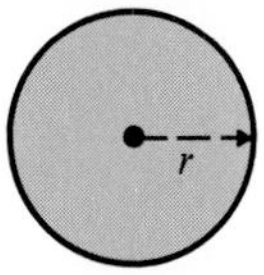 r	$V = \frac{4}{3}\pi r^3$	$s = 4\pi r^2$
Right circular cone	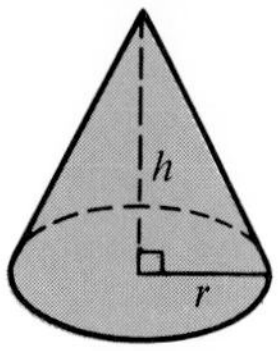 h, r	$V = \frac{1}{3}\pi r^2 h$	$s = \pi r\sqrt{r^2 + h^2}$
Square or rectangular pyramid	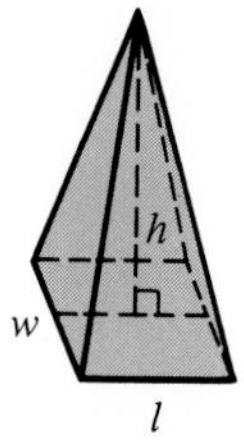 h, w, l	$V = \frac{1}{3} lwh$	

C

Squares and Square Roots

Number	Square	Square Root	Number	Square	Square Root
1	1	1.000	51	2,601	7.141
2	4	1.414	52	2,704	7.211
3	9	1.732	53	2,809	7.280
4	16	2.000	54	2,916	7.348
5	25	2.236	55	3,025	7.416
6	36	2.449	56	3,136	7.483
7	49	2.646	57	3,249	7.550
8	64	2.828	58	3,364	7.616
9	81	3.000	59	3,481	7.681
10	100	3.162	60	3,600	7.746
11	121	3.317	61	3,721	7.810
12	144	3.464	62	3,844	7.874
13	169	3.606	63	3,969	7.937
14	196	3.742	64	4,096	8.000
15	225	3.873	65	4,225	8.062
16	256	4.000	66	4,356	8.124
17	289	4.123	67	4,489	8.185
18	324	4.243	68	4,624	8.246
19	361	4.359	69	4,761	8.307
20	400	4.472	70	4,900	8.367
21	441	4.583	71	5,041	8.426
22	484	4.690	72	5,184	8.485
23	529	4.796	73	5,329	8.544
24	576	4.899	74	5,476	8.602
25	625	5.000	75	5,625	8.660
26	676	5.099	76	5,776	8.718
27	729	5.196	77	5,929	8.775
28	784	5.292	78	6,084	8.832
29	841	5.385	79	6,241	8.888
30	900	5.477	80	6,400	8.944
31	961	5.568	81	6,561	9.000
32	1,024	5.657	82	6,724	9.055
33	1,089	5.745	83	6,889	9.110
34	1,156	5.831	84	7,056	9.165
35	1,225	5.916	85	7,225	9.220
36	1,296	6.000	86	7,396	9.274
37	1,369	6.083	87	7,569	9.327
38	1,444	6.164	88	7,744	9.381
39	1,521	6.245	89	7,921	9.434
40	1,600	6.325	90	8,100	9.487
41	1,681	6.403	91	8,281	9.539
42	1,764	6.481	92	8,464	9.592
43	1,849	6.557	93	8,649	9.644
44	1,936	6.633	94	8,836	9.695
45	2,025	6.708	95	9,025	9.747
46	2,116	6.782	96	9,216	9.798
47	2,209	6.856	97	9,409	9.849
48	2,304	6.928	98	9,604	9.899
49	2,401	7.000	99	9,801	9.950
50	2,500	7.071	100	10,000	10.000

D
Common Logarithms

n	0	1	2	3	4	5	6	7	8	9
1.0	.0000	.0043	.0086	.0128	.0170	.0212	.0253	.0294	.0334	.0374
1.1	.0414	.0453	.0492	.0531	.0569	.0607	.0645	.0682	.0719	.0755
1.2	.0792	.0828	.0864	.0899	.0934	.0969	.1004	.1038	.1072	.1106
1.3	.1139	.1173	.1206	.1239	.1271	.1303	.1335	.1367	.1399	.1430
1.4	.1461	.1492	.1523	.1553	.1584	.1614	.1644	.1673	.1703	.1732
1.5	.1761	.1790	.1818	.1847	.1875	.1903	.1931	.1959	.1987	.2014
1.6	.2041	.2068	.2095	.2122	.2148	.2175	.2201	.2227	.2253	.2279
1.7	.2304	.2330	.2355	.2380	.2405	.2430	.2455	.2480	.2504	.2529
1.8	.2553	.2577	.2601	.2625	.2648	.2672	.2695	.2718	.2742	.2765
1.9	.2788	.2810	.2833	.2856	.2878	.2900	.2923	.2945	.2967	.2989
2.0	.3010	.3032	.3054	.3075	.3096	.3118	.3139	.3160	.3181	.3201
2.1	.3222	.3243	.3263	.3284	.3304	.3324	.3345	.3365	.3385	.3404
2.2	.3424	.3444	.3464	.3483	.3502	.3522	.3541	.3560	.3579	.3598
2.3	.3617	.3636	.3655	.3674	.3692	.3711	.3729	.3747	.3766	.3784
2.4	.3802	.3820	.3838	.3856	.3874	.3892	.3909	.3927	.3945	.3962
2.5	.3979	.3997	.4014	.4031	.4048	.4065	.4082	.4099	.4116	.4133
2.6	.4150	.4166	.4183	.4200	.4216	.4232	.4249	.4265	.4281	.4298
2.7	.4314	.4330	.4346	.4362	.4378	.4393	.4409	.4425	.4440	.4456
2.8	.4472	.4487	.4502	.4518	.4533	.4548	.4564	.4579	.4594	.4609
2.9	.4624	.4639	.4654	.4669	.4683	.4698	.4713	.4728	.4742	.4757
3.0	.4771	.4786	.4800	.4814	.4829	.4843	.4857	.4871	.4886	.4900
3.1	.4914	.4928	.4942	.4955	.4969	.4983	.4997	.5011	.5024	.5038
3.2	.5051	.5065	.5079	.5092	.5105	.5119	.5132	.5145	.5159	.5172
3.3	.5185	.5198	.5211	.5224	.5237	.5250	.5263	.5276	.5289	.5302
3.4	.5315	.5328	.5340	.5353	.5366	.5378	.5391	.5403	.5416	.5428
3.5	.5441	.5453	.5465	.5478	.5490	.5502	.5514	.5527	.5539	.5551
3.6	.5563	.5575	.5587	.5599	.5611	.5623	.5635	.5647	.5658	.5670
3.7	.5682	.5694	.5705	.5717	.5729	.5740	.5752	.5763	.5775	.5786
3.8	.5798	.5809	.5821	.5832	.5843	.5855	.5866	.5877	.5888	.5899
3.9	.5911	.5922	.5933	.5944	.5955	.5966	.5977	.5988	.5999	.6010
4.0	.6021	.6031	.6042	.6053	.6064	.6075	.6085	.6096	.6107	.6117
4.1	.6128	.6138	.6149	.6160	.6170	.6180	.6191	.6201	.6212	.6222
4.2	.6232	.6243	.6253	.6263	.6274	.6284	.6294	.6304	.6314	.6325
4.3	.6335	.6345	.6355	.6365	.6375	.6385	.6395	.6405	.6415	.6425
4.4	.6435	.6444	.6454	.6464	.6474	.6484	.6493	.6503	.6513	.6522
4.5	.6532	.6542	.6551	.6561	.6571	.6580	.6590	.6599	.6609	.6618
4.6	.6628	.6637	.6646	.6656	.6665	.6675	.6684	.6693	.6702	.6712
4.7	.6721	.6730	.6739	.6749	.6758	.6767	.6776	.6785	.6794	.6803
4.8	.6812	.6821	.6830	.6839	.6848	.6857	.6866	.6875	.6884	.6893
4.9	.6902	.6911	.6920	.6928	.6937	.6946	.6955	.6964	.6972	.6981
5.0	.6990	.6998	.7007	.7016	.7024	.7033	.7042	.7050	.7059	.7067
5.1	.7076	.7084	.7093	.7101	.7110	.7118	.7126	.7135	.7143	.7152
5.2	.7160	.7168	.7177	.7185	.7193	.7202	.7210	.7218	.7226	.7235
5.3	.7243	.7251	.7259	.7267	.7275	.7284	.7292	.7300	.7308	.7316
5.4	.7324	.7332	.7340	.7348	.7356	.7364	.7372	.7380	.7388	.7396

Table continued on page 496.

COMMON LOGARITHMS (*continued*)

n	0	1	2	3	4	5	6	7	8	9
5.5	.7404	.7412	.7419	.7427	.7435	.7443	.7451	.7459	.7466	.7474
5.6	.7482	.7490	.7497	.7505	.7513	.7520	.7528	.7536	.7543	.7551
5.7	.7559	.7566	.7574	.7582	.7589	.7597	.7604	.7612	.7619	.7627
5.8	.7634	.7642	.7649	.7657	.7664	.7672	.7679	.7686	.7694	.7701
5.9	.7709	.7716	.7723	.7731	.7738	.7745	.7752	.7760	.7767	.7774
6.0	.7782	.7789	.7796	.7803	.7810	.7818	.7825	.7832	.7839	.7846
6.1	.7853	.7860	.7868	.7875	.7882	.7889	.7896	.7903	.7910	.7917
6.2	.7924	.7931	.7938	.7945	.7952	.7959	.7966	.7973	.7980	.7987
6.3	.7993	.8000	.8007	.8014	.8021	.8028	.8035	.8041	.8048	.8055
6.4	.8062	.8069	.8075	.8082	.8089	.8096	.8102	.8109	.8116	.8122
6.5	.8129	.8136	.8142	.8149	.8156	.8162	.8169	.8176	.8182	.8189
6.6	.8195	.8202	.8209	.8215	.8222	.8228	.8235	.8241	.8248	.8254
6.7	.8261	.8267	.8274	.8280	.8287	.8293	.8299	.8306	.8312	.8319
6.8	.8325	.8331	.8838	.8344	.8351	.8357	.8363	.8370	.8376	.8382
6.9	.8388	.8395	.8401	.8407	.8414	.8420	.8426	.8432	.8439	.8445
7.0	.8451	.8457	.8463	.8470	.8476	.8482	.8488	.8494	.8500	.8506
7.1	.8513	.8519	.8525	.8531	.8537	.8543	.8549	.8555	.8561	.8567
7.2	.8573	.8579	.8585	.8591	.8597	.8603	.8609	.8615	.8621	.8627
7.3	.8633	.8639	.8645	.8651	.8657	.8663	.8669	.8675	.8681	.8686
7.4	.8692	.8698	.8704	.8710	.8716	.8722	.8727	.8733	.8739	.8745
7.5	.8751	.8756	.8762	.8768	.8774	.8779	.8785	.8791	.8797	.8802
7.6	.8808	.8814	.8820	.8825	.8831	.8837	.8842	.8848	.8854	.8859
7.7	.8865	.8871	.8876	.8882	.8887	.8893	.8899	.8904	.8910	.8915
7.8	.8921	.8927	.8932	.8938	.8943	.8949	.8954	.8960	.8965	.8971
7.9	.8976	.8982	.8987	.8993	.8998	.9004	.9009	.9015	.9020	.9025
8.0	.9031	.9036	.9042	.9047	.9053	.9058	.9063	.9069	.9074	.9079
8.1	.9085	.9090	.9096	.9101	.9106	.9112	.9117	.9122	.9128	.9133
8.2	.9138	.9143	.9149	.9154	.9159	.9165	.9170	.9175	.9180	.9186
8.3	.9191	.9196	.9201	.9206	.9212	.9217	.9222	.9227	.9232	.9238
8.4	.9243	.9248	.9253	.9258	.9263	.9269	.9274	.9279	.9284	.9289
8.5	.9294	.9299	.9304	.9309	.9315	.9320	.9325	.9330	.9335	.9340
8.6	.9345	.9350	.9355	.9360	.9365	.9370	.9375	.9380	.9385	.9390
8.7	.9395	.9400	.9405	.9410	.9415	.9420	.9425	.9430	.9435	.9440
8.8	.9445	.9450	.9455	.9460	.9465	.9469	.9474	.9479	.9484	.9489
8.9	.9494	.9499	.9504	.9509	.9513	.9518	.9523	.9528	.9533	.9538
9.0	.9542	.9547	.9552	.9557	.9562	.9566	.9571	.9576	.9581	.9586
9.1	.9590	.9595	.9600	.9605	.9609	.9614	.9619	.9624	.9628	.9633
9.2	.9638	.9643	.9647	.9652	.9657	.9661	.9666	.9671	.9675	.9680
9.3	.9685	.9689	.9694	.9699	.9703	.9708	.9713	.9717	.9722	.9727
9.4	.9731	.9736	.9741	.9745	.9750	.9754	.9759	.9763	.9768	.9773
9.5	.9777	.9782	.9786	.9791	.9795	.9800	.9805	.9809	.9814	.9818
9.6	.9823	.9827	.9832	.9836	.9841	.9845	.9850	.9854	.9859	.9863
9.7	.9868	.9872	.9877	.9881	.9886	.9890	.9894	.9899	.9903	.9908
9.8	.9912	.9917	.9921	.9926	.9930	.9934	.9939	.9943	.9948	.9952
9.9	.9956	.9961	.9965	.9969	.9974	.9978	.9983	.9987	.9991	.9996

Answers

CHAPTER 1

Exercise Set 1.1

1. A = {4, 5, 6, 7} **3.** C = {6, 8, 10} **5.** E = {0, 1, 2, 3, 4, 5, 6} **7.** G = { } **9.** I = {−4, −3, −2, −1, . . .}
11. $K = \{\ \}$ **13.** $\notin$ **15.** $\in$ **17.** $\notin$ **19.** $\notin$ **21.** $\subseteq$ **23.** $\not\subseteq$ **25.** $\not\subseteq$ **27.** $\not\subseteq$ **29.** $\not\subseteq$ **31.** $\subseteq$
33. $\subseteq$ **35.** $\not\subseteq$ **37.** $\subseteq$ **39.** $\not\subseteq$ **41.** $\not\subseteq$ **43.** $\subseteq$ **45.** True **47.** False **49.** True **51.** False
53. True **55.** True **57.** True **59.** False **61.** True **63.** 4 **65.** −6, 4, 0 **67.** $\sqrt{7}, \sqrt{5}$ **69.** 2, 4
71. 2, 4, −5.33, $\frac{9}{2}$, −100, −7, 4.7 **73.** 2, 4, −5.33, $\frac{9}{2}$, $\sqrt{7}$, $\sqrt{2}$, −100, −7, 4.7 **75.** $A \cup B = \{1, 2, 3, 4\}$, $A \cap B = \{2, 3\}$
77. $A \cup B = \{-1, -2, -4, -5, -6\}$, $A \cap B = \{-2, -4\}$ **79.** $A \cup B = \{0, 1, 2, 3\}$, $A \cap B = \{\ \}$
81. $A \cup B = \{2, 4, 6, 8, \ldots\}$, $A \cap B = \{2, 4, 6\}$ **83.** $A \cup B = \{0, 1, 2, 3, 4, 5, 6, 7, 8\}$, $A \cap B = \{\ \}$
85. $A \cup B = \{1, 2, 3, 4, \ldots\}$, $A \cap B = \{2, 4, 6, 8, \ldots\}$

Exercise Set 1.2

1. Commutative property of addition **3.** Distributive property **5.** Associative property of addition
7. Commutative property of addition **9.** Commutative property of multiplication **11.** Identity property of addition
13. Commutative property of addition **15.** Commutative property of addition **17.** Commutative property of addition
19. Distributive property **21.** Double negative property **23.** Identity property of multiplication
25. Inverse property of addition **27.** Inverse property of addition **29.** Inverse property of addition
31. Identity property of multiplication **33.** Inverse property of multiplication **35.** Double negative property
37. Multiplication property of 0. **39.** Double negative property **41.** $3 + x$ **43.** $x + (2 + 3)$ **45.** x **47.** x
49. x **51.** 0 **53.** $1x + 1y$ or $x + y$ **55.** 3 **57.** 0 **59.** $-4, \frac{1}{4}$ **61.** $3, -\frac{1}{3}$ **63.** $-\frac{2}{3}, \frac{3}{2}$ **65.** $6, -\frac{1}{6}$
67. $\frac{3}{7}, -\frac{7}{3}$

Exercise Set 1.3

1. $>$ **3.** $>$ **5.** $>$ **7.** $<$ **9.** $<$ **11.** $>$ **13.** $<$ **15.** $>$ **17.** $>$ **19.** $<$ **21.** $>$ **23.** $>$
25. 6 **27.** 4 **29.** 2 **31.** $\frac{1}{2}$ **33.** 0 **35.** 45 **37.** 13.84 **39.** −7 **41.** −3 **43.** −18 **45.** −8
47. $>$ **49.** $<$ **51.** $=$ **53.** $<$ **55.** $>$ **57.** $<$ **59.** $>$ **61.** $<$ **63.** $>$ **65.** $-|12|, -8, -4, 0, |-10|$
67. $|3|, |4|, |-5|, |9|, |-12|$ **69.** $-|7|, -|6|, -|5|, |-8|, |-9|$ **71.** $-9, -|7|, |-1|, 5, |15|$
73. $-|20|, -|-18|, -12, -|9|, -8$ **75.** $3, 3.1, 3.4, |-3.6|, |3.9|$ **77.** $-7.8, -|7.3|, -7.1, -7, |-7.4|$ **79.** 8
81. −10, 10 **83.** −7, 3 **85.** $a \geq 0$ **87.** { } **89.** All real numbers, $\mathbb{R}$

Exercise Set 1.4

1. 1 **3.** 10 **5.** 5 **7.** 8 **9.** -11 **11.** 39 **13.** -34 **15.** -4 **17.** -2 **19.** -31 **21.** -16 **23.** 7 **25.** 1 **27.** -12 **29.** -5 **31.** 11 **33.** -27 **35.** 1 **37.** 2 **39.** -10 **41.** -2 **43.** -1 **45.** 5 **47.** -18 **49.** -24 **51.** -48 **53.** -48 **55.** 6 **57.** -96 **59.** -3 **61.** 1 **63.** -16 **65.** -9 **67.** 1 **69.** -24 **71.** -2 **73.** 1 **75.** $-\frac{3}{16}$ **77.** $-\frac{1}{9}$ **79.** $\frac{12}{5}$ **81.** 170 ft below sea level (or -170 ft) **83.** \$4313 **85.** True **87.** False **89.** True **91.** True **93.** False **95.** False **97.** True

Just for Fun

1. -50 **2.** 84 **3.** -1 **4.** 1 **5.** 6

Exercise Set 1.5

1. 9 **3.** 8 **5.** 27 **7.** 216 **9.** -8 **11.** -1 **13.** -32 **15.** $\frac{8}{27}$ **17.** 1 **19.** 4 **21.** -3 **23.** -1 **25.** 1 **27.** 4 **29.** 8 **31.** $\frac{5}{6}$ **33.** $\frac{1}{2}$ **35.** $\frac{15}{9} = \frac{5}{3}$ **37.** 4 **39.** -2 **41.** -4 **43.** 1 **45.** 5 **47.** -6 **49.** $\frac{1}{4}$ **51.** $9, -9$ **53.** $1, -1$ **55.** $1, -1$ **57.** $9, -9$ **59.** $16, -16$ **61.** $27, -27$ **63.** $1, -1$ **65.** $-27, 27$ **67.** $-8, 8$ **69.** $64, -64$ **71.** 21 **73.** 29 **75.** 0 **77.** -19 **79.** -2 **81.** 31 **83.** 1 **85.** 7 **87.** 4 **89.** No real number when squared gives a negative number.
91. A negative number raised to an odd power is a negative number.

Just for Fun

1. (a) $2^0 + 2^1 + 2^2 + \cdots + 2^{29} = 2^{30} - 1$ cents **(b)** $(2^{30} - 1)/100 = \$10{,}737{,}418.23$ **(c)** Gains \$10,707,418.23

Exercise Set 1.6

1. 26 **3.** 19 **5.** 23 **7.** 2 **9.** -64 **11.** 29 **13.** -32 **15.** 2 **17.** 294 **19.** $\frac{4}{11}$ **21.** 2 **23.** $\frac{27}{5}$ **25.** $\frac{5}{11}$ **27.** -2 **29.** -20 **31.** 3 **33.** 17 **35.** 1 **37.** -7 **39.** 44 **41.** 75 **43.** -40 **45.** 0 **47.** 21 **49.** 33 **51.** 42 **53.** $\frac{25}{2}$ **55.** 8 **57.** $\frac{4}{9}$ **59.** $(3x + 6)^2$, 225 **61.** $6(3x + 6) - 9$, 81 **63.** $[(x + 3)/2y]^2 - 3$, 1

Just for Fun

1. $\frac{883}{48}$ **2.** 21904 **3.** $\frac{131072}{5}$ **4.** $-\frac{15}{19}$ **5.** $-\frac{3}{35}$

Review Exercises

1. $\{3, 4, 5, 6\}$ **2.** $\{0, 3, 6, 9, \ldots\}$ **3.** $\in$ **4.** $\notin$ **5.** $\notin$ **6.** $\in$ **7.** $\subseteq$ **8.** $\nsubseteq$ **9.** $\nsubseteq$ **10.** $\subseteq$ **11.** $\subseteq$ **12.** $\subseteq$ **13.** $\subseteq$ **14.** $\subseteq$ **15.** $\subseteq$ **16.** $\nsubseteq$ **17.** 4, 6 **18.** 4, 6, 0 **19.** $-3, 4, 6, 0$ **20.** $-3, 4, 6, \frac{1}{2}, 0, \frac{15}{27}, -\frac{1}{5}, 1.47$ **21.** $\sqrt{5}, \sqrt{3}$ **22.** $-3, 4, 6, \frac{1}{2}, \sqrt{5}, \sqrt{3}, 0, \frac{15}{27}, -\frac{1}{5}, 1.47$ **23.** True **24.** True **25.** False **26.** True **27.** True **28.** $A \cup B = \{1, 2, 3, 4, 5\}$, $A \cap B = \{2, 3, 4, 5\}$ **29.** $A \cup B = \{2, 3, 4, 5, 6, 7, 8, 9\}$, $A \cap B = \varnothing$ **30.** $A \cup B = \{1, 2, 3, 4, \ldots\}$, $A \cap B = \{2, 4, 6, \ldots\}$ **31.** $A \cup B = \{3, 4, 5, 6, 9, 10, 11, 12\}$, $A \cap B = \{9, 10\}$ **32.** Commutative property of addition **33.** Commutative property of addition **34.** Distributive property **35.** Commutative property of multiplication **36.** Associative property of addition **37.** Identity property of addition **38.** Associative property of multiplication **39.** Identity property of multiplication **40.** Double negative property **41.** Multiplication property of 0. **42.** Identity property of addition **43.** Identity property of multiplication **44.** Inverse property of addition **45.** Inverse property of multiplication **46.** Inverse property of multiplication **47.** Identity property of multiplication **48.** $3 + x$ **49.** $3x + 15$ **50.** $x + [6 + (-4)]$ **51.** $x \cdot 3$ **52.** $9 \cdot (x \cdot y)$ **53.** $4x - 4y + 20$ **54.** a **55.** a **56.** 0 **57.** 1 **58.** a **59.** $>$ **60.** $<$ **61.** $<$ **62.** $>$ **63.** $<$ **64.** $<$ **65.** $<$ **66.** $>$ **67.** $=$ **68.** $=$ **69.** $<$ **70.** $<$ **71.** $>$ **72.** $>$ **73.** $>$ **74.** $>$ **75.** $-5, -2, 4, |7|$ **76.** $0, \frac{3}{5}, 2, 3, |-3|$ **77.** $-2, 3, |-5|, |-7|$ **78.** $-4, -|3|, -2.1, -2$ **79.** $-4, -|-3|, 5, 6$ **80.** $-3, 0, |1.6|, |-2.3|$ **81.** 1 **82.** -63 **83.** -12 **84.** 8 **85.** -1 **86.** -2 **87.** -1 **88.** 16 **89.** 21 **90.** 21 **91.** -47 **92.** 15 **93.** 31 **94.** 6 **95.** 27 **96.** $-\frac{8}{5}$ **97.** $\frac{8}{3}$ **98.** 5 **99.** $-\frac{16}{19}$ **100.** 15 **101.** 10 **102.** -3 **103.** 39 **104.** 7 **105.** -50 **106.** -249 **107.** 204

Practice Test

1. $\{6, 7, 8, 9, \ldots\}$ **2.** $\nsubseteq$ **3.** $\subseteq$ **4.** True **5.** False **6.** True **7.** $-\frac{3}{5}, 2, -4, 0, \frac{19}{12}, 2.57, -1.92$ **8.** $-\frac{3}{5}, 2, -4, 0, \frac{19}{12}, 2.57, \sqrt{8}, \sqrt{2}, -1.92$ **9.** $A \cup B = \{5, 7, 8, 9, 10, 11, 14\}$, $A \cap B = \{8, 10\}$ **10.** $A \cup B = \{1, 3, 5, 7, \ldots\}$, $A \cap B = \{3, 5, 7, 9, 11\}$ **11.** $<$ **12.** $>$ **13.** $-|4|, -2, |3|, 6$ **14.** Distributive property **15.** Associative property of addition **16.** Commutative property of addition **17.** Inverse property of multiplication **18.** Identity property of addition **19.** 25 **20.** 23 **21.** $-\frac{1}{4}$ **22.** $-\frac{37}{22}$ **23.** $\frac{64}{5}$ **24.** 17 **25.** 39

CHAPTER 2

Exercise Set 2.1

1. Reflexive property **3.** Symmetric property **5.** Transitive property **7.** Reflexive property **9.** Addition property **11.** Multiplication property **13.** Multiplication property **15.** Addition property **17.** Addition property **19.** Multiplication property **21.** Symmetric property **23.** First **25.** Second **27.** Fifth **29.** Zero **31.** Thirteenth **33.** Twelfth **35.** $15x - 15$ **37.** $5x^2 - x - 5$ **39.** $-4x^2 - 8x + 7$ **41.** $-3y - 3$ **43.** Cannot be simplified **45.** $4xy + y^2 - 2$ **47.** $8x + 40$ **49.** $4x + 16$ **51.** $14x - 32$ **53.** $-19x - 4$ **55.** $x - 14y$ **57.** 1 **59.** $\frac{2}{5}$ **61.** 15 **63.** 9 **65.** $-\frac{2}{5}$ **67.** $\frac{55}{4}$ **69.** 0 **71.** -3 **73.** -8 **75.** 20 **77.** 1 **79.** -5 **81.** -4 **83.** 0 **85.** 8 **87.** 3 **89.** -4 **91.** $\frac{5}{2}$ **93.** No solution **95.** -5 **97.** All real numbers **99.** -2 **101.** -5 **103.** 4 **105.** $\frac{96}{5}$ **107.** -6 **109.** An equation that is true for all real numbers **111.** Equations with the same solution set **113.** $2x + 3 = 5 + 3$, $7(2x) = 7(5)$, $2x/2 = 5/2$

Just for Fun **1.** $\frac{306}{157}$ **2.** $-\frac{115}{31}$ **3.** $\frac{1524}{131}$

Exercise Set 2.2

1. $x + 5x = 24$; 4, 20 **3.** $x + (x + 1) = 51$; 25, 26 **5.** $2x - 8 = 38$; 23 **7.** $x + 3x = 48$; 12, 36 **9.** $x + (x + 2) + (x + 4) = 66$; 20, 22, 24 **11.** $10 - (3x/5) = 4$; 10 **13.** $5x = \frac{1}{2}(4x + 2) + 2$; 1, 6 **15.** $x + x + 2x = 180$; 45°, 45°, 90° **17.** $x + x + (x + 15) = 45$; 10, 10, 25 **19.** $2.89x = 37.99$; 13.15 months **21.** $16x = 236$; 14.75 days **23.** $20 + 0.18x = 38$; 100 mi **25.** $15{,}000 - 800n = 8000$; 8.75 years **27.** $x + 0.18(350) = 83$; $20 **29.** $x - 0.15x = 1275$; $1500 **31.** $250 + 20x = 400$; 7.5 **33.** $x + 0.07x = 156$; $145.79 **35.** $x + 0.0675x = 500$; $468.38 **37.** $150 + 6x = 510$; 60 days **39.** $2x + 2(2x + 2) = 40$; 6 ft, 14 ft **41.** $2(\frac{x}{2} + 1) + 2x = 20$; 6 m, 4 m **43.** $4x + 2(x + 3) = 30$; 4 ft, 7 ft **45.** $7525 + (0.87/47)x = 6069 + (0.90/29)x$; 116,258.23 miles

Just for Fun **1.**

Let n be the number.	n
Multiply by 2	$2n$
Add 33	$2n + 33$
Subtract 13	$2n + 33 - 13 = 2n + 20$
Divide by 2	$\dfrac{2n + 20}{2} = n + 10$
Subtract number started with	$n + 10 - n = 10$

Exercise Set 2.3

1. 11.2 mi **3.** 6.55 hr **5.** 2735.63 mph **7.** 38.85 ft **9.** 54.35 copies/min **11.** 6.4 cm **13.** 12.2 sec **15.** 2.5 hr **17.** 50 mph **19.** 918 ft **21.** **(a)** 78 shares Prime, 390 shares American Motors **(b)** $30 **23.** $2200 at 7%, $5800 at $5\frac{1}{4}$% **25.** $5000 at 6%, $3000 at 10% **27.** $2300 at 6%, $4200 at 9% **29.** 120 adults, 52 children **31.** 59 nickels, 24 dimes **33.** 6 hr at $4, 12 hr at $4.50 **35.** 54 lb **37.** 24 oz water **39.** 6 qt **41.** 1.2 qt soda **43.** 25 lb sand **45.** 3 oz juice, 5 oz drink **47.** $1075 for Mr., $5325 for Mrs.

Just for Fun **1.** **(a)** 123 m/hr; 61.5 m/hr; 41 m/hr **(b)** 328 m **2.** **(a)** 121.9 mph **(b)** 0.0509 hr or 183.1 sec **3.** 300,000,000 ft **4.** 82,944 stitches require 10,368 minutes or 172.8 hours **5.** 6 qt **6.** 60 miles, 7 AM

Exercise Set 2.4

1. $y = -2x + 3$ **3.** $y = (-2x + 6)/3$ **5.** $y = x - 8$ **7.** $y = (2x - 6)/4$ **9.** $y = (8x - 3)/2$ **11.** $y = 2x + 3$ **13.** $y = -x + 2$ **15.** $y = (-9x + 8)/3$ **17.** $l = A/w$ **19.** $r = d/t$ **21.** $b = 2A/h$ **23.** $t = i/pr$ **25.** $l = (P - 2w)/2$ **27.** $h = 3V/B$ **29.** $h = V/\pi r^2$ **31.** $\sigma = (x - \mu)/z$ **33.** $R = P/I^2$ **35.** $h = 2A/(b_1 + b_2)$ **37.** $m = (y - b)/x$ **39.** $m = (y - y_1)/(x - x_1)$ **41.** $n = 2S/(f + l)$ **43.** $F = \frac{9}{5}C + 32$ **45.** $z = kx/y$ **47.** $m_1 = Fd^2/km_2$ **49.** $\bar{x} = \mu + z\sigma/\sqrt{n}$

Just for Fun **1.** $R_1 = R_T R_2/(R_2 - R_T)$ **2.** **(a)** $x_1 = [x(m_1 + m_2 + m_3) - m_2x_2 - m_3x_3]/m_1$ **(b)** $m_1 = (m_2x_2 + m_3x_3 - xm_2 - xm_3)/(x - x_1)$

Exercise Set 2.5

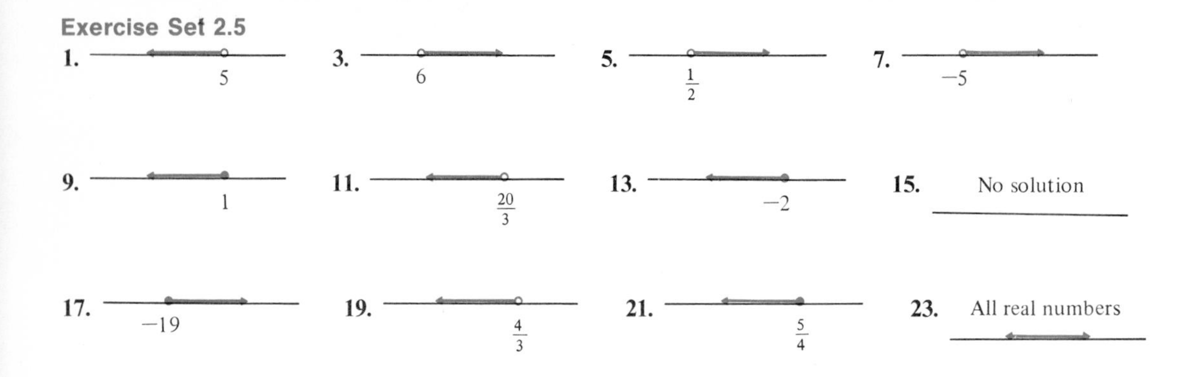

25. (1, 6) **27.** $(-\frac{3}{5}, \frac{8}{5}]$ **29.** $[\frac{7}{2}, 5)$ **31.** $(-\frac{7}{6}, \frac{2}{3})$ **33.** (0, 7) **35.** $[-\frac{56}{3}, \frac{14}{3})$ **37.** (0, 1] **39.** (−23, 2] **41.** $(-1, \frac{1}{3})$ **43.** $\{x|2 < x < 4\}$ **45.** $\varnothing$ **47.** $\{x|-3 < x < 1\}$ **49.** $\{x|x \le 2 \text{ or } x > 8\}$ **51.** $\{x|x < 4\}$ **53.** $\{x|0 \le x \le 2\}$ **55.** $\{x|x > \frac{13}{5}\}$ **57.** $\{x|x < 0 \text{ or } x > 6\}$ **59.** 11 boxes **61.** 1075 lb **63.** 7 hr **65.** 1881 books **67.** 690 pieces **69.** 82 **71.** Any value less than 3.48 **73.** $6.27 < x < 8.07$

Just for Fun **1.** $84 \le x \le 100$

Exercise Set 2.6

1. $\{-5, 5\}$ **3.** $\{-12, 12\}$ **5.** $\{0\}$ **7.** $\varnothing$ **9.** $\{-2, 8\}$ **11.** $\{-\frac{18}{5}, -\frac{12}{5}\}$ **13.** $\{-5, 1\}$ **15.** $\{\frac{3}{2}, \frac{11}{6}\}$ **17.** $\{-8, 0\}$ **19.** $\{-17, 23\}$ **21.** $\{-\frac{59}{3}, \frac{49}{3}\}$ **23.** $\{-3, 7\}$ **25.** $\{-1, \frac{11}{5}\}$ **27.** $\{y|-5 \le y \le 5\}$ **29.** $\{x|-2 < x < 8\}$ **31.** $\{x|2 < x < 12\}$ **33.** $\{z|0 \le z \le \frac{10}{3}\}$ **35.** $\{x|-1 \le x \le 6\}$ **37.** $\{x|-\frac{11}{3} < x < \frac{19}{3}\}$ **39.** $\{x|\frac{9}{2} \le x \le \frac{11}{2}\}$ **41.** $\{x|-2 \le x \le 3\}$ **43.** $\{c|-9 \le c \le 11\}$ **45.** $\{x|x < -3 \text{ or } x > 3\}$ **47.** $\{z|z \le -2 \text{ or } z \ge 2\}$ **49.** $\{x|x \le 2 \text{ or } x \ge 8\}$ **51.** $\{x|x < -\frac{5}{3} \text{ or } x > 1\}$ **53.** $\{x|x \le -4 \text{ or } x \ge -1\}$ **55.** $\{w|w \le -\frac{35}{3} \text{ or } w \ge 15\}$ **57.** $\{x|x < -\frac{1}{2} \text{ or } x > 2\}$ **59.** $\{x|x \le 0 \text{ or } x \ge \frac{4}{3}\}$ **61.** $\{x|x \le -18 \text{ or } x \ge 2\}$ **63.** $\{x|x \le -\frac{25}{3} \text{ or } x \ge \frac{65}{3}\}$

Just for Fun **1.** All real numbers **2.** All x and all y **3.** $\{-1, 9\}$ **4.** $\{-1\}$

Review Exercises

1. Tenth **2.** First **3.** Seventh **4.** Cannot be simplified **5.** $7x^2 + 2xy - 4$ **6.** 8 **7.** $2x - 6y + 6$ **8.** $\frac{49}{6}$ **9.** 20 **10.** $-\frac{13}{3}$ **11.** −3 **12.** $-\frac{9}{2}$ **13.** No solution **14.** −3 **15.** 16 **16.** sister 16, Hassan 20 **17.** 5 **18.** 20 **19.** 50 **20.** 24, 25 **21.** 13, 15 **22.** 200 **23.** 10, 19 **24.** 36 mi **25.** $25 **26.** 40°, 65°, 75° **27.** 9 **28.** $2\frac{2}{3}$ hr **29.** 50 mph **30.** $1153.85 at 12%, $13,846.15 at 5.5% **31.** 480 gal 86% octane, 720 gal 91% octane **32.** 15 lb at $4, 25 lb at $4.80 **33.** $l = A/w$ **34.** $h = A/\pi r^2$ **35.** $w = (P - 2l)/2$ **36.** $r = d/t$ **37.** $m = (y - b)/x$ **38.** $y = (2x - 5)/3$ **39.** $V_2 = P_1V_1/P_2$ **40.** $a = (2S - b)/3$ **41.** $l = (K - 2d)/2$ **42.** $t = (I - p)/pr$ **43.** $b_1 = (2A - hb_2)/h$ **44.** $t = (w + 2l)/V_0$ **45.** 7 **46.** −3 **47.** $\frac{5}{2}$ **48.** $\frac{21}{4}$ **49.** $\frac{9}{2}$ **50.** −10 **51.** $\frac{2}{5}$ **52.** $\frac{20}{9}$ **53.** 13 boxes **54.** 7 min **55.** 17 weeks **56.** (5, 11) **57.** [−3, 3) **58.** $(\frac{7}{2}, 6)$ **59.** $(\frac{8}{3}, 6)$ **60.** $[-\frac{1}{2}, \frac{23}{2})$ **61.** (2, 14) **62.** $\{x|81 \le x \le 100\}$ **63.** $\{x|-3 < x < 3\}$ **64.** $\mathbb{R}$ **65.** $\{x|x > -1\}$ **66.** $\{x|2 \le x \le \frac{5}{2}\}$ **67.** $\{x|x \le -4\}$ **68.** $\{x|x \le -4 \text{ or } x > \frac{17}{5}\}$ **69.** $\{-4, 4\}$ **70.** $\{x|-3 < x < 3\}$ **71.** $\{x|x \le -4 \text{ or } x \ge 4\}$ **72.** $\{-5, 13\}$ **73.** $\{x|x \le -3 \text{ or } x \ge 7\}$ **74.** $\{x|-11 < x < -1\}$ **75.** $\{-\frac{1}{2}, \frac{9}{2}\}$ **76.** $\{x|-2 < x < 5\}$ **77.** $\{x|x \le -\frac{1}{3} \text{ or } x \ge 3\}$ **78.** $\{-17, 23\}$ **79.** $\{-1, 4\}$ **80.** $\{x|-14 < x < 22\}$ **81.** $\{x|x \le \frac{5}{2} \text{ or } x \ge \frac{11}{2}\}$ **82.** $\{x|-\frac{7}{2} \le x \le -\frac{1}{2}\}$ **83.** $\{x|x < \frac{3}{4} \text{ or } x > \frac{13}{4}\}$ **84.** $\{x|-12 < x < 6\}$ **85.** $\{x|x \le -\frac{3}{2} \text{ or } x \ge \frac{9}{2}\}$ **86.** $\{x|x < -12 \text{ or } x > 20\}$

Practice Test

1. Sixth **2.** $\frac{27}{7}$ **3.** $-\frac{34}{5}$ **4.** 68 **5.** 23, 24 **6.** 4.13 yrs **7.** 113 mi **8.** 11.56 mi **9.** 6.25 liters **10.** $b = (a - 2c)/3$ **11.** $b_2 = (2A - hb_1)/h$ **12.** 33 **13.** (−12, 12) **14.** $\{-1, 9\}$ **15.** $\{x|x < -1 \text{ or } x > 4\}$ **16.** $\{x|\frac{1}{2} < x < \frac{5}{2}\}$

Exercise Set 3.1

1. $A(3, 1), B(-6, 0), C(2, -4), D(-2, -4), E(0, 3), F(-8, 1), G(\frac{3}{2}, -1)$ **3.**

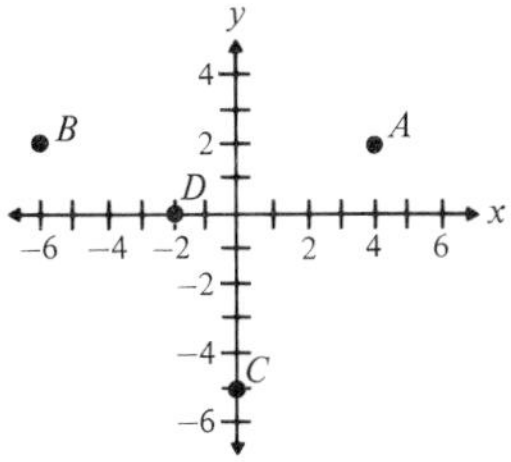

5. 3 **7.** 9 **9.** 5

11. 13 **13.** $\sqrt{90} = 3\sqrt{10}$ **15.** $\sqrt{74}$ **17.** $\sqrt{20} = 2\sqrt{5}$

19. $\sqrt{\dfrac{281}{16}} = \dfrac{\sqrt{281}}{4}$ **21.** (2, 3) **23.** (0, 0) **25.** $(-4, -5)$

27. $(-\frac{7}{2}, -5)$ **29.** $(2, -4)$ **31.** $(\frac{9}{4}, \frac{15}{4})$

33. The square of any positive number is positive.

Exercise Set 3.2

1.

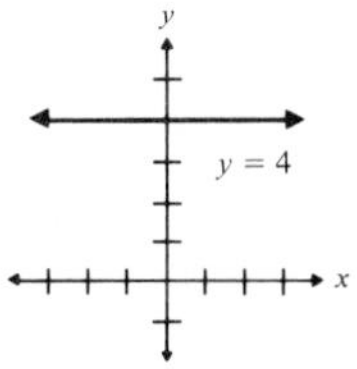

3.

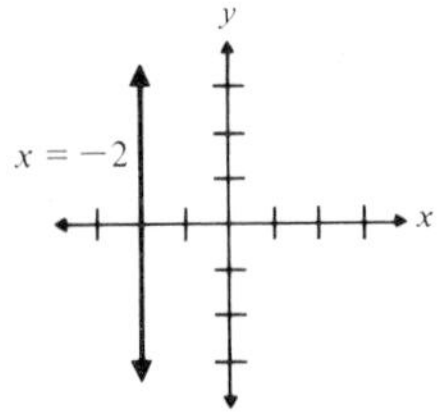

5.

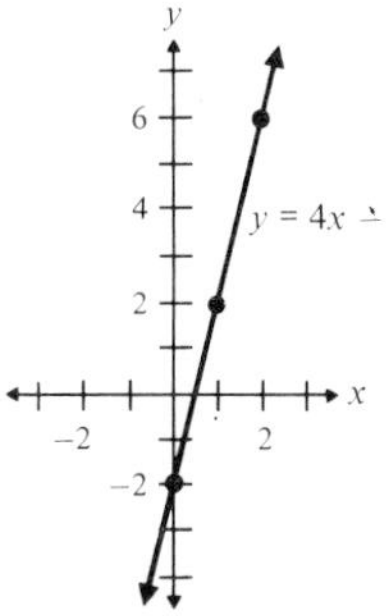

7.

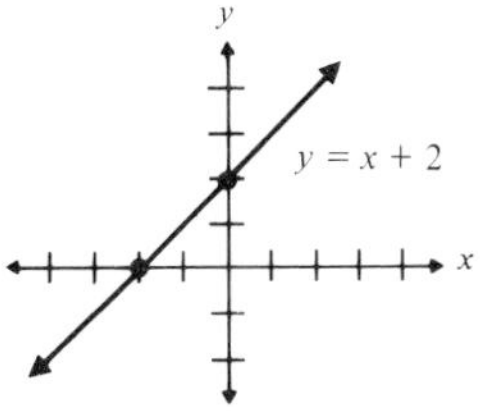

9.

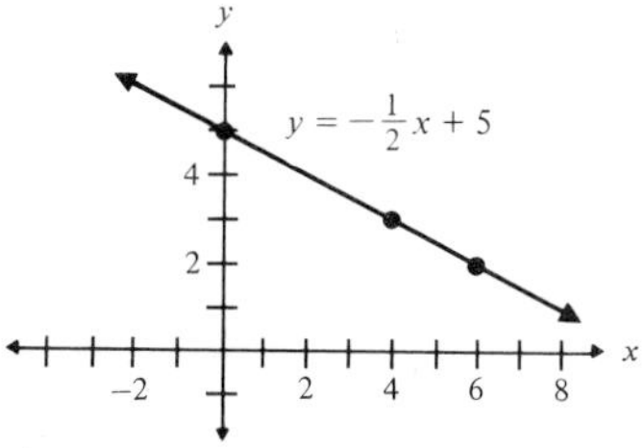

11.

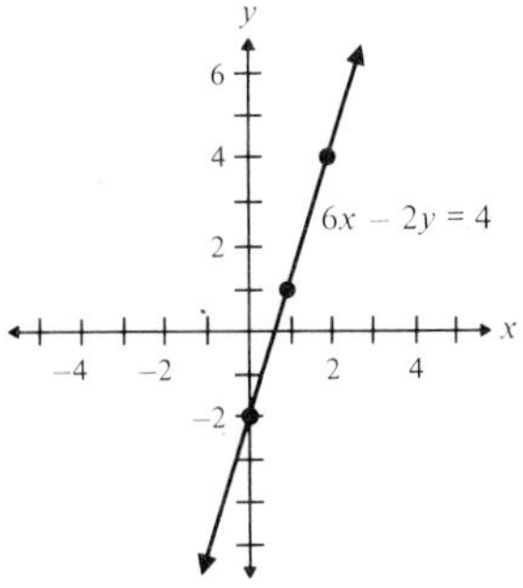

13.

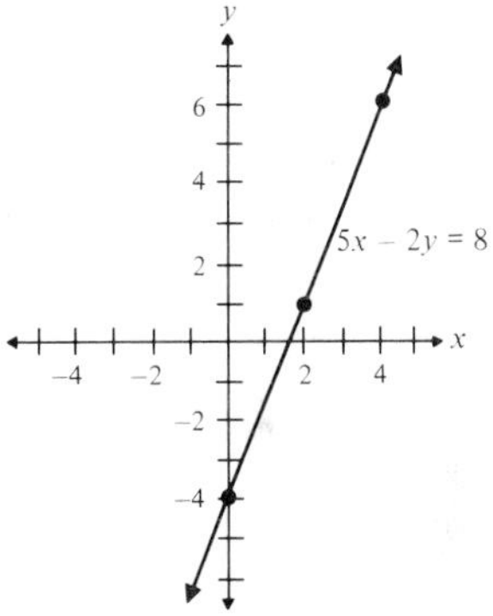

15.

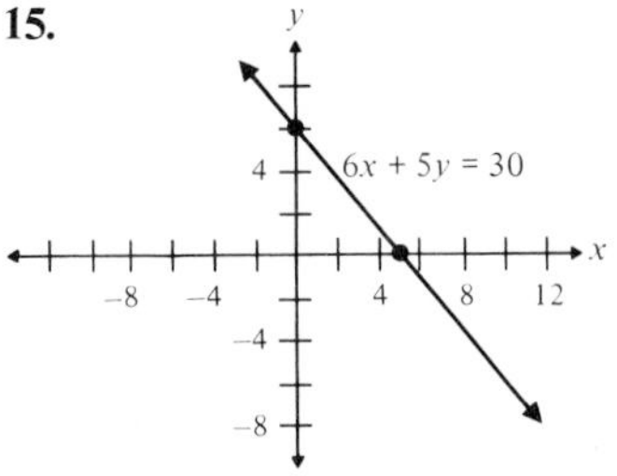

17.

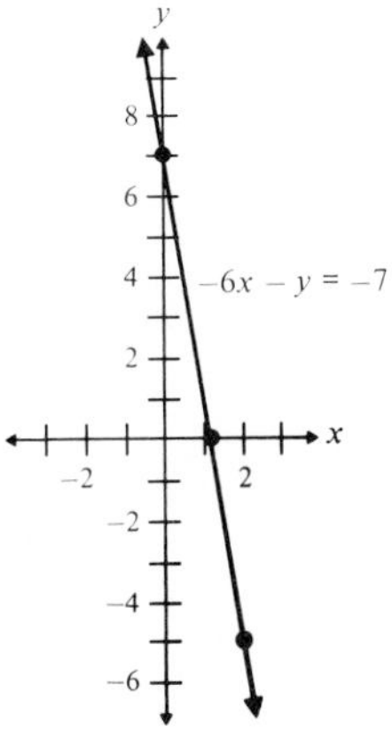

19.

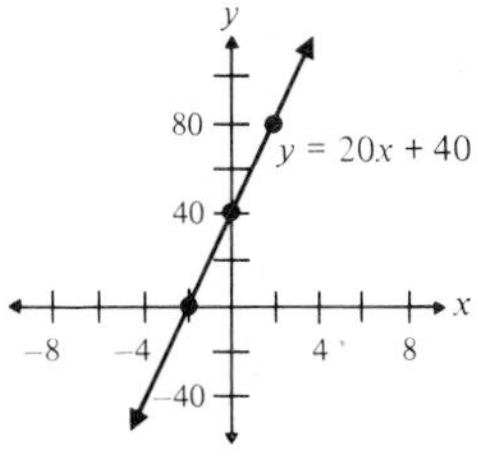

21.

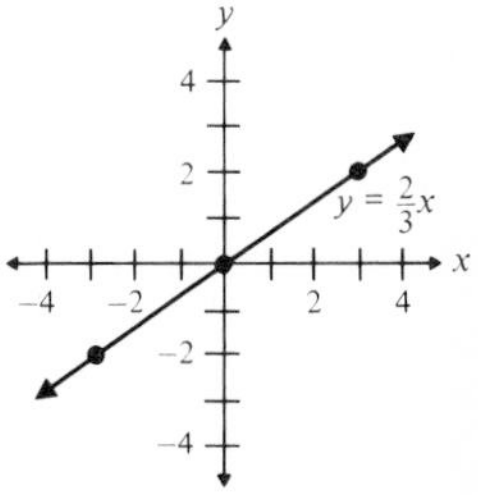

23.

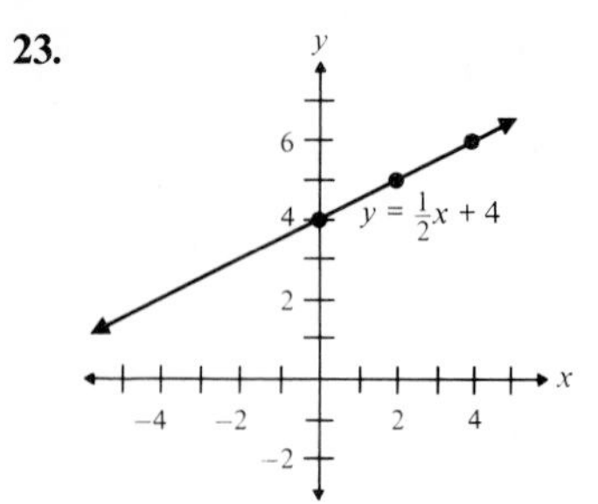

25.

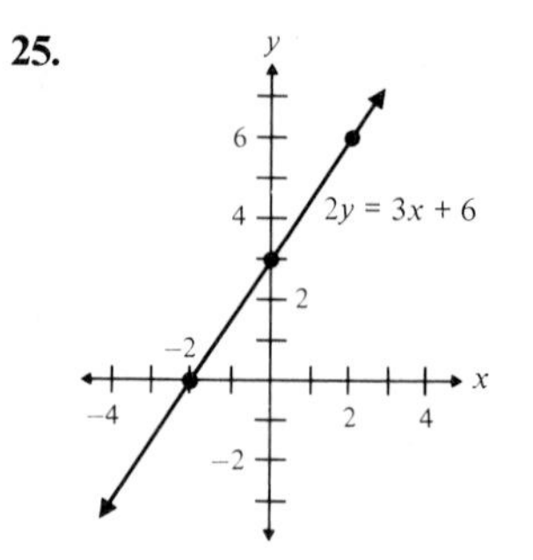

27.

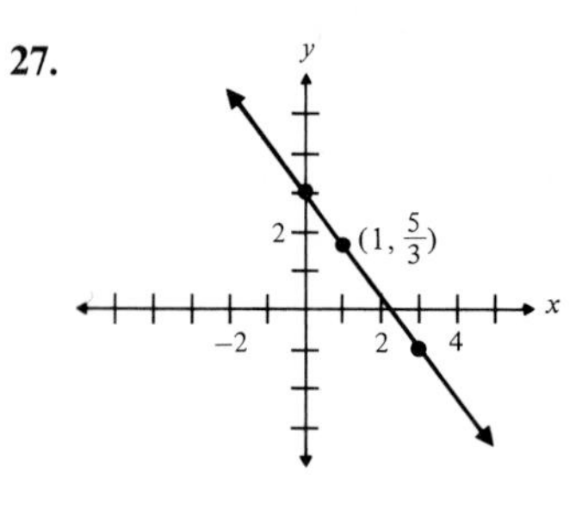

29.

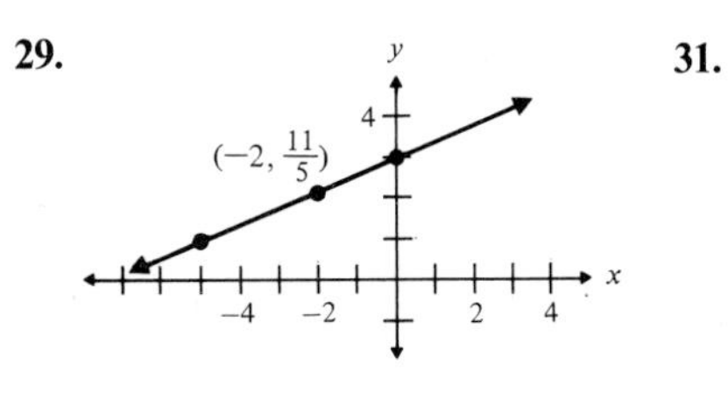

31.

33.

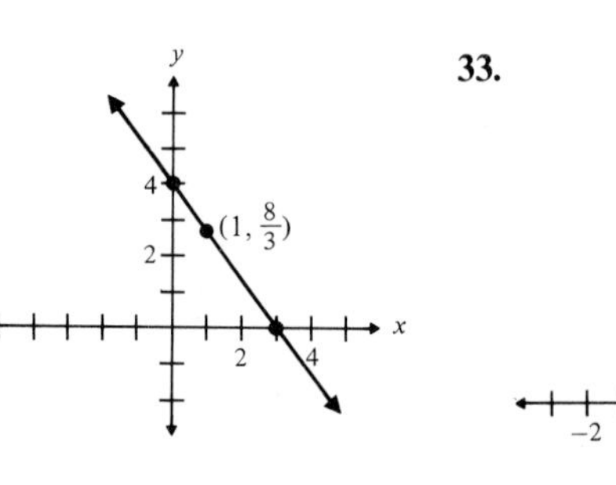

35.

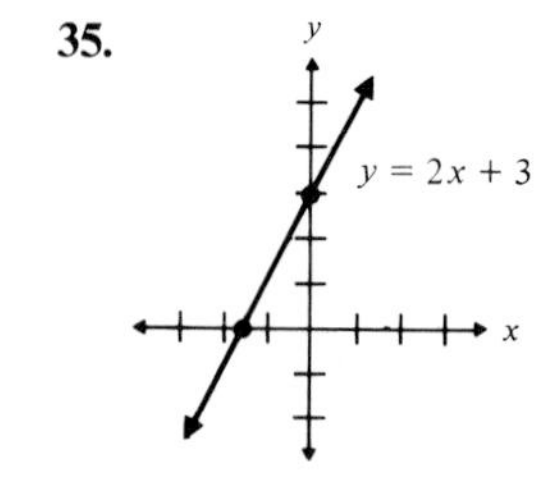

37.

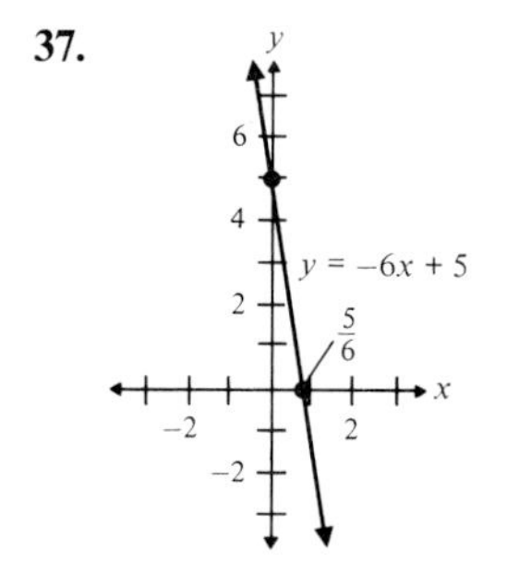

39.

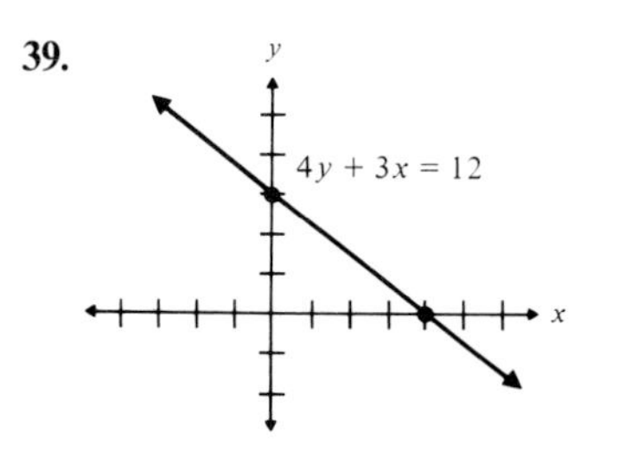

41.

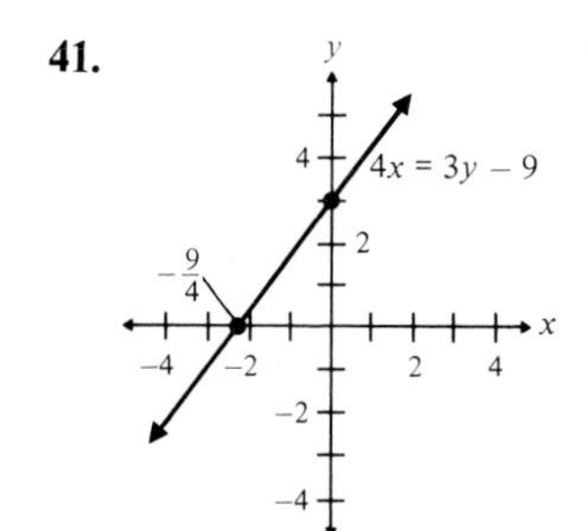

43.

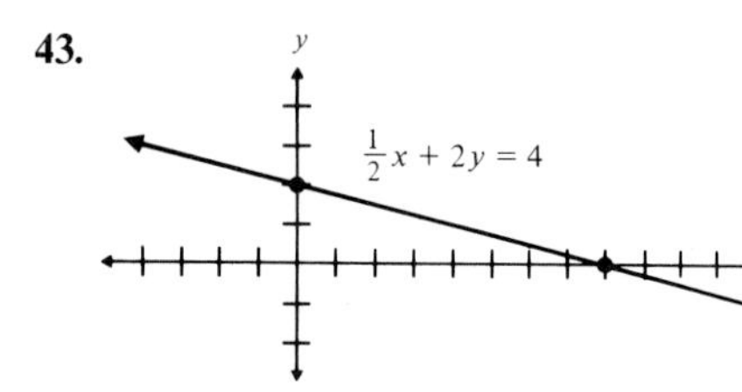

45.

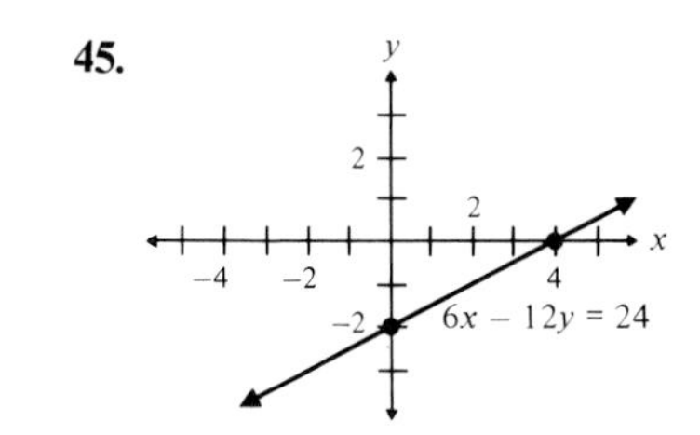

47.

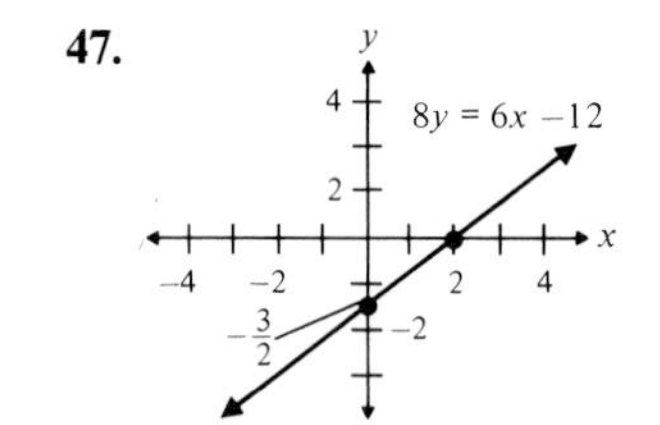

49.

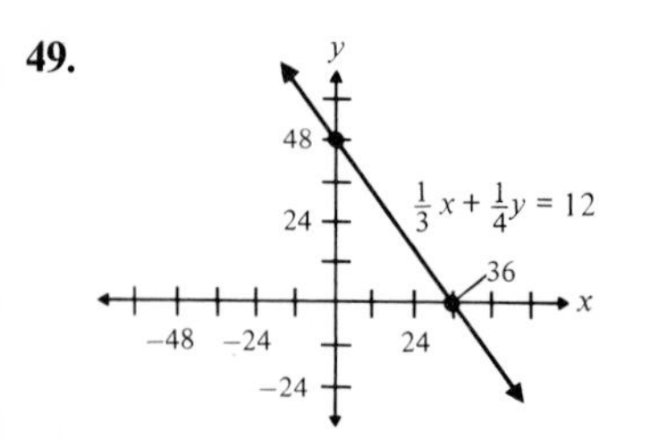

51.

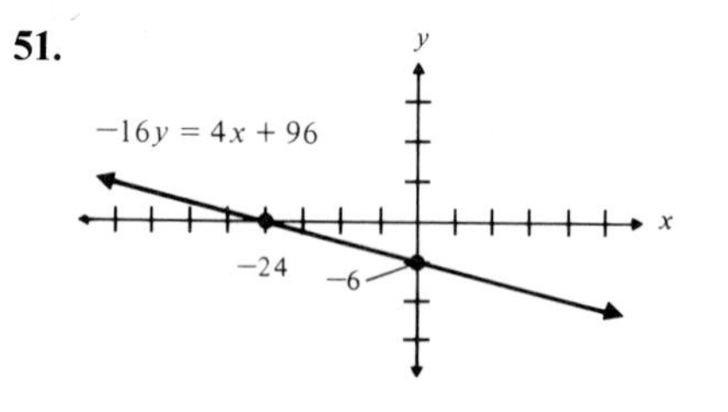

53.

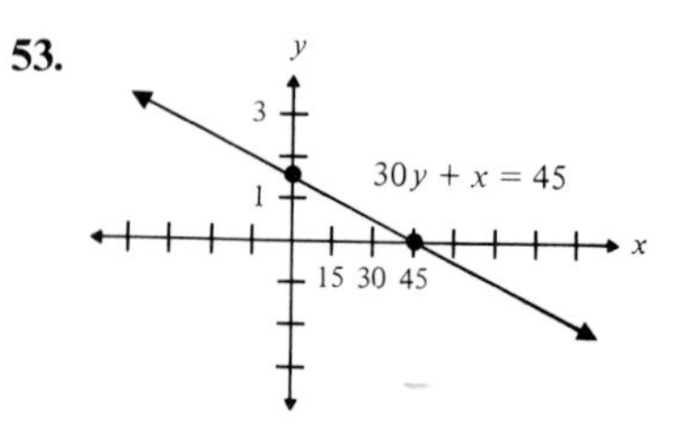

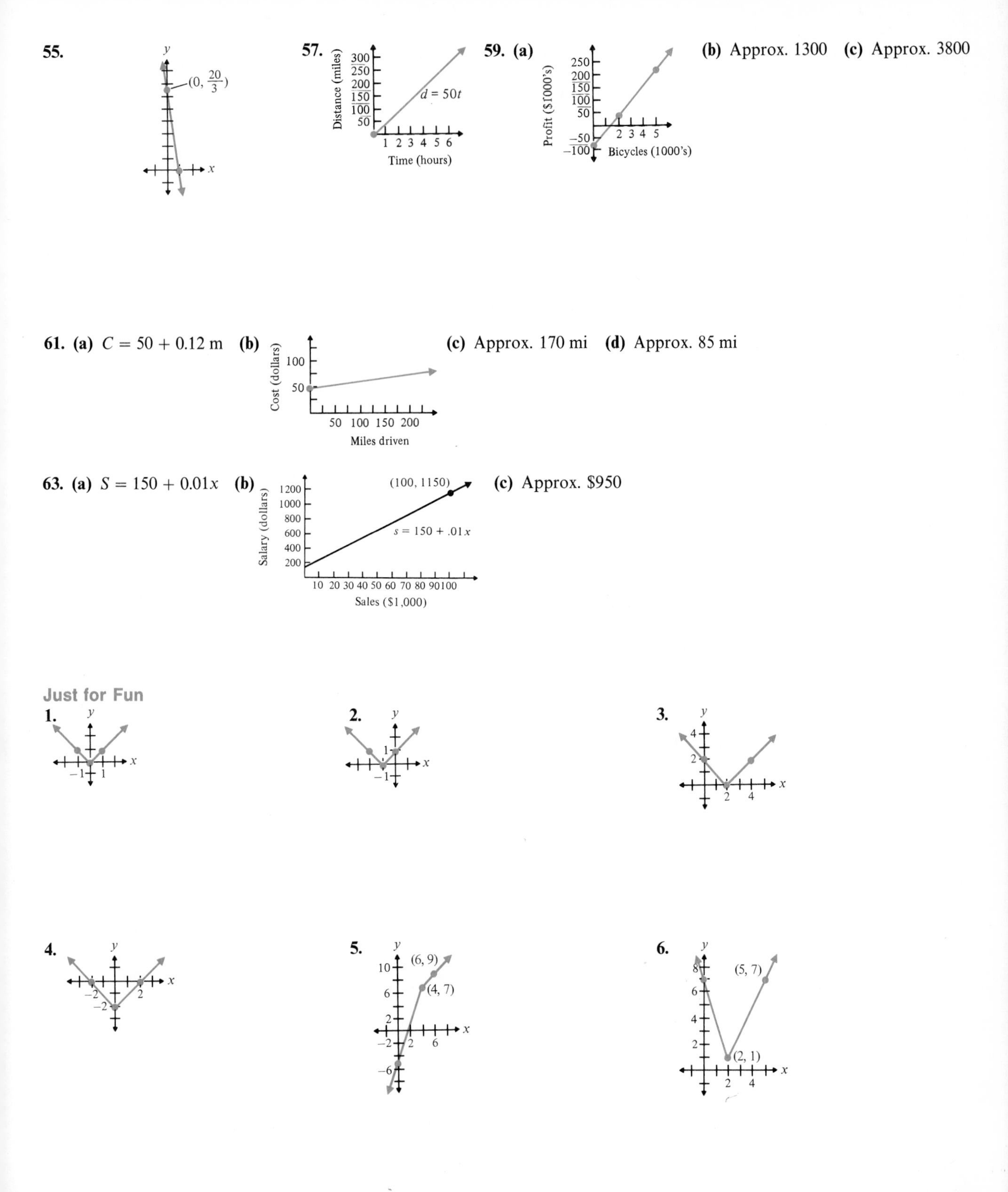
55.
y
(0, 20/3)
x
57.
300
250
200
150
100
50
Distance (miles)
d = 50t
1 2 3 4 5 6
Time (hours)
59. (a)
250
200
150
100
50
−50
−100
Profit ($1000's)
2 3 4 5
Bicycles (1000's)
(b) Approx. 1300 (c) Approx. 3800
61. (a) C = 50 + 0.12 m (b)
100
50
Cost (dollars)
50 100 150 200
Miles driven
(c) Approx. 170 mi (d) Approx. 85 mi
63. (a) S = 150 + 0.01x (b)
1200
1000
800
600
400
200
Salary (dollars)
(100, 1150)
s = 150 + .01x
10 20 30 40 50 60 70 80 90 100
Sales ($1,000)
(c) Approx. $950
Just for Fun
1.
y
x
−1
1
2.
y
x
1
−1
3.
y
4
2
x
2
4
4.
y
x
−2
2
−2
5.
y
10
6
2
−2
−6
(6, 9)
(4, 7)
x
2
6
6.
y
8
6
4
2
(5, 7)
(2, 1)
x
2
4

Exercise Set 3.3

1. -8 **3.** $-\frac{1}{2}$ **5.** -1 **7.** Does not exist **9.** -5 **11.** 0 **13.** $-\frac{1}{7}$ **15.** $\frac{1}{3}$ **17.** $\frac{1}{5}$ **19.** 0
21. Does not exist **23.** 1 **25.** Parallel **27.** Perpendicular **29.** Neither **31.** Perpendicular
33. Parallel **35.** $a = 7$ **37.** $b = 2$ **39.** $c = -4$ **41.** $x = 6$ **43.** $x = 0$
45. Select two points on the line; then find the ratio of the change in y to the change in x.
47. The line is falling going from left to right. **49.** Because the change in x is zero and you cannot divide by zero.

Exercise Set 3.4

1. $y = x + 2$ **3.** $y = -\frac{3}{2}x + 15$ **5.** Parallel **7.** Parallel **9.** Neither **11.** Perpendicular **13.** Perpendicular
15. Parallel **17.** Perpendicular **19.** $y = 4x - 5$ **21.** $y = -x + 6$ **23.** $y = -\frac{2}{3}x - \frac{8}{3}$ **25.** $y = \frac{5}{6}x + \frac{8}{3}$
27. $y = x - 3$ **29.** $y = -\frac{4}{3}x + \frac{4}{3}$ **31.** $y = 2x + 2$ **33.** $2x + 3y = 10$ **35.** $y = \frac{2}{3}x + \frac{41}{12}$ **37.** $x + 2y = 6$
39. $y = -2x - \frac{16}{3}$ **41.** $2x - 4y = 1$ **43.** $y = \frac{2}{3}x - \frac{14}{3}$ **45.** $y = -\frac{2}{3}x + 6$ **47.** $12x + 5y = 62$
49. **(a)** $-x + y = 2$ **(b)** $y = x + 2$ **(c)** $y - 2 = 1(x - 0)$

Exercise Set. 3.5

1. Function, domain $\{1, 2, 3, 4, 5\}$, range $\{1, 2, 3, 4, 5\}$ **3.** Function, domain $\{1, 2, 3, 4, 5, 7\}$, range $\{-1, 0, 2, 4, 5\}$
5. Relation, domain $\{1, 2, 3, 5\}$, range $\{-4, -1, 0, 1, 2\}$ **7.** Function, domain $\{-2, \frac{1}{2}, 0, 2, 3, 5\}$, range $\{-3, -1, 0, \frac{2}{3}, 2, 5\}$
9. Relation, domain $\{1, 2, 6\}$, range $\{-3, 0, 2, 5\}$ **11.** Relation, domain $\{0, 1, 2\}$, range $\{-7, -1, 2, 3\}$
13. Relation, domain $\{x \mid -2 \le x \le 2\}$, range $\{y \mid -2 \le y \le 2\}$ **15.** Function, domain $\mathbb{R}$, range $\{y \mid y \ge 0\}$
17. Function, domain $\{-1, 0, 1, 2, 3\}$, range $\{-1, 0, 1, 2, 3\}$ **19.** Function, domain $\mathbb{R}$, range $\mathbb{R}$
21. Relation, domain $\{-2\}$, range $\mathbb{R}$ **23.** Function, domain $\mathbb{R}$, range $\{y \mid -5 \le y \le 5\}$ **25.** **(a)** 13 **(b)** 3
27. **(a)** 4 **(b)** 9 **29.** **(a)** -1 **(b)** 2 **31.** **(a)** -1 **(b)** -11 **33.** $f(x) = 3x + 1$ **35.** $f(x) = 2x - 1$
37. $f(x) = -x - 2$ **39.** $f(x) = \frac{1}{2}x + 3$ **41.** $g(x) = 2x - 6$

43. Any set of ordered pairs. **45.** No, a function has no two distinct ordered pairs with the same first coordinate.
47. If a vertical line can be drawn at any value of x that intersects the graph more than once, then each x does not have a unique y value, and the graph is not a function.
49. The set of second coordinates of the ordered pairs.

Exercise Set 3.6

1. **3.** **5.** **7.**

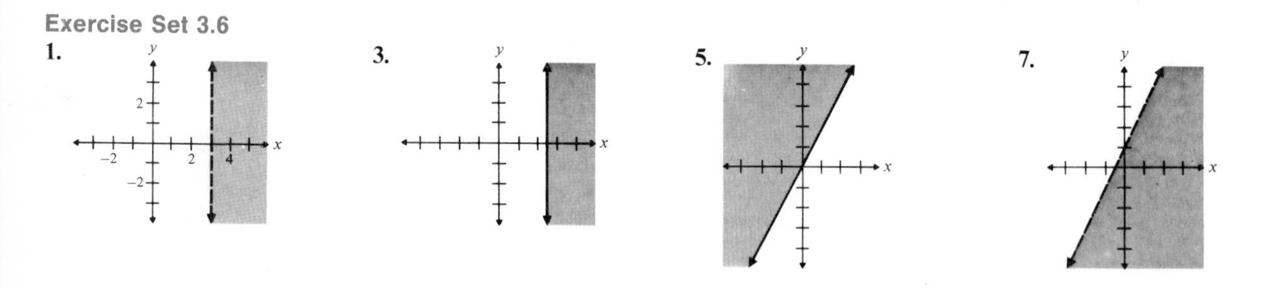

9. **11.** **13.** **15.**

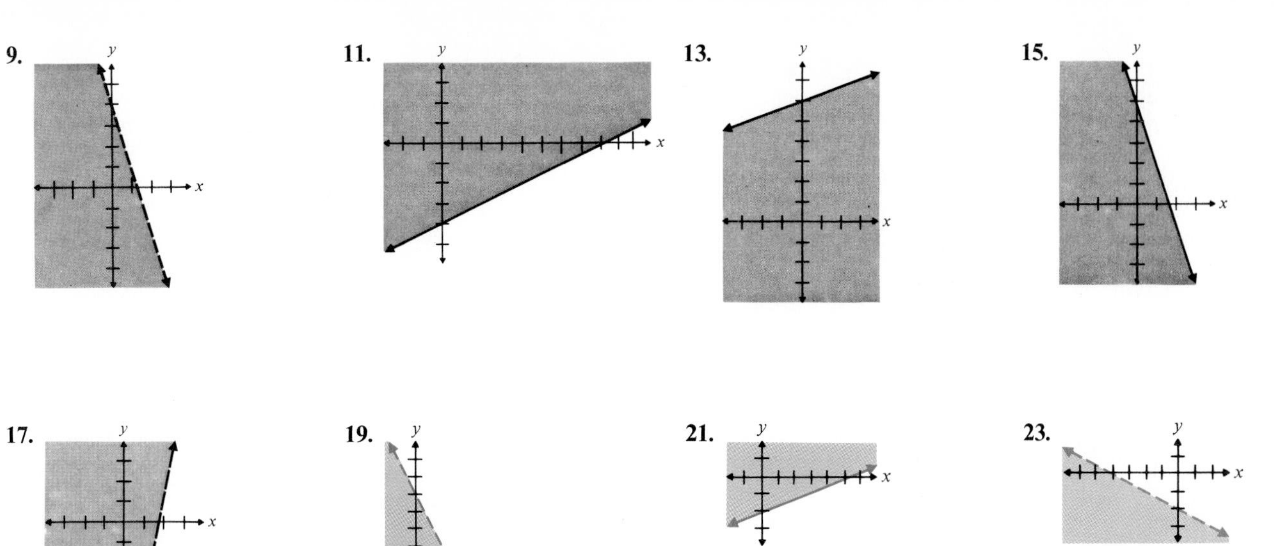

17. **19.** **21.** **23.**

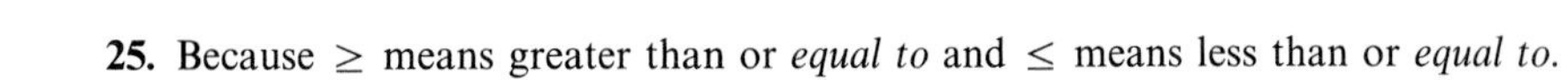

25. Because $\geq$ means greater than or *equal to* and $\leq$ means less than or *equal to*.

Just for Fun

1.

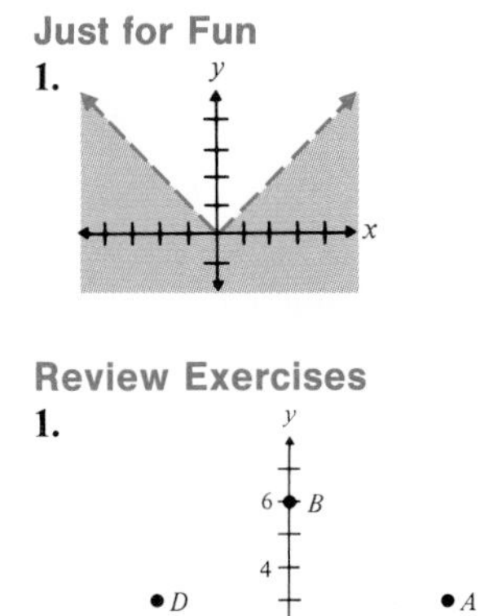

2.

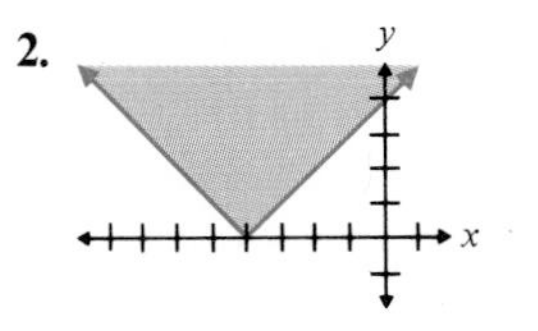

Review Exercises

1.

2. 5 **3.** 5 **4.** 13 **5.** $\sqrt{8} = 2\sqrt{2}$ **6.** 2 **7.** 13 **8.** $(\frac{5}{2}, \frac{5}{2})$ **9.** $(-\frac{1}{2}, 1)$
10. $(\frac{3}{2}, \frac{3}{2})$ **11.** $(-1, -\frac{1}{2})$ **12.** $(-4, -\frac{1}{2})$ **13.** $(-1, 6)$
14. **15.**

16.

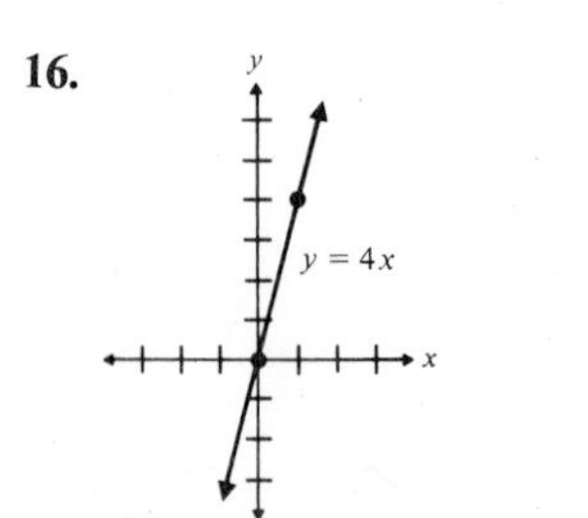

17.

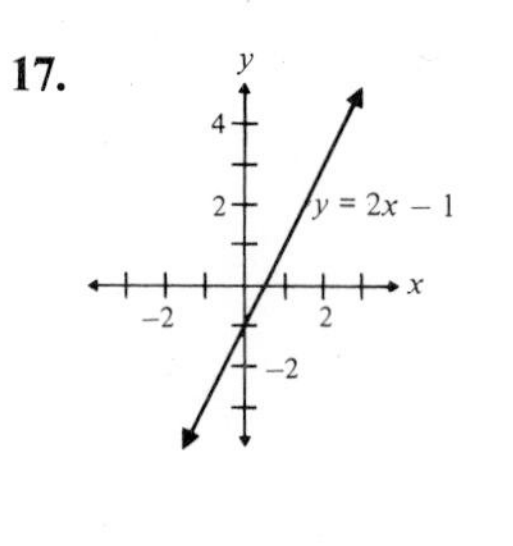

18.

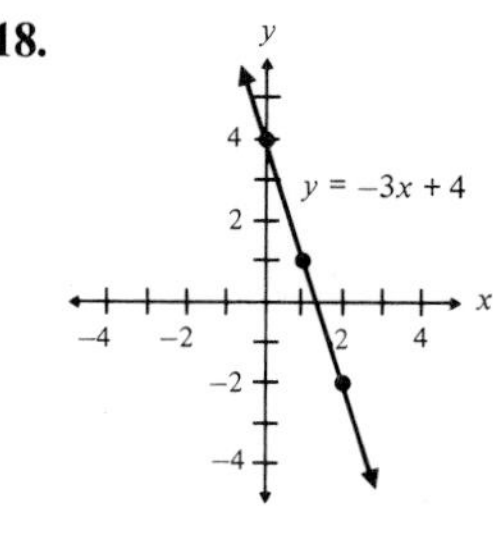

19.

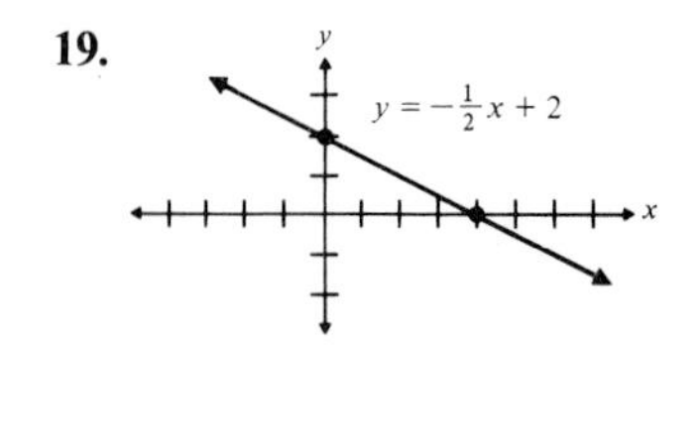

20.

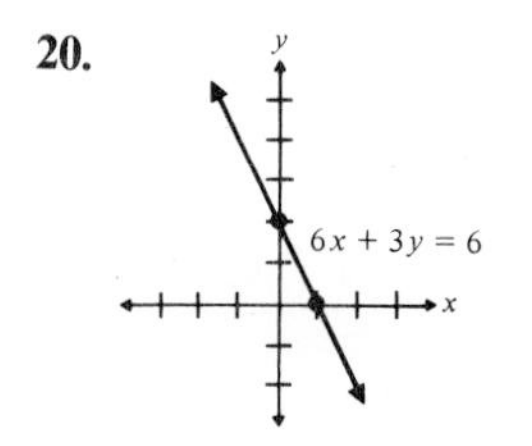

21.

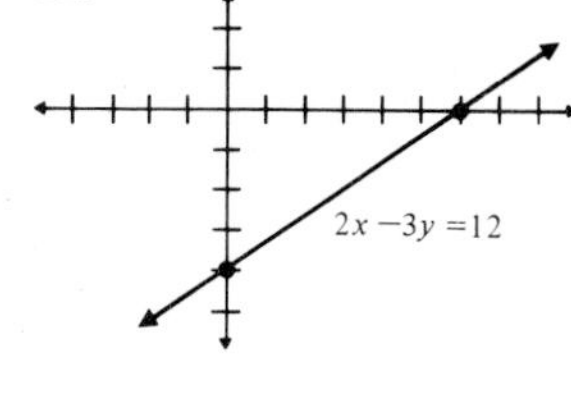

22. **23.**

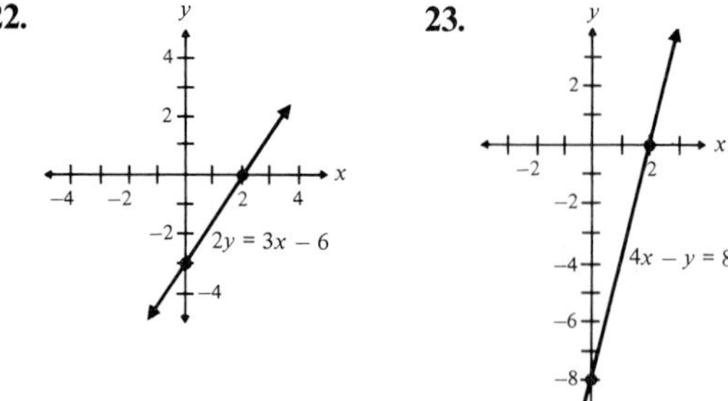

24. **25.**

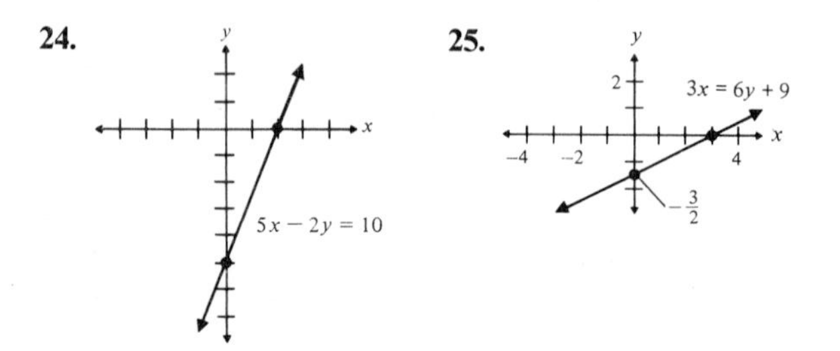

26.

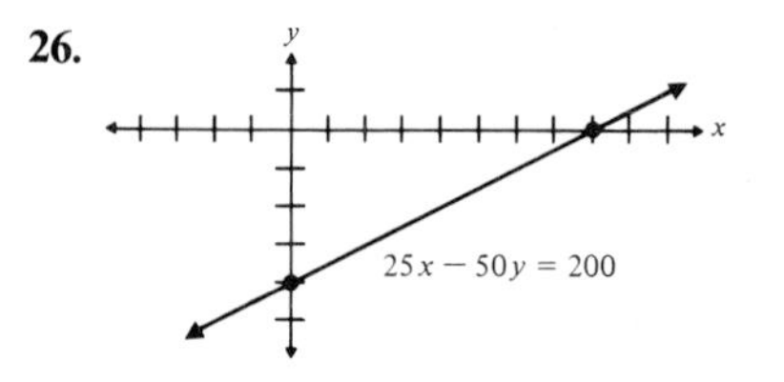

27. **28.**

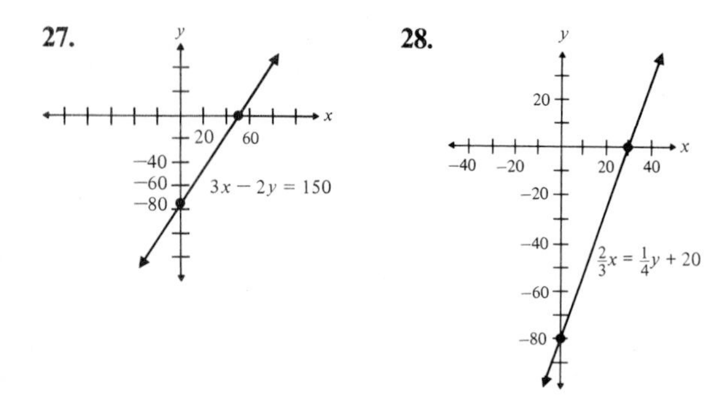

29. $m = -1, b = 5$ **30.** $m = 3, b = 5$

31. $m = -4, b = \frac{1}{2}$ **32.** $m = -\frac{2}{3}, b = 3$ **33.** $m = -\frac{1}{2}, b = \frac{3}{2}$ **34.** $m = \frac{3}{2}, b = 3$ **35.** $m = -\frac{3}{5}, b = \frac{12}{5}$
36. $m = -\frac{9}{7}, b = \frac{15}{7}$ **37.** $m = \frac{1}{2}, b = -2$ **38.** $m =$ does not exist, no y intercept **39.** $m = 0, b = 6$
40. $m = -4, b = 0$ **41.** -7 **42.** $-\frac{1}{3}$ **43.** $\frac{1}{3}$ **44.** 7 **45.** Neither **46.** Parallel **47.** Perpendicular
48. Neither **49.** $a = 3$ **50.** $x = 6$ **51.** $y = -37$ **52.** $x = 7$ **53.** $y = 3$ **54.** $x = 2$ **55.** $y = 2x + 2$
56. $y = -\frac{1}{2}x + 2$ **57.** Parallel **58.** Perpendicular **59.** Parallel **60.** Parallel **61.** Perpendicular
62. Neither **63.** $y = 2x - 2$ **64.** $y = -3x + 2$ **65.** $y = -\frac{2}{3}x + 4$ **66.** $y = x - 1$ **67.** $y = -\frac{7}{2}x - 4$
68. $y = 3x + 20$ **69.** $y = \frac{2}{5}x - \frac{18}{5}$ **70.** $y = -\frac{5}{3}x - 4$ **71.** $y = -\frac{1}{2}x + 4$

72. **(a)**

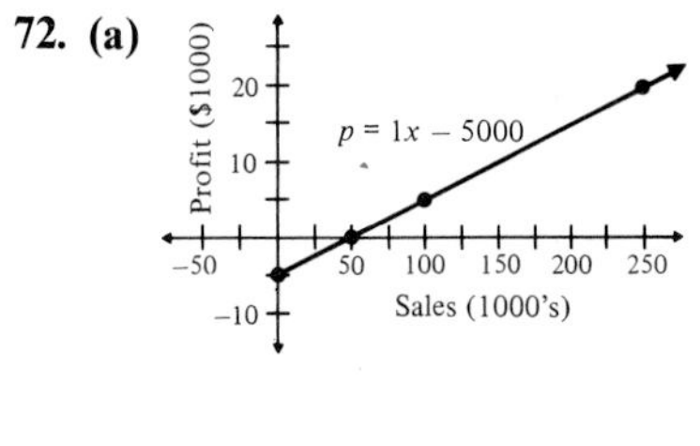

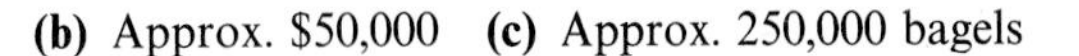
(b) Approx. $50,000 **(c)** Approx. 250,000 bagels

73.

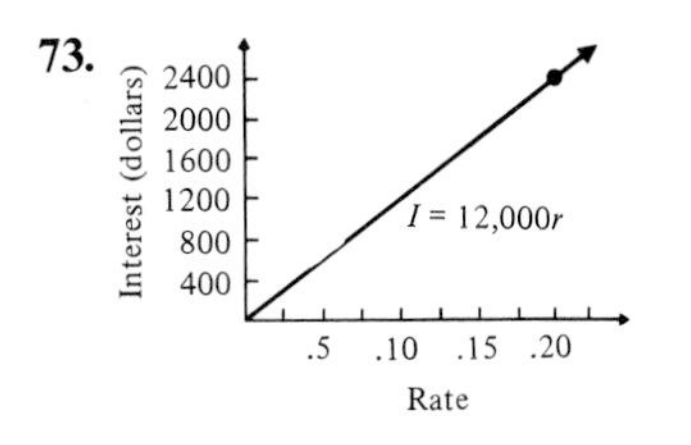

74. Domain: $\{-2, 0, 3, 6\}$, range: $\{-1, 4, 5, 9\}$

75. Domain $\{\frac{1}{2}, 2, 4, 5\}$, range: $\{-6, -1, 2, 3\}$ **76.** Domain $\{x \mid -1 \leq x \leq 1\}$, range $\{y \mid -1 \leq y \leq 1\}$
77. Domain $\{x \mid -2 \leq x \leq 2\}$, range $\{y \mid -1 \leq y \leq 1\}$ **78.** Domain $\mathbb{R}$, range $\{y \mid y \leq 0\}$ **79.** Domain $\mathbb{R}$, range $\mathbb{R}$
80. Function **81.** Function **82.** Function **83.** Not a function **84.** Function **85.** Function **86.** Function
87. Function **88.** Not a function **89.** Not a function **90.** **(a)** 10 **(b)** -5 **91.** **(a)** 7 **(b)** 4 **92.** **(a)** 2 **(b)** $\frac{5}{2}$
93. **(a)** 6 **(b)** 16 **94.** $f(x) = 2x + 4$ **95.** $f(x) = 3x - 5$ **96.** $f(x) = -4x + 2$ **97.** $f(x) = \frac{1}{2}x + 2$

98. **99.** **100.** **101.** **102.**

103.

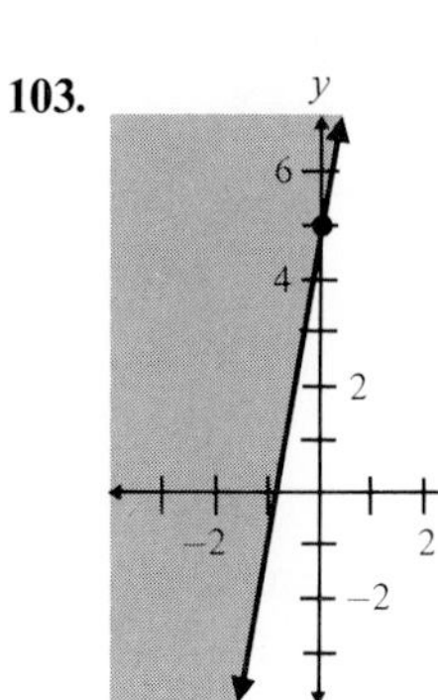

104.

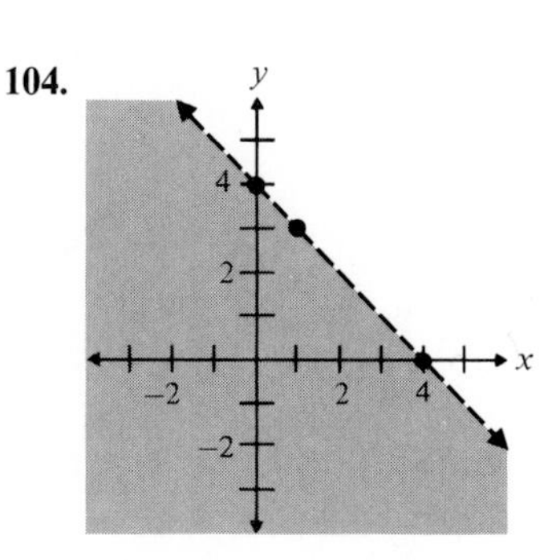

105.

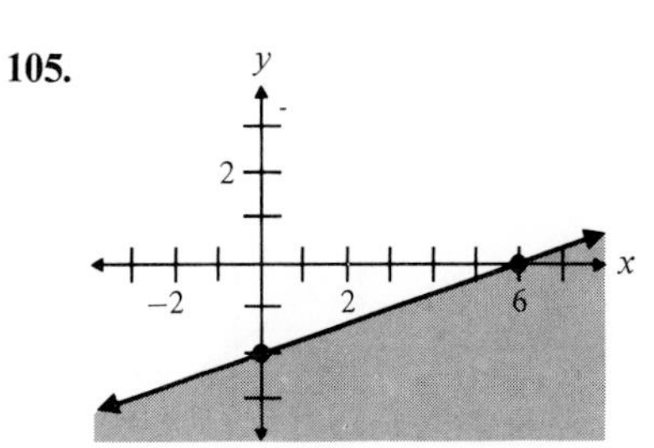

Practice Test

1. 5 **2.** $m = \frac{4}{9}$, $b = -\frac{5}{3}$ **3.** $y = 3x - 3$ **4.** $y = 4x + 7$ **5.** $y = -\frac{3}{7}x + \frac{2}{7}$ **6.** $y = \frac{3}{2}x + \frac{11}{2}$

7. **8.** **9.**

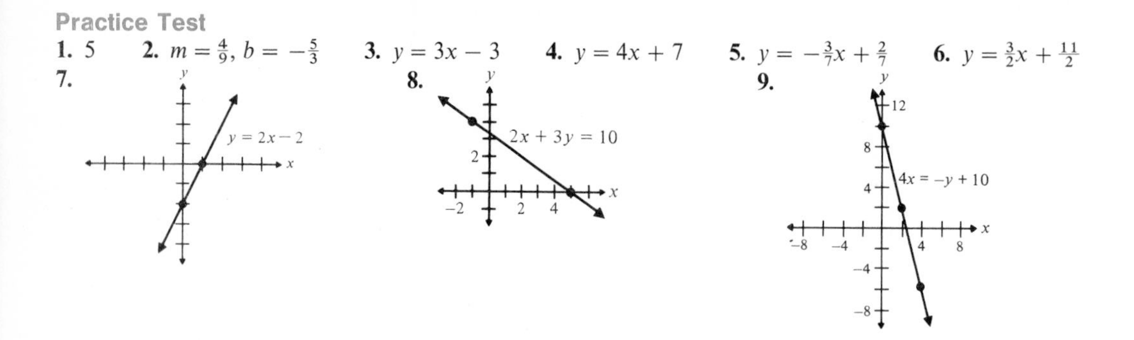

10. Domain $\{4, 2, \frac{1}{2}, 6\}$, range $\{0, -3, 2, 9\}$ **11.** *a, c* **12. (a)** 1 **(b)** −17 **13.**

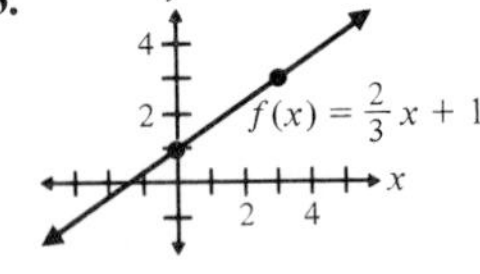

14. **15.**

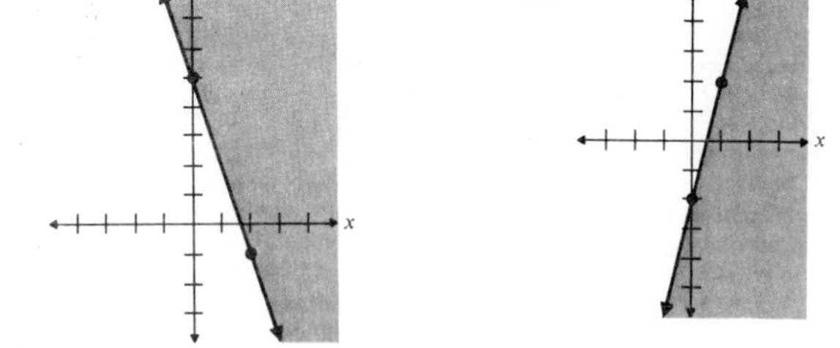

CHAPTER 4

Exercise Set 4.1

1. *c* **3.** None **5.** *b, c* **7.** *c* **9.** None **11.** $y = -\frac{1}{2}x + \frac{5}{2}$; consistent—one
$y = \frac{1}{2}x - \frac{1}{2}$

13. $y = \frac{2}{3}x + 1$; inconsistent—none
$y = \frac{2}{3}x - 2$

15. $y = \frac{2}{3}x - \frac{4}{3}$; consistent—one
$y = \frac{3}{2}x + 1$

17. $y = \frac{2}{3}x - \frac{4}{3}$; dependent—infinite number
$y = \frac{2}{3}x - \frac{4}{3}$

19. $y = \frac{3}{2}x + \frac{1}{2}$; inconsistent—no solution
$y = \frac{3}{2}x + \frac{1}{4}$

21.

23. **25.** **27.** **29.**

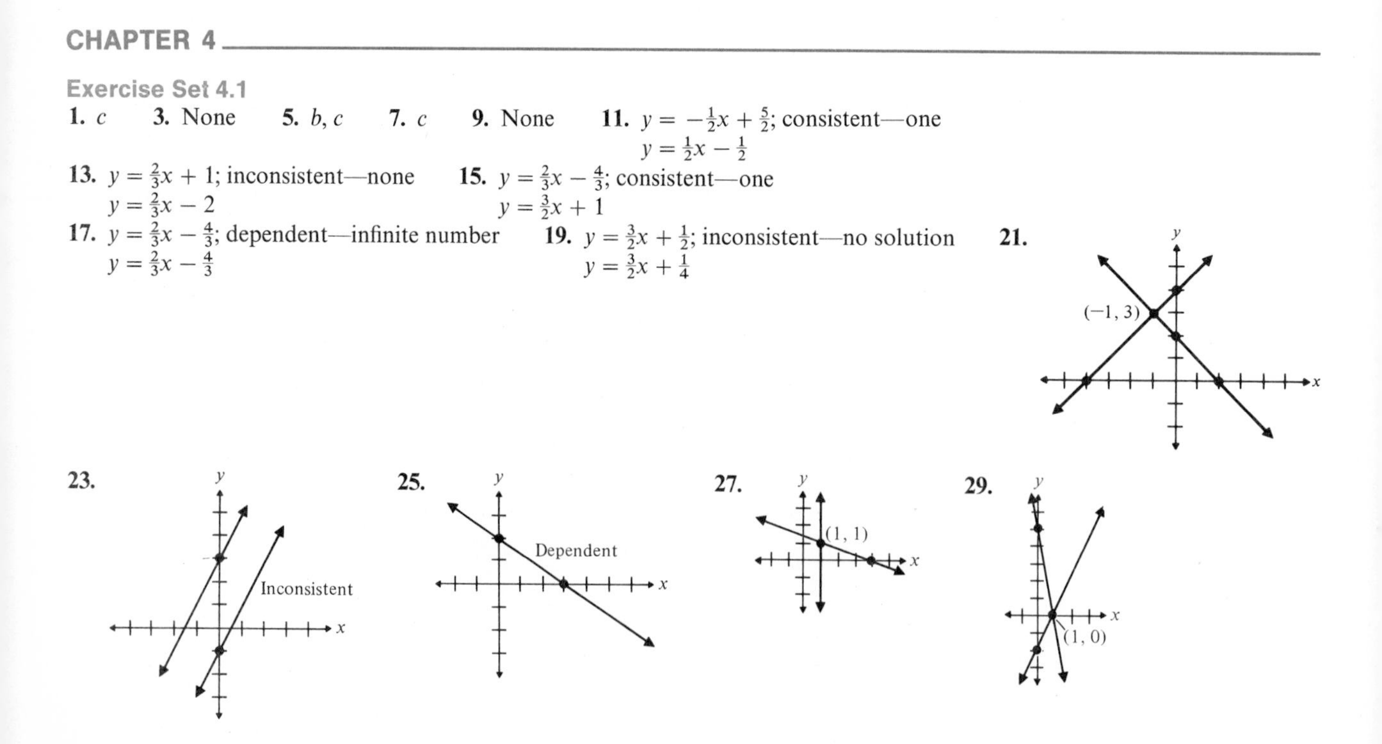

31. (5, 2) **33.** (3, 3) **35.** No solution
37. $(-1, -2)$ **39.** Infinite number of solutions
41. $(-1, 2)$ **43.** $(5, -3)$ **45.** $(-3, -4)$ **47.** $(-5, -\frac{10}{3})$ **49.** (8, 6) **51.** $(1, -3)$ **53.** $(-3, 2)$
55. $(2, \frac{9}{2})$ **57.** No solution **59.** $(4, -2)$ **61.** $(4, -1)$ **63.** Infinite number of solutions
65. $(\frac{59}{7}, \frac{60}{7})$ **67.** $(2, -1)$ **69.** $(-\frac{15}{43}, -\frac{27}{43})$ **71.** $(\frac{29}{22}, -\frac{5}{11})$ **73.** (14, 66) **75.** $(\frac{192}{25}, \frac{144}{25})$ **77.** (10, 4)
79. Write both equations in slope intercept form and compare their slopes and y intercepts.
81. Both sides of the equation will be the same.

Just for Fun **1.** $(8, -1)$ **2.** $(-\frac{105}{41}, \frac{447}{82})$ **3.** $(1/a, 5)$ **4.** $(1/a, 1/b)$

Exercise Set 4.2

1. (1, 2, 5) **3.** $(-7, -\frac{35}{4}, -3)$ **5.** (0, 3, 6) **7.** $(-1, 1, 3)$ **9.** $(2, -1, 3)$ **11.** $(-\frac{19}{2}, 10, -33)$ **13.** $(-1, -2, 3)$
15. A line, a plane **17.** None, the system is inconsistent. **19.** One

Just for Fun **1.** $(\frac{2}{3}, \frac{23}{15}, \frac{37}{15})$ **2.** $(-1, 2, 1, 5)$

Exercise Set 4.3

1. $x + y = 37$, $y = 2x + 1$, 12, 25 **3.** $x + y = 76$, $y = x + 2$, 37, 39 **5.** $x - y = 28$, $x = 3y$, 14, 42
7. $x + y = 4.5$, $x - y = 3.2$, canoe 3.85 mph, current 0.65 mph **9.** $y = 30 + 0.14x$, $y = 16 + 0.24x$, 140 mi
11. $y = 1600 + 6x$, $y = 1200 + 8x$, 200 books **13.** $x + y = 30$, $3x + 5y = 100$, 25 lb at \$3, 5 lb at \$5
15. $x + y = 100$, $0.05x + 0.00y = 0.035(100)$, 70 gal of 5%, 30 gal of skim **17.** $x + y = 8000$, $0.1x + 0.08x = 750$, \$2500 at 8%, \$5500 at 10% **19.** $c = 60{,}000 + 1500x$, $c = 20{,}000 + 3000x$, $26\frac{2}{3}$; use mini if 27 or more terminals
21. $d = 154$, $d = 70t - 42t$, 5.5 hr **23.** $x + y = 8$, $0.16x + 0.40y = 0.25(8)$, 5-oz drink, 3-oz juice
25. $d = 9t$, $d = 5t + \frac{1}{2}$, 1.125 mi **27.** $(\frac{27}{38}, -\frac{15}{38}, -\frac{6}{19})$

Just for Fun **1. (a)** $\frac{2400}{17} \approx 141.2$ cm **(b)** Pull away **2.** 50 ml – 8% sol., 20 ml – 10% sol., 30 ml – 20% sol.
3. 0 lb first alloy, 50 lb second alloy, 50 lb third alloy

Exercise Set 4.4

1. (3, 1) **3.** (3, 2) **5.** $(\frac{60}{17}, -\frac{11}{17})$ **7.** $(\frac{1}{2}, -4)$ **9.** $(2, -3)$ **11.** $(-1, 1, 3)$ **13.** $(-\frac{1}{2}, \frac{1}{2}, -\frac{3}{2})$
15. $(\frac{165}{14}, -\frac{153}{14}, -\frac{6}{7})$ **17.** $(-1, 0, 2)$ **19.** Dependent equations—infinite number of solutions
21. A square array of numbers enclosed between two vertical bars. A determinant that has two rows and two columns of elements. A determinant that has three rows and three columns of elements. **23.** It will have the opposite sign.

Exercise Set 4.5

1. **3.** 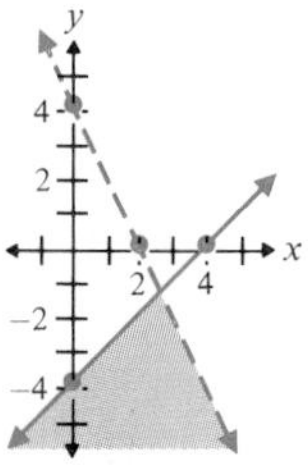 **5.** 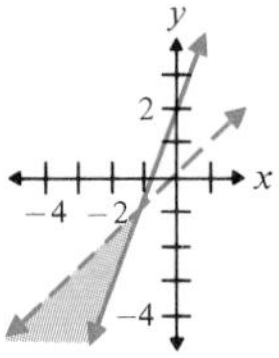 **7.** 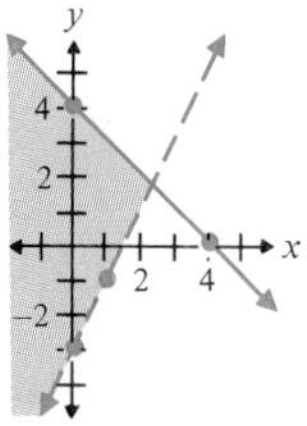 **9.** 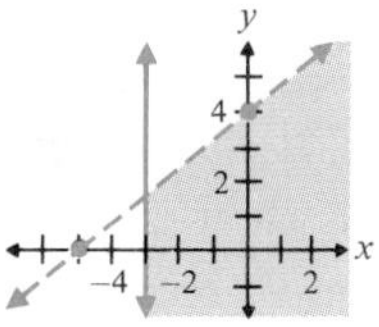 **11.** 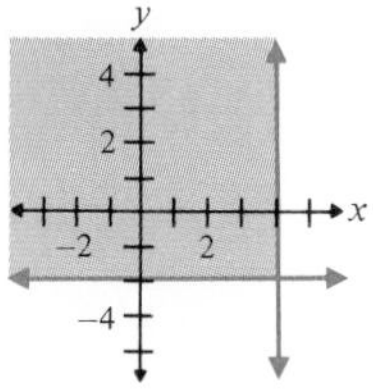 **13.** 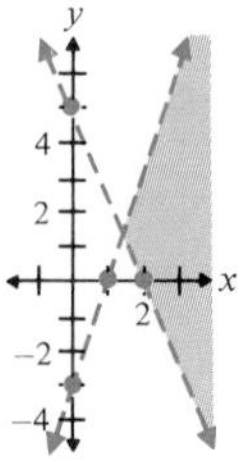 **15.** 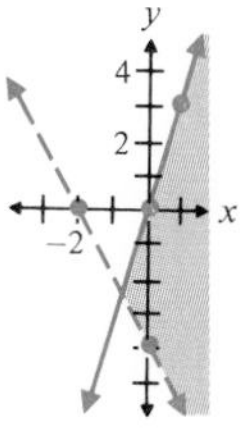 **17.** 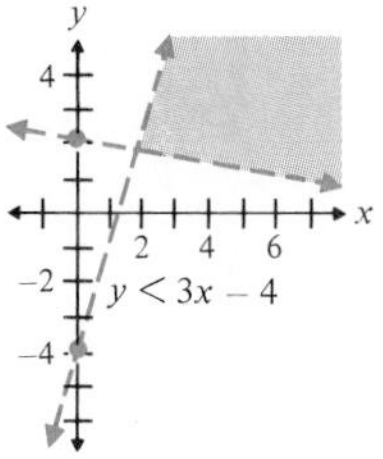 **19.**

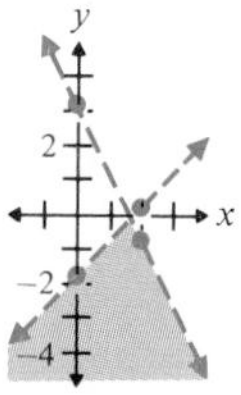

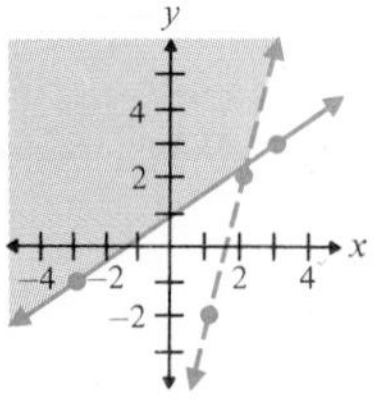

21.

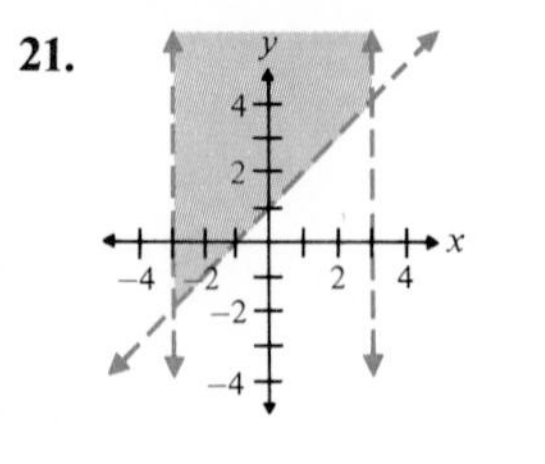

23.

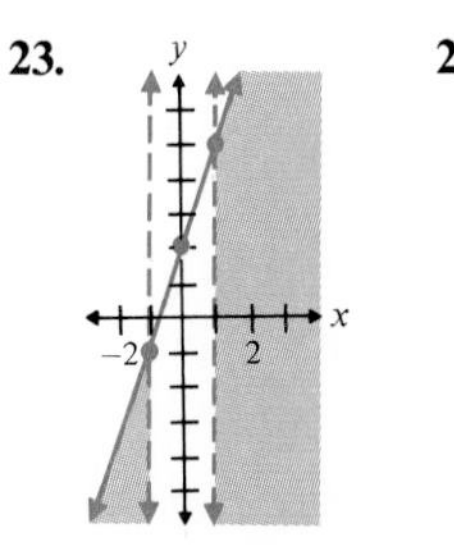

25.

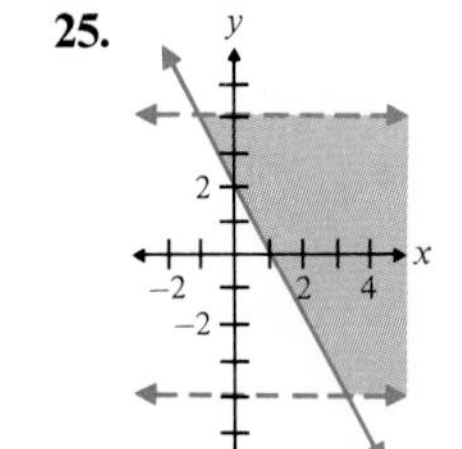

27. **29.**

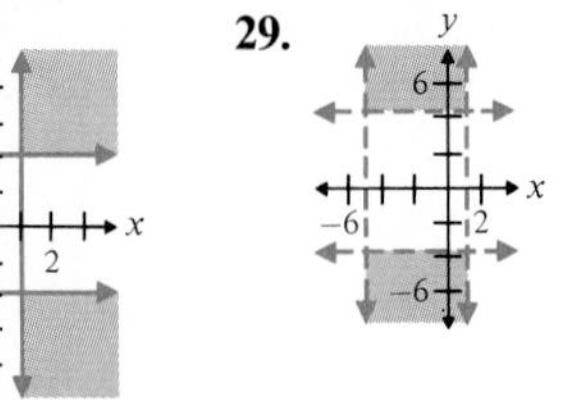

31. **33.**

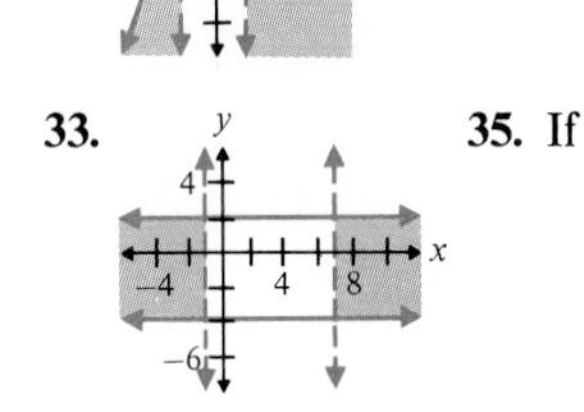

35. If the boundary lines are parallel, there may be no solution.

Just for Fun **1.** **2.** **3.** **4.**

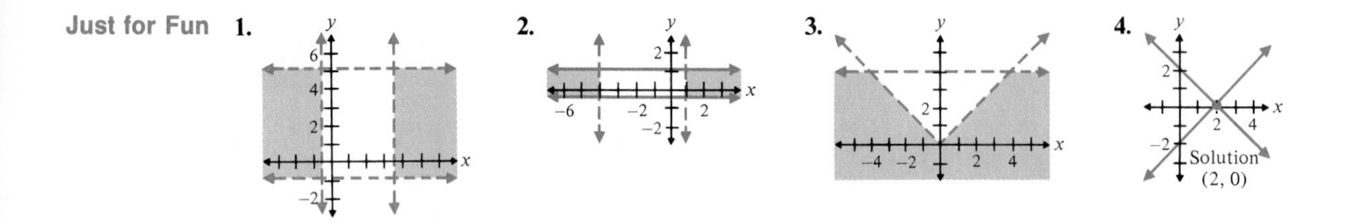

Review Exercises

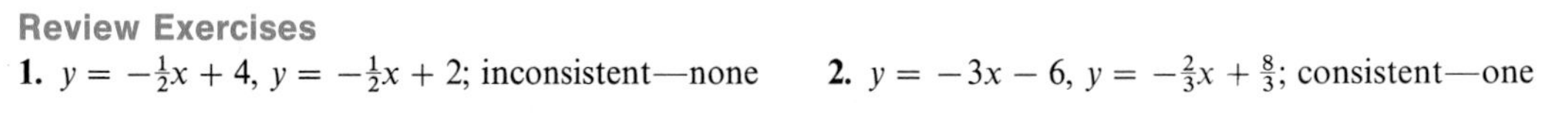

1. $y = -\frac{1}{2}x + 4$, $y = -\frac{1}{2}x + 2$; inconsistent—none **2.** $y = -3x - 6$, $y = -\frac{2}{3}x + \frac{8}{3}$; consistent—one

3. $y = \frac{1}{2}x + 4$, $y = -\frac{1}{2}x + 4$; consistent—one **4.** $y = \frac{3}{2}x + 2$, $y = \frac{2}{3}x - \frac{4}{3}$; consistent—one **5.**

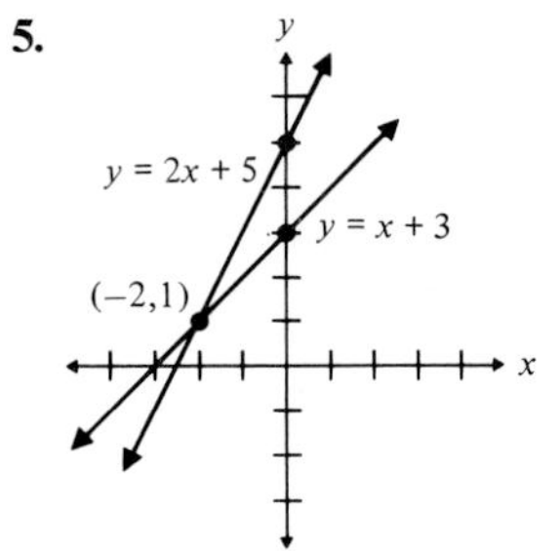

6.

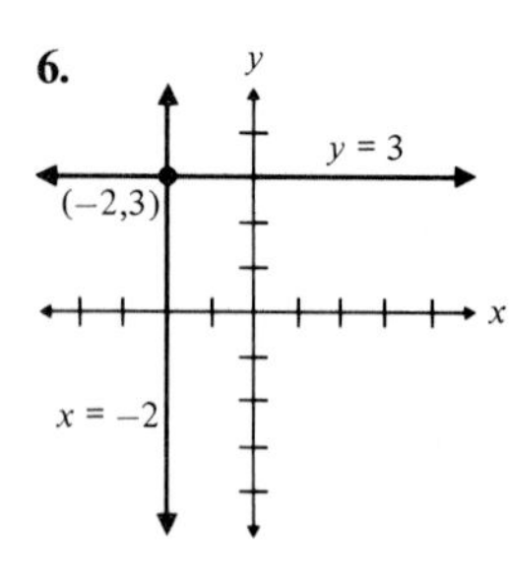

7.

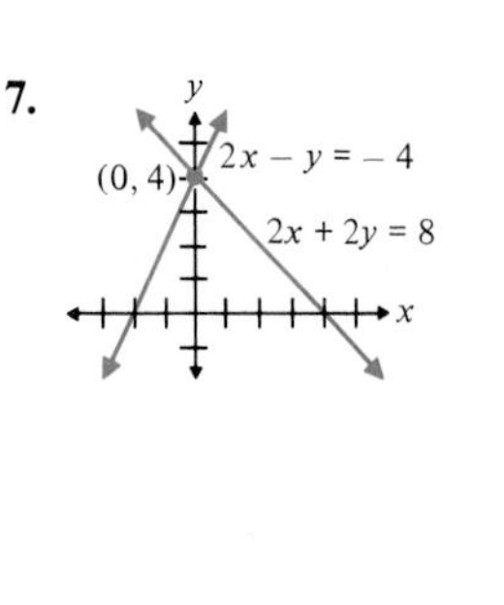

8.

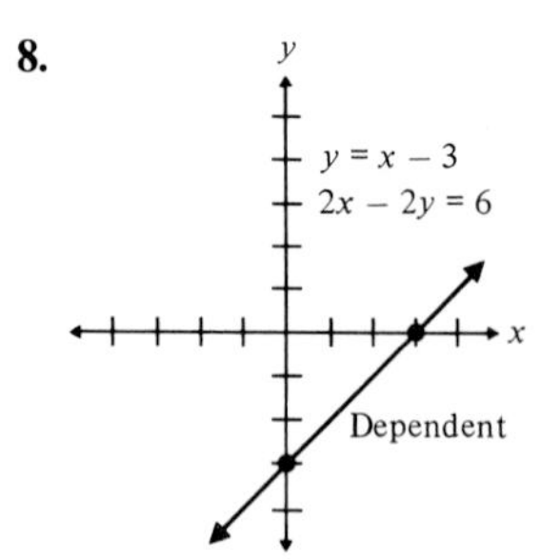

9. (3, 7) **10.** (2, 3)

11. (5, 2)

12. $(-3, 2)$ **13.** $(2, 1)$ **14.** $(5, 4)$ **15.** $(5, 2)$ **16.** $(-18, 6)$ **17.** $(8, -2)$ **18.** $(1, -2)$ **19.** $(26, -16)$
20. $(-7, 19)$ **21.** $(\frac{32}{13}, \frac{8}{13})$ **22.** $(-1, \frac{13}{3})$ **23.** $(1, 2)$ **24.** $(\frac{7}{5}, \frac{13}{5})$ **25.** $(6, -2)$ **26.** $(-\frac{78}{7}, -\frac{48}{7})$ **27.** $(2, 5, \frac{34}{5})$
28. $(5, -\frac{15}{4}, -2)$ **29.** $(1, 2, -1)$ **30.** $(3, 1, 2)$ **31.** $(\frac{8}{3}, \frac{2}{3}, 3)$ **32.** $(0, 2, -3)$ **33.** $x + y = 48$, $y = 2x - 3$, 17, 31
34. $x - y = 18$, $x = 4y$, 6, 24 **35.** $R = 0.08x$, $R = 500 + 0.03x$, \$10,000
36. $x + y = 6$, $0.3x + 0.5y = 0.4(6)$, 3 liters of each **37.** $x + y = 600$, $x - y = 530$, 565 plane, 35 wind
38. $d = 2(x + 2)$, $d = 2.5(x - 2)$, 18 mph, 40 mi **39.** $(4, -1)$ **40.** $(1, -1)$ **41.** $(-1, 2)$ **42.** $(4, 1, 3)$
43. $(\frac{3}{2}, -4, 3)$ **44.** $(-1, 5, -2)$ **45.** No solution **46.** $(1, 1, 1)$ **47.**

48. **49.** **50.**

Practice Test

1. $y = -\frac{4}{3}x - 2$, $y = \frac{4}{3}x + \frac{2}{3}$; consistent—one **2.** $y = -\frac{5}{3}x + 3$, $y = \frac{5}{3}x - 3$; consistent—one **3.**

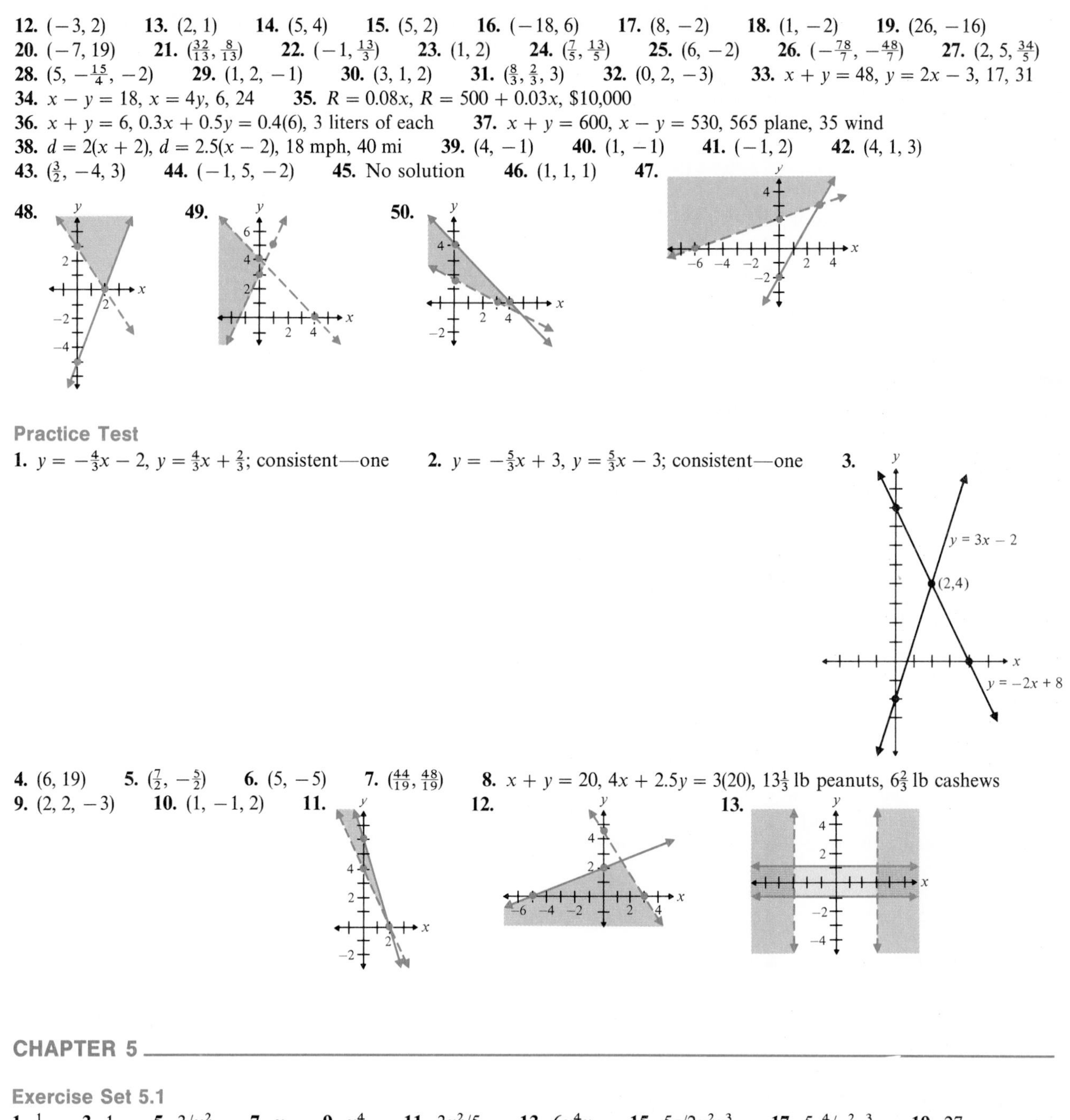

4. $(6, 19)$ **5.** $(\frac{7}{2}, -\frac{5}{2})$ **6.** $(5, -5)$ **7.** $(\frac{44}{19}, \frac{48}{19})$ **8.** $x + y = 20$, $4x + 2.5y = 3(20)$, $13\frac{1}{3}$ lb peanuts, $6\frac{2}{3}$ lb cashews
9. $(2, 2, -3)$ **10.** $(1, -1, 2)$ **11.** **12.** **13.**

CHAPTER 5

Exercise Set 5.1

1. $\frac{1}{9}$ **3.** 1 **5.** $3/x^2$ **7.** x **9.** x^4 **11.** $3y^2/5$ **13.** $6x^4y$ **15.** $5z/2x^2y^3$ **17.** $5z^4/x^2y^3$ **19.** 27
21. 2 **23.** $\frac{1}{49}$ **25.** x^6 **27.** $1/x$ **29.** $\frac{1}{25}$ **31.** $\frac{1}{49}$ **33.** x^7 **35.** $1/x^3$ **37.** $1/(2y^2)$ **39.** $6x^7$ **41.** x^2
43. $3y$ **45.** $12y^3/x^7$ **47.** $-10x^7z^5$ **49.** $8x^7y^2/z^3$ **51.** $3x^2/y^6$ **53.** $3x^3/y^5$ **55.** $3x^5$ **57.** $-x^2/y^9$
59. $2y^4z^4/3x^2$ **61.** $6x^6yz^3$ **63.** x^{7a+4} **65.** w^{7b+1} **67.** x^{w+7} **69.** x^{p+3} **71.** 3.7×10^3 **73.** 9×10^2
75. 4.7×10^{-2} **77.** 1.9×10^4 **79.** 1.86×10^{-6} **81.** 9.14×10^{-6} **83.** 5200 **85.** 40,000,000 **87.** 0.0000213
89. 0.312 **91.** 9,000,000 **93.** 535 **95.** 150,000,000 **97.** 0.0063 **99.** 320 **101.** 0.021 **103.** 20
105. 4,200,000,000,000 **107.** 0.00000045 **109.** 2000 **111.** 0.0000002 **113.** 900 **115.** 400,000 sec.

Just for Fun **1. (a)** 0 gives $10^0 = 1$, 1 gives $10^1 = 10$, 2 gives $10^2 = 100$, 3 gives $10^3 = 1000$, 4 gives $10^4 = 10{,}000$, 5 gives $10^5 = 100{,}000$, 6 gives $10^6 = 1{,}000{,}000$, 7 gives $10^7 = 10{,}000{,}000$, 8 gives $10^8 = 100{,}000{,}000$, 9 gives $10^9 = 1{,}000{,}000{,}000$, 10 gives $10^{10} = 10{,}000{,}000{,}000$ **(b)** 10,000 **(c)** $10^{1.8} \approx 63.1$
2. (a) about 5.87×10^{12} miles **(b)** 500 sec or $8\frac{1}{3}$ min. **(c)** 6.72×10^{11} sec or 21,309 years

Exercise Set 5.2

1. 1 **3.** 4 **5.** -1 **7.** -3 **9.** -1 **11.** -2 **13.** 12 **15.** x^4 **17.** 81 **19.** $\frac{1}{64}$ **21.** x^{12} **23.** 1 **25.** 1 **27.** x^2 **29.** $-1/x^3$ **31.** $3/x^8$ **33.** $27x^6$ **35.** $81x^8y^4$ **37.** $16x^4/y^4$ **39.** $1/(8x^9y^3)$ **41.** $-x^{12}/(64y^{15})$ **43.** $36x^2/y^4$ **45.** $8x^6y^{15}$ **47.** $125y^9/x^3$ **49.** $x^4y^2/9$ **51.** $1/(64x^{21}y^3)$ **53.** $125x^3y^3z^6$ **55.** $45x^{11}y^8$ **57.** $4x^7y^6/z^4$ **59.** x^2/y^{10} **61.** $4/(3x^3y^3)$ **63.** x^{5m+4} **65.** b^{5y^2+3y} **67.** m^{3-7y}

Just for Fun **1. (a)** $1^{10} = 1$, $2^{10} = 1024$, $3^{10} = 59{,}049$, $4^{10} = 1{,}048{,}576$ **(b)** $0.3^{10} = 0.0000059$, $0.2^{10} = 0.0000001$, $0.1^{10} = 0.0000000001$ **(c)** If values greater than 1 are substituted for x, $x^{10} > 1$ and x^{10} increases as x increases. When 1 is substituted for x, $x^{10} = 1$. When values less than 1 are substituted for x, $x^{10} < 1$ and x^{10} decreases as x decreases.

Exercise Set 5.3

1. Monomial **3.** Monomial **5.** Binomial **7.** Trinomial **9.** Not polynomial **11.** Binomial **13.** $-x^2 - 4x - 8$, second **15.** $10x^2 + 3xy + 6y^2$, second **17.** $2x^3 + x^2 - 3x + 5$, third **19.** In descending order, fourth **21.** $-2x^3 + 3x^2y + 5xy^2 - 6$, third **23.** $7x - 2$ **25.** $x - 6$ **27.** $-5x + 9$ **29.** $-7x + 4$ **31.** $7y - 11$ **33.** $x^2 - 8x - 2$ **35.** $3x - z + 3$ **37.** $-2x^2 + x - 12$ **39.** $2x^2 + 8x + 5$ **41.** $8y^2 - 5y - 3$ **43.** $5x^2 + x + 20$ **45.** $-7x^2 + x - 12$ **47.** $-x^2 + x + 20$ **49.** $5x^2 - 2x + 7$ **51.** $-x^3 + 3x^2y + 4xy^2$ **53.** $5xy^2 - 8$ **55.** $3x^2 - 2xy$ **57.** $4x^3 - 8x^2 - x - 4$ **59.** $7x^2y - 3x + 2$ **61.** $9x^3 - x^2 - 9$ **63.** $x^2 + x + 16$ **65.** $-x + 11$ **67.** $-3x^2 + 2x - 12$ **69.** $-3x^2 - 4x + 14$ **71.** $7x^2 + 7x - 13$ **73.** $15y^2 - 6y + 4$ **75.** $4x^2 + 7xy + 3y$ **77.** $x^3 + 4x^2 - 6x - 6$ **79.** $4x^2 + 7x - 4y^2 - 3y$ **81.** $-7x^2y + 6xy^2$

Exercise Set 5.4

1. $2x + 6$ **3.** $4x - 12$ **5.** $8x^2 - 24x$ **7.** $3x^3 + 9x^2 - 3x$ **9.** $-2x^3 + 4x^2 - 10x$ **11.** $-20x^4 + 30x^3 - 20x^2$ **13.** $6x^4y - 18x^3y^2 + 24xy^3$ **15.** $6x^5y^2 + 15x^{11} - 18x^4y$ **17.** $2xyz + \frac{8}{3}y^2z - 6y^3z$ **19.** $x^2 + x - 20$ **21.** $3x^2 - 8x - 3$ **23.** $8x^2 - 8x - 6$ **25.** $6y^2 - y - 12$ **27.** $x^2 - y^2$ **29.** $4x^3 - 6x^2 - 6x + 9$ **31.** $-2x^3 + 8x^2 - 3x + 12$ **33.** $-4x^2 + 2x + 12$ **35.** $xy + xz + y^2 + yz$ **37.** $6x^4 - 5x^2y - 6y^2$ **39.** $-12x^3 + 8x^2y^2 + 9xy - 6y^3$ **41.** $x^2 - 16$ **43.** $4x^2 - 1$ **45.** $4x^2 - 12xy + 9y^2$ **47.** $16x^2 - 40xy + 25y^2$ **49.** $16a^2 - 24ab + 9b^2$ **51.** $4r^2 + 16rs + 16s^2$ **53.** $4x^2 + 20xy + 25y^2$ **55.** $4y^4 - 20y^2w + 25w^2$ **57.** $25m^4 - 4n^2$ **59.** $9m^2 + 12m + 4 - n^2$ **61.** $25x^2 + 10x + 1 - 36y^2$ **63.** $16 - 8x + 24y + x^2 - 6xy + 9y^2$ **65.** $y^2 + 8y - 4xy + 16 - 16x + 4x^2$ **67.** $w^2 - 8w - 9x^2 + 16$ **69.** $x^3 - 7x^2 + 14x - 8$ **71.** $-2y^3 + 9y^2 - 17y + 12$ **73.** $2x^3 + 10x^2 + 9x - 9$ **75.** $5x^3 - x^2 + 16x + 16$ **77.** $-14x^3 - 22x^2 + 19x - 3$ **79.** $x^3y - x^2y^2 - 6xy^3$ **81.** $a^3 + b^3$ **83.** $a^3 + 8b^3$ **85.** $2a^3 - 7a^2b + 5ab^2 - 6b^3$ **87.** $6x^3 - x^2y - 16xy^2 + 6y^3$ **89.** $x^5 - 2x^4 - 5x^3 + 5x^2 - 2x - 12$ **91.** $x^3 + 9x^2 + 27x + 27$ **93.** $8x^3 + 36x^2 + 54x + 27$

Just for Fun **1.** $y^2 - 2y - 2xy + 2x + x^2 + 1$ **2.** $x^4 - 12x^3y + 54x^2y^2 - 108xy^3 + 81y^4$

Exercise Set 5.5

1. $3x + 4$ **3.** $2x + 1$ **5.** $3x^2 - x - 2$ **7.** $3y^2 + \dfrac{1}{y^2}$ **9.** $2x + 1 + \dfrac{1}{2x} + \dfrac{3}{2x^2}$ **11.** $2x^2 - 4xy + \dfrac{3}{2}y^2$

13. $3x^6 - x^3 + 6$ **15.** $\dfrac{z}{2} + z^2 - \dfrac{3}{2}x^2y^4z^7$ **17.** $x + 3$ **19.** $2x + 3$ **21.** $2x + 4$ **23.** $x - 2$ **25.** $x + 5$

27. $2x - 3$ **29.** $2x^2 + 3x - 1$ **31.** $2x^2 - x - 4$ **33.** $2x^2 + x - 2 - \dfrac{2}{2x - 1}$ **35.** $-x^2 - 7x - 5 - \dfrac{8}{x - 1}$

37. $2x^2 - 6x + 3$ **39.** $x + 4$

Just for Fun **1.** $2x^2 + 3xy - y^2$ **2.** $x^2 + \dfrac{2}{3}x + \dfrac{4}{9} - \dfrac{37}{9(3x - 2)}$

Exercise Set 5.6

1. $x+3$ **3.** $x-1$ **5.** $x+8+\frac{12}{x-3}$ **7.** $3x+5+\frac{10}{x-4}$ **9.** $4x^2+x+3+\frac{3}{x-1}$
11. $3x^2-2x+2+\frac{6}{x+3}$ **13.** $5x^2-11x+14-\frac{20}{x+1}$ **15.** $x^3+6x^2+11x+6+\frac{8}{x-2}$ **17.** $y^4-\frac{10}{y+1}$
19. $3x^2+3x-3$ **21.** $2x^3+2x-2$

Just for Fun **1. (a)** $3x^2-2x+5-[13/(3x+5)]$ **(b)** Because we are expressing the remainder in terms of $3x+5$ rather than $x+\frac{5}{3}$, the denominator of the remainder is altered rather than the numerator.

Exercise Set 5.7

1. (a) 2 **(b)** -10 **(c)** $3a-1$ **3. (a)** 11 **(b)** -1 **(c)** $2a+2b+3$ **5. (a)** 6 **(b)** 10 **(c)** 6
7. (a) 2 **(b)** 17 **(c)** -11 **9. (a)** 1 **(b)** -35 **(c)** $-c^2+4c-3$ **11. (a)** 23 **(b)** 15 **(c)** $3a^2+16a+28$
13. (a) 5 **(b)** $-c^2-2c+5$ **(c)** $-c^2-6c-3$ **15. (a)** 40 **(b)** $2x^2+3x+5$ **(c)** $8x^2-6x+5$
17. (a) h^2+3h-4 **(b)** $h^2+11h+24$ **(c)** $a^2+2ah+h^2+3a+3h-4$ **19. (a)** $2h^2-3h+1$
(b) $2x^2+9x+10$ **(c)** $2x^2+4xh+2h^2-3x-3h+1$ **21. (a)** 110 **(b)** 240 **23. (a)** $v=19.65$ m/s, $h=60.3$ m
(b) $v=8.45$ m/s, $h=11.125$ m **25. (a)** 34 m **(b)** 10.75 m **27. (a)** 5.6 cm **(b)** 8.988 cm **29. (a)** 91 **(b)** 204

Just for Fun **1.** $x^3+7x^2+21x+30$ **2. (a)** $2x^2+4xh+2h^2+3x+3h-4$ **(b)** $4xh+2h^2+3h$ **(c)** $4x+2h+3$
3. (a) $A=d^2-(\pi d^2/4)$ **(b)** 3.44 ft **(c)** 7.74 ft

Exercise Set 5.8

1. $x=-1$, $(-1,-8)$, up **3.** $x=2$, $(2,-2)$, down **5.** $x=1$, $(1,11)$, down **7.** $x=-1$, $(-1,-8)$, down
9. $x=\frac{1}{2}$, $(\frac{1}{2},\frac{7}{4})$, up **11.** $x=-\frac{3}{2}$, $(-\frac{3}{2},-14)$, up

13. Domain: $\mathbb{R}$
Range: $\{y \mid y \geq -1\}$

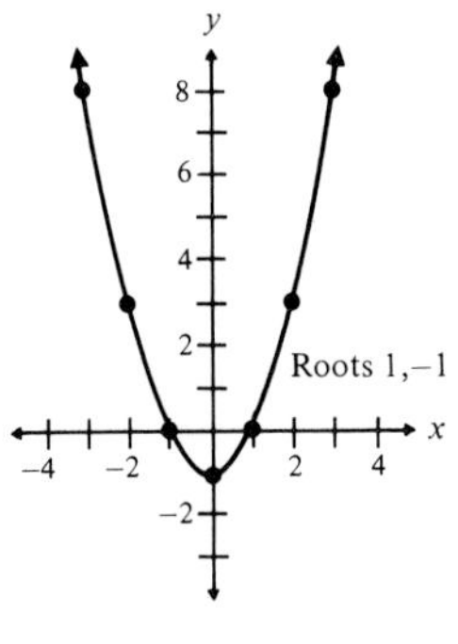

15. Domain: $\mathbb{R}$
Range: $\{y \mid y \leq 3\}$

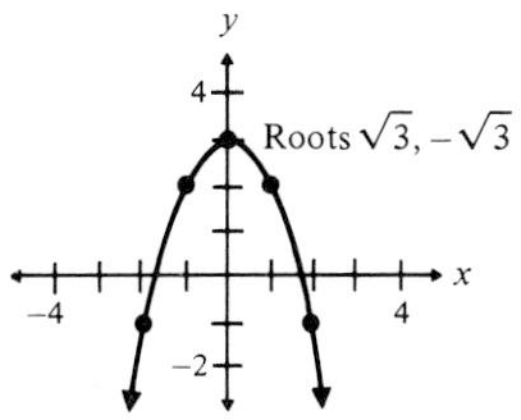

17. Domain: $\mathbb{R}$
Range: $\{y \mid y \geq 2\}$

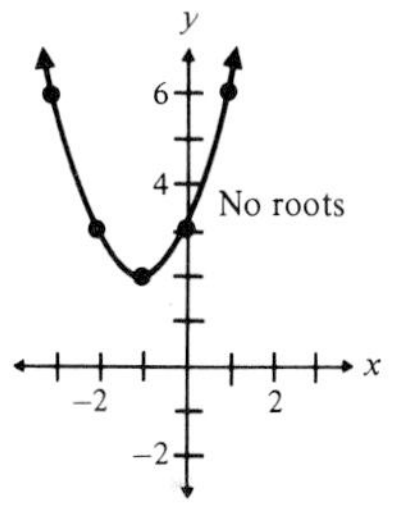

19. Domain: $\mathbb{R}$
Range: $\{y \mid y \geq -16\}$

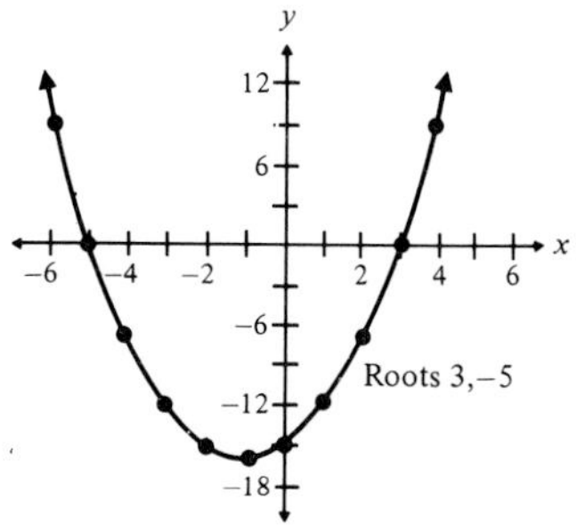

21. Domain: $\mathbb{R}$
Range: $\{y \mid y \leq -1\}$

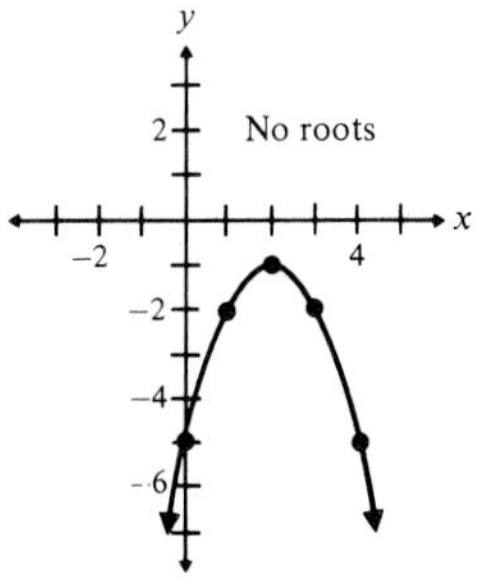

23. Domain: $\mathbb{R}$
Range: $\{y \mid y \geq -11\}$

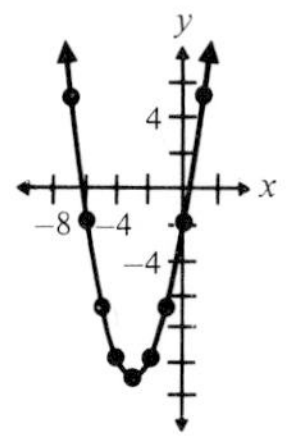

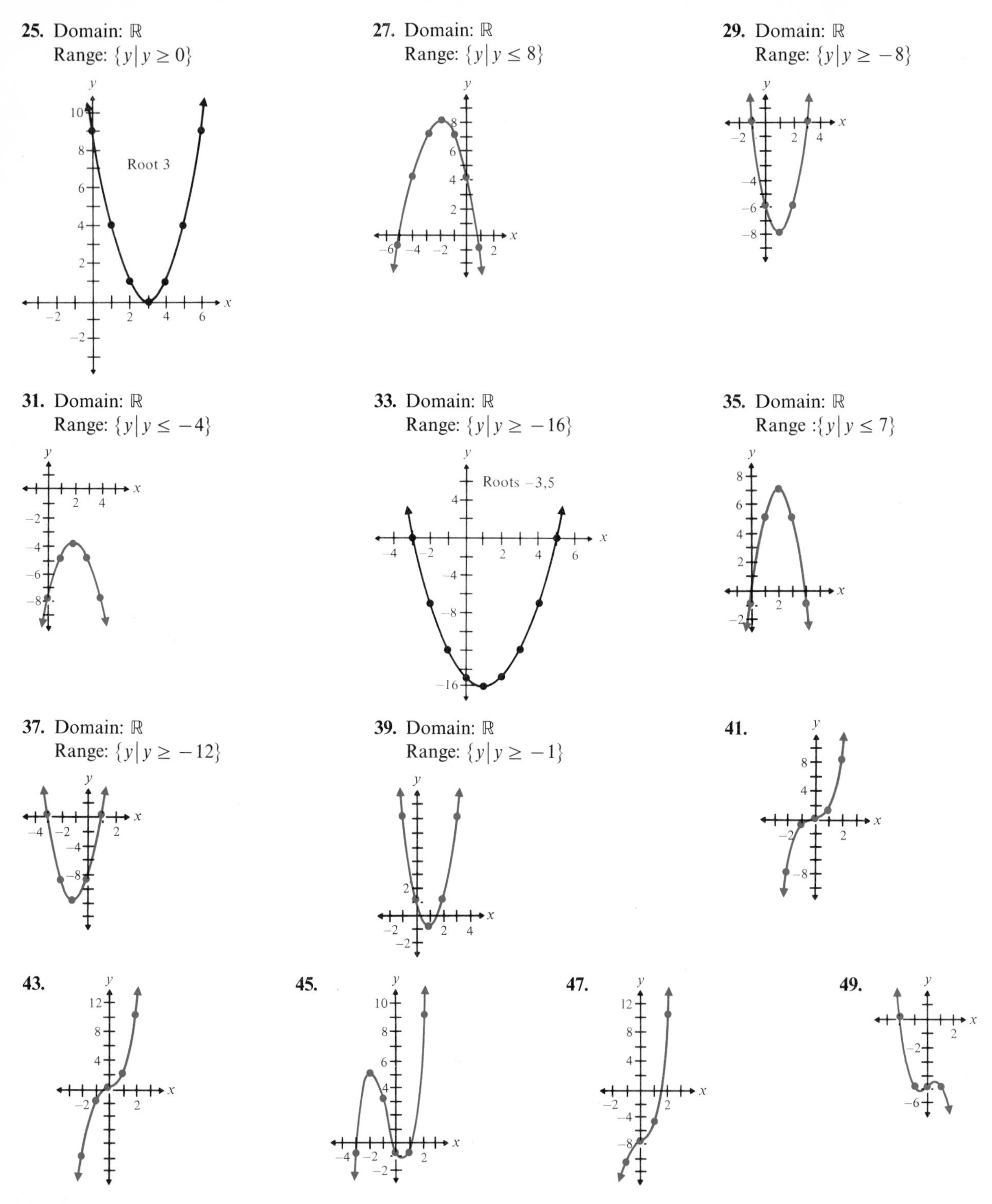

51. When the squared term is positive, the graph opens up; if negative, it opens down.

53. As x increases, y increases; as x decreases, y decreases

55. y decreases as x goes from -3 to 0; then y increases as x goes from 0 to 3. (y is a minimum when $x = 0$.)

Just for Fun

1. **2.**

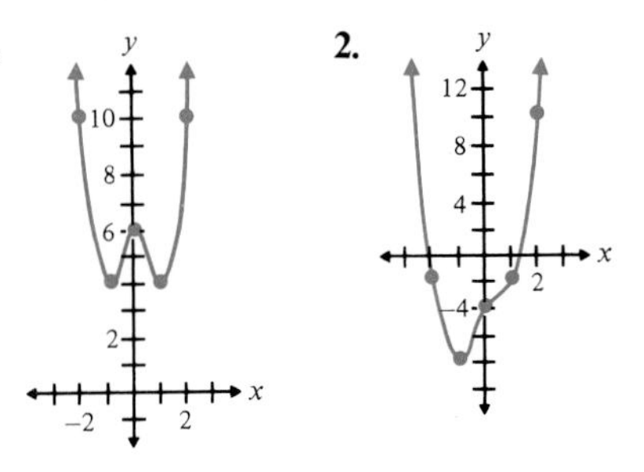

Review Exercises

1. 64 **2.** x^8 **3.** y^7 **4.** 27 **5.** x^4 **6.** y^9 **7.** $1/y$ **8.** x^7 **9.** $1/x^3$ **10.** $\frac{1}{27}$ **11.** $\frac{1}{32}$ **12.** 3
13. $9x^4$ **14.** $-12x^2y^6$ **15.** $-21x^3y^9$ **16.** $8/x^2y$ **17.** $30x^5/y^6$ **18.** $3y^7/x^5$ **19.** $3/xy^9$ **20.** $y^8/2$
21. $\dfrac{x^7}{y^8}$ **22.** $125x^3y^3$ **23.** $\dfrac{x^{10}}{9y^2}$ **24.** $\dfrac{4x^2}{y^2}$ **25.** $\dfrac{-125y^3}{x^6z^9}$ **26.** $\dfrac{z^4}{36x^2y^6}$ **27.** $\dfrac{x^9}{27}$ **28.** $\dfrac{x^{12}}{16y^8}$ **29.** $\dfrac{125}{x^9y^3}$
30. $\dfrac{8x^{10}y^{16}}{z^6}$ **31.** $\dfrac{64y^3}{x^3z^{15}}$ **32.** $\dfrac{4x^5}{z^6}$ **33.** 7.42×10^{-5} **34.** 2.6×10^5 **35.** 1.83×10^5 **36.** 1×10^{-6}
37. 248,000,000 **38.** 30,000 **39.** 0.02 **40.** 200,000,000 **41.** 0.00000004 **42.** 2000 **43.** Binomial, $-x + 5$, first
44. Trinomial, $x^2 + 5x - 3$, second **45.** Monomial, 6, zero **46.** Trinomial, $x^2 + xy - y^2$, second
47. Trinomial, $x^5y^3 + x^4y - 6xy^3$, eighth **48.** Polynomial, $2x^4 - 9x^2y + 6xy^3 - 3$, fourth **49.** Not polynomial
50. Not polynomial **51.** $10x - 5$ **52.** $7x^2 - 4x - 15$ **53.** $2x^2 + 4x + 4$ **54.** $2x^3 - 8x^2 - 9$
55. $3x^2y + 3xy - 9y^2$ **56.** $-9xy - 9x + 5y^2$ **57.** $4x^3 + 8x^2 + 12x$ **58.** $-2x^4y^2 - 2x^3y^7 + 12xy^3$
59. $12x^2 + xy - 20y^2$ **60.** $6x^2y + 9xy^2 + 2x + 3y$ **61.** $x^2 + 10x + 25$ **62.** $x^2 - 6x + 9$ **63.** $4x^2 + 20x + 25$
64. $9x^2 - 12xy + 4y^2$ **65.** $x^2 - 36$ **66.** $x^2 - y^2$ **67.** $4x^2 - 9$ **68.** $25x^2y^2 - 36$ **69.** $4x^2 - 25y^4$
70. $x^2 + 6xy + 9y^2 + 4x + 12y + 4$ **71.** $x^2 + 6xy + 9y^2 - 4$ **72.** $6x^3 - x^2 - 24x + 18$
73. $4x^4 + 12x^3 + 6x^2 + 13x - 15$ **74.** $x^3y + 6x^2y + 7xy^2 + x^2y^2 + y^3$ **75.** $5y^2 + 2$ **76.** $3x - 2y + 1$
77. $\dfrac{x^2}{2y} + x + \dfrac{3y}{2}$ **78.** $x + 4$ **79.** $2x - 3$ **80.** $2x^2 + 3x - 4$ **81.** $2x^3 + x^2 - 3x - 4$ **82.** $3x^2 + 7x + 21 + \dfrac{73}{x - 3}$
83. $2y^4 - 2y^3 - 8y^2 + 8y - 7 + \dfrac{6}{y + 1}$ **84.** $x^4 + 2x^3 + 4x^2 + 8x + 16 + \dfrac{12}{x - 2}$ **85.** $2x^2 + 2x + 6$
86. **(a)** 3 **(b)** 94 **87.** **(a)** 3 **(b)** 7 **88.** **(a)** 5 **(b)** 30 **89.** **(a)** $a^2 + 2a - 1$ **(b)** $a^2 + 6a + 7$
90. **(a)** $a^2 - a + 3$ **(b)** $a^2 + 2ab + b^2 - a - b + 3$ **91.** Domain: $\mathbb{R}$ Range: $\{y \mid y \geq 0\}$ **92.** Domain: $\mathbb{R}$ Range: $\{y \mid y \geq -1\}$

93. Domain: $\mathbb{R}$ Range: $\{y \mid y \geq 1\}$ **94.** **95.** **96.**

97. **(a)** 720 **(b)** 1500 **98.** **(a)** 84 ft **(b)** 36 ft

Practice Test

1. $y^{10}/9x^6$ **2.** $9x^6/y^{14}$ **3.** Trinomial, $-6x^4 - 4x^2y^3 + 2x$, fifth **4.** $4x^3 - 2x^2 + 2x + 8$ **5.** $4x^5 - 2y^2 + \frac{5}{x}$
6. $-6x^2 + xy + y^2$ **7.** $4x^3 + 8x^2y - 9xy^2 - 6y^3$ **8.** $x - 5 + \frac{25}{2x + 3}$ **9.** $2x^2y + 3x + 7y^2$
10. $2x^2 + 5x + 20 + \frac{53}{x - 3}$ **11.** $-6x^7y^6 + 18x^4y^7 - 9x^3y^4$ **12.** $4x^2 + 12xy + 9y^2$
13. $3x^3 + 3x^2 + 15x + 15 + \frac{79}{x - 5}$ **14.** -18 **15.** Domain: $\mathbb{R}$ Range $\{y \mid y \geq -2\}$ **16.** 6848

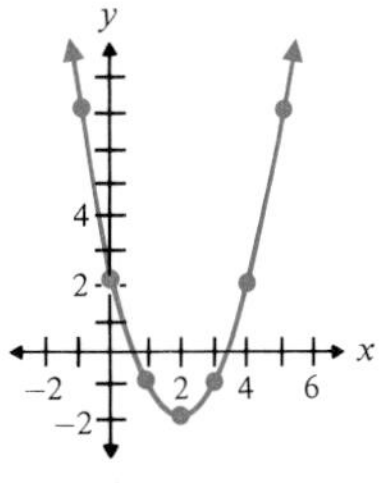

CHAPTER 6

Exercise Set 6.1

1. $8(n + 1)$ **3.** Cannot be factored **5.** $2(8x^2 - 6x - 3)$ **7.** $2p(10p^2 - 9p + 6)$ **9.** $x^3(7x^2 - 9x + 3)$
11. $3y(8y^{14} - 3y^2 + 1)$ **13.** $x(1 + 3y^2)$ **15.** Cannot be factored **17.** $4(4xy^2z + x^3y - 2)$ **19.** $8xy^2(5x + 2y^2 + 8y)$
21. $9xyz(4yz^2 + 4x^2y + x)$ **23.** $y^2z^2(14yz^3 - 28yz^4 - 9x)$ **25.** $4x^3(6x^3 + 2x - y)$ **27.** $xy(48x + 16y + 33)$
29. $(x + 2)(x + 3)$ **31.** $(4x - 3)(7x - 1)$ **33.** $(2x + 1)(8x^2 + 4x + 1)$ **35.** $(2x + 1)(4x + 1)$
37. $(x + 3)(5x^2 + 15x - 3)$ **39.** $2x^2(2x + 7)(3x^3 + 2x - 1)$ **41.** $(2x + 5)(6x^3 - 2x^2 - 1)$ **43.** $(2r - 3)^6(8pr - 12p - 3)$
45. $(x - 5)(x + 3)$ **47.** $(2x - 3)(2x + 3)$ **49.** $(3x + 1)(x + 3)$ **51.** $(2x - 1)^2$ **53.** $2(2x - 1)(2x - 5)$
55. $(a + b)(x + y)$ **57.** $(3a + 2b)(c + d)$ **59.** $(3a - 4b)(5a - 6b)$ **61.** $(x - 3y)(x + 2y)$ **63.** $(3x + y)(2x - 3y)$
65. $(2x - 5y)(5x - 6y)$ **67.** $(x - y)(x^2 + y)$ **69.** $y(2x - 3y)(x^2y^2 + 3z^2)$ **71.** $x(3x^2 + 2)(x^2 - 5)$
73. $(3p + 2q)(p^2 + q^2)$

Just for Fun **1.** $2(x - 3)(2x^4 - 12x^3 + 15x^2 + 9x + 2)$ **2. (a)** $(x + 1)[(x + 1) + 1] = (x + 1)(x + 2)$
(b) $(x + 1)^2[(x + 1) + 1] = (x + 1)^2(x + 2)$ **(c)** $(x + 1)^{n-1}[(x + 1) + 1] = (x + 1)^{n-1}(x + 2)$
(d) $(x + 1)^n[(x + 1) + 1] = (x + 1)^n(x + 2)$ **3.** $2x(x + 5)^{-2}[2 + (x + 5)^1] = 2x(x + 5)^{-2}(x + 7)$
4. $(2x - 3)^{-1/2}[x + 6(2x - 3)^1] = (2x - 3)^{-1/2}(13x - 18)$

Exercise Set 6.2

1. $(x + 6)(x + 1)$ **3.** $(p - 5)(p + 2)$ **5.** $(x - 3)(x + 2)$ **7.** $(x + 8)(x + 8)$ **9.** $(x - 32)(x - 2)$ **11.** $(a - 15)(a - 3)$
13. Cannot be factored **15.** $(x - y)(x - 3y)$ **17.** $(a - 9b)(a - 2b)$ **19.** $(x - 15y)(x + 3y)$ **21.** $5(x + 1)(x + 3)$
23. $4(x + 4)(x - 1)$ **25.** $x(x - 6)(x + 3)$ **27.** $x(x - 8)(x + 3)$ **29.** $(2x + 3)(x + 1)$ **31.** $(4w + 1)(w + 3)$
33. Cannot be factored **35.** $(x - 1)(3x - 7)$ **37.** $(2x + 5)(x + 3)$ **39.** $(5y - 1)(y - 3)$ **41.** $(3y - 5)(y + 1)$
43. $(2x + 3y)(2x - y)$ **45.** $2(4x^2 + x - 10)$ **47.** $3x(2x + 3)(2x - 5)$ **49.** $2(2a - 3b)(3a - 4b)$ **51.** $xy(x - 6)(x + 3)$
53. $ab(a - 4)(a + 3)$ **55.** $5(x + 4y)(x + y)$ **57.** $6pq^2(p - 5q)(p + q)$ **59.** $(x^2 + 3)(x^2 - 2)$ **61.** $(x^2 + 2)(x^2 + 3)$
63. $(y^2 - 10)(y^2 + 3)$ **65.** $(2a^2 + 5)(3a^2 - 5)$ **67.** $(2x + 5)(2x + 3)$ **69.** $(3r + 14)(2r + 7)$ **71.** $(2p - 7)(3p - 14)$
73. $(ab - 3)(ab - 5)$ **75.** $(3xy - 5)(xy + 1)$ **77.** $(x + 2)(x + 1)(x + 3)$ **79.** $(2a - 5)(a - 1)(5 - a)$
81. $(2x + 3)(x + 2)(x - 3)$ **83.** $2x^2 - 5xy - 12y^2$ **85.** Divide $(x^2 - xy - 6y^2)$ by $(x - 3y)$, $x + 2y$

Just for Fun **1.** Cannot divide both sides of the equation by $a - b$ since it equals 0. **2.** $(2a^n + 3)(2a^n - 5)$
3. $2(3x^ny^n - 1)(2x^ny^n + 1)$

Exercise Set 6.3

1. $(x + 9)(x - 9)$ **3.** Cannot be factored **5.** $(1 + 2x)(1 - 2x)$ **7.** $(4 + x)(4 - x)$ **9.** $(x + 6y)(x - 6y)$
11. $(5 + 4y^2)(5 - 4y^2)$ **13.** $(a^2 + 2b^2)(a^2 - 2b^2)$ **15.** $(9a^2 + 4b)(9a^2 - 4b)$ **17.** $(xy + 1)(xy - 1)$

19. $(7 + 8xy)(7 - 8xy)$ **21.** $x^2(3y + 2)(3y - 2)$ **23.** $(5 + x + y)(5 - x - y)$ **25.** $(x + y + 4)(x + y - 4)$
27. $(a + 3b + 2)(a - 3b - 2)$ **29.** $(x + 5)^2$ **31.** $(5 - t)^2$ **33.** $(y - 4)^2$ **35.** $(3y + z)^2$ **37.** $(5ab - 2)^2$
39. $(a^2 + 6)^2$ **41.** Cannot be factored **43.** $(x + y + 1)^2$ **45.** $(a + b)^2(a - b)^2$ **47.** $(x + 3 + y)(x + 3 - y)$
49. $(x - 1)(7 - x)$ **51.** $(x - 3)(x^2 + 3x + 9)$ **53.** $(x + y)(x^2 - xy + y^2)$ **55.** $(a - 5)(a^2 + 5a + 25)$
57. $(x - 4)(x^2 + 4x + 16)$ **59.** $(y + 1)(y^2 - y + 1)$ **61.** $(3 - 2y)(9 + 6y + 4y^2)$ **63.** $5(x + 2y)(x^2 - 2xy + 4y^2)$
65. $(y^2 + x^3)(y^4 - x^3y^2 + x^6)$ **67.** $5(x - 5y)(x^2 + 5xy + 25y^2)$ **69.** $(x + 2)(x^2 + x + 1)$
71. $(x - y - 3)(x^2 - 2xy + y^2 + 3x - 3y + 9)$ **73.** 16, -16 **75.** 9

Just for Fun **1. (a)** $(x + \sqrt{7})(x - \sqrt{7})$ **(b)** $(x\sqrt{2} + \sqrt{15})(x\sqrt{2} - \sqrt{15})$ **2.** $-3(2x - 13)$ **3.** $(a^n - 8)^2$

Exercise Set 6.4

1. $3(x + 4)(x - 3)$ **3.** $2xy(x - 4 + 8y)$ **5.** $4(x + 3)^2$ **7.** $2(x + 6)(x - 6)$ **9.** $5x(x^2 + 3)(x^2 - 3)$
11. $3x(x - 4)(x - 1)$ **13.** $5x^2y^2(x - 3)(x + 4)$ **15.** $x^2(x + y)(x - y)$ **17.** $x^4y^2(x - 1)(x^2 + x + 1)$
19. $x(x^2 + 4)(x + 2)(x - 2)$ **21.** $4(x^2 + 2y)(x^4 - 2x^2y + 4y^2)$ **23.** $2(a + b + 3)(a + b - 3)$ **25.** $(x + 3y)^2$
27. $x(x + 4)$ **29.** $2(5r + 9)(5r - 9)$ **31.** $(y + 5)^2$ **33.** $5a^2(3a - 1)^2$ **35.** $(x + \frac{1}{3})(x^2 - \frac{1}{3}x + \frac{1}{9})$
37. $(x + 3)(x - 3)(3x + 2)$ **39.** $ab(a + 4b)(a - 4b)$ **41.** $(3 + x + y)(3 - x - y)$

Exercise Set 6.5

1. 0, -5 **3.** 0, -9 **5.** $-\frac{5}{2}$, 3, -2 **7.** 3 **9.** 0, 12 **11.** 0, -2 **13.** -4, 3 **15.** 2, 10 **17.** 3, -6
19. -4, 6 **21.** 0, 2, -21 **23.** -5, -6 **25.** 4, -2 **27.** 32, -2 **29.** 0, 6, -3 **31.** $\frac{1}{3}$, 7 **33.** $\frac{2}{3}$, -1
35. 0, -4, 3 **37.** 0, 16 **39.** 0, 4, -4 **41.** 0, -8 **43.** -3, 5 **45.** -3, 4 **47.** $-\frac{1}{3}$, 2 **49.** 0, 1
51. 0, -3, -5, **53.** 8, 9 **55.** 9, 11 **57.** 5, 7 **59.** $w = 3$ ft, $l = 12$ ft **61.** $h = 10$ cm, $b = 16$ cm **63.** 7 m
65. 5 sec **67.** 5 sec

Just for Fun **1.** $(x^3 + 5)(x^3 - 12) = 0$, $x = \sqrt[3]{-5}$, $x = \sqrt[3]{12}$ **2.** 1, -9 **3.** ± 1, ± 2

Review Exercises

1. $4x^2$ **2.** $6xy$ **3.** $3xy^2$ **4.** $2x - 5$ **5.** $x + 5$ **6.** 1 **7.** $3(3x + 11)$ **8.** $4(3x - 2)(x + 1)$
9. $6x^4(10 + x^5 - 3xy^2)$ **10.** Cannot be factored **11.** $(x - 2)(5x + 3)$ **12.** $(2x + 1)(4x - 3)$
13. $(x - 1)(3x^2 - 3x - 2)$ **14.** $(2x - 1)(10x - 3)$ **15.** $3xy(2xy - 4x^2 - 3y^4)$ **16.** $3xy^2z^2(4y^2z + 2xy - 5x^2z)$
17. $(x + 2)(x + 3)$ **18.** $(x + 3)(x - 5)$ **19.** $(x + 7)(x - 7)$ **20.** $(x + 3)(3x + 1)$ **21.** $(5x - 1)(x + 4)$
22. $(x + 4y)(5x - y)$ **23.** $(4x + 5y)(3x - 2y)$ **24.** $(3x - 2y)(4x + 5y)$ **25.** $(4x - 1)(x + 6)$ **26.** $(3x - y)(4x - 3y)$
27. $(x + 5)(x + 3)$ **28.** $(x - 5)(x - 3)$ **29.** $(x - 9)(x + 3)$ **30.** $(x - 15)(x + 3)$ **31.** $(x - 10y)(x + 5y)$
32. $(x - 18y)(x + 3y)$ **33.** $2(x + 4)^2$ **34.** $3(x - 3)^2$ **35.** $x(x - 6)(x + 3)$ **36.** $2x(x - 7)(x - 8)$
37. $(3x + 1)(x + 4)$ **38.** $(2x - 5y)(x + 2y)$ **39.** $x(4x - 5)(x - 1)$ **40.** $x(12x + 1)(x + 5)$ **41.** $(x^2 - 5)(x^2 + 2)$
42. $(x^2 + 4)(x^2 - 5)$ **43.** $(x + 9)(x + 11)$ **44.** $(x + 6)(x - 6)$ **45.** $(x + 7)(x - 7)$ **46.** $4(x + 2y^2)(x - 2y^2)$
47. $(x^2 + 9)(x - 3)(x + 3)$ **48.** $(x - 1)(x + 5)$ **49.** $(x - 1)(x - 5)$ **50.** $(x + 4)^2$ **51.** $(2x - 3)^2$ **52.** $(3y + 4)^2$
53. $(w^2 - 8)^2$ **54.** $(x + 2)(x^2 - 2x + 4)$ **55.** $(x - 2)(x^2 + 2x + 4)$ **56.** $(2x + 3)(4x^2 - 6x + 9)$
57. $(3x - 2y)(9x^2 + 6xy + 4y^2)$ **58.** $(3x - 2y^2)(9x^2 + 6xy^2 + 4y^4)$ **59.** $(2y^2 - 5x)(4y^4 + 10xy^2 + 25x^2)$
60. $(x - 1)(x^2 + 4x + 7)$ **61.** $y^2(x + 3)(x - 5)$ **62.** $3x(x - 4)(x - 2)$ **63.** $3xy^4(x - 2)(x + 6)$
64. $3y(y^2 + 3)(y^2 - 3)$ **65.** $2y(x + 2)(x^2 - 2x + 4)$ **66.** $5x^2y(x + 2)^2$ **67.** $3x(2x + 1)(x - 4)$
68. $(x + 5 + y)(x + 5 - y)$ **69.** $3(x + 2y)(x^2 - 2xy + 4y^2)$ **70.** $(x + 4)^2(x - 1)$ **71.** $(4x + 1)(4x + 5)$
72. $(2x^2 - 1)(2x^2 + 3)$ **73.** $(x + 1)(x - 2)(x - 1)$ **74.** 5, $-\frac{2}{3}$ **75.** 0, $\frac{3}{2}$ **76.** 0, $-\frac{4}{3}$ **77.** 6, -4 **78.** -3, -5
79. -4, 2 **80.** -2, -5 **81.** 0, 2, 4 **82.** 0, 1, -2 **83.** $\frac{1}{4}$, $-\frac{3}{2}$ **84.** 2, -2 **85.** 2, -3 **86.** 5, 6
87. $w = 7$ ft, $l = 9$ ft **88.** 4 m, 11 m **89.** 5 in., 9 in. **90.** 9 sec

Practice Test

1. $4x(xy - 1)$ **2.** $(3x + 2)(x + 4)$ **3.** $4xy^3(2xy - 3 - 6x^4y^2)$ **4.** $3xy^2(3x^2 + 4xy^3 - 9y^2)$ **5.** $5(x - 2)(x + 1)$
6. $(2x + 3y)(x + 2y)$ **7.** $(x - 9)(x - 7)$ **8.** $(x - 4y)(x - 3y)$ **9.** $3x(x - 3)(x + 1)$ **10.** $(3x - 2)(2x - 1)$
11. $(5x + 2)(x + 3)$ **12.** $(9x + 4y^2)(9x - 4y^2)$ **13.** $y^6(3x - 2)(9x^2 + 6x + 4)$ **14.** $(x + 2)(x + 6)$
15. $(2x^2 + 9)(x^2 - 2)$ **16.** 5, $-\frac{2}{3}$ **17.** 3, -9 **18.** 0, -5, 1 **19.** $h = 4$ m, $b = 14$ m **20.** 7 sec

CHAPTER 7

Exercise Set 7.1

1. $\{x|x \neq -5\}$ **3.** $\{x|x \neq 3, x \neq -3\}$ **5.** $\{x|x \neq 0\}$ **7.** $\mathbb{R}$ **9.** $\{x|x \neq 2, x \neq -2\}$ **11.** $\{r|r \neq 4, r \neq -4\}$ **13.** $\{x|x \neq -1, x \neq -6\}$ **15.** $\{y|y \neq 1\}$ **17.** $\{z|z \neq 15/8\}$ **19.** $\{x|x \neq 3, x \neq 5\}$ **21.** $\{x|x \neq 4, x \neq -4\}$ **23.** $\{x|x \neq 0\}$ **25.** $\frac{3}{5}$ **27.** $1 - y$ **29.** x **31.** $x - 4x^3 = x(1 - 2x)(1 + 2x)$ **33.** $(3 - 2x^2)/9$ **35.** $2(x^2 + 2xy - 3y^2)/3$ **37.** -1 **39.** $-1/3$ **41.** $1/(x + 5)$ **43.** $-2/(x - 4)$ **45.** $x + 8$ **47.** $(x + 3)/(x - 5)$ **49.** $-(x + 5)$ **51.** $1/(x + 1)$ **53.** $(2x - 5)/2$ **55.** $(x - 8)/(x - 2)$ **57.** $(x - w)/(x + w)$ **59.** $(a^2 + ab + b^2)/(a + b)$ **61.** $\sqrt{x}$ is not a polynomial. **63.** **(a)** 1 **(b)** $\frac{1}{10}$ **(c)** $\frac{1}{100}$ **(d)** It decreases.

Just for Fun **1.** **(a)** Domain: $\{x|x \neq 2\}$ **(b)** **2.** **(a)** Domain: $\{x|x \neq 1\}$ **(b)**

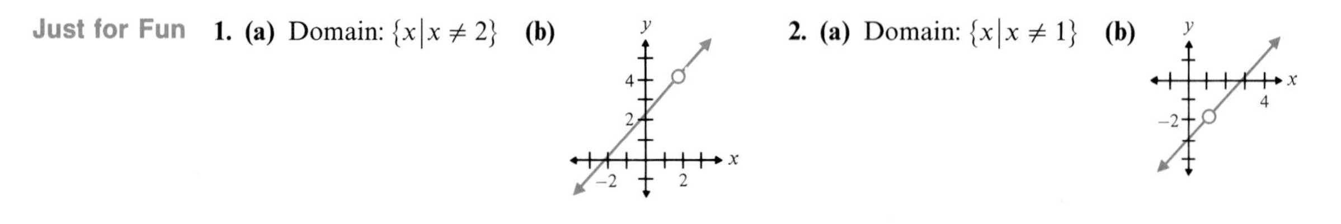

Exercise Set 7.2

1. $xy^2/4$ **3.** $20x^4/y^6$ **5.** $27x/8$ **7.** $32x^7/35my^2$ **9.** $36x^9y^2/25z^7$ **11.** $\frac{1}{2}$ **13.** x **15.** $(-3x + 2)/(3x + 2)$ **17.** 1 **19.** 1 **21.** $\dfrac{2(x + 3)}{x - 4}$ **23.** $(a + b)(a - b)$ or $a^2 - b^2$ **25.** $\dfrac{-(a + 3)}{a + 2}$ **27.** $\dfrac{x + 3}{2(x + 2)}$ **29.** 1 **31.** $\frac{1}{2}$ **33.** $\dfrac{3x + 1}{x - 1}$ **35.** $\dfrac{1}{x^2 + 2x + 4}$ **37.** $\dfrac{x^2(2x - 1)}{(x - 1)(x - 3)}$ **39.** $\dfrac{p - q}{p + q}$ **41.** $\dfrac{2x(x + 2)}{x - 5}$ **43.** $x(x^2 - 4x + 16)$ **45.** $12x^3y^2$ **47.** $\dfrac{12}{7xz}$ **49.** 1 **51.** $\dfrac{3y^2}{a^2}$ **53.** $\dfrac{z^2}{2y}$ **55.** $3x$ **57.** $\frac{2}{3}$ **59.** $\dfrac{-1}{x}$ **61.** $\dfrac{1}{x - 3}$ **63.** $\dfrac{x + 9}{x - 15}$ **65.** $\dfrac{(a + 1)^2}{9(a + b)^2}$ **67.** $\dfrac{2(x + 2)}{x - 2}$ **69.** $\dfrac{-(x + 2)}{x - 3}$ **71.** $x^2(a + b)$ **73.** $6y^3(x - y)$ **75.** 1 **77.** $(2a - 1)(a - 1)$

Exercise Set 7.3

1. $(2x - 11)/3$ **3.** $-8/x$ **5.** $1/(x - 4)$ **7.** $-(x + 2)/x$ **9.** -1 **11.** $1/(x - 2)$ **13.** $x - 5$ **15.** $(x + 1)(x + 2)$ **17.** $48x^3y$ **19.** $x - 3$ **21.** $(x - 8)(x + 3)(x + 8)$ **23.** $(x + 3)(x - 3)(x - 1)$ **25.** $(x + 1)(3x + 2)(2x + 3)$ **27.** $(x - 2)(x + 3)(x + 4)$ **29.** $10/3x$ **31.** $(3x + 12)/2x^2$ **33.** $(10y + 9)/12y^2$ **35.** $(25y^2 - 12x^2)/60x^4y^3$ **37.** $\dfrac{4 + 6y}{3y}$ **39.** $\dfrac{2x + 6}{x(x - 3)}$ **41.** $\dfrac{2x}{x - y}$ **43.** $\dfrac{2(3x - 7)}{(x - 3)(x - 1)}$ **45.** $\dfrac{4}{x - 2}$ **47.** $\dfrac{14x + 58}{(x + 3)(x + 7)}$ **49.** $\dfrac{5x^2 + 4x - 16}{(x - 4)(x + 1)}$ **51.** $\dfrac{2x^2 + 5x - 2}{(x - 5)(x + 2)}$ **53.** $\dfrac{1}{x - 5}$ **55.** $\dfrac{x + 4}{(x + 1)(x - 1)}$ **57.** $\dfrac{x^2}{y(x - y)}$ **59.** $\dfrac{x + y}{x - y}$ **61.** $\dfrac{3(x + 1)}{2x - 3}$ **63.** $\dfrac{3x - 1}{4x + 1}$ **65.** $\dfrac{2x^2 - 4xy + 4y^2}{(x - 2y)^2(x + 2y)}$ **67.** $\dfrac{12}{x - 3}$ **69.** 0 **71.** 1

Just for Fun **1.** $\dfrac{(x + 6)(x - 1)}{(x + 2)(x - 3)(x - 2)(x + 3)}$ **2.** $\dfrac{4x^2 - 9x + 27}{(x + 3)(x - 3)(x^2 - 3x + 9)}$

Exercise Set 7.4

1. $\frac{8}{11}$ **3.** $\frac{57}{32}$ **5.** $\frac{25}{1224}$ **7.** $x^3y/8$ **9.** $6xz^2$ **11.** $(xy + 1)/x$ **13.** $3/x$ **15.** $(3y - 1)/(2y - 1)$ **17.** $(x - y)/y$ **19.** $-a/b$ **21.** -1 **23.** $2(x + 2)/x^3$ **25.** $(x - 2)/8$ **27.** $(x + 1)/(1 - x)$ **29.** $-4a/(a^2 + 4)$ **31.** $(2 + a^2b)/a^2$ **33.** $ab/(b + a)$ **35.** $a + b$ **37.** $(a^2 + b)/[b(b + 1)]$ **39.** $(y - x)/(x + y)$ **41.** $(a + b)^2/ab$ **43.** $(6y - x)/3xy$ **45.** **(a)** $\frac{2}{7}$ **(b)** $\frac{4}{13}$ **47.** $R_T = R_1R_2R_3/(R_2R_3 + R_1R_3 + R_1R_2)$

Just for Fun **1.** $(4a^2 + 1)/[4a(2a^2 + 1)]$ **2.** $(x^2 + x + 1)/(x^3 + x^2 + 2x + 1)$

Exercise Set 7.5

1. 4 **3.** 48 **5.** 30 **7.** -1 **9.** 1 **11.** 4 **13.** 3 **15.** $-\frac{1}{5}$ **17.** $\frac{1}{4}$ **19.** $\frac{14}{3}$ **21.** No solution **23.** 2 **25.** $-\frac{12}{7}$ **27.** 8 **29.** $-\frac{4}{3}, 1$ **31.** -14 **33.** $-2, -3$ **35.** 4 **37.** $-\frac{5}{2}$ **39.** 5 **41.** No solution **43.** No solution **45.** $\frac{17}{4}$ **47.** 12 in. **49.** 17.14 cm **51.** Object: 16 cm, image: 48 cm **53.** $155.\overline{6}$ ohms **55.** 176.47 ohms **57.** A number obtained when solving an equation which is not a true solution to the original equation.

Just for Fun

1. 115.96 days

Exercise Set 7.6

1. $2\frac{2}{9}$ hr **3.** $3\frac{3}{7}$ days **5.** $2\frac{2}{5}$ hr **7.** $15\frac{3}{11}$ min **9.** -4 **11.** 5, 6 **13.** 3 **15.** All real numbers **17.** 1 or 2 **19.** 4 mph **21.** 8 mph **23.** 50 mph, 60 mph **25.** 9 mph

Exercise Set 7.7

1. Direct **3.** Direct **5.** Direct **7.** Direct **9.** Direct **11.** Direct **13.** Inverse **15.** Direct **17.** $C = kZ^2$; 60.75 **19.** $x = k/y$; 0.2 **21.** $L = k/P^2$; 6.25 **23.** $A = kR_1R_2/L^2$; 57.6 **25.** $x = ky$; 18 **27.** $y = kR^2$; 20 **29.** $C = k/J$; 0.41 **31.** $F = kM_1M_2/d$; 40 **33.** $S = kIT^2$; 0.2 **35.** $S = kF$; 0.7 in **37.** $I = k/d^2$; 31.25 footcandles **39.** $W = k/d^2$; 133.25 lb **41.** $R = kL/A$; 25 ohms

Review Exercises

1. $\{x|x \neq 4\}$ **2.** $\{x|x \neq -1\}$ **3.** $\mathbb{R}$ **4.** $\{x|x \neq -3\}$ **5.** $\{x|x \neq 0\}$ **6.** $\{x|x \neq 5, x \neq -2\}$ **7.** x **8.** $x - 3$ **9.** -1 **10.** $(x - 1)/(x - 2)$ **11.** $(x - 3)/(x + 1)$ **12.** $(a^2 + 2a + 4)/(a + 2)$ **13.** $6xz^2$ **14.** $-\frac{1}{2}$ **15.** $\frac{(x + y)y^2}{2x^3}$ **16.** 1 **17.** $\frac{x + y}{x}$ **18.** $\frac{x + 1}{2x - 1}$ **19.** $\frac{1}{a + 2}$ **20.** $\frac{2}{x(x + 2)}$ **21.** $\frac{32z}{x^3}$ **22.** $\frac{3}{x - y}$ **23.** $\frac{1}{3(a + 3)}$ **24.** 1 **25.** $2x(x - 5y)$ **26.** $\frac{16(x - 2y)}{3(x + 2y)}$ **27.** $\frac{x^3 + y^2}{x^3 - y^2}$ **28.** $\frac{(x + 3)(x - 1)}{4(x + 1)}$ **29.** 4 **30.** $\frac{4}{x + 10}$ **31.** $3x + 2$ **32.** $\frac{6x^2 - 3x - 16}{2x - 3}$ **33.** $\frac{-4x - 2}{2x - 3}$ **34.** $x(x + 1)$ **35.** $(x + y)(x - y)$ **36.** $(x + 7)(x - 5)(x + 2)$ **37.** $(x + 2)^2(x - 2)(x + 1)$ **38.** $\frac{3}{x}$ **39.** $\frac{24x + y}{4xy}$ **40.** $\frac{5x^2 - 12y}{3x^2y}$ **41.** $\frac{7x + 12}{x + 2}$ **42.** $\frac{5x + 12}{x + 3}$ **43.** $\frac{a^2 + c^2}{ac}$ **44.** $\frac{7x + 12}{x(x + 3)}$ **45.** $\frac{-3x^2 + 2x - 4}{3x(x - 2)}$ **46.** $\frac{x^2 - 2x - 5}{(x + 5)(x - 5)}$ **47.** $\frac{2(2x + 13)}{(x + 5)^2}$ **48.** $\frac{3(x - 1)}{(x + 3)(x - 3)}$ **49.** $\frac{4}{(x + 2)(x - 3)(x - 2)}$ **50.** $\frac{2(x - 4)}{(x - 3)(x - 5)}$ **51.** $\frac{22x + 5}{(x - 5)(x - 10)(x + 5)}$ **52.** $\frac{-1}{x - 3}$ **53.** $\frac{x + 3}{x + 5}$ **54.** $\frac{34}{9}$ **55.** $\frac{55}{26}$ **56.** $\frac{5yz}{6}$ **57.** $16x^3z^2/y^3$ **58.** $(xy + 1)/y^3$ **59.** $(xy - x)/(x + 1)$ **60.** $(4x + 2)/[x(6x - 1)]$ **61.** $2/x$ **62.** a **63.** $(2a + 1)/2$ **64.** $(x + 1)/(-x + 1)$ **65.** $x^2(3 - y)/[y(y - x)]$ **66.** 9 **67.** 1 **68.** 6 **69.** 6 **70.** 52 **71.** 20 **72.** 18 **73.** $\frac{1}{2}$ **74.** -6 **75.** -18 **76.** 120 ohms **77.** 900 ohms, 1800 ohms **78.** 3 cm **79.** 15 cm **80.** $1\frac{5}{7}$ hr **81.** $4\frac{8}{19}$ hr **82.** 2 **83.** $\frac{5}{6}$ **84.** 5 mph **85.** 50 mph, 150 mph **86.** 75 **87.** 20 **88.** 1 **89.** 20 **90.** 426.7 **91.** 5 in **92.** $119.88 **93.** 400 ft **94.** 200.96 **95.** 2.38 min

Practice Test

1. $\{x|x \neq 7, x \neq -4\}$ **2.** $-(x + 4)$ **3.** $4x^3z/3$ **4.** $a + 3$ **5.** $(x - 3y)/3$ **6.** $x^2 + y^2$ **7.** $(10x + 3)/2x^2$ **8.** $-1/[(x + 4)(x - 4)]$ **9.** $x(x + 10)/[(2x - 1)^2(x + 3)]$ **10.** $(y + x)/(y - x)$ **11.** $x^2(y + 1)/(y + x)$ **12.** 60 **13.** 12 **14.** 10 **15.** 1.125 **16.** $3\frac{1}{13}$ hr

CHAPTER 8

Exercise Set 8.1

1. 5 **3.** -3 **5.** 5 **7.** 11 **9.** -2 **11.** 12 **13.** 1 **15.** 7 **17.** 6 **19.** 1 **21.** 43 **23.** 147 **25.** 83 **27.** 179 **29.** $|y - 8|$ **31.** $|x - 3|$ **33.** $|3x + 5|$ **35.** $|6 - 3x|$ **37.** $|y^2 - 4y + 3|$ **39.** 7 **41.** 243 **43.** 4 **45.** 64 **47.** 36 **49.** 8 **51.** z^6 **53.** y^9 **55.** x **57.** z^7 **59.** x^6 **61.** x^{20} **63.** y **65.** x **67.** x^4 **69.** When n is an even integer and $x < 0$. **71.** When n is an even integer, m is an odd integer, and $x < 0$.

Exercise Set 8.2

1. $5\sqrt{2}$ **3.** $2\sqrt{10}$ **5.** $4\sqrt{2}$ **7.** $2\sqrt[3]{2}$ **9.** $2\sqrt[3]{4}$ **11.** $3\sqrt[4]{2}$ **13.** $x\sqrt{x}$ **15.** $x^5\sqrt{x}$ **17.** $b^{13}\sqrt{b}$ **19.** $y^2\sqrt[3]{y}$ **21.** $y^2\sqrt[4]{y}$ **23.** $x^3\sqrt[5]{x^4}$ **25.** $2x\sqrt{6x}$ **27.** $2x^3\sqrt{5x}$ **29.** $2x^3\sqrt[3]{10x^2}$ **31.** $xy^3\sqrt{xy}$ **33.** $5y^2\sqrt{2x}$ **35.** $3x^2y^2\sqrt[3]{3y^2}$ **37.** $3x^4y^4\sqrt[3]{2y}$ **39.** $2b^4c^2\sqrt[4]{abc}$ **41.** $2b\sqrt[5]{a^2}$ **43.** $2cw^3\sqrt[3]{4cz}$ **45.** $3x^2y^7z^{16}\sqrt[3]{3xz^2}$ **47.** 5 **49.** $10\sqrt{3}$ **51.** $2\sqrt[3]{7}$ **53.** $2\sqrt[4]{5}$ **55.** $3x^3\sqrt{10x}$ **57.** $xy^2\sqrt{42y}$ **59.** $4x^3y^2$ **61.** $2x^3y^5\sqrt{30y}$ **63.** $5xy^4\sqrt[3]{x^2y^2}$ **65.** $x^2y^2\sqrt[3]{4y^2}$ **67.** $xy^4\sqrt[3]{25x}$ **69.** $3x^3y^4\sqrt[4]{2xy^3}$ **71.** $2x^3y^3z\sqrt[3]{3xyz}$

Exercise Set 8.3

1. 3 **3.** $\frac{1}{3}$ **5.** $\frac{1}{2}$ **7.** $2\sqrt{2}$ **9.** $x/3$ **11.** $x/2$ **13.** $2x^2$ **15.** $1/2x^2$ **17.** $y^2\sqrt{5}/x$ **19.** $xy^2/2$ **21.** $\sqrt{3}/3$ **23.** $\sqrt{2}/2$ **25.** $x\sqrt{5}/5$ **27.** $x\sqrt{y}/y$ **29.** $\sqrt{3}/3$ **31.** $2\sqrt{5}/5$ **33.** $\sqrt{10}/4$ **35.** $2\sqrt{15}/5$ **37.** $x\sqrt{2}/3$ **39.** $\sqrt{3xy}/2y$ **41.** $\sqrt[3]{4}/2$ **43.** $\sqrt[3]{9}/3$ **45.** $5\sqrt[3]{x^2}/x$ **47.** $\sqrt[3]{2x^2}/2x$ **49.** $\sqrt[3]{10xy}/2y$ **51.** $5x\sqrt[4]{8}/2$ **53.** $\sqrt[4]{135x}/3x$ **55.** $2x^2\sqrt{xyz}/z$ **57.** $y^2\sqrt{10xz}/2z$ **59.** $2y^2z\sqrt{15zx}/3x$ **61.** $x^2z^3\sqrt{30xyz}/2y$ **63.** $3x^2y\sqrt{yz}/z$ **65.** $x^2y^2\sqrt[3]{15yz}/z$ **67.** 6 **69.** -3 **71.** -1 **73.** $x-25$ **75.** $y-9$ **77.** $x-y$ **79.** x^2-y **81.** $3\sqrt{2}-3$ **83.** $(-3\sqrt{6}-15)/19$ **85.** $(-4\sqrt{2}-28)/47$ **87.** $-5-\sqrt{30}$ **89.** $(\sqrt{17}+2\sqrt{2})/9$ **91.** $(5\sqrt{x}+15)/(x-9)$ **93.** $(15-5\sqrt{x})/(9-x)$ **95.** $(2x\sqrt{2x}-2\sqrt{2xy})/(x^2-y)$ **97.** **(a)** $\sqrt{2}/6$ **(b)** $\sqrt{2}/6$ **(c)** $\sqrt{2}/6$ **99.** $2/\sqrt{2} < 3/\sqrt{3}$

Exercise Set 8.4

1. $2\sqrt{3}$ **3.** $-2\sqrt[3]{7}$ **5.** $5-4\sqrt{3}$ **7.** $5\sqrt{x}$ **9.** $3\sqrt[4]{x}$ **11.** $3+2\sqrt{y}$ **13.** $x+\sqrt{x}+4\sqrt{y}$ **15.** $5-4\sqrt[3]{x}$ **17.** $2(\sqrt{2}-\sqrt{3})$ **19.** $7\sqrt{5}$ **21.** $35\sqrt{10}$ **23.** $40\sqrt{5}$ **25.** $5\sqrt{3}$ **27.** $5\sqrt{7}$ **29.** $-7x\sqrt{3}$ **31.** $14\sqrt[3]{3}$ **33.** $-6\sqrt[3]{5}$ **35.** $7\sqrt[3]{2}$ **37.** $\sqrt{3x}(4x-2)$ **39.** $5x\sqrt[3]{xy^2}$ **41.** $-x\sqrt[3]{x^2y}$ **43.** $16ay\sqrt[3]{3y}$ **45.** $3x^3y^2\sqrt{xy}$ **47.** $-x^2y^2\sqrt[4]{xy}$ **49.** $(x^3y-xy^2)\sqrt[3]{xy^2}$ **51.** $\sqrt{2}$ **53.** $2\sqrt{6}/3$ **55.** $2\sqrt{6}/3$ **57.** $15\sqrt{2}/2$ **59.** $13\sqrt{2}/2$ **61.** $2\sqrt{x}(2+1/x)$ **63.** $23+9\sqrt{3}$ **65.** $-3-5\sqrt{3}$ **67.** $16-10\sqrt{2}$ **69.** $-14+11\sqrt{2}$ **71.** $8+2\sqrt{15}$ **73.** $1-\sqrt{6}$ **75.** $19+8\sqrt{3}$ **77.** $29-12\sqrt{5}$ **79.** Approx. 5.97 **81.** $1/(\sqrt{3}+2) < 2+\sqrt{3}$

Just for Fun **1.** $\sqrt[5]{81x^3y}/x^2y^2$ **2.** $\sqrt[4]{27x^3y^2z^3}/3x^2y^2z^4$

Exercise Set 8.5

1. 25 **3.** 8 **5.** 81 **7.** 16 **9.** 8 **11.** 9 **13.** No solution **15.** No solution **17.** 1 **19.** $-\frac{1}{3}$ **21.** 5 **23.** -1 **25.** $\frac{1}{4}$ **27.** 1 **29.** -7 **31.** 2 **33.** 2 **35.** $\frac{9}{16}$ **37.** 9 **39.** 4
41. $\sqrt{x+3}$ cannot equal a negative number.
43. 0 is the only solution. For any nonzero value the left side of the equation is positive and the right side is negative.

Just for Fun **1.** 6 **2.** 6 **3.** No solution **4.** 16

Exercise Set 8.6

1. 4 **3.** $\sqrt{52} \approx 7.21$ **5.** $\sqrt{175} \approx 13.23$ **7.** $\sqrt{41} \approx 6.40$ **9.** $\sqrt{128} \approx 11.31$ **11.** $\sqrt{149} \approx 12.21$ **13.** $\sqrt{128} \approx 11.31$ **15.** 13 in. **17.** $\sqrt{18.25} \approx 4.27$ m **19.** 9 in. by 12 in. **21.** 12 in. **23.** $\sqrt{5120} \approx 71.55$ ft/sec **25.** $\sqrt{576} = 24$ in.^{2} **27.** 3.14 sec **29.** $0.2(\sqrt{149.4})^3 \approx 365.2$ days **31.** $\sqrt{1{,}000{,}000} = 1000$ lb **33.** $\sqrt{320} \approx 17.89$ ft/sec **35.** **(a)** $2, -2$ **(b)** $3, -4$ **(c)** $-4, \frac{3}{2}$ **(d)** $5, -1$

Just for Fun **1.** $\sqrt{1066.67} \approx 32.66$ ft/sec **2.** 1.59 cps **3.** **(a)** $S = 32 + 80\sqrt{\pi}$ sq in. ≈ 173.80 sq in. **(b)** $r = 4\sqrt{\pi}/\pi \approx 2.26$ in.

Exercise Set 8.7

1. $3+0i$ **3.** $3+2i$ **5.** $(6+\sqrt{3})+0i$ **7.** $0+5i$ **9.** $4+2i\sqrt{3}$ **11.** $0+3i$ **13.** $9-3i$ **15.** $0+(2-4\sqrt{5})i$ **17.** $15-4i$ **19.** $8+13i$ **21.** $18-5i$ **23.** $(4\sqrt{2}+\sqrt{3})-2i\sqrt{2}$ **25.** $19+12i\sqrt{3}$ **27.** $2\sqrt{3}-7+(7+2\sqrt{3})i$ **29.** $-6-4i$ **31.** $-1+6i$ **33.** $-12-18i$ **35.** $-4+2i\sqrt{3}$ **37.** $-6+3i\sqrt{2}$ **39.** $1+5i$ **41.** $6-22i$ **43.** $8+4\sqrt{3}+(3+2\sqrt{3})i$ **45.** $14+i\sqrt{2}$ **47.** $-5i/3$ **49.** $(3-2i)/2$ **51.** $(5-2i)/5$ **53.** $(49+14i)/53$ **55.** $(9-12i)/10$ **57.** $(3+i)/5$ **59.** $(\sqrt{2}+i\sqrt{6})/4$ **61.** $[5\sqrt{10}-2\sqrt{15}+(10\sqrt{2}+5\sqrt{3})i]/45$ **63.** True **65.** True **67.** False

Exercise Set 8.8

1. $\{x|x \geq -2\}$ **3.** $\{x|x \geq 0\}$ **5.** $\{x|x \leq 4\}$ **7.** $\{x|x \geq -4\}$ **9.** $\{x|x \geq -\frac{5}{3}\}$ **11.** $\{x|x \geq \frac{1}{14}\}$ **13.** $\{x|x \leq \frac{5}{12}\}$

15. Domain: $\{x|x \geq -4\}$
Range: $\{y|y \geq 0\}$

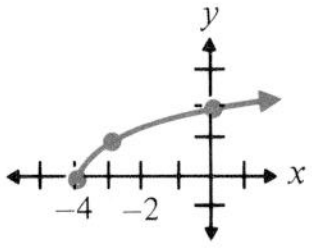

17. Domain: $\{x|x \geq 2\}$
Range: $\{y|y \geq 0\}$

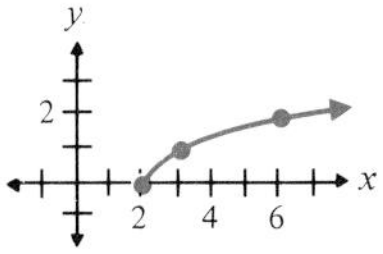

19. Domain: $\{x|x \geq -3\}$
Range: $\{y|y \leq 0\}$

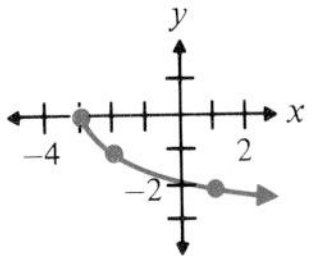

21. Domain: $\{x|x \leq 3\}$
Range: $\{y|y \geq 0\}$

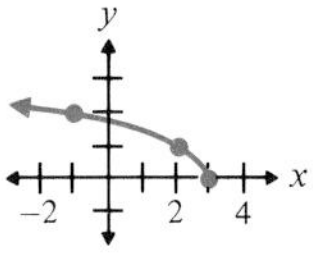

23. Domain: $\{x|x \leq 2\}$
Range: $\{y|y \leq 0\}$

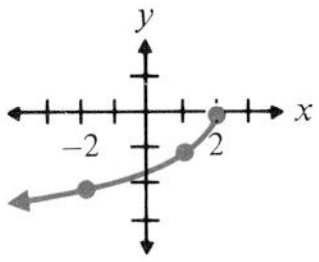

25. Domain: $\{x|x \geq 0\}$
Range: $\{y|y \geq 0\}$

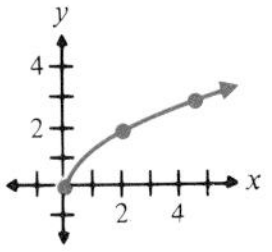

27. Domain: $\{x|x \geq -\frac{1}{2}\}$
Range: $\{y|y \geq 0\}$

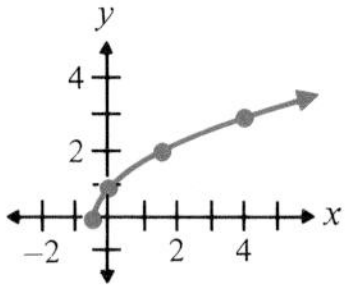

29. Domain: $\{x|x \geq -2\}$
Range: $\{y|y \geq 0\}$

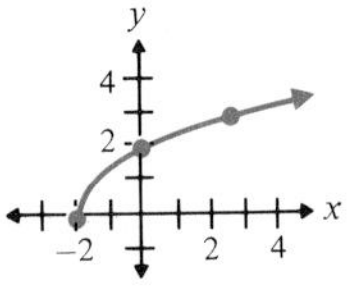

31. Domain: $\mathbb{R}$
Range: $\{y|y \geq 0\}$

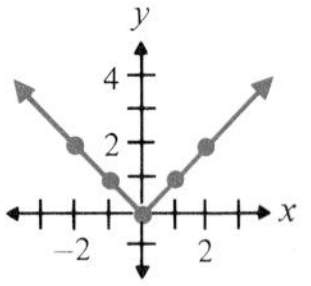

33. No, since $\sqrt{-4}$ is not a real number. **35.** $\mathbb{R}$ **37.** No, it is a cube root function.
It is a function.
Domain: $\mathbb{R}$
Range: $\mathbb{R}$

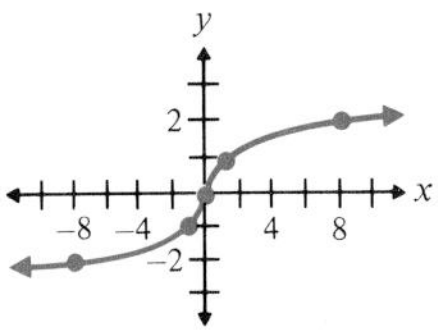

Just for Fun

1. (a) Domain: $\{x|-2 \leq x \leq 2\}$ **(b)**
(c) Range: $\{y|0 \leq y \leq 2\}$

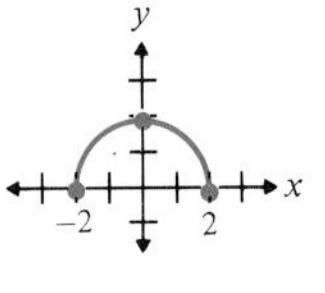

2. (a) Domain: $\{x|x \leq -2 \text{ or } x \geq 2\}$ **(b)**
(c) Range: $\{y|y \geq 0\}$

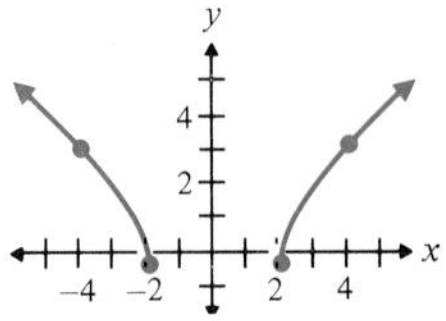

Review Exercises

1. 3 **2.** 5 **3.** -2 **4.** 4 **5.** 3 **6.** -3 **7.** 6 **8.** 19 **9.** 7 **10.** 93 **11.** $|x|$ **12.** $|x-2|$ **13.** $|a-3|$ **14.** $|x-y|$ **15.** $|2x-3|$ **16.** 9 **17.** 125 **18.** 9 **19.** 27 **20.** x^3 **21.** x^5 **22.** y^4 **23.** 81 **24.** x^5 **25.** x **26.** x **27.** $2\sqrt{6}$ **28.** $4\sqrt{5}$ **29.** $2\sqrt[3]{2}$ **30.** $3\sqrt[3]{2}$ **31.** $2y^2\sqrt{2x}$ **32.** $5xy^3\sqrt{2xy}$ **33.** $x^2y\sqrt[3]{9y^2}$ **34.** $2x^2y^3\sqrt[4]{x}$ **35.** $5x^2y^3\sqrt[3]{xy}$ **36.** $x^3y^3z^3\sqrt[4]{yz^2}$ **37.** $2x^2yz^3\sqrt[5]{x^2y^2z^2}$ **38.** 10 **39.** 2 **40.** $2x^3\sqrt{10}$ **41.** $3x^3y^4\sqrt{5x}$ **42.** $2x^2y^3\sqrt[3]{4x^2}$ **43.** $xy^2\sqrt[3]{25x}$ **44.** $2x^2y^4\sqrt[4]{x}$ **45.** $\frac{1}{2}$ **46.** $\frac{6}{5}$ **47.** $\frac{1}{3}$ **48.** $x/2$ **49.** $x/2$ **50.** $4y^2/x$ **51.** $3xy$ **52.** $5/xy$ **53.** $\sqrt{2}/2$ **54.** $\sqrt{3}/3$ **55.** $2\sqrt{3}/3$ **56.** $x\sqrt{7}/7$ **57.** $\sqrt{10}/5$ **58.** $\sqrt{2y}/y$ **59.** $\sqrt{xy}/x$ **60.** $\sqrt{15xy}/5y$ **61.** $2\sqrt[3]{x^2}/x$ **62.** $\sqrt[3]{75xy^2}/5y$ **63.** $x\sqrt{3y}/y$ **64.** $x^2y^2\sqrt{6yz}/z$ **65.** $5xy^2\sqrt{3yz}/3z$ **66.** $xy^2\sqrt{10xy}/2$ **67.** $3x^2z^2\sqrt{15yz}/5y$ **68.** $y^3z^4\sqrt{30xz}/3x^2$ **69.** 7 **70.** -2 **71.** $x-y^2$ **72.** x^2-y **73.** $28+10\sqrt{3}$ **74.** 5 **75.** $-10+5\sqrt{5}$ **76.** $(12+3\sqrt{2})/14$ **77.** $(3x-x\sqrt{x})/(9-x)$ **78.** $(x-\sqrt{xy})/(x-y)$ **79.** $6\sqrt{3}$ **80.** $-4\sqrt[3]{5}$ **81.** $2\sqrt[3]{x}$ **82.** $7\sqrt{2}$ **83.** $-\sqrt[3]{2}$ **84.** $-4\sqrt{3}$ **85.** $8-13\sqrt[3]{2}$ **86.** $69\sqrt{2}/8$ **87.** $(xy+x^3y)\sqrt{x}$ **88.** $(3x^2y^3-4x^3y^4)\sqrt{x}$ **89.** $(2x^2y^2-x+3x^3)\sqrt[3]{xy^2}$ **90.** $(4x^2-2y+1)\sqrt[4]{xy^2}$ **91.** 36 **92.** 64 **93.** 7 **94.** 8 **95.** 2 **96.** -3

97. 2 **98.** 0, 9 **99.** $\sqrt{194} \approx 13.93$ **100.** $\sqrt{56} \approx 7.48$ **101.** $\sqrt{1280} = 16\sqrt{5}$ **102.** $2\pi\sqrt{2}$ **103.** $5 + 0i$
104. $-6 + 0i$ **105.** $2 - 2i$ **106.** $3 + 4i$ **107.** $7 + i$ **108.** $1 - 2i$ **109.** $2 + 5i$ **110.** $3\sqrt{3} + (\sqrt{5} - \sqrt{7})i$
111. $12 + 8i$ **112.** $-2\sqrt{3} + 2i$ **113.** $6\sqrt{2} + 4i$ **114.** $-6 + 6i$ **115.** $17 - 6i$ **116.** $24 + 3\sqrt{5} + (4\sqrt{3} - 6\sqrt{15})i$
117. $-2i/3$ **118.** $-3i/5$ **119.** $(-2 - \sqrt{3})i/2$ **120.** $5(3 - 2i)/13$ **121.** $(5\sqrt{3} + 3i\sqrt{2})/31$
122. $[\sqrt{10} - 6\sqrt{2} + (3\sqrt{2} + 2\sqrt{10})i]/10$ **123.** $\{x|x \geq 6\}$ **124.** $\{x|x \geq -\frac{5}{3}\}$ **125.** $\{x|x \leq \frac{5}{2}\}$ **126.** $\{x|x \geq 6\}$
127. Domain: $\{x|x \geq 0\}$ Range: $\{y|y \geq 0\}$ **128.** Domain: $\{x|x \geq -2\}$ Range: $\{y|y \geq 0\}$ **129.** Domain: $\{x|x \leq 5\}$ Range: $\{y|y \geq 0\}$ **130.** Domain: $\{x|x \geq 3\}$ Range: $\{y|y \leq 0\}$

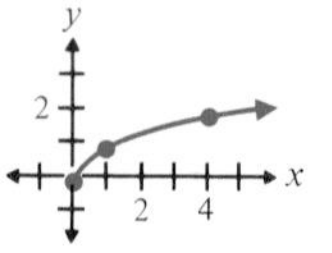

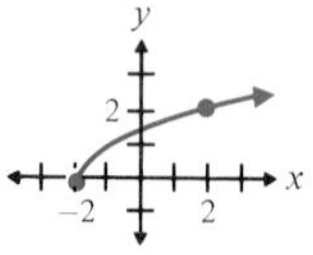

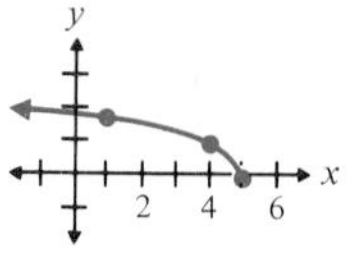

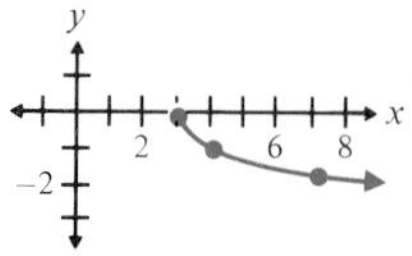

Practice Test

1. 26 **2.** $|3x - 4|$ **3.** $5x^2y^4\sqrt{2x}$ **4.** $2x^3y^3\sqrt[3]{5x^2y}$ **5.** $\sqrt{3}/3$ **6.** $x^2y^2\sqrt{yz}/2z$ **7.** $\sqrt[3]{x^2}/x$ **8.** $(2 - \sqrt{2})/2$
9. $-20\sqrt{3}$ **10.** $(2xy + 2x^2y^2)\sqrt[3]{y^2}$ **11.** $2\sqrt{5} - 2\sqrt{10} - 6 + 6\sqrt{2}$ **12.** 13 **13.** -3 **14.** 16 **15.** 12
16. $18 + 2\sqrt{2} + (6\sqrt{2} - 6)i$ **17.** $(\sqrt{5} + i\sqrt{10})/6$ **18.** Domain: $\{x|x \geq -2\}$ Range: $\{y|y \geq 0\}$

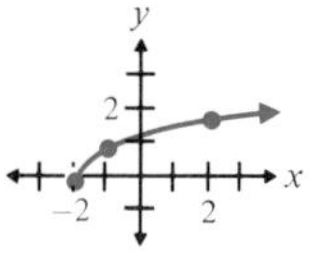

CHAPTER 9

Exercise Set 9.1

1. $1, -3$ **3.** $5, -1$ **5.** $-2, -1$ **7.** 5, 3 **9.** $-1 + i\sqrt{14}, -1 - i\sqrt{14}$ **11.** $-2, -3$ **13.** $-3, -6$ **15.** 7, 8
17. $6, -2$ **19.** $-1 + \sqrt{7}, -1 - \sqrt{7}$ **21.** $-3 + \sqrt{3}, -3 - \sqrt{3}$ **23.** $(5 + \sqrt{57})/2, (5 - \sqrt{57})/2$ **25.** $0, -2$
27. $0, \frac{1}{3}$ **29.** $1, -3$ **31.** $(-9 + \sqrt{73})/2, (-9 - \sqrt{73})/2$ **33.** $-8, -3$ **35.** $(-2 + i\sqrt{2})/2, (-2 - i\sqrt{2})/2$
37. $1 + i, 1 - i$ **39.** $(-3 + \sqrt{59})/10, (-3 - \sqrt{59})/10$ **41.** $(-1 + i\sqrt{191})/12, (-1 - i\sqrt{191})/12$ **43.** 7, 9
45. 5 ft, 12 ft **47.** $12 + 12\sqrt{2}$ ft

Just for Fun **1.** $(x + 2)^2 + (y - 3)^2 = 16$ **2.** $4(x - 6)^2 + 9(y + 4)^2 = 144$ **3.** $(x + 2)^2 - 4(y + 2)^2 = 16$

Exercise Set 9.2

1. Two real solutions **3.** No real solution **5.** Two real solutions **7.** Single real solution **9.** Two real solutions
11. No real solution **13.** Two real solutions **15.** Two real solutions **17.** 1, 2 **19.** 4, 5 **21.** $3, -8$
23. 4, 9 **25.** $5, -5$ **27.** 0, 3 **29.** $(7 + i\sqrt{159})/8, (7 - i\sqrt{159})/8$ **31.** $(7 + \sqrt{17})/4, (7 - \sqrt{17})/4$ **33.** $\frac{1}{3}, -\frac{1}{2}$
35. $(2 + \sqrt{6})/2, (2 - \sqrt{6})/2$ **37.** $(2 + i\sqrt{2})/2, (2 - i\sqrt{2})/2$ **39.** $\frac{5}{4}, -1$ **41.** $\frac{9}{2}, -1$ **43.** $3, \frac{5}{2}$
45. $(-1 + i\sqrt{23})/4, (-1 - i\sqrt{23})/4$ **47.** $0, -3$ **49.** $\sqrt{7}/2, -\sqrt{7}/2$ **51.** $(-6 + 2\sqrt{6})/3, (-6 - 2\sqrt{6})/3$
53. $(11 + \sqrt{241})/6, (11 - \sqrt{241})/6$ **55.** 2 **57.** $w = 3$ ft, $l = 7$ ft **59.** 2 in.

Just for Fun **1.** $2\sqrt{5}, -\sqrt{5}$ **2.** $-2\sqrt{6}, -3\sqrt{6}$ **3.** $1, -1, 2, -2$ **4.** 800 mph
5. $(-0.12 + \sqrt{14.3952})/1.2 \approx 3.0617$

Exercise Set 9.3

1. -2 5 **3.** -4 0 **5.** -4 4 **7.** -5 -4
9. -3 9 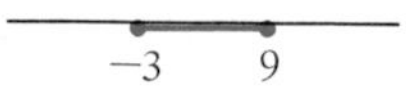 **11.** -5 7 **13.** -6 6 **15.** -3 -2

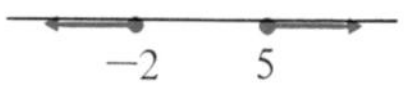

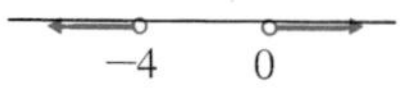

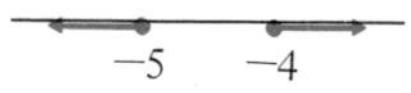

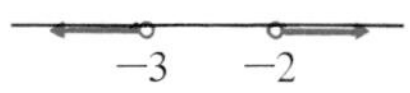

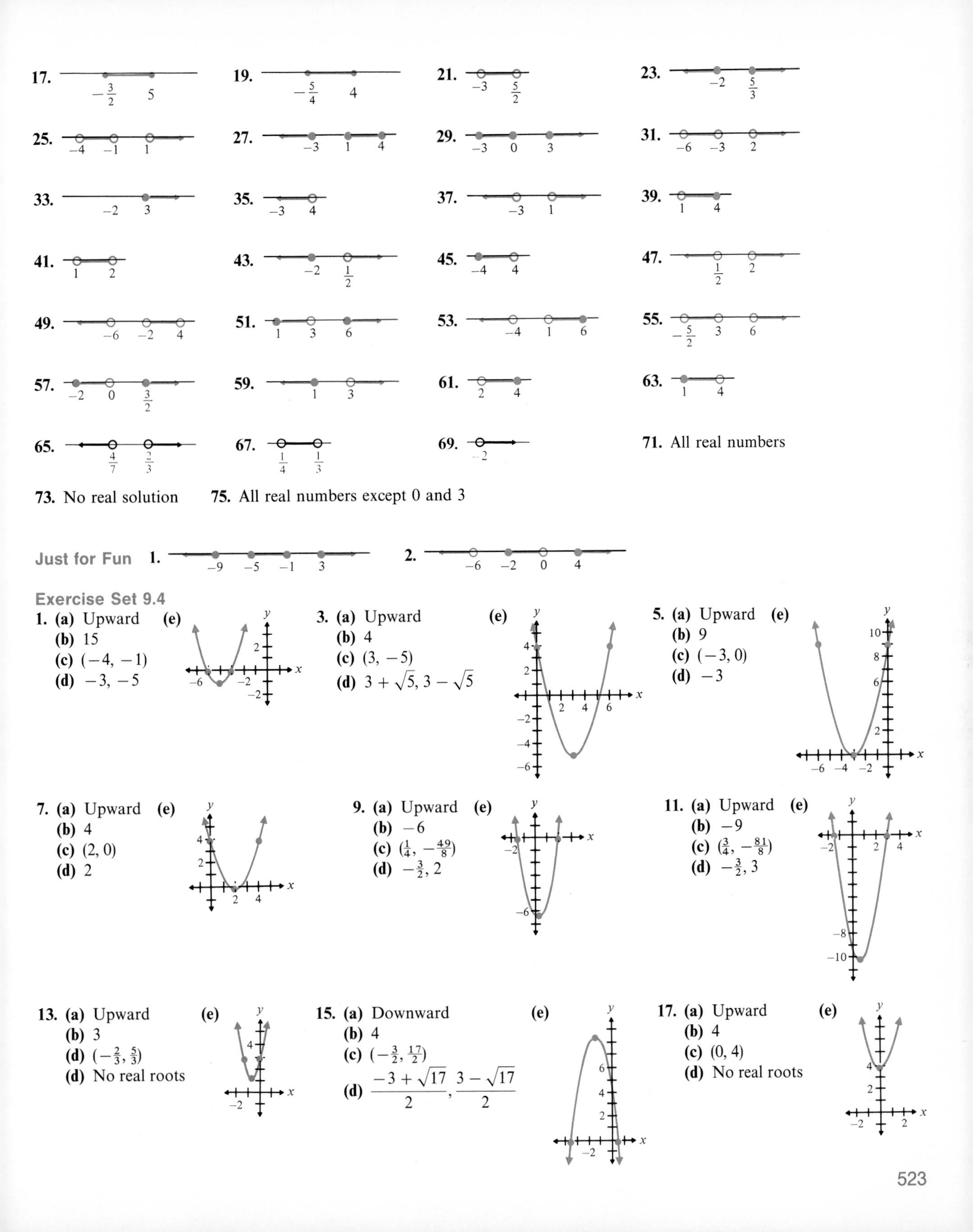

17. $-\frac{3}{2}$, 5 **19.** $-\frac{5}{4}$, 4 **21.** -3, $\frac{5}{2}$ **23.** -2, $\frac{5}{3}$

25. -4, -1, 1 **27.** -3, 1, 4 **29.** -3, 0, 3 **31.** -6, -3, 2

33. -2, 3 **35.** -3, 4 **37.** -3, 1 **39.** 1, 4

41. 1, 2 **43.** -2, $\frac{1}{2}$ **45.** -4, 4 **47.** $\frac{1}{2}$, 2

49. -6, -2, 4 **51.** 1, 3, 6 **53.** -4, 1, 6 **55.** $-\frac{5}{2}$, 3, 6

57. -2, 0, $\frac{3}{2}$ **59.** 1, 3 **61.** 2, 4 **63.** 1, 4

65. $\frac{4}{7}$, $\frac{2}{3}$ **67.** $\frac{1}{4}$, $\frac{1}{3}$ **69.** -2 **71.** All real numbers

73. No real solution **75.** All real numbers except 0 and 3

Just for Fun **1.** -9, -5, -1, 3 **2.** -6, -2, 0, 4

Exercise Set 9.4

1. **(a)** Upward **(b)** 15 **(c)** $(-4, -1)$ **(d)** $-3, -5$ **(e)**

3. **(a)** Upward **(b)** 4 **(c)** $(3, -5)$ **(d)** $3 + \sqrt{5}, 3 - \sqrt{5}$ **(e)**

5. **(a)** Upward **(b)** 9 **(c)** $(-3, 0)$ **(d)** -3 **(e)**

7. **(a)** Upward **(b)** 4 **(c)** $(2, 0)$ **(d)** 2 **(e)**

9. **(a)** Upward **(b)** -6 **(c)** $(\frac{1}{4}, -\frac{49}{8})$ **(d)** $-\frac{3}{2}, 2$ **(e)**

11. **(a)** Upward **(b)** -9 **(c)** $(\frac{3}{4}, -\frac{81}{8})$ **(d)** $-\frac{3}{2}, 3$ **(e)**

13. **(a)** Upward **(b)** 3 **(d)** $(-\frac{2}{3}, \frac{5}{3})$ **(d)** No real roots **(e)**

15. **(a)** Downward **(b)** 4 **(c)** $(-\frac{3}{2}, \frac{17}{2})$ **(d)** $\dfrac{-3 + \sqrt{17}}{2}, \dfrac{3 - \sqrt{17}}{2}$ **(e)**

17. **(a)** Upward **(b)** 4 **(c)** $(0, 4)$ **(d)** No real roots **(e)**

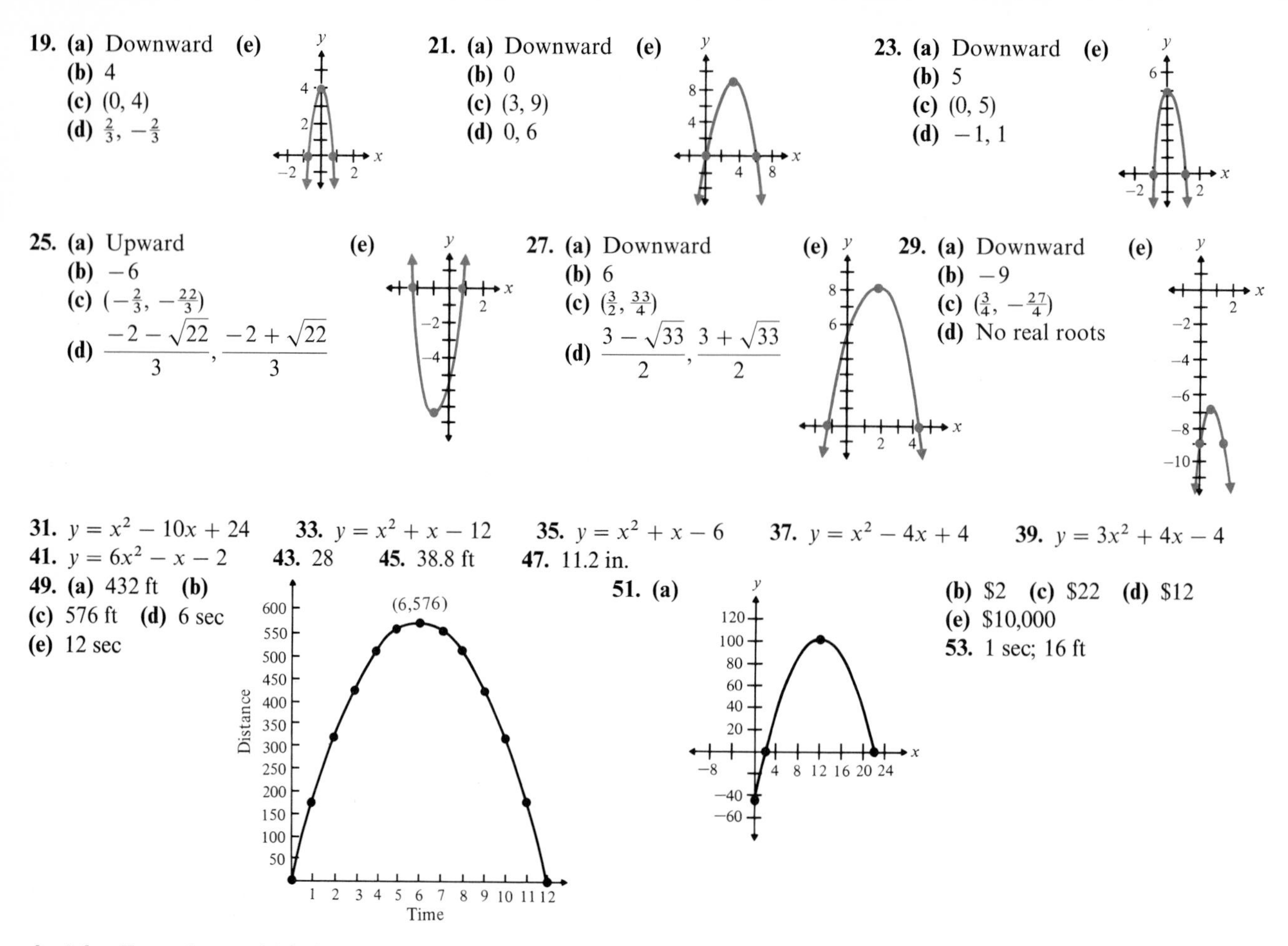

19. **(a)** Downward **(b)** 4 **(c)** (0, 4) **(d)** $\frac{2}{3}, -\frac{2}{3}$ **(e)**

21. **(a)** Downward **(b)** 0 **(c)** (3, 9) **(d)** 0, 6 **(e)**

23. **(a)** Downward **(b)** 5 **(c)** (0, 5) **(d)** $-1, 1$ **(e)**

25. **(a)** Upward **(b)** -6 **(c)** $(-\frac{2}{3}, -\frac{22}{3})$ **(d)** $\frac{-2-\sqrt{22}}{3}, \frac{-2+\sqrt{22}}{3}$ **(e)**

27. **(a)** Downward **(b)** 6 **(c)** $(\frac{3}{2}, \frac{33}{4})$ **(d)** $\frac{3-\sqrt{33}}{2}, \frac{3+\sqrt{33}}{2}$ **(e)**

29. **(a)** Downward **(b)** -9 **(c)** $(\frac{3}{4}, -\frac{27}{4})$ **(d)** No real roots **(e)**

31. $y = x^2 - 10x + 24$ **33.** $y = x^2 + x - 12$ **35.** $y = x^2 + x - 6$ **37.** $y = x^2 - 4x + 4$ **39.** $y = 3x^2 + 4x - 4$
41. $y = 6x^2 - x - 2$ **43.** 28 **45.** 38.8 ft **47.** 11.2 in.
49. **(a)** 432 ft **(b)** **(c)** 576 ft **(d)** 6 sec **(e)** 12 sec

51. **(a)** **(b)** \$2 **(c)** \$22 **(d)** \$12 **(e)** \$10,000
53. 1 sec; 16 ft

Just for Fun **1.** $w = 16$ ft, $l = 16$ ft **2.** $w = 25$ ft, $l = 50$ ft

Review Exercises

1. 2, 8 **2.** 3, 5 **3.** 1, 13 **4.** 2, -3 **5.** 9, -6 **6.** 1, -6 **7.** $-1 + \sqrt{6}, -1 - \sqrt{6}$
8. $(3 + i\sqrt{23})/2, (3 - i\sqrt{23})/2$ **9.** $2 + 2i\sqrt{7}, 2 - 2i\sqrt{7}$ **10.** 5, -3 **11.** $(1 + i\sqrt{47})/4, (1 - i\sqrt{47})/4$
12. $-3 + \sqrt{19}, -3 - \sqrt{19}$ **13.** Two real solutions **14.** No real solution **15.** No real solution
16. No real solution **17.** One real solution **18.** No real solution **19.** Two real solutions
20. Two real solutions **21.** 2, 7 **22.** 3, -10 **23.** 2, 5 **24.** 2, $-\frac{3}{5}$ **25.** -2, 9
26. $(1 + i\sqrt{119})/2, (1 - i\sqrt{119})/2$ **27.** $\frac{3}{2}, -\frac{5}{3}$ **28.** $(-2 + \sqrt{10})/2, (-2 - \sqrt{10})/2$ **29.** $(3 + \sqrt{57})/4, (3 - \sqrt{57})/4$
30. $3 + \sqrt{2}, 3 - \sqrt{2}$ **31.** $(2 + i\sqrt{14})/3, (2 - i\sqrt{14})/3$ **32.** $(3 + \sqrt{33})/3, (3 - \sqrt{33})/3$ **33.** $0, -\frac{3}{2}$ **34.** $0, \frac{5}{2}$
35. 3, 8 **36.** 7, 9 **37.** 5, -8 **38.** 3, -9 **39.** 10, -6 **40.** $(1 + i\sqrt{167})/2, (1 - i\sqrt{167})/2$
41. $(-11 + \sqrt{73})/2, (-11 - \sqrt{73})/2$ **42.** 5, -5 **43.** 0, -6 **44.** $\frac{1}{2}, -3$ **45.** $(9 + i\sqrt{39})/6, (9 - i\sqrt{39})/6$
46. $\frac{2}{3}, -\frac{3}{2}$ **47.** $(-3 + \sqrt{33})/2, (-3 - \sqrt{33})/2$ **48.** $2, \frac{5}{3}$ **49.** $1, -\frac{8}{3}$ **50.** $(3 + 3\sqrt{3})/2, (3 - 3\sqrt{3})/2$ **51.** $0, \frac{5}{2}$

52. $0, -\frac{5}{3}$ **53.** $\frac{1}{4}, -\frac{3}{2}$ **54.** $\frac{5}{2}, -\frac{5}{3}$ **55.** -5 -1 **56.** -5 3 **57.** 5 6
58. -3 0 **59.** -4 $\frac{4}{3}$ **60.** -2 2 **61.** $-\sqrt{5}$ $\sqrt{5}$

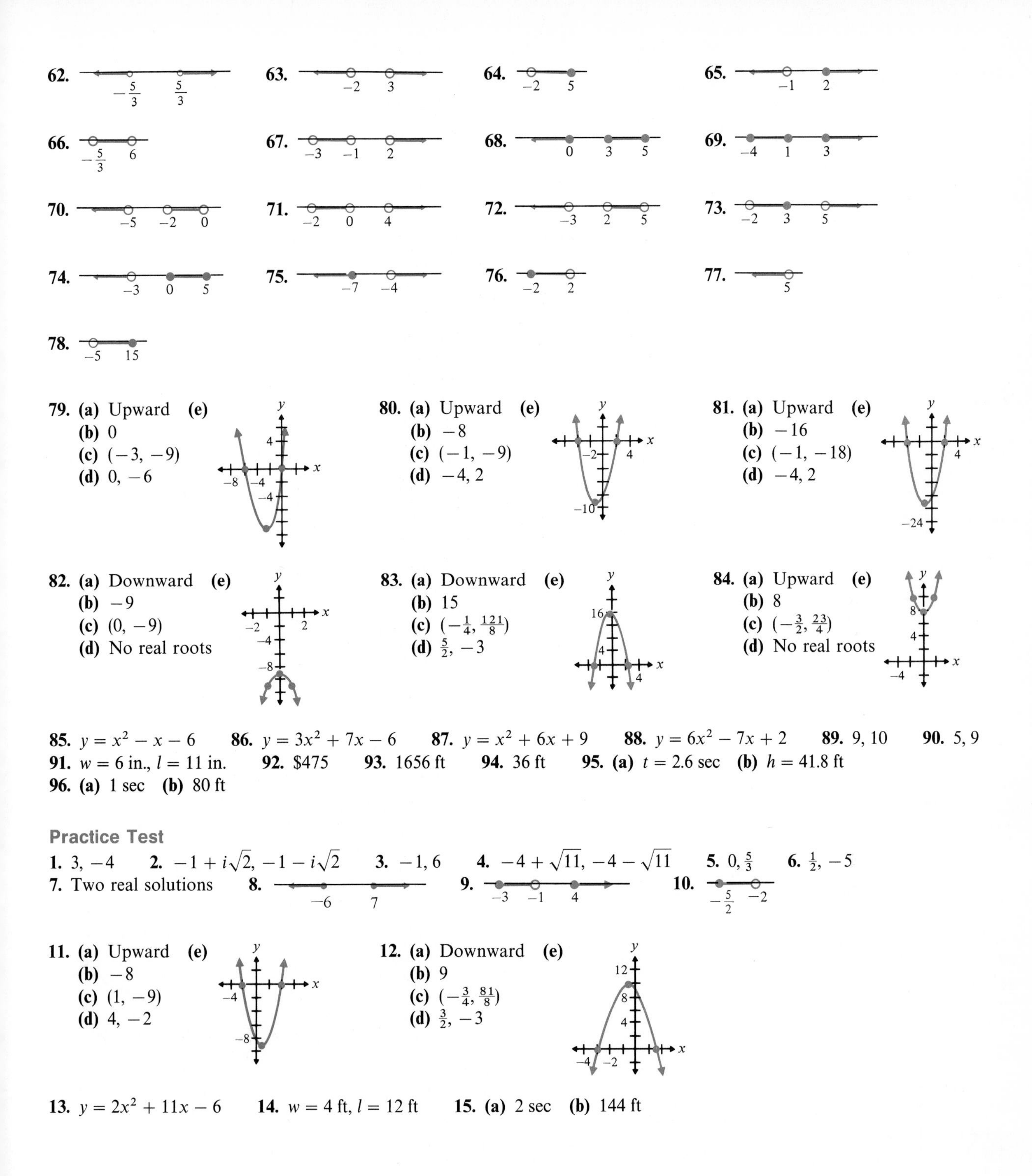

62. $-\frac{5}{3}$, $\frac{5}{3}$ **63.** -2, 3 **64.** -2, 5 **65.** -1, 2

66. $-\frac{5}{3}$, 6 **67.** -3, -1, 2 **68.** 0, 3, 5 **69.** -4, 1, 3

70. -5, -2, 0 **71.** -2, 0, 4 **72.** -3, 2, 5 **73.** -2, 3, 5

74. -3, 0, 5 **75.** -7, -4 **76.** -2, 2 **77.** 5

78. -5, 15

79. **(a)** Upward **(b)** 0 **(c)** $(-3, -9)$ **(d)** $0, -6$ **(e)**

80. **(a)** Upward **(b)** -8 **(c)** $(-1, -9)$ **(d)** $-4, 2$ **(e)**

81. **(a)** Upward **(b)** -16 **(c)** $(-1, -18)$ **(d)** $-4, 2$ **(e)**

82. **(a)** Downward **(b)** -9 **(c)** $(0, -9)$ **(d)** No real roots **(e)**

83. **(a)** Downward **(b)** 15 **(c)** $(-\frac{1}{4}, \frac{121}{8})$ **(d)** $\frac{5}{2}, -3$ **(e)**

84. **(a)** Upward **(b)** 8 **(c)** $(-\frac{3}{2}, \frac{23}{4})$ **(d)** No real roots **(e)**

85. $y = x^2 - x - 6$ **86.** $y = 3x^2 + 7x - 6$ **87.** $y = x^2 + 6x + 9$ **88.** $y = 6x^2 - 7x + 2$ **89.** 9, 10 **90.** 5, 9
91. $w = 6$ in., $l = 11$ in. **92.** \$475 **93.** 1656 ft **94.** 36 ft **95.** **(a)** $t = 2.6$ sec **(b)** $h = 41.8$ ft
96. **(a)** 1 sec **(b)** 80 ft

Practice Test

1. 3, -4 **2.** $-1 + i\sqrt{2}, -1 - i\sqrt{2}$ **3.** $-1, 6$ **4.** $-4 + \sqrt{11}, -4 - \sqrt{11}$ **5.** $0, \frac{5}{3}$ **6.** $\frac{1}{2}, -5$
7. Two real solutions **8.** -6, 7 **9.** -3, -1, 4 **10.** $-\frac{5}{2}$, -2

11. **(a)** Upward **(b)** -8 **(c)** $(1, -9)$ **(d)** $4, -2$ **(e)**

12. **(a)** Downward **(b)** 9 **(c)** $(-\frac{3}{4}, \frac{81}{8})$ **(d)** $\frac{3}{2}, -3$ **(e)**

13. $y = 2x^2 + 11x - 6$ **14.** $w = 4$ ft, $l = 12$ ft **15.** **(a)** 2 sec **(b)** 144 ft

CHAPTER 10

Exercise Set 10.1

1. $x^2 + y^2 = 9$

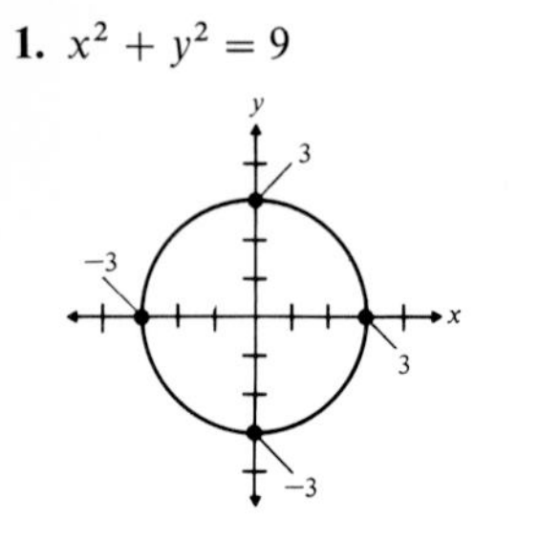

3. $(x - 3)^2 + y^2 = 1$

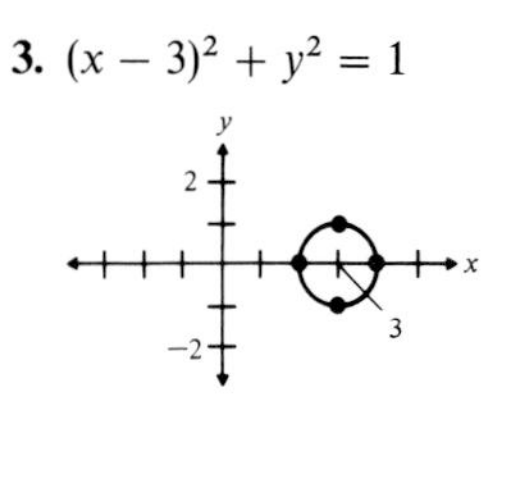

5. $(x + 6)^2 + (y - 5)^2 = 25$

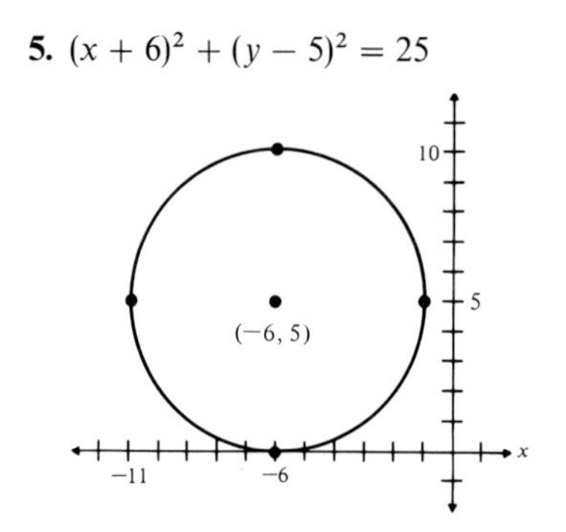

7. $(x - 4)^2 + (y - 7)^2 = 8$

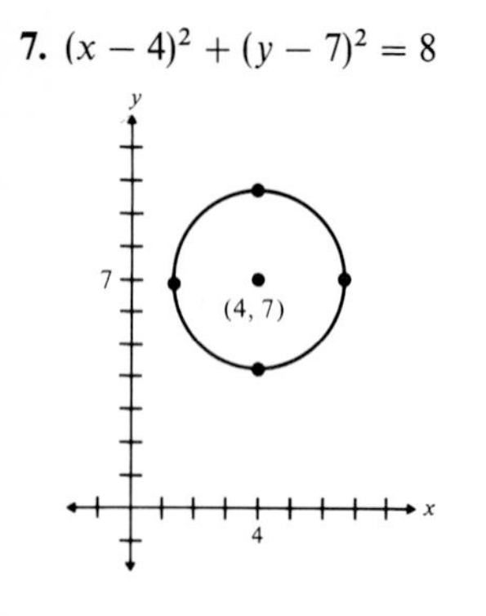

9.

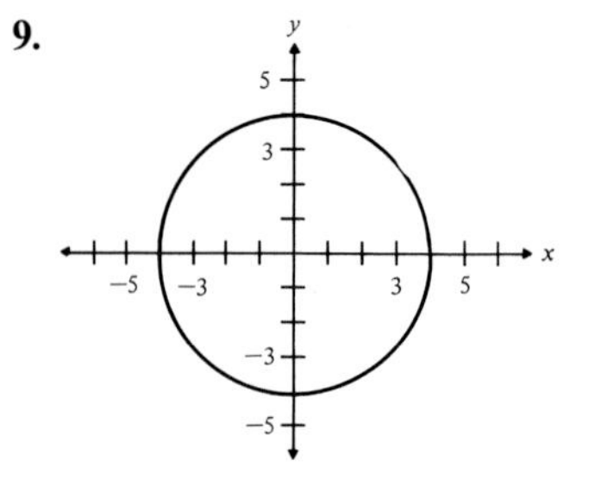

11.

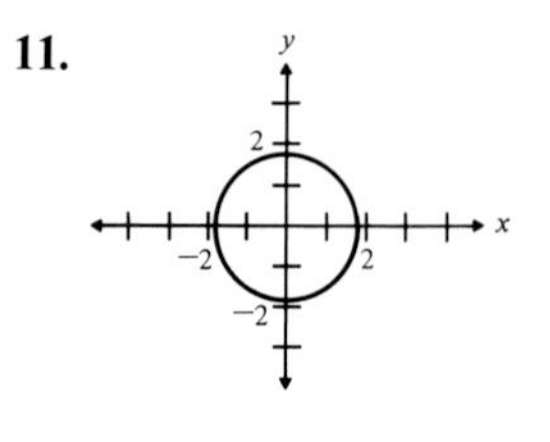

13.

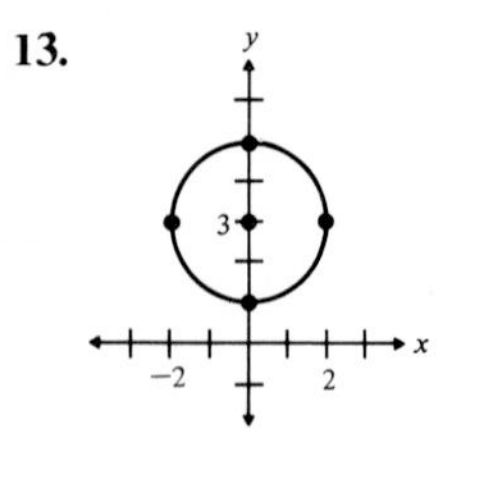

15.

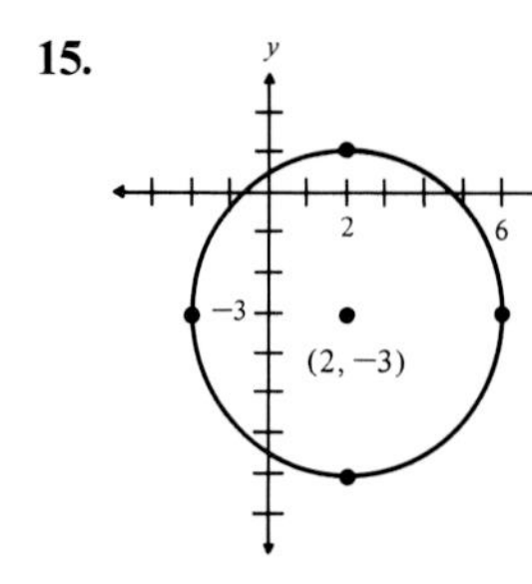

17.

31.

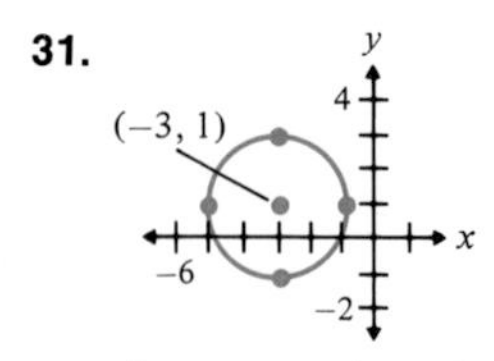

19. $x^2 + y^2 = 1$
21. $(x + 4)^2 + (y - 6)^2 = 4$
23. $(x + 5)^2 + (y + 3)^2 = 4$
25. $x^2 + (y + 5)^2 = 10^2$

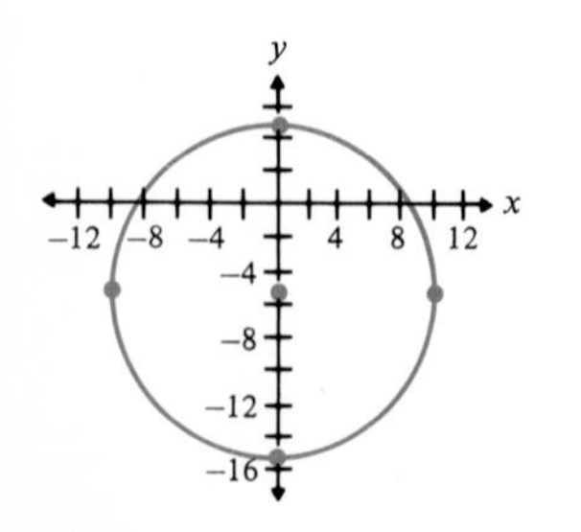

27. $(x + 4)^2 + y^2 = 5^2$

29. $(x + 1)^2 + (y - 2)^2 = 3^2$

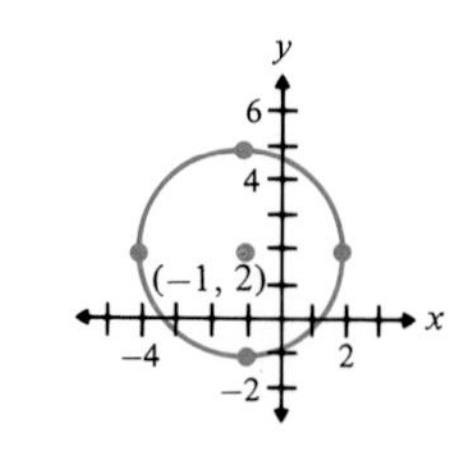

31. $(x + 3)^2 + (y - 1)^2 = 2^2$
33. $(x - 4)^2 + (y + 1)^2 = 2^2$

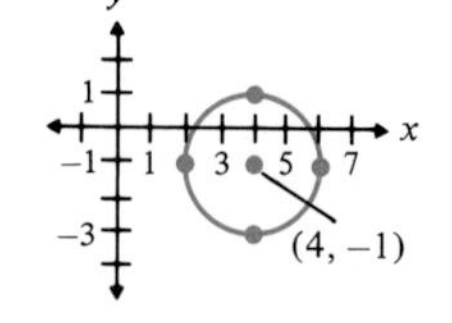

Exercise Set 10.2

1.

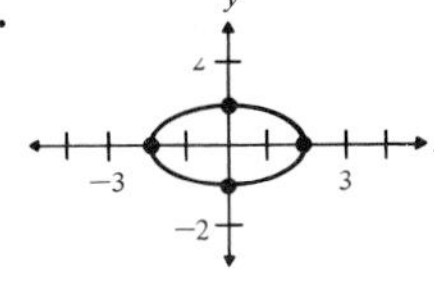

3.

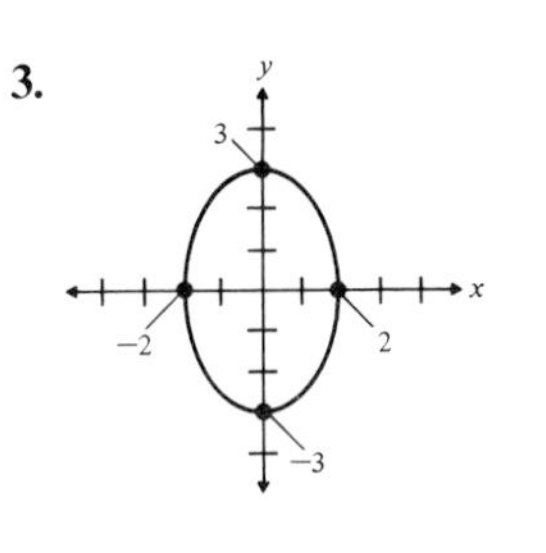

5.

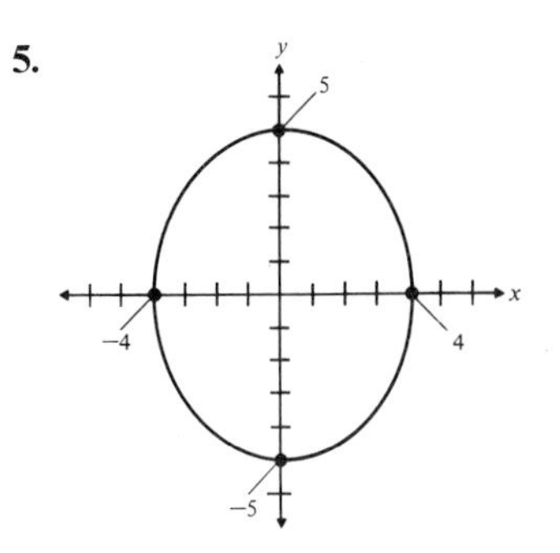

7.

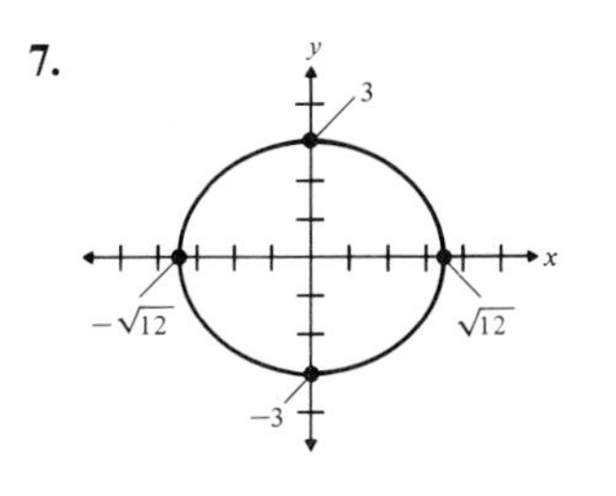

9.

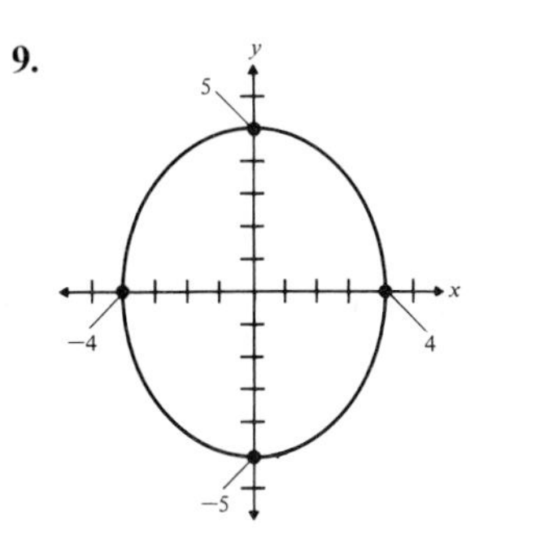

11.

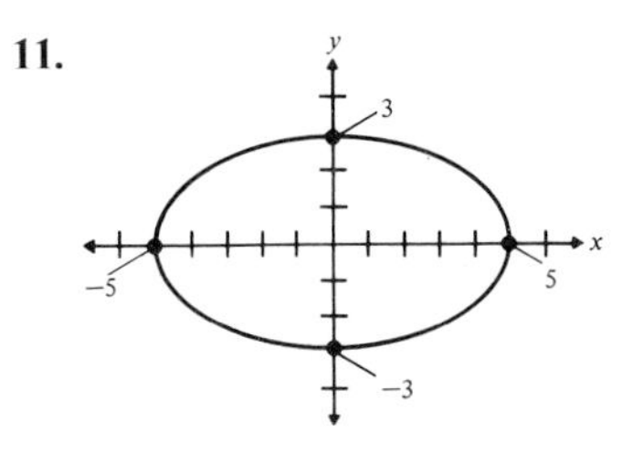

Just for Fun

1.

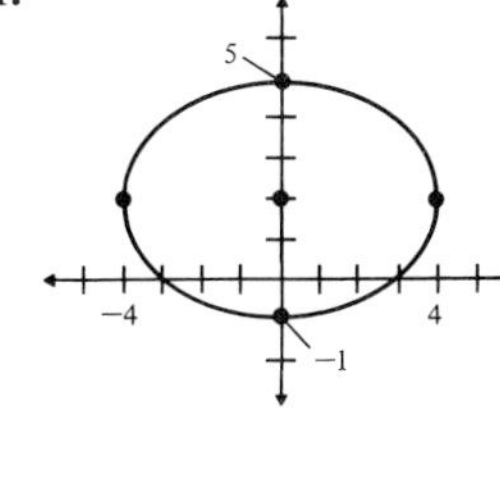

2.

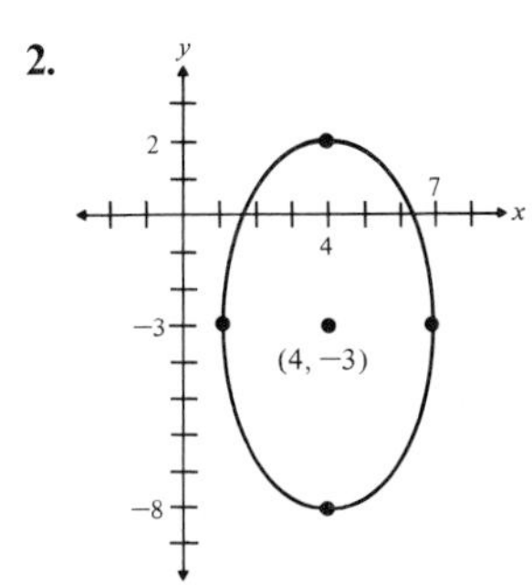

3. $\dfrac{(x-2)^2}{100} + \dfrac{(y-1)^2}{25} = 1$

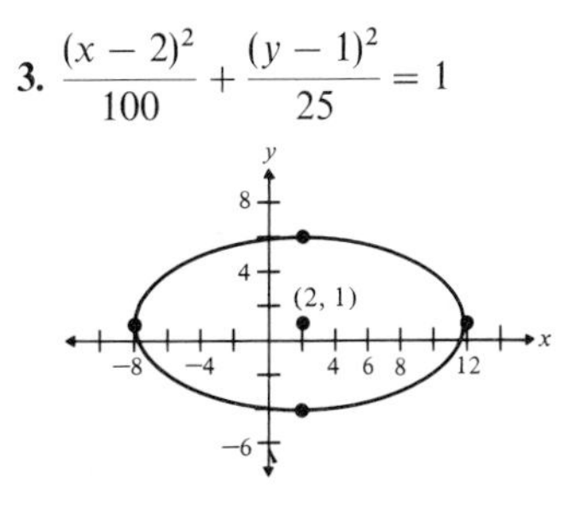

Exercise Set 10.3

1.

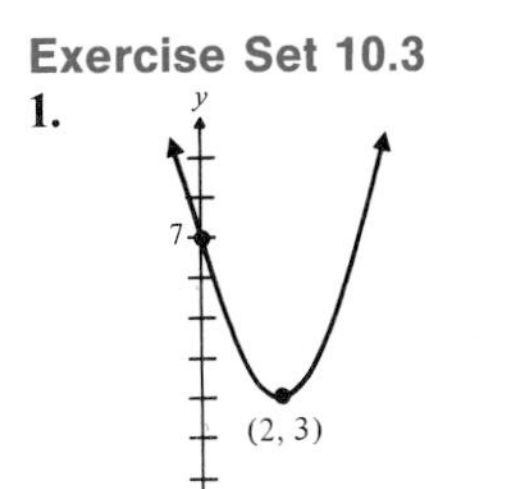

3.

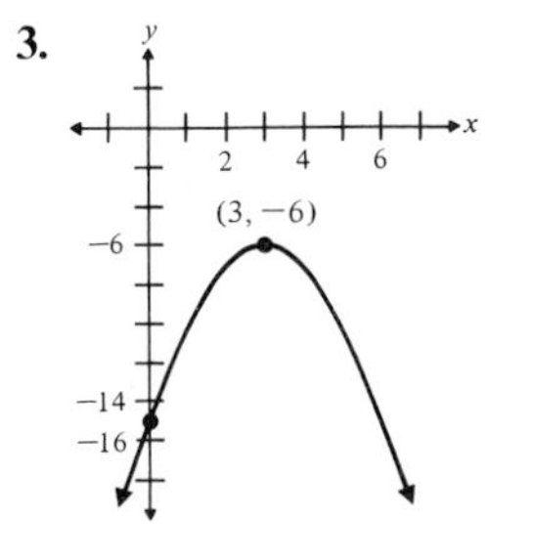

5.

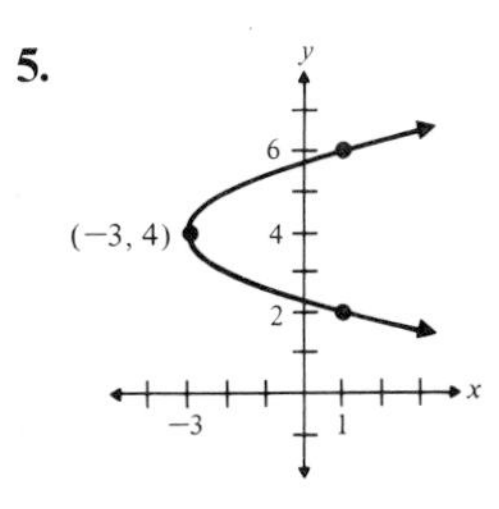

7.

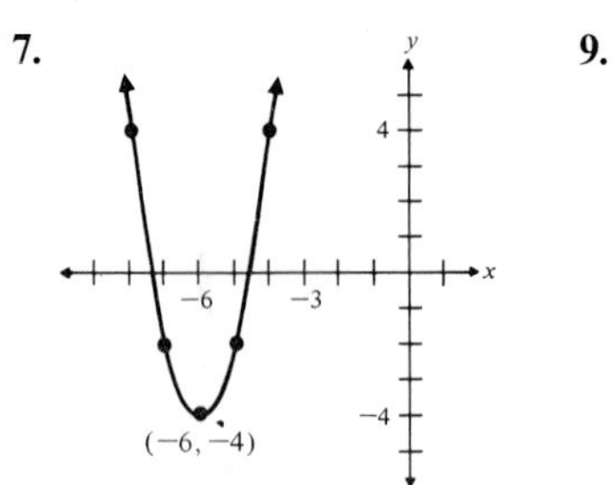

9.

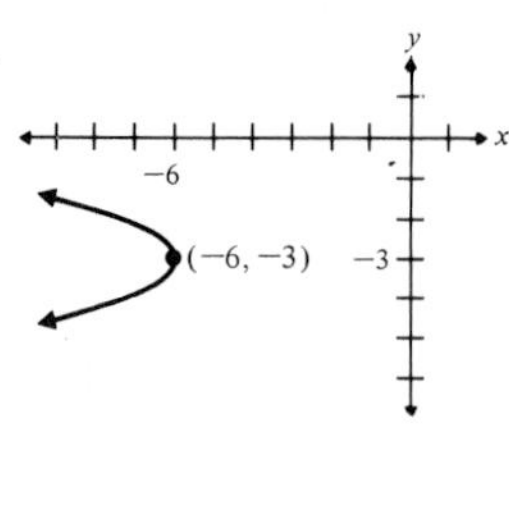

11.

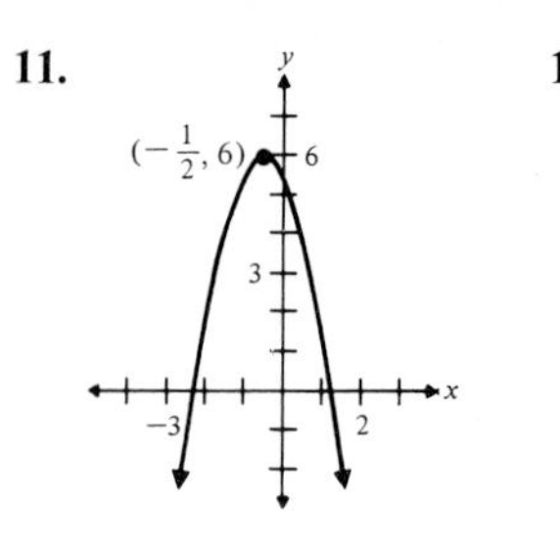

13. $y = (x + 1)^2 - 1$

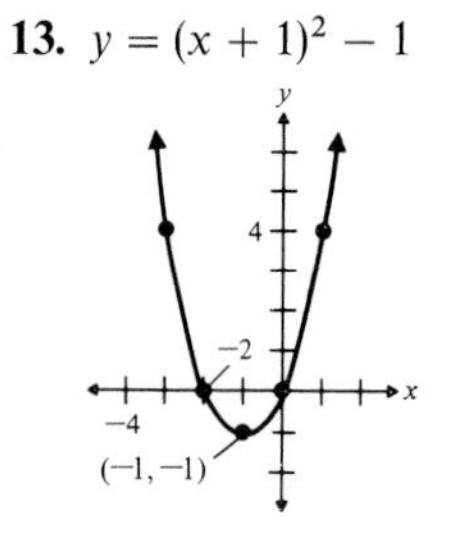

15. $x = (y + 3)^2 - 9$

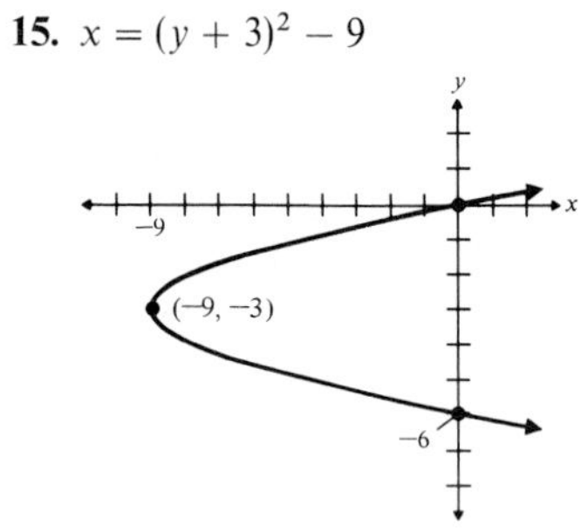

17. $y = (x + 1)^2 - 16$

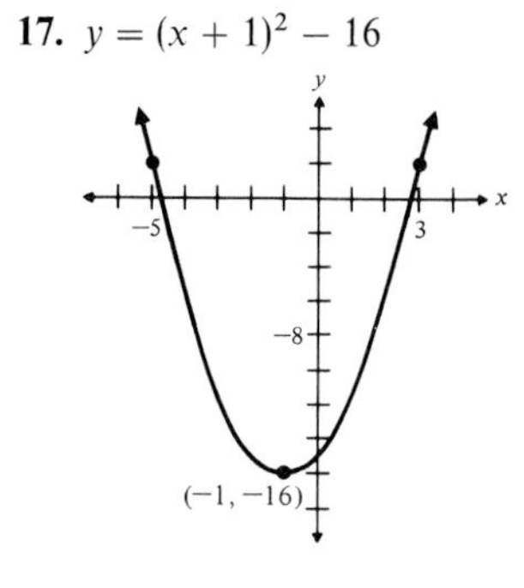

19. $x = -(y - 3)^2$

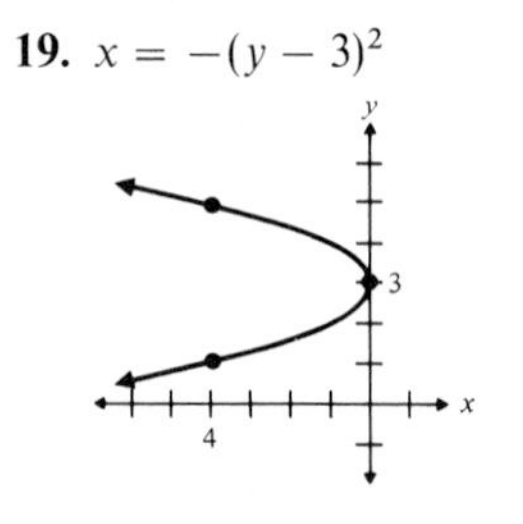

21. $y = (x + \frac{7}{2})^2 - \frac{9}{4}$

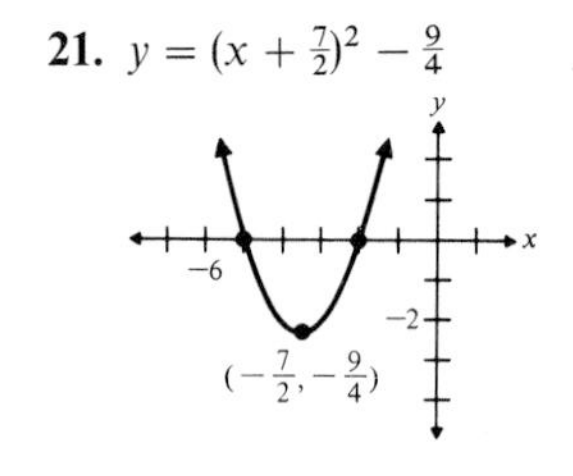

23. $x = -(y + 2)^2 + 4$

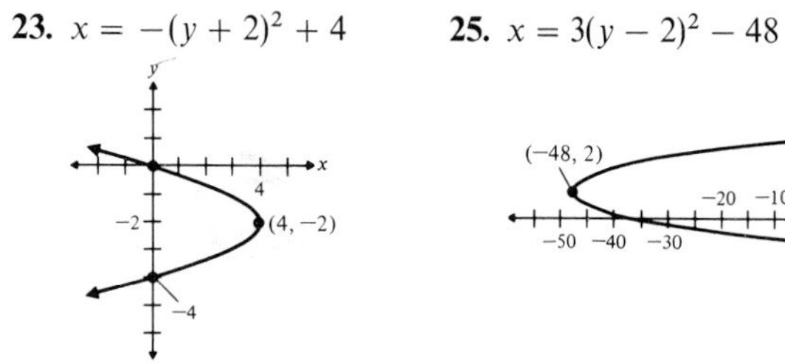

25. $x = 3(y - 2)^2 - 48$

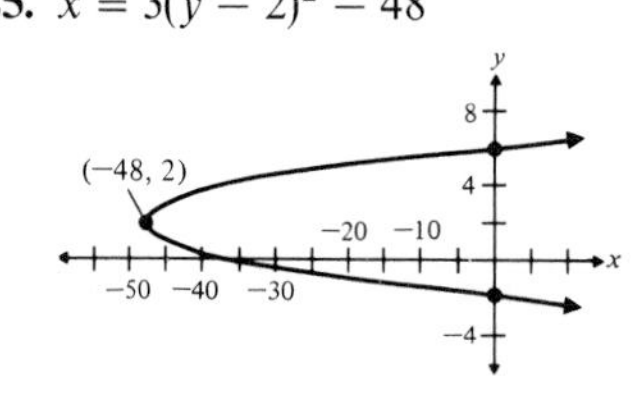

Exercise Set 10.4

1. $y = \pm\frac{1}{2}x$

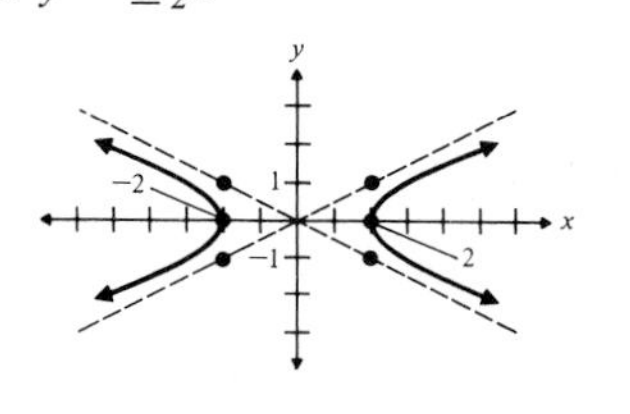

3. $y = \pm\frac{3}{4}x$

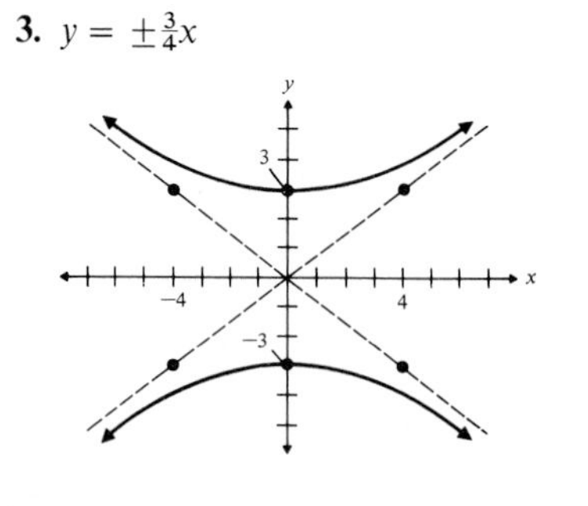

5. $y = \pm\frac{5}{6}x$

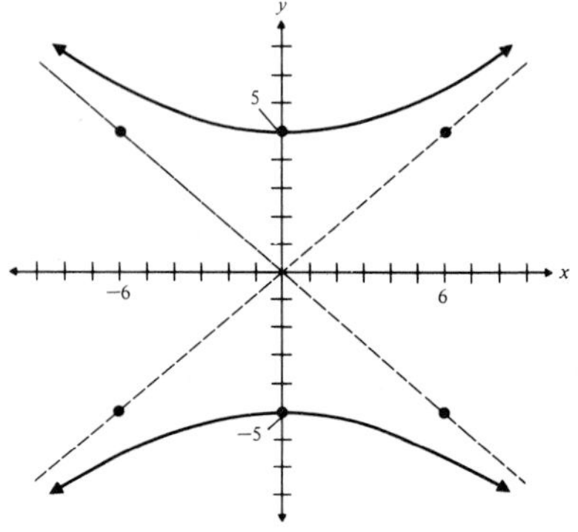

7. $y = \pm x$

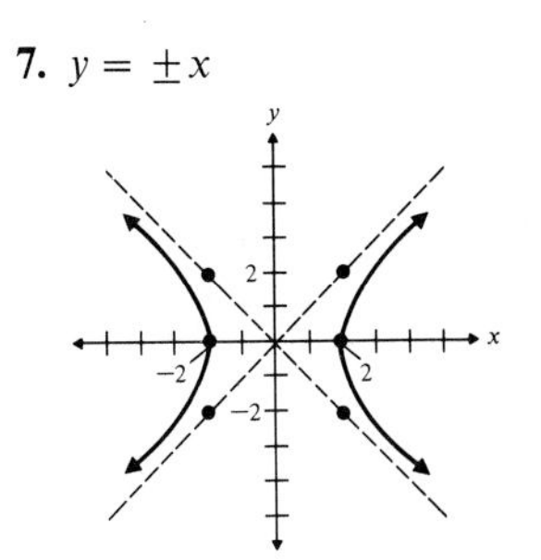

9. $y = \pm \frac{4}{9}x$

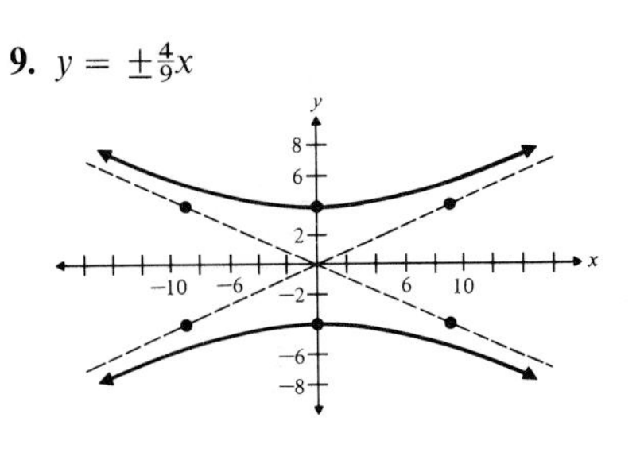

11. $y = \pm \frac{5}{4}x$

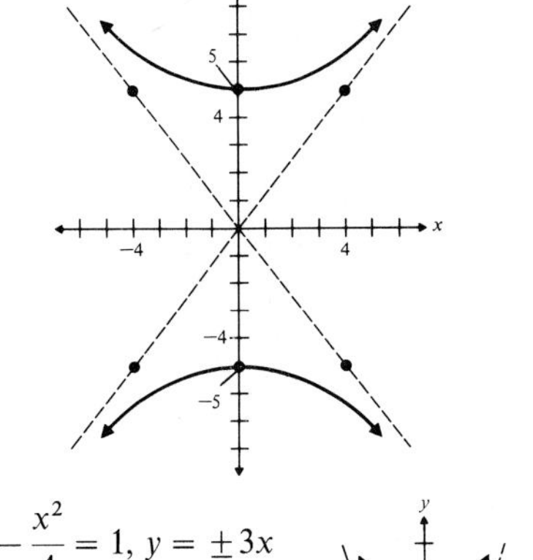

13. $\dfrac{x^2}{4} - \dfrac{y^2}{16} = 1$, $y = \pm 2x$

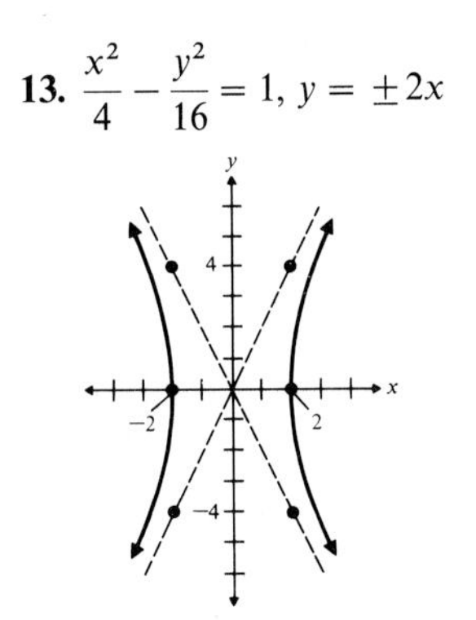

15. $\dfrac{y^2}{1} - \dfrac{x^2}{9} = 1$, $y = \pm \frac{1}{3}x$

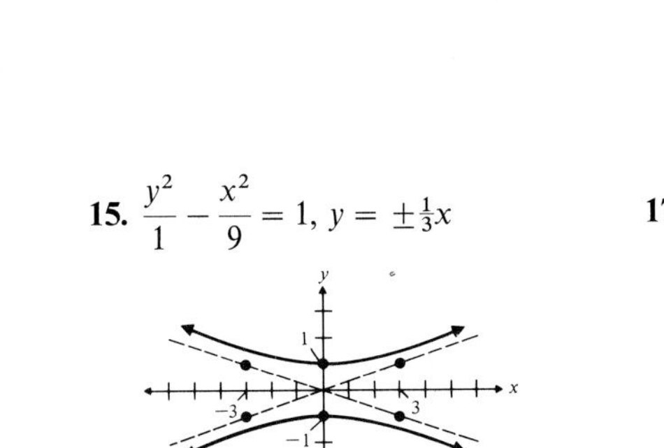

17. $\dfrac{y^2}{36} - \dfrac{x^2}{4} = 1$, $y = \pm 3x$

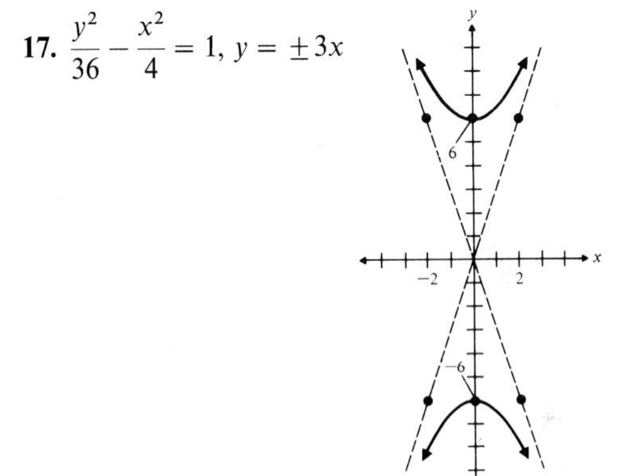

19. $\dfrac{x^2}{4} - \dfrac{y^2}{25} = 1$, $y = \pm \frac{5}{2}x$

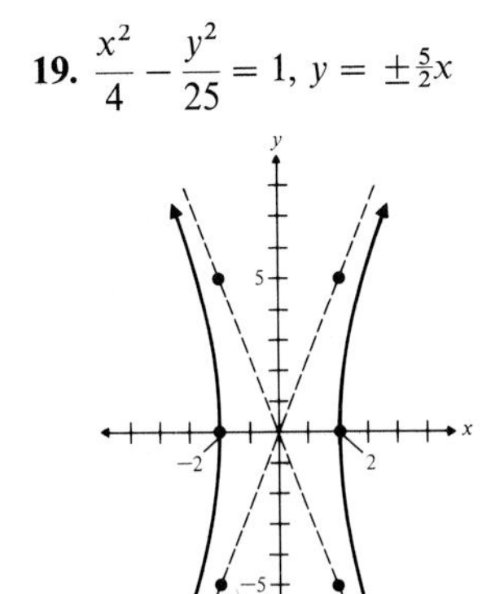

21. $\dfrac{x^2}{9} - \dfrac{y^2}{81} = 1$, $y = \pm 3x$

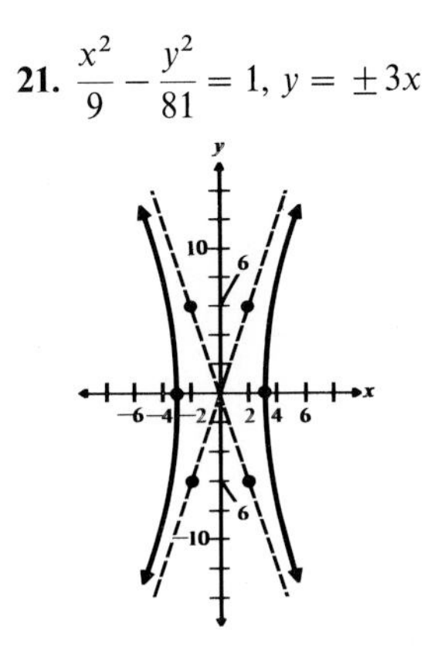

23.

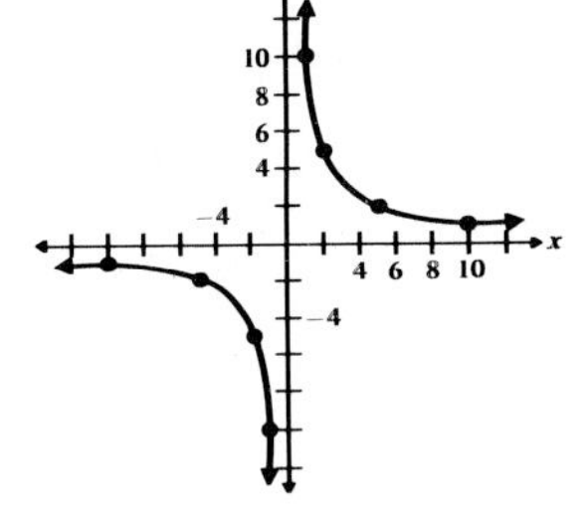

25.

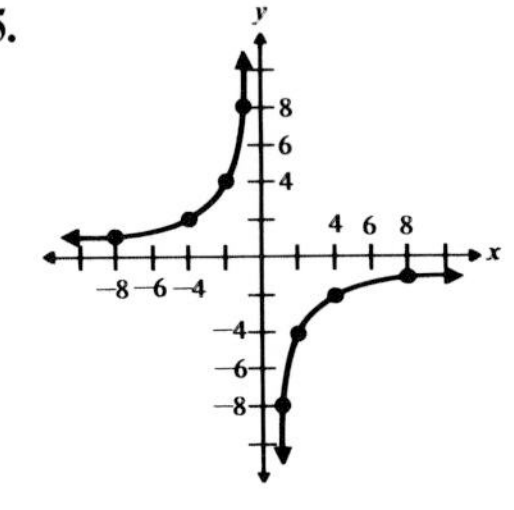

27.

29. Parabola **31.** Hyperbola **33.** Parabola **35.** Hyperbola **37.** Ellipse **39.** Parabola **41.** Circle **43.** Circle **45.** Hyperbola

Just for Fun **1.** $y + 2 = \pm\frac{2}{3}(x - 3)$ **2.** $y + 5 = \pm\frac{4}{3}(x - 1)$ **3.** $\frac{(y+1)^2}{4} - \frac{(x-1)^2}{1} = 1,\ y + 1 = \pm 2(x - 1)$

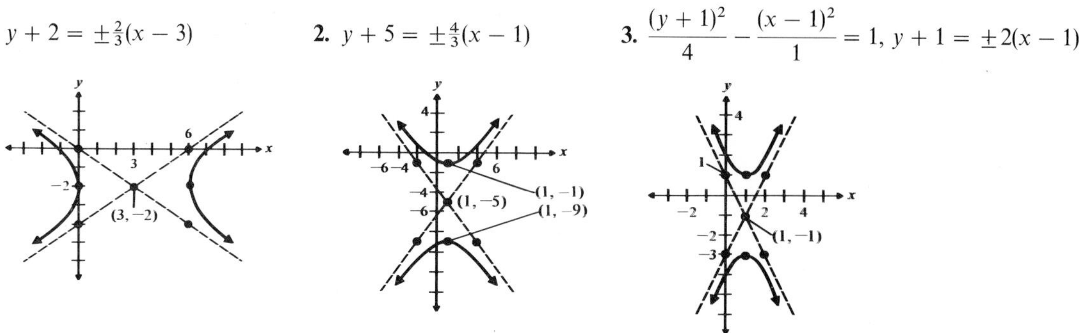

Exercise Set 10.5

1. $(0, -2), (\frac{8}{5}, -\frac{6}{5})$ **3.** $(-4, 11), (\frac{5}{2}, \frac{5}{4})$ **5.** $(2, -4), (-14, -20)$ **7.** No real solution **9.** $(\frac{4}{3}, 3), (-\frac{1}{2}, -8)$ **11.** $(0, -3), (\sqrt{5}, 2), (-\sqrt{5}, 2)$ **13.** $(2, 0), (-2, 0)$ **15.** $(3, 2), (3, -2), (-3, 2), (-3, -2)$ **17.** $(3, 0), (-3, 0)$ **19.** $(\sqrt{6}, \sqrt{3}), (\sqrt{6}, -\sqrt{3}), (-\sqrt{6}, \sqrt{3}), (-\sqrt{6}, -\sqrt{3})$ **21.** No real solution **23.** $(\sqrt{5}, 2), (\sqrt{5}, -2), (-\sqrt{5}, 2), (-\sqrt{5}, -2)$

Exercise Set 10.6

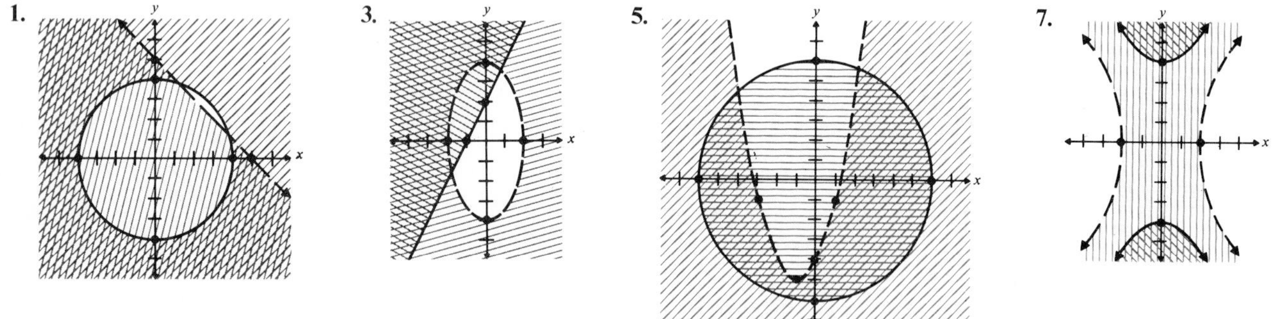

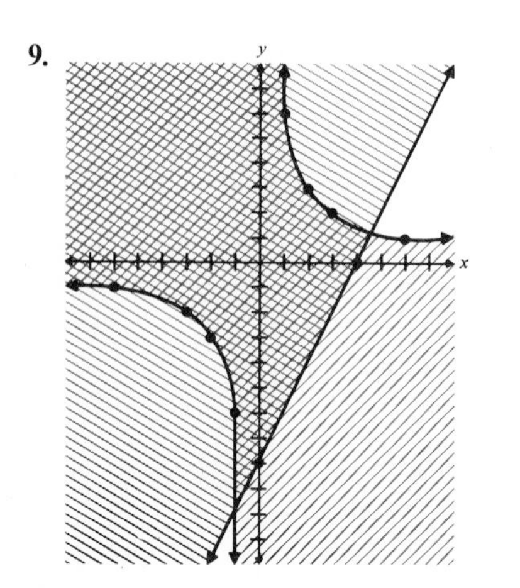

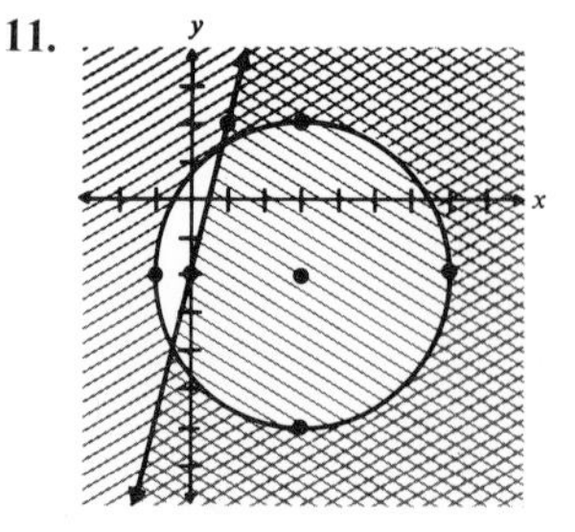

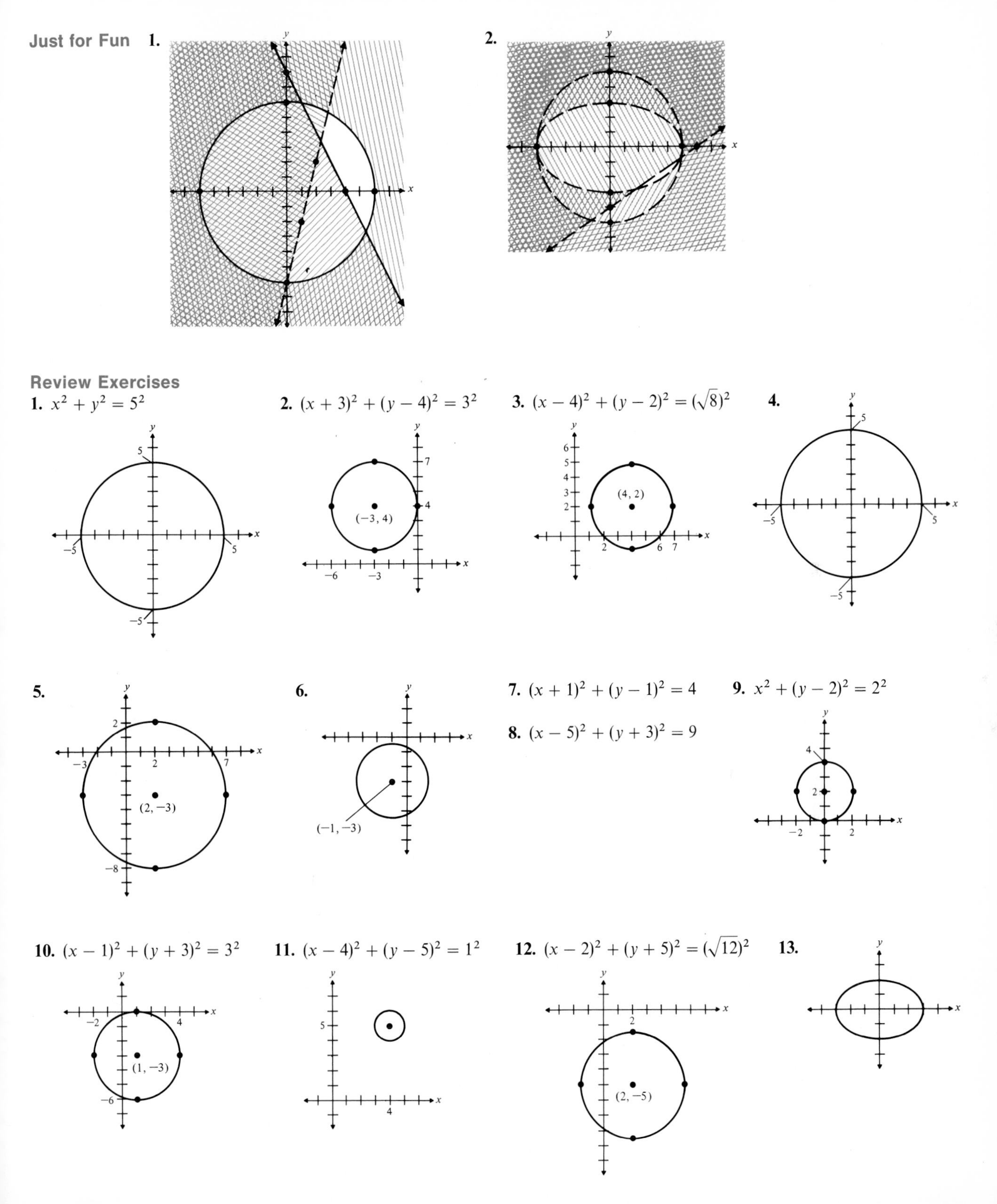
Just for Fun
1.
2.
Review Exercises
1. x² + y² = 5²
2. (x + 3)² + (y − 4)² = 3²
3. (x − 4)² + (y − 2)² = (√8)²
4.
(−3, 4)
(4, 2)
5.
6.
7. (x + 1)² + (y − 1)² = 4
8. (x − 5)² + (y + 3)² = 9
9. x² + (y − 2)² = 2²
(2, −3)
(−1, −3)
10. (x − 1)² + (y + 3)² = 3²
11. (x − 4)² + (y − 5)² = 1²
12. (x − 2)² + (y + 5)² = (√12)²
13.
(1, −3)
(2, −5)

14.

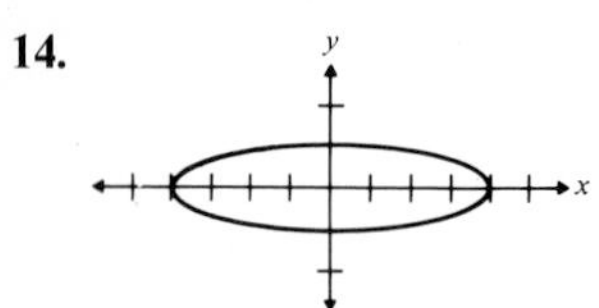

15.

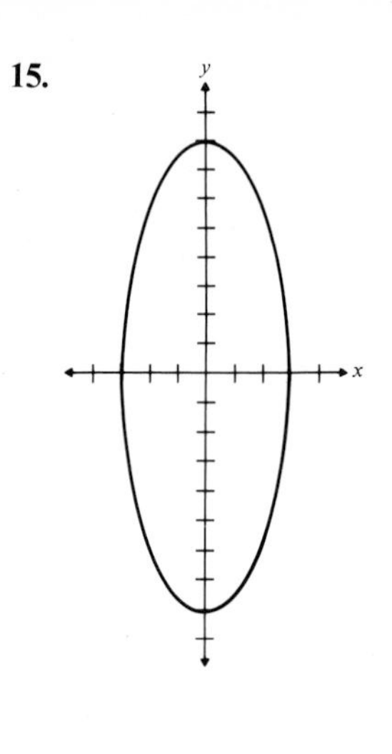

16.

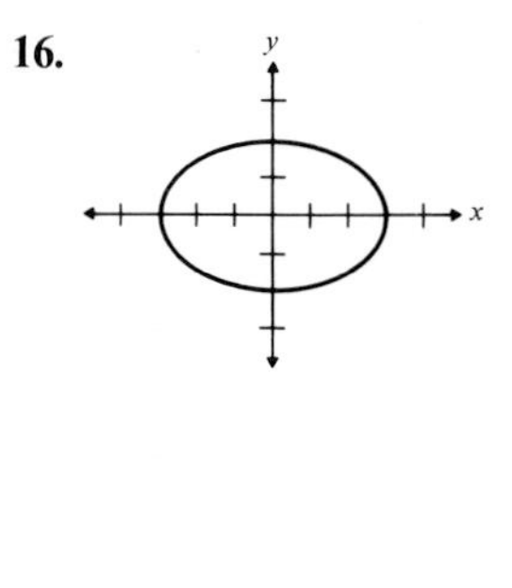

17.

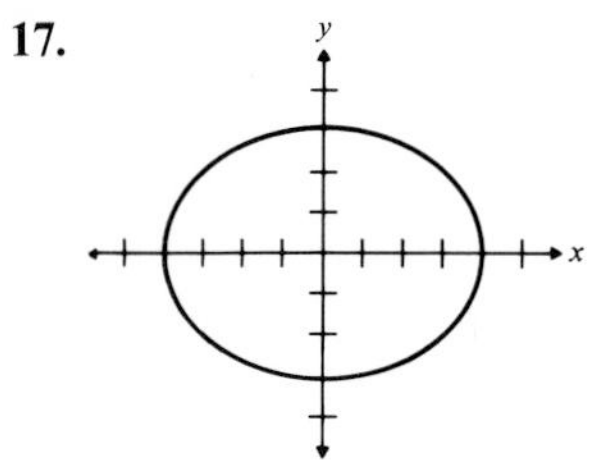

18.

19.

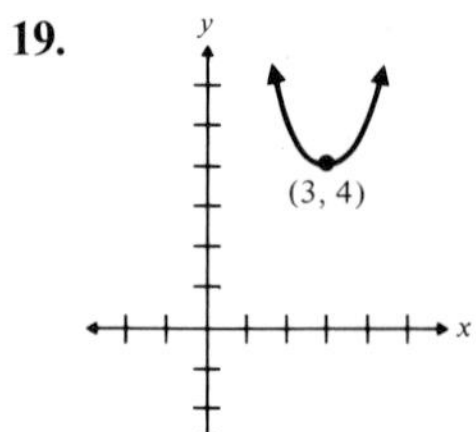

20.

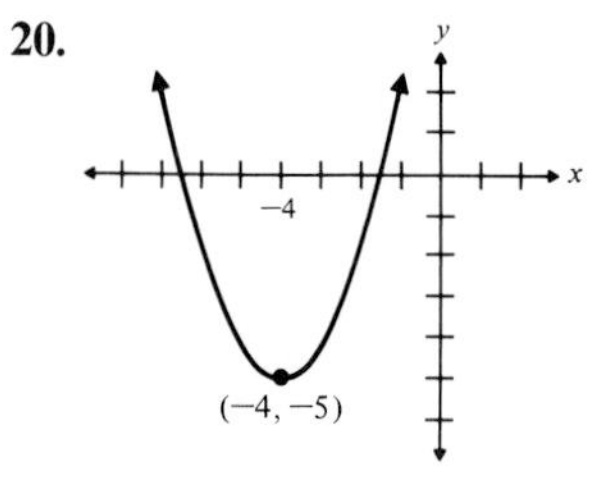

21.

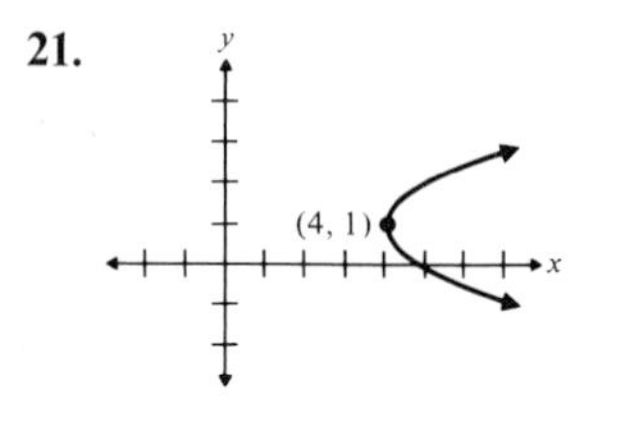

22.

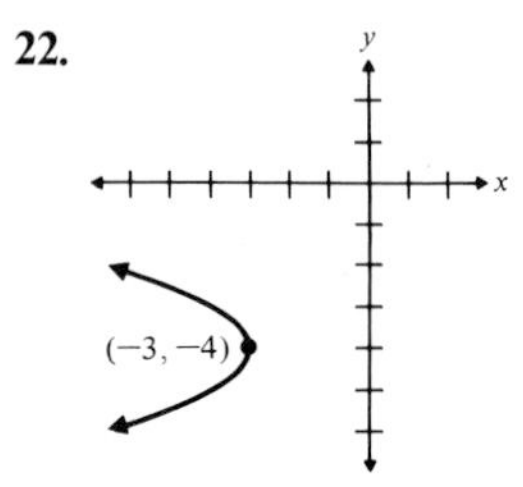

23.

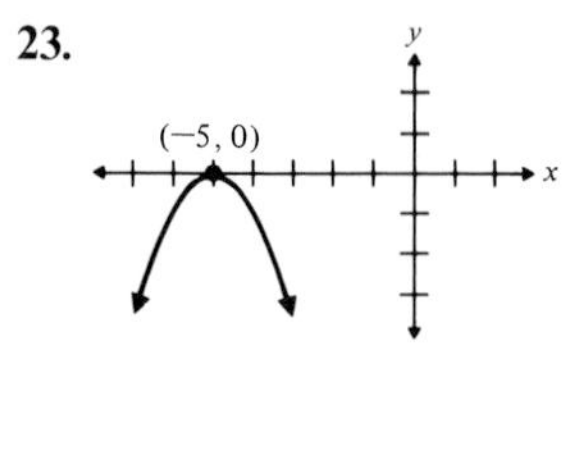

24. $y = (x - 3)^2 - 9$

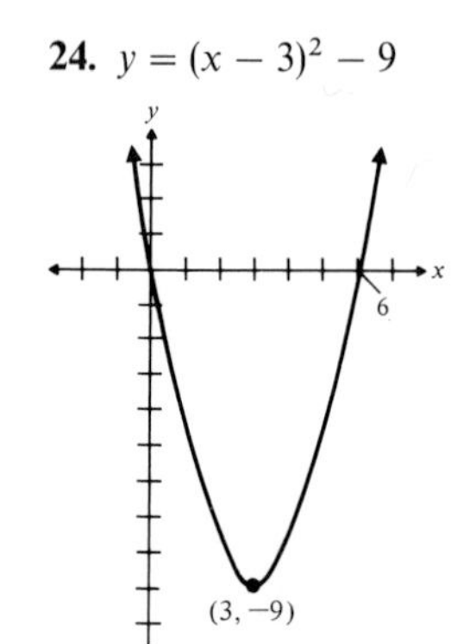

25. $y = (x - 1)^2 - 4$

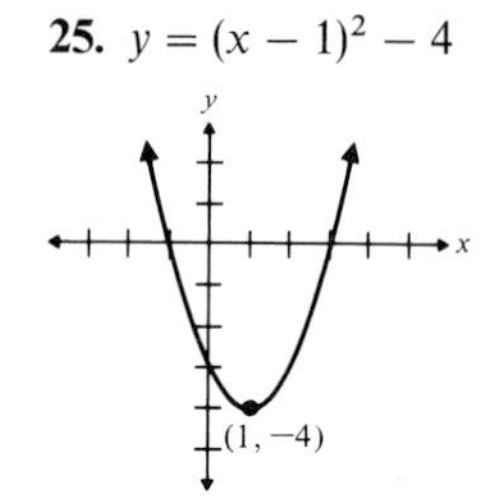

26. $x = -(y + 1)^2 + 9$

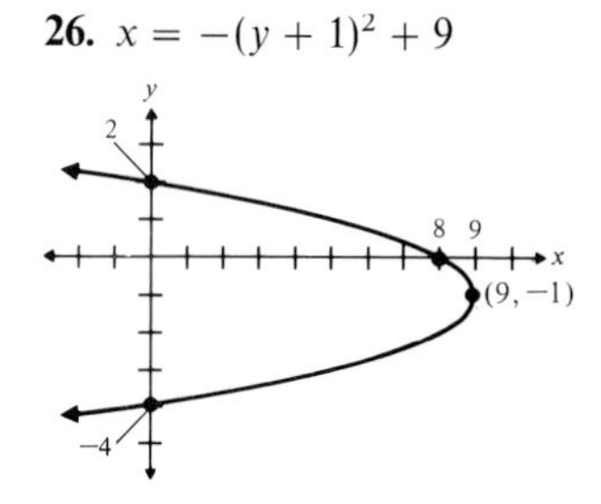

27. $x = (y + \frac{5}{2})^2 - \frac{9}{4}$

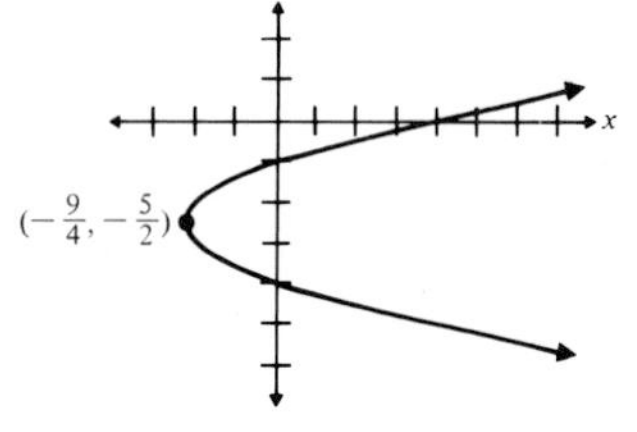

28. $y = 2(x - 2)^2 - 32$

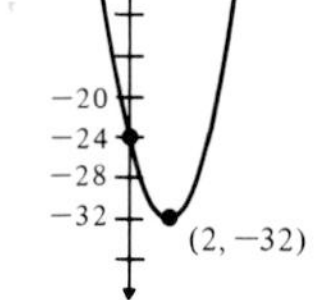

29. $y = \pm\frac{3}{2}x$

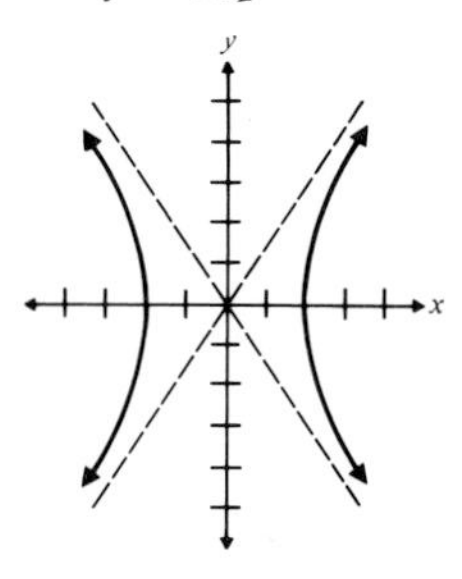

30. $y = \pm 2x$

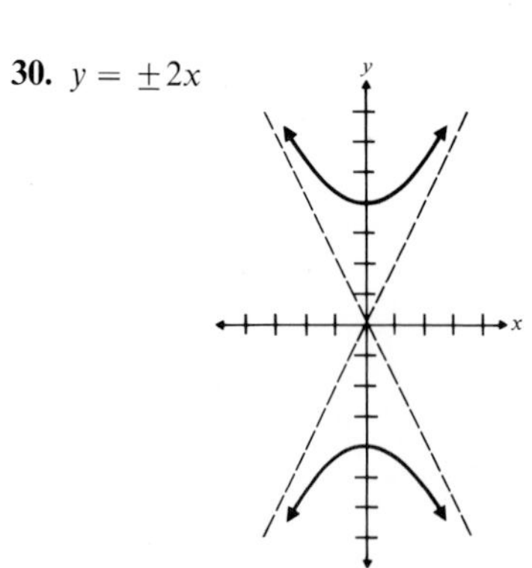

31. $y = \pm\frac{3}{5}x$

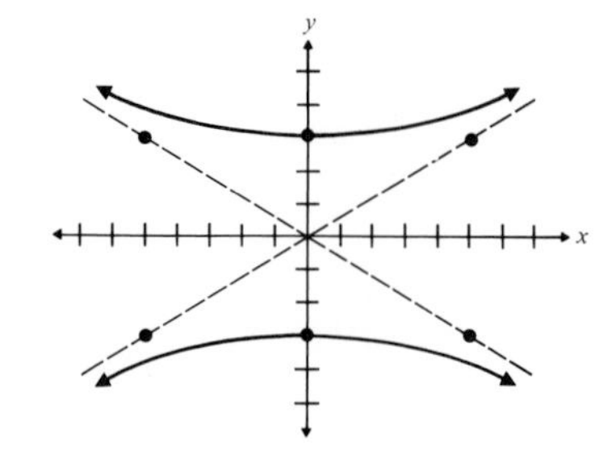

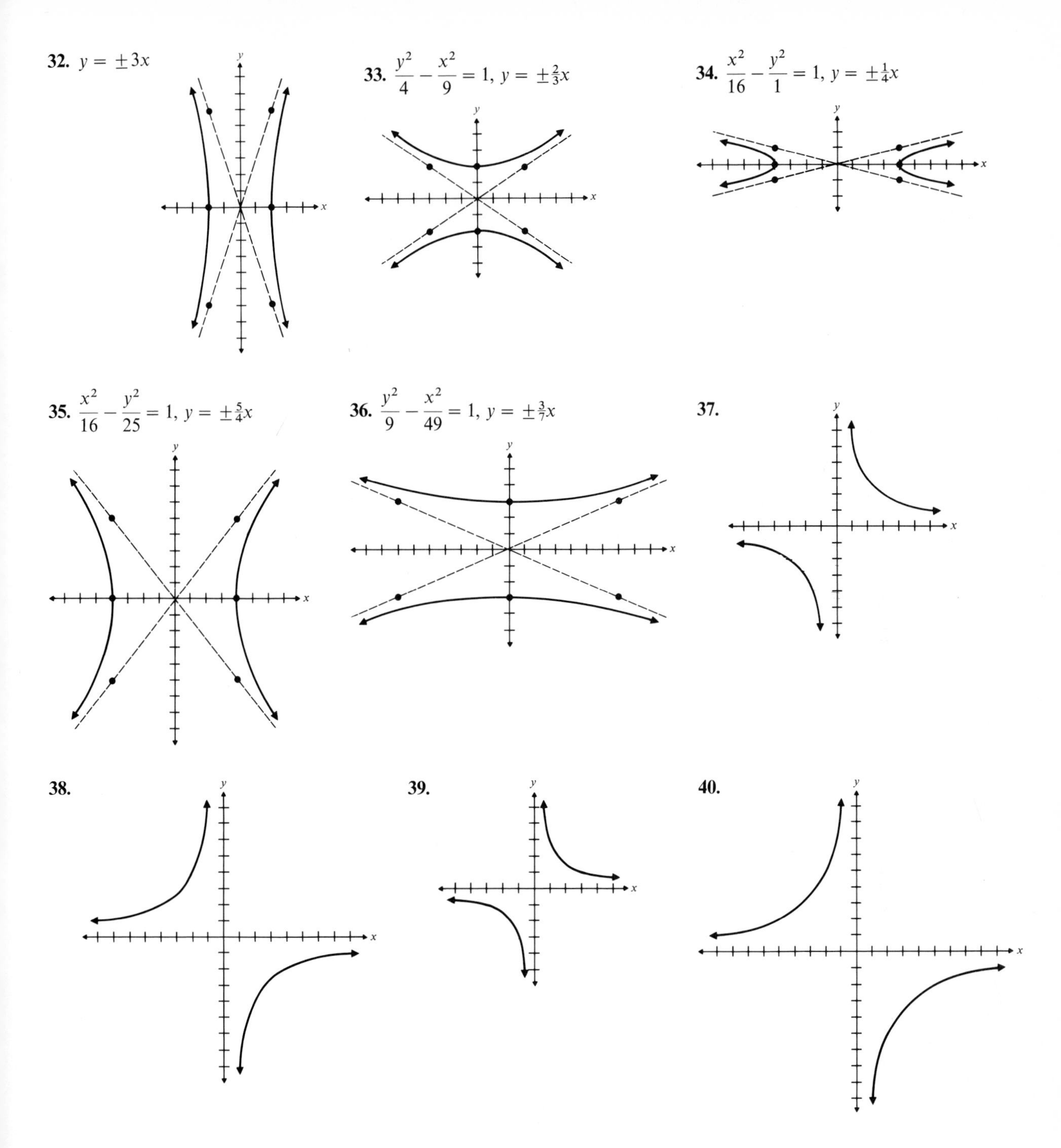

32. $y = \pm 3x$
33. $\frac{y^2}{4} - \frac{x^2}{9} = 1,\ y = \pm\frac{2}{3}x$
34. $\frac{x^2}{16} - \frac{y^2}{1} = 1,\ y = \pm\frac{1}{4}x$
35. $\frac{x^2}{16} - \frac{y^2}{25} = 1,\ y = \pm\frac{5}{4}x$
36. $\frac{y^2}{9} - \frac{x^2}{49} = 1,\ y = \pm\frac{3}{7}x$
37.
38.
39.
40.

41. Hyperbola **42.** Ellipse **43.** Circle **44.** Hyperbola **45.** Ellipse **46.** Parabola **47.** Ellipse **48.** Parabola **49.** Parabola **50.** Hyperbola **51.** $(-3, 0), (-\frac{12}{5}, \frac{9}{5})$ **52.** $(\frac{1}{3}, 15), (5, 1)$ **53.** $(2, 0), (-2, 0)$ **54.** No real solution **55.** $(4, 0), (-4, 0)$ **56.** $(4, 3), (4, -3), (-4, 3), (-4, -3)$ **57.** No real solution **58.** $(\sqrt{3}, 0), (-\sqrt{3}, 0)$

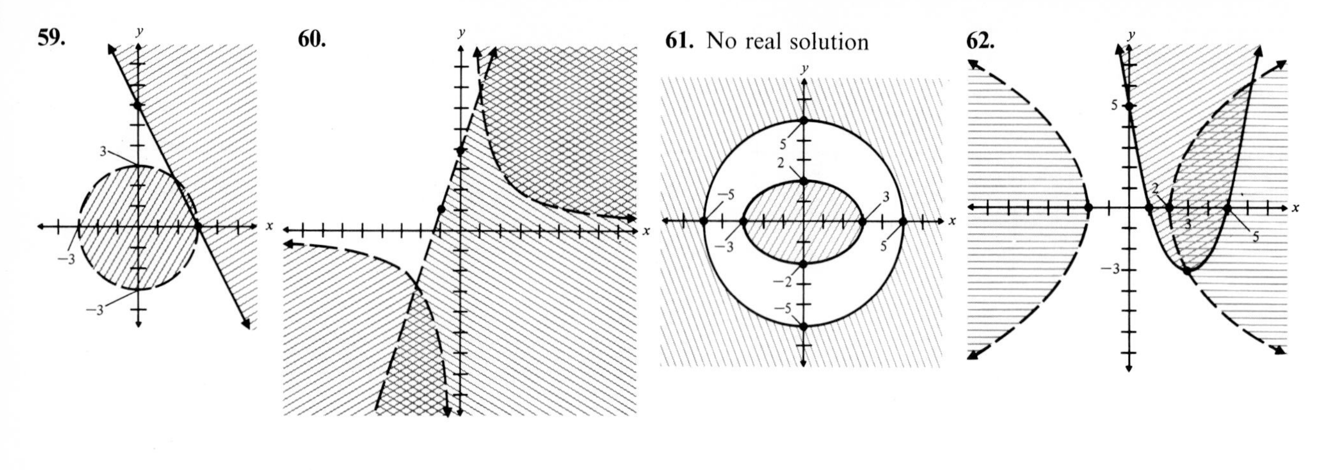

59. **60.** **61.** No real solution **62.**

Practice Test

1. $(x + 3)^2 + (y + 1)^2 = 9^2$

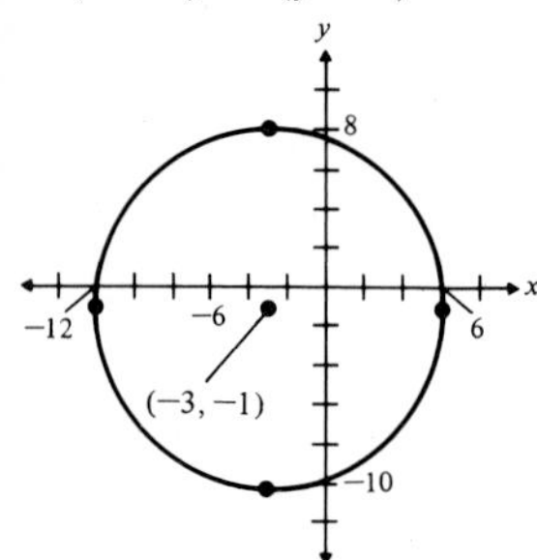

2. $(x - 1)^2 + (y - 3)^2 = 3^2$

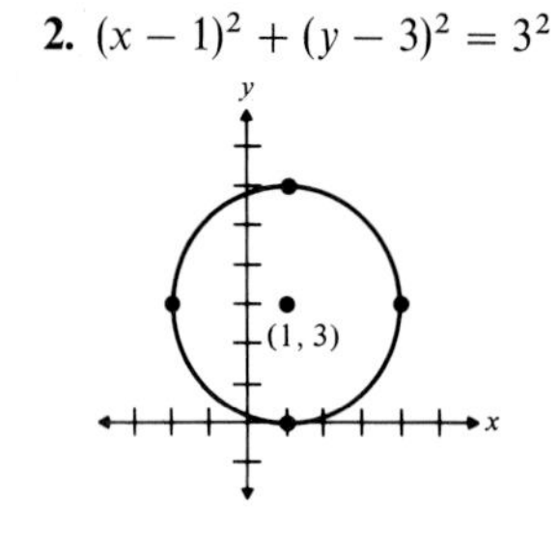

3. **4.**

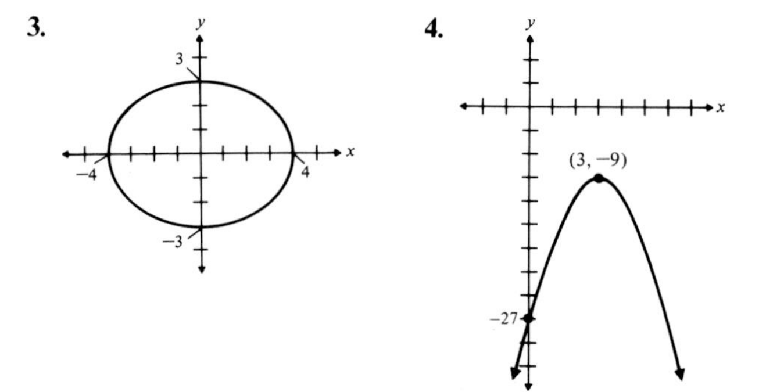

5.

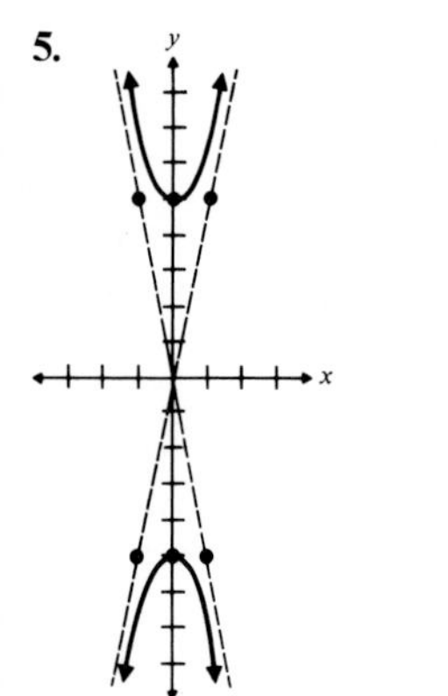

6.

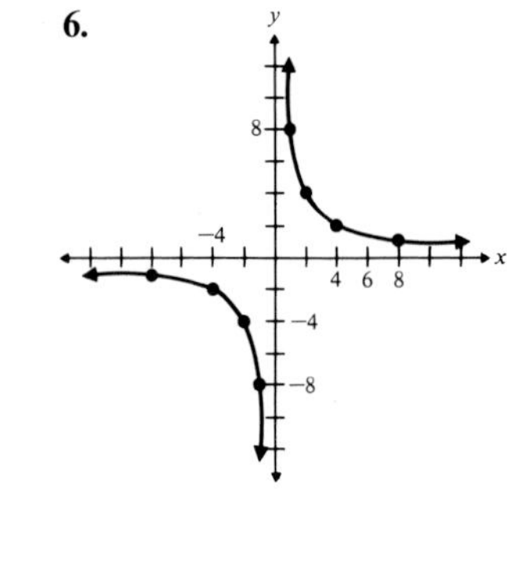

7. $(\sqrt{6}, \sqrt{10}), (-\sqrt{6}, -\sqrt{10}), (\sqrt{6}, -\sqrt{10}), (-\sqrt{6}, \sqrt{10})$

8.

CHAPTER 11

Exercise Set 11.1

1. No **3.** Yes **5.** No **7.** Yes **9.** No **11.** $f(x)$: Domain $\{-2, -1, 2, 4, 9\}$ Range $\{0, 3, 4, 6, 7\}$
$f^{-1}(x)$: Domain $\{0, 3, 4, 6, 7\}$ Range $\{-2, -1, 2, 4, 9\}$

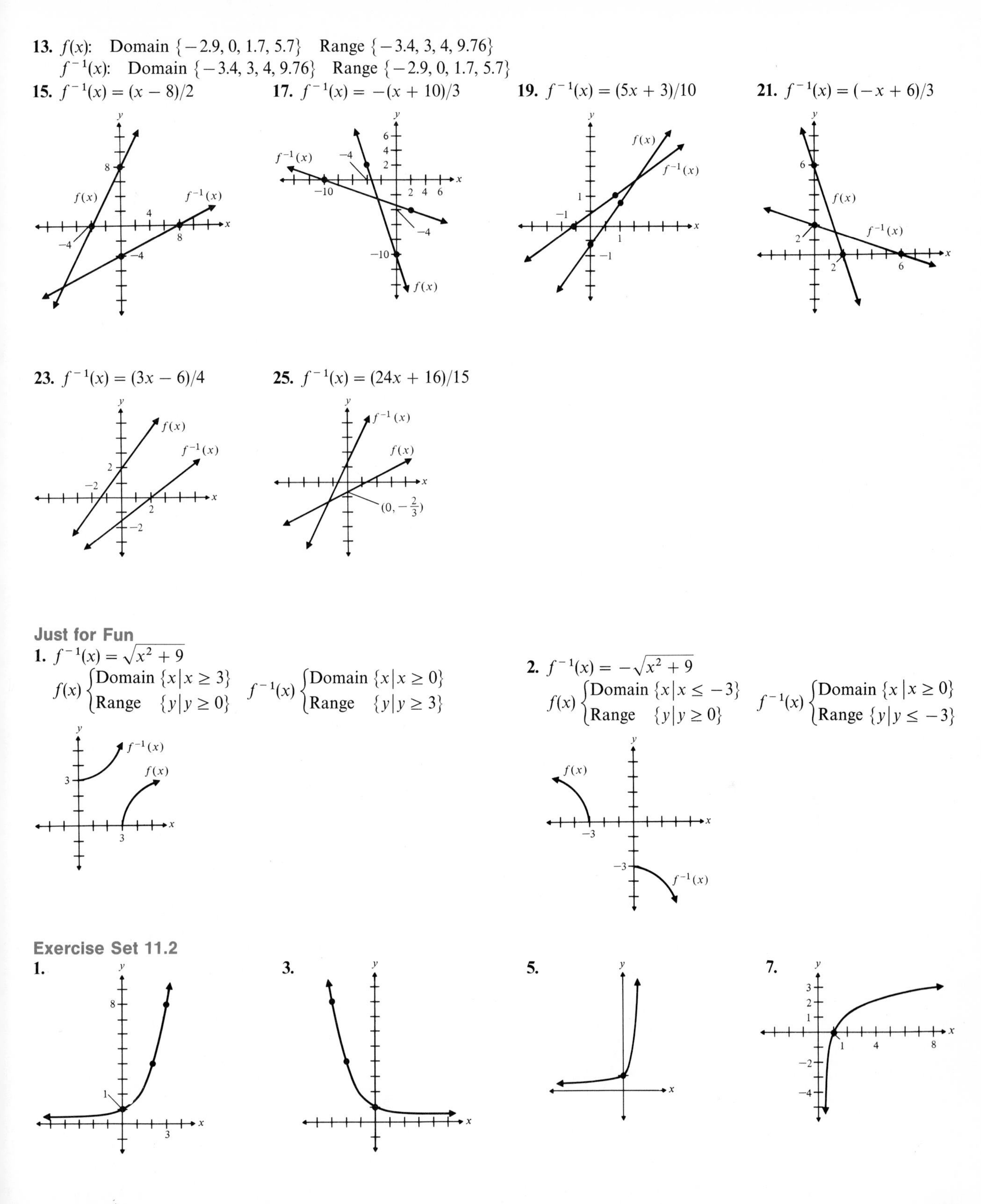

13. $f(x)$: Domain $\{-2.9, 0, 1.7, 5.7\}$ Range $\{-3.4, 3, 4, 9.76\}$
$f^{-1}(x)$: Domain $\{-3.4, 3, 4, 9.76\}$ Range $\{-2.9, 0, 1.7, 5.7\}$

15. $f^{-1}(x) = (x - 8)/2$

17. $f^{-1}(x) = -(x + 10)/3$

19. $f^{-1}(x) = (5x + 3)/10$

21. $f^{-1}(x) = (-x + 6)/3$

23. $f^{-1}(x) = (3x - 6)/4$

25. $f^{-1}(x) = (24x + 16)/15$

Just for Fun

1. $f^{-1}(x) = \sqrt{x^2 + 9}$

$f(x)\begin{cases}\text{Domain } \{x \mid x \geq 3\} \\ \text{Range } \{y \mid y \geq 0\}\end{cases}$ $f^{-1}(x)\begin{cases}\text{Domain } \{x \mid x \geq 0\} \\ \text{Range } \{y \mid y \geq 3\}\end{cases}$

2. $f^{-1}(x) = -\sqrt{x^2 + 9}$

$f(x)\begin{cases}\text{Domain } \{x \mid x \leq -3\} \\ \text{Range } \{y \mid y \geq 0\}\end{cases}$ $f^{-1}(x)\begin{cases}\text{Domain } \{x \mid x \geq 0\} \\ \text{Range } \{y \mid y \leq -3\}\end{cases}$

Exercise Set 11.2

1.

3.

5.

7.

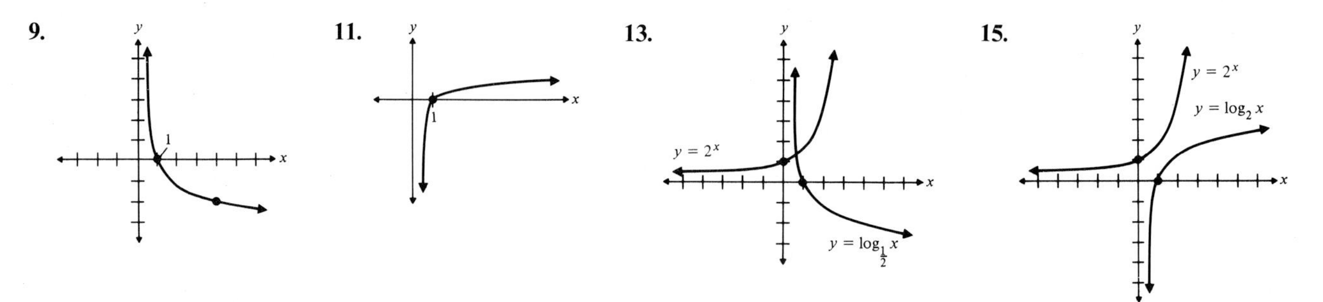

17. $\log_2 8 = 3$ **19.** $\log_3 243 = 5$ **21.** $\log_8 2 = \frac{1}{3}$ **23.** $\log_{1/4}(\frac{1}{16}) = 2$ **25.** $\log_5 (\frac{1}{25}) = -2$ **27.** $\log_4 (\frac{1}{64}) = -3$ **29.** $\log_{16} (\frac{1}{4}) = -\frac{1}{2}$ **31.** $\log_8 (\frac{1}{2}) = -\frac{1}{3}$ **33.** $2^3 = 8$ **35.** $4^3 = 64$ **37.** $(\frac{1}{2})^4 = \frac{1}{16}$ **39.** $5^{-3} = \frac{1}{125}$ **41.** $125^{1/3} = 5$ **43.** $27^{-1/3} = \frac{1}{3}$ **45.** $10^3 = 1000$ **47.** $6^y = 36, y = 2$ **49.** $2^3 = x, x = 8$ **51.** $64^y = 8, y = \frac{1}{2}$ **53.** $a^3 = 64, a = 4$ **55.** $(\frac{1}{2})^4 = x, x = \frac{1}{16}$ **57.** $a^5 = 32, a = 2$ **59.** $8^{1/3} = x, x = 2$ **61.** 256 **63.** \$14,448.89 **65.** 17.4 grams **67.** **(a)** 5 grams **(b)** 7.28×10^{-11} grams **69.** 10,000,000 **71.** **(a)** $a > 0, a \neq 1$ **(b)** $\{x|x > 0\}$ **(c)** all real numbers

Exercise Set 11.3

1. $\log_3 7 + \log_3 12$ **3.** $\log_8 7 + \log_8 (x + 3)$ **5.** $\log_4 15 - \log_4 7$ **7.** $\frac{1}{2} \log_{10} x - \log_{10} (x - 3)$ **9.** $4 \log_8 x$ **11.** $\log_{10} 3 + 2 \log_{10} 8$ **13.** $5 \log_4 x - \log_4 (x + 4)$ **15.** $4 \log_{10} x - 3 \log_{10} (x + 2)$ **17.** $\log_8 x + \log_8 (x - 6) - 3 \log_8 x$ **19.** $\log_{10} 2 + \log_{10} x - \log_{10} 3$ **21.** $\log_{10}\left(\frac{x^2}{x-2}\right)$ **23.** $\log_5 \left(\frac{x}{4}\right)^2$ **25.** $\log_{10}\left[\frac{x(x-4)}{(x+1)}\right]$ **27.** $\log_7 \left[\frac{(x-2)}{x}\right]^{1/2}$ **29.** $\log_9\left(\frac{5^2}{6^4 \cdot 3}\right)$ **31.** $\log_6\left[\frac{3^4}{(x+3)^2 x^4}\right]$

Exercise Set 11.4

1. 2.9395 **3.** 0.9031 **5.** 3.0000 **7.** 5.9330 – 10 **9.** 2.0000 **11.** 0.2405 **13.** 9.5740 – 10 **15.** 7.9405 – 10 **17.** 2.0086 **19.** 7.1072 – 10 **21.** 3.48 **23.** 209 **25.** 0.0874 **27.** 1.00 **29.** 317 **31.** 100 **33.** 0.000787 **35.** 6610 **37.** 0.0871 **39.** 15.5 **41.** 71,400 **43.** 0.0246 **45.** 4030 **47.** 17.6 **49.** 0.000515 **51.** 172 **53.** 0.040 **55.** 210,000 **57.** 44.1 **59.** 2050 **61.** 71.7

Just for Fun **1.** 0.0384 **2.** 0.001

Exercise Set 11.5

1. 5 **3.** 3 **5.** 55.50 **7.** 5.59 **9.** 3 **11.** 0, –8 **13.** $\frac{13}{2}$ **15.** $\frac{3}{2}$ **17.** 2 **19.** 2 **21.** 0.91 **23.** 20 **25.** **(a)** 31.62 km **(b)** 0.50 km **27.** \$69,914.40 **29.** 139 bacteria **31.** \$7112.10 **33.** 2.94 **35.** **(a)** 1,000,000,000,000 **(b)** 100,000

Just for Fun **1.** 2.55

Review Exercises

1. One to one **2.** One to one **3.** Not one to one **4.** Not one to one **5.** One to one **6.** Not one to one

7. $f(x)$, domain $\{-4, 0, 5, 6\}$, range $\{-3, 2, 3, 7\}$
$f^{-1}(x)$, domain $\{-3, 2, 3, 7\}$, range $\{-4, 0, 5, 6\}$

8. $f(x)$, domain $\{-3, -1, \frac{1}{2}, \sqrt{5}\}$, range $\{2, \sqrt{7}, 3, 8\}$
$f^{-1}(x)$, domain $\{2, \sqrt{7}, 3, 8\}$, range $\{-3, -1, \frac{1}{2}, \sqrt{5}\}$

9. $f^{-1}(x) = (x + 2)/4$ **10.** $f^{-1}(x) = -\frac{1}{3}(x + 5)$ **11.** $f^{-1}(x) = (3x - 5)/2$ **12.** $f^{-1}(x) = (-15x + 9)/10$

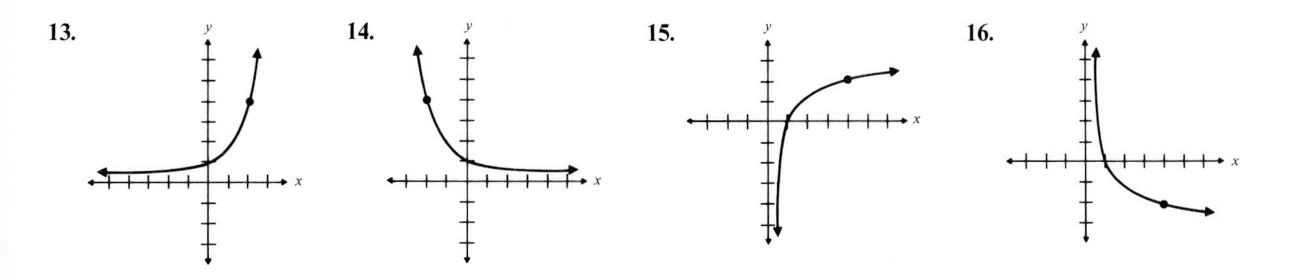

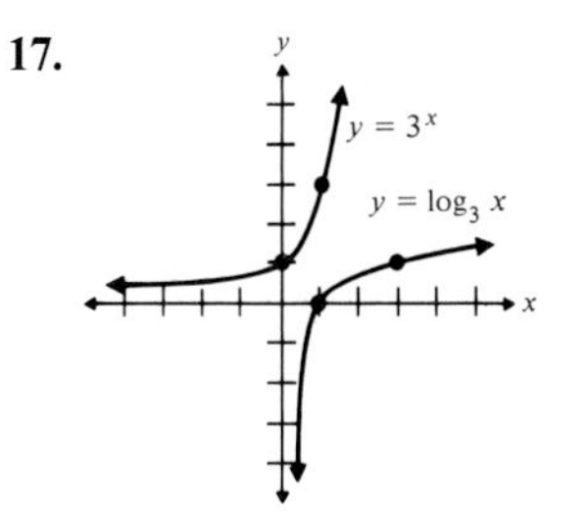

18. $\log_4 16 = 2$ **19.** $\log_8 2 = \frac{1}{3}$ **20.** $\log_6 (\frac{1}{36}) = -2$ **21.** $\log_{25} 5 = \frac{1}{2}$
22. $5^2 = 25$ **23.** $(\frac{1}{3})^2 = \frac{1}{9}$ **24.** $3^{-2} = \frac{1}{9}$ **25.** $2^5 = 32$ **26.** $4^3 = x$, 64
27. $4^2 = x$, 16 **28.** $a^3 = 8$, 2 **29.** $(\frac{1}{4})^{-3} = x$, 64 **30.** $\frac{1}{2} \log_8 12$ **31.** $5 \log (x - 8)$
32. $\log 2 + \log (x - 3) - \log x$ **33.** $4 \log x - \log 39 - \log (2x + 8)$
34. $\log_5 [(x - 2)/x^2]$ **35.** $\log [x^2/(x + 1)^3]$ **36.** $\log \left[\frac{(x \cdot 2)^3}{x}\right]$
37. $\log_8 \left[\frac{(x + 3)^2(x - 1)^4}{\sqrt{x}}\right]$ **38.** 3.9138 **39.** 6.8549 − 10 **40.** 7.2765 − 10
41. 4.2455 **42.** 56.6 **43.** 829 **44.** 0.0796 **45.** 0.0422 **46.** 214 **47.** 0.0594
48. 15.1 **49.** 1680 **50.** 6,130,000 **51.** 117 **52.** 2,610,000 **53.** 2.605
54. 4.498 **55.** 1.595 **56.** $-\frac{1}{2}$ **57.** 123 **58.** $\frac{298}{3}$ **59.** 2 **60.** 1
61. $25,700 **62.** 566.5 mg

Practice Test

1. $f^{-1}(x) = (x - 4)/2$

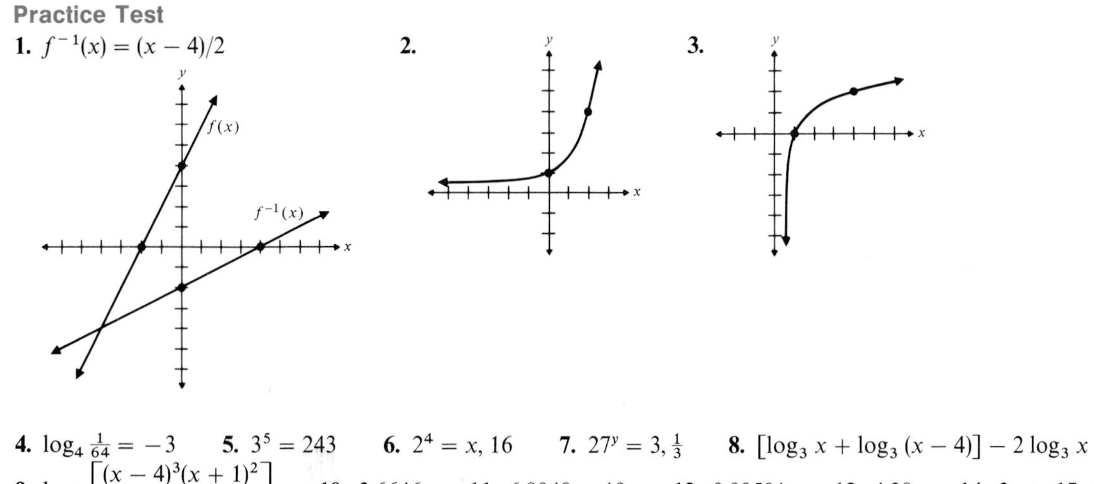

4. $\log_4 \frac{1}{64} = -3$ **5.** $3^5 = 243$ **6.** $2^4 = x$, 16 **7.** $27^y = 3, \frac{1}{3}$ **8.** $[\log_3 x + \log_3 (x - 4)] - 2 \log_3 x$
9. $\log_8 \left[\frac{(x - 4)^3(x + 1)^2}{\sqrt{x}}\right]$ **10.** 3.6646 **11.** 6.8048 − 10 **12.** 0.00501 **13.** 4.38 **14.** 3 **15.** $4660

CHAPTER 12

Exercise Set 12.1

1. 2, 4, 6, 8, 10 **3.** 6, $\frac{7}{2}, \frac{8}{3}, \frac{9}{4}$, 2 **5.** 1, $\frac{1}{2}, \frac{1}{3}, \frac{1}{4}, \frac{1}{5}$ **7.** $\frac{3}{2}, \frac{4}{3}, \frac{5}{4}, \frac{6}{5}, \frac{7}{6}$ **9.** −1, 1, −1, 1, −1 **11.** 4, −8, 16, −32, 64
13. 27 **15.** 6 **17.** 16 **19.** $\frac{81}{19}$ **21.** $s_1 = 7, s_3 = 27$ **23.** $s_1 = 3, s_3 = 17$ **25.** $s_1 = 1, s_3 = 3$
27. $s_1 = \frac{1}{2}, s_3 = 7$ **29.** 64, 128, 256 **31.** 15, 17, 19 **33.** $\frac{1}{6}, \frac{1}{7}, \frac{1}{8}$ **35.** 1, −1, 1 **37.** $\frac{1}{81}, \frac{1}{243}, \frac{1}{729}$
39. $\frac{1}{16}, -\frac{1}{32}, \frac{1}{64}$ **41.** −25, −33, −41

Exercise Set 12.2

1. 3, 7, 11, 15, 19; $a_n = 3 + 4(n-1)$ **3.** $-5, -3, -1, 1, 3$; $a_n = -5 + 2(n-1)$ **5.** $\frac{1}{2}, 2, \frac{7}{2}, 5, \frac{13}{2}$; $a_n = \frac{1}{2} + \frac{3}{2}(n-1)$
7. 100, 95, 90, 85, 80; $a_n = 100 - 5(n-1)$ **9.** 22 **11.** -23 **13.** 13 **15.** 2 **17.** 9 **19.** 6
21. $s_{10} = 100, d = 2$ **23.** $s_8 = \frac{52}{5}, d = \frac{1}{5}$ **25.** $s_5 = 20, d = \frac{4}{5}$ **27.** $s_{11} = 407, d = 6$
29. $-4, -6, -8, -10$; $a_{10} = -22, s_{10} = -130$ **31.** $-8, -13, -18, -23$; $a_{10} = -53, s_{10} = -305$
33. 100, 93, 86, 79; $a_{10} = 37, s_{10} = 685$ **35.** $\frac{9}{5}, \frac{12}{5}, \frac{15}{5}, \frac{18}{5}$; $a_{10} = \frac{36}{5}, s_{10} = 45$ **37.** $n = 26, s_n = 442$
39. $n = 17, s_n = \frac{153}{2}$ **41.** $n = 9, s_n = 315$ **43.** $n = 29, s_n = 1479$ **45.** 500,500 **47.** 267 **49.** 1 ft **51.** 210 logs

Exercise Set 12.3

1. 5, 15, 45, 135, 405 **3.** 90, 30, 10, $\frac{10}{3}, \frac{10}{9}$ **5.** $-15, 30, -60, 120, -240$ **7.** $3, \frac{9}{2}, \frac{27}{4}, \frac{81}{8}, \frac{243}{16}$ **9.** 160 **11.** 13,122
13. $\frac{1}{64}$ **15.** 6144 **17.** 1023 **19.** 10,160 **21.** $\frac{2565}{128}$ **23.** $-\frac{9279}{625}$ **25.** $r = \frac{1}{2}, a_n = 5(\frac{1}{2})^{n-1}$
27. $r = -3, a_n = 2(-3)^{n-1}$ **29.** $r = 3, a_n = -1(3)^{n-1}$ **31.** $r = 2, a_1 = 7$ or $r = -2, a_1 = 7$ **33.** $r = 3, a_1 = 5$
35. \$572,503.52 **37. (a)** 3 days **(b)** 1.172 grams **39. (a)** 437.25 million **(b)** 11.9 years **41. (a)** $\frac{1}{2}, \frac{1}{4}, \frac{1}{8}, \frac{1}{16}, \frac{1}{32}$
(b) $a_n = a_1 r^{n-1} = \frac{1}{2}(\frac{1}{2})^{n-1} = (\frac{1}{2})^n$ **(c)** amount remaining $= \frac{1}{128}$ or 0.78% **43.** $(\frac{2}{3})^{10}$ or 1.7%
45. (a) 7840, 6272, 5017.6 **(b)** $a_n = 7840(\frac{4}{5})^{n-1}$ **(c)** \$3211.26

Just for Fun **1. (a)** \$3211.26 **(b)** \$3211.26 **2.** $n = 21, s = 2{,}097{,}151$

Exercise Set 12.4

1. 12 **3.** $\frac{25}{3}$ **5.** $\frac{5}{3}$ **7.** $\frac{81}{10}$ **9.** 2 **11.** 24 **13.** -45 **15.** -15 **17.** $\frac{3}{11}$ **19.** $\frac{5}{9}$ **21.** $\frac{17}{33}$ **23.** 40 ft

Just for Fun **1.** 190 ft **2.** 4 ft

Exercise Set 12.5

1. $x^3 + 12x^2 + 48x + 64$ **3.** $a^4 - 4a^3b + 6a^2b^2 - 4ab^3 + b^4$ **5.** $243a^5 - 405a^4b + 270a^3b^2 - 90a^2b^3 + 15ab^4 - b^5$
7. $16x^4 + 16x^3 + 6x^2 + x + \frac{1}{16}$ **9.** $(x^4/16) - (3x^3/2) + (27x^2/2) + 54x + 81$ **11.** $x^{10} + 10x^9y + 45x^8y^2 + 120x^7y^3$
13. $2187x^7 - 5103x^6y + 5103x^5y^2 - 2835x^4y^3$ **15.** $x^{16} - 24x^{14}y + 252x^{12}y^2 - 1512x^{10}y^3$ **17.** 1.217 **19.** 1.072

Just for Fun **1.** 10.013 **2.** 2.991 **3.** 0.924

Review Exercises

1. 3, 4, 5, 6, 7 **2.** $1, \frac{1}{2}, \frac{1}{3}, \frac{1}{4}, \frac{1}{5}$ **3.** 2, 6, 12, 20, 30 **4.** $\frac{1}{5}, \frac{2}{3}, \frac{9}{7}, 2, \frac{25}{9}$ **5.** 25 **6.** 2 **7.** $\frac{16}{81}$ **8.** 88
9. $s_1 = 5, s_3 = 24$ **10.** $s_1 = 2, s_3 = 28$ **11.** $s_1 = \frac{4}{3}, s_3 = \frac{227}{60}$ **12.** $s_1 = -3, s_3 = -4$ **13.** 16, 32, 64; $a_n = 2^{n-1}$
14. $-\frac{1}{2}, \frac{1}{4}, -\frac{1}{8}$; $a_n = -8(-\frac{1}{2})^{n-1}$ **15.** $\frac{32}{3}, \frac{64}{3}, \frac{128}{3}$; $a_n = \frac{2}{3}(2)^{n-1}$ **16.** $-3, -6, -9$; $a_n = 9 - 3(n-1) = 12 - 3n$
17. $-1, 1, -1$; $a_n = (-1)^n$ **18.** 30, 36, 42; $a_n = 6 + 6(n-1) = 6n$ **19.** 5, 7, 9, 11, 13
20. $\frac{1}{2}, -\frac{3}{2}, -\frac{7}{2}, -\frac{11}{2}, -\frac{15}{2}$ **21.** $-12, -\frac{25}{2}, -13, -\frac{27}{2}, -14$ **22.** $-100, -\frac{499}{5}, -\frac{498}{5}, -\frac{497}{5}, -\frac{496}{5}$ **23.** 26
24. -15 **25.** -4 **26.** $\frac{1}{2}$ **27.** 6 **28.** 7 **29.** $s = 112, d = 2$ **30.** $s = -210, d = -6$ **31.** $s = \frac{63}{5}, d = \frac{2}{5}$
32. $s = -42, d = -\frac{1}{3}$ **33.** 2, 6, 10, 14; $a_{10} = 38, s_{10} = 200$ **34.** $-8, -11, -14, -17$; $a_{10} = -35, s_{10} = -215$
35. $\frac{5}{6}, \frac{9}{6}, \frac{13}{6}, \frac{17}{6}$; $a_{10} = \frac{41}{6}, s_{10} = \frac{115}{3}$ **36.** $-80, -76, -72, -68$; $a_{10} = -44, s_{10} = -620$ **37.** $n = 11, s_n = 308$
38. $n = 9, s_n = 36$ **39.** $n = 11, s_n = \frac{231}{10}$ **40.** $n = 19, s_n = 760$ **41.** 5, 10, 20, 40, 80 **42.** $-12, -6, -3, -\frac{3}{2}, -\frac{3}{4}$
43. $20, -\frac{40}{3}, \frac{80}{9}, -\frac{160}{27}, \frac{320}{81}$ **44.** $-100, -20, -4, -\frac{4}{5}, -\frac{4}{25}$ **45.** $\frac{4}{243}$ **46.** 6400 **47.** -2048 **48.** $\frac{160}{6561}$ **49.** 3060
50. $\frac{37{,}969}{1215}$ **51.** $-\frac{4305}{64}$ **52.** $\frac{172{,}539}{256}$ **53.** $r = 2, a_n = 6(2)^{n-1}$ **54.** $r = \frac{1}{3}, a_n = 8(\frac{1}{3})^{n-1}$ **55.** $r = 5, a_n = -4(5)^{n-1}$
56. $r = \frac{2}{3}, a_n = \frac{9}{5}(\frac{2}{3})^{n-1}$ **57.** 14 **58.** -6 **59.** -15 **60.** $\frac{49}{10}$ **61.** 4 **62.** $\frac{21}{2}$ **63.** -36 **64.** $\frac{25}{6}$
65. $\frac{52}{99}$ **66.** $\frac{125}{333}$ **67.** $81x^4 + 108x^3y + 54x^2y^2 + 12xy^3 + y^4$ **68.** $8x^3 - 36x^2y^2 + 54xy^4 - 27y^6$
69. $x^9 - 18x^8y + 144x^7y^2 - 672x^6y^3$ **70.** $256a^{16} + 3072a^{14}b + 16{,}128a^{12}b^2 + 48{,}384a^{10}b^3$ **71.** 15,150
72. (a) 30,000, 31,000, 32,000, 33,000, 34,000 **(b)** $a_n = 30{,}000 + (n-1)1000$ **(c)** 39,000 **73.** \$102,400
74. \$1070.05 **75.** 100 ft

Practice Test

1. $3, 1, \frac{5}{9}, \frac{3}{8}, \frac{7}{25}$ **2.** $s_1 = 5, s_3 = \frac{23}{2}$ **3.** $a_n = \frac{1}{3} + \frac{1}{3}(n-1) = \frac{1}{3}n$ **4.** $a_n = 5(2)^{n-1}$ **5.** 12, 9, 6, 3 **6.** $\frac{5}{8}, \frac{10}{24}, \frac{20}{72}, \frac{40}{216}$
7. 16 **8.** -32 **9.** 13 **10.** $\frac{512}{729}$ **11.** $\frac{39{,}063}{5}$ **12.** $r = \frac{1}{2}, a_n = 12(\frac{1}{2})^{n-1}$ **13.** 9
14. $x^4 + 8x^3y + 24x^2y^2 + 32xy^3 + 16y^4$

Index